b1145b711

AN INTRODUCTION TO THE INTEL FAMILY OF MICROPROCESSORS

A Hands-On Approach Utilizing the 80x86 Microprocessor Family

Third Edition

JAMES L. ANTONAKOS
Broome Community College

Prentice Hall

Upper Saddle River, Columbus, Ohio

Library of Congress Cataloging-in-Publication Data

Antonakos, James L.
   An introduction to the Intel family of microprocessors : a hands
-on approach utilizing the 80x86 microprocessor family/ James L. Antonakos.
   — 3rd ed.
      p.   cm.
   Includes index.
   ISBN 0-13-893439-8
   1. Intel 80x86 (Microprocessor)   I. Title.
QA76.8.I2924A58   1999
004.165—dc21

97-46893
CIP

Cover photo: © Karen Guzak
Editor: Charles E. Stewart, Jr.
Production Supervision: Custom Editorial Productions, Inc.
Design Coordinator: Karrie M. Converse
Cover Designer: Linda Fares
Production Manager: Deidra M. Schwartz
Illustrations: Custom Editorial Productions, Inc.
Marketing Manager: Ben Leonard

This book was set in Times Roman and Helvetica by Custom Editorial Productions, Inc. and was printed and bound by Courier/Westford, Inc. The cover was printed by Phoenix Color Corp.

© 1999, 1996 by Prentice-Hall, Inc.
Simon & Schuster/A Viacom Company
Upper Saddle River, New Jersey 07458

Earlier edition © 1993 by Macmillan Publishing Company

Printed in the United States of America

10 9 8 7 6 5 4 3 2

ISBN 0-13-893439-8

Prentice-Hall International (UK) Limited, *London*
Prentice-Hall of Australia Pty. Limited, *Sydney*
Prentice-Hall Canada Inc., *Toronto*
Prentice-Hall Hispanoamericana, S. A., *Mexico*
Prentice-Hall of India Private Limited, *New Delhi*
Prentice-Hall of Japan, Inc., *Tokyo*
Simon & Schuster Asia Pte. Ltd., *Singapore*
Editora Prentice-Hall do Brasil, Ltda., *Rio de Janeiro*

This one is for Heather and Jeff,
with all my love.

PREFACE

The rapid advancement of microprocessors into our everyday affairs has both simplified and complicated our lives. Whether we use a computer in our everyday job, or come in contact with one elsewhere, most of us have used a computer at one time or another. Most people know that a microprocessor is lurking somewhere inside the machinery, but what a microprocessor is, and what it does, remains a mystery.

PURPOSES OF THE BOOK

This book is intended to help remove the mystery concerning the Intel 80x86 microprocessor family through detailed coverage of its hardware and software and examples of many different applications. Some of the more elaborate applications are visible to us. A large collection of personal computers use 80x86-based architecture, as do some popular commercial video games, electronic engravers, and speech recognition systems. Industry and the government have adopted the 80x86 for many commercial and military applications as well.

The book is intended for 2- or 4-year electrical engineering, engineering technology, and computer science students. Professional people, such as engineers and technicians, will also find it a handy reference. The material is intended for a one-semester course in microprocessors. Prior knowledge of digital electronics, including combinational and sequential logic, decoders, memories, Boolean algebra, and operations on binary numbers, is helpful. This presumes knowledge of standard computer-related terms, such as RAM, EPROM, TTL, and so forth. Appendix D is included as a reference on binary numbers and arithmetic for those students who would like a quick review.

ADDITIONS TO THE THIRD EDITION

The majority of new material added to the second edition to create the third edition of this textbook is found in two new chapters. Chapters 15 and 16 cover, respectively, Pentium architecture and protected mode operation. While the textbook is strongly based in the world of the 8088 microprocessor (especially the single-board computer project in Chapter 14), coverage of the advanced Intel processors has been increased, with attention paid to the

Pentium. Keep in mind that many of the concepts developed for the 8088 apply to the entire 80x86 family (via real mode on the advanced processors).

The former Chapter 3 covered the entire 8088 instruction set. It has been split into two chapters (the new Chapters 3 and 4), with added material now covering the entire 80x86 instruction set, up through the Pentium. Simple methods are shown to allow the processor's real mode to access the new 32-bit registers available beginning with the 80386.

The term "80x86" is used throughout the text to refer to the entire family of compatible Intel processors, the 8088, 8086, 80286, 80386, 80486, and Pentium. Please think of "80x86" as a single machine as well, one processor having many abilities.

In all the programming chapters, new, simplified assembler directives such as .MODEL, .CODE, and .DATA are being used. Anyone requiring the older versions should look for information on my homepage by clicking the 8088's textbook image.

In addition, many new exercises, questions, and figures have been added. Also, a new section covering troubleshooting techniques has been added to each chapter. These techniques should help eliminate some of the problems encountered working with assembly language and microprocesssor interfacing.

The old Chapter 5 on DEBUG has been remade as Appendix H, with material added on the Codeview utility. Coverage of some material has been rearranged (during the Chapter 3 split), and, last, a fond farewell was made to the old Appendix D, which covered the 8085 microprocessor.

OUTLINE OF COVERAGE

For those individuals who have no prior knowledge of microprocessors, **Chapter 1, Microprocessor-Based Systems,** is a good introduction to the microprocessor, how it functions internally, and how it is used in a small system. Chapter 1 is a study of the overall operation of a microprocessor-based system.

Chapter 2, An Introduction to the 80x86 Microprocessor Family, highlights the main features of the 80x86. Data types, addressing modes, and instructions are surveyed. All processors in the 80x86 family are examined as well.

Chapter 3, 80x86 Instructions, Part 1, and **Chapter 4, 80x86 Instructions, Part 2,** introduce the entire real-mode instruction set of the 80x86. These instructions include data transfer, string, arithmetic, logical, bit manipulation, program transfer, and processor control instructions. Addressing modes, flags, and the structure of a source file are also covered. Over 90 examples are provided to help the student grasp the material.

Chapter 5, Interrupt Processing, covers the basic sequence of an interrupt, as well as multiple interrupts, special interrupts, and interrupt service routines. The instructor may choose to cover this chapter after Chapter 6 to get right to the programming examples.

Chapter 6, An Introduction to Programming the 80x86, contains the first real programming efforts. Numerous programming examples show how the 80x86 performs routine functions involving binary and BCD mathematics, string operations, data-table manipulation, and number conversions. The concept of a software *driver* program is developed.

Chapter 7, Programming with DOS and BIOS Function Calls, opens the power of the personal computer to the student. DOS interrupt 21H and other interrupts are included to show how a real-mode program running on a personal computer accesses DOS and the hardware of the machine. The keyboard, video display, speaker, and printer are all used in applications. Sixteen programs are included to give the student sufficient exposure to the required programming techniques.

Chapter 8, Advanced Programming Applications, introduces the student to many advanced concepts, such as linking multiple object files, instruction execution time, interrupt handling, memory management, 80x87 coprocessor programming, and macro usage. Fifteen more programming applications are included to support the new concepts.

Chapter 9, Using Disks and Files, introduces the student to the operation of the disk system and the organization of a diskette. The different structures used with all disks, such as the boot sector and FAT, are covered, as well as the many different interrupt functions that support disk access. A number of example programs are included to illustrate how data can be read from and written to a diskette.

The hardware operation of the 8088 is covered in **Chapter 10, Hardware Details of the 8088.** All CPU pins are discussed, as are timing diagrams, the difference between minmode and maxmode operation, the Personal Computer Bus Standards, and two chips that are essential to 8088-based systems: the 8284 clock generator and the 8288 bus controller.

Chapter 11, Memory System Design, covers the details needed to design an operational RAM and EPROM-based memory system for the 8088. Static and dynamic RAMs, EPROMs, DMA, and full and partial address decoding are covered.

The I/O system is covered in **Chapter 12, I/O System Design.** In this chapter the difference between the processor's memory space and port space is covered, as are the techniques needed to design port address decoders. Two 8085-based peripherals are covered: the 8255™ parallel interface adapter and the 8251 UART™, with 8088 interfacing and programming examples provided.

Programming with 80x86 Peripherals is the subject of **Chapter 13.** In this chapter, three peripherals designed to interface with the 80x86 are examined. These peripherals implement interval timing, interrupt control, and floating-point operations. Programming and interfacing are discussed, with specific examples.

Many textbooks rarely cover hardware and software with an equal amount of detail. This book was written to give equal treatment to both, culminating in a practical exercise: *building and programming your own single-board computer!* **Chapter 14, Building a Working 8088 System,** is included to give students a chance to design, build, and program their own 8088-based computers. The system contains 8KB of EPROM, 8KB of RAM, a serial I/O device, a parallel I/O device, and 8-bit D/A and A/D converters. Future memory expansion is built in. The hardware is designed first, followed by design of the software monitor program.

Some books choose to explain the operation of a commercial system, such as the SDK-86™. This approach is certainly worthwhile, but does not give the student the added advantage of knowing *why* certain designs were used. The hardware and software designs in Chapter 14 are sprinkled with many questions, which are used to guide the design toward its final goal.

The single-board computer presented in Chapter 14 can be easily wire-wrapped in a short period of time (some students have constructed a working computer in seven days), directly from the schematics provided in the chapter. It is reasonable to say that most students can build a working system in one semester.

The hardware operation of the Pentium is covered in **Chapter 15, Hardware Details of the Pentium.** Descriptions of each CPU pin are included, as are explanations of the various methods employed by the Pentium to access data over its buses. The operation of the Pentium's superscaler arechitecture, internal pipelining, branch prediction, instruction and data caches, and floating-point unit are discussed.

Finally, **Chapter 16, Protected Mode Operation,** presents the details associated with protected mode operation. The virtual memory techniques made possible by the use of segments and paging are explained. Important issues such as protection, exceptions, multitasking, and input/output are also discussed. Virtual-8086 mode is also covered.

USES OF THE BOOK

Due to the information presented, some chapters are much longer than others. Even so, it is possible to cover certain sections of selected chapters out of sequence, or to pick and choose sections from various chapters. Chapters 3 and 4 could be covered in this way, with emphasis placed on additional addressing modes or groups of instructions at a rate deemed appropriate by the instructor. Also, some sections in Chapter 13 may be skipped, depending on the instructor's preference for peripherals. Some instructors may wish to cover hardware (Chapters 10 through 12) before programming (Chapters 3 through 9). There is no reason this cannot be done.

To aid the instructor, answers to selected odd-numbered end-of-chapter study questions are included in the text and are also provided in a detailed solutions manual. The solutions manual is designed in such a way that solutions to all odd-numbered questions are grouped together, followed by solutions to all even-numbered questions. This allows the instructor to release selected odd-only or even-only answers to students, while retaining others for testing purposes.

The appendixes to this text are used to present a full list of 8088 instructions, their allowed addressing modes, flag usage, and instruction times. In addition, the appendixes contain data sheets for three 80x86-based peripherals, the 8259™ interrupt controller, the 8254™ interval timer, and the 8087™ floating point coprocessor. This avoids the need for second references. Also included are detailed information on DOS and BIOS interrupts, a review of binary numbers, and keyboard scan codes.

In summary, over 250 illustrations and 70 different applications are used to give the student sufficient exposure to the 80x86. The added benefit of Chapter 14, where a working system is developed, makes this book an ideal choice for a student wishing to learn about microprocessors. The new material in Chapters 15 and 16 lays the groundwork for more advanced hardware and software development. Furthermore, even though this book deals only with the 80x86 family, the serious microprocessor student should also be exposed to other CPUs as well. But to try to cover two or more different microprocessors in one text does not do either microprocessor justice. For this reason, all attention is paid to the 80x86 family, and not to other CPUs.

THE COMPANION DISKETTE

The diskette included with the book contains all of the source files presented in the book. The files are stored in separate directories related to their specific chapters. In addition, the object code library NUMOUT.LIB from Chapter 8 and the various binary and executable files related to the single-board computer from Chapter 14 are also included. The diskette contains an executable file called README.COM, which explains the diskette contents in detail.

ACKNOWLEDGMENTS

I would like to thank my editor, Charles Stewart, and his assistants, Kate Linsner and Kimberly Yehle, for their help while I was putting this book together. I would also like to thank the many students and instructors who used the second edition and contacted me via e-mail with questions and comments. In addition, I would like to thank my copyeditor, Cindy Lanning, as well as Jim Reidel, who managed the book through production.

The following individuals provided many useful comments during the rewrite, and I am grateful for their advice: Shelton Houston, University of Southern Mississippi and Dimitri Kagaris, Southern Illinois University.

James L. Antonakos
antonakos_j@sunybroome.edu
http://www.sunybroome.edu/~antonakos_j

BRIEF CONTENTS

CONTENTS

PART 1

Introduction

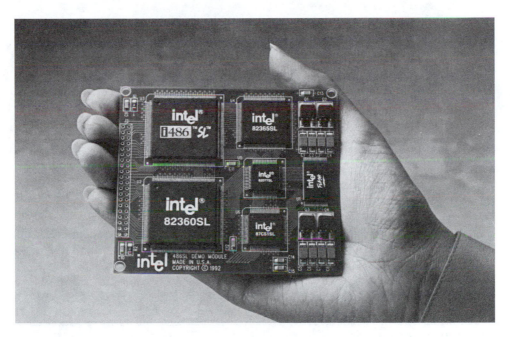

An example of an Intel486™ SL Microprocessor-based system, at just 3.25 × 3.75 inches, fits in the palm of your hand.

CHAPTER 1

Microprocessor-Based Systems

OBJECTIVES

In this chapter you will learn about:

* The block diagram of a microprocessor-based system and the function of each section
* The processing cycle of a microprocessor
* The way software is used to initialize hardware and peripherals
* The history of the microprocessor and of the different generations of computers
* The technique used to create software for the personal computer
* Some typical errors encountered during program development

1.1 INTRODUCTION

The invention of the microprocessor has had a profound impact on many aspects of our lives, since today even the most mundane chores are being accomplished under its supervision—something that allows us more time for other productive endeavors. Even a short list of the devices using the microprocessor shows how dependent we have become on it:

1. Pocket calculators
2. Digital watches (some with calculators built in)
3. Automatic tellers (at banks and food stores)
4. Smart telephones
5. Compact disk players
6. Home security and control devices
7. Realistic video games
8. Talking dolls and other toys
9. VCRs
10. Personal computers

The purpose of this chapter is to show how a microprocessor is used in a small system and to introduce the operation of the personal computer. We will see what types of hardware may be connected to the microprocessor, and why each type is needed. We will also see how software is used to control the hardware, and how that software can be developed.

Section 1.2 shows how the microprocessor has evolved over time, from the initial 4-bit machines to today's 32-bit processors. Section 1.3 covers the block diagram of a typical microprocessor-based system and explains each functional unit. Section 1.4 explains the basic operation of a microprocessor. Section 1.5 discusses the hardware and software requirements of a small microprocessor control system. Section 1.6 brings the material of the first five sections together in a technical description of the personal computer. Section 1.7 shows how software is developed for, and used by, the personal computer. Finally, section 1.8 introduces the first of a series of troubleshooting techniques.

1.2   EVOLUTION OF MICROPROCESSORS

We have come a long way since the early days of computers, when ENIAC (for Electronic Numerical Integrator and Computer) was state of the art and occupied thousands of feet of floor space. Constructed largely of vacuum tubes, it was slow, prone to breakdowns, and performed a limited number of instructions. Even so, ENIAC ushered in what was known as the **first generation** of computers.

Today, thanks to advances in technology, we have complete computers that fit on a piece of silicon no larger than your fingernail and that far outperform ENIAC.

When the transistor was invented, computers shrank in size and increased in power, leading to the **second generation** of computers. **Third-generation** computers came about with the invention of the integrated circuit, which allowed hundreds of transistors to be packed on a small piece of silicon. The transistors were connected to form logic elements, the basic building blocks of digital computers. With third-generation computers we again saw a decrease in size and increase in computing power. Machines like the 4004™ and 8008™ by Intel® found some application in simple calculators, but they were limited in power and addressing capability. When improvements in integrated circuit technology enabled us to place *thousands* of transistors on the same piece of silicon, computers really began to increase in power. This new technology, called large-scale integration (LSI), is even faster than the previous medium- and small-scale integration (MSI and SSI, respectively) technologies, which dealt with only tens or hundreds of transistors on a chip. LSI technology has created the **fourth generation** of computers that we use today. An advanced form of LSI technology, VLSI, meaning very large scale integration, is now being used to increase processing power.

The first microprocessors that became available with third-generation computers had limited instruction sets and thus restricted computing abilities. Although they were suitable for use in electronic calculators, they simply did not have the power needed to operate more complex systems, such as guidance systems or scientific applications. Even some of the early fourth-generation microprocessors had limited capabilities because of the lack of addressing modes and instruction types. Eight-bit machines like the 8080™, Z80™, and 6800™ were indeed more advanced than previous microprocessors, but they still did not possess multiply and divide instructions. How frustrating and time consuming to have to write a program to do these operations!

Within the last decade, microprocessor technology has improved tremendously. Thirty-two-bit processors can now multiply and divide, operate on many different data types (4-, 8-, 16-, 32-bit numbers), and address *billions* of bytes of information. Processors of the 1970s were limited to 64KB, a small amount of memory by today's standards.

Each new microprocessor to hit the market boasts a fancier instruction set and faster clock speed, and indeed our needs for faster and better processors keep growing. A new technology called RISC (for Reduced Instruction Set Computer) has recently gained acceptance. This technology is based on the fact that most microprocessors use only a small portion of their entire instruction set. By designing a machine that uses only the more common types of instructions, processing speed can be increased without the need for a significant advance in integrated circuit technology. The Pentium microprocessor, manufactured by Intel, uses many of the architectural techniques employed by RISC machines.

Why the need for superfast machines? Consider a microprocessor dedicated to displaying three-dimensional color images on a video screen. Rotating the three-dimensional image around an imaginary axis in real time (in only a few seconds or less) may require millions or even billions of calculations. A slow microprocessor would not be able to do the job.

Eventually we will see fifth-generation computers. The whole artificial intelligence movement is pushing toward that goal, with the desired outcome being the production of a machine that can think. Until then, we will have to make the best use of the technology we have available.

1.3   SYSTEM BLOCK DIAGRAM

Any microprocessor-based system must, of necessity, have some standard elements such as memory, timing, and input/output (I/O). Depending on the application, other exotic circuitry may be necessary as well. Analog-to-digital (A/D) converters and their counterparts, digital-to-analog (D/A) converters, interval timers, math coprocessors, complex interrupt circuitry, speech synthesizers, and video display controllers are just a few of the special sections that may also be required. Figure 1.1 depicts a block diagram of a system containing some standard circuitry and functions normally used.

As the figure shows, all components communicate via the **system bus.** The system bus is composed of the processor address, data, and control signals. The **central processing unit (CPU)** is the heart of the system, the master controller of all operations that can be performed. The CPU executes instructions that are stored in the **memory** section. For the sake of future expansion, the system bus is commonly made available to the outside world (through a special connector). Devices may then be added easily as the need arises. Commercial systems have predefined buses that accomplish this. All devices on the system bus must communicate with the processor, usually within a tightly controlled period of time. The **timing** section governs all system timing and thus is responsible for the proper operation of all system hardware. The timing section usually consists of a crystal oscillator and timing circuitry (counters designed to produce the desired frequencies) set up to operate the processor at its specified clock rate. Using a high-frequency crystal oscillator and dividing it down to a lower frequency provides for greater stability.

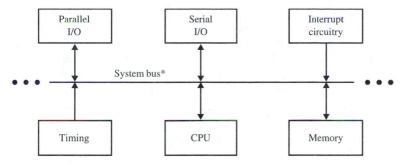

* Data, address, and control signals

FIGURE 1.1   Standard block diagram of a microprocessor-based system

The CPU section consists of a microprocessor and the associated logic circuitry required to enable the CPU to communicate with the system bus. These logic elements may consist of data and address bus drivers, a bus controller to generate the correct control signals, and possibly a math coprocessor. **Coprocessors** are actually microprocessors themselves; their instruction set consists mainly of simple instructions for transferring data, and complex instructions for performing a large variety of mathematical operations. Coprocessors perform these operations at very high clock speeds with a great deal of precision (80-bit results are common). In addition to the basic add/subtract/multiply/divide operations, coprocessors are capable of finding square roots, logarithms, a host of trigonometric functions, and more.

The actual microprocessor used depends on the complexity of the task that will be controlled or performed by the system. Simple tasks require nothing more complicated than an 8-bit CPU. A computerized cash register would be a good example of this kind of system. Nothing more complicated than binary coded decimal (BCD) arithmetic—and possibly some record keeping—is needed. But for something as complex as a flight control computer for an aircraft or a digital guidance system for a missile, a more powerful 16- or 32-bit microprocessor must be used.

The **memory** section usually has two components: **read-only memory (ROM)** and **random access memory (RAM).** Some systems may be able to work properly without RAM, but all require at least a small amount of ROM. The ROM is included to provide the system with its intelligence, which is ordinarily needed at start-up (power-on) to configure or initialize the peripherals, and sometimes to help recover from a catastrophic system failure (such as an unexpected power failure). Some systems use the ROM program to download the main program into RAM from a larger, external system, such as a personal computer (PC) or a mainframe computer. In any event, provisions are usually made for adding additional ROM as the need arises.

There are three types of RAM. For small systems that do not process a great deal of data, the choice is static RAM. Static RAM is fast and easy to interface, but comes in small sizes (as little as 16 bytes per chip). Larger memory requirements are usually met by using dynamic RAM, a different form of memory that has high density (256K bits per chip or more), but that unfortunately requires numerous refreshing cycles to retain the stored data. Even so, dynamic RAM is the choice when large amounts of data must be stored, as in a system gathering seismic data at a volcano or receiving digitized video images from a satellite.

Both static and dynamic RAM lose their information when power is turned off, which may cause a problem in certain situations. Previous solutions involved adding battery backup circuitry to the system to keep the RAMs supplied with power during an outage. But batteries can fail, so a better method was needed; thus came the invention of **nonvolatile memory (NVM),** which is memory that retains its information even when power is turned off. NVM comes in small sizes and therefore is used to store only the most important system variables in the event of a power outage.

Another type of storage media is the floppy or hard disk. Both types of disks provide the system with large amounts of storage for programs and data, although the data are accessed at a much slower rate than that of RAM or ROM. Floppy and hard disks also require complex hardware and software to operate, and are not needed in many control applications.

With a microprocessor used in control applications, there will be times when the system must respond to special external circumstances. For example, a power failure on a computer-controlled assembly line requires immediate attention by the system, which must contain software designed to handle the unexpected event. The event actually *interrupts* the processor from its normal program execution to service the unexpected event. The system software is designed to handle the power-fail interrupt in a certain way and then return to the main program. An interrupt, then, is a useful way to grab the processor's attention, get it to perform a special task, and then resume execution from the point where it left off.

Not all types of interrupts are unexpected. Many are used to provide systems with useful features, such as real-time clocks, multitasking capability, and fast I/O operations.

The interrupt circuitry needed from system to system will vary depending on the application. A system used for keeping time has to use only a single interrupt line connected to a timing source. A more complex system, such as an assembly line controller that may need to monitor multiple sensors, switches, and other items, may require many different prioritized interrupts and would therefore need more complex interrupt circuitry.

Some systems may require serial I/O for communication with an operator's console or with a host computer. In Figure 1.2 we see how a small system might communicate with

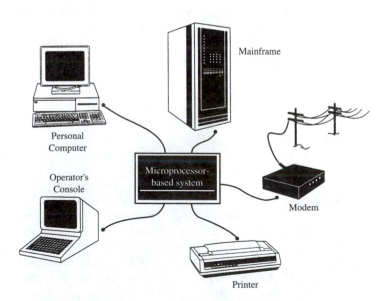

FIGURE 1.2   Serial communication possibilities in a small system

other devices or systems via serial communication. Although this type of communication is slow, it has the advantage of simplicity: only two wires (for receive and transmit) plus a ground are needed. Serial communication is easily adapted for use in fiber-optic cables. Parallel I/O, on the other hand, requires more lines (at least eight), but has the advantage of being very fast. A special parallel operation called **direct memory access (DMA)** is used to transfer data from a hard disk to a microcomputer's memory. Other uses for parallel I/O involve reading switch information, controlling indicator lights, and transferring data to A/D and D/A converters and other types of parallel devices.

All of these sections have their uses in a microprocessor-based system. Whether they are actually used depends on the designer and the application.

1.4   MICROPROCESSOR OPERATION

No matter how complex microprocessors become, they will still follow the same pattern of operations during program execution: endless fetch, decode, and execute cycles. During the fetch cycle, the processor loads an instruction from memory into its internal instruction register. Some advanced microprocessors load more than one instruction into a special buffer to decrease program execution time. The idea is that while the microprocessor is decoding the current instruction, other instructions can be read from memory into the instruction **cache,** a special type of internal high-speed memory. In this fashion, the microprocessor performs two jobs at once, thus saving time.

During the decode cycle, the microprocessor determines what type of instruction has been fetched. Information from this cycle is then passed to the execute cycle. To complete the instruction, the execute cycle may need to read more data from memory or write results to memory.

While these cycles are proceeding, the microprocessor is also paying attention to other details. If an interrupt signal arrives during execution of an instruction, the processor usually latches onto the request, holding off on interrupt processing until the current instruction finishes execution. The processor also monitors other signals such as WAIT, HOLD, or READY inputs. These are usually included in the architecture of the microprocessor so that slow devices, such as memories, can communicate with the faster processor without loss of data.

Most microprocessors also include a set of control signals that allows external circuitry to take over the system bus. In a system where multiple processors share the same memory and devices, these types of control signals are necessary to resolve **bus contention** (two or more processors needing the system bus at the same time). Multiple-processor systems are becoming more popular now as we continue to strive toward faster execution of our programs. **Parallel processing** is a term often used to describe multiple-processor systems and their associated software.

Special devices called **microcontrollers** are often used in simple control systems because of their many features. Microcontrollers are actually souped-up microprocessors with built-in features such as RAM, ROM, interval timers, parallel I/O ports, and even A/D converters. Microcontrollers are not used for large systems, however, because of their small instruction sets. Unfortunately, we have yet to get everything we want on a single chip!

1.5   HARDWARE/SOFTWARE REQUIREMENTS

We saw earlier that it is necessary to have at least some ROM in a system to take care of peripheral initialization. What type of initialization is required by the peripherals? The serial device must have its baud rate, parity, and number of data and stop bits programmed. Parallel devices must be configured because most of them allow the direction (input or output) of their I/O lines to be programmed in many different ways. It is then necessary to set the direction of these I/O lines when power is first applied. For a system containing a D/A converter it may be important to output an initial value required by the external hardware. Because we can never assume that correct conditions exist at power-on, the microprocessor is responsible for establishing them.

Suppose a certain system contains a video display controller. Start-up software must select the proper screen format and initialize the video memory so that an intelligent picture (possibly a menu) is generated on the screen of the display. If the system uses light-emitting diode (LED) displays or alphanumeric displays for output, they must be properly set as well. High-reliability systems may require that memory be tested at power-on. While this adds to the complexity of the start-up software and the time required for initialization, it is a good practice to follow. Bad memory devices will certainly cause a great deal of trouble if they are not identified.

Other systems may employ a special circuit called a **watchdog monitor.** The circuit operates like this: during normal program execution, the watchdog monitor is disabled. Should the program veer from its proper course, the monitor will automatically reset the system. A simple way to make a watchdog monitor is to use a binary counter, clocked by a known frequency. If the counter is allowed to increment up to a certain value, the processor is automatically reset. The software's job, if it is working correctly, is to make sure that the counter never reaches this count. A few simple logic gates can be used to clear the counter under microprocessor control, possibly whenever the CPU examines a certain memory location.

For flexibility, the system may have been designed to download its main program from a host system. If this is the case, the system software will be responsible for knowing how to communicate with the host and place the new program into the proper memory locations. To guarantee that the correct program is loaded, the software should also perform a running test on the incoming data, requesting the host to retransmit portions of the data whenever it detects an error.

Sometimes preparing for a power-down is as important as doing the start-up initialization. A power supply will quite often supply voltage in the correct operating range for a few milliseconds after the loss of AC power. During these few milliseconds the processor must execute the shutdown code, saving important system data in nonvolatile RAM or doing whatever is necessary for a proper shutdown. If the system data can be preserved, it may be possible to continue normal execution when power is restored.

For systems that will be expanded in the future, the system bus must be made available to the outside world. To protect the internal system hardware, all signals must be properly buffered. This involves using tri-state buffers or similar devices to isolate the internal system bus from the bus available to the external devices. Sometimes optoisolators are used to completely separate the internal system signals from the external ones. The only connection in optoisolators is a beam of light, which makes them ideal when electrical isolation is required.

Figure 1.3 sums up all of these concepts with an expanded block diagram of a microprocessor-based control system. Notice once again that all devices in the system communicate with the CPU via the system bus.

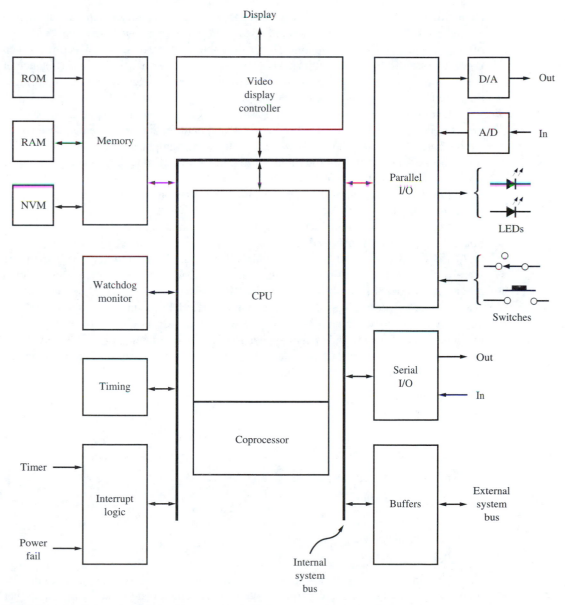

FIGURE 1.3   Expanded block diagram of a microprocessor-based system

1.6   THE PERSONAL COMPUTER

All of the material in this chapter, up to this point, has dealt with general microprocessor-based systems. In this section, we will see how a specific microprocessor-based system, the PC, uses many of the hardware features already described. Although the PC has been around for many years and has evolved into a powerful machine containing very advanced technology, such as CD-ROM drives, it began as a much simpler machine constructed around the 8/16-bit Intel 8088 microprocessor. The 8088 came out in the late 1970s and offered a higher level of computing power than the 8-bit processors of the time. When IBM® chose the 8088 for use in its new PC, it paved the way for worldwide acceptance of the new processor. Many companies began copying the architecture of the PC and offered their own compatible 8088-based computer systems. Thus began the PC market.

One reason the PC market grew as fast as it did was due to the usefulness of the features the PC offered. The initial PC contained a keyboard for entering commands and data, a monochrome video display for viewing text and simple graphics, one or two floppy disk drives for storing information and running programs, and a memory large enough for many useful applications. It also came equipped with a software program called **DOS,** for Disk Operating System, which made it possible to access files on the disk drives and run programs with the use of simple commands.

Most of the electronics within the PC were contained on a single printed circuit board called the **motherboard.** Memory chips, timing circuitry, interrupt logic, the 8088 microprocessor, and other hardware all resided on the motherboard. Included were a number of **expansion slots,** plastic connectors with metal fingers into which other circuit boards could be plugged. The PC's system bus was wired to each expansion slot, so any card plugged into an expansion slot had the power of the entire machine available to it. Expansion cards were used to add new features to the basic machine, such as a color video display, a hard disk, or additional memory. Today, there are hundreds of different expansion cards available. A small sample of them shows the wide variety of hardware applications:

Modem/Fax

LAN controller

Data acquisition

Sound/speech synthesis

High-resolution color graphics

Image processing

CD-ROM drive

Hand-held and flatbed scanner

Serial/parallel I/O

Clearly, with the right number and type of expansion cards, the PC can be configured to do just about anything. For our purposes, we will concentrate on the hardware that comes with a base machine, with a few add-ons, namely the hard disk and color-display cards.

Let us now take a detailed look at the inside of the personal computer. Figure 1.4 is the block diagram for a typical PC motherboard. As shown, all communication is through the system bus. The microprocessor may be an 8088 (as found on the original PC), or one of the newer 32-bit processors from Intel, the 80386™, 80486™, or Pentium. A nice feature of the advanced Intel microprocessors is that they all execute programs written for the 8088. So, even if your machine is new, all of the software presented in this textbook will run on it.

If a motherboard contains an 8088 microprocessor, there is usually a socket provided for an additional chip, the 8087 Floating-Point Coprocessor. This device is capable of performing mathematical calculations much faster than the 8088, and is designed to work parallel with the processor. Motherboards based on the 80486 and Pentium do not contain this socket because the coprocessor is built into the processor itself.

For high-speed data transfers involving memory, the motherboard contains an 8237™ DMA Controller. This device can be easily programmed to move large chunks of data without assistance from the processor.

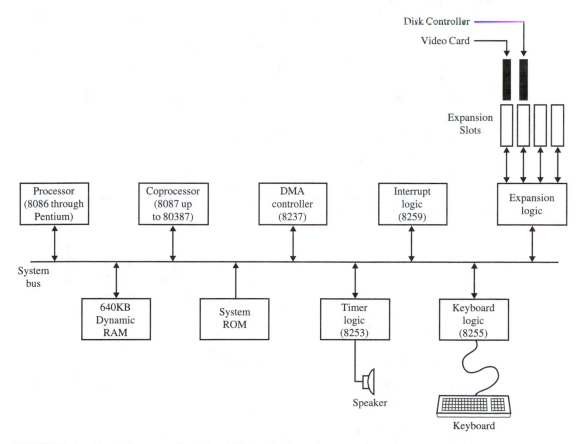

FIGURE 1.4   Block diagram of a typical PC motherboard

The PC has many features that require the use of the interrupt system. An 8259™ Programmable Interrupt Controller is included to handle the interrupts generated by the PC's time-of-day clock, keyboard, serial and parallel I/O devices, and disk drives.

Today, it is rare to find a PC that does not contain at least 640KB of RAM. All application programs written for the many versions of DOS use RAM during their execution. The memory is contained in a handful of dynamic RAM chips. Special timing circuitry is used to ensure that each memory chip gets refreshed and accessed properly.

The motherboard contains a small amount of ROM as well. This ROM is referred to as **system ROM** and is used to control the PC when it is turned on. The system ROM is responsible for checking and initializing all peripherals and devices on the motherboard, and for starting up the disk drive to load DOS.

As mentioned before, the PC maintains a time-of-day clock. This clock is a combination of software and hardware. A special timing device, the 8253™ Programmable Interval Timer, is used to generate timing pulses at regular intervals. These pulses interact with the interrupt logic and DOS to simulate the passage of time. The 8253 also controls the PC's speaker. With proper programming, it is possible to make the speaker beep and generate other sounds.

A parallel I/O device, the 8255 Programmable Peripheral Interface, is used to monitor and read the PC's keyboard and motherboard option switches.

Finally, expansion logic is used to drive the system bus signals on the expansion slots. This makes it possible for circuitry on an expansion card to access every device on the motherboard.

We do not have to know a great deal about the hardware just described to write programs for the personal computer. We will, however, examine the methods used to access the motherboard devices and thus directly control such things as the keyboard, speaker, and disk drives.

1.7   DEVELOPING SOFTWARE FOR THE PERSONAL COMPUTER

To get the most use out of the power of the PC, it is necessary to understand how to control and use the hardware on the motherboard and the software capabilities of the processor. We will explore the software architecture of the 80x86 family in great detail in the following chapters, first by examining the instruction set and then by looking at programming examples. What we will see is that the 80x86 processors speak a different language than we do.

Machine Language vs. Assembly Language

Our language is one of words and phrases. The 80x86 language is a string of 1s and 0s. For example, the instruction

```
ADD AX,BX
```

contains a word, **ADD,** that means something to us. Apparently, we are adding **AX** and **BX** together, whatever they are. So, even though we might be unfamiliar with the 80x86 instruction set, the instruction ADD AX,BX means something to us.

If we were instead given the binary string

```
0000 0001 1101 1000
```

or the hexadecimal equivalent

```
01 D8
```

and asked its meaning, we might be hard-pressed to come up with anything. We associate more meaning with ADD AX,BX than we do with 01 D8, which *is* the way the instruction is actually represented. All programs for the 80x86 will simply be long strings of binary numbers.

Because of the processor's internal decoders, different binary patterns represent different instructions. Here are a few examples to illustrate this point:

```
01 D8    ADD    AX,BX    ;add BX to AX, result in AX
29 D8    SUB    AX,BX    ;subtract BX from AX, result in AX
21 D8    AND    AX,BX    ;AX equals AX AND BX
40       INC    AX       ;add 1 to AX
4B       DEC    BX       ;subtract 1 from BX
8B C3    MOV    AX,BX    ;copy BX into AX
```

Can you guess the meaning of each instruction just by reading it? Do the hexadecimal codes for each instruction mean anything to you? What we see here is the difference between **machine language** and **assembly language.** The machine language for each instruction is represented by the hexadecimal codes. This is the binary language of the machine. The assembly language is represented by the wordlike terms that mean something to us. Putting groups of these wordlike instructions together is how a program is constructed. In the next few sections we will see how an assembly language program is written, converted into machine language, and executed.

NUMOFF: Our First Machine Language Program

When the personal computer is first turned on, instructions in the start-up software turn the NUM-LOCK indicator on. This indicator is located near the NUM-LOCK button on the keyboard. Pushing NUM-LOCK manually every time the PC is turned on (or even rebooted) is annoying. Luckily, there is a single bit stored in a specific memory location used by DOS that controls the state of the NUM-LOCK indicator. We are about to see that it is possible to write an 80x86 program to manipulate the NUM-LOCK status bit.

Creating an Executable Program

Using a PC-based word processor, enter the following text file exactly as you see it. Save the file under the name of NUMOFF.ASM.

```
;NUMOFF.ASM: Turn NUM-LOCK indicator off.
;
        .MODEL SMALL
        .STACK
        .CODE
        .STARTUP
        MOV     AX,40H      ;set AX to 0040H
        MOV     DS,AX       ;load data segment with 0040H
```

```
        MOV     SI,17H      ;load SI with 0017H
        AND     BYTE PTR [SI],0DFH   ;clear NUM-LOCK bit
        .EXIT
        END
```

The twelve lines of code constitute a **source file,** the starting point of any 80x86-based program. Thus, NUMOFF.ASM is a source file.

To convert NUMOFF.ASM into a group of hexadecimal bytes that represents the corresponding machine language, we make use of two additional programs: ML and LINK. ML is a **macro assembler,** a program that takes a source file as input and determines the machine language for each source statement. ML creates two additional files. These are the *list* and *object* files. The list file contains all of the text from the source file, plus additional information, as we will soon see. The object file contains only the machine language.

To assemble NUMOFF.ASM, enter the following command at the DOS prompt:

```
ML /c /Fl NUMOFF.ASM <cr>
```

where **<cr>** indicates a carriage return. This instructs ML to assemble NUMOFF.ASM and create NUMOFF.LST (the list file) and NUMOFF.OBJ (the object file). The list file created by ML looks like this:

```
                    ;NUMOFF.ASM: Turn NUM-LOCK indicator off.
                    ;
                            .MODEL SMALL
                            .STACK
0000                        .CODE
                            .STARTUP
0017   B8 0040      MOV     AX,40H      ;set AX to 0040H
001A   8E D8        MOV     DS,AX       ;load data segment with 0040H
001C   BE 0017      MOV     SI,17H      ;load SI with 0017H
001F   80 24 DF     AND     BYTE PTR [SI],0DFH   ;clear NUM-LOCK bit
                            .EXIT
                            END
```

Here, it is obvious that ML has determined the machine language for each source statement. The first column is the set of memory locations where the instructions are stored. The second column is the group of machine language bytes that represent the actual 80x86 instructions.

To make the object file executable, we run the **LINK** program, which converts NUMOFF.OBJ into NUMOFF.EXE. The DOS command to do this is:

```
LINK NUMOFF;<cr>
```

Right now we have our first working 80x86 program, NUMOFF.EXE! To test it, press the NUM-LOCK button on the PC's keyboard until the NUM-LOCK light goes on. Then execute NUMOFF.EXE by entering:

```
NUMOFF.EXE<cr>
```

at the DOS prompt. The NUM-LOCK light should go off. This is what NUMOFF.EXE does. If you want to automatically turn the NUM-LOCK indicator off each time your PC is powered up (or rebooted), add one statement to your AUTOEXEC.BAT file:

```
NUMOFF.EXE
```

Many of the programs in later chapters will be assembled and linked in this fashion.

Generating Machine Code with DEBUG

An alternate technique for generating and executing machine code is through the use of the **DEBUG** program. Let us briefly look at how DEBUG can be used to perform the same task as the NUMOFF.EXE program. We'll take a detailed look at DEBUG later.

From the DOS prompt, start up DEBUG with:

```
DEBUG<cr>
```

You will get a minus sign (–) as a prompt, which helps to distinguish the DEBUG environment from the DOS environment.

Now enter all of the following text shown in bold. You will almost surely see different addresses on your machine than those that appear here, but this will not affect what we are trying to do.

```
-a<cr>
7F2D:0100 mov ax,40<cr>
7F2D:0103 mov ds,ax<cr>
7F2D:0105 mov si,17<cr>
7F2D:0108 and byte ptr [si],df<cr>
7F2D:010B <cr>
-
```

The <cr> on the last line gets you out of the 'a' mode. Since 'a' stands for *assemble,* DEBUG has assembled each statement entered by the user and placed the corresponding machine code into memory. This is indicated by the way addresses are incrementing in the address field (0100 to 0103 to 0105, etc.).

Now push the NUM-LOCK button on your keyboard so that the NUM-LOCK indicator is on. To execute the program statement by statement, enter a single 't' at DEBUG's prompt:

```
-t<cr>
```

This is the *trace* command, and it is used to single-step through an 80x86 program. Each time you enter 't' DEBUG will execute one instruction of your program and display the results.

If you hit 't' three more times you should see the NUM-LOCK indicator go off. This completes the program. To exit from DEBUG back to DOS, enter:

```
-q<cr>
```

This shows that DEBUG is also a useful way of executing machine language programs. We will write and execute a number of programs this way, and learn much more about how DEBUG works.

Programming Exercise 1.1: Can you think of a way to do the opposite of NUMOFF? That is, can you change NUMOFF so that it turns NUM-LOCK on?

Hint: Use an OR operation.

Programming Exercise 1.2:   Can you think of a way to toggle the NUM-LOCK indicator? This would cause NUM-LOCK to alternate between on and off.

Hint: Use an XOR operation.

1.8   TROUBLESHOOTING TECHNIQUES

You may think it premature to begin discussing troubleshooting techniques, when we have been exposed to so little of the 80x86 family architecture. Even so, we have already seen a number of places where errors can occur, and it would be worthwhile to discuss them. For example, the NUMOFF.ASM source file could have contained one or more *typographical* errors, such as a misspelled instruction (MVO versus MOV), or a missing comma, or a comma where a semicolon was expected. Generally, when errors such as these are present in a source file, the assembler will report them with a brief error message.

Even if the source file does not have any typographical errors, we could still run into trouble. We could enter the command to invoke ML or DEBUG incorrectly, or not use the correct options.

When the source file correctly assembles and links, and an executable program has been created, there is still the possibility of a *run-time* error in the program. Run-time errors are typically caused by incorrect sequences of instructions and incomplete or faulty logical thinking.

To avoid a loss of time and effort, it is good to keep these common stumbling blocks in mind. Paying attention to the details will really pay off, as you learn to create a working program with a minimum of time and effort.

SUMMARY

In this chapter we have examined the operation of microprocessor-based systems. We saw that the complexity of the hardware, and thus of the software, is a function of the type of application. Through the use of many different types of peripherals, such as parallel and serial devices, analog-to-digital converters, and others, a system can be tailored to perform almost any job. We also reviewed the basic fetch, decode, and execute cycle of a microprocessor, and examined the other duties the CPU performs, one of which was interrupt handling.

We also covered the initialization requirements of peripherals used in a microprocessor-based system, and why it is necessary to perform initialization in the first place. Other types of hardware and software requirements were also examined, such as the use of a watchdog monitor and a nonvolatile memory.

Four different generations of computers were presented and their differences highlighted. Current computing trends dealing with parallel processing and artificial intelligence were also introduced.

This was followed by an introduction to the motherboard hardware of a typical personal computer. Any *compatible* PC must use the same hardware. Because software is needed to control the hardware, we finished with a quick look at two techniques for creating and executing machine language programs. The first technique used ML and LINK, and the second technique used DEBUG. Both methods will be covered in detail in following chapters.

STUDY QUESTIONS

1. Make a list of ten additional products containing microprocessors that we use every day.
2. a) The *cycle time* of a microprocessor is the time for one complete clock cycle. For example, if the clock frequency of a microprocessor is 2 million cycles per second (2 MHz), then each cycle takes 500 ns (500 billionths of a second). Compare the cycle time of a microprocessor running at 2 MHz with one running at 50 MHz.
   b) If a typical 80x86 instruction requires four clock cycles to execute, how long does the instruction take to execute if the processor clock speed is 25 MHz?
3. Speculate on the uses for timing signals in the serial I/O, memory, and interrupt sections.
4. Why do coprocessors enhance the capabilities of an ordinary CPU?
5. Draw a block diagram for a computerized cash register. The hardware should include a numerical display, a keyboard, and a compact printer.
6. What kind of initialization software would be required for the cash register of Question 5?
7. What would be the difference in system RAM requirements for two different cash registers, one without record keeping and one with?
8. What type of information should be stored in NVM during a power failure in a system designed to control navigation in an aircraft?
9. What types of interrupts may be required in a control system designed to monitor all doors, windows, and elevators in an office complex?
10. Name some advantages of downloading the main program into a microprocessor-based system. Are there any disadvantages?
11. Suppose that a number of robots making up a portion of an automobile assembly line are connected to a master factory computer. What kinds of information might be passed between the factory computer and the microprocessors controlling each robot?
12. A certain hard disk transfers data at the rate of 8 million bits per second. Explain why the CPU would not be able to perform the transfer itself, thus requiring the use of a DMA controller.
13. How does a plug-in card use motherboard hardware?
14. What additional kinds of plug-in cards can you think of?
15. Suppose that three microprocessors are used in the design of a new video game containing color graphics and complex sounds. How might each microprocessor function?

16. Why did processing speed increase with each new generation of computers?
17. List five different applications that might need the fast computing power of a RISC-based machine.
18. One reason 32-bit processors are faster than 8-bit machines is because they operate on four times as many data bits at the same time. Why doesn't everyone using an 8-bit machine just switch over to a 32-bit processor?
19. An upward-compatible microprocessor is one that can execute instructions from earlier models. How would a designer of the new CPU implement upward compatibility?
20. Which of the statements in the NUMOFF.ASM source file actually generated code?
21. It takes a certain amount of time to execute each instruction in NUMOFF. Which instruction do you think takes the longest?
22. What are all the NUMOFF files created in Section 1.7?
23. How are DEBUG statements different from statements in NUMOFF.ASM?
24. What advantages does a microcontroller have over a microprocessor? What disadvantages does it have?
25. Why do typographical errors prevent a source file from being assembled?

CHAPTER 2

An Introduction to the 80x86 Microprocessor Family

OBJECTIVES

In this chapter you will learn about:

- Real-mode and protected-mode operation
- The register set of the 80x86 family
- The addressing capabilities and data types that may be used
- The different addressing modes and instruction types available
- The usefulness of interrupts
- Some of the differences between the 8086 and the 80286, 80386, 80486, and Pentium microprocessors

2.1   INTRODUCTION

Since its arrival on the computer scene, the 80x86 microprocessor family has quickly taken the lead in the personal computer market. 80x86 processors are used everywhere, from dedicated control systems to superfast network file servers. Today, the Pentium offers compatibility with all previous 80x86 machines, with many new architectural improvements.

In this chapter we will examine the features of the 80x86 microprocessor family. Only basic material will be covered, leaving the hardware and software details for upcoming chapters. From reading this chapter you should become aware that the 80x86 is a machine with many possibilities.

Section 2.2 discusses the differences between real-mode and protected-mode operation. Section 2.3 covers the software model of the 80x86 family. Section 2.4 introduces the numerous registers contained within the processor, followed by a brief discussion of data organization, instruction types, and addressing modes in Sections 2.5, 2.6, and 2.7, respectively. Interrupts are the subject of Section 2.8. Sections 2.9 through 2.13 deal with the upward-compatible architectures found in the 8086, 80286, 80386, 80486, and Pentium microprocessors, respectively. Troubleshooting techniques are presented in Section 2.14.

2.2   REAL-MODE AND PROTECTED-MODE OPERATION

The first processor in the 80x86 family was the 16-bit 8086, which was capable of addressing 1MB of memory, a significant improvement over the 8-bit machines available in the late 1970s. Twenty address lines were provided on the processor to access the 1MB of memory (2^{20} equals 1,048,576, or 1 MB). The advanced processors that followed the 8086, beginning with the 80286, all contained additional address lines. The 80386, 80486, and Pentium all contain 32 address lines, giving them the ability to access 2^{32}, or 4096MB, of memory. This large addressing space allows the advanced Intel processors to perform many operating system chores—such as multitasking—that are difficult, or even impossible, on the 8086.

Beginning with the 80286, the advanced Intel processors all contained the ability to operate in two different modes of operation, *real mode* and *protected mode*. In real mode, the advanced processors, including the Pentium, simply operate like very fast 8086s, with the associated 1MB memory limit. Real mode operation is automatically selected upon power-up. So a Pentium-based PC that boots up into DOS is operating in real mode (DOS is a real-mode operating system).

In protected mode, the full 4096MB (4 *gigabytes*) of memory is available to the processor, as are special privileged instructions and many other architectural goodies, including support for multitasking, virtual memory addressing, memory management and protection, and control over the internal data and instruction cache. The Windows operating system runs in protected mode to take advantage of these improvements. Writing programs that run in protected mode requires special background knowledge of operating systems theory. For this reason, the majority of the examples presented in this book are developed for real-mode operation. We will, however, in the final chapter, explore the details of protected mode.

2.3   THE SOFTWARE MODEL OF THE 80X86 FAMILY

The 80x86 family of microprocessors contains four data registers referred to as AX, BX, CX, and DX. All are 16 bits wide and may be split up into two halves of 8 bits each. Figure 2.1 shows how each half is referred to by the programmer. Five other 16-bit registers are available for use as pointer or index registers. These registers are the *stack pointer* (SP), *base pointer* (BP), *source index* (SI), *destination index* (DI), and *instruction pointer* (IP). None of the five may be divided up in a manner similar to the data registers. AX, BX, CX, DX, BP, SI, and DI are referred to as *general purpose* registers.

Beginning with the 80386, the register sizes were extended to 32 bits. These new *extended* registers are referred to as EAX, EBX, ECX, and EDX. All are 32 bits wide. The

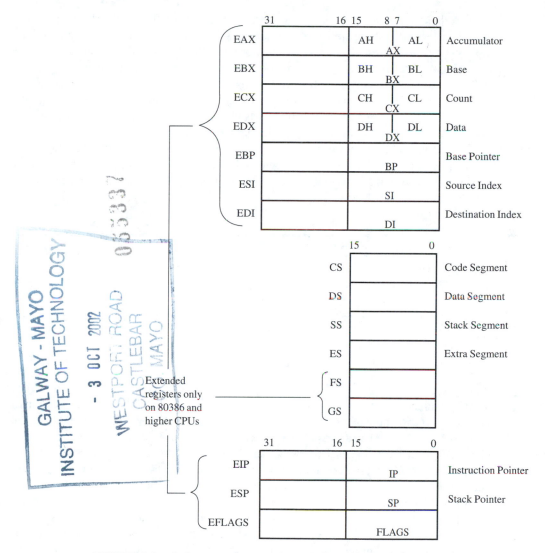

FIGURE 2.1   Software model of the 80x86 microprocessor family

lower 16 bits of each register are the original AX, BX, CX, and DX registers, and may still be split up into halves of 8 bits each. Pointer and index registers are also extended. Two additional segment registers (FS and GS) were also added.

Normally the 32-bit registers are not fully available when the processor is operating in real mode. To maintain compatibility with earlier 80x86 machines, only the 16-bit registers AX, BX, CX, DX, BP, SI, and DI are available for use. A special technique can be used to utilize a 32-bit register (such as EAX) on an instruction-by-instruction basis. This will be demonstrated in Section 2.5.

A major difference between the 80x86 and many other CPUs on the market has to do with the next group of registers, the *segment registers*. Four segment registers are used by the processor to control all accesses to memory and I/O, and must be maintained by the programmer. The *code segment* (CS) is used during instruction fetches, the *data segment* (DS) is most often used by default when reading or writing data, the *stack segment* (SS) is used during stack operations such as subroutine calls and returns, and the *extra segment* (ES) is used for anything the programmer wishes. All segment registers are 16 bits long. Section 2.4 explains in more detail how the segment registers are used.

Finally, a 32-bit *flag register* is used to indicate the results of arithmetic and logical instructions. Included are zero, parity, sign, and carry flags, as well as flags that are only used in protected mode. Together, these sixteen registers make an impressive set!

2.4 PROCESSOR REGISTERS

The 16-bit real-mode registers introduced in Section 2.3 are combined in an interesting way to form the necessary 20-bit address required by memory. If you recall, there were no 20-bit registers shown in the software model. How then does the processor generate 20-bit addresses in real mode?

Segment Registers

The six segment registers, CS, DS, SS, ES, FS, and GS are all 16-bit registers controlled by the programmer. A real-mode segment, as defined by Intel, is a 64KB block of memory starting on any 16-byte boundary. Thus, 00000, 00010, 00020, 20000, 8CE90, and E0840 are all examples of valid segment addresses. The information contained in a segment register is combined with the address contained in another 16-bit register to form the required 20-bit address. Figure 2.2 shows how this is accomplished. In this example, the code segment register contains A000 and the instruction pointer contains 5F00. The processor forms the 20-bit address A5F00 in the following way: first, the data in the code segment register is shifted 4 bits to the left. This has the effect of turning A000 into A0000. Then the contents of the instruction pointer are added, giving A5F00. All external addresses are formed in a similar manner, with one of the four segment registers used in each case. As we will see in Chapter 3, each segment

FIGURE 2.2 Generating a
20-bit address in real mode

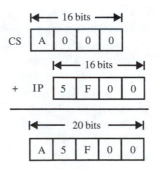

register has a default usage. The processor knows which segment register to use to form an address for a particular application (instruction fetch, stack operation, and so on). The processor also allows the programmer to specify a different segment register when generating some addresses.

In protected mode, the segment registers are used as selectors that point to predefined *segment descriptors,* which are described in the last chapter.

General Purpose Registers

The seven general purpose registers (AX, BX, CX, DX, BP, SI, and DI) available to the programmer in real mode can be used in many different ways, and they also have some specific roles assigned to them. For instance, the accumulator (AX) is used in multiply and divide operations and also in instructions that access I/O ports. The count register (CX) is used as a counter in loop operations, providing up to 65,536 passes through a loop before termination. The lower half of CX, the 8-bit CL register, is also used as a counter in shift/rotate operations. Data register DX is used in multiply and divide operations and also as a pointer when accessing I/O ports. The last two registers are the source index and destination index (referred to as SI and DI, respectively). These registers are used as pointers in string operations.

Even though these registers have specific uses, they may be used in many other ways simply as general purpose registers, allowing for many different 16-bit operations.

Recall that the general purpose registers are actually 32 bits wide. However, when the processor operates in real mode, it defaults to the original 16-bit 8086 register sizes. It is possible to take advantage of the 32-bit extended registers while running in real mode. A special 1-byte code called the *operand size prefix* is inserted before each instruction that uses a 16-bit register. For these instructions, the processor will use the full 32-bit register length. This 1-byte code has the value 66 hexadecimal. Let us examine an actual problem to solve using the 32-bit registers.

Example 2.1: The distance from the Earth to the Sun is approximately 93 million miles. The speed of light is roughly 186,000 miles per second. How can we determine the time required for a ray of light to reach the Earth?

Solution: First, convert the input numbers into hexadecimal:

```
93,000,000      =      058B1140
186,000         =      0002D690
```

Both numbers require more than 16 bits of storage and will not fit into any of the 16-bit general purpose registers. If the operand size prefix code is used, both numbers are easily stored in the extended registers. We can use the DEBUG program to solve the problem. Here are the contents of a short text file called SPEED.SCR:

```
a
db 66
sub dx,dx
db 66
mov ax,1140
dw 058b
```

```
db 66
mov bx,d690
dw 0002
db 66
div bx

r
t 4
q
```

The first line contains DEBUG's **a** (assemble) command. This command instructs DEBUG to begin assembling real-mode instructions. The second line contains the **db** (define byte) command, and is used to supply the single-byte operand size prefix. Note the use of db 66 before each of the four instructions. The three lines

```
db 66
mov ax,1140
dw 058b
```

will execute as if they were written like this:

```
mov eax,058b1140
```

The **dw** (define word) command is used to supply the upper 16 bits of the extended register value. This is how the distance value is set up in register EAX, with the processor running in real mode. A similar method is used to initialize the speed of light in register EBX.

The blank line after the DIV instruction is used to terminate the assemble command. The **r** (register) command on the next line causes DEBUG to display the initial register values. Then comes the **t** (trace) command, which traces the execution of the four instructions entered with the assemble command. The last line contains the **q** (quit) command, to quit DEBUG and return to DOS.

All of these commands can be entered manually from the keyboard after starting DEBUG. Since they are all contained in the SPEED.SCR file, redirecting the DOS input device with the command:

```
C> DEBUG < SPEED.SCR
```

will cause DEBUG to read the commands from the SPEED.SCR file instead.

Either way, the final register display from DEBUG looks like this:

```
AX=01F4  BX=D690  CX=0000  DX=1140  SP=FFEE  BP=0000  SI=0000  DI=0000
DS=229B  ES=229B  SS=229B  CS=229B  IP=0112  NV UP EI NG NZ NA PO NC
229B:0112 C6F70A          MOV    BH,0A
```

When the DIV instruction executes, the value 058B1140 is divided by 2D690, giving a final result in AX of 01F4 (which indicates a time of 500 seconds, or 8 minutes and 20 seconds).

Try running DEBUG with the db 66 statements removed. The results are drastically different.

Flag Register

Figure 2.3 shows the eleven flag assignments within the lower 16 bits of the flag register. The flags are divided into two groups: **control** flags and **status** flags. The control

15														0
-	NT	IOPL	OF	DF	IF	TF	SF	ZF	-	AF	-	PF	-	CF

FIGURE 2.3 Lower word of flag register

flags are IF (*interrupt enable flag*), DF (*direction flag*), and TF (*trap flag*). The status flags are CF (*carry flag*), PF (*parity flag*), AF (*auxiliary carry flag*), ZF (*zero flag*), SF (*sign flag*), OF (*overflow flag*), NT (*nested task*), and IOPL (*input/output privilege level*). Most of the instructions that require the use of the ALU affect the flags. Remember that the flags allow ALU instructions to be followed by conditional instructions.

The content/operation of each flag is as follows:

CF: Contains carry out of MSB of result

PF: Indicates if result has even parity

AF: Contains carry out of bit 3 in AL

ZF: Indicates if result equals zero

SF: Indicates if result is negative

OF: Indicates that an overflow occurred in result

IF: Enables/Disables interrupts

DF: Controls pointer updates during string operations

TF: Provides single-step capability for debugging

IOPL: Priority level of current task

NT: Indicates if current task is nested

The upper 16 bits of the flag register are used for protected-mode operation. See the final chapter for details.

2.5   DATA ORGANIZATION

The 80x86 has the capability of performing operations on many different types of data. In this section, we will examine what some of the more common data types are and how they are represented and used by the processor.

Bits, Bytes, and Words

The processor contains instructions that directly manipulate single bits and other instructions that use 8-, 16-, and even 32-bit numbers. By common practice, 8-bit binary numbers are referred to as **bytes.** Processor register halves AL, BH, and CL are examples of where bytes might be stored and used.

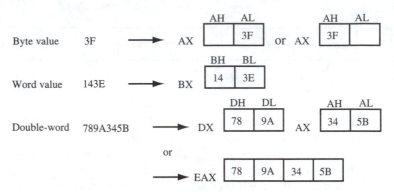

FIGURE 2.4   Storing different data types in registers

Sixteen-bit numbers are known as **words** and require an entire processor register for storage. Registers DX, BP, and SP are used to hold word data types. In register DX, DH will contain the upper 8 bits of the number, and DL will hold the lower 8 bits.

Some instructions (particularly multiply and divide) allow the use of 32-bit numbers. These data types are called **double-words** (or long-words). In this case, the 32-bit number is stored in registers DX and AX, with DX holding the upper 16 bits of the number. Even though an extended register, such as EAX, may be used to store a 32-bit number, it is difficult to work with extended registers in real mode. These data types are illustrated in Figure 2.4.

It is important to keep track of the data type being used in an instruction, because incorrect or undefined types may lead to incorrect program assembly or execution.

One of the differences between the Intel line of microprocessors and those made by other manufacturers is Intel's way of storing 16-bit numbers in memory. A method that began with the 8-bit 8080 and has been used on all upgrades from the 8085 to the Pentium is a technique called **byte-swapping.** This technique is sometimes confusing for those unfamiliar with it, but becomes clear after a little exposure. When a 16-bit number must be written into the system's byte-wide memory, the low-order 8 bits are written into the first memory location and the high-order 8 bits are written into the second location. Figure 2.5 shows how the 2 bytes that make up the 16-bit hexadecimal number 2055 are written into locations 18000 and 18001, with the low-order 8 bits (55) going into the first location (18000). This is what is known as byte-swapping. The lower byte is always written first, followed by the high byte. Byte-swapping is one of the most significant differences between Intel processors and other machines, such as Motorola's 68000™ (which does not swap bytes).

Reading the 16-bit number out of memory is performed automatically by the processor with the aid of certain instructions. The processor knows that it is reading the lower byte first and puts it in the correct place. Programmers who manipulate data in memory must remember to use the proper practice of byte-swapping or discover that their programs do not give the correct, or expected, results. It should be easy to remember the mechanics of byte-swapping because the lower byte is always read/written to the lower memory address.

FIGURE 2.5 Intel byte-swapping

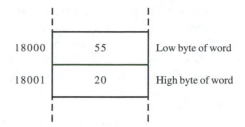

| 18000 | 55 | Low byte of word |
| 18001 | 20 | High byte of word |

Assembler Directives DB, DW, DUP, and EQU

Representing data types in a source file requires the use of special *assembler directives* designed to create the required data and perform all appropriate byte-swapping. Consider the following sample portion of a list file:

```
0000                           .DATA

0000 03                        NUM1    DB      3
0001 64                        NUM2    DB      100
0002 00                        NUM3    DD      ?
0003 0F 30 C8 3A CE            NUMS    DB      15,48,200,3AH,0CEH
0008 48 65 6C 6C 6F 24         MSG     DB      'Hello$'

000E 0006                      WX      DW      6
0010 03E8                      WY      DW      1000
0012 1234 ABCD                 WZ      DW      1234H,0ABCDH
0016 0000                      TEMP    DW      ?

0018  000A [00]                SCORES  DB      10 DUP(0)
0022  0007 [0000]              TIMES   DW      7 DUP(?)

= 000D                         TOP     EQU     13
= 157C                         MORE    EQU     5500
```

In this example, byte and word data types are defined through the use of the **DB (define byte)** and **DW (define word)** assembler directives. As you can see, DB and DW both allow one or more numbers in their data field or no numbers at all, as is the case with the ? in the data field. DB and DW will convert any numbers in their data fields into hexadecimal and store the numbers in the appropriate place within the object file. If an ASCII character string is found in the data field (as in 'Hello$') the ASCII byte associated with each character is generated.

Because each DB or DW statement in the source file may generate different lengths of data, the assembler keeps track of a *program counter* to indicate the starting address of each data group. For example, the bytes for the fourth DB statement begin at address 0003 within the data segment, and the word for the second DW statement is located at address 0010 within the data segment. In many cases, we supply a *label* with a particular DB or DW statement to use with instructions found elsewhere in the source file. The labels are assigned the value of the starting address in each DB or DW statement. Examine the sample data again. Do you see why the address associated with MSG is 0008?

DB and DW statements can be used to simply reserve byte or word space by using ? in the data field. When many reserved bytes or words are needed, the **DUP (duplicate)** assembler directive is used. Although it is legal to use DB   ?,?,?,?,? in a source file, a simple DB   5 DUP(?) should be used instead. This saves the programmer the effort of having to count the number of question marks entered. DUP has an additional advantage. Suppose that 2000 reserved bytes are needed. The statement DB   2000 DUP(?) is clearly more desirable than two hundred DB   ?,?,?,?,?,?,?,?,?,? statements. In our example, DUP is used to create both byte (SCORES) and word (TIMES) data tables.

One last assembler directive is very useful because it defines a value that can be used in other source statements but does *not* generate any code. This is the **EQU (equate)** directive. Notice how EQU is used to assign the value 13 to TOP and the value 5500 to MORE. Any instruction in the source program that uses TOP and MORE will automatically use the values 13 and 5500, respectively. From a practical viewpoint, suppose you have written a 5000-line source program that contains the instruction

```
MOV  AL,77
```

in many different places. If, for some reason, you had to change all the 77s to 88s, you might be in for a long editing session. A simple solution would be to define the value 77 with an EQU statement, as in

```
VAL   EQU   77
```

then use the EQU label in each instruction, like so

```
MOV  AL,VAL
```

If it is necessary to change 77 to 88, only the EQU statement has to be edited.

This brief introduction to data types should help us when we examine the 80x86 instruction set.

2.6   INSTRUCTION TYPES

The 80x86 instruction set is composed of six main groups of instructions. A discussion of instruction specifics will be postponed until Chapters 3 and 4. Examining the instructions briefly here, however, will give a good overall picture of the capabilities of the processor. All instructions are part of the original 8086 instruction set, unless otherwise indicated.

Data Transfer Instructions

Data transfer instructions are used to move data among registers, memory, and the outside world. Also, some instructions directly manipulate the stack, while others may be used to alter the flags.

The data transfer instructions are:

IN	Input byte or word from port
LAHF	Load AH from flags
LDS	Load pointer using data segment

LEA	Load effective address
LES	Load pointer using extra segment
MOV	Move to/from register/memory
OUT	Output byte or word to port
POP	Pop word off stack
POPF	Pop flags off stack
PUSH	Push word onto stack
PUSHF	Push flags onto stack
SAHF	Store AH into flags
XCHG	Exchange byte or word
XLAT	Translate byte

Additional 80286 instructions are:

INS	Input string from port
OUTS	Output string to port
POPA	Pop all registers
PUSHA	Push all registers

Additional 80386 instructions are:

LFS	Load pointer using FS
LGS	Load pointer using GS
LSS	Load pointer using SS
MOVSX	Move with sign extended
MOVZX	Move with zero extended
POPAD	Pop all double (32-bit) registers
POPD	Pop double register
POPFD	Pop double flag register
PUSHAD	Push all double registers
PUSHD	Push double register
PUSHFD	Push double flag register

Additional 80486 instruction is:

BSWAP	Byte swap

New Pentium instruction is:

MOV	Move to/from control register

Arithmetic Instructions

These instructions make up the arithmetic group. Byte and word operations are available on almost all instructions. A nice addition are the instructions that multiply and divide. Previous 8-bit microprocessors did not include these instructions, forcing the programmer to write subroutines to perform multiplication and division when needed. Addition and subtraction of both binary and BCD operands are also allowed.

The arithmetic instructions are:

AAA	ASCII adjust for addition
AAD	ASCII adjust for division
AAM	ASCII adjust for multiply
AAS	ASCII adjust for subtraction
ADC	Add byte or word plus carry
ADD	Add byte or word
CBW	Convert byte or word
CMP	Compare byte or word
CWD	Convert word to double-word
DAA	Decimal adjust for addition
DAS	Decimal adjust for subtraction
DEC	Decrement byte or word by one
DIV	Divide byte or word (unsigned)
IDIV	Integer divide byte or word
IMUL	Integer multiply byte or word
INC	Increment byte or word by one
MUL	Multiply byte or word (unsigned)
NEG	Negate byte or word
SBB	Subtract byte or word and carry
SUB	Subtract byte or word

Additional 80386 instructions are:

CDQ	Convert double-word to quadword
CWDE	Convert word to double-word

Additional 80486 instructions are:

CMPXCHG	Compare and exchange
XADD	Exchange and add

New Pentium instruction is:

CMPXCHG8B	Compare and exchange 8 bytes

Bit Manipulation Instructions

Instructions capable of performing logical, shift, and rotate operations are contained in this group. Many common Boolean operations (AND, OR, NOT) are available in the logical instructions. These, as well as the shift and rotate instructions, operate on bytes or words. Single-bit operations are available on processors from the 80386 and up.

The bit manipulation instructions are:

AND	Logical AND of byte or word
NOT	Logical NOT of byte or word
OR	Logical OR of byte or word
RCL	Rotate left through carry byte or word
RCR	Rotate right through carry byte or word
ROL	Rotate left byte or word
ROR	Rotate right byte or word
SAL	Arithmetic shift left byte or word
SAR	Arithmetic shift right byte or word
SHL	Logical shift left byte or word
SHR	Logical shift right byte or word
TEST	Test byte or word
XOR	Logical exclusive-OR of byte or word

Additional 80386 instructions are:

BSF	Bit scan forward
BSR	Bit scan reverse
BT	Bit test
BTC	Bit test and complement
BTR	Bit test and reset
BTS	Bit test and set
SETcc	Set byte on condition
SHLD	Shift left double precision
SHRD	Shift right double precision

String Instructions

String operations simplify programming whenever a program must interact with a user. User commands and responses are usually saved as ASCII strings of characters, which may be processed by the proper choice of string instruction.

The string instructions are:

CMPS	Compare byte or word string
LODS	Load byte or word string
MOVS	Move byte or word string
MOVSB (MOVSW)	Move byte string (word string)

REP	Repeat
REPE (REPZ)	Repeat while equal (zero)
REPNE (REPNZ)	Repeat while not equal (not zero)
SCAS	Scan byte or word string
STOS	Store byte or word string

Program Transfer Instructions

This group of instructions contains all jumps, loops, and subroutine (called **procedure**) and interrupt operations. The great majority of jumps are **conditional,** testing the processor flags before execution.

The program transfer instructions are:

CALL	Call procedure (subroutine)
INT	Interrupt
INTO	Interrupt if overflow
IRET	Return from interrupt
JA (JNBE)	Jump if above (not below or equal)
JAE (JNB)	Jump if above or equal (not below)
JB (JNAE)	Jump if below (not above or equal)
JBE (JNA)	Jump if below or equal (not above)
JC	Jump if carry set
JCXZ	Jump if CX equals zero
JE (JZ)	Jump if equal (zero)
JG (JNLE)	Jump if greater (not less or equal)
JGE (JNL)	Jump if greater or equal (not less)
JL (JNGE)	Jump if less (not greater or equal)
JLE (JNG)	Jump if less or equal (not greater)
JMP	Unconditional jump
JNC	Jump if no carry
JNE (JNZ)	Jump if not equal (not zero)
JNO	Jump if no overflow
JNP (JPO)	Jump if no parity (parity odd)
JNS	Jump if no sign
JO	Jump if overflow
JP (JPE)	Jump if parity (parity even)
JS	Jump if sign
LOOP	Loop unconditional
LOOPE (LOOPZ)	Loop if equal (zero)

| LOOPNE (LOOPNZ) | Loop if not equal (not zero) |
| RET | Return from procedure (subroutine) |

Additional 80286 instructions are:

BOUND	Check index against array bounds
ENTER	Enter a procedure
LEAVE	Leave a procedure

Additional 80386 instructions are:

| IRETD | Interrupt return |
| JECXZ | Jump if ECX is zero |

Processor Control Instructions

This last group of instructions performs small tasks that sometimes have profound effects on the operation of the processor. Many of the instructions manipulate the flags.

The processor control instructions are:

CLC	Clear carry flag
CLD	Clear direction flag
CLI	Clear interrupt enable flag
CMC	Complement carry flag
ESC	Escape to external processor
HLT	Halt processor
LOCK	Lock bus during next instruction
NOP	No operation
STC	Set carry flag
STD	Set direction flag
STI	Set interrupt enable flag
WAIT	Wait for $\overline{\text{TEST}}$ pin activity

Additional 80286 instructions (protected mode only) are:

ARPL	Adjust requested privilege level
CLTS	Clear task switched flag
LAR	Load access rights
LGDT	Load global descriptor table
LIDT	Load interrupt descriptor table
LLDT	Load local descriptor table
LMSW	Load machine status word
LSL	Load segment limit
LTR	Load task register

SGDT	Store global descriptor table
SIDT	Store interrupt descriptor table
SLDT	Store local descriptor table
SMSW	Store machine status word
STR	Store task register
VERR	Verify segment for reading
VERW	Verify segment for writing

Additional 80486 instructions are:

INVD	Invalidate cache
INVLPG	Invalidate TLB entry
WBINVD	Write back and invalidate cache

New Pentium instructions are:

CPUID	CPU identification
RDMSR	Read from model-specific register
RDTSC	Read from time stamp counter
RSM	Resume from system management mode
WRMSR	Write to model-specific register

In Chapter 3 we will begin examining each of the 80x86 instructions in detail, and see many examples of how they are used.

2.7   ADDRESSING MODES

The 80x86 offers the programmer a wide number of choices when referring to a memory location. Many people believe that the number of **addressing modes** contained in a microprocessor is a measure of its power. If that is so, the 80x86 should be counted among the most powerful processors. Many of the addressing modes are used to generate a **physical address** in memory. Recall from Figure 2.2 that a 20-bit address is formed by the sum of two 16-bit address values. One of the four segment registers will always supply the first 16-bit address. The second 16-bit address is formed by a specific addressing mode operation. The resulting 20-bit address points to one specific location in the processor's 1MB real-mode addressing space. We will see that there are a number of different ways the second part of the address may be generated. Protected-mode addressing will be covered in the final chapter.

Real-Mode Addressing Space

All addressing modes eventually create a physical address that resides somewhere in the 00000 to FFFFF addressing space of the processor. Figure 2.6 shows a brief memory map of the real-mode addressing space, which is broken up into 16 blocks of 64KB each. Each

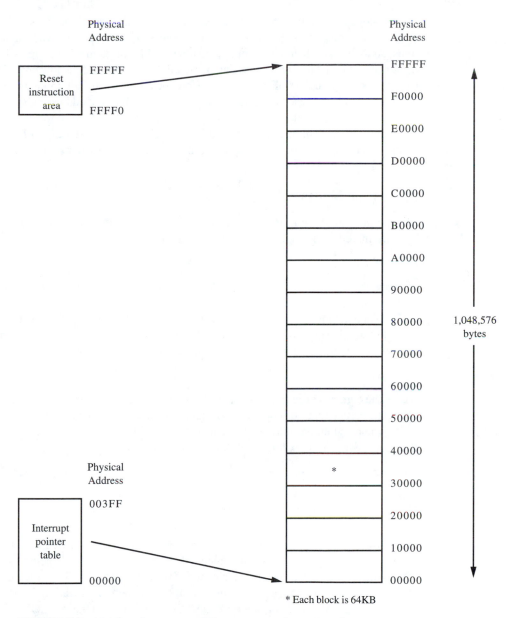

FIGURE 2.6 Addressing space of the processor in real mode

64KB block is called a **segment.** A segment contains all the memory locations that can be reached when a particular segment register is used. For example, if the data segment contains 0000, then addresses 00000 through 0FFFF can be generated when using the data segment. If, instead, register DS contains 1800, then the range of addresses becomes 18000 through 27FFF. It is important to see that a segment can begin on *any* 16-byte boundary. So, 00000, 00010, 00020, 035A0, 10800, and CCE90 are all acceptable starting addresses for a segment.

Altogether, 1,048,576 bytes can be accessed by the processor. This is commonly referred to as 1 **megabyte.** Small areas of the addressing space are reserved for special operations. At the very high end of memory, locations FFFF0 through FFFFF are assigned the role of storing the initial instruction used after a RESET operation. At the low end of memory, locations 00000 through 003FF are used to store the addresses for all 256 interrupts (although not all are commonly used in actual practice). This dedication of addressing space is common among processor manufacturers, and may force designers to conform to specific methods or techniques when building systems around the 80x86. For instance, EPROM is usually mapped into high memory, so that the starting execution instructions will always be there at power-on.

Addressing Modes

The simplest addressing mode is known as **immediate.** Data needed by the processor is actually included in the instruction. For example:

```
MOV  CX,1024
```

contains the immediate data value 1024. This value is converted into binary and included in the code of the instruction.

When data must be moved between registers, **register** addressing is used. This form of addressing is very fast, because the processor does not have to access external memory (except for the instruction fetch). An example of register addressing is:

```
ADD   AL,BL
```

where the contents of registers AL and BL are added together, with the result stored in register AL. Notice that both operands are names of internal registers.

The programmer may refer to a memory location by its specific address by using **direct** addressing. Two examples of direct addressing are:

```
MOV  AX,[3000]
```

and

```
MOV  BL,COUNTER
```

In each case, the contents of memory are loaded into the specified registers. The first instruction uses square brackets to indicate that a memory address is being supplied. Thus, 3000 and [3000] are allowed to have two different meanings. 3000 means the number 3000, whereas [3000] means the number stored at memory location 3000. The second instruction uses the symbol name COUNTER to refer to memory. COUNTER must be defined somewhere else in the program for it to be used this way.

When a register is used within the square brackets, the processor uses **register indirect** addressing. For example:

```
MOV  BX,[SI]
```

instructs the processor to use the 16-bit quantity stored in the SI (source index) register as a memory address. A slight variation produces **indexed** addressing, which allows a small offset value to be included in the memory operand. Consider this example:

```
MOV  BX,[SI + 10]
```

The location accessed by the instruction is the sum of the SI register and the offset value 10.

When the register used is the base pointer (BP), **based** addressing is employed. This addressing mode is especially useful when manipulating data in large tables or arrays. An example of based addressing is:

```
MOV  CL,[BP + 4]
```

Including an index register (SI or DI) in the operand produces **based-indexed** addressing. The address is now the sum of the base pointer and the index register. An example might be:

```
MOV  [BP + DI],AX
```

When an offset value is also included in the operand, the processor uses **based-indexed with displacement** addressing. An example is:

```
MOV  DL,[BP + SI + 2]
```

Obviously, the 80x86 allows the base pointer to be used in many different ways.

Other addressing modes are used when string operations must be performed.

The processor is designed to access I/O ports, as well as memory locations. When **port** addressing is used, the address bus contains the address of an I/O port instead of a memory location. I/O ports may be accessed two different ways. The port may be specified in the operand field, as in:

```
IN  AL,80H
```

or indirectly, via the address contained in register DX:

```
OUT  DX,AL
```

Using DX allows a port range from 0000 to FFFF, or 65,536 individual I/O port locations. Only 256 (00 to FF) posts are allowed when the port address is included as an immediate operand.

All of these addressing modes will be covered again in detail in Chapter 3.

32-Bit Addressing Modes

The Pentium architecture supports an additional method of generating addresses, especially designed to take advantage of protected-mode operation. In protected mode, addresses are 32 bits wide, spanning a 4-gigabyte range. The 32-bit addresses are generated as indicated in Figure 2.7.

As usual, a segment register is used as part of the address calculation. Unlike real-mode addressing, any general purpose register may be used as a base register or index register. The only exception is ESP, which can be used as a base register but not an index register.

A scale factor is also included to multiply the contents of the index register by 1, 2, 4, or 8. This is very useful when dealing with arrays of data composed of bytes, words, double-words, or quad-words. The MOV instruction shown in Figure 2.7 multiplies index register ECX by a scale factor of four.

Though designed for protected mode applications, the 32-bit addressing modes may be used while running in real mode by using an *address size prefix* byte before the instruction that uses the 32-bit addressing mode. This is illustrated in Example 2.2.

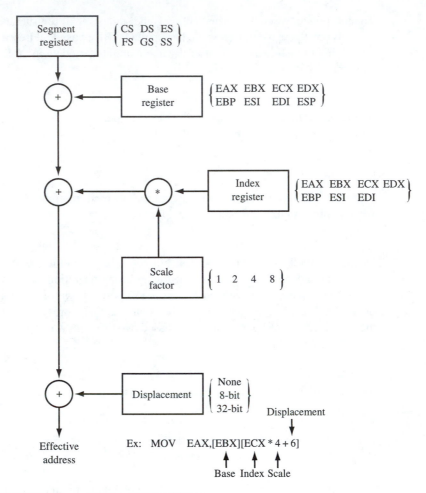

FIGURE 2.7   Generating a 32-bit address

Example 2.2:   Examine the following portion of a list file for a real-mode program with instructions using 32-bit registers or addressing modes:

```
0010 B4 09                          MOV     AH,9
0012 8D 16 0000 R                   LEA     DX,TABLE
0016 CD 21                          INT     21H
0018 B4 09                          MOV     AH,9
001A 66 8D 1E 0000 R                LEA     EBX,TABLE
001F 66 BA 00000002                 MOV     EDX,2
0025 67 8D 14 93                    LEA     DX,[EBX][EDX*4]
0029 CD 21                          INT     21H
```

The second LEA instruction and following MOV instruction both contain extended registers in their operand fields. Since real-mode code is being generated, the default register size is 16 bits. Thus, the operand size prefix byte 66H is used prior to the code for the LEA and MOV instructions to switch to 32-bit mode.

In a similar fashion, the address size prefix byte 67H is used before the third LEA instruction to allow processing of the 32-bit addressing mode specified by EBX, EDX, and the scale factor. As with the operand size prefix, the address size prefix works for a single instruction only, and must precede all instructions that utilize 32-bit addressing.

2.8   INTERRUPTS

An interrupt is an event that occurs while the processor is executing an instruction. The instruction might be part of a group of instructions in a *main* program, such as a word processing application. The interrupt temporarily suspends execution of the main program in favor of a special routine that services the interrupt. When interrupt processing is complete, the processor is returned back to the exact place in the main program where it left off. For example, a **timer** interrupt might occur while the word processing application is in the middle of a spell checking procedure. Spell checking is suspended and the timer interrupt service routine takes over. The routine might simply increment the seconds counter on the time-of-day clock. When the interrupt is finished, spell checking resumes.

Let us take a look at how interrupts are implemented by the 80x86 and how one particular interrupt is used to control the computer when DOS is running.

Hardware and Software Interrupts

The processor is capable of responding to 256 different types of interrupts. These interrupts are generated in a number of different ways. External hardware interrupts are caused by activating the processor's NMI and INTR signals. NMI is a **nonmaskable interrupt** and cannot be ignored by the CPU. INTR is a maskable interrupt that the processor may choose to ignore depending on the state of an internal interrupt enable flag.

Internal interrupts are caused by execution of an INT instruction. INT is followed by an interrupt number from 0 to 255, giving the programmer the option of generating any number of specific interrupts during program execution. We will see later that machines based on the 80x86 that contain a software disk operating system (DOS) have very specific functions assigned to certain interrupts (INT 21H for example) that allow the user to read the keyboard, write text to the screen, control disk drives, and so forth.

Some interrupts are generated internally by the processor itself. *Divide error* is one example. This interrupt is caused when division by zero is detected in the execution unit (during execution of the IDIV or DIV instructions). The processor can also generate single-step interrupts at the end of every instruction if a certain flag called the trap flag is set. Another internal interrupt is INTO (interrupt on overflow).

The Interrupt Vector Table

All interrupts use a dedicated table in memory for storage of their interrupt service routine (ISR) addresses. The table is called an **interrupt pointer table** (or interrupt vector table) and is 1,024 bytes long, enough storage space for 256 4-byte entries. Because an ISR address occupies 4 bytes of storage, the table holds addresses for all 256 interrupts. Each ISR address is composed of a 2-byte CS value and a 2-byte instruction pointer address. Thus, if the table entry for a type-0 interrupt (divide error) were CS:0100 and IP:0400, the divide error ISR code would have to be located at physical address 01400.

Interrupt processing will be covered in detail in Chapter 5. For now, let us examine one particularly useful interrupt.

A Brief Look at DOS Interrupt 21H

One of the most useful DOS interrupts is number 21H. This interrupt was chosen as the entry point into DOS for programmers writing their own DOS applications. Although there are many other interrupts assigned to specific functions by DOS, INT   21H is loaded with so many different functions we rarely need to use others.

It is possible to use DEBUG to determine where the code for DOS's INT   21H routine (or any other interrupt) is located. Because each interrupt vector (CS:IP) requires 4 bytes, the offset into memory for INT   21H's vector is four times 21H, or address 00084H. The following DEBUG session uses this address to find out where the ISR for INT   21H is located:

```
C> debug
-d 0:80    L 10
0000:0080 94 10 16 01 B4 16 26 07-4F 03 FB 0A 8A 03 FB 0A
-q
```

The CS:IP addresses for INT   21H are shown in bold. In this example, CS = 0726H and IP = 16B4H (remember that words are *byte-swapped*). This gives a 20-bit memory address of 08914H. To examine the actual instructions within the INT   21H service routine, use DEBUG's unassemble command with the CS:IP addresses shown (e.g., –u 726:16b4).

The way INT   21H is used is simply a matter of loading specific registers with data and issuing an INT   21H. For example, to read the computer's time we use the following two instructions:

```
MOV  AH,2CH ;get system time function number
INT  21H    ;DOS call
```

DOS will return the time as follows:

$$CH = \text{hours}$$
$$CL = \text{minutes}$$
$$DH = \text{seconds}$$
$$DL = \text{100ths of seconds}$$

Many of the programs we will examine in later chapters use INT   21H and other interrupts.

2.9   THE 8086: THE FIRST 80x86 MACHINE

This section and those that follow describe the historical evolution of the 80x86 family. This is done to gain an appreciation for the improvements that have been made to the 80x86 architecture prior to the design of the Pentium. We begin with the 8086.

The 8086 microprocessor is a 16-bit machine with a 16-bit data bus and a 20-bit address bus. This allows for 2^{20}, or 1MB of addressing space. The instruction set and addressing modes presented in this chapter first became available with the 8086, which operates in one mode only: real mode. Every processor in the 80x86 family to come after the 8086 supports the initial instruction set.

When the 8086 is reset (or first turned on), the processor fetches its first instruction from address FFFF0H. On the PC, this address enables the motherboard's system ROM, which begins the process of booting DOS. All of the 80x86 machines, even the Pentium, follow this mechanism when reset.

A slightly reengineered version of the 8086 was the 8088 microprocessor, which is identical to the 8086 except for the use of an external data bus that is only 8 bits wide. This forces the 8088 to access memory twice as often as the 8086, resulting in a slight perfor mance penalty in terms of execution speed. Both the 8086 and the 8088 were used in the first PCs to hit the market.

Figure 2.8 compares the pin assignments of the 8088 with those of the 8086. As you can see, the majority of pins are the same for both processors. Differences exist on address lines A_8 through A_{15}. On the 8088, these lines are merely address lines. On the 8086, these are **multiplexed address/data** lines AD_8 through AD_{15}. This is where the other half of the 16-bit data bus comes in. The 8086 is capable of reading or writing 16-bits of data at once and for that reason has a slight speed advantage over the 8088.

Internally, both processors are almost identical. Any program written for the 8088 will run without change on the 8086, and vice versa. This means that the instruction sets of both machines are identical. The 8-bit version was a popular choice among manufacturers of personal computers and is therefore featured in this text, but keep in mind that much of what will be said about the 8088 also applies to the 8086.

Both processors have the capability of operating in **minimum mode** or **maximum mode.** Both modes have different ways of dealing with the external buses; thus we see (again in Figure 2.8) that some pins on the processors have two functions. For example, pin 29 on both machines is $\overline{WR}$ in minimum mode, and $\overline{LOCK}$ in maximum mode. The use of maximum mode requires some additional hardware outside the CPU to generate bus control signals.

It is useful to understand that the 8088 and 8086 are not two entirely different microprocessors. You may wish to think of the 8088 as an 8086 on the inside, with an 8-bit data bus on the outside.

A souped-up version of the 8086, called the 80186, contained special hardware such as programmable timers, counters, interrupt controllers, and address decoders. The 80186 was never used in the PC, but was ideal for systems that required a minimum of hardware.

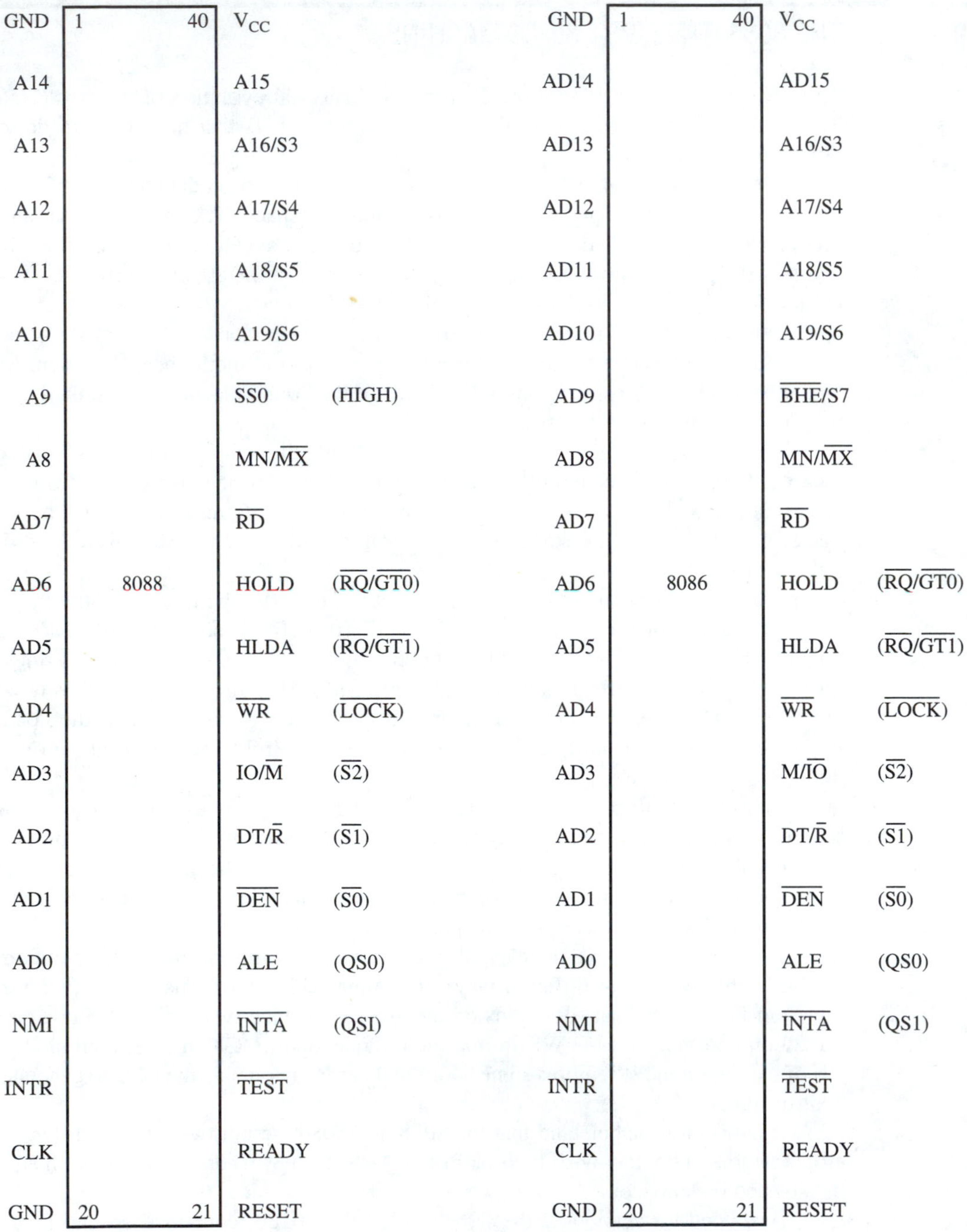

GND	1	40	V$_{CC}$			GND	1	40	V$_{CC}$	
A14			A15			AD14			AD15	
A13			A16/S3			AD13			A16/S3	
A12			A17/S4			AD12			A17/S4	
A11			A18/S5			AD11			A18/S5	
A10			A19/S6			AD10			A19/S6	
A9			$\overline{SS0}$	(HIGH)		AD9			$\overline{BHE}$/S7	
A8			MN/$\overline{MX}$			AD8			MN/$\overline{MX}$	
AD7			$\overline{RD}$			AD7			$\overline{RD}$	
AD6	8088		HOLD	($\overline{RQ/GT0}$)		AD6	8086		HOLD	($\overline{RQ/GT0}$)
AD5			HLDA	($\overline{RQ/GT1}$)		AD5			HLDA	($\overline{RQ/GT1}$)
AD4			$\overline{WR}$	($\overline{LOCK}$)		AD4			$\overline{WR}$	($\overline{LOCK}$)
AD3			IO/$\overline{M}$	($\overline{S2}$)		AD3			M/$\overline{IO}$	($\overline{S2}$)
AD2			DT/$\overline{R}$	($\overline{S1}$)		AD2			DT/$\overline{R}$	($\overline{S1}$)
AD1			$\overline{DEN}$	($\overline{S0}$)		AD1			$\overline{DEN}$	($\overline{S0}$)
AD0			ALE	(QS0)		AD0			ALE	(QS0)
NMI			$\overline{INTA}$	(QSI)		NMI			$\overline{INTA}$	(QS1)
INTR			$\overline{TEST}$			INTR			$\overline{TEST}$	
CLK			READY			CLK			READY	
GND	20	21	RESET			GND	20	21	RESET	

Note: ( ) denotes a MAX mode signal

FIGURE 2.8   8088 and 8086 pin assignments

2.10   A SUMMARY OF THE 80286

The next major improvement in Intel's line of microprocessors was the 80286 High-Performance Microprocessor with Memory Management and Protection™. The 80286 does not contain the internal DMA controllers, timers, and other enhancements. Instead, the 80286 concentrates on the features needed to implement **multitasking,** an operating system environment that allows many programs or tasks to run seemingly simultaneously. In fact, the 80286 was designed with this goal in mind. A 24-bit address bus gives the processor the capability of accessing 16MB of storage. The internal memory management feature increases the storage space to 1 **gigabyte** of virtual address space. That's over 1 billion locations of virtual memory! Virtual addressing is a concept that has gained much popularity in the computing industry. Virtual memory allows a large program to execute in a smaller physical memory. For example, if a system using the 80286 contained 8MB of RAM, memory management and virtual addressing permits the system to run a program containing 12MB of code and data, or even multiple programs in a multitasking environment, *all of which* may be larger than 8MB.

To implement the complicated addressing functions required by virtual addressing, the 80286 has an entire functional unit dedicated to address generation. This unit is called the **address unit.** It provides two modes of addressing: 8086 real address mode and protected virtual address mode. The 8086 real address mode is used whenever an 8086 program executes on the 80286. The 1MB addressing space of the 8086 is simulated on the 80286 by the use of the lower 20 address lines. Processor registers and instructions are totally upward-compatible with the 8086.

Protected virtual address mode uses the full power of the 80286, providing memory management, additional instructions, and protection features, while at the same time retaining the ability to execute 8086 code. The processor switches from 8086 real address mode to protected mode when a special instruction sets the protection enable bit in the machine's status word. Addressing is more complicated in protected mode, and is accomplished through the use of **segment descriptors** stored in memory. The segment descriptor is the device that really makes it possible for an operating system to control and protect memory. Certain bits within the segment descriptor are used to grant or deny access to memory in certain ways. A section of memory may be write protected or made to execute only by the setting of proper bits in the access rights byte of the descriptor. Other bits are used to control how the segment is mapped into virtual memory space and whether the descriptor is for a code segment or a data segment. Special descriptors, called **gate descriptors,** are used for other functions. Four types of gate descriptors are call gates, task gates, interrupt gates, and trap gates. They are used to change privilege levels (there are four), switch tasks, and specify interrupt service routines.

The instruction set of the 80286 is identical to that of the 8086, with additional instructions thrown in to handle the new features. Many of the instructions are used to load and store the different types of descriptors found in the 80286. Other instructions are used to manipulate task registers, change privilege levels, adjust the machine status word, and verify read/write accesses. Clearly, the 80286 differs greatly from the 8086 in the services it offers, while at the same time filling a great need for designers of operating systems.

2.11   A SUMMARY OF THE 80386

Intel continued its upward-compatible trend with the introduction of the 386 High Performance 32-bit CHMOS Microprocessor with Integrated Memory Management™. Software written for the 8088, 8086, 80186, and 80286 will also run on the 386. A 132-pin Grid Array™ package houses the 386, which offers a full 32-bit data bus and 32-bit address bus. The address bus is capable of accessing over 4 gigabytes of physical memory. Virtual addressing pushes this to over 64 *trillion* bytes of storage.

The register set of the 386 is compatible with earlier models, including all eight general purpose registers plus the four segment registers. Although the general purpose registers are 16 bits wide on all earlier machines, they can be extended to 32 bits on the 386. Their new names are EAX, EBX, ECX, and so on. Two additional 16-bit data segment registers are included, FS and GS. Like the 80286, the 386 has two modes of operation: real mode and protected mode. When in real mode, segments have a maximum size of 64KB. When in protected mode, a segment may be as large as the entire physical addressing space of 4 gigabytes. The new extended flags register contains status information concerning privilege levels, virtual mode operation, and other flags concerned with protected mode. The 386 also contains three 32-bit control registers. The first, machine control register, contains the machine status word and additional bits dealing with the coprocessor, paging, and protected mode. The second, page fault linear address, is used to store the 32-bit address that caused the last page fault. In a virtual memory environment, physical memory is divided up into a number of fixed size **pages.** Each page will at some time be loaded with a portion of an executing program or other type of data. When the processor determines that a page it needs to use has not been loaded into memory, a **page fault** is generated. The page fault instructs the processor to load the missing page into memory. Ideally, a low page-fault rate is desired.

The third control register, page directory base address, stores the physical memory address of the beginning of the page directory table. This table is up to 4KB in length and may contain up to 1024 page directory entries, each of which points to another page table area, whose information is used to generate a physical address.

The segment descriptors used in the 80286 are also used in the 386, as are the gate descriptors and the four levels of privilege. Thus, the 386 functions much the same as the 80286, except for the increase in physical memory space and the enhancements involving page handling in the virtual environment.

The computing power of each of the processors that have been presented can be augmented with the addition of a floating-point coprocessor. All sorts of mathematical operations can be performed with the coprocessors with 80-bit binary precision. The 8087 coprocessor is designed for use with the 8088 and 8086, the 80287 with the 80286, and the 80387 with the 386.

2.12   A SUMMARY OF THE 80486

This processor is the next in Intel's upward-compatible 80x86 architecture. Surprisingly, there are only a few differences between the 80486 and the 80386, but these differences create a significant performance improvement.

Like the 80386, the 80486 is a 32-bit machine containing the same register set as the 80386 and all of the 80386's instruction set with a few additional instructions. The 80486 has a similar 4-gigabyte addressing space using the same addressing features.

The first improvement over the 80386 is the addition of an 8KB **cache** memory. A cache is a very high-speed memory, with an access time usually ten times faster than that of conventional RAM used for external processor memory. The 80486's internal cache is used to store both instructions and data. Whenever the processor needs to access memory, it will first look in the cache for it. If the data is found in the cache, they are read out much faster than if they had to come from external RAM or EPROM. This is known as a *cache hit*. If the data is not found in the cache, the processor must then access the slower external memory. This is called a *cache miss*. The processor tries to keep the cache's hit ratio as high as possible. Consider the following example:

$$\text{RAM Access Time} = 70 \text{ ns}$$

$$\text{Cache Access Time} = 10 \text{ ns}$$

$$\text{Hit Ratio} = 0.85$$

$$\text{Average Memory Access Time} = 0.85 \times (10 \text{ ns}) \text{ Hit}$$

$$+ (1 - 0.85) \times (10 \text{ ns} + 70 \text{ ns}) \text{ Miss}$$

$$= 20.5 \text{ ns}$$

The average memory access time for a hit ratio of 0.85 is less than 21 ns! This is due to the following reasoning: If data is found in the cache (85% of the time), the access time is only 10 ns. If data is not found (15% of the time), the access time equals 80 ns (the cache access time plus the RAM access time), because the processor had to read the cache to find out the data was *not* there.

If you consider that a large portion of a program (or even an *entire* program) might fit within the 8KB cache, you will agree that the program will execute very quickly, because most instruction fetches will be for code already in the cache. This architectural improvement significantly increases the processing speed of the 80486. Some of the new 80486 instructions are included to help maintain the cache.

The 80486 has two other improvements. Although it executes the same instruction set as the 80386, the 80486 does so with a redesigned internal architecture. This new design allows many 80486 instructions to execute with fewer clock cycles than those required by the 80386. This reduction in clock cycles adds additional speed to the 80486's execution. Also, the 80486 comes with an on-chip coprocessor. You might recall that the 80386 can be connected to an external 80387 coprocessor to enhance performance. The 80486 has the equivalent of an 80387 built right into it! And because the coprocessor is closer to the CPU, data is transferred quicker, which leads to another performance boost.

Thus, although the 80386 and 80486 share many similarities, the 80486's differences create a much more powerful processor.

2.13   A SUMMARY OF THE PENTIUM

The newest, and fastest, chip in the Intel high-performance microprocessor line is the Pentium. As usual, upward compatibility has been maintained. The Pentium will run all programs written for any machine in the 80x86 line, though it does so at speeds many times that of the fastest 80486. And the Pentium does so with a radically new architecture!

There are two major computer architectures in use: CISC and RISC. CISC stands for Complex Instruction Set Computer. RISC stands for Reduced Instruction Set Computer. All of the 80x86 machines prior to the Pentium can be considered CISC machines. The Pentium itself is a mixture of both CISC and RISC technologies. The CISC aspect of the Pentium provides for upward compatibility with the other 80x86 architectures. The RISC aspects lead to additional performance improvements. Some of these improvements are separate 8KB data and instruction caches, dual integer pipelines, and branch prediction.

As Figure 2.9 shows, the Pentium processor is a complex machine with many interlocking parts. At the heart of the processor are the two integer pipelines, the U pipeline and the V pipeline. These pipelines are responsible for executing 80x86 instructions. A floating point unit is included on the chip to execute instructions previously handled by the external 80x87 math coprocessors. During execution, the U and V pipelines are capable of executing two integer instructions at the same time, under special conditions, or one floating point instruction.

The Pentium communicates with the outside world via a 32-bit address bus and a 64-bit data bus. The bus unit is capable of performing *burst* reads and writes of 32 bytes to memory, and through bus cycle pipelining, allows two bus cycles to be in progress simultaneously.

An 8KB instruction cache is used to provide quick access to frequently used instructions. When an instruction is not found in the instruction cache, it is read from the external data bus and a copy placed into the instruction cache for future references. The branch target buffer and prefetch buffers work together with the instruction cache to fetch instructions as fast as possible. The prefetch buffers maintain a copy of the next 32 bytes of prefetched instruction code, and can be loaded from the cache in a single clock cycle, due to the 256-bit-wide data output of the instruction cache.

The Pentium uses a technique called *branch prediction* to maintain a steady flow of instructions into the pipelines. To support branch prediction, the branch target buffer maintains a copy of instructions in a different part of the program located at an address called the *branch target*. For example, the branch target of a CALL XYZ instruction is the address of the subroutine XYZ. Certain instructions, such as CALL, may cause the processor to jump to an entirely different program location, instead of simply proceeding to the instruction in the next location. So, just in case the code from the target address is needed, the branch target buffer maintains a copy of it and feeds it to the instruction cache.

A separate 8KB data cache stores a copy of the most frequently accessed memory data. Since memory accesses are significantly longer than processor clock cycles, it pays to keep a copy of memory data in a fast-reading cache. The data and instruction caches may both be enabled/disabled with hardware or software. Both also employ the use of a translation lookaside buffer, which converts logical addresses into physical addresses when virtual memory is employed.

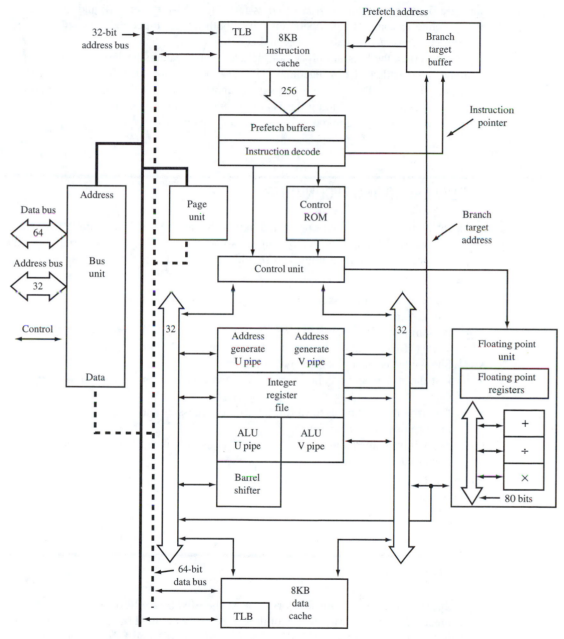

FIGURE 2.9   Pentium architecture block program

The floating-point unit of the Pentium maintains a set of floating-point registers and provides 80-bit precision when performing high-speed math operations. This unit has been completely redesigned from the one used inside the 80486 and is also pipelined. The floating-point unit uses hardware in the U and V pipelines to perform the initial work during a floating-point instruction (such as fetching a 64-bit operand), and then uses its own pipeline to complete the operation. Since both integer pipelines are used, only one floating-point instruction may be executed at a time.

Altogether, the Pentium processor includes many features designed to increase performance over earlier 80x86 machines. This was possible by blending CISC and RISC technology together. The benefit to us as programmers lies in the fact that all Intel processors from the 8086 up, including the Pentium, run the same basic instruction set that we will learn about beginning in the next chapter, but they do it faster and faster.

2.14   TROUBLESHOOTING TECHNIQUES

A great amount of material was presented in this chapter regarding the Intel 80x86 architecture. It would be helpful to really commit some of the most basic material about the 80x86 to memory. As a minimum, you should be able to do the following without much thought:

- Name all of the processor registers, their bit sizes, and whether they can be split into 8-bit halves.
- Be familiar with several architectural features, such as the processor's addressing space (1MB in real mode), interrupt mechanism, and I/O mechanism.
- Show what is meant by Intel byte-swapping.
- List the names and meanings of the most common flags, such as zero, carry, and sign.
- Describe the method used to form a 20-bit address in real mode (combining a segment register with an offset).
- Show how an instruction is composed of an operation, a set of operands, and a particular addressing mode.

Knowing these basics thoroughly will assist you in mastering the 80x86 instruction set that we will examine in Chapters 3 and 4.

SUMMARY

This chapter has taken an introductory look at the 80x86 family of upward-compatible microprocessors. The software model of the 80x86 was examined first, showing all the 16-bit general purpose registers (AX, BX, CX, DX, SI, DI, BP, and SP) and the four 16-bit segment registers (CS, DS, SS, and ES), as well as the 32-bit extended registers beginning with the 80386.

Although 80x86 real mode contains only 16-bit registers, its architecture allows the generation of 20-bit physical addresses, giving the processor a 1-megabyte addressing space. One of the segment registers is always involved in a memory access. The general purpose registers were shown to have specific tasks assigned to them by default, such as the use of AX in multiply and divide operations. Techniques to use 32-bit registers and addressing modes were also demonstrated.

A technique called Intel byte-swapping was also introduced, which accounts for the way a 16-bit number is stored in memory (low byte first). This technique is rarely seen on other microprocessors.

The entire instruction set was presented to give you a feel for the type of operations the 80x86 is capable of performing. This discussion was followed by a brief explanation of what a segment is, and what addressing modes are available. Some examples were shown to illustrate the use of different addressing modes. This was followed by an explanation of the processor's interrupt structure, and a summary of the entire 80x86 family.

In the next chapter we will take a detailed look at the software operation of the 80x86.

STUDY QUESTIONS

1. Name all of the general purpose registers and some of their special functions.
2. How are the segment registers used to form a 20-bit address?
3. a) If CS contains 03E0H and IP contains 1F20H, from what address is the next instruction fetched?
   b) If SS contains 0400H and SP contains 3FFEH, where is the top of the stack located?
   c) If a data segment begins at address 24000H, what is the address of the last location in the segment?
4. Explain what the instruction and data caches are used for.
5. Are the U and V pipelines identical in operation?
6. a) Show the DB statement needed to define a list of numbers called FACTORS that contains all the integer factors of the decimal value 50.
   b) Show how a DW statement can be written to reserve 250 words of the value 7.
7. What is a segment?
8. Two memory locations, beginning at address 3000H, contain the bytes 34H and 12H. What is the word stored at location 3000H? See Figure 2.10 for details.

FIGURE 2.10 For Question 2.8

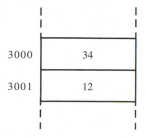

9. What is Intel byte-swapping?
10. Count the number of different instructions available. How many are there?
11. How many addressing modes are provided?
12. What is a physical address?
13. Why is register addressing so fast?
14. What do square brackets mean when they appear in an operand (e.g., MOV AX,[3000])?
15. What is the difference between MOV   AX,1000H and MOV   AX,[1000H]?
16. How does port addressing differ from memory addressing?
17. What is an interrupt?
18. Name one instruction that can cause an interrupt.
19. How many interrupts does the 80x86 support?
20. What are some of the differences between real mode and protected mode?
21. List the important features of the 80286.
22. What is one advantage of virtual memory?
23. What is a page fault?
24. Compare two 386 systems, one containing 512KB of RAM, the second containing 4 megabytes. How would the number of page faults compare when:
    a) a 220KB application is executed on both machines?
    b) a 6-megabyte application is executed on both machines?
25. Which has the greater effect on the number of page faults, physical memory size or the size of the program being executed?
26. Why would we resist building a complete physical memory for the 386? Does the reason apply to the 8086?
27. Why would anyone possibly need 4.3 billion bytes for a program? Can you think of any applications that may require this much memory?
28. List three differences between the 80286 and the 80386.
29. List three differences between the 80386 and the 80486.
30. Compute the average memory access time from the following information:

RAM Access Time = 80 ns

Cache Access Time = 10 ns

Hit Ratio = 0.92

31. What makes the Pentium so different from other 80x86 CPUs?
32. Use DEBUG to find the address of INT   21H on your DOS machine.

PART 2

Software Architecture

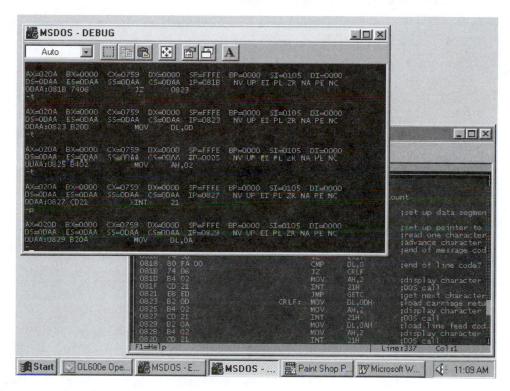

Microsoft® Windows® 95 desktop showing a program under development.

CHAPTER 3

80x86 Instructions, Part 1: Addressing Modes, Flags, Data Transfer, and String Instructions

OBJECTIVES

In this chapter you will learn about:

*   The style of source files written in 80x86 assembly language
*   The different addressing modes of the 80x86
*   The operation and use of the processor flags
*   Data transfer and string instructions

3.1   INTRODUCTION

This chapter is intended to introduce you to the first part of the instruction set of the 80x86 microprocessor family, and the ways that different addressing modes and data types can be used to make the instructions do the most work for you. The combination of instructions and addressing modes found in the 80x86 makes the job of writing code much easier and more efficient than before. In this chapter we will examine the various addressing modes, flags, and conventions used when representing data. We will take a detailed look at the data transfer and string instructions, leaving the remainder of the instruction set for Chapter 4.

Section 3.2 introduces the conventions followed when writing 80x86 assembly language source code. Section 3.3 explains the different instruction types available; this is followed by coverage of the 80x86's addressing modes in Section 3.4. Processor flags are detailed in Section 3.5 to set the stage for the first two instruction groups, data transfer and string instructions, which are presented in Sections 3.6 and 3.7, respectively. Troubleshooting techniques are presented in Section 3.8.

3.2   ASSEMBLY LANGUAGE PROGRAMMING

Program execution in any microprocessor system consists of fetching binary information from memory and decoding that information to determine the instruction represented. The information in memory may have been programmed into an EPROM or downloaded from a separate system. But where did the program come from and how was it written? As humans, we have trouble handling many pieces of information simultaneously and thus have difficulty writing programs directly in **machine code,** the binary language understood by the microprocessor. It is much easier for us to remember the mnemonic SUB AX,AX than the corresponding machine code 2BC0. For this reason, we write **source files** containing all the instruction mnemonics needed to execute a program. The source file is converted into an **object file** containing the actual binary information the machine will understand by a special program called an **assembler.** Some assemblers allow the entire source file to be written and assembled at one time. Other assemblers, called **single-line assemblers,** work with one source line at a time and are restricted in operation. These kinds of assemblers are usually found on small microprocessor-based systems that do not have disk storage and text editing capability.

The assembler discussed here is not a single-line assembler. Instead it is a **cross-assembler.** Cross-assemblers are programs written in one language, such as C, that translate source statements into a second language: the machine code of the desired processor. Figure 3.1 shows this translation process. The source file in the example, TOTAL.ASM, is presented as input to the assembler. The assembler will convert all source statements into the correct binary codes and place these into the object file TOTAL.OBJ. Usually, the object file contains additional information concerning program relocation and external references, and thus is not yet ready to be loaded into memory and executed. A second file created by the assembler is the **list file,** TOTAL.LST, which contains all the original source file text plus the additional code generated by the assembler. The list file may be displayed on the screen, or printed. The object file may not be printed or displayed, since it is just code.

FIGURE 3.1   Source program assembly

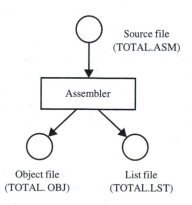

A Sample Source File

Let us look at a sample source file, a subroutine designed to find the sum of 16 bytes stored in memory. It is not important at this time that you understand what each instruction does. We are simply trying to get a feel for what a source file might look like and what conventions to follow when we write our own programs.

```
        ORG  8000H
TOTAL:  MOV  AX,7000H    ;load address of data area
        MOV  DS,AX       ;init data segment register
        MOV  AL,0        ;clear result
        MOV  BL,16       ;init loop counter
        MOV  SI,0        ;init data pointer
ADDUP:  ADD  AL,[SI]     ;add data value to result
        INC  SI          ;increment data pointer
        DEC  BL          ;decrement loop counter
        JNZ  ADDUP       ;jump if counter not zero
        MOV  [SI],AL     ;save sum
        RET              ;and return
        END
```

The first line of source code contains a command that instructs the assembler to load its program counter with 8000H. The ORG (for origin) command is known as an assembler **pseudo-opcode,** a fancy name for a mnemonic that is understood by the assembler but not by the microprocessor. ORG does not generate any source code; it merely sets the value of the assembler's program counter. This is important when a section of code must be loaded at a particular place in memory. The ORG statement is a good way to generate instructions that will access the proper memory locations when the program is loaded into memory.

Hexadecimal numbers are followed by the letter H to distinguish them from decimal numbers. This is necessary since 8000 decimal and 8000 hexadecimal differ greatly in magnitude. For the assembler to tell them apart, we need a symbol that shows the difference. Some assemblers use $8000; others use &H8000. It is really a matter of whose software you purchase. All examples in this book will use the 8000H form.

The second source line contains the major components normally used in a source statement. The label TOTAL is used to point to the address of the first instruction in the subroutine. ADDUP is also a label. Single-line assemblers do not allow the use of labels.

The opcode is represented by MOV and the operand field by AX,7000H. The order of the operands is <destination>, <source>. This indicates that 7000H is being MOVed into AX. So far we have three fields: label, opcode, and operand. The fourth field, if it is used, usually contains a comment explaining what the instruction is doing. Comments are preceded by a semicolon (;) to separate them from the operand field. In writing source code, you should follow the four-column approach. This will result in a more understandable source file.

The final pseudo-opcode in most source files is END. The END statement informs the assembler that it has reached the end of the source file. This is important, because many assemblers usually perform two passes over the source file. The first pass is used to determine the lengths of all instructions and data areas, and to assign values to all symbols (labels) encountered. The second pass completes the assembly process by generating the machine code for all instructions, usually with the help of the symbol table created in the first pass. The second pass also creates and writes information to the list and object files. The list file for our example subroutine looks like this:

```
 1   8000                           ORG    8000H
 2   8000   B8   0070   TOTAL:      MOV    AX,7000H
 3   8003   8E   D8                 MOV    DS,AX
 4   8005   B0   00                 MOV    AL,0
 5   8007   B3   10                 MOV    BL,16
 6   8009   BE   0000               MOV    SI,0
 7   800C   02   04     ADDUP:      ADD    AL,[SI]
 8   800E   46                      INC    SI
 9   800F   FE   CB                 DEC    BL
10   8011   75   F9                 JNZ    ADDUP
11   8013   88   04                 MOV    [SI],AL
12   8015   CB                      RET
13                                  END
```

Normally the comments would follow the instructions, but they have been removed for the purposes of this discussion.

The first column of numbers represents the original source line number.

The second column of numbers represents the memory addresses of each instruction. Notice that the first address matches the one specified by the ORG statement. Also notice that the ORG statement does not generate any code.

The third column of numbers is the machine code generated by the assembler. The machine codes are intermixed with data and address values. For example, B8 00 70 in line 2 represents the instruction MOV AX,7000H, with the MOV instruction coded as B8 and the immediate word 7000H coded in byte-swapped form as 00 70. In line 3, the machine codes 8E D8 represent MOV DS,AX. Neither of these 2 bytes are data or address values, as they were in line 2. Look for other data values in the instructions on lines 4 through 6. Line 5 makes an important point: the assembler will convert decimal numbers into hexadecimal (the 16 in the operand field has been converted into 10 in the machine code column).

Finally, another look at the list file shows that there are 1-, 2-, and 3-byte instructions present in the machine code. When an address or data value is used in an instruction, chances are good that you will end up with a 2- or 3-byte instruction (or possibly even more).

Following the code on each line is the text of the original source line. Having all of this information available is very helpful during the debugging process.

Assembler Directives ORG, SEGMENT, ENDS, ASSUME, and END

The actual form of an 80x86 source file is much more complicated than the simple example we have just examined. When writing 80x86 source files, we separate code areas from data areas. We may even have a separate area reserved for the stack. The new source file for TOTAL, which includes separate areas called **segments**, contains many new pseudocodes to help the assembler generate the correct machine code for the object file.

```
DATA      SEGMENT   PARA 'DATA'
          ORG       7000H
POINTS    DB        16 DUP(?)          ;save room for 16 data bytes
SUM       DB        ?                  ;save room for result
DATA      ENDS

CODE      SEGMENT   PARA 'CODE'
          ASSUME CS:CODE,DS:DATA

          ORG       8000H
```

```
TOTAL:    MOV     AX,DATA          ;load address of data segment
          MOV     DS,AX            ;init data segment register
          MOV     AL,0             ;clear result
          MOV     BL,16            ;init loop counter
          LEA     SI,POINTS        ;init data pointer
ADDUP:    ADD     AL,[SI]          ;add data value to result
          INC     SI               ;increment data pointer
          DEC     BL               ;decrement loop counter
          JNZ     ADDUP            ;jump if counter not zero
          MOV     SUM,AL           ;save sum
          RET                      ;and return

CODE      ENDS
          END     TOTAL
```

In this new source file we use a DATA segment and a CODE segment. These areas are defined by the **SEGMENT** pseudocode and terminated by **ENDS** (for end segment). Note that the SEGMENT assembler directive does *not* affect the value in any segment register. The TOTAL subroutine is placed inside the CODE segment. In the DATA segment, 16 bytes of storage are reserved with the DB   16 DUP(?) instruction. DB stands for define byte, and DUP for duplicate. The ? means we do not know what value to put into memory and is a way of telling the assembler that we do not care what byte values are used in the reserved data area. We could easily use DB   16 DUP(0) to place 16 zeros into the reserved area. Words can also be defined/reserved in a similar fashion, using, for example, DW   8 DUP(0) or DW   8 DUP(?). The ORG   7000H statement tells the assembler where to put the data areas. It is not necessary for this ORG value to be smaller than the ORG of the subroutine. ORG   87C0H would have also worked in place of ORG   7000H. It is all a function of where RAM exists in your system.

The addition of the TOTAL label in the END statement informs the assembler that TOTAL, not SUM, is the starting execution address. This information is also included in the object file.

Most assemblers now accept simplified segment directives and automatically generate the necessary code to manage segments. The TOTAL source file, rewritten with simplified segment directives, now looks like this:

```
          .MODEL  SMALL
          .DATA
          ORG     7000H
POINTS    DB      16 DUP(?)        ;save room for 16 data bytes
SUM       DB      ?                ;save room for result

          .CODE
          ORG     8000H
TOTAL:    MOV     AX,7000H         ;load address of data area
          MOV     DS,AX            ;init data segment register
          MOV     AL,0             ;clear result
          MOV     BL,16            ;init loop counter
          LEA     SI,POINTS        ;init data pointer
ADDUP:    ADD     AL,[SI]          ;add data value to result
          INC     SI               ;increment data pointer
          DEC     BL               ;decrement loop counter
          JNZ     ADDUP            ;jump if counter not zero
          MOV     SUM,AL           ;save sum
          RET                      ;and return
          END     TOTAL
```

The first directive, **.MODEL,** instructs the assembler that the type of program being created falls into a category called *small*. Small programs are programs that contain one code segment and one data segment. All of the programs in this book are small programs. Other models allow for multiple code segments, multiple data segments, or both, as indicated in Table 3.1.

The **.DATA** directive indicates the beginning of a data segment. There is no need to indicate the end of the data segment. This is automatically assumed when the **.CODE** directive is encountered, which begins the code segment. Clearly, it is easier for the programmer to use these simplified segment directives, rather than get bogged down with the minute details of assembler syntax. The remaining programs in the book will utilize simplified segment directives.

When a large program must be written by a team of people, each person will be assigned a few subroutines to write. They must all assemble and test their individual sections to ensure the code executes correctly. When all portions of the program (called **modules,** after a technique called modular programming) are assembled and tested, their object files are combined into one large object file via a program called a **linker.** Figure 3.2 represents this process. The linker examines each object file, determining its length in bytes, its proper place in the final object file, and what modifications should be made to it.

In addition, a special collection of object files is sometimes available in a **library** file. The library may contain often used subroutines, or patches of code. Instead of continuously reproducing these code segments in a source file, a special pseudocode is used to instruct the assembler that the code must be brought in at a later time (by the linker). This helps keep the size of the source file down, and promotes quicker writing of programs.

When the linker is through, the final code is written to a file called the **load** module. Another program called a **loader** takes care of loading the program into memory. Usually the linker and loader are combined into a single program called a **link-loader.**

So, writing the source file is actually only the first step in a long process. But even before a source file can be written, the programmer must understand the instructions that will be used in the source file. The remaining sections will begin coverage of this important topic.

TABLE 3.1 Predefined .MODEL types

Memory Model	Data Segments	Code Segments	Special Features
TINY	one	one	Only used for .COM file. CS and DS combined.
SMALL	one	one	Smallest .EXE file model.
MEDIUM	one	multiple	
COMPACT	multiple	one	
LARGE	multiple	multiple	
HUGE	multiple	multiple	Uses normalized addresses.
FLAT	one	one	32-bit addressing.

FIGURE 3.2 Linking
multiple object files together

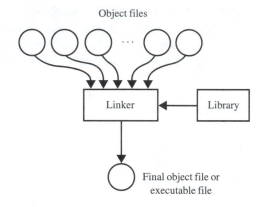

3.3   INSTRUCTION TYPES

For purposes of discussion in this section, the instruction set of the 80x86 microprocessor is divided into seven different groups:

1. Data transfer
2. Strings
3. Arithmetic
4. Bit manipulation
5. Loops and jumps
6. Subroutine and interrupt
7. Processor control

The data transfer group contains instructions that transfer data from memory to register, register to register, and register to memory. Instructions that perform I/O are also included in this group. Data may be 8, 16, or 32 bits in length, and all of the processors registers may be used (including the stack pointer and flag register).

The next group deals with strings. A string is simply a collection of bytes stored sequentially in memory, whose length can be up to 64KB. Instructions are included in this group to move, scan, and compare strings.

The arithmetic group provides addition and subtraction of 8-, 16-, and 32-bit values, signed and unsigned multiplication and division of 8-, 16-, 32-, and 64-bit numbers, and special instructions for working with BCD and ASCII data.

The bit manipulation group is used to perform AND, OR, XOR (Exclusive -OR), and NOT (1's complement) operations on 8-, 16-, and 32-bit data contained in registers or memory. Shift and rotate operations on bytes and words are also included, with single or multi-bit shifts or rotates possible.

Loops and jumps are contained in the next group. Many different forms of instructions are available, with each one testing a different condition based on the state of the processor's flags. The loop instructions are designed to repeat automatically, terminating when certain conditions are met.

The subroutine and interrupt group contains instructions required to call and return from subroutines and handle interrupts. The processor stack can be manipulated by a spe-

cial form of the return instruction, and two classes of subroutines, called near and far procedures, are handled by these instructions.

The final group of instructions is used to directly control the state of some of the
flags, enable/disable the 80x86 interrupt mechanism, and synchronize the processor with
external peripherals.

Many different addressing modes can be used with most instructions, and in the next
section we will examine these addressing modes in detail.

3.4   ADDRESSING MODES

The power of any instruction set is a function of both the types of instructions implemented
and the number of addressing modes available. Suppose that a microprocessor could not
directly manipulate data in a memory location. The data would have to be loaded into a
processor register, manipulated, and written back into memory. If an addressing mode were
available that could directly operate on data in memory, the task could be done more effectively. In this section we will examine the many different addressing modes available in the
80x86, and see how each is used. The examples presented make use of the MOV instruction, which has the following syntax: MOV   <destination>,<source>. It will be obvious in
the examples what is being accomplished with each MOV instruction, so a detailed description is not included here. Also, whenever the contents of a memory location or an address
or data register is referred to, assume that the value or address is hexadecimal.

Creating an Effective Address

Chapter 2 introduced the concept of segmented memory, where a segment is a 64KB block
of memory accessed with the aid of one of the six segment registers. Whenever *any* real-
mode address is generated by the 80x86, the 20-bit value that appears on the processor's
address bus is formed by some combination of segment register and an additional numerical offset. This address is referred to as the **effective address.** Many of the data transfer
instructions use the data segment register by default when forming an effective address.
As we saw in Chapter 2, instruction fetches generate effective addresses via the addition
of the code segment register and the instruction pointer. Stack operations automatically
use the stack segment register and the stack pointer. Thus, from a programming standpoint,
almost all instructions require proper initialization of a segment register for correct execution and generation of effective addresses.

One way to initialize a segment register is to first place the desired segment address
into the AX register and then copy the contents of AX into the corresponding segment register. For example, to initialize the data segment register to 1000H, we would use the following instructions:

```
MOV   AX,1000H
MOV   DS,AX
```

Remember that the direction of data transfer in the operand field is from right to left, so
1000H is MOVed into AX, and then the contents of AX are MOVed into DS, as you can
see in Figure 3.3. (Unfortunately, MOV   DS,l000H is an illegal instruction. We are *not
allowed* to move a numerical value directly into a segment register!)

FIGURE 3.3   Initializing the
data segment register

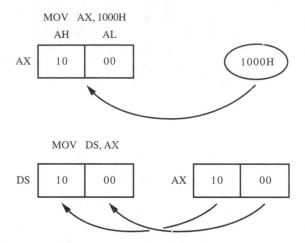

All of the examples used to explain the processor's addressing modes assume that
the corresponding segment register has already been initialized.

Immediate Addressing

We often use **immediate data** in the operand field of an instruction. Immediate data represents a constant that we wish to place somewhere, by MOVing it into a register or a
memory location. In MOV   AX,1000H, we are placing the immediate value 1000H into
register AX. Immediate data is represented by 8-, 16-, or 32-bit numbers that are coded
directly into the instruction. For example, MOV   AX,l000H is coded as B8 00 10, and
MOV   AL,7 is coded as B0 07. Notice in the first instruction that the 16-bit value 1000H
has its lower byte (00) and its upper byte (10) reversed in the actual machine code. As we
saw in Chapter 2, this technique is commonly referred to as *Intel byte-swapping,* and it is
the way all 16-bit numbers are stored.

An important question at this time is "How did the assembler know that the 7 in
MOV   AL,7 should be coded as the single byte value 07, and not the 2-byte value 07 00
(in byte-swapped form)?" The answer is that the assembler looked at the size of the other
operand in the instruction (AL). Because AL is the lower half of the AX register, the assembler knows that it may contain only 8 bits of data. Knowing this, do you see why MOV
AL,9C8H would be an illegal instruction? If you cannot answer this question, think about
how many bits are needed to represent the hexadecimal number 9C8. Because 12 bits are
needed, we cannot possibly store 9C8H in the 8-bit AL register.

Let's apply what we have just seen in an example.

Example 3.1:   What is the result of MOV   CX,7?

Solution: Because register CX is specified, the immediate value 7 is coded as the 2-
byte number 00 07. Figure 3.4 shows how this 2-byte number is placed into CX.

The machine code for MOV   CX,7 is B9 07 00.

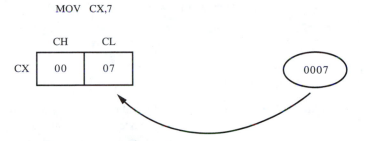

FIGURE 3.4 An example of immediate addressing

Here are some additional examples of immediate addressing:

```
MOV  AL,20    ;put 20 decimal into AL
ADD  BH,20H   ;add 20 hexadecimal to BH
SHL  DX,1     ;shift DX left one bit
AND  AH,40H   ;and AH with 40 hexadecimal
SUB  EAX,1    ;subtract 1 from EAX
```

Register Addressing

The operand field of an instruction in many cases will contain one or more of the internal registers. Register addressing is the name we use to refer to operands that contain one of these registers. The MOV DS,AX instruction from the previous section is an example of register addressing, because we are using only processor registers in the operand field. Instructions of this form often execute very quickly, because they do not require any memory accesses beyond the initial instruction fetch cycles. The machine code corresponding to MOV DS,AX is 8E D8, a 2-byte instruction containing all the information necessary for the processor to perform the desired operation. DEC DX (decrement the DX register) is another example of register addressing, and has 4A as its corresponding machine code.

Example 3.2: If register AX contains the value 1000H and register BL contains 80H, what is the result of MOV AL,BL?

Solution: The contents of the BL register are copied into the AL register, leaving AH undisturbed. The final value in AX is 1080H. Notice that the contents are *copied* during the MOV. MOV can also be interpreted as "make a copy of." So, when we move data from one place to another, we actually are just making a second copy of the source data.

The machine code for MOV AL,BL is 88 D8.

It is important to note that only the code segment register is needed to execute the instructions used in Examples 3.1 and 3.2.

Here are some additional examples of register addressing:

```
PUSH   AX        ;save copy of AX on stack
ADD    BH,CL     ;add CL to BH, result in BH
XCHG   BX,CX     ;exchange BX and CX
MUL    DL        ;multiply AL by DL
DIV    EBX       ;divide EDX:EAX by EBX
```

Direct Addressing

In this addressing mode the effective address is formed by the addition of a segment register and a displacement value that is coded directly into the instruction. The displacement is usually the address of a memory location *within a specific segment.* One way to refer to a memory location in the operand field is to surround the memory address with square brackets. The instruction MOV   [7000H],AX illustrates this concept. The 7000H is not thought of as immediate data here, but rather as address 7000H within the segment pointed to by the DS register. DS is the segment register used by the processor whenever brackets ([ ]) are used in the operand field (unless we override the use of DS by specifying a different segment register). A detailed example should serve to explain this addressing mode more clearly.

Example 3.3:   What is the result of MOV   [7000H],AX? Assume that the DS register contains 1000H, and AX contains 1234H.

Solution: Examine Figure 3.5 very carefully. Remember that when the 80x86 uses a segment register to form an effective address, it shifts the segment register 4 bits to the left

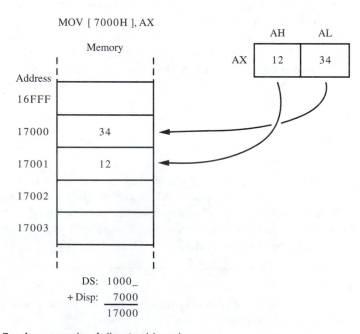

FIGURE 3.5   An example of direct addressing

(turning 1000H into 10000H) and then adds the specified 16-bit displacement (or offset). Thus, the effective address generated by the processor is 17000H. Because register AX is used in the operand field, 2 bytes will be written into memory, with the lower byte (34) going into the first location (17000H) and the upper byte (12) going into the second location (17001H). Once again, we can see that the processor has byte-swapped the data as they were written into memory.

The machine code for MOV   [7000H],AX is A3 00 70.

Another form of direct addressing requires the use of a label in the operand field. The sample program TOTAL that we examined in Section 3.2 used direct addressing in the MOV   SUM,AL instruction. Another look at the TOTAL source file should convince you that the SUM label represents a displacement value within the segment pointed to by the DS register.

The automatic use of the DS register for memory accesses can be overridden by the programmer by specifying a different segment register within the operand field. If we wish to use the extra segment register in the instruction of Example 3.3, we would write MOV   ES:[7000H],AX. The machine code required to allow the use of the ES register in this instruction is 26 A3 00 70. Note the similarity to the machine code in Example 3.3.

The first byte in the instruction, 26H, is called a *segment override prefix*. This byte instructs the processor to use the extra segment instead of the default data segment. Table 3.2 lists the override prefixes for each segment.

Here are some additional examples of direct addressing:

```
ADD   AL,[4000]    ;add contents of location 4000 to AL
OR    [440H],BL    ;or contents of location 440H with BL
DEC   COUNT        ;decrement value stored at COUNT
ADD   SPIN,6       ;add 6 to value stored at SPIN
```

Register Indirect Addressing

In this addressing mode we have a choice of four registers to use within the square brackets to specify a memory location. The four registers to choose from are the two base registers (BX and BP) and the two index registers (SI and DI). The 16-bit contents of the specified register are added to the DS register in the usual fashion (or SS if BP is used).

TABLE 3.2 Segment override prefix bytes

Segment	Prefix
CS	2EH
DS	3EH
ES	26H
FS	64H
GS	65H
SS	36H

Example 3.4: What is the effective address generated by the instruction MOV DL,[SI], if register SI contains 2000H, and the DS register contains 800H?

Solution: Shifting the DS register 4 bits to the left and adding the contents of register SI produces an effective address of 0A000H. Figure 3.6 illustrates this process.

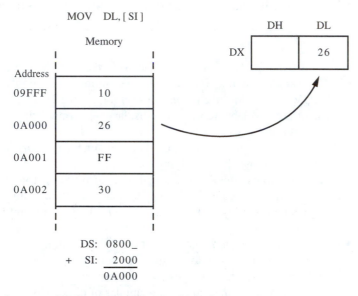

FIGURE 3.6   An example of register indirect addressing

The machine code for MOV   DL,[SI] is 8A 14.

Using a register in the operand field to point to a memory location is a very common programming technique. It has the advantage of being able to generate any address within a specific segment simply by changing the value of the specified register.

Here are some additional examples of register indirect addressing:

```
MOV   ES:[DI],AL      ;put copy of AL into location pointed to
                       by DI in the extra segment
ADD   CX,[BP]         ;add word at location pointed to by BP (in
                       stack segment) to CX
XOR   [BX],DH         ;exclusive-or value at location pointed to
                       by BX with DH
JMP   [BP]            ;jump indirect to address stored at
                       location pointed to by BP
```

Based Addressing

This addressing mode uses one of the two base registers (BX or BP) as the pointer to the desired memory location. It is very similar to register indirect addressing, the difference being that an 8- or 16-bit displacement may be included in the operand field. This displacement is also added to the appropriate segment register, along with the base register, to generate the effective address. The displacement is interpreted as a signed, 2's complement number. The 8-bit displacement gives a range of −128 to +127. The 16-bit displacement gives a range of −32,768 to 32,767. So, the signed displacement allows us to point forward and backward in memory, a handy tool when accessing data tables stored in memory.

Example 3.5: What is the effective address generated by the instruction MOV AX,[BX+4]? Assume that the DS register contains 100H and register BX contains 600H.

Solution: Figure 3.7 shows how the DS register, the BX register, and the displacement value are added together to create the effective address 1604H. This address is then used to read the word out of locations 1604H and 1605H into the AX register.

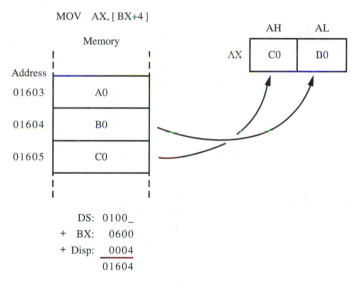

FIGURE 3.7 An example of based addressing

The machine code for MOV   AX,[BX+4] is 8B 47 04.

When the BP register is used, the stack segment is used in place of the data segment. Here are some additional examples of based addressing:

```
SUB   [BP+10],BH      ;subtract BH from value stored at location
                      pointed to by BP plus 10 in the stack segment
MOV   DX,CS:[BX-2]    ;put a copy of the value stored at
                      location pointed to by BX minus 2 (in the
                      code segment) into DX
AND   [BP+4000H],CH   ;and CH with value stored at location
                      pointed to by BP plus 4000H
CMP   AL,[BX+100]     ;compare AL with value stored at
                      location pointed to by BX plus 100
```

Indexed Addressing

Like based addressing, indexed addressing allows the use of a signed displacement. The difference is that one of the index registers (SI or DI) must be used in the operand field.

Example 3.6:   What is the effective address generated by the instruction MOV [DI–8],BL? Assume that the DS register contains 200H and that register DI contains 30H.

Solution: Figure 3.8 shows how the negative displacement is combined with the DI register. Notice that although the DI register points to address 30H within the data segment, the actual location accessed is 8 bytes behind.

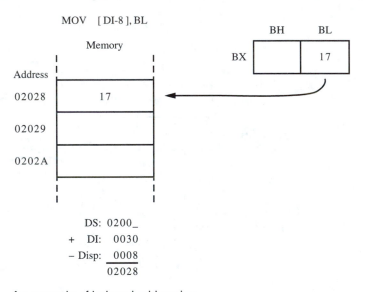

FIGURE 3.8   An example of indexed addressing

The machine code for MOV   [DI–8],BL is 88 5D F8. As in the previous example, the third byte in the machine code is the displacement value. In this case, the displacement of –8 is coded as F8 hexadecimal, which is the 2's complement of 8.

Here are some additional examples of indexed addressing:

```
MOV    AL,ES:[SI]        ;put a copy of the value stored in
                         location to by SI (in the extra segment) into AL
CALL   [SI+2]            ;call subroutine whose address is
                         stored at location pointed to by SI plus 2
XOR    [DI-80H],DX       ;exclusive-OR value stored at location
                         pointed to by DI minus 80H with DX
ADD    [DI],BX           ;add BX to value stored at location
                         pointed to by DI
```

Based Indexed Addressing

This addressing mode combines the features of based and indexed addressing but does not allow the use of a displacement. Two registers must be used as the source or destination operand, one from BX/BP and the other from SI/DI. The contents of both registers are *not* interpreted as signed numbers; therefore, each register may have a range of values from 0 to 65535. Once again, use SS instead of DS when BP is used.

Example 3.7:   What is the effective address generated by MOV   [BP+SI],AH? Assume the SS register contains 2000H, register BP contains 4000H, and register SI contains 800H.

Solution: Shifting the SS register 4 bits to the left and adding the contents of BP and SI gives an effective address of 24800H. See Figure 3.9 for an illustration of this. Also note that a different form of the instruction is used in the figure. MOV   [BP][SI],AH is a different way of saying MOV   [BP+SI],AH. Both methods of specifying the operand are acceptable.

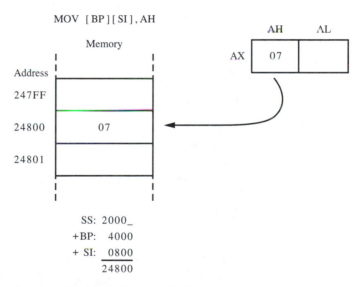

FIGURE 3.9   An example of based indexed addressing

The machine code for MOV   [BP+SI],AH is 88 22.

Here are some additional examples of based indexed addressing:

```
MOV    AX,[BX+SI]         ;put copy of value stored in location
                          pointed to by BX plus SI into AX
ADD    [BP+DI],BL         ;add BL to value stored at location
                          pointed to by BP plus DI in the stack segment
OR     CL,ES:[BX][DI]     ;or CL with value stored at location
                          pointed to by BX plus DI (in the extra segment)
PUSH   [BP][SI]           ;push value stored at location
                          pointed to by BP plus SI in the stack segment
```

Note that [BP+SI] and [BP][SI] are both acceptable operand formats.

Based Indexed with Displacement Addressing

This addressing mode combines all of the features of the addressing modes we have been examining. As with all the others, it still accesses memory locations only within one 64KB segment. It does, however, give the programmer the option of using two registers to access stored information, and the addition of the signed displacement makes this addressing mode the most flexible of all.

Example 3.8: What memory location is accessed by MOV   CL,[BX+DI+2080H]? Assume that the DS register contains 300H, register BX contains 1000H, and register DI contains 10H.

Solution: Figure 3.10 shows how the three registers and the displacement are added to create the memory address 06090H. The byte stored at this location is copied into register CL.

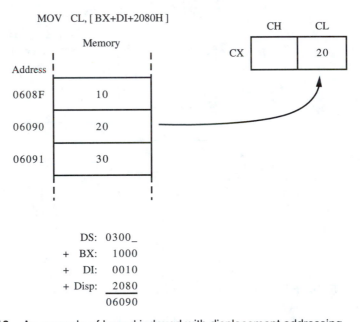

FIGURE 3.10   An example of based indexed with displacement addressing

The machine code for MOV   CL,[BX+DI+2080H] is 8A 89 80 20. In this case the last 2 bytes are the byte-swapped displacement.

Here are some additional examples of based indexed with displacement addressing:

```
ADD    AL,[BP+DI-50]       ;add the value stored in location pointed to by BP plus
                            DI minus 50 (in the stack segment) to AL
XOR    [BX+SI-4E00H],DX    ;exclusive-OR DX with the value stored in location pointed to
                            by BX plus SI minus 4E00H
```

```
JMP     [BP+SI+1000]            ;jump to address stored in location
                                pointed to by BP+SI+1000 in the stack segment
MOV     ES:[BX+DI+4],AX         ;put copy of AX into location pointed
                                to by BX plus DI plus 4 (in the extra
                                segment)
```

String Addressing

The instructions that handle strings do not use the standard addressing modes covered in the previous examples. So, when we look at a string instruction we will not see any registers listed in the operand field. For example, consider the string instruction MOVSB (move byte string). No processor registers are shown in the instruction, but the processor knows that it should use both SI and DI during execution of the instruction (as well as DS and ES). All of the string-based instructions assume that the SI register points to the first element in the source string (which might be either a byte value or a word value) and that the DI register points to the first element of the destination string. The 80x86 will automatically adjust the contents of SI and DI during execution of the string instruction.

We will examine string operations in more detail in Section 3.7.

Port Addressing

Intel brand microprocessors differ from other processors on the market in their implementation of the use of **I/O ports** for data communication between the processor and the outside world. One way to get information into the processor is to read it from memory. Another way is to read an **input port.** When sending data out of the CPU, we can direct it into a memory location or send it to an **output port.** The 80x86 provides the programmer with up to 65,536 input and output ports (although many useful designs rarely use more than a handful of I/O ports). An I/O port is accessed in much the same way that memory is accessed, by placing the address of the I/O port onto the address bus and enabling certain control signals. The address of the I/O port can be coded directly into an instruction, as in:

```
IN   AL,40H
```

or

```
OUT  80H,AL
```

In this case, the port address must be between 00 and FFH, a total of 256 different I/O ports. Notice that the AL register is used to receive the port information. We could also use AX to receive 16 bits of data from an input port (as in IN   AX,38H).

A second method of addressing I/O ports requires that the port address be placed into the DX register. The corresponding instructions are:

```
IN   AL,DX
```

and

```
OUT  DX,AL
```

Because we are now using register DX to store the port address, our choices range from 0000 to FFFFH, a total of 65,536 I/O locations. Note that IN   AX,DX and OUT DX,AX are also allowed.

I/O ports are very useful for communication with peripherals connected to the processor, such as serial and parallel I/O devices, video display chips, clock/calendar chips, and many others.

Using 32-Bit Addressing in Real Mode

Recall from Chapter 2 that the 80x86 provides an additional method of generating addresses that produces 32-bit effective addresses. In the real mode, these 32-bit addresses must fall within the familiar 64KB range (0000H to FFFFH) used within a segment. The advantage is that any extended register may be used as a base register or as an index register (or both in the same instruction). The only exception is the ESP register, which may only be used as a base register.

A typical instruction using 32-bit addressing might look like this:

```
MOV    AX,[EBX][ECX*4]
```

In this example, EBX is the base register and ECX is the index register. Index registers may be *scaled* (multiplied) by a factor of 1, 2, 4, or 8. This allows easy access of 1-, 2-, 4-, and 8-byte quantities.

When a single register is used to specify the 32-bit address, it is automatically treated as a base register unless a scale factor is included. The following two instructions illustrate this important difference:

```
MOV    SI,[EDX]       ;EDX is a base register
MOV    DI,[EDX*8]      ;EDX is an index register
```

The same register may also be used as both a base and index register, as in:

```
MOV    BX,[EAX][EAX*2]
```

An optional displacement of 8 or 32 bits may also be included. This is useful for specifying the starting address of a block of data, such as an array.

When the base register is EBP or ESP, the stack segment is used for the memory access. Otherwise, the data segment is used by default.

Keep in mind that it is not possible to mix 16- and 32-bit registers in the same address operand. So, an instruction like:

```
MOV    AL,[EBX][CX]
```

causes an error during assembly, since EBX and CX may not be used together.

Take the time now to review the addressing modes we have just examined, because they are crucial to understanding the operation of the instructions we will begin looking at in Section 3.6.

3.5   THE PROCESSOR FLAGS (CONDITION CODES)

The 80x86 has a number of status indicators that are referred to as *condition codes* or *flags*. Both terms convey useful information. The "condition" part of condition codes refers to information about the most recently executed instruction. Did the last DEC   AL instruction produce a 0 in AL? The "zero condition" is an example of a condition code. From another point of view, when we see a flag waving, we often know that some new event has just occurred. In this sense, the **zero flag** is a way for the processor to wave at the programmer when a zero condition has occurred. Figure 3.11 shows how the zero flag changes from a 0 to a 1 when the AL register is finally decremented to zero. The flag is really a single bit in the flag register. This bit can be only a 0 or a 1. There are many instructions that look at the zero flag and make a decision based on its contents (e.g., jump if the zero flag is not 0).

The processor has five main flags that are commonly tested during the execution of a program and others that we will examine later. The five main flags and their bit position within the processor's flag register are shown in Figure 3.12(a). The flags and their meaning are as follows.

FIGURE 3.11 Operation of the zero flag

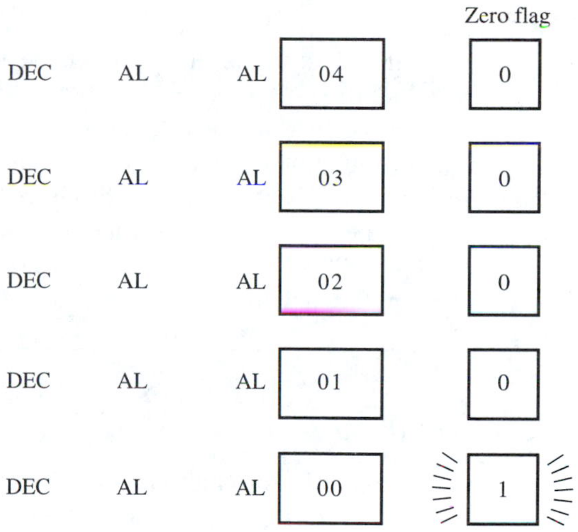

FIGURE 3.12 Flag register, lower word

7	6	5	4	3	2	1	0
S	Z	-	A	-	P	-	C

(a) Condition code half

15	14	13	12	11	10	9	8
-	NT	IOPL		O	D	I	T

(b) Additional flags

Sign (Bit 7)

When we make use of signed binary data in a program, there are times when we wish to know if the last addition or subtraction produced a positive or negative result. Remember that when we use 2's complement format, the most significant bit in the data is used as the sign bit. This would be bit 7 for a byte value and bit 15 for a word value. The processor examines this bit and adjusts its sign flag accordingly. When the sign flag is 0, the processor is saying that the number produced by the last arithmetic or logical instruction is positive. When the sign flag is a 1, the number can be interpreted as a negative number in 2's complement notation.

Example 3.9:   Consider the following two pairs of instructions:

```
MOV   AL,3FH      MOV   AL,7FH
INC   AL          INC   AL
```

In each case, the final value in the accumulator (AL) will be interpreted as a *signed* binary number.

In the first pair of instructions, the accumulator value 3FH is incremented to 40H, which is 01000000 in binary. Notice that the MSB (bit 7) is 0, indicating that 40H is a *positive* number. The sign flag will be cleared in this case.

In the second pair of instructions, the accumulator value 7FH is incremented to 80H, which is 10000000 in binary. The MSB is now 1, indicating that 80H is a *negative* number. The sign flag will be set by this result.

Zero (Bit 6)

We were briefly introduced to the zero flag in the beginning of this section. The zero flag is set (made equal to 1) or cleared (made equal to 0) after execution of many arithmetic and logical instructions. In Figure 3.11 we saw that the zero flag was set when the AL register was finally decremented to 0. The zero flag is often used in programs to determine when a match has been found in a compare operation, when a register or memory location contains 0, and when a loop should be terminated. It is not difficult to see that the zero flag is set when a "zero" is created by an instruction.

Example 3.10:   From the hardware perspective, it is not difficult to determine if a group of bits are all 0. As Figure 3.13 shows, a NOR gate is used to generate an output signal called ZERO. A NOR gate outputs a 1 when all of its inputs are 0. Think of the eight NOR inputs as if they were connected to the individual bits of AL. Thus, when AL becomes 00000000, ZERO will go high. Any other 8-bit pattern in AL will cause ZERO to be low (indicating a *not-zero* condition). Can you imagine what is needed to check for 0 in a 16- or 32-bit register?

FIGURE 3.13 Hardware generation of zero condition

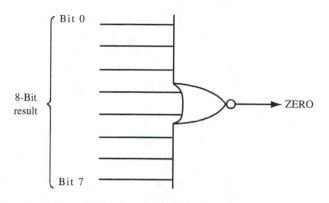

Carry (Bit 0)

Suppose that register BL contains FEH. If we increment BL we will get FFH. This represents an 8-bit number containing all 1s (or 255 as an unsigned decimal). If we increment BL again, what do we get? The correct answer is not 100H, but 00 *with a 1 in the carry flag.* Figure 3.14 shows this concept in graphical form. Since 100H requires 9 bits for representation, we cannot store 100H in register BL. We can store only the lower 8 bits, which are all low. The carry flag is used to store the ninth bit (or the 17th bit in a 16-bit operand and a 33rd bit in a 32-bit operand).

Example 3.11: The carry flag is also used to indicated a *borrow* as a result of a subtraction. Consider these two pairs of instructions:

```
MOV  AL,6      MOV  AL,6
SUB  AL,1      SUB  AL,9
```

Remember that subtraction in binary is found by adding the 2's complement of one number to another.

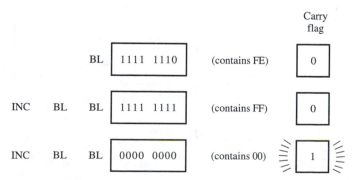

FIGURE 3.14 Operation of the carry flag

In the first pair of instructions, subtracting 1 from 6 leaves 5. The carry flag is cleared in this case, because we subtracted a smaller number from a larger one.

In the second pair of instructions, subtracting 9 from 6 gives FDH (or 11111101 in binary, the 2's complement representation of −3). Because we subtracted a larger number from a smaller one, the carry flag will be set, indicating a borrow.

Auxiliary Carry (Bit 4)

The operation of this flag is very similar to that of the carry flag except that the carry is out of bit 3 instead of bit 7 (or 15 or 31). In other words, the auxiliary carry flag indicates a carry out of the lower 4 bits. The main reason for including this flag is to aid in the execution of the processor's decimal adjust instructions. There are no instructions that directly test the state of this flag as there are for the sign, zero, and carry flags.

Parity (Bit 2)

The parity bit is used to indicate the presence of an even number of 1s in the lower 8 bits of a result. For example, if a logical instruction produced the bit pattern 11010010 in register AL, the parity flag would be set, because there are an even number of 1s (four actually) in AL. If the pattern had been 11110010, the parity flag would be cleared, because five 1s is not even.

Example 3.12:   The state of the parity flag is easily determined through the use of exclusive-NOR gates. As shown in Figure 3.15, seven exclusive NOR gates are used to determine the parity of an 8-bit result. An exclusive NOR gate outputs a 1 when its inputs are identical (00 or 11) and a 0 when its inputs are different. From Figure 3.15 it is clear that the input number 11010010 generates a 1 on the PARITY output, indicating even parity.

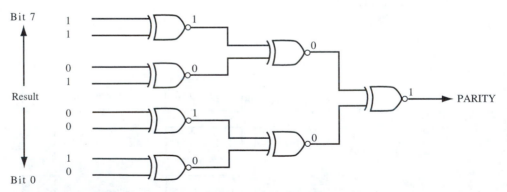

FIGURE 3.15   Parity generation with exclusive NOR gates

Of the five flags just examined, the first three (sign, zero, and carry) are the most often used. We must be familiar with the processor flags to understand the results produced by many of the instructions we will begin examining in the next section.

Other Flags

A number of flags are found in bits 8–15 of the flag register (see Figure 3.12(b) for these). The trace, direction, interrupt-enable, overflow, I/O priority level, and nested task flags will all become important to us later, as will the protected mode flags in bits 16–31 of the flag register. For now, simply keep in mind that they are also part of the flag register.

The flags, or condition codes, contain valuable information concerning the operation of a program on an instruction-by-instruction basis. Thus, we should make good use of the flags when writing programs. Employ the conditional jump instructions where possible, and pay attention to how the flags are affected by all instructions in the program. Sometimes a well-written program that appears completely logical in its method will still yield incorrect results because a flag condition was overlooked.

Keep the condition codes in mind as you study the remaining sections, and be sure to use Appendix B as you examine each new instruction.

3.6   DATA TRANSFER INSTRUCTIONS

As with any microprocessor, a detailed presentation of available instructions is important. Unless you have a firm grasp of what can be accomplished with the instructions you may use, your programming will not be efficient. Indeed, you may even create problems for yourself.

Still, there is no better teacher than experience. You should experiment with these instructions and examine their results. Compare what you see with the manufacturer's data. A difficult concept often becomes clear in practice.

Each of the following sections deals with a separate group of instructions. Information about the instruction, how it works, how it is used, what its mnemonic looks like, how it affects the condition codes, and more, will be presented for each instruction. Even so, it is strongly suggested that you constantly refer to Appendix B as you read about each new instruction. Most of the material in this appendix, such as allowable addressing modes and condition code effects, is not reproduced here.

The instructional groups are organized in such a way that the more commonly used instructions are presented first, followed by less frequently used instructions. For example, in the group dealing with data movement instructions, the MOV instruction is presented first due to its wide range of applications. Although the LDS and LES instructions come first alphabetically, they have restricted functions and are not needed in many programming applications. So, they do not get to steal the spotlight from MOV by appearing first.

Hopefully, covering the instructions in this fashion will allow you to study the important instructions first (in each group), and the other instructions as the need arises.

In most cases the machine code for the instruction will be included. This is done simply to compare instruction lengths and explain new features about the 80x86.

Examples will also be given for each instruction. Let us begin our coverage of the instruction set with the data transfer instructions.

Moving Data

This group of instructions makes it possible to move (copy) data around inside the processor and between the processor and its memory and I/O systems. Remember that the more popular instructions are covered first.

MOV   Destination,Source (Move Data).   We have already seen many examples of the MOV instruction in its role of explaining the addressing modes. So, instead of repeating those examples here, we will examine other aspects of the MOV instruction. When we use MOV to load immediate data into a register, the assembler will look at the size of the specified register in the operand field to determine if the immediate data is a 1-, 2-, or 4-byte number. For example:

```
MOV    AL,30H
```

and

```
MOV    AX,30H
```

are two different instructions and the immediate data 30H is interpreted differently in both of them. In the first instruction the 30H is coded as a byte value because it is being MOVed into AL. In the second instruction the 30H is coded as a word value because it is being MOVed into AX. This is clearly shown by the resulting code for both instructions. MOV AL,30H is coded as B0 30. MOV   AX,30H is coded as B8 30 00. Note that the second two bytes represent the byte-swapped value 0030H.

This much we have already seen. But what happens when the assembler does not have any way of determining how large the operands should be? For example, in MOV   [SI],0 the processor does not know if the 0 should be coded as a byte value, word value, or as a double-word value. This instruction would then produce an error during assembly. For cases like this, include some additional information in the instruction's operand field. If you wish to MOV a byte value into memory, use MOV   BYTE PTR [SI],0. Word values require MOV   WORD PTR [SI],0 and double-word values require MOV   DWORD PTR [SI],0. The **byte ptr, word ptr,** and **dword ptr** assembler directives stand for "byte pointer," "word pointer," and "double-word pointer." The corresponding code for MOV   BYTE PTR [SI],0 is C6 04 00. For MOV   WORD PTR [SI],0 it is C7 04 00. For MOV   DWORD PTR [SI],0, we get the machine code 66 C7 04 00000000. Notice the operand-size prefix byte. This pointer feature of the assembler can be applied to many 80x86 instructions.

MOVSX Destination,Source (Move with Sign Extended).   When working with signed binary values, it is common to convert 8- or 16-bit numbers into 16- or 32-bit *sign-extended* numbers. Recall that the MSB of a signed number is used to represent the sign (+/−) of the number. It is commonly called the **sign bit.** Negative numbers have a sign bit equal to one. The MOVSX instruction examines the state of the sign bit when extending the source value. When the source is an 8-bit operand, the upper byte of the destination will be set to 00H for a positive source value or FFH for a negative source value. Thus, the sign of the source is extended through the upper 8 bits of the destination.

When the source is a 16-bit operand, the sign is extended through the upper 16 bits of the destination, resulting in 0000H or FFFFH in the upper two bytes. Example 3.13 demonstrates the operation of MOVSX.

Example 3.13: Suppose that register AL contains 36H and register BX contains C3EEH. What are the results of MOVSX   AX,AL and MOVSX   EBX,BX?

Solution: When MOVSX   AX,AL executes, the result in AX equals 0036H. The sign bit of AL is zero, causing the upper 8 bits of AX to be set to zero.

MOVSX   EBX,BX causes the sign bit of BX (a one) to be extended through the upper 16 bits of EBX, giving FFFFC3EEH as the 32-bit result. Each destination register has the same sign as its corresponding source register.

The machine code for MOVSX   AX,AL is 0F BE C0. The machine code for MOVSX   EBX,BX is 66 0F BF DB.

MOVZX Destination,Source (Move with Zero Extended). This instruction is used to convert 8- and 16-bit values into 16- and 32-bit values by adding leading zeros to the source operand. Thus, a value like 89H becomes 0089H, and 3700H becomes 00003700H. This is similar to what MOVSX does, except that the sign bit is ignored. MOVZX should not be used when working with signed numbers.

Example 3.14: Repeat Example 3.13 for the same register values, except use MOVZX instead of MOVSX. What are the final results?

Solution: Register AX contains 0036H, and EBX contains 0000C3EEH. Note the difference in EBX from the FFFFC3EEH value of Example 3.13. The upper 16 bits of EBX are not sign-extended by MOVZX as they are by MOVSX.

The machine code for MOVZX   AX,AL is 0F B6 C0. The machine code for MOVZX   EBX,BX is 66 0F B7 DB.

PUSH   Source (Push Data onto Stack). It is often necessary to save the contents of a register so that it can be used for other purposes. The saved data may be copied into another register or written into memory. If we are interested only in saving the contents of one or two registers we could simply reserve a few memory locations and then directly MOV the contents of each register into them. This practice is limiting, however, because we cannot easily modify our program in the future (say, to save additional registers in memory) without making a significant number of changes. Instead, we will use a special area of memory called the **stack.** The stack is a collection of memory locations pointed to by the stack pointer register and the stack segment register. When we wish to write data into the stack area we use the PUSH instruction. We commonly refer to this as pushing data onto the stack. The processor will automatically adjust the stack pointer during the push in such a way that the next item pushed will not interfere with what is already there. Example 3.15 shows the operation of PUSH in more detail.

Example 3.15: The stack segment register has been loaded with 0000 and the stack pointer register with 2000H. If register AX contains 1234H, what is the result of PUSH   AX?

Solution: The stack pointer (presently containing 2000H) points to a location referred to as the top of the stack. Whenever we push an item onto the stack, the stack pointer is decremented by 2. This is necessary because all pushes involve 2 bytes of data (usually the contents of a register). Decrementing the stack pointer during a push is a standard way of implementing stacks in hardware. Figure 3.16 shows the new contents of memory after PUSH AX has executed. The data contained in memory locations 01FFF and 01FFE is replaced by the contents of register AX. Notice that the contents have been byte-swapped during the write (34 comes before 12) and that the new stack pointer value is 1FFE. Remember that the stack *builds* toward 0. As more items are pushed, the address within the stack pointer gets smaller by 2 each time. Also notice that the contents of register AX remain unchanged.

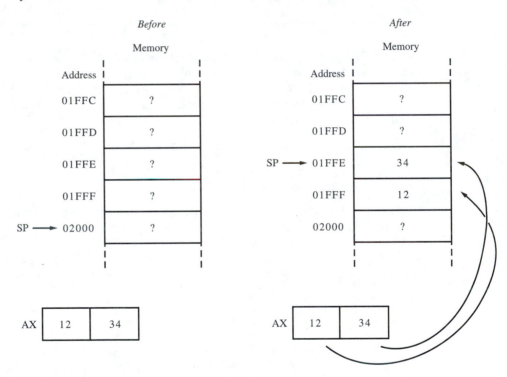

FIGURE 3.16   Execution of PUSH   AX

The machine code for PUSH   AX is 50.

When the SP register is pushed, the value written to the stack is the value of SP before the push.

PUSHW/PUSHD Source (Push Word/Double-Word Data onto Stack).   The operation of PUSHW and PUSHD is similar to PUSH. PUSHW is used to push immediate word values onto the stack, as in:

```
PUSHW      1000H
```

PUSHD is used to push double-word values onto the stack. PUSHD may be used to push the 32-bit processor registers onto the stack, as in:

```
PUSHD       EAX
```

PUSHW decrements the stack pointer by 2 and PUSHD decrements SP by 4. Note that PUSH EAX and PUSHD EAX are equivalent.

Example 3.16: Figure 3.17 illustrates the changes made to stack memory and the SP during execution of these two instructions:

```
PUSHW       1000H
PUSHD       EAX
```

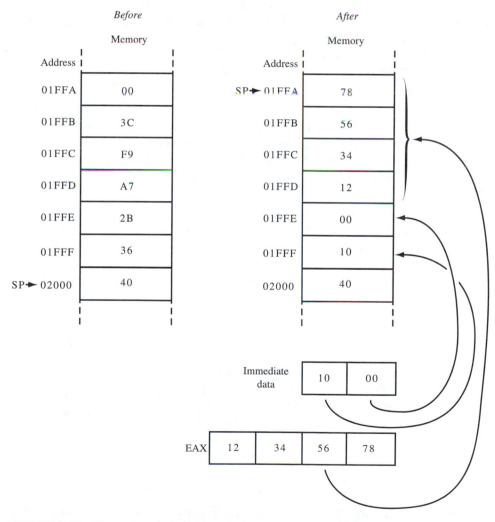

FIGURE 3.17 Execution of PUSHW 1000H and PUSHD EAX

TABLE 3.3   Order of registers pushed with PUSHA/PUSHAD

When Pushed	PUSHA	PUSHAD
First	AX	EAX
	CX	ECX
	DX	EDX
	BX	EBX
	SP	ESP
	BP	EBP
	SI	ESI
Last	DI	EDI

Register EAX contains 12345678H. The stack pointer is decremented by 6 during execution of both instructions.

    The machine code for PUSHW   1000H is 68 00   10. The machine code for PUSHD EAX is 66   50.

PUSHA/PUSHAD   (Push All Registers/Push All Double-Registers).   These instructions automatically push all general purpose registers and the stack pointer onto the stack. PUSHA pushes all 16-bit registers and PUSHAD pushes all 32-bit registers. Table 3.3 indicates the order in which the registers are pushed.

Example 3.17:   If the stack pointer initially contains 2000H, what is its value after PUSHAD executes?

Solution: Since eight 32-bit registers are pushed, the SP is decremented by 4 a total of eight times. This gives a final SP value of 1FE0H.

    The machine code for PUSHAD is 66 60.

POP   Destination (Pop Word off Stack).   The POP instruction is used to perform the reverse of a PUSH. The stack pointer is used to read 2 bytes of data and copy them into the location specified in the operand field. The stack pointer is automatically incremented by 2. All registers except CS and IP may be popped. The destination operand may be a memory location.

Example 3.18:   Assume the contents of the stack segment register and the stack pointer are 0000 and 2000H, respectively. What is the result of POP   BX?

Solution: Figure 3.18 shows a snapshot of memory contents in the stack area. The contents of location 2000 (20) are copied into the lower byte of BX. The stack pointer is incremented and the contents of location 2001 (30) are copied into the upper half of BX. The old

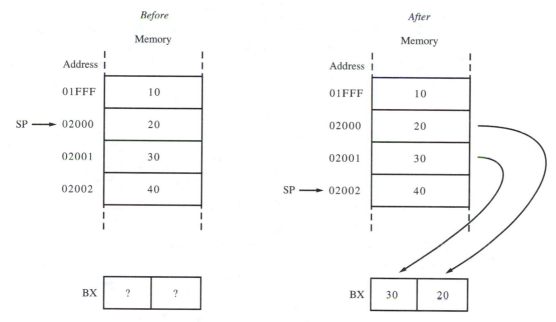

FIGURE 3.18 Execution of POP   BX

contents of BX are lost. The stack pointer is then incremented a second time. Compare the operation of PUSH and POP and you will see that they complement one another.

The machine code for POP   BX is 5B.

POPA/POPAD Destination (Pop All Registers/Pop All Double-Registers). These instructions complement the PUSHA/PUSHAD instructions. All general purpose registers (16-bit for POPA, 32-bit for POPAD) are popped from the stack in the order indicated in Table 3.4. It is important to note that the contents of the SP (or ESP) are not loaded with the data popped off the stack. This is necessary to prevent the stack from changing locations halfway through the execution.

TABLE 3.4 Order of registers popped with POPA/POPAD

When Popped	POPA	POPAD
First	DI	EDI
	SI	ESI
	BP	EBP
	SP	ESP
	BX	EBX
	DX	EDX
	CX	ECX
Last	AX	EAX

Note: The value popped for SP/ESP is discarded.

Example 3.19:   Figure 3.19 shows the result of executing the POPA instruction. Initially, the SP contains 1FF0H. After POPA executes, the SP contains 2000H, *not* the value of 8877H popped off the stack. The memory data is shown in byte-swapped format, which accounts for the difference of 2 between each address.

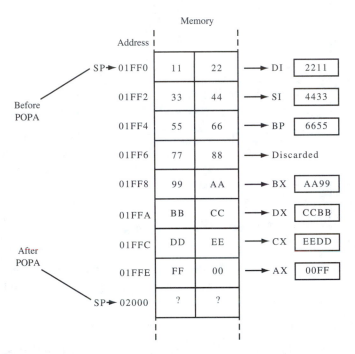

FIGURE 3.19   Loading all general purpose registers with POPA

The machine code for POPA is 61.

IN   Accumulator,Port (Input Byte or Word from Port).   This is another instruction that was briefly introduced during our examination of addressing modes. The processor has an I/O space that is separate from its memory space. There are 65,536 possible I/O ports available for use by the programmer. In almost all cases a machine designed around the 80x86 would use only a handful of these ports.

The input port is actually a hardware device connected to the processor's data bus. When executing the IN instruction, the processor will output the address of the input port on the address bus. The selected input port will then place its data onto the data bus to be read by the processor. Data read from an input port *always ends up in the accumulator.*

The processor allows two different forms of the IN instruction. If a full 16-bit port address must be specified, the port address is loaded into register DX, and IN   AL,DX or IN   AX,DX is used to read the input port. If the port number is between 00 and FFH, a different form of IN may be used. In this case, to input from port 80H we would use IN AL,80H or IN   AX,80H. Using AL in the operand field causes 8 bits of data to be read. Two bytes can be input by using AX in the operand field.

Example 3.20: What is the result of IN AL,80H if the data at input port 80 is 22?

Solution: The byte value 22 is copied into register AL.
The machine code for IN AL,80H is E4 80.

It may be helpful for you to remember the expression "All I/O is through the accumulator" when working with the I/O instructions. Input ports can be used to read data from keyboards, A/D converters, DIP switches, clock chips, UARTs, and other peripherals that may be connected to the CPU.

INS Destination,DX (Input String from Port). This instruction is very similar to IN, except that the destination is a memory location instead of the accumulator. The memory location is pointed to by the DI register and is located in the extra segment. It is common to use the notation ES:DI to represent a segment:offset pair. So, ES:DI is the memory location where the input port (address in DX) data will be written to. The destination operand specified in the instruction is only used to indicate the size of the data transfer: byte, word, or double-word.

After the input port data has been written into memory, the DI register is automatically incremented or decremented by 1, 2, or 4, depending on the state of the processor's direction flag. This is a feature of string operations, which we will examine more closely in Section 3.7.

Three simplified instructions are equivalent to INS, one for each data size. INSB inputs bytes, INSW inputs words, and INSD inputs double-words. All three use DX as the port address and ES:DI as the destination, but do not require any operands, as INS does.

Example 3.21: The following instructions (and associated machine code) perform the same job:

```
6C       INSB   or   INS   BYTE PTR ES:[DI],DX
6D       INSW   or   INS   WORD PTR ES:[DI],DX
66 6D    INSD   or   INS   DWORD PTR ES:[DI],DX
```

OUT Port,Accumulator (Output Byte or Word to Port). This instruction is a complement to the IN instruction. With OUT, we can send 8 or 16 bits of data to an output port. The port address may be loaded into DX for use with OUT DX,AL or OUT DX,AX, or specified within the instruction, as in OUT 80H,AL or OUT 80H,AX.

Example 3.22: What happens during execution of OUT DX,AL if AL contains 7C and DX contains 3000?

Solution: The port address stored in register DX is output on the address bus, along with the 7C from AL on the data bus. The output port circuitry must recognize address 3000 and store the data.

The machine code for OUT   DX,AL is EE.

OUTS DX,Source (Output String to Port).   This instruction reads data from string memory located at DS:SI and outputs it to the port specified by register DX. The size of the source operand controls the size of the data output to the port. The SI register is automatically incremented or decremented (depending on the state of the direction flag) after the data transfer is made.

Three simplified forms of OUTS are recognized by the assembler. They are OUTSB, OUTSW, and OUTSD. Each outputs a byte, word, or double-word, respectively, to the port specified by register DX. No operands are necessary, since the data will automatically be read using the SI register and the data segment.

Example 3.23:   Using a source operand to specify the size of the data transfer may be done like this:

```
6E          OUTS    DX,BYTE PTR [SI]
6F          OUTS    DX,WORD PTR [SI]
66  6F      OUTS    DX,DWORD PTR [SI]
```

The same result may be accomplished by using one of the simplified forms:

```
6E          OUTSB
6F          OUTSW
66  6F      OUTSD
```

Remember that the SI register is automatically incremented or decremented by 1, 2, or 4 after each operation.

String operations, including another look at OUTS, will be covered in detail in Section 3.7.

LEA   Destination,Source (Load Effective Address).   This instruction is used to load the offset of the source memory operand into one of the processor's registers. The memory operand may be specified by any number of addressing modes. The destination may not be a segment register.

Example 3.24:   What is the difference between MOV   AX,[40H] and LEA   AX,[40H]?

Solution: In MOV   AX,[40H] the processor is directed to read 2 bytes of data from locations 40H and 41H and place the data into register AX. In LEA   AX,[40H] the processor simply places 40H into register AX. Modifying these instructions slightly should help to further define the difference just presented. Suppose that a label called TIME has been defined in a program, and has the memory address value of 40H associated with it. Using MOV   AX,TIME will cause the data at locations TIME and TIME+1 (40H and 41H) to be read from memory and stored in AX. Using LEA   AX,TIME will cause the effective address of the label (which is 40H) to be copied into AX. The memory location

TIME is never accessed in LEA   AX,TIME. All operands dealing with memory locations
are assumed to be found in a 64KB block pointed to by the data segment register.

What does LEA   AX,[SI] do? It *does not* read the contents of the memory locations
pointed to by SI. Instead, the value of SI at execution time is loaded into AX.

The machine code for LEA   AX,[40H] is 8D 06 40 00. The machine code for LEA
AX,[SI] is 8D 04.

PUSHF/PUSHFD   (Push Flags onto Stack).   There are times when it is necessary to save
the state of each flag in the processor's flag register. Usually this is done whenever the
processor is interrupted. Saving the flags and restoring them at a later time, along with the
processor registers, is a proven technique for resuming program execution after an interrupt. PUSHF pushes the lower 16 bits of the flag register onto the stack. Use PUSHFD to
push the entire 32-bit flag register.

Example 3.25:   Assume that the stack segment register and the stack pointer have been
loaded with addresses 0000 and 2000H, respectively, and that the flag register contains
00C3H. What is the result of PUSHF?

Solution: PUSHF writes the contents of the flag register into stack memory. The operation of PUSHF is similar to PUSH, as you can see in Figure 3.20. The stack pointer is
decremented by 2. Then the lower byte of the flag register (C3) is written into stack memory, followed by the upper byte (00). As usual, we see that the 16-bit flag register has been
byte-swapped as it was written into memory.

The machine code for PUSHF is 9C. The machine code for PUSHFD is 66 9C.

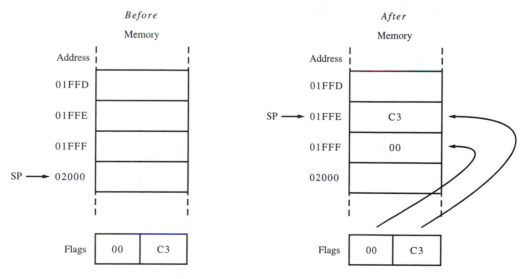

FIGURE 3.20   Operation of PUSHF

POPF/POPFD   (Pop Flags off Stack).   This instruction reverses the operation of PUSHF/PUSHFD, popping 2 or 4 bytes off the stack and storing them in the flag register. The operation is similar to POP, with the stack pointer increased by 2 or 4 at completion.

    The machine code for POPF is 9D. The machine code for POPFD is 66 9D.

XCHG   Destination,Source (Exchange Data).   This instruction is used to swap the contents of two 8-, 16-, or 32-bit operands. One operand must be a processor register (excluding the segment registers). The other operand may be a register or a memory location. If a memory location is used as an operand it is assumed to be within a data segment.

Example 3.26:   Registers AL and BL contain 30 and 40, respectively. What is the result of XCHG   AL,BL?

Solution: After execution, AL contains 40 and BL contains 30.

    The machine code for XCHG   AL,BL is 86 C3. It may be interesting to note that the machine code for XCHG   BL,AL (which performs the same operation as XCHG AL,BL) is 86 D8.

BSWAP   Destination (Byte-Swap).   This instruction swaps bytes in a 32-bit general purpose register. The upper and lower bytes switch places, as do the middle 2 bytes. BSWAP is useful for converting 32-bit numbers from *little endian* format into *big endian* format, and vice versa. A number stored in little endian format has its lower byte in the lowest memory location. Big endian format places the upper byte in the lowest memory location. Big endian numbers are found in Motorola processors, such as the 680x0 series. Intel machines use little endian numbers (Intel byte-swapping).

Example 3.27:   The contents of register EAX are swapped with BSWAP as indicated in Figure 3.21. Note that executing two BSWAP instructions in a row with the same register restores the register to its original value.

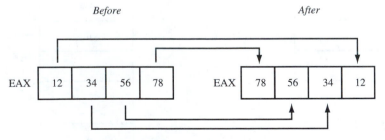

FIGURE 3.21   Result of BSWAP EAX

    The machine code for BSWAP EAX is 66 0F C8.

XLAT Translate-Table (Translate Byte). Some programming applications require quick translation from one binary value to another. For example, in an image processing system, binary video information defining brightness in a gray-level image can be falsely colored by translating each binary pixel value into a corresponding color combination. The process is easily accomplished with the aid of a color look-up table. XLAT is one instruction that is useful for implementing such a look-up table. XLAT assumes that a 256-byte data table has been written into memory at the starting address contained in register BX. The number in register AL at the beginning of execution is used as an index into the translation table. The byte stored at the address formed by the addition of BX and AL is then copied into AL. The translation table is assumed to be in the data segment.

Example 3.28: A translation table resides in memory with a starting address of 0800H. How does XLAT know where the table is? If register AL contains 04, what is the result of XLAT?

Solution: XLAT uses register BX as the pointer to the beginning of the translation table, so it is necessary to place the address 0800H into BX before executing XLAT (assume here that the DS register contains 0000). This can be easily done with MOV BX,0800H. Figure 3.22 shows the result of executing XLAT with AL equal to 04. The byte at address 0804H is copied into register AL, giving it a final value of 44.

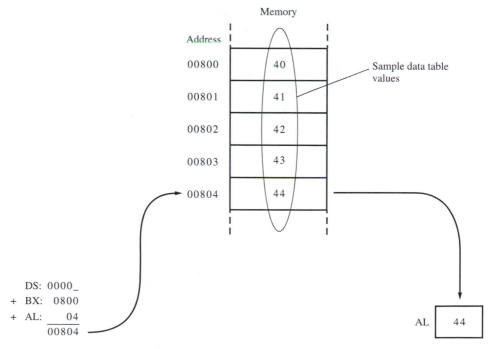

FIGURE 3.22 Execution of XLAT

The machine code for XLAT is D7.

A very useful application of XLAT is encryption. Suppose that the 26 letters of the alphabet are scrambled and then written into memory at a starting address found in register BX. If register AL is restricted to numbers from 1 to 26 (1 representing A, 26 representing Z), executing XLAT will map one letter of the alphabet into a different letter. In this fashion, we can encrypt text messages one character at a time. A second table would then be needed to translate the encrypted message back into correct form.

Another application involves PC keyboards. Because each key on a keyboard generates a unique 8-bit code, it is possible to create different keyboard "layouts," where the layout is actually a translation table that contains user-defined keyboard codes. The original keyboard codes are used as addresses within the translation table to generate the correct code.

LDS   Destination,Source (Load Pointer Using DS).   This instruction is used to load *two* 16-bit registers from a 4-byte block of memory. The first 2 bytes are copied into the register specified in the destination operand of the instruction. The second 2 bytes are copied into the DS register. This instruction will come in handy when working with source strings, which we will look at in the string instructions section.

Example 3.29:   Assume that the DS register contains 0100 and register SI contains 0020. What is the result of LDS   DI,[SI]?

Solution: Figure 3.23 shows how the SI register is used to access the data that will be copied into DI and DS. Note that the addresses read out of memory are byte-swapped. After execution, the DS register contains 9080 and register DI contains F030.

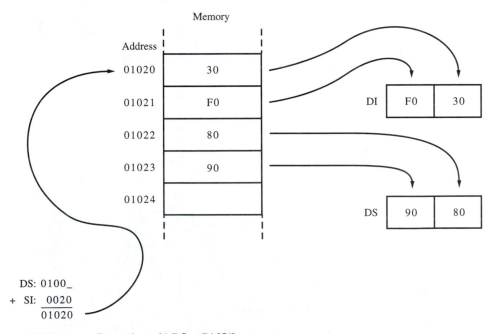

FIGURE 3.23   Execution of LDS   DI,[SI]

The machine code for LDS   DI,[SI] is C5 3C.

LES/LFS/LGS/LSS Destination,Source (Load Pointer Using ES/FS/GS/SS).   This instruction is nearly identical to LDS. The difference is that the second address read out of memory is written into the indicated segment register instead of the DS register.

Example 3.30:   All of the load segment/register instructions may be used with 32-bit extended registers as well. Here are a few examples with accompanying machine code:

```
67 66 C5 06            LDS     EAX,FWORD PTR [ESI]
66 C4 1E 1000          LES     EBX,FWORD PTR TEMP
67 66 0F B4 0C 97      LFS     ECX,FWORD PTR [EDI][EDX*4]
66 0F B2 26 2000       LSS     ESP,FWORD PTR NEWSTACK
```

The labels TEMP and NEWSTACK are located at addresses 1000H and 2000H, respectively. The FWORD PTR directive is used to indicate the 32-bit size of the offset in the address operand.

   The last instruction is particularly useful because it is able to completely change the working stack address during its execution.

LAHF   (Load AH Register from Flags).   One way to determine the state of the flags is to load a copy of them into a register. Then individual bits within the register can be manipulated or tested by the programmer. LAHF can be used to copy the lower byte of the flag register into register AH.

Example 3.31:   The lower byte of the flag register contains 83H. What is the result of LAHF and what is the state of each flag?

Solution: LAHF copies 83 into register AH. Refer to Figure 3.12 for information on the flag positions. Because the binary equivalent of 83H is 10000011, we see that the sign and carry flags are currently set, and the zero, auxiliary carry, and parity flags are cleared.

   The machine code for LAHF is 9F.

SAHF   (Store AH Register into Flags).   This instruction is used to load a new set of flags into the flag register. The contents of register AH are copied into the lower byte of the flag register, giving new values to all five main processor flags.

   The machine code for SAHF is 9E.

Assembler Directives OFFSET, BYTE PTR, WORD PTR, DWORD PTR, FWORD PTR, and SEG

There are times when we need to provide a small amount of assistance to the assembler so that it can figure out how to code an instruction. For example, the instruction:

```
MOV   AL,[SI]
```

contains an operand reference to 8-bit register AL. This indicates to the assembler that the memory location pointed to by [SI] is a *byte* location. By a similar method, the instruction:

```
MOV   [SI],AX
```

contains a reference to the 16-bit register AX. What happens when an instruction contains no reference to size, as in:

```
MOV   [SI],5
```

In this case, the assembler will give an error saying that the operand must have size specified. The assembler does not know if we are trying to write the byte 05, the word 0005, or even the double-word value 00000005 into memory.

To get around instances such as this, we make use of the **BYTE PTR (byte pointer), WORD PTR (word pointer),** and **DWORD PTR (double-word pointer)** assembler directives. In terms of our example, to move the byte value 05 into memory we would use:

```
MOV   BYTE PTR [SI],5
```

and to write the word 0005 we would use:

```
MOV   WORD PTR [SI],5
```

Double-word values are specified like this:

```
MOV DWORD PTR [SI],5
```

Situations like these arise only when we have not used the DB, DW, or DD directives to define the size of a data area.

It is often necessary to load the address of a label into a register. In particular, DOS INT   21H, Function 09H, requires that the address of an ASCII text string be loaded into register DX prior to its call. There are two ways to accomplish this. In the first method, we use the LEA instruction:

```
B4 09          MOV   AH,9
8D 16 000E     LEA   DX,MESSAGE
CD 21          INT   21H
```

In the second method, we use the assembler directive **OFFSET** to perform the same chore:

```
B4 09          MOV   AH,9
BA 000E        MOV   DX,OFFSET MESSAGE
CD 21          INT   21H
```

Can you spot the advantage of using the OFFSET directive? Notice that the LEA   DX instruction requires 4 bytes of machine code. The equivalent MOV   DX instruction needs only 3 bytes of code and fewer clock cycles as well, which results in faster execution. This might not seem like a big savings, but consider that a source file might be hundreds or even thousands of lines long. An instruction like LEA   DX,MESSAGE might appear quite often in the source file. If it appears 100 times, then 100 bytes are saved using MOV   DX with OFFSET. Writing efficient programs, programs that both run quickly and require little space in memory, is important. Using programming techniques like the ones just shown can help shorten the length of an executable program.

The previous instructions that set up register DX for INT 21H's function 09H assumed that the DS register had already been initialized to its proper value. But what if this has not been done? Then it is the programmer's responsibility to initialize the data segment to the address assigned by DOS when the program was loaded into memory. For example, how can we determine the segment address for the MESSAGE string? This is accomplished with the **SEG** directive. Two statements are needed to initialize a segment register with SEG, such as:

```
MOV    AX, SEG MESSAGE
MOV    DS,AX
```

The SEG directive determines the segment where the MESSAGE data is located and places the segment address into AX. This address is then loaded into the DS register with the second MOV instruction. Normally, when only one data area is used in a program, it is not necessary to use the SEG operator, since the single data area is automatically assigned to the data segment. When multiple data areas are required, it may be necessary to initialize the ES, FS, and GS registers. The SEG directive will come in handy in this case.

Last, if 32-bit addressing modes are used, it may be necessary to declare a 48-bit pointer variable. Pointer variables contain a 16-bit segment portion and a 16- or 32-bit offset portion. So, pointers are either 32 bits or 48 bits in length. A 48-bit pointer can be declared like this (in the .DATA section of the source file):

```
BIGPTR    DF    ?
```

where the **DF (Define Far Word)** directive automatically reserves 6 bytes of storage for the pointer. Any instruction that uses BIGPTR as an operand, such as:

```
LDS    EDX,BIGPTR
```

will access the 6 bytes correctly.

When it is not possible to use a far label (such as BIGPTR), the **FWORD PTR (Far Word Pointer)** directive can be used to specify the proper operand size. Example 3.30 shows examples of this directive.

3.7   STRING INSTRUCTIONS

A particularly nice feature of the 80x86 is its ability to handle strings. A string is a collection of bytes, words, or long-words that can be up to 64KB in length. An example of a string might be a sequence of ASCII character codes that constitute a password, or the ASCII codes for "YES." A string could also be the 7 bytes containing your local telephone number. No matter what kind of information is stored in a string, there are a number of common operations we find useful to perform on them. During the course of executing a program, it may become necessary to make a copy of a string, compare one string with another to see if they are identical, or scan a string to see if it contains a particular byte or word. The processor has instructions designed to do this automatically. A special instruction called the **repeat prefix** can be used to repeat the copy, compare, or scan operations. Register CX has an assigned role in this process. It contains the repeat count necessary for the repeat prefix. CX is decremented during the string operation, which terminates when CX reaches 0. The SI and DI registers are also vital parts of all string operations. The processor assumes that the SI register points to the first element of the source string, which must be located in the data segment. The destination string is located in a similar way via the DI register and must reside in the extra segment. A special flag called the **direction flag** is used to control the way SI and DI are adjusted during a string instruction. They are automatically incremented or decremented by 1, 2, or 4, based on the value of the direction flag and the size of the string elements.

Figure 3.24 gives an example of two text strings stored in memory. The first string spells out SHOPPER, and is followed by a blank (20H) and a carriage return code (0DH).

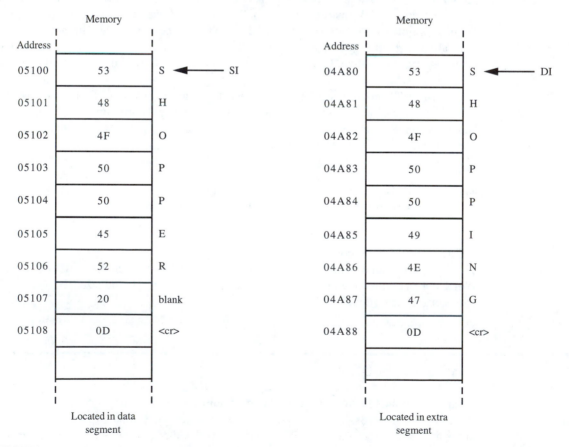

FIGURE 3.24   Two sample text strings

The second string spells out SHOPPING and is followed by only a carriage return code. Although the strings are the same length (9 bytes) they are different after the fifth character. We will be able to use these strings in our examples to understand the operation of the string instructions.

Initializing the String Pointers

Before we can use any string instruction we have to set up the SI, DS, DI, and ES registers. There are a number of ways this can be done. The source string (SHOPPER) in Figure 3.24 could be pointed to by these instructions:

```
MOV   AX,510H   ;string segment-address
MOV   DS,AX
MOV   SI,0      ;string offset within segment
```

When the contents of SI and DS are combined to form an effective address, 05100H will be the first byte accessed in the data segment. A similar technique is used to initialize the destination string (SHOPPING):

```
MOV   AX,4A8H   ;string segment-address
MOV   ES,AX
MOV   DI,0      ;string offset within segment
```

Remember that we cannot move immediate data directly into a segment register, hence our use of the accumulator in the first instruction.

Another way to initialize the string pointers is through the LDS and LES instructions. In this way, both strings can be initialized with only two instructions:

```
LDS   SI,SRCSTR
LES   DI,DSTSTR
```

where SRCSTR and DSTSTR are the labels of two 4-byte pointer fields that contain the string offset and segment values. Refer to Example 3.29 for a review of LDS. (LES works in a similar fashion.)

Using String Instructions

REP/REPE/REPZ/REPNE/REPNZ. These five mnemonics are available for use by the programmer to control the way a string operation is repeated, if at all. REP *(repeat)*, REPE *(repeat while equal)*, REPZ *(repeat while zero)*, REPNE *(repeat while not equal)* and REPNZ *(repeat while not zero)* are all recognized by the assembler as prefix instructions for string operations. MOVS *(move string)* and STOS *(store string)* make use of the REP prefix. When preceded by REP, these string operations repeat until CX decrements to 0. REPE and REPZ operate the same way, but are used for SCAS *(scan string)* and CMPS *(compare string)*. Here, an additional condition is needed to continue the string operation. Each time SCAS or CMPS completes its operation, the zero flag is tested and execution continues (repeats) as long as the zero flag is set. This makes sense, because a compare operation involves an internal subtraction, and the subtraction produces a 0 result when both operands match. The zero flag is set in the case of matching operands and cleared for different ones.

REPNE and REPNZ also repeat as long as CX does not equal 0, but require that the zero flag be cleared to continue. So, we have three ways to repeat string operations:

1. Repeat while CX does not equal 0.
2. Repeat while CX does not equal 0 and the zero flag is set.
3. Repeat while CX does not equal 0 and the zero flag is cleared.

If necessary, combinations of string operations can be used to perform a specific kind of string function.

MOVS   Destination-String,Source-String (Move String).   This instruction is used to make a copy of the source string in the destination string. The names of the strings must be used in the operand fields so that the processor knows whether they are byte strings or word strings. A byte string might be defined like this:

```
STRINGX   DB    'SHOPPER ',0DH
```

and a word string like this:

```
STRINGY   DW    1000H,2000H,3000H,4000H
```

The assembler will associate the DB or DW directive used in the source line to set the type of string being defined. Thus, the assembler will know what kind of strings it is working with when it encounters the operands in MOVS. The operands in MOVS are only used to set the string size. The DS:SI and ES:DI registers must already be initialized to the starting address of each string.

When MOVS executes with the REP prefix, CX must be initialized to the proper count. The state of the direction flag will determine which way the strings are copied. If it is cleared, SI and DI will auto-increment. If the direction flag is set, SI and DI will auto-decrement. The direction flag can be cleared with the CLD instruction and set with the STD instruction.

MOVSB/MOVSW/MOVSD (Move String).   These three mnemonics can be used in place of MOVS and cause identical execution. Because they explicitly inform the assembler of the string size, there is no need to include the string operands in the instruction.

Example 3.32:   What instructions are necessary to make a copy of the SHOPPER string from Figure 3.24? We want the destination string to have a starting address of 3000H, and the index registers should auto-increment during the string operation.

Solution: Because the SHOPPER string is 9 bytes long, we must initialize CX to 9. The direction flag must be cleared to get the copy performed in the correct manner, and REP must be used to copy bytes from the source string until CX is decremented to 0. One way to do all this would be:

```
B8 10 05    MOV    AX,510H     ;source string segment-address
8E D8       MOV    DS,AX
29 F6       SUB    SI,SI       ;source string offset
B8 00 03    MOV    AX,300H     ;destination string segment-address
```

```
8E C0      MOV    ES,AX
29 FF      SUB    DI,DI      ;destination string offset
FC         CLD               ;auto-increment
F3         REP               ;repeat while CX <> 0
A4         MOVSB             ;copy string
```

Note the alternate method used to place 0000H in SI and DI. The SUB instructions require 2 bytes of code each. MOV   SI,0 and MOV   DI,0 would require 3 bytes each. The code is included for each instruction for your interest.

CMPS   Destination-String,Source-String (Compare String).   This instruction is used to compare two strings. The compare operation, as we have already seen, is accomplished by an internal subtraction of the destination and source operands. So, in this case, a byte or word from the destination string is subtracted from the corresponding element of the source string. If the two elements are equal, the zero flag will be set. Different elements cause the zero flag to be cleared. The REPZ prefix will allow strings to be checked to see if they are identical. This is a very handy tool when writing interactive programs (programs that require a response from the user). For example, if a user enters "ZOOM IN" when asked for a command, the program can check this string against all legal command strings to see if it matches any of them. String comparisons are also employed in **spell checkers,** programs that automatically find misspelled words in a text file. (The text for this book was run through a spell checker in a relatively short period of time.)

Because the flags are adjusted during execution of CMPS, we know if the two strings matched by examining the zero flag at the end of execution. JZ   MATCH (jump to MATCH if the zero flag is set) can be used to detect matching strings.

Example 3.33:   Assume that DS:SI and ES:DI have been initialized to the starting addresses of the two strings from Figure 3.24. If REPZ   CMPS STRINGA,STRINGB is executed with CX equal to 4, do the strings match? Do they match if CX equals 8? What state must the direction flag be in?

Solution: The direction flag must be cleared so that SI and DI auto-increment during the compare. When CX equals 4, the processor compares only the first 4 bytes of each string. Because each string begins with "SHOP", the zero flag remains set throughout the compare and we get a match. When CX equals 8, CMPS will repeat until SI and DI point to the sixth byte in each string. Then the comparison fails due to the "E" in "SHOPPER" and the "I" in "SHOPPING." The zero flag is then cleared and the instruction terminates, even though CX has not yet decremented to 0. This indicates that the strings are different.

The machine code for CMPS is A6 for byte strings, A7 for word strings, and 66 A7 for double-word strings.

The assembler allows the use of CMPSB, CMPSW, and CMPSD as alternate forms of the instruction. Once again, no operands are needed with these instructions.

SCAS   Destination-String (Scan String).   This instruction is used to scan a string by comparing each string element with the value saved in AL, AX, or EAX. AL is used for byte strings AX for word strings, and EAX for double-word strings. The string element pointed to by ES:DI is internally subtracted from the accumulator and the flags adjusted accordingly. Once again the zero flag is set if the string element matches the accumulator, and cleared otherwise. The accumulator and the string element remain unchanged. If REPNZ is used as the prefix to SCAS, the string is effectively searched for the item contained in the accumulator. Alternately, if REPZ is used, a string can be scanned until an element *differs* from the accumulator. This is especially handy when working with text strings. Suppose a text string contains a number of leading blanks (ASCII code 20H) before the actual text begins. SCAS can be used with 20H in AL and REPZ as the prefix to skip over the leading blanks. When the first nonblank (non-20H byte) character is encountered, the scan will terminate with DI pointing to the nonblank character.

Example 3.34:   Suppose that the ES and DI registers are initialized to point to the starting address of the "SHOPPING" string in Figure 3.24. What is the result of REPNZ SCAS if CX contains 9 and AL contains 4EH?

Solution: With CX set to 9 the processor will be able to scan each element of the "SHOPPING" string. However, the REPNZ prefix will allow the scan to continue only as long as the current string element does *not* match the byte stored in AL. There is no match between the accumulator until ES:DI points to address 04A86H. At that point, the accumulator matches the contents of memory and the scan terminates with the zero flag set.

  The machine code for SCAS is AE for a byte string, AF for a word string, and 66 AF for a double-word string.

The assembler recognizes SCASB, SCASW, and SCASD as alternate forms of SCAS.

LODS   Source-String (Load String).   This instruction is used to load the current string element into the accumulator and automatically advance the SI register to point to the next element. It is assumed that DS:SI have already been set up prior to execution of LODS. The direction flag determines whether SI is incremented or decremented.

Example 3.35:   Refer to Figure 3.24 once again. Assume that DS:SI currently point to address 05105H and that the direction flag is set. What is the result of executing LODS with a byte-size operand?

Solution: LODS copies the byte from the current address indicated by DS:SI (which is 45H) into AL and then *decrements* SI by 1 to point to the next element. SI is decremented because the direction flag is set. The next string element is at address 05104H.

The machine code for LODS is AC for byte strings, AD for word strings, and 66 AD for double-word strings.

The assembler accepts LODSB, LODSW, and LODSD as explicit uses of LODS.

STOS   Destination-String (Store String).   This instruction is used to write elements of a string into memory. The contents of AL, AX, or EAX are written into memory at the address pointed to by ES:DI and then DI is adjusted accordingly depending on the state of the direction flag and the size of the string elements.

Example 3.36:   We wish to modify the "SHOPPING" string of Figure 3.24 by adding the word "MALL" to it. The carriage return code 0DH will be replaced by a blank, and the character codes for "MALL" and another carriage return will be added. How should STOS be used to accomplish this modification?

Solution: Because the modification represents 6 bytes to write into memory, we will use the word version of STOS to write two codes into memory at a time. First, we will write the blank code and the letter "M". Then the codes for "A" and "L", and finally the code for "L" and the carriage return. The direction flag will be cleared to allow auto-incrementing of DI, and ES:DI must be initialized to address 04A88H to begin. The necessary instructions are as follows:

```
B8 A8 04    MOV    AX,4A8H    ;destination string segment-address
8E C0       MOV    ES,AX
BF 08 00    MOV    DI,8       ;destination offset
FC          CLD               ;auto-increment
B8 20 4D    MOV    AX,'M '    ;code for BLANK and 'M'
AB          STOSW
B8 41 4C    MOV    AX,'LA'    ;code for 'A' and 'L'
AB          STOSW
B8 4C 0D    MOV    AX,0D4CH   ;code for 'L' and CR
AB          STOSW
```

The first three instructions are needed to initialize ES:DI to address 04A88H. Because the 04A8H number in ES will become 4A80H when the processor forms the effective address, DI must be initialized to 0008 to get the right starting address for the modification. Because the 8088 will byte-swap the string words as they are written into memory, it is necessary to place them into AX in reverse order. For example, in MOV   AX,'M ' the ASCII codes for M and blank become the word 4D20H (when the assembler looks up the ASCII codes). The lower byte of 4D20H is 20H, the first character we wish to write into string memory. The upper byte gets written into memory next, which places the code for "M" *after* the code for blank. Figure 3.25 indicates this operation. Three STOSWs are needed to write the six new string characters.

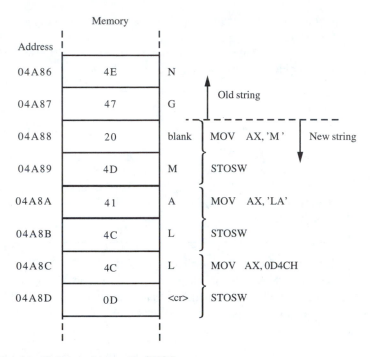

FIGURE 3.25   Modifying a string with STOS

The machine code for STOS is AA for byte strings, AB for word strings, and 66 AB for double-word strings.

As with the other string instructions, the assembler recognizes STOSB, STOSW, and STOSD as explicit forms of STOS.

Another Look at INS/OUTS

Recall that there are two string instructions, INS and OUTS, that transfer data between an I/O port and string memory. The source string DS:SI supplies data to the OUTS instruction, and INS writes its data to the destination string ES:DI. As usual with string operations, the index register is incremented or decremented by 1, 2, or 4 after each operation.

Example 3.37:   The following registers contain the indicated hexadecimal values:

```
DS = 0400    ES = 0600    CX = 0003
SI = 0100    DI = 0200    DX = 03E0
```

Furthermore, the direction flag is clear. What are the results of executing these instructions?

```
REP INSB    REP OUTSW
```

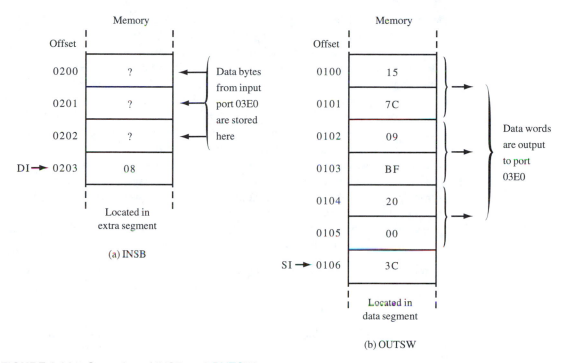

FIGURE 3.26   Operation of INSB and OUTSW

Solution: The INSB instruction is repeated three times. This causes the processor to read 3 bytes from input port 03E0 and store them in string locations 0200, 0201, and 0202, as indicated by Figure 3.26(a). The final value in DI is 0203.

The OUTSW instruction also executes three times. Each word read from string memory is sent to output port 03E0. The words are located at addresses 0100, 0102, and 0104. The final value in SI is 0106. Figure 3.26(b) illustrates OUTSW's execution.

3.8   TROUBLESHOOTING TECHNIQUES

With a large portion of the instruction still left to cover, it would be good to stop here and review some key points to remember when working with 80x86 instructions.

- Examine the relationship between the instruction mnemonic and the resulting machine code. You will begin to see patterns. These patterns will help you discover errors such as leaving the H off a hexadecimal number like 16H in the instruction MOV   BL,16. The assembler will convert the 16 into 10H and generate the machine code B3 10 instead of B3 16. Knowing what to expect is a good place to start.
- Know your data sizes. It is frustrating to look at the instruction MOV   AL,500 and wonder why it does not assemble. It looks OK, does it not? But remember that AL is

the lower 8-bit half of AX, and can only handle integers as large as 255. Knowing your data sizes will help you avoid problems like this.

- Do not make the assembler guess. For example, the instruction MOV [SI],5 looks OK, but actually generates an error because the assembler does not know if the "5" is an 8-bit 5, a 16-bit 5, or a 32-bit 5. This affects how many memory locations are accessed by [SI] and could lead to trouble if the assembler does not use the correct size. To force a particular size, such as 16 bits, use MOV WORD PTR [SI],5.

- Learn how to use several of the most basic addressing modes, which segment registers are involved, and how they form the physical address.

- Once again, always be aware of how an instruction uses, or affects, the processor flags. Many programs do not work simply because the flags are ignored by the programmer. This will be especially true in Chapter 4 when we examine the arithmetic and logical instructions.

- Before using any string instructions, make sure the direction flag is set to the appropriate value, so that the index registers update properly.

These points, and others we will see in Chapter 4, should go a long way toward eliminating many of the common errors encountered when working with assembly language.

SUMMARY

In this chapter we began coverage of the 80x86 instruction set. This included detailed looks at each addressing mode and flag available to the programmer in the real mode. Data transfer and string instructions were explained by example, including operations allowed on the 32-bit extended registers. The important concept of a stack was introduced, to illustrate how to save data in memory.

The structure of source and list files was also covered, with attention paid to the assembler details necessary for writing proper code. All of the information presented here will be useful as you complete coverage of the instruction set in Chapter 4.

STUDY QUESTIONS

1. Explain the use of the ORG, DB, DW, and END assembler directives.
2. What happens when a source file is assembled?
3. What two files are created by the assembler?
4. What are the opcode, data type, and operand(s) in this instruction:

   ```
   MOV    AH,7
   ```

5. What is meant by byte-swapping?
6. Which of the assembler directives produce data: ORG, DB, DW, DD, DF, SEG, END?
7. What does a linker do?

8. List the seven basic instruction groups.
9. Identify the source and destination addressing mode in each of these instructions:
   a) MOV   AX,BX
   b) MOV   AH,7
   c) MOV   [DI],AL
   d) MOV   AX,[BP]
   e) MOV   AL,[SI+6]
   f) JNZ   XYZ
   g) CBW
10. Why are the flags so important in a control-type program?
11. What does this two-instruction sequence do?

```
XCHG      AX,BX
XCHG      BX,CX
```

12. What does this instruction accomplish?

```
BSWAP EBX
```

13. Memory locations 00490H through 00493H contain, respectively, 0A, 9C, B2, and 78. What does AX contain after each instruction? (Assume that SI contains 00490H and that BP contains 0002.)
    a) MOV   AX,[SI]
    b) MOV   AX,[SI+1]
    c) MOV   AX,[SI][BP]
14. Registers AX, BX, CX, and DX contain, respectively, 1111H, 2222H, 3333H, and 4444H. What are the contents of each register after this sequence of instructions?

```
PUSH      AX
PUSH      CX
PUSH      BX
POP       DX
POP       AX
POP       BX
```

15. What is the difference between LDS and LES? When should each be used?
16. Show the instructions needed to scan a 200-byte string for the byte 25H.
17. Repeat Question 16 for the word value 9A25H.
18. Redo Example 3.32 for auto-decrement copying.
19. Redo Example 3.36 by using byte operation instead.
20. Explain how the processor switches from 16-bit operands to 32-bit operands, when operating in the real mode.
21. What are the simplified segment directives discussed in this chapter?
22. If EAX contains 00000200H, EBX contains 00000003H, and the data segment contains 1000H, what is the effective address generated by these instructions?
    a) MOV   ECX,[EAX]
    b) MOV   ECX,[EBX][EAX]
    c) MOV   ECX,[EAX][EBX*8]
    d) MOV   ECX,[ESI][EDI]
23. Which registers in Question 22 are base registers? Which are index registers?

24. Can the stack pointer be used as an index register when using 32-bit addressing?
25. What scale values may be used in 32-bit addressing?
26. What is a segment override prefix? Show an example of when it may be used.
27. If an I/O port address is greater than FFH, what must be done to use IN or OUT (or INS/OUTS)?
28. What is the difference between IN and INS?
29. If BX contains 3000H, what are the results of these two instructions?
    a) MOVSX EBX,BX
    b) MOVZX EBX,BX
30. Repeat Question 29 for BX equal to 9A00H.
31. How can all general purpose registers be pushed or popped from the stack with a single instruction?
32. How is register AL used during execution of XLAT?
33. The segment address of the word variable ABC is not known. Show how the SEG directive can be used to initialize the extra segment so that ABC is accessible via ES. Also show the instruction needed to read ABC into register DX.
34. What are the physical addresses actually used during the string operations shown in Figure 3.26?
35. What are the word values output to the port in Figure 3.26?

CHAPTER 4

80x86 Instructions, Part 2: Arithmetic, Logical, Bit Manipulation, Program Transfer, and Processor Control Instructions

OBJECTIVES

In this chapter you will learn about:

- The arithmetic, logical, and bit manipulation instructions
- The program transfer (loop/jump and subroutine/interrupt) instructions
- The processor control instructions
- How an assembler generates machine code
- The special properties of relocatable code

4.1   INTRODUCTION

Chapter 3 introduced the flags, addressing modes, data transfer, and string instructions. This chapter completes coverage of the 80x86 instruction set. As before, each instruction is presented with an accompanying example and its associated machine code.

In addition, we will examine the properties of the relocatable machine code used by the processor, and see how it is generated by an assembler.

Section 4.2 covers the arithmetic instructions. These are followed by the logical and bit manipulation instructions in Sections 4.3 and 4.4, respectively. Program transfer instructions (loops, jumps, subroutines) are explained in Section 4.5. The details of the processor control instructions are examined in Section 4.6. Section 4.7 shows how an assembler generates 80x86 machine code. The relocation properties of the processor's machine code are discussed in Section 4.8. A final set of troubleshooting techniques for the 80x86 instruction set is provided in Section 4.9.

4.2   ARITHMETIC INSTRUCTIONS

This group of instructions provides the 80x86 with its basic integer math skills. Floating-point math operations are handled by a separate group of instructions, which we will examine in Chapter 13. Addition, subtraction, multiplication, and division can all be performed on different sizes and types of numbers. You will need to refer to Appendix B often to understand fully the effects of each instruction on the processor's flags.

The Instructions

ADD   Destination,Source (Add Byte, Word, or Double-Word).   This instruction is used to add 8-, 16-, or 32-bit operands together. The sum of the two operands replaces the destination operand. All flags are affected.

Example 4.1:   If AX and BX contain 1234H and 2345H, respectively, what is the result of ADD   AX,BX? How would this answer compare with ADD   BX,AX?

Solution: Considering ADD   AX,BX, adding 1234H to 2345H gives 3579H, which replaces the contents of AX. BX still contains 2345H. ADD   BX,AX generates the same sum, but the contents of BX are changed instead.

The machine code for ADD   AX,BX is 01 D8. For ADD BX,AX we get 01 C3.

We can also add constants (immediate data) to registers or memory. For example, ADD CX,7 would add 7 to the number stored in register CX. The machine code for this instruction is 83 C1 07. The 80x86 will sign-extend the immediate value before it is used.

ADC   Destination,Source (Add Byte, Word, or Double-Word with Carry).   The operation of ADC is similar to ADD; however, in addition to the source operand the processor also adds the contents of the carry flag. The carry flag is then updated based on the size of the result. Other flags are also affected. ADC is commonly used to add multibyte operands together (such as 128-bit numbers).

Example 4.2:   Using the same register values from Example 4.1, what is the result of ADC   AX,BX if the carry flag is set?

Solution: Figure 4.1 shows how AX, BX, and the carry flag are added together to get 357AH, which replaces the contents of register AX. The carry flag is then cleared, because 357AH fits into a 16-bit register.

The machine code for ADC   AX,BX is 11 D8.

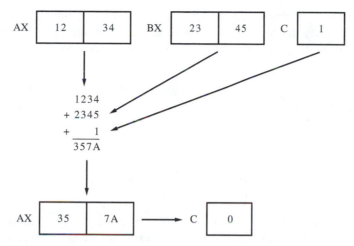

FIGURE 4.1 Execution of ADC AX,BX

INC Destination (Increment Byte, Word, or Double-Word by 1). There are many times when we need to add only 1 to a register or to the contents of a memory location. We could do this using the ADD instruction, but it is simpler to use INC. Also, in many cases the use of INC generates less machine code than the corresponding ADD instruction, resulting in faster execution. All flags except the carry flag are affected. Use ADD destination,1 to update the carry flag.

Example 4.3: The two instructions ADD AX,1 and INC AX both increase the value of register AX by 1. The machine code for each differs greatly, with ADD AX,1 assembling into 3 bytes (05 01 00) and INC AX only requiring 1 byte (40). Because 3 bytes take longer to fetch from memory than 1 byte, INC AX executes faster than ADD AX,1.

SUB Destination,Source (Subtract Byte, Word, or Double-Word). SUB can be used to subtract 8-, 16-, or 32-bit operands. If the source operand is larger than the destination operand, the resulting borrow is indicated by setting the carry flag.

Example 4.4: If AL contains 00 and BL contains 01, what is the result of SUB AL,BL?

Solution: Subtracting 01 from 00 in 8-bit binary results in FFH, which is the 2's complement representation of −1. So, the contents of AL are replaced with FFH, and both the carry and sign flags are set to indicate a borrow and a negative result.

SBB   Destination,Source (Subtract Byte, Word, or Double-Word with Borrow).   SBB executes in much the same way as SUB, except the contents of the carry flag are also subtracted from the destination operand. The contents of the carry flag are updated at completion of the instruction.

DEC   Destination (Decrement Byte, Word, or Double-Word by 1).   DEC provides a quick way to subtract 1 from any register or the contents of any memory location. All flags except the carry flag are affected.

Example 4.5:   What is the result of DEC   **byte ptr** [200H]? Assume that the DS register contains 0500H.

Solution: Because we are decrementing the contents of a memory location, the assembler must be informed of the operand size. This is accomplished with the **byte ptr** directive. Figure 4.2 shows the change made to location 05200H when the instruction executes.

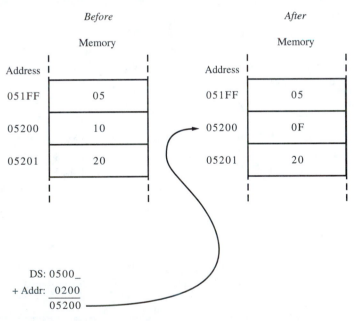

FIGURE 4.2   Execution of DEC   BYTE PTR [200H]

The machine code for DEC   **byte ptr** [200H] is FE 0E 00 02. Notice that the last 2 bytes are the byte-swapped offset 0200H.

CMP   Destination,Source (Compare Byte, Word, or Double-Word).   This very useful instruction allows the programmer to perform comparisons on byte, word, and double-word values. Comparisons are employed in search algorithms and whenever range checking needs to be done on input data. For example, it may be beneficial to use CMP to check a register for a 0 value prior to multiplication or division. The internal operation of CMP is actually a subtraction of the destination and source operands without any modification to either. The flags are updated based on the results of the subtraction. The zero flag is set after a CMP if the destination and source operands are equal, and cleared otherwise. All other flags are also affected. Immediate data is sign-extended to the size of the destination before the comparison is performed.

Example 4.6:   What CMP instruction is needed to determine if the accumulator (AX) contains the value 3B2EH?

Solution: We should use CMP   AX,3B2EH to do the checking. As we will see later, we should follow the CMP instruction with some kind of conditional jump (as in JZ MATCH or JNZ   NOMATCH). We could also test each half of AX individually, using CMP   AH,3BH and CMP   AL,2EH, but this results in more code and slower execution time.
        The machine code for CMP   AX,3B2EH is 3D 2E 3B.

Example 4.7:   Let us examine the operation of CMP when both signed and unsigned numbers are used as operands. Consider the following instruction sequence:

```
MOV   AL,20
CMP   AL,10
CMP   AL,30
```

When the first CMP instruction executes, both the sign and carry flags will be cleared because the accumulator (20) was larger than the immediate data (10). The internal subtraction 20 − 10 gives a positive result.
        When the second CMP instruction executes, the internal subtraction performed is 20 − 30. In this case, the sign and carry flags are both set (due to the negative result).
        All three numbers used (10, 20, and 30) represent *positive* numbers. What happens when *negative* numbers are used? The three instructions are now as follows:

```
MOV   AL,90H     ;represents −112
CMP   AL,80H     ;represents −128
CMP   AL,0A0H    ;represents −96
```

Remember that negative numbers are represented using 2's complement notation.
        The sign and carry flags are both cleared during execution of the first CMP instruction, because −112 minus −128 results in a positive value. When the second CMP instruction is executed, the sign and carry flags are both set as a result of the internal subtraction −112 minus −96.

When both positive and negative numbers are compared, as in:

```
MOV   AL,10
CMP   AL,90H
```

the sign and carry flags are affected differently. In this case, the sign flag is cleared and the carry flag is set. Can you determine why?

CMPXCHG   Destination,Source (Compare and Exchange).   This instruction compares the destination operand with the accumulator (AL, AX, or EAX, depending on the size of the destination). The flags are set accordingly. If the accumulator equals the destination, the source operand is copied into the destination. If the accumulator and destination operands are different, the accumulator is replaced by the value in the destination.

Example 4.8:   If AL, BL, and CL contain the respective values 10H, 20H, and 30H, what is the result of CMPXCHG   BL,CL? What is the result if BL initially equals 10H?

Solution: Figure 4.3 shows the results of each CMPXCHG execution.

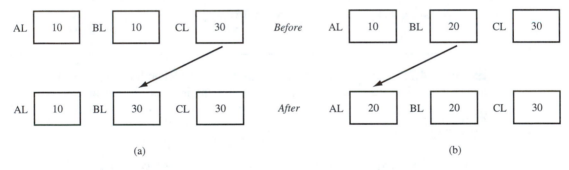

(a)   (b)

FIGURE 4.3   Execution of CMPXCHG   BL,CL (a) when AL equals BL and (b) when AL does not equal BL

In both cases, data is rapidly exchanged between two registers. This type of data exchange is especially useful in operating system software that supports multiple processes through the use of **semaphores.** A semaphore is essentially a counter whose value decides which process is able to execute next. It is often necessary to update the semaphore value with a single instruction. When multiple instructions are used, such as:

```
              CMP   AL,BL
              JNZ   NOTEQUAL      ;jump to NOTEQUAL if AL <> BL
              MOV   BL,CL
              JMP   NEXT
NOTEQUAL      MOV   AL,BL
NEXT          ---
```

it is possible to incorrectly update the semaphore due to unexpected interrupts that switch between the two or more processes accessing the semaphore at the same time. The

CMPXCHG instruction solves this problem by performing the equivalent work of the five example instructions.

CMPXCHG8B Destination (Compare and Exchange 8 Bytes). This instruction is similar to CMPXCHG except that the comparison is fixed at 64 bits. The EDX and EAX registers (EDX:EAX) specify the 64-bit operand (EDX being the upper 32 bits) that is compared with the destination. If equal, the destination receives a copy of the 64-bit number in ECX:EBX (ECX holding the upper 32 bits).

If the comparison is not equal, the destination value is copied into EDX:EAX. Only the zero flag is affected by the result of the comparison.

Example 4.9: What are the results of the CMPXCHG8B instruction given this sequence of instructions:

```
MOV        EDX,12345678H
MOV        EAX,9ABCDEF0H
MOV        ECX,01020304H
MOV        EBX,05060708H
CMPXCHG8B  VAL64BIT
```

VAL64BIT is a 64-bit (8-byte) memory operand, defined as follows:

```
VAL64BIT   DQ    123456789ABCDEF0H
```

Solution: Since the 64-bit memory operand VAL64BIT is equal to EDX:EAX, the contents of VAL64BIT are replaced by ECX:EBX. Its new value is 0102030405060708H.

Note the use of the DQ (define quadword) directive used to reserve room for VAL64BIT.

MUL Source (Multiply Byte, Word, or Double-Word Unsigned). This unsigned multiply instruction treats byte, word, and double-word numbers as unsigned binary values. This gives an 8-bit number a range of 0 to 255, a 16-bit number the range 0 to 65,535, and a 32-bit number the range 0 to 4,294,967,296. The source operand specified in the instruction is multiplied by the accumulator. If the source is a byte, AL is used as the multiplier, with the 16-bit result replacing the contents of AX. If the source is a word, AX is used as the multiplier, and the 32-bit result is returned in registers DX:AX, with DX containing the upper 16 bits of the result. For double-word source values, the multiplier is EAX. The 64-bit result is stored in EDX:EAX. All flags are affected.

Example 4.10: What is the result of MUL CL if AL contains 20H and CL contains 80H? What is the result of MUL AX if AX contains A064H?

Solution: The decimal equivalents of 20H and 80H are 32 and 128, respectively. The product of these two numbers is 4096, which is 1000H. Upon completion, AX will contain 1000H. The decimal equivalent of A064H is 41,060. MUL   AX multiplies the accumulator by itself, giving 1,685,923,600 as the result, which is 647D2710H. The upper half of this hexadecimal number (647DH) will be placed into register DX. The lower half (2710H) will replace the contents of AX.

The machine code for MUL   CL is F6 E1. The machine code for MUL   AX is F7 E0.

IMUL   Source (Integer Multiply Byte, Word, or Double-Word).   This multiply instruction is very similar to MUL except that the source operand is assumed to be a *signed* binary number. This gives byte operands a range of –128 to 127, word operands a range of –32,768 to 32,767, and double-word operands a range of –2,147,483,648 to 2,147,483,647. Once again the operand size determines whether the result will be placed into AX, DX:AX, or EDX:EAX.

Example 4.11:   What is the result of IMUL   CL if AL contains 20H and CL contains 80H?

Solution: Although this example appears to be exactly like Example 4.10, it actually is not, because the 80H in register CL is interpreted as –128. The product of –128 and 32 is –4096, which is F000H in 2's complement notation. This is the value placed into AX upon completion.

The machine code for IMUL   CL is F6 E9.

Example 4.12:   IMUL also accepts two or three operands under certain restrictions. When two operands are used, the first operand must be a 16/32-bit register. The second operand may be a register, a memory operand, or immediate data. The immediate data will be sign-extended to 16 or 32 bits. The result must be 16 or 32 bits wide. Examples are:

```
0F AF C3          IMUL       AX,BX         ;AX = AX * BX
0F AF 0D          IMUL       CX,[DI]       ;CX = CX * [DI]
6B DB 32          IMUL       BX,50         ;BX = BX * 50
6B C9 F4          IMUL       CX,-12        ;CX = CX * -12
66 0F AF C3       IMUL       EAX,EBX       ;EAX = EAX * EBX
```

When three operands are used, the first operand must be a 16/32-bit register. The second operand may be a register or memory operand. The third operand must be an immediate value. Examples of this form are as follows:

```
6B C3 05          IMUL       AX,BX,5          ;AX = BX * 5
6B CA F6          IMUL       CX,DX,-10        ;CX = DX * -10
66 69 1C 000007D0 IMUL       EBX,[SI],2000    ;EBX = [SI] * 2000
66 69 1C FFFFF830 IMUL       EBX,[SI],-2000   ;EBX = [SI] * -2000
```

Notice that the 32-bit size of EBX in the last two instructions cause the immediate data to be sign-extended to 32 bits.

DIV   Source (Divide Byte, Word, or Double-Word Unsigned).   In this instruction the accumulator is divided by the value represented by the source operand. All numbers are interpreted as *unsigned* binary numbers. If the source is a byte, the quotient is placed into AL and the remainder into AH. If the source is a word, it is divided into the 32-bit number represented by DX:AX (with DX holding the upper 16 bits). The 16-bit quotient is placed into AX and the 16-bit remainder into DX. For double-word operands, the 64-bit value in EDX:EAX is divided by the source. The 32-bit quotient is placed in EAX. The 32-bit remainder is saved in EDX. If the quotient is too large to fit in the destination, a type-0 interrupt is generated. All flags are affected.

Example 4.13:   What is the result of DIV   BL if AX contains 2710H and BL contains 32H?

Solution: The decimal equivalents of 2710H and 32H are 10,000 and 50, respectively. The quotient of these two numbers is 200, which is C8H. This is the byte placed into AL. Because the division had no remainder, AH will contain 00.

The machine code for DIV   BL is F6 F3.

IDIV   Source (Integer Divide Byte, Word, or Double-Word).   As with MUL and IMUL, we also see similarities between DIV and IDIV. In IDIV the operands are treated as signed binary numbers. Everything else remains the same.

Example 4.14:   What is the result of IDIV   CX if AX contains 7960H, DX contains FFFEH, and CX contains 1388H?

Solution: Because a word operand is specified, the 32-bit number represented by DX and AX will be divided by CX. Combining DX and AX gives FFFE7960H, which is −100,000. CX represents 5,000. Dividing −100,000 and 5,000 gives −20, which is FFECH. This value is placed into AX, and 0000 is placed into DX because the division has no remainder.

The machine code for IDIV   CX is F7 F9.

Example 4.15:   A simple random-number generator uses a 64-bit pattern to represent its random value. A new random number is generated by dividing the current random number by a 32-bit generator pattern. The quotient and remainder are combined to form the new random pattern, with the remainder making up the upper 32 bits. One way to do this is as follows:

```
Address and data in data segment:
0000 12345678     RANDHI   DD     12345678H
0004 9ABCDEF0     RANDLO   DD     9ABCDEF0H
0008 5A5A5A5A     GTERM    DD     5A5A5A5AH
```

```
Random-number generator:
66  8B 16 0000      MOV     EDX,RANDHI
66  A1 0004         MOV     EAX,RANDLO
66  8B 1E 0008      MOV     EBX,GTERM
66  F7 FB           IDIV    EBX
66  89 16 0000      MOV     RANDHI,EDX
66  A3 0004         MOV     RANDLO,EAX
```

When the six instructions are executed, the new values of RANDHI and RANDLO are 5296DBCCH and 33944A55H, respectively. This may certainly be viewed as a random pattern, although significant testing is required to determine how well the generator works when used repeatedly.

NEG   Destination (Negate Byte, Word, or Double-Word).   This instruction is used to find the signed 2's complement representation of the number in the destination. This is accomplished by subtracting the destination operand from 0. All flags are affected.

Example 4.16:   What is the result of NEG   AX if AX contains FFECH?

Solution: Subtracting FFECH from 0 gives 0014H, which is +20. Take another look at Example 4.14 and you will see that we have just proved the results of the IDIV instruction.
    The machine code for NEG   AX is F7 D8.

CBW   (Convert Byte to Word).   This instruction is used to extend a signed 8-bit number in AL into a signed 16-bit number in AX. This is sometimes necessary before performing an IDIV or IMUL. No flags are affected.

Example 4.17:   What does AX contain after execution of CBW if AL initially contains 37H? What if AL contains B7H?

Solution: Because 37H is a positive signed number, the result in AX is 0037H. Note that the most significant bit is still a 0. B7H, however, is a negative signed number, resulting in AX becoming FFB7H. Note here that the MSB is a 1, indicating a negative result.
    The machine code for CBW is 98.

CWD/CWDE   (Convert Word to Double-Word).   CWD extends the sign of the number stored in AX through all 16 bits of register DX. This results in a 32-bit signed number in DX and AX. CWD is useful when preparing for IMUL and IDIV. CWDE also extends the sign of AX, but does it in the upper 16 bits of EAX. The 32-bit result is in one register. No flags are affected.

Example 4.18: What is in DX after CWD executes with AX containing 4000H? What if CWDE is executed?

Solution: Because 4000H is positive (the MSB is 0), DX will contain 0000. When CWDE executes, EAX will contain 00004000H.

The machine code for CWD is 99. The machine code for CWDE is 66 98.

CDQ (Convert Double-Word to Quadword). This instruction is similar to CWD. The sign bit of EAX is extended through EDX. This gives a 64-bit result in EDX:EAX.

Example 4.19: What are the results of CDQ if EAX contains C8000000H?

Solution: Since the MSB of EAX is a one, the contents of EDX are set to FFFFFFFFH. The machine code for CDQ is 66 99.

DAA (Decimal Adjust for Addition). When a hexadecimal number contains only the digits 0–9 (as in 07H, 35H, 72H, and 98H), the number is referred to as a **packed decimal number**. So, instead of a range of 00 to FFH we get 00 to 99H. When these types of numbers are added together, they do not always produce the correct packed decimal result. Consider the addition of 15H and 25H. Straight hexadecimal addition gives 3AH as the result. Because we are interpreting our numbers as packed decimal numbers, 3AH seems to be an illegal answer. The processor is able to correct this problem with the use of DAA, which will modify the byte stored in AL so that it looks like a packed decimal number. Numbers of this type are also referred to as **binary coded decimal (BCD)** numbers.

All flags are affected.

Example 4.20: If ADD   AL,BL is executed with AL containing 15H and BL containing 25H, what is the resulting value of AL? What does AL contain if DAA is executed next?

Solution: Adding 15H to 25H gives 3AH, as we just saw. When DAA executes it will examine AL and determine that it should be corrected to 40H to give the correct packed decimal answer. It is interesting that the processor adds 06H to AL to convert it into a packed number.

The machine code for DAA is 27.

DAS (Decimal Adjust for Subtraction). This instruction performs the same function as DAA except it is used after a SUB or SBB instruction. All flags are affected.

Example 4.21:   If AL contains 10H and CL contains 02H, what is the result of SUB AL,CL? What happens if DAS is executed next?

Solution: Subtracting 02H from 10H will result in AL containing 0EH. Following the SUB instruction with DAS will cause AL to be converted into 08H, the correct packed decimal answer.

   The machine code for DAS is 2F.

AAA   (ASCII Adjust for Addition).   Occasionally the need arises to perform addition on ASCII numbers. The ASCII codes for 0–9 are 30H through 39H. Addition of two ASCII codes unfortunately does not result in a correct ASCII or decimal answer. For example, adding 33H (the ASCII code for 3) and 39H (the ASCII code for 9) gives 6CH. Using DAA to correct this result will give 72H, which is the correct packed decimal answer, but not the answer we are looking for. After all, 3 plus 9 equals 12, which must be represented by the ASCII characters 31H and 32H. AAA is used to perform this correction by using it after ADD or ADC. AAA will examine the contents of AL, adjusting the lower 4 bits to make the correct decimal result. Then it will clear the upper 4 bits, and add 1 to AH if the lower 4 bits of AL were initially greater than 9. Only the carry and auxiliary carry flags are directly affected.

Example 4.22:   Register AX is loaded with 0033H. Register BL is loaded with 39H. ADD   AL,BL is executed, giving AL a new value of 6CH. What happens if AAA is executed next?

Solution: AAA will see the C part of 6CH and correct it to 2 (by adding 6). Next, the upper 4 bits of AL will be cleared. Now AL contains 02. Because the lower 4 bits of AL (prior to execution of AAA) were greater than 9, 1 will be added to AH, making its final value 01. We end up with AX containing 0102H. If we now add 30H to each byte in AX, we will get 3132H, the two ASCII digits that represent 12.

   The machine code for AAA is 37.

AAS   (ASCII Adjust for Subtraction).   This instruction performs the same correction procedure that AAA does, except it is used after SUB or SBB to modify the results of a subtraction. Also, if the number in the lower 4 bits of AL is greater than 9, 1 will be *subtracted* from AH.

Example 4.23:   AX is loaded with 0037H and BL is loaded with 32H. SUB   AL,BL is executed and the processor replaces the contents of AL with 05H. What happens if AAS is now executed? What are the results if BL was initially loaded with 39H?

Solution: Because the number in the lower 4 bits of AL (5) is not greater than 9, AAS does not change it. The upper 4 bits of AL are cleared. Because no change was needed there is no need to add 1 to AH. The final value of AX is 0005H. Adding 30H to each byte in AX gives 3035H, the two ASCII codes for the number 05.

If BL is initially loaded with 39H the result of SUB   AL,BL is FEH, which is placed in AL. When this value is examined by AAS the E in FEH will be changed to 8 (by subtracting 6) and the upper 4 bits will be cleared. AL now contains 08H. Note that 7 minus 9 is –8. AL now contains the 8 part of the answer. Because AL required modification, AH will be decremented. The final value of AX is FF08H. The FFH in AH indicates that a borrow occurred (and so does the carry flag). Adding 30H to AL results in the correct ASCII code for 8 (which is 38H).

The machine code for AAS is 3F.

AAM   (ASCII Adjust for Multiply).   This instruction is used to adjust the results of a MUL instruction so that the results can be interpreted as ASCII characters. Both the operands used in the multiplication must be less than or equal to 9 (their upper 4 bits must be 0). This ensures that the upper byte of the result placed in AX will be 00, a necessary requirement for proper operation of AAM. AAM will convert the binary product found in AL into two separate digits. One digit replaces the contents of AL and the other digit replaces the contents of AH. Software can then be used to add 30H to each value to obtain the ASCII codes for the correct product. Only the parity, sign, and zero flags are directly affected.

Example 4.24:   AL and BL are given initial values of 7 and 6, respectively. MUL BL produces 002AH in AX. Note that 7 times 6 is 42 and that 2AH is equal to 42. What are the contents of AX after AAM is executed?

Solution: Because the upper 4 bits of each operand used in the multiplication were 0 and the lower 4 bits contained numbers less than or equal to 9, AAM will be able to correctly convert the product in AX. AAM converts the 2AH in AL into the digits 4 and 2 and writes them in AH and AL, respectively. The final contents of AX are 0402H. If we now add 30H to each byte in AX we get 3432H, the two ASCII codes for 42.

The machine code for AAM is D4 0A.

AAD   (ASCII Adjust for Division).   This instruction is executed *before* DIV to change an unpacked two-digit BCD number in AX into its binary equivalent in AL. AH must be 00 after the change to produce the correct results. The number in AL can then be divided (via DIV) to get the correct binary quotient in AL. Only the parity, sign, and zero flags are directly affected.

Example 4.25:   Register AX is loaded with 0600H. These 2 bytes correspond to the unpacked BCD number 60, not its decimal equivalent of 1,536. If BL is loaded with 04H we expect that the use of AAD and DIV will produce the correct answer of 15 (which is 60 divided by 4). What are the contents of AX after AAD and after DIV   BL?

Solution: AAD examines AX and finds that it contains a valid unpacked BCD number. The value 600H is converted to 003CH, the correct binary equivalent of 60. DIV   BL then divides this number by 4 to produce 000FH in AX. Remember that AL contains the quotient and AH contains the remainder. The quotient 0FH equals 15 and the remainder is 0, so AAD correctly adjusted AX before the division.

The machine code for AAD is D5 0A.

XADD   Destination,Source (Exchange and Add Byte, Word, or Double-Word).   In this instruction, the source operand is a register and the destination is a register or memory location. The source and destination operands are added together, with the sum replacing the contents of the destination. The original destination is copied into the source register.

All flags are affected.

Example 4.26:   What is the result of XADD   AX,BX if AX contains 2000H and BX contains 3000H?

Solution: The sum of AX and BX is 5000H. This becomes the new value stored in AX. BX is loaded with 2000H.

The machine code for XADD is 0F C1 D8.

4.3   LOGICAL INSTRUCTIONS

The instructions in this group are used to perform Boolean (logical) operations on binary data.

The Instructions

NOT   Destination ("NOT" Byte, Word, or Double-Word).   This instruction finds the complement of the binary data stored in the destination operand. All 0s are changed to 1s, and all 1s to 0s. No flags are affected.

Example 4.27:   What is the result of NOT   AL if AL contains 3EH?

Solution: The binary equivalent of 3EH is 00111110. When these bits are complemented we get 11000001, which is C1H. This is the final value in AL.

The machine code for NOT   AL is F6 D0.

AND   Destination,Source ("AND" Byte, Word, or Double-Word).   This instruction performs the logical AND operation on the destination and source operands. AND is a useful way of turning bits off (making them 0). Refer to Figure 4.4 for a review of the truth table for a two-input AND gate. Note that it takes two 1s to make a 1 on the output of the AND gate. The

FIGURE 4.4   Truth table for
an AND gate

A	B	f
0	0	0
0	1	0
1	0	0
1	1	1

processor ANDs the bits in each operand together to produce the final result. This logical operation is very useful for masking off unwanted bits during many types of comparison operations. All flags are affected.

Example 4.28:   What is the result of AND   AL,[345H]? Assume that the DS register contains 2000H and that AL contains BCH.

Solution: Figure 4.5 shows how the processor accesses location 20345H and ANDs its contents (27H = 00100111) with that of the AL register (BCH = 10111100). The result is 24H, which is placed into AL.

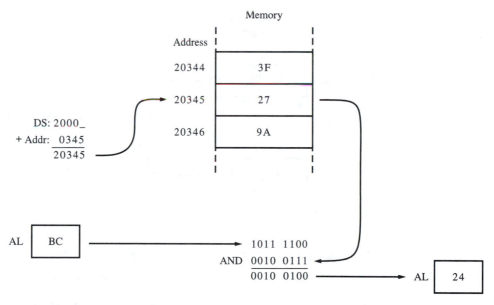

FIGURE 4.5   Operation of AND   AL,[345H]

The machine code for AND   AL,[345H] is 22 06 45 03.

OR   Destination,Source ("OR" Byte, Word, or Double-Word).   This instruction performs a logical OR of each bit in the source and destination operands. OR can be used to turn bits on (make them 1). Figure 4.6 shows the truth table for a two-input OR gate. The output of the OR gate will be high if a 1 is present on either (or both) inputs. Respective bits in each operand are ORed together to produce the final result. All flags are affected.

FIGURE 4.6   Truth table for
an OR gate

A	B	f
0	0	0
0	1	1
1	0	1
1	1	1

Example 4.29: Register BX contains 2589H. What is the result of OR BX,0C77CH?

Solution: As a matter of habit, whenever we use a hexadecimal number in an instruction, we precede it with a leading 0 if the first hexadecimal digit is greater than 9. So, the correct way to specify C77CH in the OR instruction is to write 0C77CH. The purpose for doing this is to prevent the assembler from thinking that C77CH is the name of a label somewhere in the program. Putting a leading 0 first does not change the value of the hexadecimal number, but does tell the assembler that it is working with a number and not a name.

So, in Figure 4.7 we see how the OR instruction operates on BX and the immediate data to produce the final result of E7FDH in register BX.

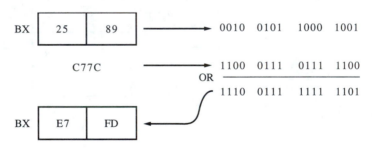

FIGURE 4.7   Operation of OR   BX,0C77CH

The machine code for OR   BX,0C77CH is 81 CB 7C C7.

XOR   Destination,Source (Exclusive-OR Byte, Word, or Double-Word).   XOR is a special form of OR in which the inputs *must be different* to get a 1 out of the logic gate. Figure 4.8 shows the truth table for a two-input XOR gate. Clearly, the output is high only when the inputs are different. The XOR function can be used to toggle bits to their opposite states. For example, 011 XORed with 010 gives 001. Note that the second bit in the first group has been changed to 0. If we XOR 001 with 010 again, we get 011. The two XOR operations simply toggle the second bit back and forth. This is due to the 1 in the second group of bits. It may be interesting to note that XOR   AL,0FFH does the same job as NOT AL. All flags are affected by XOR.

FIGURE 4.8 Truth table for an Exclusive-OR gate

A	B	f
0	0	0
0	1	1
1	0	1
1	1	0

Example 4.30: What is the result of XOR AX,CX if AX contains 1234H and CX contains 4567H?

Solution: Figure 4.9 shows how the two 16-bit binary numbers in AX and CX are XORed together. The result is placed in AX.

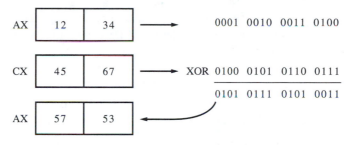

FIGURE 4.9 Operation of XOR AX,CX

The machine code for XOR AX,CX is 31 C8.

TEST Destination,Source ("TEST" Byte, Word, or Double-Word). TEST can be used to examine the state of individual bits, or groups of bits. Internally, the processor performs an AND operation on the two operands without changing their contents. The flags are updated based on the results of the AND operation. In this way, testing individual bits is easily accomplished by putting a 1 in the correct bit position of one operand and examining the zero flag after TEST executes. If the bit being tested was high, the zero flag will be cleared (a *not-zero* condition). If the bit was low, the zero flag will be set.

Example 4.31: What instruction is needed to test bit 13 in register DX?

Solution: We must determine the immediate data needed to perform the desired test. Figure 4.10 shows how the 1 needed to perform the test requires the immediate word 2000H. Because immediate data cannot be used as a destination operand, we end up with the instruction TEST DX,2000H. It will be necessary to examine the flags (possibly only the zero flag) to decide what to do with the results of the test.

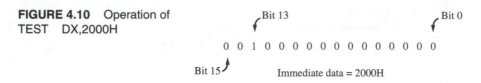

FIGURE 4.10 Operation of
TEST   DX,2000H

The machine code for TEST   DX,2000H is F7 C2 00 20.

BSF/BSR   Destination,Source (Bit Scan Forward/Reverse). The BSF instruction scans the source operand for the first bit that equals 1, beginning with the LSB. The bit position (index) of the first 1 found is saved in the destination. The destination must be a 16- or 32-bit register. The source may be a 16- or 32-bit register or memory operand.

BSR operates in a similar manner, except for the direction of the scan, which begins with the MSB instead.

Example 4.32:   Register EBX contains 4CE00000H. What is the result of BSF EAX,EBX? What happens if BSR is used?

Solution: As indicated in Figure 4.11, BSF begins scanning for a 1 at bit 0. The first 1 is found at bit 21. Register EAX is loaded with 15H (21 decimal).

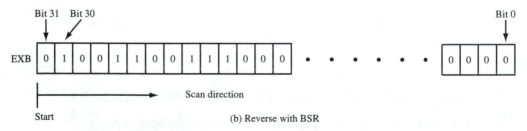

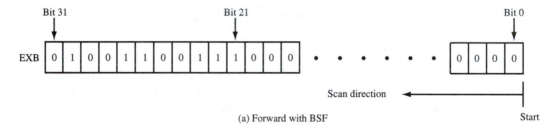

FIGURE 4.11 Scanning bits in EBX

When BSR   EAX,EBX is used, the result in EAX is 1EH (30 decimal), since scanning begins with the MSB. This indicates that the first 1 found is bit 30.

The machine code for BSF   EAX,EBX is 66 0F BC C3. The machine code for BSR   EAX,EBX is 66 0F BD C3.

Example 4.33: A very useful application of BSF/BSR is edge detection in an image-processing application. A special image called a *binary image* that contains only black and white colors has the property that an edge exists anywhere there is a zero to one transition. For example, consider this 32-bit sample from a binary image (zero is black, one is white):

00000000001111111100000000000000000

The edges of the seven-pixel wide bar occur at bit positions 16 and 22. These positions may be easily found using BSF and BSR.

BT/BTC/BTS/BTR Destination,Source (Bit Test and Complement/Set/Reset). These four instructions are used to determine the value of a specific bit in the 16- or 32-bit destination operand (register or memory). The bit to be tested is indicated by the source operand, which may be a register or an immediate value. Bits are numbered from 0 to 31.

The state of the bit that is tested is copied into the carry flag. Thus, if the bit is zero, the carry flag will be zero.

After testing the bit, the BTC instruction complements it. The destination is modified. The BTS instruction tests and then sets the indicated bit. BTR tests and resets the chosen bit.

Example 4.34: The contents of register AX and the carry flag are shown after each instruction executes:

```
B8 5555          MOV    AX,5555H     ;AX = 5555, carry = 0
0F BA E0 0A      BT     AX,10        ;AX = 5555, carry = 1
0F BA E0 0B      BT     AX,11        ;AX = 5555, carry = 0
BB 000A          MOV    BX,10        ;AX = 5555, carry = 0
0F A3 D8         BT     AX,BX        ;AX = 5555, carry = 1
43               INC    BX
0F A3 D8         BT     AX,BX        ;AX = 5555, carry = 0
0F BA F8 0F      BTC    AX,15        ;AX = D555, carry = 0
0F BA E8 01      BTS    AX,1         ;AX = D557, carry = 0
0F BA F0 00      BTR    AX,0         ;AX = D556, carry = 1
```

In control applications, a single bit is often used to operate a device (open/close a relay, turn an indicator light on/off) or sense a specific input condition (door open/closed, control switches). The bit test instructions provide a simple way to perform all the necessary operations on a single bit. This makes the job of writing the control system code quicker and easier.

SETcc Destination (Set Byte on Condition). This instruction tests the specified condition and sets the destination byte to 01H if true. If the condition is false, the destination is set to zero. The true/false condition is based on the state of one or more processor flags. Table 4.1 shows all the conditions that may be tested with SETcc.

TABLE 4.1   Conditions used with SETcc

Instruction	Meaning	Instruction	Meaning
SETA	Set byte if above	SETNE	Set byte if not equal
SETAE	Set byte if above or equal	SETNG	Set byte if not greater
SETB	Set byte if below	SETNGE	Set byte if not greater or equal
SETBE	Set byte if below or equal	SETNL	Set byte if not less
SETC	Set byte if carry	SETNLE	Set byte if not less or equal
SETE	Set byte if equal	SETNO	Set byte if not overflow
SETG	Set byte if greater	SETNP	Set byte if not parity
SETGE	Set byte if greater or equal	SETNS	Set byte if not sign
SETL	Set byte if less	SETNZ	Set byte if not zero
SETLE	Set byte if less or equal	SETO	Set byte if overflow
SETNA	Set byte if not above	SETP	Set byte if parity
SETNAE	Set byte if not above or equal	SETPE	Set byte if parity even
SETNB	Set byte if not below	SETPO	Set byte if parity odd
SETNBE	Set byte if not below or equal	SETS	Set byte if sign
SETNC	Set byte if not carry	SETZ	Set byte if zero

Example 4.35:   What does SETNZ AL do?

Solution: SETNZ checks the state of the zero flag. If the zero flag indicates a NZ condition (zero flag is clear), register AL is set to 01H. Otherwise, AL is set to 00H.
The machine code for SETNZ AL is 0F 95 C0.

4.4   BIT MANIPULATION INSTRUCTIONS

This group of instructions is used to shift or rotate bits left or right in register or memory operands. These operations are very useful when converting data from one form to another, or for manipulating specific patterns, such as a single bit that moves through each position of a register.

The Instructions

SHL/SAL   Destination,Count (Shift Logical/Arithmetic Left Byte, Word, or Double-Word).   This is the first of a number of different instructions that we will encounter that go by two names. The programmer may have personal reasons for using SHL instead of SAL, or vice versa, but to the assembler both names are identical and generate the same machine code.

The count operand indicates how many bits the destination operand is to be shifted to the left. Shifts may be from 1 to 31 bits, and may be specified as an immediate value, as in SHL   AX,7. For multi-bit shifts, the count may also be placed in register CL. The corresponding instruction for a multi-bit shift would then be SHL   AL,CL. All bits in the destination operand are shifted left, with a 0 entering the LSB each time. Bits that shift out of the MSB position are placed into the carry flag. Other flags are modified according to the final results.

Example 4.36:   What are the results of SHL   AL,CL if AL contains 75H and CL contains 3?

Solution: Figure 4.12 shows the state of register AL after each shift is performed. A 0 is shifted in from the right each time. The last bit shifted out of the MSB position is the final state of the carry flag. The final contents of AL are A8H. Could the same results be obtained by executing SHL   AL,1 three times? The answer is yes, at the expense of generating more code and consuming more execution time.

FIGURE 4.12   Execution of SHL AL,CL

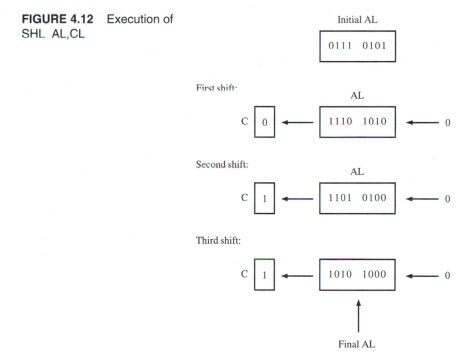

This type of shift instruction is very useful for computing powers of 2, because each time a binary number is shifted left it doubles in value.

The machine code for SHL   AL,CL is D2 E0. The machine code for SHL   AL,1 is D0 E0.

SHR   Destination,Count (Shift Logical Right Byte, Word, or Double-Word).   This instruction has the opposite effect of SHL, with bits shifting to the right in the destination operand. Zeros are shifted in from the left and the bits that fall out of the LSB position end up in the carry flag. Figure 4.13 details this operation. Keep in mind that shifting right 1 bit is equivalent to dividing by 2.

FIGURE 4.13   Operation of
SHR

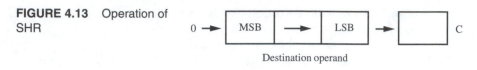

Destination operand

Example 4.37:   What instruction (or instructions) is needed to divide the number in BX by 32?

Solution: Because 32 equals 2 raised to the fifth power, shifting BX to the right five times is equivalent to dividing it by 32. The shift count 5 must be placed into register CL before the shift instruction SHR   BX,CL is executed. An alternate solution would involve execution of SHR   BX,5.

    The machine code for SHR   BX,CL is D3 EB. The machine code for SHR   BX,5 is C1 EB 05.

SAR   Destination,Count (Shift Arithmetic Right Byte, Word, or Double-Word).   SAR is very similar to SHR with the important difference being that the MSB is shifted back into itself. This serves to preserve the original sign of the destination operand. When we deal with signed numbers, the MSB is used as the sign bit. If SHR is used to shift a signed negative number to the right, the 0 that gets shifted into the MSB position forces the computer to then interpret the result as a positive number. SAR prevents this from happening. Negative numbers stay negative and positive numbers stay positive, and each gets smaller with every shift right.

Example 4.38:   What are the results of SAR   CL,1 if CL initially contains B6H?

Solution: B6H is the signed 2's complement representation of –74. Figure 4.14 shows how the contents of register CL are shifted right 1 bit, with the MSB coming back in from the left to preserve the negative sign. The final value in CL is DBH, which corresponds to –37, exactly one-half of the original number. A 0 is shifted out of the LSB into the carry flag.

FIGURE 4.14   Operation of
SAR   CL,1

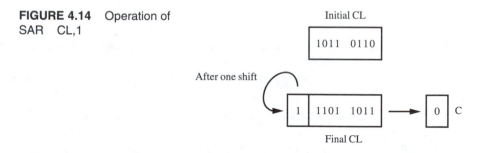

    The machine code for SAR   CL,1 is D0 F9.

SHLD/SHRD   Destination,Source,Count (Shift Left/Right Double Precision).   These instructions are similar to SHL and SHR, with the exception that instead of simply shifting zeros into the destination from left or right, bits from the source operand are shifted in. The destination operand may be a 16- or 32-bit register or memory location. The source operand must be a 16- or 32-bit register. The count may be specified by the value in register CL, or by an immediate value.

SHLD shifts the destination left the number of bits specified by the count. Bits from the source operand, beginning with the MSB, are shifted into the LSB of the destination.

SHRD shifts the destination right the number of bits specified by the count. Source operand bits, beginning with the LSB, are shifted into the MSB of the destination.

The source operand is not affected.

Example 4.39:   Registers AX, BX, and CX contain the following values, respectively: 1234H, 5678H, and 9ABCH. What are the results of these instructions?

```
SHLD    AX,BX,4
SHRD    BX,CX,8
```

Solution: As indicated in Figure 4.15(a), the SHLD instruction shifts 4 bits from **BX** into AX. The final result in AX is 2345H. SHRD shifts 8 bits from **CX** into BX, giving BX a final value of BC56H. This is illustrated in Figure 4.15(b).

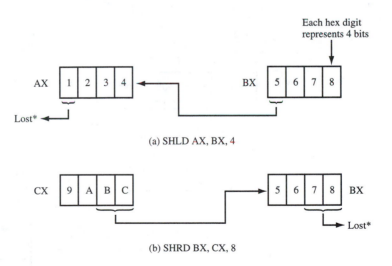

(a) SHLD AX, BX, 4

(b) SHRD BX, CX, 8

*Except carry is set to last bit shifted out

FIGURE 4.15   Double-precision shifts

The machine code for SHLD AX,BX,4 is 0F A4 D8 04. The machine code for SHRD BX,CX,8 is 0F AC CB 08.

ROL   Destination,Count (Rotate Left Byte, Word, or Double-Word).   The difference between ROL and SHL is that bits that get rotated out of the LSB position get rotated back into the MSB. Thus, data inside the destination operand is never lost, only circulated within itself. A copy of the bit that rotates out of the LSB is placed into the carry flag. The only other flag affected is the overflow flag.

Example 4.40:   What is the result of ROL   **byte ptr** [SI],1? Assume that the DS register contains 3C00H and that register SI contains 0020H.

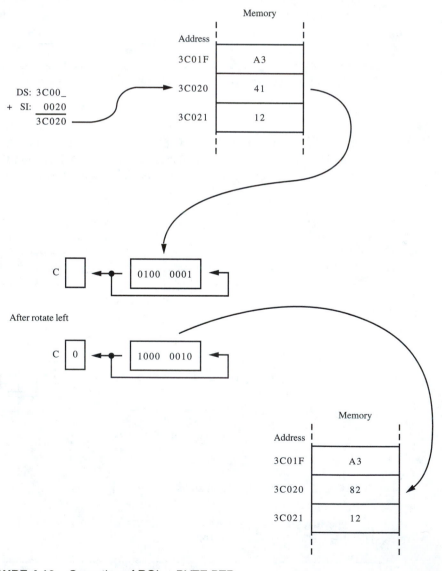

FIGURE 4.16   Operation of ROL   BYTE PTR

Solution: In Figure 4.16 the DS and SI registers combine to form an effective address of 3C020H. The byte stored at this location is rotated 1 bit left, resulting in a final value of 82H. The 0 rotated out of the MSB is placed in the carry flag.

    The machine code for ROL   **byte ptr** [SI],1 is D0 04.

ROR   Destination,Count (Rotate Right Byte, Word, or Double-Word).   This instruction has the opposite effect of ROL, with bits moving to the right within the destination operand. The bit that rotates out of the LSB goes into the carry flag and also into the MSB.

Example 4.41:   What is the result of executing ROR   AX,1 twice? AX initially contains 3E95H.

Solution: Figure 4.17 shows the results of each rotate right operation. The final value in AX is 4FA5H. The carry flag is cleared as a result of the second shift.

FIGURE 4.17   Operation of two ROR   AX,1 instructions

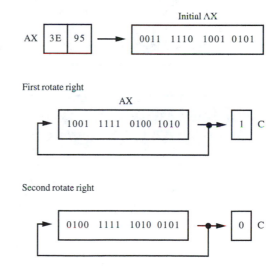

    The machine code for ROR   AX,1 is D1 C8.

RCL   Destination,Count (Rotate Left Through Carry Byte, Word, or Double-Word).   The operation of RCL is similar to ROL except for how the carry flag is used. ROL is an 8-, a 16-, or a 32-bit rotate. RCL is a 9-, 17-, or 32-*bit* rotate. The bit that gets rotated out of the MSB goes into the carry flag and the bit that was in the carry flag gets rotated into the LSB. By controlling the carry flag we can place new data into the destination operand, if that is our desire.

Example 4.42:   What is the result of executing RCL   DL,1? Assume that the carry flag is initially cleared and that DL contains 93H.

Solution: The contents of DL are rotated left 1 bit through the carry. The 0 in the carry flag rotates into the LSB of DL. The 1 in the MSB of DL rotates out of the register into the carry flag. This is shown in Figure 4.18. The final value in DL is 26H.

FIGURE 4.18   Operation of RCL   DL,1

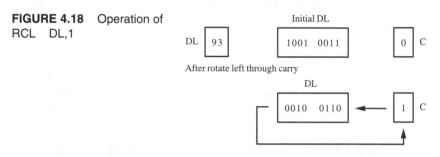

The machine code for RCL   DL,1 is D0 D2.

RCR   Destination,Count (Rotate Right Through Carry Byte, Word, or Double-Word).   This instruction complements the operation of RCL, rotating data bits right within the destination operand. The bit rotated out of the LSB goes into the carry flag. The bit that comes out of the carry flag goes into the MSB.

Example 4.43:   What is in AX after execution of RCR   AX,CL if CL contains 2 and AX contains ABCDH? The carry flag is initially cleared.

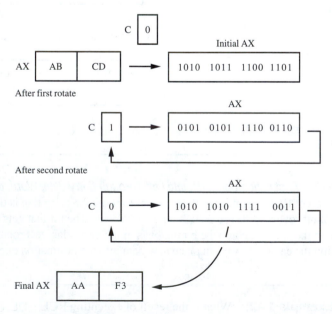

FIGURE 4.19   Operation of RCR   AX,CL

Solution: Figure 4.19 shows the result of each rotate right operation. The final value in AX is AAF3H. The carry flag is cleared after the second rotate.
    The machine code for RCR   AX,CL is D3 D8.

4.5   PROGRAM TRANSFER INSTRUCTIONS

This group of instructions allows the programmer to choose where the processor fetches its next instruction from. There are two main ways to change the execution path of a program. In general, the programmer may cause the program to *jump* to a new instruction location, or change its path temporarily by the use of a *subroutine*. Let us see how this is done.

Loop and Jump Instructions

Normally, the processor fetches instructions from sequential memory locations and places them into an instruction queue. Many instructions simply perform a specific operation on one of the processor's registers or on the data contained in a memory location. There are, however, many times when the results of an instruction need to be interpreted. Did the last instruction set the zero flag? Was the result of the last subtraction negative? Did a 0 rotate into bit 7 of register BL? There may be many different choices to make during execution, and these choices require a number of different portions of machine code. The program needs to be able to change its execution path to respond appropriately. For example, in a program designed to play tic-tac-toe it makes a big difference who goes first and whether you put an X in the center or not. The machine language responsible for handling these differences may need to be executed in a different order each time the game is played. For this reason we need to be able to change the path of program execution by forcing the processor to fetch its next instruction from a new location. This is commonly referred to as a **jump.** A jump alters the contents of the processor's instruction pointer (and sometimes its CS register as well). Remember that the CS and IP registers are combined to determine the address of the next instruction fetch. If we change the contents of these registers, the next instruction is fetched from a new address. The CS and IP registers define a 64KB segment of memory. Any instruction that jumps to an address within this segment performs an **intrasegment transfer.** For example, an instruction located within a code segment beginning at address 10000H can perform an intrasegment transfer to address 1C000H. But what about instructions located in other areas of the system memory? Suppose that a certain 80x86-based system has 32KB of EPROM located at address 48000H. It is not possible for an instruction located in the code segment at 10000H to do an intrasegment jump to one at 48000H. In this case an **intersegment transfer** is needed. The difference between intrasegment transfers and intersegment transfers is that an intrasegment transfer requires a change only in the IP register, whereas the intersegment transfer requires both the CS and IP registers to be modified. For example, if the CS register contains 1000H, then all values of the IP from 0000 to FFFFH generate addresses 10000H to 1FFFFH. No other addresses are possible. If the CS register instead contained 4800H, then the range of reachable addresses is from 48000H to 57FFFH. Do you see why an intersegment transfer needs also to modify the CS register?

When the assembler examines a jump instruction it is often able to determine what kind of jump is required, but many times it needs specific guidance from the programmer to generate the correct code. Assembler directives **near** and **far** are used to indicate intrasegment and intersegment transfers and are included directly in the source statement. For instance, JMP   FAR PTR TESTBIT indicates an intersegment transfer to the address associated with the label TESTBIT. The machine code generated for this instruction would contain information for both the CS and IP registers. JMP   NEXTITEM, on the other hand, is assumed to contain a near label. NEXTITEM must be contained within the current code segment.

JMP   Target (Unconditional Jump to Target).   The word unconditional in the description of this instruction means that no condition has to be met for the processor to jump. JMP *always* causes a program transfer. The target operand may be a near or far variable and must be indicated as such for the assembler to produce machine code. In a **direct** jump the target address is given in the instruction. In an **indirect** jump the target address might be contained in a register or memory location. A special form of the direct jump, called the *short* jump, is used whenever the target address is within +127 or –128 bytes of the current IP address. Instead of specifying the actual target address within the instruction, a **relative offset** is used instead. In the case of the short jump, the relative offset is coded in a single byte, which is interpreted as a signed binary number. Thus, short jumps are capable of moving forward up to 127 locations and backward up to 128 locations. When the target address is outside of this range, but still within the same segment, a near jump is used and the target address is coded as a 2-byte offset within the segment. Far jumps require 4 bytes for addresses: 2 for the new IP address and 2 for the new CS address.

No flags are affected.

Example 4.44:   Let us examine the code for three different types of JMP instructions. A short jump JMP   120H is located at address 100H. The corresponding machine code is EB 1E. The EB part of the machine code means "short jump." The 1E part is the relative offset of the target address. This offset is added to the current IP value, which is 102H, the first available location following JMP   120H. If we add 1EH to 102H we get 120H, the desired target address to jump to. Now let us assume that a second short jump JMP   0C0H is located at address 102H. The machine code for this instruction is EB BC. Because the target address is smaller than the current instruction pointer address of 104H, the relative offset will be negative. The offset value BC is a negative signed number equal to the difference between C0H and 104H. In both of these instructions, the current IP value was *not* the address of the JMP instruction itself, but the address of the *next* instruction. Because JMP   120H is located at 100H and only required 2 bytes, the next instruction begins at 102H.

Now consider a near jump JMP   3000H located at address 200H. The assembler knows that it should use a near jump because the difference between the 3000H target and the current IP was greater than 127. The machine code for JMP   3000H is E9 00 30. The E9 part indicates a near jump and 00 30 is the byte-swapped target address.

A far jump JMP   FAR PTR XYZ requires 5 bytes of code. The first byte is EA, which indicates a far jump. The next 2 bytes are the new IP address, and the last 2 bytes are the new CS value. Where does EA 00 10 00 20 take the processor? The new IP address from this machine code is 1000H. The new CS address is 2000H. The processor will perform a far jump to address 21000H.

Other legal forms of JMP involve using registers or the contents of memory locations to supply the target. For example, JMP   AX indicates a near jump to the address contained in register AX. If AX contains 4500H, then the processor will jump to address 4500H within the code segment. JMP   [SI] causes the processor to read the word stored in the memory locations pointed to by the SI register, and use that word as the target address. Figure 4.20 shows this instruction in more detail. The [SI] operand indicates an indirect jump and requires that the memory word pointed to by SI be read and placed into the IP register. In this fashion, JMP   [SI] sets the processor up to fetch its next instruction from address 1234H within the current code segment.

The machine code for JMP   AX is FF E0. The machine code for JMP   [SI] is FF 24.

Conditional Jumps.   The 80x86 has a large variety of conditional jumps, all of which depend on specific flag states to enable their operation. Many of these conditional jumps test only one specific flag. For example, the JC and JNC instructions examine only the state of the carry flag to determine their next step. Others, such as JA and JNA, perform a Boolean test of more than one flag. Furthermore, none of the flags are affected by the jumps. As we have previously seen, a number of instructions go by more than one name. Such is the case

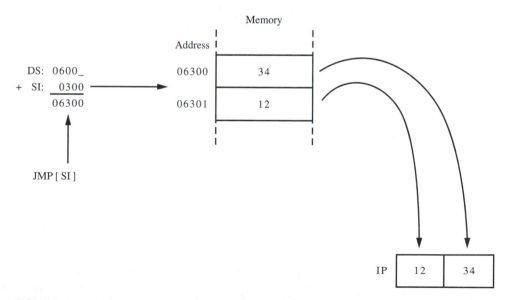

FIGURE 4.20   Operation of an indirect JMP

for the unconditional jumps. The instructions, their required flag conditions, and their interpretations are as follows:

Instruction	Flag Tested	Explanation
JNC	$CF = 0$	jump if no carry
JAE	"	jump if above or equal
JNB	"	jump if not below
JC	$CF = 1$	jump if carry
JB	"	jump if below
JNAE	"	jump if not above or equal
JNZ	$ZF = 0$	jump if not zero
JNE	$ZF = 0$	jump if not equal
JZ	$ZF = 1$	jump if zero
JE	"	jump if equal
JNS	$SF = 0$	jump if no sign
JS	$SF = 1$	jump if sign
JNO	$OF = 0$	jump if no overflow
JO	$OF = 1$	jump if overflow
JNP	$PF = 0$	jump if no parity
JPO	"	jump if parity odd
JP	$PF = 1$	jump if parity
JPE	"	jump if parity even
JA	$(CF \text{ or } ZF) = 0$	jump if above
JBNE	"	jump if not below or equal
JBE	$(CF \text{ or } ZF) = 1$	jump if below or equal
JNA	"	jump if not above
JGE	$(SF \text{ xor } OF) = 0$	jump if greater or equal
JNL	"	jump if not less
JL	$(SF \text{ xor } OF) = 1$	jump if less
JNGE	"	jump if not greater or equal
JG	$((SF \text{ xor } OF) \text{ or } ZF) = 0$	jump if greater
JNLE	"	jump if not less or equal
JLE	$((SF \text{ xor } OF) \text{ or } ZF) = 1$	jump if less or equal
JNG	"	jump if not greater

Note that you must supply the target address in the jump instruction. Unlike JMP, we are not allowed to use register names or other fancy addressing mode operands. The nice thing about relative jumps is that they always jump to the correct location within the code segment, no matter where the code segment appears in memory.

Example 4.45:   Let us look at a few ways conditional instructions may be used to test information.

Consider the following group of instructions, which determine if the AL register contains a lowercase ASCII letter.

```
CMP   AL,'a'
JB    NOTLOWER
CMP   AL,'z'+1
JNB   NOTLOWER
```

If AL does not contain the ASCII code of a lowercase letter (61H to 7AH for 'a' or 'z'), the processor will jump to NOTLOWER. If AL does contain a valid code, then neither of the conditional jumps will be taken, and the processor will execute the first instruction following JNB   NOTLOWER. Remember that JB and JC are equivalent instructions, as are JNB and JNC. Thus, the same lowercase test can be performed like this:

```
CMP   AL,'a'
JC    NOTLOWER
CMP   AL,'z'+1
JNC   NOTLOWER
```

Next, suppose that you are working with a portion of code that adds small increments to a register, as in:

```
ADD   BL,8
```

If BL contains a value in the range 00 to 77H, the result of the ADD will be in the range 08 to 7FH. If, instead, BL contained a value in the range 78H to 7FH, the result after adding eight would be 80H to 87H. In terms of signed numbers, BL has gone through a sign change. Why has this happened? The value 78H corresponds to 120 decimal. Adding eight to this gives us 128, which is represented by 80H, or 10000000 binary. Note that the MSB is high, which indicates the *negative* value −128 from the 2's complement point of view. If we simply interpret 80H as an *unsigned* integer, the 80H means +128. This unintentional sign change will be indicated by the processor's **overflow** flag. A similar result is obtained when

```
SUB   BL,8
```

is used, with the overflow flag being set when the 80H barrier is passed. To test for this condition, use the JO and the JNO instruction, as in:

```
ADD   BL,8
JO    OVERFLOW
```

Note that the overflow flag tells us something different than the carry flag does. There is no carry when BL goes from 78H to 80H, but there is an overflow.

Many conditional jumps test more than one flag. For instance, JG uses three flags (sign, overflow, and zero) to determine its outcome. Examine the following three groups of code. Which JG instructions are taken?

```
MOV   AL,30H       MOV   AL,30H       MOV   AL,30H
CMP   AL,20H       CMP   AL,40H       CMP   AL,90H
JG    XXX          JG    YYY          JG    ZZZ
```

The JG to XXX is taken because 30H is greater than 20H. The JG to YYY is not taken because 30H is not greater than 40H. The JG to ZZZ *is taken* because the 90H value is interpreted by the processor as the negative number –112 (due to its MSB being high). Thus, the CMP instruction determines that 30H is greater than –112 (90H), and so the JG is taken. Note that if you want 90H to be interpreted as the *unsigned* value 144, a different conditional jump (such as JC or JNC) should be used.

Example 4.46:   Is it possible to perform a conditional jump that has a target address outside of the relative range of +127/–128 bytes?

Solution: Yes. Two instructions are needed to synthesize a conditional jump of this type. Consider the following code:

```
         JNZ    SKIPJMP
         JMP    NEWPLACE
SKIPJMP:  ---
```

Here, the conditional jump JNZ is used to jump over a near JMP to NEWPLACE whenever the zero flag is cleared. When the zero flag is set, the JNZ will *not* jump to SKIPJMP, but will simply continue execution with the next instruction in memory, which is the JMP   NEWPLACE instruction. So, the zero flag must be set to perform a near JMP to NEWPLACE.

LOOP   Short-Label (Loop).   This instruction will decrement register CX and perform a short jump to the target address if CX does not equal 0. LOOP is very useful for routines that need to repeat a specific number of times. CX must be loaded prior to entering the section of code terminated by LOOP.

Example 4.47:   In Figure 4.21(a) a LOOP instruction is used to execute a short section of code. The number placed into CX at the beginning of the loop (10 decimal in this case) determines how many times the loop repeats. Notice that the LOOP BACK instruction

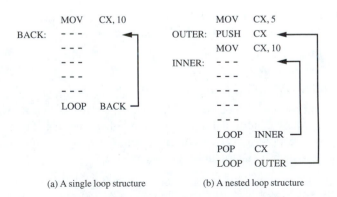

(a) A single loop structure   (b) A nested loop structure

FIGURE 4.21   Using LOOP instructions

does not contain a reference to CX. The processor knows it should automatically decrement CX while executing this instruction. LOOP BACK actually performs a short jump to BACK as long as CX does not equal 0.

In some cases it becomes necessary to perform a "loop within a loop," or *nested* loop as it is more commonly known. Figure 4.21(b) shows an example of a nested loop. The outer loop executes five times and the inner loop ten times. Instructions in the body of the loop execute fifty times. The PUSH and POP instructions are used to preserve the contents of the outer loop counter.

A single LOOP can be repeated up to 65,535 times using CX as the counter. How many loops are possible with a nested loop? The answer is over *4 billion!*

LOOPE/LOOPZ Short-Label (Loop if Equal, Loop if Zero). This instruction is similar to LOOP except for a secondary condition that must be met for the jump to take place. In addition to decrementing CX, LOOPZ also examines the state of the zero flag. If the zero flag is set and CX does not equal 0, LOOPZ will jump to the target. If CX equals 0, or if the zero flag gets cleared within the loop, the loop will terminate. As always, when a condi tional instruction does not have the correct flag condition, execution continues with the next instruction. LOOPE is an alternate name for this instruction.

LOOPNE/LOOPNZ Short-Label (Loop if not Equal, Loop if not Zero). LOOPNZ is the opposite of LOOPZ. The zero flag must be *cleared* to allow further looping. LOOPNE has the same function.

JCXZ/JECXZ Short-Label (Jump if CX/ECX = 0). JCXZ jumps to the target address if register CX contains 0. JECXZ does the same if ECX equals 0. CX and ECX are not adjusted in any way. There are times when JCXZ or JECXZ should be used at the beginning of a section of LOOP code to skip over the code when CX or ECX equals 0.

Example 4.48: In Figure 4.22 a JCXZ instruction is used to skip over the AGAIN loop whenever CX equals 0. If the JCXZ instruction were absent, how many times would the loop execute?

FIGURE 4.22 Operation of JCXZ

```
DOLOOP:   JCXZ   SKIPLOOP
AGAIN:    ADD    AX, 3
          LOOP   AGAIN
SKIPLOOP: - - -
```

Solution: If JCXZ is removed and the loop is entered with CX equal to 0, it will repeat 65,536 times! This is because LOOP decrements CX before it tests for 0. If CX is decremented from 0 it becomes FFFFH, which means 65,535 more decrements are needed to get back to 0. Note that the code (JCXZ included or not) works properly for every other starting value of CX.

Subroutine and Interrupt Instructions

Although subroutines and interrupts are called or generated in different ways, they share some common ground. Both require the use of the stack for proper operation. Both have a means of returning to where they came from (via information returned from the stack).

A **subroutine** is a collection of instructions that is CALLed from one or many other different locations within a program. At the end of a subroutine is an instruction that tells the processor how to RETurn (go back) to where it was called from. Figure 4.23 shows how a subroutine called ALPHA can be called from many different places within a program. The RET instruction at the end of ALPHA always causes the processor to go to the instruction immediately following the last CALL. It does this by popping a return address off the stack, which was placed there by the corresponding CALL. For example, when the first CALL is encountered, the address of the instruction immediately following the CALL is pushed onto the stack. When the RET instruction at the end of ALPHA executes, the return address is popped off the stack and placed into the IP register. Using the stack in this way guarantees that RET will go back to the right place no matter which of the three CALL instructions has been used.

Interrupts operate in a similar manner, but require more use of the stack. In addition to a return address, the 80x86 interrupt mechanism also pushes the processor flag register onto the stack. An interrupt is generated through software by the use of the INT or INTO instructions, which are to interrupts what CALL is to subroutines. An **interrupt service routine,** or ISR for short, is a common way of referring to the code used to handle an interrupt. Because the stack is used differently, RET is not used to return from an ISR. IRET is used instead to ensure that the stack is popped correctly. The processor has many different types of interrupts, which we will cover in detail in Chapter 5.

CALL   Procedure-Name (Call Procedure).   This instruction is used to call a subroutine. The subroutine may be located in the current code segment or in a different one. Like the intrasegment and intersegment jumps, we also have intrasegment and intersegment CALLs. We think of these as calls to near or far *procedures.* A procedure is another way of referring to a subroutine and has to do with the assembler directives PROC and ENDP used in the source file. The direct and indirect types of JMPs that we have previously seen also apply to CALL. In a direct CALL, the address of the procedure is specified in the instruction (usually via the label naming the procedure). In an indirect CALL, the procedure address may be contained in a processor register or memory location. CALL pushes the address of the instruction immediately following

FIGURE 4.23   Calling a subroutine from many different places

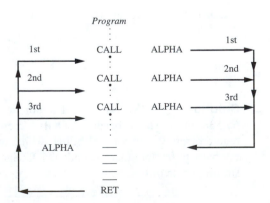

itself onto the stack. In the case of a near CALL, this address is just one word of data. In the case of a far CALL, two pushes are performed: one for the instruction offset and the other for the code segment address. CALL then jumps to the address of the procedure and resumes execution.

Example 4.49: Figure 4.24 shows the contents of the stack after near and far CALLs to the same procedure. For the near call, only the offset of the next instruction is pushed. For the far call, the instruction offset and the code segment value are pushed. If both instructions were located at offset 100H within a code segment located at 5000H (e.g., CS equals 0500H), the stack would be filled with the data you see in the figure. The near call to XYZ will cause only the return address 103H to be pushed. Because the near call requires 3 bytes of machine code, address 103H is the first available instruction address.

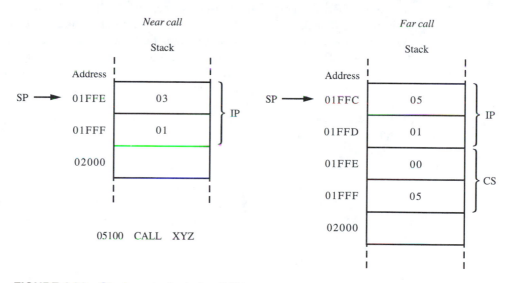

FIGURE 4.24   Stack contents during CALL

The far call is a 5-byte instruction and needs to push more information onto the stack. First the contents of the CS register are pushed (0500H). Then the instruction offset is pushed. This offset is 105H now because of the length of the far call instruction. The procedure must be careful not to modify the contents of these stack locations. It is also important to guarantee that the SP register points to the correct address before executing RET. It is okay to push items onto the stack while inside the procedure, but these items must all be popped off before returning.

RET Optional-Pop-Value (Return from Procedure).   This instruction is used to provide an orderly exit from a procedure that has been called. It is important to understand that RET does not know *anything* about the procedure it terminates except for its type being far or near. All RET does is pop the appropriate information off the stack and place it into the assigned registers. For a near procedure, RET pops one word off the stack and writes it into the IP register.

For a far procedure, RET pops two words. The first goes into the IP register, the second into the CS register. Once these pops are made (for either type of return), the optional-pop-value is used to pop additional words off the stack, if any at all. The optional-pop-value is added to the stack pointer, advancing it past information that was pushed before the procedure was called. This value is assumed to be 0 when no pop-value is included.

Example 4.50:   The return addresses of two different subroutines are contained in the stack. The subroutine called first was a far procedure. The instruction that called it pushed the return address 0350:1C00, where 0350H is the segment address and 1C00H is the offset. The second subroutine called was a near procedure with a return address of 4080H. The second procedure was called from inside the first procedure. What are the top three items on the stack?

Solution: Figure 4.25 shows the stack created by the return addresses for each procedure. When the second procedure finishes, its near RET will pop 4080H off the stack and put it into the IP register. This gets us back into the first procedure (which called the second). The far RET in this procedure pops 1C00H off the stack, puts it into the IP register, and then pops 0350H off the stack and places it into the CS register. Any number of procedure return addresses can be nested in this way, as long as the correct RETs are always used and you do not run out of stack memory.

FIGURE 4.25   Nested return addresses

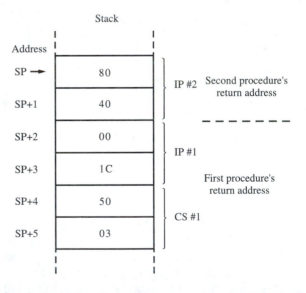

The machine code for a near RET is C3. The machine code for a far RET is CB. When pop-values are used, 2 additional bytes are added to the instructions. They contain the pop-value. The machine code for a near RET 2 is C2 02 00.

INT   Interrupt-Type (Interrupt).   The 80x86 supports 256 software interrupts, which initially operate in a manner similar to CALL. When INT is encountered, the processor will first push the flags onto the stack. This is an important step, because the flags tell us part of the processor's internal state at the time it was interrupted. Then it clears two special flags called the **trace flag** and the **interrupt-enable flag.** It does this to prevent additional software interrupts from occurring while it processes the first.

Next, the processor pushes the current CS register onto the stack and follows that with the contents of the IP register. The last two pushes resemble the operation of a far call. The *number* of the interrupt, rather than its address, is coded into the instruction, and is now used to find the correct place in the processor's **interrupt vector table,** a 1KB block of memory located in the beginning of the processor's address space. Its range of address is from 00000 to 003FFH. The interrupt number is called the **interrupt-type,** and is multiplied by 4 to get the address within the vector table that contains the interrupt service routine address. All ISR addresses are 4 bytes long and in the standard CS:IP format. Once the vector address is known, the processor reads the ISR address out of the table, places the information into the CS and IP registers, and resumes program execution at the new location.

Example 4.51:   What is the sequence of events for INT 08H if it generates a CS:IP return address of 0100:0200? The flag register contains 0081H.

Solution: The flags are pushed first. The return CS and return IP addresses are pushed next. Multiplying interrupt type 8 by 4 gives 32, which is 00020H. The ISR address stored at this vector address (see Figure 4.26) is read into CS:IP and execution continues at address 05810H.

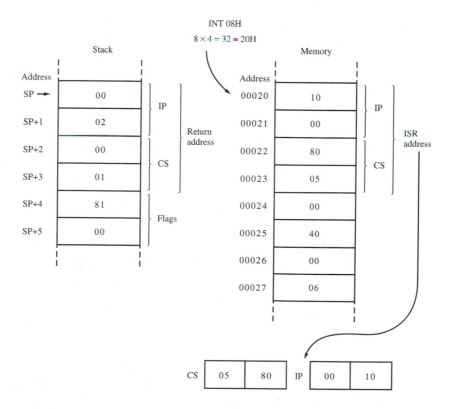

FIGURE 4.26   Operation of INT 08H

IRET/IRETD   (Interrupt Return).   Because the operation of an interrupt requires different handling of the stack, an IRET instruction must be used at the end of an ISR. A normal RET instruction will not pop the flags off and will most likely result in incorrect program execution. IRET pops the CS and IP return addresses and then restores the flag register by popping its value too. Thus, the flags will resemble their state at the exact moment of the interrupt. When 32-bit addressing is in effect, IRETD is used to pop 32-bit addresses off the stack.

Example 4.52:   What is the return address and the contents of the flag register when IRET uses the stack information from Figure 4.27?

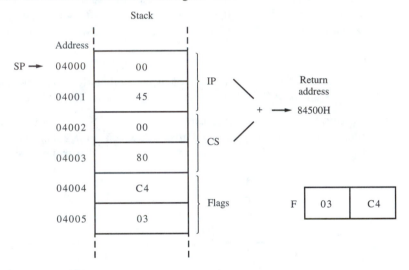

FIGURE 4.27   Operation of IRET

Solution: The word 4500H is popped into the IP register. Then, 8000H is popped into the CS register. Together, these values indicate a return address of 84500H. Next, 03C4H is popped into the flags. The new value of the stack pointer is 04006H.

The machine code for IRET is CF.

INTO   (Interrupt on Overflow).   If the overflow flag is set when INTO is encountered, it will generate a type-4 interrupt, with all of the pushes and operations performed by INT. The interrupt vector address of INTO is 00010H. If the overflow flag is cleared when INTO executes, no interrupt is generated and execution resumes with the next instruction. This interrupt is useful for determining when the results of an arithmetic operation have gone out of bounds.

BOUND   Index,Range (Check Array Index Against Bounds).   This instruction tests the value of an index register against a predefined allowed range of indexes. If the array index is within range, no further action is taken. If the array index is out of range, an INT 05H is generated. The index operand may be any 16- or 32-bit register. The range operand specifies the address of two 16- or 32-bit numbers that indicate the allowed index range.

Example 4.53: The following excerpt from a source file shows how the BOUND instruction may be used. The index contained in the SI register is checked against the range 0 to 9.

```
        .DATA
RANGE   DW      0,9
ARRAY   DB      10 DUP(?)

        .CODE
        .
        .
        BOUND   SI,RANGE
        MOV     AL,[SI]
        .
        .
```

If the SI register has an index value between 0 and 9, the MOV instruction will be executed. If not, an INT 05H is generated. Note that MS-DOS uses INT 05H to print the contents of the screen, so using BOUND in be real mode will require you to change the operation of INT 05H's ISR.

ENTER/LEAVE   (Enter/Leave a Procedure).   These instructions are used to manage the **stack frame** used by high-level languages such as C. A stack frame stores pointers and variables used by the current function or procedure being executed in the high-level language. By convention, the BP register is used as a **frame pointer,** and indicates the beginning of the previous stack frame. The first thing the ENTER instruction does is to push the BP onto the stack. Thus, the current stack frame now points back to the previous stack frame. This value will be popped by the LEAVE instruction before the function RETurns, restoring the old frame pointer.

Next, the new value of the SP is copied into BP. This sets the frame pointer to the beginning of the current stack frame.

Finally, the amount of dynamic storage space to reserve, in bytes, is subtracted from the SP. Dynamic storage space is used by the procedure for storage of local variables.

The ENTER instruction takes two operands. The first operand specifies the amount of dynamic storage space in bytes. This value must be a multiple of 2. The second operand indicates the **lexical nesting level** of the new procedure being entered. The lexical nesting level (or simply level) is determined by how many subprocedures are currently active. For instance, if the main program calls Procedure A, A's level is 1. If Procedure A then calls Procedure B, Procedure B's level is 2. When the level is 0, ENTER is equivalent to:

```
PUSH    BP
MOV     BP,SP
SUB     SP,dsp
```

where dsp specifies the amount of dynamic storage space.

When the specified level is not 0, ENTER also pushes pointers onto the stack that references the stack frames of all previous procedures.

The LEAVE instruction restores the original (previous) values of the BP and SP registers. Equivalent instructions are:

```
MOV     SP,BP
POP     BP
```

We will examine how ENTER and LEAVE are used in the C programming language in Chapter 8.

Example 4.54:   The BP and SP registers contain the offsets 2008H and 2000H, respectively. The stack segment is at 0000H. What is the result of executing ENTER   4,0?

Solution: Figure 4.28 shows the changes made to the stack when ENTER   4,0 executes. First, the BP is pushed, which decrements the SP to 1FFEH. This value is then copied into the BP register. Last, 4 is subtracted from the SP, giving it a final value of 1FFAH.

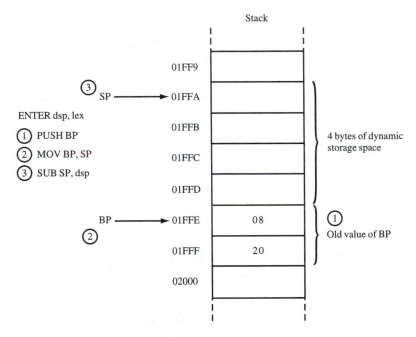

FIGURE 4.28   Operation of ENTER   4,0

The corresponding LEAVE instruction will restore the value of the BP to 2008H and the SP to 2000H.

The machine code for ENTER 4,0 is C8 0004 00. The machine code for LEAVE is C9.

Assembler Directives PROC and ENDP

As mentioned previously, a subroutine or procedure must be of a specific type. The two types, near and far, are specified by the PROC statement that begins the subroutine. In the following example, two procedures, TEST and DELAY, are defined with different PROC statements. Once again, this is necessary to ensure that the stack is used properly.

```
TEST    PROC   FAR            DELAY  PROC   NEAR
         .                            .
         .                            .
        CALL   DELAY                  .
         .                           RET
         .                    DELAY  ENDP
TEST    ENDP
```

In this example, DELAY is located somewhere within the source file. It would be accept-able to write the code this way also:

```
TEST    PROC   FAR
         .
         .
        CALL   DELAY
         .
         .
         .
DELAY   PROC   NEAR
         .
         .
         .
DELAY   ENDP
TEST    ENDP
```

Note that the prodecures are now *nested*. The assembler still requires matching ENDP statements for each PROC statement.

   If a FAR procedure needs to be CALLed, the assembler requires the statement:

```
CALL    FAR PTR <name>
```

In many cases, this statement prevents unwanted assembly errors.

4.6 PROCESSOR CONTROL INSTRUCTIONS

This last group consists of instructions used to make changes to specific bits within the flag register and to externally synchronize the processor. Also included is an instruction found on practically all microprocessors: the NOP.

The Instructions

CLC (Clear Carry Flag).   The instruction clears the carry flag. Its machine code is F8.

STC (Set Carry Flag).   This instruction sets the carry flag. Its machine code is F9.

CMC (Complement Carry Flag).   This instruction inverts the carry flag. Its machine code is F5. These three carry-flag instructions are useful when dealing with the rotate instructions and with multi-bit mathematical routines.

CLD (Clear Direction Flag).   The direction flag is cleared by this instruction. Its machine code is FC. The direction flag is used by the processor to determine if index registers should be incremented or decremented during string operations.

STD   (Set Direction Flag).   The direction flag is set by this instruction. Its machine code is FD.

CLI   (Clear Interrupt-Enable Flag).   This instruction clears the interrupt-enable flag. This prevents the processor from responding to a hardware INTR request. No other interrupts are affected. Its machine code is FA.

STI   (Set Interrupt-Enable Flag).   Setting the interrupt-enable flag allows the processor to acknowledge an interrupt signal on the INTR input. Its machine code is FB. If a request was made to INTR while the interrupt-enable flag was cleared (which we call a *pending* interrupt), that interrupt request would be acknowledged after completion of the next instruction after STI.

HLT   (Halt Until Interrupt or Reset).   This instruction terminates program execution. Its machine code is F4. It places the processor in a halt state, in which it will remain until it receives a hardware reset, or an interrupt request on NMI or INTR. If the request is on INTR, interrupts must be enabled to leave the halt state.

NOP   (No Operation).   Although this instruction does not do anything at all (no flags or registers are affected), it is still quite useful in many ways. Specifically, one application of NOP is its use within timing delay loops. There are applications that require specific time delays to be generated (for slowing down a graphics display to see the action, for example). Sometimes it is useful to throw a couple of NOPs into the timing loop, to increase the loop time. After all, it takes a few microseconds to fetch, decode, and execute NOP. It is nice to be able to use an instruction that takes up a little time but does nothing. Its machine code is 90.

LOCK   (Lock Bus During Next Instruction).   Executing LOCK causes the processor's $\overline{\text{LOCK}}$ output to go low for the duration of the next instruction. Thus, LOCK is used as a prefix to another instruction. Locking the bus is a way of giving the processor some privacy while it executes the subsequent instruction, which cannot be interrupted until it completes. The prefix code for LOCK is F0.

We will examine other processor control instructions in Chapter 16.

4.7   HOW AN ASSEMBLER GENERATES MACHINE CODE

During assembly, source statements are read one line at a time and examined. If they contain legal instructions, the assembler can determine the required machine code by filling in missing bits in a basic opcode format. For example, the instructions MOVSB and MOVSW both have the same basic opcode format: 1010010w, where *w* is a variable that takes on 0 for a byte operation and 1 for a word operation. Thus, if the assembler reads in a source statement that contains MOVSB, once it recognizes the MOVS part of the instruction, it will then look for a "B" or a "W" to determine the correct value of *w*. It is easy to see that MOVSB has A4 as its machine code and that MOVSW has A5 as its machine code. If the assembler does not see a "B" or a "W" after MOVS, it will generate an error message.

Other instructions are more complicated to assemble than MOVS. One form of the multiplication instruction MUL can have two sizes of operands (byte and word), and the operands can be registers or memory locations. Clearly, the basic opcode format for MUL must be more complicated. In fact, the format is 1111011w mod 100 r/m. This strange-looking string of symbols is used to specify all possible opcodes for MUL.

Consider the instruction MUL   CX. The *w* bit works as it did in the MOVS example. Because CX is a word-size register, *w* will be set to 1. *Mod* is a 2-bit variable that is set to 11 by the assembler when the operand is a processor register. The register specified in the instruction is indicated by the 3-bit variable *r/m*. Register CX has the code 001 associated with it. So, MUL   CX becomes 11110111 11100001, or F7 E1 in machine code. MUL   SI would be F7 E6, since SI's internal code is 110.

Suppose an instruction has two operand fields. In XCHG   AL,BL, the operands are both byte-size registers. Each register will have its own internal code. The destination register AL is coded as 000. The source register BL is coded as 011. The basic opcode format for XCHG is 1000011w mod reg r/m. The new variable *reg* represents the 3-bit code for the destination operand. *R/m* indicates the source operand. When exchanging registers, *w* will be set to 11. Inserting these 2 bits and the 6 bits that are used to specify AL and BL, we get 10000110 11000011. This is the machine code 86 C3. Similarly, XCHG   CX,DX has 87 CA for its machine code. Can you determine the 3-bit codes associated with CX and DX? (You should get 001 and 010, respectively.)

So, we see that the process of assembling a program involves a great deal of time picking the right combination of bits to place into opcode patterns. Let us keep this in mind when we assemble our own programs later in the book. When the assembler indicates an error, it's not because it picked the wrong bits; it is because the information supplied was incorrect, missing, or of the wrong type.

4.8   THE BEAUTY OF RELOCATABLE CODE

In the early days of microprocessors few addressing modes were available. The limited ability of the early microprocessor to access its own memory required, in most cases, the generation of specific *absolute* addresses within instructions. For example, if an instruction located at address 1000H needed to examine the data in memory location 106FH, the address 106FH would have to be coded in the instruction as a 2-byte absolute address. The relative mode of addressing gets rid of the absolute address portion of the instruction and replaces it with a single byte of relative address information. Now only the offset 6FH must be saved within the instruction's machine code. You may think that saving a single byte here and there will make little difference. In short programs this is true. But in longer programs, which are more likely to access temporary values or data tables stored in memory or perform a significant number of jumps, a substantial savings in space results.

A second, more important feature of relative addressing is its built-in **relocatability.** A relocatable program is a program that can be loaded into memory at any address and then executed normally. Early microprocessors did not generate relocatable code and thus required their programs to load into memory at a specific location. For instance, a program ORGed at address 3C00H had to load into memory at address 3C00H in order to run properly. The same program loaded at 5000H, or anywhere else, would contain absolute addresses (such as the address 106FH with the example instruction at the beginning of this section) that did not look at the correct locations. For the program to run correctly, each absolute address would have to be adjusted depending on the new load address. For example, the 106FH address might be changed to 506FH. While this type of program modification is possible, programs will require additional load time when we try to execute them. Relocatable code does not have this disadvantage. An instruction that accesses the location "30 locations forward" of its own address will get the address right no matter where it is loaded. The 80x86's ability to use relocatable code is an attractive feature.

4.9   TROUBLESHOOTING TECHNIQUES

Now that we have seen the entire 80x86 instruction set, let us spend a moment on some tips designed to keep the software development process running smoothly.

- Periodically review the entire instruction set. Programmers tend to fall in love with certain instructions and methods of writing code, and it is easy to forget that the processor already contains an instruction to fill your need.
- Understanding the binary number system and its associated operations is just as important as knowing the addressing modes and instruction types. It is difficult to perform any kind of interfacing, design, or programming without a good background knowledge of 1s and 0s. Take the time to review Appendix D if you have not already done so; it may save you some effort in the future.
- The arithmetic, logical, and bit manipulation instructions all rely heavily on the flags. As we have seen again and again, the flags are very important. Get into the habit of checking out an instruction to see how it affects the flags. For example, you may write a short loop like this:

```
BACK:      DEC    BL
           ADD    AL,7
           JNZ    BACK
```

Logically, everything seems in order. AL and BL are both changed in the loop. The problem is that the JNZ instruction is meant to test the NZ condition created by the DEC instruction, but the ADD instruction, which also updates the zero flag, changes the outcome. It is best to place the instruction that affects a flag as close as possible to the instruction that uses it.
- When using subroutines, or any group of instructions that manipulate the stack, it helps to go through the code to verify that the same number of pushes and pops are made. It is very easy to add an extra push or pop when writing stack-based code for the first time. Generally, the DOS environment on the personal computer is very unforgiving when its stack is abused. If the machine locks up or reboots when you run your stack-based program, it would be good to begin your troubleshooting by examining what the stack is doing.

It may be a good idea to keep a logbook of the problems you encounter working with assembly language. Until you have seen the same problem enough times, or learn how to avoid it, keeping track of the problem and its solution will save a lot of time and effort should you encounter it again.

SUMMARY

In this chapter we examined the remainder of the 80x86 instruction set. The arithmetic, logical, bit manipulation, program transfer, and processor control instructions were all covered in detail. The basic operation of an assembler was covered, with examples given to show how bit fields within the instruction are filled in. The ability of relocatable code to execute at any load address was also discussed.

Use this chapter, and Chapter 3, for reference when you begin studying programming applications in Chapter 6.

STUDY QUESTIONS

1. If AX contains 1234H, what is the result of ADD   AL,AH?
2. What is the state of each flag after ADD   AL,AH in the previous question?
3. Memory location 2000H has the word 5000H stored in it. What does each location contain after INC BYTE PTR [2000H]?
4. Repeat Question 3 for DEC WORD PTR [2000H].
5. What is the state of the zero flag after CMP   CL,30H if CL does not contain 30H?
6. What instruction is needed to check whether the upper bytes of AX and BX are equal?
7. Show the instructions needed to multiply AX by 25. Assume the results are unsigned.
8. If DX contains 00EEH and AX contains 0980, what is the result of:

```
MOV    BX,0F0H
DIV    BX
```

9. Repeat Question 8 for the instruction IDIV   BX.
10. What instruction is needed to find the signed 2's complement of CX?
11. What are the results of CBW if AL contains 30H? What if AL contains 98H?
12. What is the final value of AL in this series of instructions?

```
MOV    AL,27H
MOV    BL,37H
ADD    AL,BL
DAA
```

13. If DX contains 7C9AH, what is the result of NOT   DX?
14. If AL contains 55H and BL contains AAH, what is the result of:
    a) AND   AL,BL
    b) OR   AL,BL
    c) XOR   BL,AL

15. What does SHL   AL,1 do if AL contains 35H?
16. What is the data in BX after SHR   BX,CL if CL contains 6?
17. Memory location 1000H contains the byte 9F. What instructions are needed to rotate it 1 bit right?
18. Explain the difference between short and near jumps.
19. What conditional jump instruction should be used after CMP   AL,30H to jump when AL equals 30H?
20. What instruction is needed in Question 19 to jump when AL is less than 30H?
21. The lower byte of the flag register contains 95H. Which of the following instructions will actually jump?
    a) JZ
    b) JNC
    c) JP
    d) JA
    e) JGE
    f) JLE
22 How many times does the NOP instruction execute in the following sequence?

```
            MOV     CX,20H
    XYZ:    PUSH    CX
            MOV     CX,9
    ABC:    NOP
            LOOP    ABC
            POP     CX
            LOOP    XYZ
```

23. A near CALL generates a return IP of 1036H. What are the contents of stack memory after the call if SP initially contains 2800H?
24. Repeat Question 23 for a far CALL located in code segment 0400H.
25. What are the final SP values in Questions 23 and 24?
26. Explain the operation of near and far RETurns.
27. What are all of the activities following an INT   16H?
28. What does INTO do in this sequence of instructions?

```
    MOV     AL,30H
    MOV     BL,0E0H
    ADD     AL,BL
    INTO
```

29. What instructions are needed to add AL, BL, and DL together, and place the result in CL? Do not destroy BL or DL.
30. What is the difference between MOV   AX,0 and SUB   AX,AX? There may be more than one difference to comment on.
31. Multiply the contents of AX by 0.125. Because fractional multiplication is not available, you must think of an alternate way to solve this problem. Assume that AX is unsigned.
32. What does this sequence of instructions do?

```
    MUL     BL
    DIV     CL
```

33. What are the largest two decimal numbers that may be multiplied by MUL? Repeat for IMUL.

34. Find the volume of a cube whose length on one side has been placed into BL. The volume should be in DX when finished.
35. Write the appropriate AND instruction to preserve bits 0, 3–9, and 13 of register BX, and clear all others.
36. What OR instruction is needed to set bits 2, 3, and 5 of AL?
37. Show how ROL can be used to rotate a 32-bit register composed of AX and BX, with AX containing the upper 16 bits.
38. Why is it important for SAL to preserve the value of the MSB?
39. Why must a subroutine contain RET as the final instruction?
40. Modify the data summing example so that bytes are added together instead of words.
41. Write a subroutine to compute the area of a right triangle whose side lengths are stored in AL and BL. Return the result in AX.
42. Write a subroutine that will compute the factorial of the number contained in DL. Store the result in word location FACTOR.
43. A data byte at location STATUS controls the calling of four subroutines. If bit 7 is set, ROUT1 is called. If bit 5 is clear, ROUT2 is called. ROUT3 is called when bits 2 and 3 are high, and ROUT4 is called if bit 0 is clear and bit 1 is set. These conditions may all exist at one time, so prioritize the routines in this way: ROUT1, ROUT3, ROUT2, and ROUT4.
44. Write a routine to swap nibbles in AL. For example, if AL contains 3E, then it will contain E3 after execution.
45. Write a subroutine that will increment the packed-decimal value stored in COUNT and reset it to 00 each time it reaches 60.
46. Write a subroutine that sets the zero flag if AH is in the range 30 to 212, and clears the zero flag otherwise.
47. Show the instructions needed to count the number of 1s found in AL. For example, if AL contains 10110001, the number of 1s is 4.
48. Write a routine that determines if BH contains a *palindrome*. A palindrome (in binary) is a pattern of 0s and 1s that reads the same forward and backward. For example, 11000011 is a palindrome, and 11001000 is not.
49. What instructions are necessary to determine the largest number contained in BX, CX, and DX? The number should be placed in AX when found.
50. Repeat Question 49 for the smallest number in AX, BX, CX, and DX.
51. Show the instructions needed to solve this equation:

$$AX = (5BX + CX/3)/DX$$

52. Registers AX, BX, and CX contain the respective values 2000H, 1000H, and 3000H. What is the result of CMPXCHG BX,CX?
53. What does IMUL BX,CX,12 do?
54. What is the result of dividing the decimal number 100,000,000 by 1024? Show how this can be done using IDIV.
55. What is the result of executing CDQ if EAX contains 8F005000H?
56. What does XADD BX,CX do if BX contains 000AH and CX contains 1000H?
57. Register ECX contains the value 07E98600H. What happens when these instructions are executed?
    a) BSF EBX,ECX
    b) BSR EDX,ECX

58. The AX register contains the value ABCDH. What are the results of executing the following?
    a) BT AX,1
    b) BTC AX,5
    c) BTS AX,10
    d) BTR AX,14
59. If the zero flag is set, what does SETNZ CL do?
60. If AX and BX contain 2468H and 1357H, respectively, what are the contents of each register after execution of the following?
    a) SHLD AX,BX,6
    b) SHRD BX,AX,6
61. Since the BOUND instruction causes problems with INT 05H in the MS-DOS environment, can you devise an equivalent set of instructions that tests the bounds of an array index? If the index is out of bounds, jump to the label BADINDEX.
62. How is the stack changed when ENTER 2,0 executes? The initial stack pointer is at offset 1E00H and BP contains 1E20H.

CHAPTER 5

Interrupt Processing

OBJECTIVES

In this chapter you will learn about:

- The differences between hardware and software interrupts
- The differences between maskable and nonmaskable interrupts
- Interrupt processing procedures
- The vector address table
- Multiple interrupts and interrupt priorities
- Special function interrupts
- The general requirements of all interrupt handlers

5.1   INTRODUCTION

The 80x86 provides a very flexible method for recovering from what are known as *cata-strophic* system faults. Through the same mechanism, external and internal interrupts may be handled and other events not normally associated with program execution may be taken care of. The method that does all of this for us is the 80x86 interrupt handler. In this chapter we will see that there are many kinds of interrupts. Some of these deal with issues that have always plagued programmers (such as the divide-by-zero operation), while still others may be defined by the programmer. The emphasis in this chapter is on the definition of the numerous interrupts available. Actual programming examples designed to handle interrupts will be covered in the next chapter.

Section 5.2 explains the differences between hardware and software interrupts. Section 5.3 gives details on the processor's interrupt vector table. The entire process of how an interrupt is handled is presented in Section 5.4, followed by a description of multiple interrupts in Section 5.5. Section 5.6 explains the special interrupts incorporated into the 80x86, such as the divide-by-zero interrupt. Examples of actual interrupt service routines in Section 5.7 are presented next, with troubleshooting techniques completing the chapter in Section 5.8.

5.2   HARDWARE AND SOFTWARE INTERRUPTS

An **interrupt** is an event that causes the processor to stop its current program execution and perform a specific task to service the interrupt. We are all interrupted many times during the day. A ringing telephone or doorbell, a knock on the door, or a question from a friend all indicate the need to communicate with you. The situation is much the same with the microprocessor. The interrupt is used to get the processor's attention. Interrupts may be used to inform the processor in an alarm system that a fire has started or a window has been opened. In a personal computer, interrupts are used to keep accurate time, read the keyboard, operate the disk drives, and access the power of the disk operating system.

Two kinds of interrupts are available: hardware and software interrupts. Hardware interrupts are generated by changing the logic levels on either of the processor's hardware interrupt inputs. These inputs are **NMI (nonmaskable interrupt)** and **INTR (interrupt request).** INTR can be enabled and disabled through software with the use of the STI and CLI instructions. This means that INTR can be *masked* (disabled). NMI gets its name from the fact that its operation cannot be disabled. NMI *always* causes an interrupt sequence when it is activated. Chapter 10 shows that a rising edge is needed on NMI to trigger the interrupt mechanism, and that INTR is level sensitive, requiring a high logic level to interrupt the processor.

The processor automatically generates a type-2 interrupt when NMI is activated. This number refers to the entry in the processor's interrupt vector table that is reserved for NMI. INTR (when enabled) initiates an interrupt acknowledge cycle, which is used to read an interrupt vector number or type from the processor's data bus.

Software interrupts are generated directly by an executing program. These types of interrupts are also called *exceptions*. Some instructions, like INT and INTO, initiate interrupt processing when they are executed. Other instructions are capable of generating an interrupt when a certain condition is met. DIV and IDIV, for example, cause a type-0 interrupt (divide error) whenever division by 0 is attempted.

What happens when a software and hardware interrupt occur at the same time? The processor has a technique for handling this situation; it requires that the interrupts be *prioritized*. Table 5.1 shows the interrupt priority scheme used by the 80x86.

Interrupts with the highest priority are divide-error, INT, and INTO. NMI and INTR have lower priorities, with single-step having the lowest. If both hardware interrupts are activated simultaneously, NMI will be serviced first, with INTR **pending** until it gets its chance to be recognized by the processor. If a divide-error and NMI occur simultaneously, divide-error will be recognized first, followed by NMI.

TABLE 5.1   Interrupt priorities

Interrupt	Priority
Divide-error	Highest
INT, INTO	
NMI	
INTR	
Single-step	Lowest

In the next section we will examine the details of the interrupt vector table, which is accessed when any type of interrupt is initiated.

5.3   THE INTERRUPT VECTOR TABLE

All types of interrupts, whether hardware or software generated, point to a single entry in the processor's interrupt vector table. This table is a collection of 4-byte addresses (two for CS and two for IP) that indicate where the processor should jump to execute the associated interrupt service routine. Because 256 interrupt types are available, the interrupt vector table is 1,024 bytes long. The lKB block of memory reserved for the table is located in the address range 00000 to 003FFH. Some earlier processors automatically loaded their first instruction from address 0000 after a reset. The 80x86 fetches its first address from location FFFF0H. This indicates that we are able to begin program execution without first having to place values into the interrupt vector table. If we plan on using any interrupts, it will be necessary to initialize the required vectors within the table. This can be done easily with a few MOV instructions.

Figure 5.1 shows the organization of the interrupt vector table. Each 4-byte entry consists of a 2-byte IP register value followed by a 2-byte CS register value. This indicates that interrupt service routines are considered far routines. Notice that some of the vectors are predefined. Vector 0 has been chosen to handle division-by-zero errors. Vector 1 helps to implement single-step operation. Vector 2 is used when NMI is activated. Vector 3 (breakpoint) is normally used when troubleshooting a new program. Vector 4 is associated with the INTO instruction. Vector 5 is used when the BOUND instruction reports an out-of-range index. Vectors 6 through 18 perform various housekeeping duties. Some of these vectors (10, 11, 14) are operational only in protected or virtual-8086 mode. Table 5.2 shows the associated vector assignments. Vectors 19 through 31 are reserved by Intel for use in their products. This does not mean that these interrupt vectors are unavailable to us, but that we should refrain from using them in an Intel machine unless we know how they have been assigned.

Vectors 32 through 255 are unassigned and free for us to use. To initialize an interrupt vector we must write the 4 bytes of the interrupt service routine address into the table locations reserved for the interrupt. A short example shows one way this can be done.

FIGURE 5.1   Interrupt vector
table

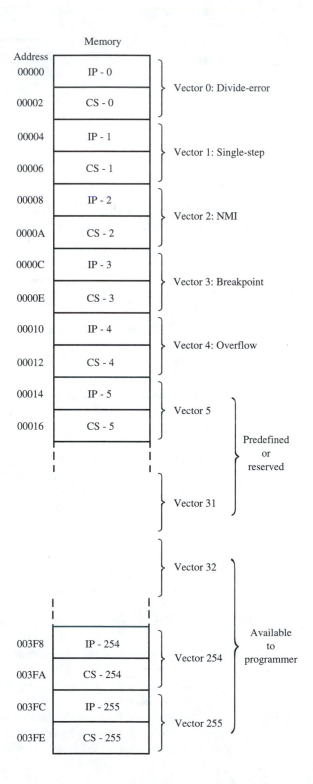

TABLE 5.2 Interrupt/exception vectors

Vector	Description
0	Divide error
1	Debugger call (single-step)
2	NMI
3	Breakpoint
4	INTO
5	BOUND range exceeded
6	Invalid opcode
7	Device not available
8	Double fault
9	Reserved
10	Invalid task state segment
11	Segment not present
12	Stack exception
13	General protection
14	Page fault
15	Reserved
16	Floating-point error
17	Alignment check
18	Machine check
19–31	Reserved
32–255	Maskable interrupts (nonreserved)

Example 5.1: The interrupt service routine for a type-40 interrupt is located at address 28000H. How is the interrupt vector table set up to handle this interrupt?

Solution: An easy way to determine the address within the interrupt vector table that is used by an interrupt is to multiply the interrupt number by 4. Multiplying 40 by 4 and converting into hexadecimal gives 000A0H as the starting address of the vector for INT 40. The interrupt service routine address 28000H can be generated by many different combinations of CS and IP values. If CS is loaded with 2800H and IP with 0000, we get the correct address of 28000H. Thus, it is necessary to write these two address values into memory starting at 000A0H. The short section of code shown here is one way to do this:

```
PUSH    DS                      ;save current DS address
MOV     AX,0                    ;set new DS address at 0000
MOV     DS,AX
MOV     DI,00A0H                ;offset for INT  40 vector
MOV     WORD PTR [DI],0         ;store IP address
MOV     WORD PTR [DI + 2],2800H ;store CS address
POP     DS                      ;get old DS address back
```

The PUSH DS instruction is used to save the current value of the DS register on the stack. Because we need to access memory in the 00000 to 003FFH range, it is convenient to load DS with 0000 and use DI as the offset into the vector table. Notice that the first word written is 0000, which goes into locations 000A0H and 000A1H. Then the CS value 2800H is written into locations 000A2H and 000A3H. Figure 5.2 provides a snapshot of

FIGURE 5.2   ISR address
for INT 40H and INT 41H

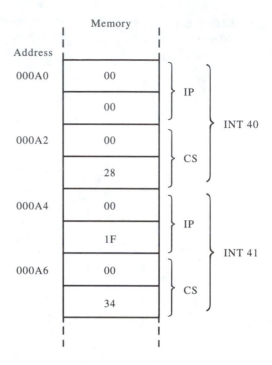

memory after these instructions execute. Now, when INT   40 executes, it will cause a jump to the interrupt service routine located at address 28000H. The POP   DS instruction restores the old value of the DS register.

Figure 5.2 also shows the contents of memory for the vector associated with INT 41. What is the address of the ISR? As usual, the two words have been byte-swapped. The word at address 000A4H is 1F00H. The word at address 000A6H is 3400H. The effective address created by the addition of these two words is 35F00H, which is the address of the ISR for INT   41.

The machine code for INT   40 is CD 28 (note that 28H equals 40 decimal). The machine code for INT   41 is CD 29. All INT instructions begin with CD as their first byte and have the interrupt number as the second byte. The only exception to this rule is INT 3, *breakpoint,* which has only the byte CC as its opcode.

In the examples presented later we will see other ways in which the interrupt vector table can be initialized.

5.4   THE INTERRUPT PROCESSING SEQUENCE

In Chapter 4 we were introduced to the interrupt process in the coverage of the INT and INTO instructions. These software interrupts initiate a sequence of steps in which the flags and return address are saved prior to loading CS and IP with the ISR address. The same process is followed for hardware interrupt NMI, which automatically generates a type-2 interrupt. The sequence initiated by INTR (when interrupts are enabled) is slightly different, since the processor must first read the interrupt number from the data bus. When interrupts are enabled, INTR causes the processor to perform two **interrupt acknowledge cycles.** In the 8088 external devices recognize these cycles by examining the state of the $\overline{\text{INTA}}$ output, which goes low during each cycle. The first low-going pulse on $\overline{\text{INTA}}$ is used to indicate to other devices on the system bus that the processor is beginning an interrupt acknowledge cycle. In minimum mode this indicates that the processor will not acknowledge a hold request until the interrupt acknowledge cycle completes. In maximum mode the processor activates its $\overline{\text{LOCK}}$ output to prevent a system bus takeover during the interrupt acknowledge cycle.

The second low pulse on $\overline{\text{INTA}}$ indicates that the interrupt number should be placed onto the lower byte of the processor's data bus. A special peripheral designed to respond to the 8088's interrupt acknowledge cycle is the 8259A programmable interrupt controller, which we will cover in Chapter 13.

In the Pentium, output signals $M/\overline{IO}$, $D/\overline{C}$, $W/\overline{R}$, and $\overline{\text{ADS}}$ all go low to indicate an interrupt acknowledge cycle.

Once the interrupt number is read from the data bus, the processor performs all of the steps that we are familiar with. Let us review the overall process once more.

Get Vector Number

The processor obtains the interrupt number in one of three ways. First, the interrupt number may be specified directly using one of the INT instructions. Second, the processor may automatically generate the number, as it does for INTO, NMI, and divide-error. Third, it may have to read the interrupt number from the data bus (after receiving INTR).

Once the interrupt number is obtained, it is used to form the location within the interrupt vector table that contains the requested ISR address.

Save Processor Information

Once the interrupt vector is known, the processor pushes the flag register onto the stack. This is done to preserve some of the processor's internal state at the time of the interrupt (a very necessary step if we are to resume normal execution later). Once the flags are pushed, the processor clears the interrupt enable and trace flags to disable INTR while interrupt processing is taking place. Next, the IP and CS values at the time of the interrupt are pushed onto the stack. Figure 5.3 shows how the stack is used when an interrupt occurs. The contents of the flag register at the time of the interrupt were 06C2H. The address of the instruction that was about to be fetched when the interrupt occurred was 1109:3C00 (in CS:IP format).

Fetch New Instruction Pointer

Once the return address has been pushed, the processor can fetch the new values of IP and CS out of the interrupt vector table and begin execution of the interrupt service routine. The address generated by the interrupt number is used to read the two ISR address words out of the table.

One word of caution: Because the stack contains the information needed to return to the interrupt point, we must be careful not to change the contents of stack memory or alter

FIGURE 5.3   Stack contents
during an interrupt

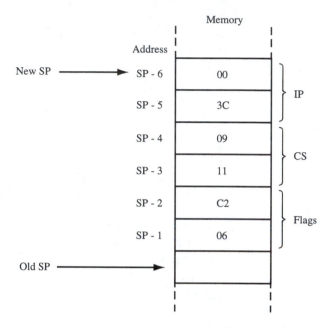

the stack pointer in any way that would prevent the correct information from being popped off. The processor will not remember anything about the interrupt and relies only on the data popped off the stack for a proper return.

5.5   MULTIPLE INTERRUPTS

In the course of program execution, chances are good that eventually two interrupts might request the processor's attention at the same time. For example, just as division-by-zero is attempted in an executing program, NMI is also activated. The processor needs to "break the tie" when this happens and recognize one of the two interrupts first. When we examined Table 5.1, we saw that divide-error has a higher priority than NMI. So, in our current example, divide-error will be recognized first, and the following sequence of steps will occur:

1. Divide-error is recognized.
2. The flags are pushed.
3. The return address (CS and IP) is pushed.
4. The interrupt-enable and trace flags are cleared.
5. NMI is recognized.
6. The new flags are pushed.
7. The new return address is pushed.
8. The interrupt-enable and trace flags are cleared.
9. The NMI ISR is executed.
10. The second return address is popped.
11. The second set of flags are popped.
12. The divide-error ISR is executed.

13. The first return address is popped.
14. The first set of flags are popped.
15. Execution resumes at the instruction following the one that initiated the divide-error.

It is easy to see that the stack plays an important role during this process.

A more common occurrence of a multiple interrupt is seen when the processor's trace flag is set. The trace flag, when set, puts the processor into single-step mode, where a type-1 interrupt is generated after completion of every instruction. If the current instruction is INT or INTO, you can see that two interrupts will need servicing: the INT or INTO interrupt and the single-step interrupt. Single-step has the lowest priority of all interrupts and thus gets recognized last. We will see how single-stepping with the trace flag can be a useful tool when debugging a program.

5.6 SPECIAL INTERRUPTS

We will now examine the specific operation of a few selected interrupts. Some may be very useful to implement, while others may never be needed in a program for proper operation. Even so, you are better off knowing how each one works and when to use it.

Divide-Error

Figure 5.4 shows the contents of four memory locations that contain the code for these two instructions:

```
B3 00   MOV   BL,0
F6 F3   DIV   BL
```

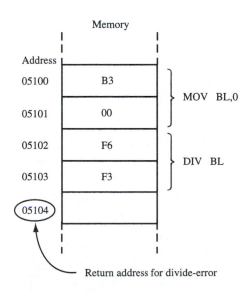

FIGURE 5.4 Instruction sequences causing a divide-error interrupt

The first instruction is located at address 05100H. The second instruction is located at address 05102H. Because DIV   BL is a 2-byte instruction, the next instruction must be located at address 05104H. This address becomes the return address when DIV   BL generates a divide-error interrupt.

The interrupt service routine for divide-error can do anything the user wishes to recover from the error. One programmer may wish simply to load the accumulator with 0 or some other number, while another may want to display an error message on the user's display screen.

Since divide-error is a type-0 interrupt, its address vector is stored in memory locations 00000 through 00003.

Single-Step

This interrupt relies on the setting of the trace flag in the flag register. When the trace flag is set, the processor will generate a type-1 interrupt after each instruction executes. Remember from the interrupt processing sequence that after the flags are pushed the processor clears the trace flag, disabling single-stepping while it executes the trace ISR. An extremely useful debugging tool can be written and used within the trace ISR. This single-step debugger may be programmed to display the contents of each processor register, the state of the flags, and other useful information after execution of each instruction in a user program. Because the trace flag is cleared before the ISR is called, we need not worry about single-stepping through the trace ISR.

You may remember from our coverage of 80x86 instructions in Chapters 3 and 4 that there are no instructions available that directly affect the state of the trace flag. There are other techniques that can be used to do this. For instance, a copy of the flags can be loaded into AX by first pushing a copy of the flag register onto the stack and then popping them into AX. Then an OR instruction can be used to set the trace flag. Once this is done, the flags are restored by pushing AX back onto the stack and then popping the flags. The instructions needed to accomplish this are:

```
PUSHF
POP   AX
OR    AX,100H
PUSH AX
POPF
```

Once this is done, the processor will enter and remain in single-step mode until the trace flag is cleared. This can be done with the same set of instructions, replacing the OR with AND   AX,0FEFFH.

Example 5.2:   Assume that the trace flag is set and that the trace ISR displays the contents of AX on the screen after each instruction executes. What do we expect to see when this group of instructions executes?

```
MOV   AX,1234H
INC   AL
DEC   AH
NOT   AX
```

Solution: Because the trace flag is set, a single-step interrupt will be generated after each of the four instructions. When the first instruction completes, the trace ISR will display AX=1234. The second instruction will increase AL to 35, causing the ISR to display AX=1235 next. The third instruction will decrease AH to 11. The trace ISR will then display AX=1135. Finally, after all bits in AX are inverted, the trace ISR will display AX=EECA. We will see that the personal computer has built-in routines capable of displaying messages and data on the screen, so writing a trace ISR is not as complicated a task as it appears.

The ISR address vector for single-step must be stored in memory locations 00004 through 00007.

NMI

Because NMI can never be ignored by the processor, it finds useful application in events that the computer absolutely must respond to. One such event is the disastrous **power-fail.** The processor unfortunately forgets the contents of its registers and flags when power is turned off and thus has no chance of getting back to the correct place in a program if its power is interrupted. One way to prevent this from happening and provide a way for the processor to resume execution is to use NMI to interrupt the processor at the beginning of a power-fail. Because the computer's power supply will continue to supply a stable voltage for a few milliseconds after it loses AC, the processor has plenty of time to execute the necessary instructions. Suppose that a certain system contains a small amount of **non-volatile memory.** This type of memory retains its data after it loses power and acts like RAM when power is applied. So, in the event of a power-fail, the NMI ISR should store the contents of each processor register in the NVM. These values can then be reloaded when power comes back up. In this fashion we can recover from a power-fail without loss of intelligence.

The ISR address vector for NMI is stored in memory locations 00008 through 0000BH.

Breakpoint

This interrupt is a type-3 interrupt, but is coded as a single byte for reasons of efficiency. Breakpoint aids in debugging in the following way: a program being debugged will have the first byte of one of its instructions replaced by the code for breakpoint (CC). When the processor gets to this instruction, it will generate a type-3 interrupt. The ISR associated with breakpoint is similar to the trace ISR and should be capable of displaying the processor register contents and also the address at which the breakpoint occurred. Before the ISR exits, it will replace the breakpoint byte with the original first byte of the instruction. Figure 5.5 shows how the breakpoint routine makes a copy of the first byte in the NOT AL instruction stored at location 06200H. The first byte of the instruction (F6H) is copied into a temporary location, and then replaced by the breakpoint instruction code CC. Some people like to refer to this as *setting the breakpoint*. Once the breakpoint is set, a fetch from address 06200H will initiate a breakpoint interrupt.

FIGURE 5.5 Setting a
breakpoint

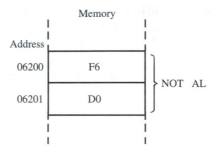

(a) Original instruction code

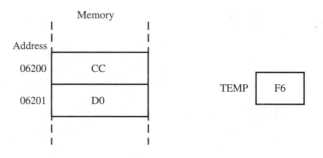

(b) First instruction byte is replaced by INT 3 (breakpoint)

Clearing the breakpoint is accomplished by copying the instruction byte from the temporary location back into its original location.

Example 5.3: A programmer wishes to find out if a conditional jump takes place. Where should the programmer place the breakpoint instruction? The code being tested looks like this:

```
        CMP   AL,0
        JNZ   XYZ
        NOT   AL
XYZ:    INC   AL
```

Solution: The programmer has two choices for placement of the breakpoint instruction. It could be placed in the location occupied by NOT   AL, which would cause a breakpoint when the JNZ does *not* jump to XYZ. It could also be placed in the location occupied by INC   AL, activating when the JNZ *does* take place. Either way, the programmer will know the results of the CMP and JNZ instructions (by the presence or absence of a breakpoint).

The ISR address vector for breakpoint is stored in memory locations 0000CH through 0000FH.

Overflow

This type-4 interrupt is initiated only when the INTO instruction is executed with the overflow flag set. Its applications, like divide-error, tend to be of a corrective nature. You may think of overflow as the watchdog for multi-bit addition and subtraction operations, much like divide-error watches out for division-by-zero. If the overflow flag is cleared, INTO will not generate an interrupt.

Example 5.4: Will the following sequence of code generate an overflow interrupt?

```
MOV AL,70H
MOV BL,60H
ADD AL,BL
INTO
```

Solution: Yes. Although the numbers in AL and BL can both be interpreted as positive signed numbers, the sum (D0H) looks like a negative signed number. In this case the overflow flag is set and INTO will generate an interrupt.

The ISR address vector for overflow is stored in memory locations 00010H through 00013H.

INTR

Up to now we have only discussed the basics of this hardware interrupt signal. Let us take a closer look. No interrupt is generated by INTR unless the interrupt-enable flag is set. This can easily be accomplished with the STI instruction. INTR must remain high until sampled by the processor, unlike NMI, which is a rising-edge triggered input. It is therefore necessary when using INTR to allow it to remain high only until an interrupt acknowledge cycle begins. For 8088 systems, the 8259A programmable interrupt controller, which we will cover in Chapter 13, automatically interfaces with INTR and $\overline{\text{INTA}}$. If the power of this peripheral is not needed, then custom interrupt circuitry must be designed. First, we need the INTR connection. The circuit shown in Figure 5.6 uses a flip-flop to condition the INTR input. A D-type flip-flop is used to convert the

FIGURE 5.6 INTR conditioning circuitry

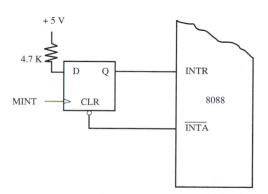

high-level requirement of INTR into an edge-sensitive request. A rising edge on MINT (maskable interrupt) will clock a 1 through the flip-flop, placing a high level on INTR. When the 8088 recognizes INTR and begins its interrupt acknowledge cycle, $\overline{\text{INTA}}$ will go low. This will clear the flip-flop and remove the INTR request. MINT must go low and back high again for another interrupt to be recognized. Once again, the Pentium uses different signals to indicate an interrupt acknowledge cycle. Figure 5.7 shows how the signals are decoded to generate an $\overline{\text{INTAK}}$ signal.

During the second low-going pulse on $\overline{\text{INTA}}$ the processor will expect an interrupt number to be placed onto the data bus. The additional circuitry of Figure 5.8 uses a tri-state

FIGURE 5.7 Decoding an interrupt acknowledge cycle on the Pentium

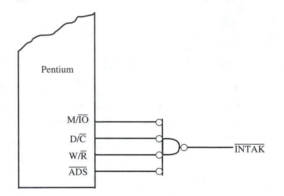

FIGURE 5.8 Circuitry needed to place an 8-bit interrupt number onto data bus

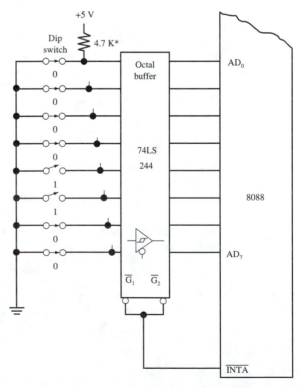

* All switches must be pulled up with individual resistors.

buffer to jam the 8-bit interrupt number onto the data bus when $\overline{INTA}$ goes low. You may notice that the interrupt number will appear on the data bus twice, once for each low-going transition of $\overline{INTA}$. The processor will ignore the first appearance because it tri-states the data bus during the first transition of $\overline{INTA}$. The 8-bit interrupt number will appear on the data bus but will be ignored by the processor until the second transition of $\overline{INTA}$. The DIP switch allows any of the 256 interrupt codes to be used. The DIP switch is currently set to produce interrupt code 30H. On the Pentium, the interrupt vector is applied to data bus signals D0 through D7.

For systems that require additional interrupts but still do not require the use of the 8259A, a few additional parts are needed to expand this simple interrupt circuit into a more complex one with more interrupts and a prioritization scheme. Figure 5.9 shows an interrupt circuit that allows up to eight levels of prioritized interrupts. The 74LS148 priority encoder will output a 3-bit binary number whenever any of its eight inputs go low. Also, only the highest priority input is recognized. So, if $\overline{INT}_0$ and $\overline{INT}_4$ are both

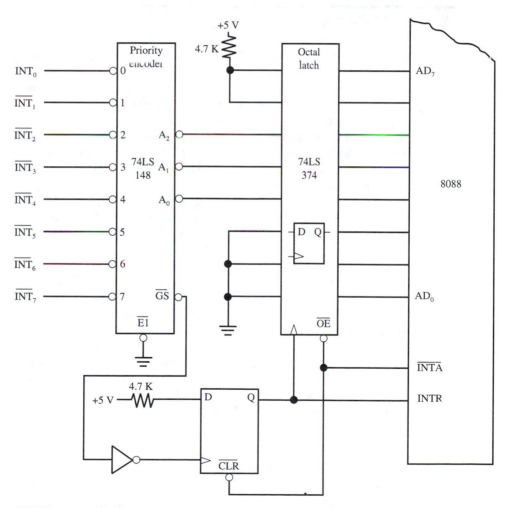

FIGURE 5.9   Prioritized interrupt circuitry

low, $\overline{INT}_4$ is recognized and the output becomes 011. Note that the output is actually the inverted binary value of the input that is active. When any input is grounded, the $\overline{GS}$ output will go low. This causes a 1 to be clocked through the D-type flip-flop, signaling an INTR and latching the priority encoder's output in the 74LS374 octal flip-flop. When $\overline{INTA}$ goes low in acknowledgment of the INTR signal, the D-type flip-flop is cleared (removing the INTR request) and the output of the octal flip-flop is enabled, placing the interrupt number onto the data bus. A similar circuit may be used with the Pentium. The bit pattern stored in the octal flip-flop is 11⟨???⟩000 for any interrupt. Specifically, $\overline{INT}_0$ causes the outputs of the 74LS148 to become 111, giving us the interrupt number F8H. $\overline{INT}_1$ generates 110 at the 148's output, for an interrupt code of F0H. Take the time to find the other six interrupt numbers right now. Do you see a pattern emerge?

5.7   INTERRUPT SERVICE ROUTINES

The interrupt service routine referred to many times in this chapter is the actual section of code that takes care of processing a specific interrupt. The ISR for a divide-error is necessarily different from one designed to handle breakpoint or NMI interrupts.

Even though these interrupt service routines are written to accomplish different goals, there are portions of each that, for the sake of good programming, look and operate the same. Recall that any time an interrupt occurs, the processor pushes the flags and return address onto the stack before vectoring to the address of the ISR. Clearly, we must see that the ISR will change the data in various registers while it is processing the interrupt. Because we desire to return to the same point in our program where we left off *before* the interrupt occurred and resume processing, we insist that all prior conditions exist upon return. This means that we must return from the ISR with the state of all registers preserved. It is now the responsibility of the ISR to preserve the state of any registers that it alters. Figure 5.10 shows how this is done. In this example, DIV BL causes a divide-error interrupt. The first thing the ISR does is save the processor registers on the stack. These registers are saved with the PUSH instruction. You will need a PUSH for each register you need to save (or you can push all the registers with PUSHA). The contents of all processor registers, including the flags, are often called the **environment** or **context** of the machine. Putting a copy of everything onto the stack saves the environment that existed at the time of the interrupt.

When the body of the ISR finishes execution, it is necessary to reload the registers that were saved at the beginning of the routine. The POP instruction is used for this purpose. POPs must be done in the reverse order of the PUSHes. An ISR that uses AX, BX, and CX would look something like this:

```
ANISR:  PUSH  AX        ;save registers
        PUSH  BX
        PUSH  CX
        ;
        ;body of ISR
        ;
```

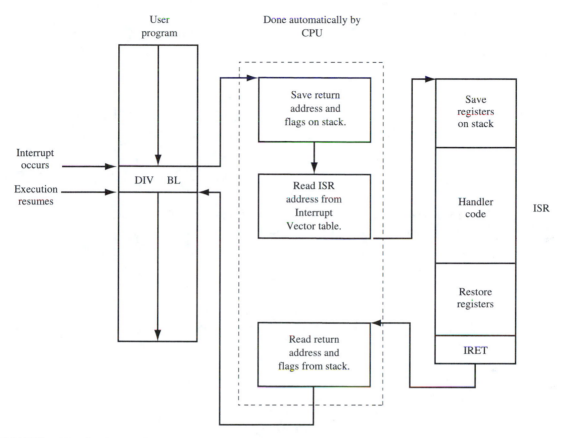

FIGURE 5.10   Storing environment during interrupt processing

```
POP   CX          ;get registers back
POP   BX
POP   AX
IRET              ;return from interrupt
```

Saving all registers is preferable, and will save you much heartache in the future, when you find that saving one or two registers was insufficient as the needs of the routine became more complex. In this case, use POPA to restore all registers.

A few examples of actual interrupt service routines will prepare you for writing your own later. Try to find similarities between each routine.

An NMI Time Clock

Figure 5.11 shows a simple way to provide the processor with some timekeeping intelligence. In this application, the processor's NMI input is connected to a 60-Hz clock source. Thus, the processor gets interrupted 60 times per second. The only task the NMITIME ISR needs to perform is to decrement a counter until it reaches 0, and then call the far routine ONESEC. ONESEC is then called once every second. The counter is decremented once

FIGURE 5.11   Interrupt circuit for NMITIME

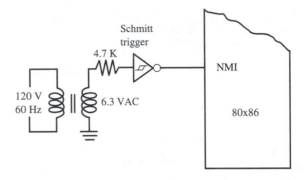

for each NMI signal. Other code is needed to initially set the count to 60, with NMITIME resetting it automatically on each 0 count.

```
NMITIME: DEC   COUNT                ;decrement 60th's counter
         JNZ   EXIT                 ;did we go to 0?
         MOV   COUNT,60             ;yes, reset counter and
         CALL  FAR PTR ONESEC       ;call ONESEC
EXIT:    IRET
```

The user program must reserve room for the COUNT location in its data segment area. COUNT DB 60 is all that is needed to reserve the byte location. Initialization software is also needed to load 60 into COUNT and place the ISR address for NMITIME into the interrupt vector table. One way to do this would be:

```
MOV   COUNT,60       ;init 60th's counter
PUSH  DS             ;save current DS address
SUB   AX,AX          ;set new DS address to 0000
MOV   DS,AX
LEA   AX,NMITIME     ;load address of NMITIME ISR
MOV   [8],AX         ;store IP address
MOV   AX,CS          ;and CS address
MOV   [0AH],AX
POP   DS             ;get old DS address back
```

The offset of NMITIME within the code segment is stored in locations 00008 and 00009, and the CS value in locations 0000AH and 0000BH. The value of the DS register remains unchanged after execution of the initialization software.

Notice that NMITIME does not destroy a single register. In this example we can get by without having to save anything on the stack.

ONESEC can be used to do a number of things. Most likely, it will update a second set of count locations that keep track of the seconds, minutes, and hours for a 12- or 24-hour clock. Software with access to these counters will be able to use the passage of time in an accurate way.

A Divide-Error Handler

This routine is used to handle a division-by-zero. Because the AX and DX registers may be undefined as a result of division by 0, DIVERR will load AX with 0101H and DX with 0. This guarantees that 8- or 16-bit division always ends up with a non-zero result. DIVERR also calls a special routine called DISPMSG, which is used to output an ASCII text message on the display screen of the computer. The ASCII message must end with a '$' character, and the address of the first byte of the message must be loaded into the SI register prior to calling DISPMSG.

```
DIVERR:  PUSH SI                   ;save current SI value
         MOV  AX,101H              ;load result with default
         SUB  DX,DX                ;clear DX
         LEA  SI,DIVMSG            ;init pointer to error message
         CALL FAR PTR DISPMSG      ;output error message
         POP  SI                   ;get old SI value back
         IRET
```

PUSH and POP are used to preserve the contents of SI. The error message must be located in the program's data segment, and look similar to this:

```
DIVMSG   DB   'Division by zero attempted!$'
```

The first byte of the text message ("D") will be located in the address associated with the label DIVMSG. The last byte of the message is the required "$" character.

An ISR with Multiple Functions

This interrupt service routine will be used to perform one of four different functions when it is executed. The ISR is called with an INT   20H instruction. Register AH is examined upon entry into ISR20H to determine what should be done. If AH equals 0, AL and BL will be added together, with the result placed in AL. If AH equals 1, the registers will be subtracted. Multiplication occurs when AH equals 2, and division when AH equals 3. Any other values of AH cause ISR20H to return without changing either register. Using AH in this way lets us do more than one thing with INT   20H. This technique is commonly used when we write programs for 80x86-based PCs running a disk operating system. One interrupt might have many different functions, all of which interface with a disk drive, display device, or printer connected to the computer. We will see specific examples of these kinds of special-function interrupts in the next chapter. The code for this ISR looks like this:

```
ISR20H: CMP     AH,4  ;AH must be 0-3 only
        JNC     EXIT
        CMP     AH,0  ;is AH 0?
        JZ      ADDAB
        CMP     AH,1  ;is AH 1?
        JZ      SUBAB
        CMP     AH,2  ;is AH 2?
        JZ      MULAB
        DIV     BL    ;AH is 3, use divide function
        EXIT:   IRET
        ADDAB:  ADD   AL,BL ;add function
                IRET
        SUBAB:  SUB   AL,BL ;subtract function
                IRET
        MULAB:  MUL   BL    ;multiply function
                IRET
```

The method of using a register value to select an interrupt function is also used by MS-DOS programs that require the use of operating system functions. For example, the instructions

```
LEA     DX,MESSAGE
MOV     AH,9
INT     21H
```

select the MS-DOS **Display Message** function of INT   21H. The function number is passed to INT   21H through register AH. The address of the ASCII message to display must be placed in DX prior to the interrupt.

We will make extensive use of the MS-DOS interrupts in Chapters 6 through 9.

The initialization code required for ISR20H is:

```
PUSH DS             ;save old DS address
SUB  AX,AX          ;set new DS address to 0000
MOV  DS,AX
LEA  AX,ISR20H      ;load address of ISR20H
MOV  [80H],AX       ;store IP address
MOV  AX,CS          ;and CS address
MOV  [82H],AX
POP  DS             ;get old DS address back
```

It is easy to verify that 4 times 20H is 80H, the interrupt vector table address required.

5.8   TROUBLESHOOTING TECHNIQUES

It pays to remember the details of interrupt processing when troubleshooting a program. Many times the fault of erratic execution in an 80x86-based system is a poorly written or incomplete interrupt handler. Reviewing the basic principles can help eliminate some of the more obvious problems.

- A valid stack must exist to save all of the information required to support the interrupt handler.
- A typical interrupt pushes the current flags and return address (CS:IP).
- Vector addresses are equal to four times the vector number.
- In the interrupt vector table, the handler address is stored in byte-swapped form, with IP as the first word, and CS as the second.
- It may be necessary to save and restore registers (via PUSH/POP) in the interrupt handler.
- Use IRET (return from interrupt) to return from an interrupt handler. RET does not work properly with interrupts.
- For an interrupt to work, its vector must be loaded with the starting address of the handler, and the handler code must be in place as well.

These points are a minimal set of programming tips. You will discover others as you begin writing your own interrupt handlers.

SUMMARY

We have seen that there is a fixed process used by the 80x86 to implement and process an interrupt. The CPU, when interrupted, saves the flag register and program counter on the stack, clears the trace and interrupt-enable flags, and loads the interrupt service routine address from the interrupt vector table. The interrupt vector table occupies memory locations 00000 through 003FFH and contains pairs of words that represent the execution addresses for each of the 256 interrupts. These pairs correspond to IP and CS values of each ISR. The interrupt number used to access the table may be internally generated by the processor, or may be supplied by external hardware during an interrupt acknowledge cycle. The processor has only two hardware interrupts: NMI and INTR. NMI cannot be disabled, but INTR can.

Interrupts are caused through software or by an external hardware request. The software interrupt may be generated intentionally by the programmer via INT and INTO, or by accident, via a run-time error such as division-by-zero. All interrupt service routines should preserve the state of any registers used to allow a proper return.

Three examples of actual interrupt service routines were also presented, followed by a short set of troubleshooting tips.

STUDY QUESTIONS

1. What is the processor's environment? Why is it important to save the environment during interrupt processing?
2. Explain the different ways interrupts are generated.
3. How is INTR disabled? How is it enabled?
4. What is the interrupt vector table address for an INT 21H?
5. The address of the ISR for INT 25H is 03C0:9AE2 (in CS:IP format). Show how this address is stored within the interrupt vector table.
6. What is the effective address of the ISR in Question 5?
7. Write the instructions necessary to place the ISR address of Question 5 into its proper place in the interrupt vector table.
8. Show the contents of stack memory after an interrupt has been initiated. Assume that the stack pointer is at address 3C00H prior to the interrupt and that CS, IP, and the flag register contain 0400H, 1890H, and 0182H, respectively.
9. What interrupt number has a vector table address range of 00280H to 00283H?
10. Which interrupt is recognized first, NMI or single-step?
11. Repeat Example 5.2 with an initial AX value of E03FH.
12. What high-priority event might require the use of NMI in a computer designed for aircraft engine control?
13. An analysis of a computer power failure showed that the computer had valid voltage levels for 2.5 milliseconds. Suppose that the microprocessor had an average instruction execution time of 0.2 microseconds. How many instructions can be executed during the power failure?
14. Explain why an orderly software shutdown is possible in Question 13. Assume that the shutdown involves pushing all processor registers onto a stack in NVM.
15. In Example 5.4 what is the highest AL value that will not cause an INTO interrupt?
16. Design an INTR circuit that has only two inputs: $\overline{XINT}$ and $\overline{YINT}$. Both signals are active low. $\overline{XINT}$ should generate interrupt number 90H, and $\overline{YINT}$, 91H.

17. What are all eight interrupt numbers for the circuit in Figure 5.9?
18. Modify the interrupt circuit of Figure 5.9 so that interrupt numbers 48H through 4FH are produced.
19. If the ONESEC procedure called by NMITIME took more than 18 milliseconds to execute, would any problems arise?
20. What changes must be made to NMITIME if the frequency of the NMI clock is 1800 Hz? We still want to call ONESEC once per second.
21. What is the result of executing INT   20H with AX containing 0303H and BL containing 04? Refer to ISR20H in Section 5.7 for details.
22. Rewrite ISR20H so that four new functions are added. These functions are:

```
AH=05H : NOT AL
AH=10H : AND AL,BL
AH=20H : OR  AL,BL
AH=80H : XOR AL,BL
```

23. Figure 5.12 shows the contents of a few locations within the interrupt vector table. What will the new program counter be when the interrupt that uses these locations is processed?
24. What INT instruction is required in Question 23?
25. How is the single-step interrupt useful for examining the operation of a running program?
26. During an interrupt acknowledge cycle, 30H is placed on the data bus. Where is the ISR address fetched from?
27. The flag register contains 0346H. Will INTO generate an interrupt? Is trace enabled? Are interrupts enabled?

FIGURE 5.12   For Question 23

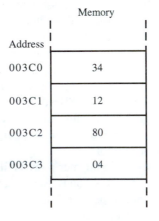

28. What do you imagine are some of the problems with these two interrupt service routines:

```
ISR1:   PUSH   AX              ISR2:   PUSH   AX
        PUSH   BX                      PUSH   CX
        ;                              ;
        ;body                          ;body
        ;                              ;
        POP    AX                      POP    BX
        IRET                           POP    AX
                                       RET
```

29. Write an interrupt service routine that will multiply the contents of AX by 7. If the new value of AX is greater than 8400H, call the far routine OVERSCAN. *Note:* OVERSCAN destroys AX, BX, and DI.

30. What instructions are necessary to load a copy of the return address (CS:IP) into registers AX and BX, from *inside* the interrupt service routine? What stack operations may be used?

31. Use DEBUG to execute these three instructions:

```
MOV   DL,41
MOV   AH,2
INT   21
```

What are the results? Remember to use "p" and not "t" to execute INT   21.

PART 3

Programming

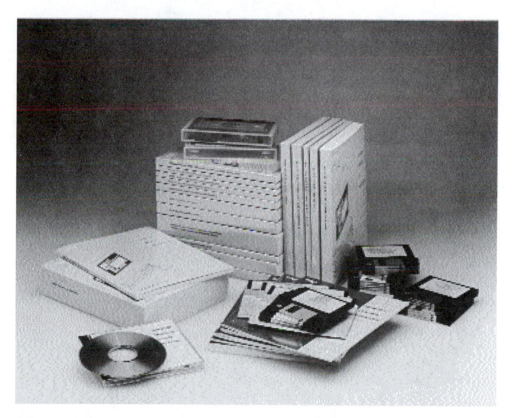

Software utilities and documentation play an important role in program development

CHAPTER 6

An Introduction to Programming the 80x86

OBJECTIVES

In this chapter you will learn about:

- Breaking down a large program into small tasks
- How to write a software driver
- Reading character strings from a keyboard
- Packing BCD digits into a byte
- Item search and lookup in a data table
- Comparison of data strings
- Sorting algorithms
- The use of condition flags to return results from a routine
- Binary and BCD math
- Writing a routine to perform a complex mathematical function
- Open- and closed-loop control systems (simplified theory)
- How to convert binary numbers to decimal numbers, and vice versa
- Insertion of an item into a linked-list
- The operation of stacks and queues

6.1 INTRODUCTION

Getting the most use out of your microprocessor requires expertise both in designing the hardware around it and in writing the code that it will execute. The purpose of this chapter is to familiarize you with some of the standard programming principles as they apply to the 80x86. We will limit ourselves to writing straight code in this chapter, using only the power of the 80x86 instruction set. Hopefully, you will see that many complex tasks may be performed in this way, without the use of external peripherals, which will be covered in Chapters 12 and 13.

Section 6.2 explains how large programming jobs are broken down into smaller tasks. Section 6.3 shows how a software driver program is written to test a new code module. Sections 6.4 and 6.5 deal, respectively, with collecting data and simple search techniques for use with data tables (or arrays). Section 6.6 details the various string operations. Integer sorting is covered in Section 6.7 and is followed by a detailed treatment of binary and BCD mathematical routines in Section 6.8. Section 6.9 shows two examples of the processor in control applications. Section 6.10 gives examples of binary and decimal conversion techniques. Section 6.11 discusses three basic data structures: linked-lists, stacks, and queues. The chapter concludes with another set of troubleshooting tips in Section 6.12.

6.2   TACKLING A LARGE PROGRAMMING ASSIGNMENT

Writing a large, complex program from scratch is a difficult job, even for the most seasoned programmers. Even if this could be done easily, other considerations exist to complicate matters. The final program must be tested to ensure correct operation. It is a rare occurrence for a new program to work perfectly the first time.

For these reasons, a more sensible approach is to break down the large program into smaller tasks. Each task may be thought of as a subroutine to be called, when needed, by the main program. The subroutines will each perform a single task, and thus will be easier to individually test and correct, as necessary. The technique of writing a large program in this way is often referred to as **structured programming.** We will not concern ourselves with all the details of structured programming. Instead, we will study a sample programming assignment and use the techniques previously mentioned to break down the assignment into smaller jobs.

The Assignment

The programming assignment is presented to us in the form of a **specification.** The specification describes the job that must be performed by the program. It also contains information concerning any input and output that may need to be performed, and sometimes a limit on the amount of time the program may take to execute.

Consider the following specification:

Specification: Subroutine WORDCOUNT

Purpose: To generate a data table of all different words contained in a paragraph of text, and a second data table containing the frequency of occurrence of each word.

Restrictions: Do not distinguish between uppercase and lowercase characters. Ignore punctuation, except where it defines the end of a word. No words will appear more than 255 times.

Input: A data table, headed by symbol TEXT, that contains the para-
                  graph to be analyzed, represented by ASCII codes. The length of
                  the paragraph text is undefined, but the last character in the text
                  will always be '$'. This character will not appear anywhere else in
                  the text.

Output: A data table, headed by symbol WORDS, that contains a list of all
                  different words encountered in the paragraph text. Each word ends
                  with ".", and the entire table ends with '$'. A second data table,
                  headed by symbol COUNTS, containing the frequency counts for
                  each entry in WORDS.

There is sufficient detail in the specification for us to determine what must be done. How
to do it is another matter.

Breaking Down the Program into Modules

Once we understand what is required of the program through information presented in the
specification, the next step is to break down the program into smaller modules. This means
that subroutine WORDCOUNT will actually become a main subroutine, which calls other
subroutines. We must identify the other subroutines needed. This step of the process
requires skill and practice. When you have given it enough thought, you might agree that
these subroutines are required:

INITIALIZE	Initialize all pointers, counters, and tables needed.
GETWORD	Get the next word from the paragraph text.
LOOKUP	Search WORDS to see if it contains the present word.
INSERT	Insert new word into WORDS.
MAKECOUNT	Make a new entry in COUNTS.
INCREASE	Increase frequency count in COUNTS for a word found by LOOKUP.

There may, of course, be other required routines, depending on who is writing the code.
The idea is to create a subroutine to accomplish only *one* task. None of the identified routines performs more than one task.

Once the input and output for each subroutine are identified, the code can be written
for each one and the subroutines tested.

Testing the Modules

Testing of each subroutine module is done separately through a special program called a
driver. The driver supplies the subroutine with sample input data and examines the subroutine's output for correctness. It is up to the programmer to select the type and quantity
of the sample data. We will look at an example of a software driver in Section 6.3.

When all modules have been tested and verified for proper operation, they can be
combined into one large module—WORDCOUNT in our example—and this module can
be tested also.

Creating the Final Module

WORDCOUNT, as mentioned before, will consist of calls to the subroutines identified in the section on breaking down the program into modules. **Pseudocode,** a generic programming language, can be used to determine the structure of WORDCOUNT (and of the other subroutines as well). The following pseudocode is one way WORDCOUNT may be implemented:

```
subroutine WORDCOUNT
  INITIALIZE
  repeat
    GETWORD
    if no word found then
      return
    LOOKUP
    if word found then
    INCREASE
    else
    INSERT
    MAKECOUNT
  forever
end WORDCOUNT
```

WORDCOUNT is implemented as an infinite loop because the length of the paragraph text is unknown. The only way out of the loop is to have GETWORD fail to find a new word in the text (that is, by reaching the '$'). This approach satisfies another requirement of structured programming: routines should contain one entry point and one exit point. Many of the routine examples that we will study in this chapter will be written in this fashion. The programmer decides how the REPEAT-FOREVER and IF-THEN-ELSE statements are implemented.

Standard Control Structures

The **IF-THEN** statement can be coded in many different ways. The actual structure is IF <condition> THEN <action>. The *condition* must be satisfied for the *action* to take place. In the WORDCOUNT example, the first IF statement causes the subroutine to return if GETWORD did not find a new word in the paragraph text. Let us assume that GETWORD returns FFH in AL if it did find a word, and 00 if it did not. One way to code the IF statement might look like this:

```
        CALL    GETWORD
        CMP     AL,0
        JNZ     NEXT
        RET
NEXT:   --
```

The CALL to GETWORD will adjust the value of AL accordingly. The CMP instruction is used to determine if AL contains 0. If it does not, a jump to NEXT is performed. This will cause execution to continue (with a CALL to LOOKUP as the next instruction). If AL does contain 0, the JNZ will not take place and the RET instruction will execute instead.

IF-THEN-ELSE statements are very similar, with coding somewhat like this:

```
          CALL    LOOKUP
          CMP     AL,0
          JZ      THENCODE
ELSECODE: --
```

Other pseudocode structures include the REPEAT-UNTIL and WHILE-DO. The REPEAT-UNTIL structure looks like this:

```
repeat
    <statement>
until <condition>
```

Coding the REPEAT-UNTIL structure depends on the type of condition being tested. A sample structure and its associated code may look like this:

```
initialize counter to 100              MOV    BX,100
repeat
    GETDATA                   AGAIN:    CALL   GETDATA
    PROCESSDATA                         CALL   PROCESSDATA
    decrement counter                   DEC    BX
until counter = 0                       JNZ    AGAIN
```

One important point about using loop counters is that the loop-count register (BX in this example) must be altered during execution of the statements within the loop.

The WHILE-DO structure is slightly different, performing the condition test at the *beginning* of the loop instead of the end. One example of a WHILE-DO is:

```
while char <> 'A' do
    <statements>
end-while
```

The corresponding machine instructions for this loop might look like this:

```
WHILE:    CMP    AL,'A'
          JZ     NEXT
          <loop instructions>
          JMP    WHILE
NEXT:     --
```

Here, it is important to modify the loop variable (AL in this case) somewhere within the loop, to avoid getting stuck inside it.

Another programming structure that is useful is the **CASE** statement. The structure of the CASE statement may look like this:

```
case <item> of
    <item 1> : <statement 1>
    <item 2> : <statement 2>
        .
        .
    <item x> : <statement x>
```

where each item (1...x) is checked for a match with the item at the beginning of the CASE statement. Only the statement associated with the matching item is executed. The number of items to match is not limited. An example of a CASE statement with three choices is as follows:

```
case selvalue of
    0 : clear counter
    1 : increment counter
    2 : decrement counter
```

In this example, the *selvalue* variable must contain a value from 0 to 2 to select one of the three statements. The corresponding assembly language for this CASE statement is:

```
          CMP    AL,0     ;is it 0?
          JNZ    C1
          MOV    BL,0     ;clear counter
```

```
            JMP    NEXT
C1:         CMP    AL,1    ;is it 1?
            JNZ    C2
            INC    BL      ;increment counter
            JMP    NEXT
C2:         CMP    AL,2    ;is it 2?
            JNZ    NEXT    ;no matches
            DEC    BL      ;decrement counter
NEXT:       --
```

In this example the selvalue variable is stored in AL and the counter is represented by BL. Note that the conditional jump JNE may be used in place of JNZ if desired, because they are equivalent. The same is true for conditional jumps JZ and JE.

Some programmers use a modified form of the CASE structure that allows execution of a statement when no match is made with any item. The pseudocode for this structure is:

```
case <item> of
     <item 1>   : <statement 1>
     <item 2>   : <statement 2>
        .
        .
     <item x>   : <statement x>
     otherwise  : <otherwise-statement>
```

It is a simple matter to execute the otherwise-statement when no matches are found by performing a JNZ   OTHER after the last CMP instruction.

Remember that there are no fixed methods for converting pseudocode into machine instructions. Use your imagination and you will undoubtedly come up with your own techniques.

6.3   WRITING A SOFTWARE DRIVER

The programs presented in the remaining sections of this chapter are written as subroutines (or procedures) that must be called to perform their functions. As we saw in the previous section, there may be many subroutines combined in a single application, with each subroutine possibly written by a different person. Thus, each programmer must test (and correct if necessary) the subroutine he or she has written. In this section we see how a new programming module is tested with a *software driver* program. The driver executes the new module with data supplied by the programmer and verifies that the module performs the associated task correctly.

The following procedure was written to solve the quadratic equation $Y = 5X^2 - 2X + 6$, where the value for X is stored in AL and the result of the equation (Y) is returned in AX:

```
QUAD   PROC   NEAR
       MOV    BL,AL    ;save copy of input value
       MUL    BL       ;compute X^2
       MOV    CX,5
       MUL    CX       ;compute 5X^2
       XCHG   DX,AX    ;save temporary result
       MOV    AL,2
       MUL    BL       ;compute 2X
```

```
            SUB    DX,AX      ;compute 5X^2 - 2X
            XCHG   DX,AX      ;get current result into AX
            ADD    AX,6       ;compute 5X^2 - 2X + 6
            RET
QUAD        ENDP
```

The driver program must pass an X value into the procedure and check the returned value for accuracy. Multiple test cases are preferable because they will show how the routine performs over a range of input values. This requires the programmer to first determine what the correct results should be. Consider these input and output pairs:

X Input	Y Output
0	6
1	9
10	486
100	49806

The software driver presented here will send each X-input value to the procedure one at a time and check for a match with the expected Y-output value each time. If all four tests pass, the driver assumes the new routine is acceptable. If any one test fails, an error message is output.

```
;Program TESTQUAD.ASM: Software driver program for QUAD procedure.
;
            .MODEL SMALL
            .DATA
X1    DB    0             ;test case 1
Y1    DW    6
X2    DB    1             ;test case 2
Y2    DW    9
X3    DB    10            ;test case 3
Y3    DW    486
X4    DB    100           ;test case 4
Y4    DW    49806
PASS  DB    'Procedure passes.',0DH,0AH,'$'
FAIL  DB    'Procedure fails.',0DH,0AH,'$'

            .CODE
            .STARTUP
            MOV    AL,X1      ;load first test value
            CALL   QUAD       ;compute result
            CMP    AX,Y1      ;look for match
            JNZ    BAD
            MOV    AL,X2      ;load second test value
            CALL   QUAD       ;compute result
            CMP    AX,Y2      ;look for match
            JNZ    BAD
            MOV    AL,X3      ;load third test value
            CALL   QUAD       ;compute result
            CMP    AX,Y3      ;look for match
            JNZ    BAD
            MOV    AL,X4      ;load fourth test value
            CALL   QUAD       ;compute result
            CMP    AX,Y4      ;look for match
            JNZ    BAD
            LEA    DX,PASS    ;set up pointer to pass message
```

```
        JMP    SEND        ;go output message
BAD:    LEA    DX,FAIL     ;set up pointer to fail message
SEND:   MOV    AH,9        ;display string function
        INT    21H         ;DOS call
        .EXIT


QUAD    PROC   NEAR
        MOV    BL,AL       ;save copy of input value
        MUL    BL          ;compute X^2
        MOV    CX,5
        MUL    CX          ;compute 5X^2
        XCHG   DX,AX       ;save temporary result
        MOV    AL,2
        MUL    BL          ;compute 2X
        SUB    DX,AX       ;compute 5X^2 - 2X
        XCHG   DX,AX       ;get current result into AX
        ADD    AX,6        ;compute 5X^2 - 2X + 6
        RET
QUAD    ENDP

        END
```

If more test cases are needed, the test data should be arranged as a data table so that a loop can be used to step through each test case.

Some procedures may require only a single test case to determine whether they function correctly. In any case, if the new procedure should fail, it may be necessary to use DEBUG to single-step through the driver program until the error is found.

Writing driver programs for the routines presented in the remaining sections should be a rewarding programming experience.

6.4   DATA GATHERING

When a microprocessor is used in a control application, one of its most important tasks is to gather data from the external process. These data may be composed of inputs from different types of sensors, parallel or serial information transmitted to the system from a separate source, or simply keystrokes from the user's keyboard.

Usually, a section of memory is set aside for the storage of the accumulated data, so the processor can alter or examine it at a later time. The rate at which new data arrive, as in keystrokes from a keyboard, may be very slow, with a new item arriving every few milliseconds or so. When the data rate is slow, the processor will waste valuable execution time waiting for the next new data item. Therefore, an efficient solution is to store the data as they arrive, and process them only when all items have been stored. We will examine two examples of gathering data in this section. The first example deals with keyboard buffering and the second example deals with packing BCD numbers.

The Keyboard Buffer

One of the first things anyone involved with computers learns is that nothing happens until you hit return. All keystrokes up to return must be saved for processing after return is hit. The subroutine presented here, KEYBUFF, is used to store these keystrokes in a buffer until return is hit. The processor will then be free to examine the contents of the keyboard buffer at a later time. KEYBUFF makes use of an INT   21H function, which is used to get a keystroke from the keyboard. The ASCII code for the key is returned by INT   21H in the lower byte of AL. INT   21H takes care of echoing the key back to the user's display. An important point to keep in mind is that INT   21H will not return a value in AL until a key is struck.

The ASCII codes for the keys entered are saved in a buffer called KEYS, which is limited to 128 characters. No code is provided to prevent more than this number of keystrokes. Can you imagine what problems occur when the 129th key is entered?

```
In the current data segment...
KEYS      DB     128 DUP(?)
.
.
.
KEYBUFF   PROC   FAR
          LEA    DI,KEYS    ;DI points to start of buffer
NEXTKEY:  MOV    AH,1
          INT    21H        ;read keystroke and echo to screen
          MOV    [DI],AL    ;save key in buffer
          INC    DI         ;point to next buffer location
          CMP    AL,0DH     ;continue until return is seen
          JNZ    NEXTKEY
          RET
KEYBUFF   ENDP
```

An important feature missing in this example is the use of special codes for editing. No means are provided for editing mistaken keys entered by the user. At the very least, the user should be able to enter a backspace to correct a previous error. You are encouraged to solve this problem, and the other one dealing with limiting the number of keystrokes, yourself.

Packing BCD Numbers

Any program that deals with numbers must use one of two approaches to numeric processing. The program must treat the numbers either as binary values or as BCD values. The use of binary operations provides for large numbers with a small number of bits (integers over 16 million can be represented with only 24 bits), but is limited in accuracy when it comes to dealing with fractions. The use of BCD provides for greater accuracy, but requires software to support the mathematical routines, and this software greatly increases the execution time required to get a result. Even so, BCD numbers have found many uses, especially in smaller computing systems. The example we will study here is used to accept a multidigit BCD number from a keyboard and store it in a buffer called BCDNUM. The trick is to take the ASCII codes that represent the numbers 0 through 9 and convert them into BCD numbers. Two BCD numbers at a time are packed into a byte as shown in Figure 6.1. BCDNUM will be limited to 6 bytes, thus making 12-digit BCD numbers possible. The subroutine PACKBCD will take care of packing the received BCD numbers into bytes and storing them in BCDNUM. No error checking is provided to ensure that no more than 12 digits are entered, or that the user has entered a valid digit. If the number entered is less than 12 digits long, the user hits return to complete the entry. All numbers will be right justified when saved in BCDNUM. This means that numbers less than 12 digits long will be filled with leading zeros.

FIGURE 6.1 Packing two BCD digits together

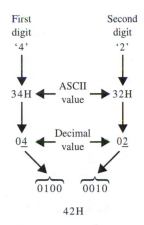

```
In current data segment . . .
BCDNUM       DB    6 DUP(?)
    .
    .
    .
PACKBCD      PROC  FAR
             LEA   DI,BCDNUM        ;point to beginning of buffer
             MOV   CX,6             ;init loop counter
CLEARBUFF:   MOV   BYTE PTR [DI],0  ;clear all bytes in BCDNUM
             INC   DI               ;with this loop
             LOOP  CLEARBUFF
             DEC   DI               ;move to end of buffer
GETDIGIT:    MOV   AH,1             ;get a number from the user
             INT   21H
             CMP   AL,0DH           ;done?
             JZ    DONE
             SUB   AL,30H           ;remove ASCII bias
             MOV   BL,AL            ;save first digit
             MOV   AH,1             ;get another number
             INT   21H
             CMP   AL,0DH           ;done?
             JZ    SAVEIT
             SUB   AL,30H           ;remove ASCII bias
             MOV   CL,4             ;prepare for 4-bit shift
             SHL   BL,CL            ;move BCD digit into upper nybble
             OR    AL,BL            ;pack both digits into AL
             MOV   [DI],AL          ;save digits in buffer
             DEC   DI               ;decrement pointer
             JMP   GETDIGIT
SAVEIT:      MOV   [DI],BL          ;save last digit in buffer
DONE:        RET
PACKBCD      ENDP
```

The loop at the beginning of PACKBCD writes zeros into all 6 bytes of BCDNUM. This is done to automatically place all leading zeros into the buffer before any digits are accepted. Notice also that DI has been advanced to the end of the buffer when the loop has finished. We need DI to start at the end of BCDNUM because we decrement it to store the digits as they are entered. INT 21H is used to get a BCD number from the user (assuming that no invalid digits are entered). Subtracting 30H from the ASCII values returned by INT 21H converts the ASCII character code (35H for '5') into the correct BCD value. The SHL and OR instructions perform the packing of two BCD digits into a single byte. The input number 12345 is stored as 00 00 00 05 34 12. You should experiment with other formats.

Programming Exercise 6.1:   Modify the keyboard buffer routine KEYBUFF so that the user may not enter more than 128 keys. KEYBUFF should automatically return if 128 keys are entered.

Programming Exercise 6.2:   Modify KEYBUFF to allow for two simple editing features. If a backspace key is entered (ASCII code 08H), the last key entered should be deleted. (What problem occurs, though, when backspace is the first key entered?) The second editing feature is used to cancel an entire line. If the user enters a Control-C (ASCII code 03H), the contents of the entire buffer are deleted.

Programming Exercise 6.3:   Modify KEYBUFF to include a count of the number of keys entered, including the final return key. This number should be stored in COUNT on return from KEYBUFF.

Programming Exercise 6.4:   Modify KEYBUFF so that all lowercase letters (a–z) are converted to uppercase letters (A–Z) before being placed in the buffer. All other ASCII codes should remain unchanged.

Programming Exercise 6.5:   Modify KEYBUFF so that the contents of the buffer are displayed (using display string from INT   21H) if Control-R is entered, and the buffer is cleared if Control-C is entered.

Programming Exercise 6.6:   Modify the PACKBCD routine so that a maximum of 12 digits may be entered. PACKBCD should automatically return after processing the 12th digit.

Programming Exercise 6.7:   Modify the PACKBCD routine to scan the buffer after the entire number has been entered and eliminate leading zeros.

6.5   SEARCHING DATA TABLES

In this section we will see a few examples of how a block of data may be searched for single or multibyte items. This technique is a valuable tool that has many applications. In a large database, information about many individuals may be stored. Their names, addresses, social security numbers, phone numbers, and many other items of importance may be saved. Finding out if a person is in the database by searching for any of the items just mentioned requires an extensive search of the database. In an operating system, information about users may be stored in a special access table. Their user names, account numbers, and passwords might be included in this table. When a user desires to gain access to the system, his or her entry in the table must be located by account number or name and the password checked and verified. Once on the system, the user will begin entering commands. The commands entered must be checked against an internal list to see if they exist before processing can take place. In a word processing program, a special feature might exist that allows a search of the entire document for a desired string. Every occurrence of this string must be replaced by a second string. For example, the author may notice that every occurrence of "apples" must be changed to "oranges." If only one or two of these strings exist, the author will edit them accordingly. But if "apples" occurs in 50 different places, it becomes very time consuming and inefficient to do this manually. Let us now look at a few examples of how a data table may be searched.

Searching for a Single Item

The first search technique we will examine involves searching for a single item. This item might be a byte or a word value. The following subroutine searches a 100-element data table for a particular byte value. Upon entry to the subroutine, the byte to be searched for is stored in ITEM. The item may or may not exist within the data table. To account for these two conditions, we will need to return an indication of the result of the search. The carry flag is used to do this. If the search is successful, we will return with the carry flag set. If the search fails, we return with the carry flag cleared.

```
In current data segment...
VALUES      DB      100 DUP(?)
ITEM        DB      ?
   .
   .
   .
FINDBYTE    PROC    FAR
            LEA     SI,VALUES    ;init data pointer
            MOV     CX,100       ;init loop counter
            MOV     AL,ITEM      ;load AL with search item
COMPARE:    CMP     AL,[SI]      ;compare item with data in table
            JZ      FOUND
            INC     SI           ;point to next item
            LOOP    COMPARE      ;continue comparisons
            CLC                  ;clear carry flag, search failed
            RET
FOUND:      STC                  ;set carry flag, item found
            RET
```

Notice how STC and CLC have been used to directly modify the carry flag, depending on the results of the search. Using the carry flag in this manner allows the programmer to write much simpler code. For example, only two instructions are needed to determine the result of the search:

```
CALL    FINDBYTE
JC      SUCCESS
```

Of course, other techniques may be used to indicate the results. The nice thing about using the flags is that they require no external storage and can be used whenever a binary condition (true/false) is the result.

Searching for the Highest Integer

When working with data it often becomes necessary to find the largest value in a given set of numbers. This is useful for finding the range of the given set and also has an application in sorting. MAXVAL is a subroutine that will search an array called NUMBERS for the largest positive byte integer. No negative numbers are allowed at this time. The result of the search is passed back to the caller in the lower byte of BX.

```
In current data segment . . .
NUMBERS     DB      128 DUP(?)
.
.
.
MAXVAL      PROC    FAR
            LEA     SI,NUMBERS      ;init data pointer
            MOV     BL,0            ;assume 0 is largest to begin with
            MOV     CX,128          ;init loop counter
CHECKIT:    CMP     [SI],BL         ;compare current value with new data
            JC      NOCHANGE        ;jump if new value is not larger
            MOV     BL,[SI]         ;load new maximum value
NOCHANGE:   INC     SI              ;point to next byte
            LOOP    CHECKIT         ;continue until all bytes checked
            RET
MAXVAL      ENDP
```

Using JC after the compare operation treats all bytes as unsigned integers. Other forms of the conditional jump will allow signed numbers to be detected as well.

Programming Exercise 6.8:   Modify the FINDBYTE data search subroutine so that the length, in bytes, of the data table is passed via LENGTH. The maximum length of the data is 1,024 items.

Programming Exercise 6.9:   Modify FINDBYTE so that the position of ITEM within VALUES is returned in POSITION, if the search is successful. For example, if ITEM is the first element, POSITION should be 0. If ITEM is the 11th element, POSITION should be 000A.

Programming Exercise 6.10:   Modify the MAXVAL subroutine so that negative numbers (represented in 2's complement notation) may be included in the data.

Programming Exercise 6.11:   Modify the MAXVAL subroutine in two ways: (1) The memory address of the maximum value is returned in DI, and (2) comparison of positive *word* values is performed if AL equals 00 upon entry to MAXVAL, and comparison of byte values is performed otherwise.

6.6   STRING OPERATIONS

As previously defined, a string is a collection of bytes or words that represent information. For example, the display string function (AH = 9) of DOS's INT   21H requires strings of the form:

```
ANYSTRING   DB    'This is a text string.$'
```

in which the end of the string is indicated by the '$' character. What is required to process a display string? Assume that register DX has been loaded with the starting address of ANYSTRING (via LEA   DX,ANYSTRING). The SENDOUT routine shown here reads string characters one at a time and outputs them to the display until the '$' character is seen.

```
SENDOUT   PROC    FAR
          MOV     SI,DX       ;use SI as string pointer
GETCHAR:  MOV     DL,[SI]     ;read a string character
          CMP     DL,'$'      ;end of string?
          JZ      EXIT        ;jump if match
          MOV     AH,2        ;display character function
          INT     21H         ;DOS call
          INC     SI          ;point to next string character
          JMP     GETCHAR     ;and repeat
EXIT:     RET
SENDOUT   ENDP
```

One disadvantage of the SENDOUT routine is that it is not possible to output the '$' character to the display, because it is the end-of-string marker. One way to fix this would be to use a byte value of 0 to terminate the string, as in:

```
NEWSTRING   DB    'This string ends differently.',0
```

The CMP instruction must be changed to CMP   DL,0 to use this new format.

There are many other uses for text strings. They can specify a DOS file name (and path), as in:

```
RUNFILE   DB    'C:\PROGRAMS\RUNME.COM'
```

or a list of abbreviated days of the week:

```
DAYS   DB    'MonTueWedThuFriSatSun'
```

In this section we will examine a number of techniques that use text strings to perform useful operations.

Comparing Strings

A very important part of any program that deals with input from a user involves recognizing the input data. Consider the password required by most users of large computing systems. The user must enter a correct password or be denied access to the system. Because the password may be thought of as a string of ASCII characters, some kind of string comparison operation is needed to see if the user's password matches the one expected by the system. The following subroutine compares two strings of 10 characters each, returning with the carry flag set if the strings are exactly the same. If you think of one string as the password entered by the user and the other as the password stored within the system, you will see how they are compared.

```
In current data segment . . .
STRINGA      DB      'alphabetic'
STRINGB      DB      'alphabet  '
.
.
.
CHKSTRING    PROC    FAR
             MOV     SI,0              ;init character pointer
             MOV     CX,10             ;init loop counter
CHECKCHAR:   MOV     AL,STRINGA[SI]    ;get character from STRINGA
             CMP     AL,STRINGB[SI]    ;compare with STRINGB character
             JNZ     NOMATCH           ;even one difference causes failure
             INC     SI                ;point to next character
             LOOP    CHECKCHAR         ;check all elements
             STC                       ;strings match
             RET
NOMATCH:     CLC                       ;strings are different
             RET
CHKSTRING    ENDP
```

The two strings used in the example are not identical because the last two characters are different.

A Command Recognizer

Consider a small single-board computer system that allows you to do all of the following:

1. Examine/alter memory (EXAM)
2. Display memory (DUMP)
3. Execute a program (RUN)
4. Terminate program execution (STOP)
5. Load a program into memory (LOAD)

Each of the five example commands has a specific routine address within the memory map of the system. For example, the DUMP command is processed by the code beginning at address 04A2C. The **command recognizer** within the operating system of the small computer must recognize that the user has entered the DUMP command, and jump to address 04A2C. This requires that both a string-compare operation and a table lookup be performed. The following routine is one way this may be accomplished:

```
In current data segment . . .
COMMANDS     DB      'EXAM'
             DB      'DUMP'
             DB      'RUN '
             DB      'STOP'
             DB      'LOAD'
JUMPTABLE    DW      DOEXAM
             DW      DODUMP
             DW      DORUN
             DW      DOSTOP
             DW      DOLOAD
COMBUFF      DB      4 DUP(?)
```

```
         .
         .
         .
RECOGNIZE    PROC    FAR
             LEA     BX,COMMANDS          ;point to command table
             MOV     DI,0                 ;init index within JUMPTABLE
             MOV     CX,5                 ;init loop counter
NEXTCOM:     PUSH    CX                   ;save loop counter
             MOV     CX,4                 ;prepare for command matching
             MOV     SI,0
CHKMATCH:    MOV     AL,[BX + SI]         ;get a table character
             CMP     COMBUFF[SI],AL       ;and compare it with command
             JNZ     NOMATCH
             INC     SI                   ;point to next character
             LOOP    CHKMATCH             ;continue comparison
             POP     CX                   ;match found, fix stack
             JMP     JUMPTABLE[DI]        ;jump to command routine
NOMATCH:     POP     CX                   ;get loop counter back
             ADD     BX,4                 ;point to next command text
             ADD     DI,2                 ;and next routine address
             LOOP    NEXTCOM              ;go check next command
             JMP     COMERROR             ;command not found
RECOGNIZE    ENDP
```

The set of valid commands begins at COMMANDS. The addresses for each command routine begin at JUMPTABLE. The command entered by the user is saved in the 4 bytes beginning at COMBUFF. The purpose of RECOGNIZE is to compare entries in COMMANDS with COMBUFF. Every time a match is not found, a pointer (DI) is advanced to point to the next routine address in JUMPTABLE. When a match is found, DI will point to the start of the routine address saved in memory. This routine address is then used by JMP. If none of the commands match the user's, a jump is made to COMERROR (possibly a routine that will output an error message saying "Illegal command").

Accessing a Simple Database

In this example we will see how a simple database is defined and accessed through the use of string operations. The database is composed of predefined *records* that contain useful information about employees at a fictitious business. A sample record from the database looks like this:

```
DB    'Jennifer Indigo, CS/AT, 2676, A7',0DH
```

which corresponds to the following record format:

```
<First Name>blank<Last Name>comma<dept>comma<phone>comma<room><cr>
```

where each <item> is referred to as a *field* and is terminated by a specific character (such as blank or comma). As you can see, a record is terminated with the code for carriage return.

Any number of records may be strung together to make up the database. In this example the following database is used:

```
DBASE    DB    'James Antonakos, EET, 2356, B20',0DH
         DB    'Mike Fisher, RWA, 2376, A19',0DH
         DB    'Dave Guza, MPC, 2389, B26',0DH
         DB    'Jennifer Indigo, CS/AT, 2676, A7',0DH
         DB    'William Robinson, LIS, 2300, J2',0DH
         DB    'Michele Tanner, ILY, 2143, B45',0DH
         DB    0      ;end of database
```

Notice that the length of each record is different. This is due to the varying length of each first name, last name, department, and room number. For simplicity, records are limited to a maximum length of 64 characters, which allows for lengthy names. Even so, as the database shows, the different length of each record prevents us from assuming that the starting character in each field is in a known position. For this reason we must *search* the database for information when we need it. The procedure shown here, LASTNAME, scans the current database record (pointed to by DI) for the individual's last name and displays it. The processor's SCAS instruction is used to find the first blank character in the record, which indicates the end of the first name. SCAS automatically adjusts DI during the scan, so that it points to the first letter of the last name when SCAS completes. A similar search method is used to find the end of the record after the last name is printed out.

Upon entry to LASTNAME, register DI must point to the beginning of a record in the database. Upon exit, DI will point to the beginning of the next record. The last name is displayed on the screen as it is read. It is assumed that the DS and ES registers have been set up accordingly.

```
LASTNAME    PROC    NEAR
            MOV     AL,20H      ;character to find
            MOV     CX,64       ;max length of record
            CLD                 ;set auto increment
            REPNZ   SCASB       ;skip to beginning of last name
DISPNAM:    MOV     DL,[DI]     ;load string character
            CMP     DL,','      ;past end of last name?
            JZ      FIXEND      ;jump if so
            MOV     AH,2        ;otherwise, output character to display
            INT     21H         ;DOS call
            INC     DI          ;advance to next character
            JMP     DISPNAM     ;and repeat
FIXEND:     MOV     AL,0DH      ;character to find
            MOV     CX,64       ;max length of record
            REPNZ   SCASB       ;skip to end of record
            RET
LASTNAME    ENDP
```

If LASTNAME is called repeatedly, each last name in the database will be found. A driver program to do this, and to format the output with carriage return and line feed codes, might display these results:

Antonakos

Fisher

Guza

Indigo

Robinson

Tanner

Similar routines can be written to access and display any group of items from each record. These routines can also be used to create other databases. For example, LASTNAME can be modified to copy just the last name into a new area of memory where a database of last names is being constructed. In this case, other string instructions, such as MOVS, might be useful.

Programming Exercise 6.12:   Modify the SENDOUT routine so that a carriage return and a line feed are output when the '$' character is encountered at the end of the string.

Programming Exercise 6.13:   Rewrite the CHKSTRING procedure using the CMPSB instruction. Remember that both strings are located in the current data segment and adjust the ES register accordingly.

Programming Exercise 6.14:   The CHECKSTR subroutine is limited for two reasons. First, the starting addresses of the two strings are set when the subroutine is entered. Second, the length of the two strings is fixed at 10 characters apiece. Modify CHECKSTR so that registers SI and DI are loaded from the addresses stored in locations STRING1 and STRING2 and the string length is loaded from LENGTH.

Programming Exercise 6.15:   Using a modified version of CHECKSTR, write a routine that will count the number of occurrences of the word "the" in the block of text. The text block begins at address 3000 and ends at address 37FF.

Programming Exercise 6.16:   Rewrite the RECOGNIZE procedure using the CMPSW instruction. Remember that both strings are located in the current data segment and adjust the ES register accordingly.

Programming Exercise 6.17:   The command recognizer RECOGNIZE works only with uppercase commands. Rewrite the code so that uppercase and lowercase commands may be recognized. For example, "DUMP" and "dump" should be identical in comparison.

Programming Exercise 6.18:   Write a command recognizer that will recognize single-letter commands. The commands may be either uppercase or lowercase, and have the following addresses associated with them:

> A: 20BE
> B: 3000
> C: 589C
> D: 2900

Programming Exercise 6.19:   Modify the LASTNAME procedure so that the last name is copied into a buffer, defined as:

```
LNAME   DB   33 DUP(?)
```

The last name written into the LNAME buffer should be terminated with a '$' character.

Programming Exercise 6.20:   Write a routine called PHONELIST that scans the current database record for the first name and phone fields and displays them on the screen, as in:

James   2356

Mike   2376

etc.

Be sure that each phone number begins in the same column.

6.7   SORTING

It is often necessary to sort a group of data items into ascending (increasing) or descending order. On average, the search time for a sorted list of numbers is smaller than that of an unsorted list. Many different sorting algorithms exist, with some more efficient than others. The sorting algorithm covered here is called a **bubble sort.** A bubble sort consists of many passes over the elements being sorted, with comparisons and swaps of numbers being made during each pass. A short example should serve to introduce you to the technique of the bubble sort. Consider this group of numbers:

7 10 6 3 9

It is necessary to perform only four comparisons to determine the highest number in the group. We will repeatedly compare one element in the group with the next element, starting with the first. If the second element is larger than the first, the two numbers will be swapped. By this method we guarantee that after four comparisons, the largest number is at the end of the array. Check this for yourself. Initially, 7 and 10 are compared and not swapped. Then 10 and 6 are compared and swapped because 10 is greater than 6. The new array looks like this:

7 6 10 3 9

Next, 10 and 3 are compared and swapped. Then 10 and 9 are compared and swapped. At the end of the first pass, the array is:

7 6 3 9 10

It is not necessary now to ever compare any of the elements in the array with the last one, since we know it to be the largest. The next pass will only compare the first four numbers, giving this array at the end of the second pass:

6 3 7 9 10

The third pass will produce:

3 6 7 9 10

and you may notice now that the array is sorted. However, this is due to the original arrangement of the numbers; for completeness, a final pass must be performed on the first two numbers. Note that the five numbers being sorted required four passes. In general, N numbers will require $N - 1$ passes. The subroutine SORT presented here implements a bubble sort. DX is used as the pass counter, registers AL and BL are used for swapping elements, and CX is used as a loop counter. The number of elements to be sorted is saved as a word count in NVAL. The appropriately sized DUP statement is needed for VALUES, with only sixteen locations reserved in this example. Also, only positive integers may be sorted (because of the use of JNC in the comparison).

```
In current data segment. . .
VALUES    DB     16 DUP(?)
NVALS     DW     ?
.
.
.

SORT      PROC   FAR
          MOV    DX,NVALS              ;get number of data items
          DEC    DX                    ;subtract 1 to start
DOPASS:   MOV    CX,DX                 ;init loop counter
          MOV    SI,0                  ;init data pointer
CHECK.    MOV    AL,VALUES[SI]         ;get first element
          CMP    VALUES[SI + 1],AL     ;compare with second element
          JNC    NOSWAP
          MOV    BL,VALUES[SI + 1]     ;swap elements
          MOV    VALUES[SI + 1],AL
          MOV    VALUES[SI],BL
NOSWAP:   INC    SI                    ;point to next element
          LOOP   CHECK                 ;continue with pass
          DEC    DX                    ;decrement pass counter
          JNZ    DOPASS                ;until finished
          RET
SORT      ENDP
```

By advancing SI in steps of one, memory references VALUES[SI] and VALUES[SI+1] always access the next two elements in the VALUES array. When it is necessary to swap them, temporary variable BL is used to hold the contents of the second location while it is being replaced by the contents of the first location. The stack may be used for this purpose also, but at the expense of additional execution time. The use of different conditional jump instructions will allow for negative numbers to be sorted as well.

Programming Exercise 6.21   Modify the SORT routine so that signed and unsigned numbers are allowed.

Programming Exercise 6.22:   Modify the SORT routine so that it exits as soon as an entire pass fails to produce a single swap.

Programming Exercise 6.23:   Modify the SORT routine so that it sorts only array elements that lie between the addresses contained within SI (starting address) and DI (ending address). For example, if SI contains 03A7 and DI contains 03B2, only the twelve numbers in the address range 03A7 to 03B2 are sorted.

6.8   COMPUTATIONAL ROUTINES

This section covers examples of how the 80x86 performs standard mathematical functions. Because the processor has specific instructions for both binary and BCD operations, we will examine sample routines written around those instructions. Math processing is a major part of most high-level languages and the backbone of specialized application programs, such as spreadsheets and statistical analysis packages. Most processors, however, are limited in their ability to perform complicated math. When complex functions such as SIN(X) or LOG(Y) are needed, the programmer is faced with a very difficult task of writing the code to support them. Even after the code is written and judged to be correct, it will most likely be very lengthy and slow in execution speed. For this reason, some systems are designed with math coprocessor chips. These chips are actually microprocessors themselves whose instruction sets contain only mathematical instructions. Adding a coprocessor eliminates the need to write code to perform the math function. SIN(X) is now an instruction executed by the coprocessor. The main CPU simply reads the result from the coprocessor. The coprocessor available for the 80x86 is the 80x87 floating-point coprocessor, which we will examine in Chapter 13.

The examples we will see in this section deal only with addition, subtraction, multiplication, and division. We will, however, also look at a few ways these simple operations can be applied to simulate more complex ones.

Binary Addition

Binary addition is accomplished with ADD and ADC. Both perform addition on registers and/or memory locations. The example presented here is used to find the signed sum of a set of data. The data consists of signed 8-bit numbers. Because it is possible for the sum to exceed 127, we use 16 bits to represent the result.

```
In current data segment...
SCORES    DB       200 DUP(?)
SUM       DW       ?
  .
  .
  .
TOTAL     PROC     FAR
          LEA      SI,SCORES      ;init pointer to data
          MOV      CX,200         ;init loop counter
          MOV      BX,0           ;clear result
ADDEM:    MOV      AL,[SI]        ;load AL with value
          CBW                     ;sign extend into 16 bits
          ADD      BX,AX          ;add new value to result
          INC      SI             ;point to next data item
          LOOP     ADDEM          ;do all values
          MOV      SUM,BX         ;save result in memory
          RET
TOTAL     ENDP
```

Even though the data consists of signed 8-bit numbers, we can perform 16-bit additions if we first use CBW to extend the signs of the input numbers (from 8 to 16 bits).

Binary Subtraction

Binary subtraction is implemented by SUB and SBB. Both instructions work with memory locations and/or registers. The example presented here shows how two blocks of memory may be subtracted from each other. One application in which this technique is useful involves digitally encoded waveforms. Suppose that two analog signals, sampled at an identical rate, must be compared. If the difference is computed by subtracting the binary representation of each waveform and the resultant waveform is displayed by sending the new data to a digital-to-analog converter, we will see a straight line at the output if the waveforms are identical. WAVE1 and WAVE2 are labels associated with the 2K word blocks of memory that must be subtracted. Because of the addressing mode used, the resulting data will overwrite the data saved in WAVE2's area.

```
In current data segment...
WAVE1       DW      2048 DUP(?)
WAVE2       DW      2048 DUP(?)
.
.
.
SUBWAVE     PROC    FAR
            LEA     SI,WAVE1        ;init pointer to beginning of WAVE1
            LEA     DI,WAVE2        ;init pointer for WAVE2
            MOV     CX,2048         ;init loop counter
SUBEM:      MOV     AX,[SI]         ;get sample from WAVE1
            SUB     [DI],AX         ;subtract and replace WAVE2 sample
            ADD     SI,2            ;advance WAVE1 pointer
            ADD     DI,2            ;advance WAVE2 pointer
            LOOP    SUBEM           ;do all samples
            RET
SUBWAVE     ENDP
```

Binary Multiplication

Two instructions are available for performing binary multiplication. MUL (unsigned multiply) is used to multiply 8-, 16-, or 32-bit operands, one of them contained in the accumulator. IMUL (signed multiply) generates a signed result using signed operands of 8, 16, or 32 bits each. In both cases, the results are stored in AX/EAX or AX/EAX and DX/EDX. When 64-bit precision is not enough, we must turn to an alternate method to perform the math.

The example presented here is used to perform 64- by 32-bit multiplication on unsigned integers. The 96-bit result represents a significant increase over the 64 bits the processor is limited to. The method used to perform the multiplication is diagrammed in Figure 6.2. The 64-bit operand is represented by two 32-bit halves, A and B. The 32-bit operand is represented by C. Multiplying B by C will yield a 64-bit result. The same is true for A and C, except that A is effectively shifted 32 bits to the left, making its actual value much larger. To accommodate this, 32 zeros are placed into the summing area in such a way that they shift the result of A times C the same number of positions to the left. This is analogous to writing down a 0 during decimal multiplication by hand. The lower 32 bits of the result are the same as the lower 32 bits of the BC product. The middle 32 bits of the result are found by adding the upper 32 bits of the BC product to the lower 32 bits of the AC product. The upper 32 bits of the result equal the upper 32 bits of the AC product, plus any

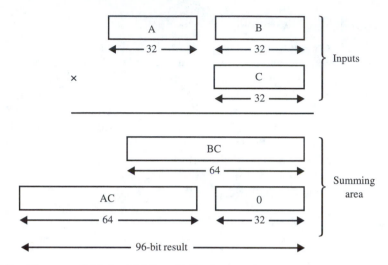

FIGURE 6.2   Diagram of 64- by 32-bit multiplication

carry out of the middle 32 bits. In the following routine, registers EDX and EAX contain the
64-bit value we know as AB (with EDX holding the upper 32 bits). Register EBX contains
the 32-bit multiplier C.

```
In current data segment...
LOWER      DD        ?
MIDDLE     DD        ?
UPPER      DD        ?
    .
    .
    .
MULTIPLY   PROC      FAR
           PUSH      EDX                    ;save a copy of EDX (A) on stack
           MUL       EBX                    ;do B times C
           MOV       LOWER,EAX              ;save partial results
           MOV       MIDDLE,EDX
           POP       EAX                    ;pop stack (A) into EAX
           MUL       EBX                    ;do A times C
           ADD       MIDDLE,EAX             ;generate middle 32-bits of result
           ADC       EDX,0                  ;increment EDX if carry present
           MOV       UPPER,EDX              ;save upper 32-bits of result
           RET
MULTIPLY   ENDP
```

It should be possible to relate the code of this example to Figure 6.2. Generating the individual AC and BC products is easily done via MUL. Adding the upper 32 bits of the BC
product to the lower 32 bits of the AC product is accomplished by using the ADD instruction. Any overflow out of the middle 32 bits will be placed into the carry flag. This carry is
then added to the upper 32 bits of the AC product to complete the operation.

Binary Division

The 80x86 supports binary division with its DIV and IDIV (unsigned and signed division)
instructions. Both instructions can divide a 64-bit quantity by a 32-bit quantity. The 64-bit
result is composed of a 32-bit quotient and a 32-bit remainder. When the quotient is too

large for the destination register, a type-0 interrupt will be generated on completion of the instruction. Many applications exist for the division operation. It can be used to find averages, probabilities, factors, and many other items that are useful when we are working with sets of data. The following subroutine is used to find a factor of a given number when supplied with another factor. For example, FACTOR will return 50 as a factor, when 6 and 300 are supplied as input (because 300 divided by 6 equals 50 exactly). FACTOR will return 0 if no factor exists (for example, 300 divided by 7 gives 42.857143, which is not an integer; thus, the two numbers cannot be factors).

```
In current data segment...
NUMBER      DQ      ?               ;64-bit input number
FACTOR1     DD      ?               ;32-bit input factor
FACTOR2     DD      ?               ;32-bit output factor
.
.
.

FACTOR      PROC    FAR
            LEA     ESI,NUMBER      ;point to input number
            MOV     EAX,[ESI]       ;load input number into EDX:EAX
            MOV     EDX,[ESI+4]
            DIV     FACTOR1         ;divide by input factor
            CMP     EDX,0           ;was division even?
            JNZ     NOFACTOR
            MOV     FACTOR2,EAX     ;save output factor
            JMP     EXIT
NOFACTOR:   MOV     FACTOR2,0       ;clear output factor
EXIT:       RET
FACTOR      ENDP
```

Because the remainder appears in EDX, we examine it for 00000000 to see if the division was even.

BCD Addition

In the binary number system, we use 8 bits to represent integer numbers in the range 0 to 255 (00 to FF hexadecimal). When the same 8 bits are used to store a binary coded decimal (BCD) number, the range changes. Integers from 0 to 99 may now be represented, with the 10s and 1s digits using 4 bits each. If we expand this reasoning to 16 bits, we get a 0 to 65,535 binary integer range, and a 0 to 9999 BCD range. Notice that the binary range has increased significantly. This is always the case and represents one of the major differences between binary and BCD numbers. Even so, we use BCD to solve a nasty problem encountered when we try to represent some numbers using binary. Consider the fractional value 0.7. It is impossible to exactly represent this number using a binary string. We end up with 0.101100110011.... The last four bits (0011) keep repeating. So, we can get very close to 0.7 this way (0.699999...), but never actually get 0.7. When we use this binary representation in a calculation, we will automatically generate a roundoff error. The purpose of BCD is to eliminate the roundoff error (at the cost of a slower computational routine).

For the purposes of this discussion we will use a BCD representation that consists of 4 bytes stored in consecutive memory locations. The first byte is the most significant byte. The fourth byte is least significant. All BCD numbers stored this way (0 to 99999999) will be right justified. Examine the following two numbers and their memory representations to see what is meant by right justification:

$$34298: \quad 00\ 03\ 42\ 98$$
$$7571364: 07\ 57\ 13\ 64$$

We can increase the range of numbers by adding more bytes of storage per number. Each new byte gives two additional BCD digits. Furthermore, we could also add an additional byte to store the exponent of the number. A single byte could represent exponents from 127 to −128 if we used signed binary numbers. Standards exist that define the format of a BCD number (and of binary numbers as well, for use with floating-point units), but we will not cover them at this time.

The example presented here shows how two BCD numbers (each stored in memory at NUMA and NUMB) can be added together. The DAA (decimal adjust for addition) instruction is used together with ADC to perform the BCD addition. ADC will add 2 bytes together, each containing two BCD digits. The result will be corrected by DAA, with the carry flag containing any carry out of the most significant digit. For example, if 37 and 85 are added, the carry will equal 1 and the result operand will contain 22. Because we have defined the 4-byte storage array for a BCD number to be right justified, it is necessary to begin adding with the least significant byte in the array. The result is stored in NUMB, overwriting the BCD number already saved.

```
In current data segment...
NUMA        DB      4 DUP(?)
NUMB        DB      4 DUP(?)
.
.
.
ADDBCD      PROC    FAR
            MOV     SI,3            ;init pointer to LSB
            MOV     CX,4            ;init loop counter
            CLC                     ;clear carry to start
DECIADD:    MOV     AL,NUMA[SI]     ;get first BCD number
            ADC     AL,NUMB[SI]     ;add second BCD number
            DAA                     ;correct result into BCD
            MOV     NUMB[SI],AL     ;save result
            DEC     SI              ;point to next pair of digits
            LOOP    DECIADD         ;do all digits
            RET
ADDBCD      ENDP
```

Upon return from the subroutine, the carry flag will contain any carry out of the MSB.

BCD Subtraction

BCD subtraction is implemented in much the same way as BCD addition, and the subroutine presented here uses the same 4-byte BCD number definition covered in the previous section. The difference in this routine is that the addresses of the two BCD numbers are assumed to be contained in registers SI and DI upon entry. Assuming that SI points to NUMA and DI to NUMB, two different subtractions are possible.

```
In current data segment. . .
NUMA        DB      4 DUP (?)
NUMB        DB      4 DUP (?)
.
.
.
```

```
AMINUSB      PROC   FAR
             XCHG   SI,DI        ;swap pointers
BMINUSA:     MOV    CX,4         ;init loop counter
             ADD    SI,3         ;adjust pointer to end of BCD number
             ADD    DI,3
             CLC                 ;clear carry flag to start
DECISUB:     MOV    AL,[DI]      ;get first number
             SBB    AL,[SI]      ;subtract second number
             DAS                 ;adjust result into BCD
             MOV    [DI],AL      ;store result
             DEC    SI           ;adjust pointers
             DEC    DI
             LOOP   DECISUB      ;do all digits
             RET
AMINUSB      ENDP
```

Upon return, the carry flag will indicate any borrow from the MSB. If the carry flag is set upon return, the result of the subtraction is negative. The result will replace the contents of NUMB when BMINUSA is the entry point to the subroutine. Entering at AMINUSB will cause the result to replace NUMA.

BCD Multiplication

Because BCD multiplication is not directly implemented on the 80x86, there are at least two ways it can be simulated. One method is to convert both BCD numbers into their binary equivalents and then use IMUL to find the result. Of course, the binary result will have to be converted back into BCD. A procedure to accomplish this method is as follows:

```
BCDMUL1      PROC   FAR
             LEA    SI,NUM1      ;point to first BCD number
             CALL   TOBINARY     ;convert into binary
             MOV    BX,AX        ;save result here
             LEA    SI,NUM2      ;point to second BCD number
             CALL   TOBINARY     ;and convert it to binary
             IMUL   BX           ;find the signed product
             CALL   TOBCD        ;convert result into BCD
             RET
BCDMUL1      ENDP
```

The two conversion routines, TOBINARY and TOBCD, operate as follows: TOBINARY converts the BCD number pointed to by SI into a signed binary number and returns it in AX. TOBCD converts the signed binary number in DX:AX into a BCD number and saves it in memory at BCDNUM.

A second approach is to do all the math in BCD. This will require a number of repetitive additions to generate the answer. The need for this looping will, unfortunately, slow down the execution speed. This disadvantage is overcome by the ability to multiply large BCD numbers. BCDMUL2 will multiply two 2-digit BCD numbers stored in the lower byte of registers AX and BX. The BCD result will be placed in AX. Further programming easily extends the input numbers into additional digits.

The multiplication performed by BCDMUL2 is detailed in Figure 6.3. As the figure shows, the product resulting from the 10s digits of the multiplier is shifted left one BCD digit, to simulate the result of multiplying by 10.

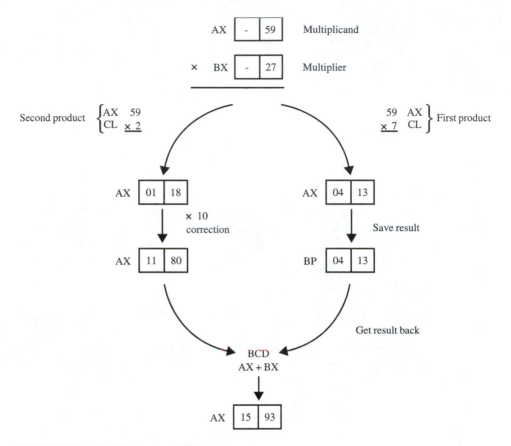

FIGURE 6.3   Multiplying two BCD numbers

```
BCDMUL2   PROC    FAR
          MOV     AH,0        ;fix AX
          PUSH    AX          ;save copy of AL on stack
          MOV     CL,BL       ;find the 1s product
          AND     CL,0FH
          CALL    FAR PTR ALTIMESCL
          MOV     BP,AX       ;save 1s product for later
          POP     AX          ;get AL back
          MOV     CL,4        ;prepare for shift
          SHR     BL,CL       ;move upper BCD digit into lower half
          MOV     CL,BL       ;find the 10s product
          CALL    FAR PTR ALTIMESCL
          MOV     CL,4        ;prepare for shift
          SHL     AX,CL       ;multiply BCD value by 10
          MOV     BX,BP       ;get 1s product back
          ADD     AL,BL       ;form lower half of result
          DAA
          ADC     AH,BH       ;form upper half of result
          XCHG    AL,AH       ;DAA only works on AL
          DAA
          XCHG    AL,AH
          RET
BCDMUL2   ENDP
```

```
ALTIMESCL   PROC    FAR
            MOV     CH,0        ;fix loop counter
            MOV     DL,AL       ;save AL
            MOV     AX,0        ;clear result
AGAIN:      ADD     AL,DL       ;perform repetitive additions
            DAA                 ;until product is found
            ADC     AH,0
            LOOP    AGAIN
            RET
ALTIMESCL   ENDP
```

Notice how a subroutine is used to make the overall process easier to code and read. Although we have not shown it in this example or in previous examples, it is assumed that a valid stack pointer has been assigned to save the subroutine return address and other information. Note also that neither BCDMUL1 or BCDMUL2 checks for multiplication by 0.

BCD Division

All of the BCD operations we have examined so far have ignored treatment of exponents. A collection of subroutines that perform BCD math must have methods of dealing with exponents or be very limited in its applications. As previously mentioned, we can add a single byte to our BCD format to include exponents in the calculations. A single byte gives a signed integer range from -128 to 127. This slightly changes the format of the BCD numbers represented, and requires **normalization** of the numbers before conversion. Normalization is necessary because we have no way of storing a decimal point within the binary data we use to represent a number. Through normalization, we end up with a standard representation by altering the mantissa and adjusting the exponent accordingly. For example, 576.4 and 5.764E2 are equal, as are 23497.28 and 2.349728E4. In these examples, both numbers have been normalized so that the first digit of the mantissa is always between 1 and 9. This method works for fractional numbers as well. Here, we have 0.0035 equaling 3.5E–3. The addition of an exponent byte to our format, together with the new technique of normalization, will require that we now left justify our BCD numbers. Representing these numbers in our standard format gives

576.4:	02	57	64	00	00
23497.28:	04	23	49	72	80
0.0035:	FD	35	00	00	00

where the first byte is used to represent the signed binary exponent. Notice the 2's complement representation of the exponent –3 in the third set of data bytes.

Adding exponent capability to our BCD format complicates the routines we have already seen. The addition routine (as well as subtraction) will give valid results only when we are adding two numbers whose exponents are equal. Because this is rarely the case, we need to adjust the exponent of one number before doing the addition. For instance, if we wish to add 5027 and 394, we must first normalize both numbers:

5027:	03	50	27	00	00
394:	02	39	40	00	00

Because the exponents are different, we have to adjust one of the numbers to correctly add them. If we adjust the number with the higher exponent, we may lose accuracy in our answer. It is much safer to adjust the smaller number. This gives us

$$5027: \quad 03 \quad 50 \quad 27 \quad 00 \quad 00$$
$$394: \quad 03 \quad 03 \quad 94 \quad 00 \quad 00$$

It is clear now that BCD addition of the 4 trailing bytes will give the correct answer. Notice that we have not changed the value of the second number, only its representation.

BCD multiplication and division also require the use of exponents for best results. Unfortunately, it is not a simple matter of adding exponents for multiplication and subtracting them for division. Special rules are invoked when we multiply or divide two negative numbers. In any case, we must take all rules into account when writing a routine that will handle exponents.

The BCD division routine presented here keeps track of exponents during its calculations. The subroutine NUMALIGN adjusts the dividend so that it is always 1 to 9 times greater than the divisor. NUMALIGN modifies the exponent of the dividend as well. Subroutine MSUB performs multiple subtractions. The number of times (0 to 9) the divisor is subtracted from the dividend is returned in the lower 4 bits of AL. Both routines use SI and DI as pointers to the memory locations containing the BCD representations of the dividend and divisor. Register BX accumulates the individual results from MSUB into a 4-digit BCD result. The exponent is generated by the EXPONENT subroutine, which uses the initial exponent values plus the results of ALIGN to calculate the final exponent, which is returned in the lower byte of AX.

```
In current data segment. . .
DIVIDEND   DB      5 DUP (?)               ;reserve 5 bytes (one for exponent)
DIVISOR    DB      5 DUP (?)
.
.
.
BCDDIV     PROC    FAR
           LEA     SI,DIVIDEND             ;init pointer to dividend
           LEA     DI,DIVISOR              ;init pointer to divisor
           MOV     AL,0                    ;clear exponent accumulator
           MOV     CX,4                    ;init loop counter
DIVIDE:    CALL    FAR PTR NUMALIGN        ;align numbers
           CALL    FAR PTR MSUB            ;perform multiple subtractions
           PUSH    CX                      ;save loop counter
           MOV     CL,4
           SHL     BX,CL                   ;shift result one digit left
           POP     CX                      ;get loop counter back
           AND     AL,0FH                  ;mask out result from MSUB
           OR      BL,AL                   ;save result in BL
           LOOP    DIVIDE                  ;continue for more precision
           CALL    FAR PTR EXPONENT        ;generate final exponent
           RET
BCDDIV     ENDP
```

BCDDIV does not check for division by 0, but this test could be added easily with a few instructions.

Deriving Other Mathematical Functions

Once subroutines exist for performing the basic mathematical functions (addition, subtraction, multiplication, and division), these subroutines may be used to derive more complex functions. The examples presented here show how existing routines can be combined to simulate higher level operations. All of the examples to be presented assume that the following multiprecision subroutines exist:

Routine	Operation
ADD	$(BP) = (SI) + (DI)$
SUBTRACT	$(BP) = (SI) - (DI)$
MULTIPLY	$(BP) = (SI) \times (DI)$
DIVIDE	$(BP) = (SI) / (DI)$

In all cases, SI and DI point to the two input numbers upon entry to the subroutine and BP points to the result. Thus, (SI) means the number pointed to by SI, not the contents of SI. By defining the routines in this way, we can avoid discussion about whether the numbers are binary or BCD.

The first routine examined is used to raise a number to a specified power (for example, 5 raised to the 3rd power is 125). This routine uses the binary number in AL as the power. The number raised to this power is pointed to by SI. The final result is pointed to by BP.

```
POWER       PROC    FAR
            CALL    FAR PTR COPY        ;make a copy of the input number
MAKEPOW:    CALL    FAR PTR MULTIPLY    ;compute next power result
            XCHG    BP,SI               ;use result as next input
            DEC     AL                  ;continue until done
            JNZ     MAKEPOW
            XCHG    BP,SI               ;final result is pointed to by BP
            RET
POWER       ENDP
```

POWER is written such that the power must be 2 or more. Negative powers and powers equal to 0 or 1 are not implemented in this routine (they are left as an exercise). COPY is a subroutine that makes a copy of the input number pointed to by SI. The copied number is pointed to by DI.

The next routine is used to generate factorials. A factorial of a number (for example, 5! or 10! or 37!) is found by multiplying all integers up to and including the input number. For instance, 5! equals $1 \times 2 \times 3 \times 4 \times 5$. This results in 5! equaling 120. Do a few factorial calculations yourself, and you will see that the result gets very large, very quickly! FACTORIAL will compute the factorial of the integer value stored in AL. The result is pointed to by BP.

```
FACTORIAL   PROC    FAR
            LEA     SI,ONE              ;init sequence counter
            LEA     DI,ONE              ;init first multiplier
NEXTNUM:    CALL    FAR PTR MULTIPLY    ;compute partial factorial
            XCHG    BP,DI               ;use result as next input
            CALL    FAR PTR INCREMENT   ;increment sequence counter
            DEC     AL                  ;continue until done
            JNZ     NEXTNUM
            XCHG    BP,DI               ;result pointed to by BP
            RET
FACTORIAL   ENDP
```

INCREMENT is a subroutine that performs a specific task: add 1 to the number pointed to by SI. We use INCREMENT to generate the sequence of integers that get multiplied

FIGURE 6.4 Finding square roots by iteration

$$\text{Estimate} = \frac{\dfrac{\text{number}}{\text{estimate}} + \text{estimate}}{2}$$

Example: Find square root of 42
Initial Estimate: 21

Number of iterations	Estimate
0	21
1	11.5
2	7.57608
3	6.55992
4	6.48121
5	6.4807

$(6.4807)^2 = 41.999$

together. The symbol ONE refers to a predefined storage area in memory that contains the value 1 in standard format.

The next routine, ROOT, computes square roots. The formula, and an example of how it works, is presented in Figure 6.4. This type of formula is **iterative.** This means that we must run through the formula a number of times before getting the desired result. Notice in Figure 6.4 how each new application of the square root formula brings the estimate of the answer closer to the correct value. After applying the formula only five times, we have a result that comes very close to the square root. A few more iterations will increase the accuracy of the result even more. Fewer iterations are needed when the initial estimate is close to the desired value. For instance, if the original estimate used in Figure 6.4 was 7 instead of 21, fewer iterations would have been needed to get to 6.4807. The routine presented here implements the formula of Figure 6.4.

```
In current data segment. . .
NUMBER     DB     5 DUP (?)
ESTIMATE   DB     5 DUP (?)
   .
   .
   .
ROOT       PROC   FAR
           LEA    SI,NUMBER         ;point to input number
           LEA    DI,TWO            ;predefined constant 2
           CALL   FAR PTR DIVIDE    ;calculate original estimate
           LEA    BX,ESTIMATE       ;save estimate
           CALL   FAR PTR SAVE
           MOV    CX,10             ;prepare for 10 iterations
ITERATE:   LEA    SI,NUMBER
           LEA    DI,ESTIMATE
           CALL   FAR PTR DIVIDE    ;number / estimate
           XCHG   BP,SI             ;use result in following addition
           CALL   FAR PTR ADDER     ;(number / estimate) + estimate
           XCHG   BP,SI             ;use result in following division
```

```
          LEA     DI,TWO
          CALL    FAR PTR DIVIDE      ;entire formula implemented now
          LEA     BX,ESTIMATE         ;save new estimate
          CALL    FAR PTR SAVE
          LOOP    ITERATE
          RET
ROOT      ENDP
```

The subroutine SAVE is used to make a copy of the number pointed to by BP. The copy is stored in memory starting at the location pointed to by BX. The XCHG instruction is used to swap pointers, thus making the results of ADDER and DIVIDE available for the next operation.

The last example we will examine is used to computer powers of **base e.** From calculus, it can be shown that an infinite series of terms can be used to generate the result of raising **e** (2.7182818) to any power, as Figure 6.5 illustrates. Notice that only the first seven terms are needed to get a reasonable amount of accuracy. Many complex functions can be represented by an infinite series, which we can then implement in software using a loop operation. The following routine generates the first ten terms of the exponential series, using the POWER and FACTORIAL routines already discussed. We assume, however, that POWER and FACTORIAL give valid results for all input values (including 0 and 1).

```
In current data segment. . .
X         DB      5 DUP (?)
TEMP      DB      5 DUP (?)
ETOX      DB      5 DUP (?)
.
.
.
EPOWER        PROC    FAR
```

FIGURE 6.5 Generation of e^x by infinite series

$$e^x = \sum_{n=0}^{\infty} \frac{x^n}{n!}$$

$$= \frac{x^0}{0!} + \frac{x^1}{1!} + \frac{x^2}{2!} + \frac{x^3}{3!} + \frac{x^4}{4!} + \ldots$$

$$= 1 + x + \frac{x^2}{2} + \frac{x^3}{6} + \frac{x^4}{24} + \ldots$$

Example: Find e^1

Number of terms	Result
1	1
2	2
3	2.5
4	2.66666
5	2.70833
6	2.71666
7	2.71805

$(e^1 = 2.7182818)$

```
                   LEA    SI,ZERO              ;predefined constant 0
                   LEA    DI,ETOX
                   CALL   FAR PTR COPY         ;clear result
                   MOV    CX,10                ;init loop counter
         NEXTERM:  LEA    SI,X                 ;compute numerator
                   MOV    AL,CL
                   CALL   FAR PTR POWER
                   LEA    BX,TEMP              ;save numerator
                   CALL   FAR PTR SAVE
                   MOV    AL,CL                ;compute denominator
                   CALL   FAR PTR FACTORIAL
                   LEA    SI,TEMP              ;divide to generate term
                   XCHG   BP,DI
                   CALL   FAR PTR DIVIDE
                   XCHG   BP,DI                ;add current term to result
                   LEA    SI,ETOX
                   CALL   FAR PTR ADDER
                   LEA    BX,ETOX              ;save result
                   CALL   FAR PTR SAVE
                   LOOP   NEXTERM
                   RET
         EPOWER    ENDP
```

Again, XCHG is used to redirect output results back into the math routines. XCHG is also used to swap pointers for storing results in memory. ETOX contains the final result when EPOWER finishes execution.

These examples should serve to illustrate the point that complex mathematical functions can be implemented with a small amount of software. Once a library of these routines has been defined and tested, even more complex equations and functions may be implemented. All that is needed is a CALL to the appropriate subroutine (or collection of subroutines).

Programming Exercise 6.24: Modify the TOTAL routine so that 32-bit numbers are added together. Return the sum in registers DX and AX, with DX containing the upper 16 bits of the sum.

Programming Exercise 6.25: Write a subroutine called BIGMUL that will compute the 128-bit result obtained by multiplying two 64-bit integers. The two input numbers should be in registers EAX and EBX on entry to BIGMUL. Use the MULTIPLY subroutine in your code to implement a process similar to that shown in Figure 6.2.

Programming Exercise 6.26: Use the FACTOR subroutine to find all factors of the number saved in a new variable, INVALUE. Place the factors into a data array called FACTORS.

Programming Exercise 6.27: Write a subroutine called BIGADD that will perform a BCD addition of all 32 bits in registers EAX and EBX. Place the result in ECX.

Programming Exercise 6.28: Write a subroutine called TOBINARY that will convert the BCD number pointed to by DI into an unsigned binary number. The result should be returned in AX.

Programming Exercise 6.29: Write a subroutine called TOBCD that converts the unsigned 16-bit binary number in DX into a BCD number. The result should be returned in DX also.

Programming Exercise 6.30: Write a subroutine that performs BCD division by first converting the BCD numbers to binary. DIV should be used to perform the division. The result should be converted back into BCD. Use TOBCD and TOBINARY in your subroutine.

Programming Exercise 6.31: Modify the POWER subroutine so that any integer power can be used, including negative powers and 0.

Programming Exercise 6.32: Change FACTORIAL so that factorials of 70 or more are not allowed. Return 0 as the result in these cases.

6.9   CONTROL APPLICATIONS

In this section we will examine two examples of how the 80x86 may be used in control applications. Control systems are designed in two different ways: open-loop and closed-loop systems. Figure 6.6 shows two simple block diagrams outlining the main difference between these two types of control systems. An open-loop control system uses its input data to effect changes in its outputs. A closed-loop system contains a feedback path where data concerning the present output conditions is sampled and supplied along with the external inputs. A burglar alarm is an example of an open-loop control system. The system may be designed to monitor sensors at various windows and doors. It may also include circuitry to digitize readings from temperature sensors. When any of the sensors detects an abnormal condition (for example, a window opening), the computer may be directed to dial an emergency phone number and play a recorded help message.

A typical application of a closed-loop control system involves the operation of a motor. Suppose we want to control the speed of the motor by making adjustments to an input voltage to the system. The speed of the motor is proportional to the input voltage and increases as the input voltage increases. We cannot simply apply the input voltage to the motor's windings, for it may not be large enough to operate the motor. Usually an amplifier is involved that is capable of driving the motor. But a problem occurs when the motor encounters a load (for example, by connecting the motor shaft to a pump). The increased load on the motor will cause the motor speed to decrease. To maintain a constant speed in the motor at this point, we need an increase in the input voltage. We cannot hope or expect the operator to constantly watch the motor and adjust the input voltage accordingly. For this reason, we add a feedback loop, which is used to sample the motor speed and generate an equivalent voltage. An *error* voltage is generated by

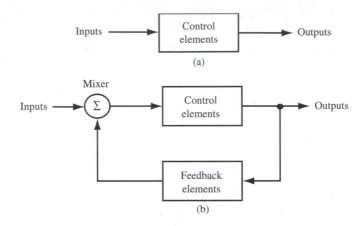

FIGURE 6.6   Control system block diagram: (a) open-loop and (b) closed-loop

comparing the actual speed of the motor (the voltage generated by the feedback circuit) with the desired speed (set by the input voltage). The motor speed voltage may be generated by a tachometer connected to the output of the motor. The error voltage is used to increase or decrease the speed of the motor until it is operating at the proper speed.

Let us look at how the processor might be used to implement the two control systems just described.

A Computerized Burglar Alarm

In this section, we will use the 80x86 to monitor activity on 100 windows and doors in a small office building. The office building consists of four floors, with fifteen doors and ten windows on each floor. The alarm console consists of an electronic display containing a labeled light-emitting diode for each window and door and a serial data terminal capable of displaying ASCII information. The operation of the system consists of two tasks: (1) illuminating the appropriate LED for all open doors and windows, and (2) sending a message to the terminal whenever a door or window opens or closes. It is necessary to continuously scan all of the windows and doors to detect any changes. The circuitry used to monitor the doors and windows and drive the LED display is connected to the processor's system bus so that all I/O can be done by reading and writing to ports. Figure 6.7 shows the assignments of all input and output devices for the first floor of the office building.

As the figure shows, fifteen door and ten window inputs are assigned for the first floor. Whenever a door or window is open, its associated bit will be low. To sample the bits, the processor must do an I/O read from the indicated port (7000 to 7003). Floors 2, 3, and 4 are assigned the same way, with the following port addresses:

Floor 2: 7004–7007

Floor 3: 7008–700B

Floor 4: 700C–700F

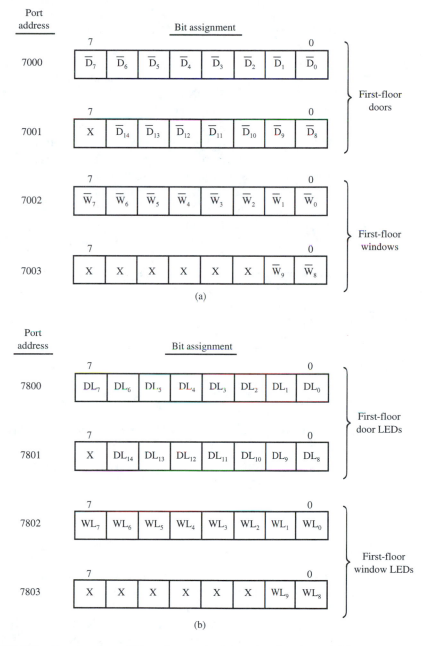

FIGURE 6.7 Burglar alarm I/O assignments: (a) system inputs and (b) system outputs

The door and window LEDs for the first floor are illuminated when their respective bits are high. The processor must do an I/O write to ports 7800 through 7803 to activate LEDs for the first floor. The other-floor LEDs work the same way, with these port addresses assigned to them:

Floor 2: 7804–7807

Floor 3: 7808–780B

Floor 4: 780C–780F

The serial device used by the system to communicate with the ASCII terminal is driven by a subroutine called CONSOLE. The 7-bit ASCII code in the lower byte of register AL is sent to the terminal when CONSOLE is called.

Knowing these definitions, we can design a system to constantly monitor all 100 doors and windows. The technique we will use is called **polling.** Each input port will be read and examined for any changes. If a door or window has changed state since the last time it was read, a message will be sent to the terminal, via CONSOLE, indicating the floor and door/window number. Because we need to remember the last state of each door and window, their states must be saved. A block of memory, called STATUS, will be used for this purpose. STATUS points to a 16-byte block of memory, which we will think of as four blocks of 4 bytes each. Each 4-byte block will store the bits for all doors and windows on a single floor.

When the program first begins operations, the state of each door and window is unknown. For this reason, we initialize STATUS by reading all system inputs when the program starts up. The code to perform the initialization is contained in a subroutine called INIT, and is as follows:

```
In current data segment...
STATUS    DB      16 DUP (?)

INIT      PROC    FAR
          LEA     DI,STATUS       ;init pointer to STATUS
          MOV     DX,7000H        ;init pointer to first input port
          MOV     CX,16           ;init loop counter
SYSREAD:  IN      AL,DX           ;read system information
          MOV     [DI],AL         ;save it in memory
          NOT     AL              ;complement input data
          ADD     DX,800H         ;set up output port address
          OUT     DX,AL           ;update display
          SUB     DX,800H         ;generate next input port address
          INC     DX
          INC     DI              ;point to next STATUS location
          LOOP    SYSREAD
          RET
INIT      ENDP
```

When INIT completes execution, the display has been updated to show the state of all 100 doors and windows, and STATUS has been loaded with the same information.

Once the initial states are known, future changes can be detected by using an Exclusive OR operation. Remember that Exclusive OR produces a 1 when both inputs are different. Figure 6.8 shows how state changes can be detected with Exclusive OR. To incorporate this into the program, XOR is used during updates to detect changes. Note that up to sixteen changes at once can be detected by XORing entire words. It is then a matter of scanning the individual bits to determine if any state changes occurred. A subroutine called DETECT will do this for us. DETECT will sense any state changes and send the appropriate message (for example, first floor: door 12 opened) to the terminal. When DETECT is called, data registers AX, BX, and DX will be interpreted as follows:

FIGURE 6.8 Detecting state changes with XOR

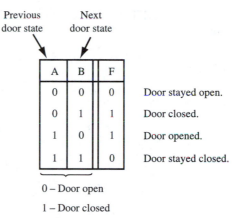

0 – Door open

1 – Door closed

AX: Contains current door or window states.

BX: Contains door or window state changes.

DX: Bit 8 cleared means AX contains door information.

   Bit 8 set means AX contains window information.

   Bits 0 and 1 contain the floor number (0—first, 1—second, 2—third, 3—fourth)

We will not cover the code involved in getting DETECT to do its job. You are encouraged to write this routine yourself, preferably using a rotate or shift instruction to do the bit testing. Because DETECT will have to output ASCII text strings (for example, "First floor," "Second floor," "opened"), the following code may come in handy:

```
In current data segment...
MSG1    DB      'First floor $'
MSG2    DB      'Second floor $'
MSG3    DB      'Third floor $'
MSG4    DB      'Fourth floor $'
MSG5    DB      'door $'
MSG6    DB      'window $'
MSG7    DB      'opened $'
MSG8    DB      'closed $'
.
.
.
SEND    PROC    FAR
        MOV     AL,[SI]             ;get a message character
        CMP     AL,'$'              ;end of message character?
        JZ      EXIT
        CALL    FAR PTR CONSOLE     ;send character to terminal
        INC     SI                  ;point to next character
        JMP     SEND
EXIT:   RET
SEND    ENDP
```

The SEND subroutine must be entered with register SI pointing to the address of the first character in the text string to be sent. SEND could be a subroutine called by DETECT during its analysis of the door and window states.

Using DETECT, the code to poll all doors and windows in the office building becomes:

```
        CALL FAR PTR INIT         ;get initial states and update display
BEGIN:  MOV  CX,0                 ;start with first floor doors
        MOV  DX,7000H             ;point to first input port
        LEA  SI,STATUS            ;point to STATUS information
NEWFLOOR: IN  AL,DX               ;get door data
        NOT  AL
        ADD  DX,800H              ;update display
        OUT  DX,AL
        MOV  AH,AL                ;save first eight door states
        SUB  DX,800H              ;point to next door
        INC  DX
        IN   AL,DX                ;get remaining door data
        NOT  AL
        ADD  DX,800H              ;update display
        OUT  DX,AL
        XCHG AL,AH                ;correct AX for DETECT
        NOT  AX
        MOV  BL,[SI]              ;get past door status
        MOV  BH,[SI + 1]
        ADD  SI,2                 ;advance to next status group
        XOR  BX,AX                ;compute state changes
        XCHG CX,DX                ;get floor number into DX
        CALL FAR PTR DETECT       ;find doors that have changed
        XCHG CX,DX                ;get port address back
        SUB  DX,800H              ;point to window data
        INC  DX
        IN   AL,DX                ;get window data
        NOT  AL
        ADD  DX,800H              ;update display
        OUT  DX,AL
        MOV  AH,AL                ;save first eight window states
        SUB  DX,800H              ;point to next window
        INC  DX
        IN   AL,DX                ;get remaining window data
        NOT  AL
        ADD  DX,800H              ;update display
        OUT  DX,AL
        XCHG AL,AH                ;correct AX for DETECT
        NOT  AX
        MOV  BL,[SI]              ;get past window status
        MOV  BH,[SI + 1]
        ADD  SI,2                 ;advance to next floor
        XOR  BX,AX                ;compute state changes
        OR   CX,100H              ;set bit-8 in CX
        XCHG CX,DX                ;get floor number into DX
        CALL FAR PTR DETECT       ;find windows that have changed
        XCHG CX,DX                ;get port address back
        AND  CH,0                 ;clear bit-8
        SUB  DX,800H              ;point to next floor
        INC  DX
        INC  CX                   ;increment floor counter
        CMP  CX,4                 ;done?
        JZ   REPEAT
        JMP  NEWFLOOR             ;both JMPs are needed since
REPEAT: JMP  BEGIN               ;relative range has been exceeded
```

While you write DETECT, do not forget that the main routine uses a number of registers and that these registers should not be altered. The stack would be a good place to store them for safekeeping.

A Constant-Speed Motor Controller

In this section we will see how the 80x86 may be used in a closed-loop control system to maintain constant speed in a motor. The schematic of the system is shown in Figure 6.9. The speed control is a potentiometer whose output voltage varies from 0 to some positive voltage. This voltage is digitized by an 8-bit analog-to-digital converter, such that 0 volts is 00H and the most positive voltage is FF. The processor reads this data from port 8000. For a purely digital speed control system, this circuitry is eliminated and the speed is set directly by software.

The motor speed is controlled by the output of an 8-bit digital-to-analog converter (with appropriate output amplifier, capable of driving the motor). The motor's minimum speed, 0 RPM, occurs when the computer outputs 00 to the D/A (by writing to port 8020). The motor's maximum speed occurs when FF is sent to the D/A. A tachometer is connected to the motor shaft through a mechanical coupling. The output of the tachometer is digitized also. Again, the minimum and maximum tachometer readings correspond to 00 and FF. A purely digital system would use a digital shaft encoder instead of a tachometer and A/D. The tach is read from input port 8010H.

Each converter is calibrated with respect to a common reference. In theory, a 17 from the SPEED A/D causes a 17 to be sent to the MOTOR D/A, which in turn causes the TACH A/D to read 17 when at the proper speed. In practice the relationship is not so linear, due to external effects of deadband, friction, and other losses in the motor.

The purpose of the program is to operate the motor at a constant speed by comparing the SPEED data with the TACH data. When SPEED equals TACH the motor is turning at the desired speed. When SPEED is less than TACH, the motor is spinning too fast. When SPEED is greater than TACH, the motor is rotating too slowly. The idea is to subtract the TACH value from the SPEED value. The difference determines how much the motor speed should be increased or decreased.

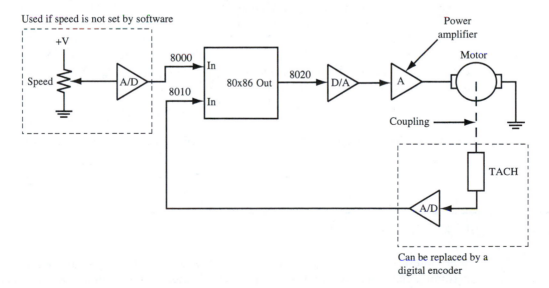

FIGURE 6.9 Constant-speed motor controller

```
SERVO:       MOV    AL,0              ;initial motor speed is 0 RPM
             MOV    DX,8020H
             OUT    DX,AL
GETSPEED:    MOV    DX,8000H          ;read new speed value
             IN     AL,DX
             MOV    AH,AL             ;save speed here
             ADD    DX,10H            ;and new tachometer value
             IN     AL,DX
             CMP    AH,AL             ;SPEED minus TACH
             JG     INCREASE
             JZ     GETSPEED          ;no change
             XCHG   AL,AH
INCREASE:    SUB    AH,AL             ;compute error value
             MOV    AL,AH
             CALL   FAR PTR GAIN
             MOV    DX,8020H          ;output new motor speed
             OUT    DX,AL
             JMP    GETSPEED
```

Because the motor's speed will not change instantly from very slow to very fast, or vice versa, the program will loop many times before the motor gets to the proper speed. For safety or functional reasons, it may not be desirable to try to change the motor speed from slow to fast instantly. Instead, the program should *ramp up* to speed gradually by limiting the size of the error voltage presented to the D/A during speed increases. Subroutine GAIN is used for this purpose, to alter the contents of AL before AL is output to the motor D/A. The ramp up/down speed of the motor, and therefore the response of the closed-loop system, will be a function of the operation of GAIN.

Programming Exercise 6.33:   Write the DETECT subroutine used by the computerized burglar alarm.

Programming Exercise 6.34:   An office complex consisting of sixty-four offices and sixteen hallways is to have its lighting controlled by a computer. Each office has one switch to control its light. Hallways have a switch at each end. Each switch is assigned a bit position in a particular memory location that can be read by the computer, and a closed switch represents a 0. Each light (think of all lights in a hallway as a single light) is also assigned a certain bit in a memory location that the computer can write to. A logic 1 is needed to turn on any light. How many byte locations are needed for all I/O? Write a program that will constantly monitor all switches and adjust the complex lighting as necessary.

Programming Exercise 6.35:   Modify the SERVO program so that the motor's speed will ramp up during periods where a large speed increase is desired.

6.10   NUMBER CONVERSIONS

When information is exchanged between a human and a computer, it is often necessary to perform a conversion from one number system to another. For example, when numeric digits are entered on a keyboard, their ASCII codes must first be converted into decimal values and then the individual values must be combined into a single, equivalent binary number. In general, this process is referred to as *decimal-to-binary* conversion. Likewise, when the computer generates a binary result, the result is often converted into the corresponding ASCII codes for output to the display screen or for storage in a text file. The technique used to perform this transformation is called *binary-to-decimal* conversion. In this section, we will look at examples of each type of conversion.

Decimal to Binary

This conversion takes a three-digit sequence of ASCII digits and converts them into an equivalent unsigned 8-bit number. For example, the ASCII digits '1' — '7' — '2' are converted into the hexadecimal number ACH (10101100 binary). The three-digit ASCII sequence can specify any number from 0 to 255 decimal. The DTOB procedure presented here uses the ASCII digits stored in memory locations HUN, TEN, and ONE as the decimal input number. The ASCII bias of 30H is subtracted from each number to get an actual decimal value. The HUN value is multiplied by 100 and the TEN value is multiplied by 10. These values are combined with the value from ONE to get the final unsigned 8-bit result, which is stored in BINVAL. Note that negative numbers are not allowed.

```
In current data segment...
BINVAL  DB      ?
HUN     DB      ?
TEN     DB      ?
ONE     DB      ?
  .
  .
  .
DTOB    PROC    FAR
        MOV     AL,HUN       ;get hundreds digit
        SUB     AL,30H       ;remove ASCII bias
        MOV     BL,100       ;multiply by 100
        MUL     BL
        MOV     CX,AX        ;save temp result
        MOV     AL,TEN       ;get tens digit
        SUB     AL,30H       ;remove ASCII bias
        MOV     BL,10        ;multiply by 10
        MUL     BL
        ADD     CX,AX        ;add to temp result
        MOV     AL,ONE       ;get ones digit
        SUB     AL,30H       ;remove ASCII bias
        ADD     CL,AL        ;add to get final result
        MOV     BINVAL,CL    ;save conversion value
        RET
DTOB    ENDP
```

Input values that exceed 255 (256 to 999) will not produce a correct result since only the lower byte of CX is saved.

Binary to Decimal

This conversion technique takes an unsigned 8-bit quantity and converts it into a three-digit sequence of ASCII digits. For example, the hexadecimal number 81H converts to the ASCII codes 31H, 32H, and 39H, which represent the decimal number 129. The unsigned 8-bit input number can specify any decimal value from 0 to 255.

Examine the BTOD procedure. The unsigned 8-bit input value is stored in BINVAL. This input number is converted into decimal in two steps. First, it is divided by 100 to get the hundreds digit. Then, the remainder is divided by 10 to get the tens and ones digits. Each digit is converted into ASCII by adding 30H to it. The three-digit ASCII result is stored in three memory locations: HUN, TEN, and ONE. HUN will take on only the ASCII codes for '0', '1', or '2'. TEN and ONE can be any ASCII code from '0' to '9'. Placing the result in memory allows further processing (possibly leading 0 suppression) before the result is sent to the display or output in some other way.

```
In current data segment...
BINVAL  DB    ?
HUN     DB    ?
TEN     DB    ?
ONE     DB    ?
  .
  .
  .
BTOD    PROC  FAR
        MOV   AL,BINVAL    ;load binary input value
        SUB   AH,AH        ;prepare for division by 100
        MOV   BL,100
        DIV   BL           ;get hundreds digit
        ADD   AL,30H       ;convert into ASCII digit
        MOV   HUN,AL       ;and save
        XCHG  AL,AH        ;get remainder
        SUB   AH,AH        ;prepare for division by 10
        MOV   BL,10
        DIV   BL           ;get tens digit
        ADD   AL,30H       ;convert into ASCII digit
        MOV   TEN,AL       ;and save
        ADD   AH,30H       ;convert ones digit into ASCII
        MOV   ONE,AH       ;and save
        RET
BTOD    ENDP
```

What other types of conversion might be useful? Ideally, we would like to have all of the common bases represented: binary, octal, decimal, and hexadecimal. Conversions between any two of these bases are straightforward. Additionally, we might want to allow negative integers as well and increase the range of integers to 65,535 or higher.

Programming Exercise 6.36: Modify the DTOB procedure so that negative integers are included. The range of *signed* integers represented with 8 bits is −128 to +127.

Programming Exercise 6.37: Modify the DTOB procedure so that the ASCII digits are entered from the keyboard. Allow a positive integer range from 0 to 65,535.

Programming Exercise 6.38: Modify the BTOD procedure so that leading zeros are replaced by blanks. For example, if the result is '0' — '5' — '9', the leading 0 in the hundreds position gets replaced by a blank, giving ' ' — '5' — '9'.

Programming Exercise 6.39: Modify the BTOD procedure so that the ASCII digits are output to the display screen. Allow a signed range of integers from –32,768 to +23,767.

6.11 DATA STRUCTURES

In this section we will examine a number of different data structures. Data structures are organized groups of data that must be accessed in a certain way. The organization of the data within a structure is up to the programmer. The data structures covered here (linked-lists, stacks, and queues) are often used by programmers to solve many different types of programming problems.

Linked-Lists

A linked-list is a collection of data elements called **nodes** that is created dynamically. Dynamic creation means that the size of the linked-list is not fixed when it is first created. As a matter of fact, it is empty when first created. As an example, if we want to reserve enough room in memory for 100 integer bytes, we use

```
DATA    DB    100 DUP (?)
```

This assembler directive is used because we already know how many numbers are going to be used. The beauty of a linked-list is that its size can be changed as necessary, either increased or decreased, with a maximum size limited only by the amount of free memory available in the system. This method actually saves space in memory, because it does not dedicate entire blocks of RAM for storing numbers. Rather, a small piece of memory is allocated each time a node is added to the linked-list. A node is most commonly represented by a pair of items. The first item is usually used for storing a piece of data. The second item is a pointer; it is used to point to the next node in the linked-list. Figure 6.10 contains an example of a three-node linked-list. Each node in the list stores a single ASCII character. The beginning of the linked-list, the first node, is pointed to by P. The nodes are linked together via pointers from one node to another. The last node in the list, node 3, contains 0 in its pointer field. We will interpret this as a pointer to nowhere (and thus the last node). The empty pointer is commonly called **nil** or **null.**

The actual representation of the node on a particular system can take many forms. Because the linked-list must reside in memory, it makes sense to assign one or more locations for the data part (or data field), and two locations for the pointer part (also called the pointer field). Why two locations for the pointer field? Because all nodes re-

FIGURE 6.10 A sample linked-list

side in memory. To point to a certain node, we must know its address, which occupies 16 bits.

For this discussion, assume that all nodes consist of 4 data bytes and 2 address bytes. Consider a subroutine called GETNODE that can be called every time a new node is added to the linked-list. GETNODE must find 6 bytes of contiguous (sequential) memory to allocate the node. When it finds them, it will return the starting address of the 6-byte block in register SI. Let us take another look at our example linked-list, only this time addresses have been added to each node. Figure 6.11 shows how the pointer field of each node contains the address of the next node in the list. The address in the pointer field of node 3 indicates the end of the list. The pointer P may be a register containing 1000, the address of the first node in the list. To generate this list, GETNODE has been called three times. GETNODE returned different addresses each time it was called. First came 1000, then 7800, and finally 2052. Linked-lists do not have to occupy a single area of memory. Rather, they may be spread out all over the processor's address space and still be connected by the various pointer fields.

To add a node to the linked-list, a simple procedure is followed. First, a new node is allocated by calling GETNODE. Register SI holds the address of this new node. The pointer field address of this new node, which will have to be modified to add it to the list, starts at SI plus 4 (because the data field occupies the first 4 bytes). To add the new node to the existing list, a copy of the pointer P is written into the pointer field of the new node. Then, to make the new node the first node in the list, P is changed to the address of the new node. Figure 6.12 shows this step-by-step process, assuming that register DI is used to store P. Once the new node has been inserted, its data field, now pointed to by DI, may be loaded with new data. Assume the new data comes from register AL. Note from Figure 6.12 that insertion of the new node into the beginning of the

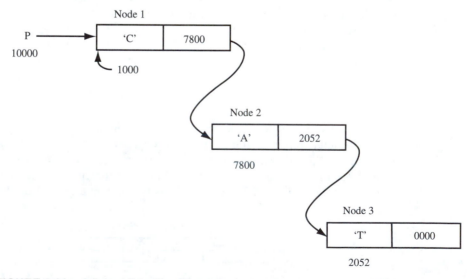

FIGURE 6.11 A linked-list with address assignments

linked-list has changed its contents from 'CAT' to 'SCAT'. The code to perform the insertion described in Figure 6.12 is as follows:

```
INSERT PROC FAR
          CALL FAR PTR GETNODE   ;get a new node from storage pool
          MOV  [SI + 4],DI       ;load pointer field with P
          MOV  DI, SI            ;update pointer P to new node
          MOV  [DI],AL           ;load data field with AL
          MOV  [DI + 1],20H      ;pad rest of data field with blanks
          MOV  [DI + 2],20H
          MOV  [DI + 3],20H
          RET
INSERT ENDP
```

ASCII blank codes (20H) are used to fill the remaining three data bytes in the data field.

1. GETNODE returns new node.

2. Pointer field of new node is loaded with P.

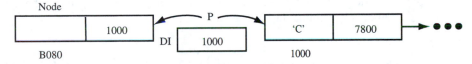

3. Pointer P to list is changed.

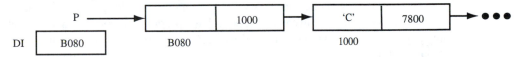

4. Data field of new node is loaded.

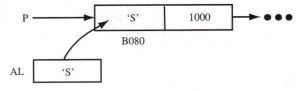

ASCII code for 'S' in lower byte of AX

FIGURE 6.12 Adding a node to a linked-list

Linked-lists are ordinarily used to represent arrays in memory. The data field may be used to store ASCII characters (as in this example), integers, Boolean data, and even pointers to other linked-lists. Linked-lists are very useful tools employed in the functions of operating systems. They are also supported by computer languages such as C.

Many computations are greatly simplified by the use of a software-controlled stack or queue. Expression evaluation and round-robin selection algorithms are just two examples of where a stack and a queue are used. The method of implementation is not critical; either may be designed as a special form of a linked-list or simply as fixed-size memory structures. The latter approach will be used here, with address registers pointing to the stack and queue memory structures.

Stacks

A stack is an area in memory reserved for reading and writing special data items, such as return addresses and register values. For example, a CALL instruction automatically pushes a return address onto the stack (using SP). Registers may be pushed onto the stack (written into stack memory) with a PUSH instruction, as in:

```
PUSH    AX
```

where the entire contents of AX are written into the stack area pointed to by SP. SP is automatically decremented by 2 during execution.

Items previously pushed onto the stack can be popped off the stack (read out of memory) in a similar fashion, as in:

```
POP    AX
```

where stack memory is read out into AX, and SP is automatically incremented by 2 during execution. Thus, we see that using a stack requires manipulation of a stack-pointer register.

One characteristic of a stack is that the last item pushed is always the first item popped. For this reason, stacks are commonly referred to as a **LIFO** (last in first out) structure.

It is possible, and often necessary, for a programmer to design a custom stack area for use within a program. For instance, suppose that a programmer requires a stack that allows only eight words to be pushed onto it. The 80x86 has no mechanism for limiting the amount of pushes (or pops) onto a stack. If this is necessary, a set of stack procedures must be written. The following routines implement a stack that allows a maximum of eight pushes. The SPUSH routine is used to place data onto the stack. The data pushed must be in AX. If the push is successful, a success code of 00H will be returned in the lower byte of BX. If more than eight pushes are attempted, the routine returns with error code 80H in the lower byte of BX, without pushing any data. The SPOP routine is used to remove an item from the stack. If a pop is attempted on an empty stack, error code 0FFH is returned in BX. Successful pops return data in AX. The stack-pointer register is SP and is assigned the address of a free block of memory from the storage pool by a routine called MAKESTACK. MAKESTACK must be called before the stack can be used. MAKESTACK also assigns addresses to SI and DI, which are used by SPUSH and SPOP to determine when a stack operation is possible. The structure of the stack is indicated in Figure 6.13. DI points to the bottom of the stack structure and SI points to the top. SP points to the stack location that will be used for the next push or pop.

FIGURE 6.13 Software-controlled stack structure

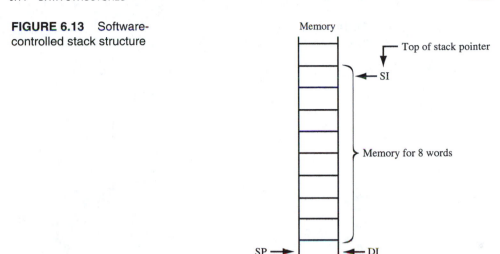

Examine the following routines to see how SI and DI are used to check for legal pushes and pops.

```
SPUSH      CMP   SP,SI        ;ok to push?
           JZ    STKFULL      ;no, go return error code
           PUSH  AX           ;push AX onto stack
           SUB   BL,BL        ;indicate a successful push
           JMP   NEXT
STKFULL    MOV   BL,80H       ;stack full error code
           JUMP  NEXT
SPOP       CMP   SP,DI        ;ok to pop?
           JZ    STKEMPTY     ;no, go return error code
           POP   AX           ;pop AX off stack
           SUB   BL,BL        ;indicate a successful pop
           JUMP  NEXT
STKEMPTY:  MOV   BL,0FFH      ;stack empty error code
NEXT:      ---
```

Note that multiple stacks can be maintained by saving the contents of SP, SI, and DI and loading new addresses into each register.

Queues

Queues are also memory-based structures, but their operation is functionally different from that of a stack. In a queue, the first item loaded is the first item to be removed. For this reason, queues are referred to as **FIFO** (first in first out) structures. Figure 6.14 shows a diagram of a queue that has had the data items 'A', 'B', and 'C' loaded into it. 'A' was loaded first, 'C' last. When we begin removing items from the queue, the 'A' will come out first (unlike the stack structure, which would have popped 'C' first).

A routine called MAKEQUEUE is used to assign a block of memory from the storage pool. MAKEQUEUE initializes four registers whose use and meanings are as follows:

FIGURE 6.14 Software-controlled queue structure

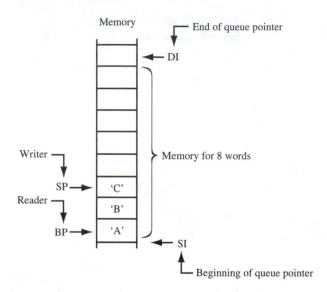

Register	Use/Meaning
SP	Pointer for write operation
BP	Pointer for read operation
DI	Contains end-of-queue address
SI	Contains beginning-of-queue address

Initially, SP, SI, and BP are all loaded with the same value. The address placed into DI is determined by the desired size of the queue.

Before an item can be written into the queue, SP must be compared with DI. If SP equals DI, it is necessary to reload SP with SI's address. This allows SP to *wrap around* the end of the queue (a similar technique is used to wrap BP around when reading). Next, SP is used to write the data item into memory. Then SP is decremented by 2 to prepare for the next write.

Data is removed from the queue by the read-pointer register BP. After the data is read, BP is decremented by 2. Note that serious data errors result when BP reads a location that has not been written into by SP yet. For this reason it is necessary to pay special attention to the positions of SP and BP within the queue. This part of the queue software is left for you to devise on your own. The routines presented here perform write and read operations with wrap around, but with no error checking.

```
INQUEUE:   CMP    SP,DI         ;need to wrap around?
           JNZ    NOADJSP       ;no
           MOV    SP,SI         ;yes, reload SP
NOADJSP:   PUSH   AX            ;write data into queue memory
           JMP    NEXT
OUTQUEUE:  CMP    BP,DI         ;need to wrap around?
           JNZ    NOADJBP       ;no
           MOV    BP,SI         ;yes, reload BP
NOADJBP:   MOV    AX,[BP]       ;read data out of queue memory
           SUB    BP,2          ;adjust read pointer
NEXT:      ---
```

In both routines, AX is used as the queue data register.

Programming Exercise 6.40:   Write a subroutine called SEARCH that will search the data fields of a linked-list for a certain piece of data. The data item to be located is saved in DL.

Programming Exercise 6.41:   Modify the INSERT subroutine so that new nodes are added to the end of the linked-list, not the beginning.

Programming Exercise 6.42:   Show how two stacks can be maintained by saving copies of all stack-related registers.

Programming Exercise 6.43:   Show how a queue can be controlled without the use of PUSH and POP instructions.

6.12   TROUBLESHOOTING TECHNIQUES

This chapter dealt mainly with individual programming modules that performed a single chore. In each case, the subroutine was already developed and tested for you; limitations of the routines were also mentioned.

How is a subroutine created from scratch? The answer is: lots of different ways. Every programmer will use his or her own individual techniques to create a new set of instructions to solve a problem. Some may prefer to use flowcharts or pseudocode, others feel comfortable writing the instructions from scratch. Let us look at one method to create a new subroutine.

First comes the problem. It may be something like this:

> "Write a FAR subroutine called FIND7 that counts the number of times the value 7 appears in an array of 100 word-size integers. The starting address of the array is the label VALUES. Return the result in AL."

This specification is very clear about what needs to be done. We begin by framing the subroutine:

```
FIND7      PROC   FAR
  .
  .
  .
           RET
FIND7      ENDP
```

Next, we add the instructions that will allow 100 passes through the loop.

```
FIND7      PROC   FAR
           MOV    CX,100
AGAIN:
```

```
          .
          .
          .
          LOOP   AGAIN
          RET
FIND7     ENDP
```

Using register CX does not interfere with the requirement that we return the result in register AL, and has the added advantage of being automatically used by the LOOP instruction to control the number of passes.

Now, in addition to initializing CX, we must initialize AL and a pointer to the VALUES array. AL must be initialized to zero, since it represents a count (there really may be zero 7s in the data).

```
FIND7     PROC   FAR
          SUB    AL,AL
          LEA    SI,VALUES
          MOV    CX,100
AGAIN:
          .
          .
          .
          LOOP   AGAIN
          RET
FIND7     ENDP
```

All that remains is the actual loop code. What we want to do, 100 times, is examine a data item from the VALUES array, compare it with 7, and, if equal, increment the AL register. This is done as follows:

```
FIND7     PROC      FAR
          SUB       AL,AL
          LEA       SI,VALUES
          MOV       CX,100
AGAIN:    CMP       BYTE PTR [SI],7
          JNZ       NOTEQ
          INC       AL
NOTEQ:
          .
          .
          .
          LOOP      AGAIN
          RET
FIND7     ENDP
```

Is there anything left to do? Yes, it is very important to adjust the pointer register SI with each pass through the loop, so that the CMP instruction always reads a new data item from the VALUES array. Since the size of each data item in VALUES is a word, SI must be incremented by two each pass.

The final FIND7 subroutine looks like this:

```
FIND7     PROC      FAR
          SUB       AL,AL
          LEA       SI,VALUES
          MOV       CX,100
AGAIN:    CMP       BYTE PTR [SI],7
          JNZ       NOTEQ
```

```
             INC      AL
NOTEQ:       ADD      SI,2
             LOOP     AGAIN
             RET
FIND7        ENDP
```

FIND7 is now ready to be combined with a driver program for testing.

   This step-by-step approach, or one you develop on your own, will become very auto-matic with practice. It requires familiarity with the most basic programming chores: initialization, counting, looping, and comparing. It is worthwhile to invent your own programming problems as well, and then try to write the instructions to solve them.

SUMMARY

In this chapter we examined a number of different applications the 80x86 is capable of per-forming. These applications find widespread use in industrial and commercial settings. In addition, we covered many different techniques, such as code conversion, table lookup, sorting, and mathematical processing with both binary and BCD numbers. The overall idea is to get a sense of how the instructions can be combined to perform any task that we imag-ine. Many more applications are possible and we have only scratched the surface here, but the routines presented in this chapter should serve as a foundation on which to build when you try to write an application of your own.

STUDY QUESTIONS

1. What is an advantage of writing programs with pseudocode?
2. Is it possible for a software driver to fully test a new piece of code?
3. How must the DS and ES segment registers be initialized for the routines presented in Section 6.6?
4. Suppose that the number of integers to sort is fixed to a maximum of 32. Is there a different sorting technique that will sort the input numbers with fewer loops/passes than a bubble sort? If so, explain your technique.
5. How must the format for storing BCD digits be changed to allow for fractional numbers like 0.783 or 457.05?
6. How might exponents be processed in numbers like 35E3 or 2.6E-7?
7. How are round-off errors eliminated by using BCD?
8. Consider the normalization of two numbers, one positive and one negative. Are the exponents adjusted in the same way for each number?
9. In the treatment of BCD numbers, the proposed format contained no provision for representing negative numbers. How might the format be changed to include them?
10. The 5-byte BCD format discussed in this chapter uses 1 byte for a signed exponent and 4 bytes for the mantissa. What is the largest positive integer that can be represented?

11. Why are field terminator characters in a database record useful?
12. When converting from one base to another, do you find any similarities in technique?
13. What does a specification for a subroutine contain?
14. Describe how to develop a subroutine.
15. What is polling?

ADDITIONAL PROGRAMMING EXERCISES

1. Modify the TESTQUAD driver program so that ten values are used to test the QUAD routine. Use a data table in your new program.
2. Write a subroutine called VALDIGIT, which will determine if the ASCII code contained in the lower byte of AL is a digit from '0' to '9'. If so, simply return. If not, jump to ERROR.
3. Modify PACKBCD to include signed BCD numbers. If the first character entered by the user is a minus sign, place 80H into SIGN. If the first character is a number or a plus sign, place 0 into SIGN.
4. Write a driver program for the MAXVAL procedure. Use the following data to test the routine:

DB   5,50,20,77,6,3,22

5. Write a subroutine that will convert the BCD number stored in BCDNUM into a binary number. Return the result in AX.
6. Write a subroutine called STRSIZE that returns the number of characters in a text string. The text string is terminated with an ASCII return character (0DH). The string begins at address TEXTLINE. Return the character count in BX.
7. Write a subroutine to find the average of a block of words that starts at location SAMPLES and whose length, in words, is saved in SIZE. Place the average in AVERAGE.
8. Write a driver program for the SORT routine. Use the following data to test the procedure:

DB   9,1,8,2,7,3,6,4,5

9. Write a subroutine called JUSTIFY that will left justify any BCD number stored in the 4-byte array called BCDIN. For example, if BCDIN contains 00 05 37 19, JUSTIFY must replace BCDIN with 53 71 90 00.
10. Two sample routines for performing BCD multiplication were presented in this chapter. Both routines had limited precision. Build on these two routines by extending their precision. For example, BCDMUL2 was limited to multiplying two-digit numbers. Write a routine that will multiply four-digit BCD numbers, using BCDMUL2 as a subroutine.
11. Write a subroutine called EXPONENT that will return a signed 8-bit exponent in register BL. Inputs to EXPONENT are SI, DI, and AX. SI and DI both point to the exponent byte of the two numbers being divided (with DI pointing to the divisor exponent). AX initially contains an exponent adjustment value (in signed 8-bit format) that must always be added to the generated exponent value.

12. Write a subroutine called ZEROCHECK that examines the BCD number pointed to by register BP and returns with the zero flag set if the BCD number is equal to 0.

13. Write a subroutine called HYPOT that computes the hypotenuse of a triangle whose sides have lengths pointed to by SI and DI. Return the result address in BP. The lengths are stored as words.

14. Find the infinite series for SIN(X) in a calculus book and implement it in a subroutine called SIN. Use EPOWER as an example of how to do this.

15. Implement COS(X) via subroutine COS. Use the following formula as a guide and solve it for COS(X) before writing any code:

$$SIN^2(X) + COS^2(X) = 1$$

Make use of ROOT and SIN in your subroutine.

16. Consider the office complex for Programming Exercise 6.34 and its associated definitions. Suppose for reasons of efficiency that no light may be on continuously for more than 30 minutes after 5 P.M. For example, if an office light is turned on at 6:17 P.M., the computer automatically shuts if off at 6:47. Use the PC's system clock interrupt function to control an automatic shutoff routine for the lights. *Note:* The light may be immediately turned on again (for another 30 minutes) if someone in the office hits the switch again. The automatic shutoff feature ends at 6 A.M.

17. Write a routine called BTOB that takes the 8-bit number in AL and displays its binary equivalent on the screen. For example, if AL contains 4EH, '01001110' is output to the display.

18. Write a routine that converts a Roman-numeral input string (such as "MCMLXXI") and determines the decimal equivalent. Return the result in AX.

19. Repeat Exercise 18 for the opposite conversion, decimal-to-Roman numeral.

20. Write a subroutine that returns with the zero flag set if the input coordinates in AX are inside the rectangle defined by coordinates in BX and CX. Register BX indicates the upper left corner (BH = row, BL = column), and CX the lower right corner.

CHAPTER 7

Programming with DOS and BIOS Function Calls

OBJECTIVES

In this chapter you will learn about:

- The use of DOS and BIOS function calls
- How to read the PC's keyboard
- How to send text to the video display
- The operation of the PC's speaker
- How to control the printer
- The structure of the program segment prefix
- How to use DOS's command-line interface

7.1   INTRODUCTION

The purpose of this chapter is to show you how the hardware components of the PC can be accessed and controlled through the use of DOS and BIOS interrupts. These interrupts require specific registers to be preloaded with various types of data to obtain the desired result. Each interrupt covered has the capability of performing multiple functions, so it is necessary to indicate which function of a particular interrupt is desired. In all cases, the number of the interrupt function to be selected must be placed into register AH *prior* to the interrupt call. For example, to read the computer's system date, AH must be loaded with 2AH, and then an INT   21H issued, as in:

```
MOV  AH,2AH   ;read system date function
INT  21H      ;DOS call
```

As you will see, only a few instructions are needed to read the keyboard or write a character to the video display, because the DOS or BIOS interrupt actually performs all the work. Sample programs are included to show how each interrupt covered may be used in an application. It is important to remember that all sample programs provided in this chapter leave room for improvement. Only the code necessary to show how the interrupt is used and to make the application work is provided. For instance, in some programs that require keyboard input from the user, multiple blanks will not be ignored and, in some cases, only uppercase letters will be accepted. The point of each program is to show you how the interrupt function is used. Programming exercises are included to encourage understanding and modification of each program.

Section 7.2 shows how the PC's keyboard is read. Section 7.3 details the operation of the video display. Creating simple tones with the PC's speaker is the subject of Section 7.4. This is followed by a few simple printer applications in Section 7.5. Section 7.6 discusses the structure of DOS's Program Segment Prefix and how a portion of it is used to obtain information from the DOS command line. A few additional interrupts and their associated applications are presented in Section 7.7. Section 7.8 covers some important troubleshooting techniques.

7.2   USING THE KEYBOARD

If it were possible for you to open the case on your computer's keyboard, you might be surprised at how little circuitry you would find. As a matter of fact, most keyboards manufactured today use a single integrated circuit called a *microcontroller* to scan all the keys and generate the specific codes. For example, when the 'A' key is pressed, the keyboard's internal microcontroller generates the code for 'A' and transmits it to the computer's motherboard in a format known as serial data transmission. This means that the multi-bit key code is sent to the motherboard one bit at a time. Circuitry on the motherboard assembles the serial data back into the proper format, which is commonly referred to as **parallel** data. The motherboard circuitry then generates a specific hardware interrupt to inform DOS that a key has been pressed. DOS contains an interrupt service routine designed to store and manipulate key codes received from the keyboard. This complicated process is transparent to us and performed automatically by DOS. All we have to do as users of the keyboard's information is issue a certain interrupt whenever we want to read the last key pressed. Several different keyboard interrupts are provided for our use.

DOS INT   21H, Function 01H: Wait for Keyboard Input

This interrupt waits for the user to press a key on the keyboard and returns the ASCII code for the key in register AL. The character associated with the pressed key is echoed to the video display. To issue this interrupt, register AH must be loaded with 01H.

Example 7.1:   Trace the following instructions using DEBUG. Because INT   21H is used, use the **p** command to trace each instruction.

```
MOV  AH,1
INT  21
```

You will notice that when you trace the INT   21 instruction, the system will hang until you press a key on the keyboard. Press the '5' key. DEBUG should immediately return with a register display. The contents of AL should be 35H, the ASCII code for '5'.

Some keys on the keyboard do not generate ASCII codes. Examples of these keys are the four arrow keys and the function keys (F1 through F10). These keys generate *extended* ASCII codes (refer to Appendix I). In this case, the value 0 will be returned in AL. To get the extended ASCII key code you must issue a second interrupt. Consider this segment of code:

```
MOV  AH,01H    ;read keyboard with echo
INT  21H
CMP  AL,0      ;extended key?
JNZ  GOTKEY    ;no, plain ASCII in AL
MOV  AH,01H    ;read keyboard again to get extended key code
INT  21H       ;extended key code is returned in AL also
```

Experiment with this interrupt to determine the codes for keys you commonly use.

DOS INT   21H, Function 08H: Console Input Without Echo

This interrupt is similar to the previous one except that the character associated with the pressed key is not echoed to the screen. This interrupt is useful in applications where secret information, such as a password, must be entered. To use this interrupt, register AH must be loaded with 08H. The ASCII key code is returned in register AL as previously discussed.

Example 7.2:   Repeat Example 7.1 using these instructions:

```
MOV  AH,8
INT  21
```

When you press the '5' key this time you will not see it displayed on the screen.

BIOS INT   16H, Function 00H: Read Keyboard Input

This interrupt provides an alternate method of reading the keyboard. To issue the interrupt, register AH must be loaded with 00H. If an ASCII key is pressed, the ASCII code for the key is returned in AL. If an extended key is pressed, a 0 is returned in AL and the code for the extended key is returned in AH. There is no need to call this interrupt twice to get the extended key code. Also, the character for the key is *not* echoed to the video display.

Example 7.3:   Execute the following code using DEBUG's **p** command:

```
MOV   AH,0
INT   16
MOV   AH,0
INT   16
```

At the first INT   16, press the '5' key. You will notice that the key is not echoed to the screen. The register display should show 35H in AL. On the second INT   16, press the up arrow. You should see 4800H in AX. This indicates that AL equals 00 and that the scan code for the up arrow is 48H.

BIOS INT   16H, Function 01H: Read Keyboard Status

This interrupt is useful for determining the state of the keyboard buffer, such as finding out if any keys have been pressed. There are times when we do not want to wait for a key to be pressed on the keyboard, and other times when we do not care what key is pressed (as in *Hit any key to continue . . .*). Register AH must be loaded with 01H to use this interrupt. The values returned are the same as in Function 00H, except for the following: if no key has been pressed (the buffer is empty), the zero flag will be set. If a key has been pressed (the buffer is not empty), the zero flag will be cleared.

The keyboard buffer is used to store the last fifteen keystrokes. You may have noticed that it is possible to *type ahead* of the current DOS command. For instance, put a floppy disk into drive A. Enter DIR A: and press enter. Then immediately enter VER and press enter. Because it takes a few moments for the disk to get up to speed and be read by DOS, the VER command is not echoed to the screen. It is saved in the keyboard buffer. When the DIR A: command completes, the VER command is read out of the buffer and processed. This interrupt actually indicates whether the keyboard buffer is empty or not empty. If there is at least one keystroke stored in the keyboard buffer, the zero flag will be cleared. It is then necessary to use Function 00H to remove the indicated key from the buffer. If the buffer is empty, the zero flag will be set.

BIOS INT   16H, Function 02H: Return Shift Flag Status

This interrupt is useful for determining what control keys have been pressed in the past and the current state of the keyboard. Register AH must be loaded with 02H to use this interrupt. Upon return, register AL contains the following information:

    Bit 7: Insert Active (when high)

    Bit 6: Caps Lock Active (when high)

    Bit 5: Num Lock Active (when high)

    Bit 4: Scroll Lock Active (when high)

    Bit 3: Alt Pressed (when high)

    Bit 2: Ctrl Pressed (when high)

    Bit 1: Left Shift Pressed (when high)

    Bit 0: Right Shift Pressed (when high)

This information is useful when the keyboard is used interactively. Some game programs use keyboard keys to control the game action (for example, left- and right-shift move a starship left and right, respectively, on the screen) and this interrupt provides easy access to a number of keys.

Example 7.4:   Turn Num Lock and Caps Lock on and trace the following instructions with DEBUG's **p** command:

```
MOV  AH,2
INT  16
```

The contents of AL should be 60H, indicating that bits 6 and 5 are set.

Keep in mind that there are many other DOS and BIOS interrupts designed to access and control the keyboard. The five presented here should get you started with your applications.

PASSWORD: A Keyboard Application

The PASSWORD program presented here uses DOS INT 21H, Function 08H to obtain a secret password from the user. The password is not echoed to the screen. If the user enters the password correctly, the program exits to DOS; otherwise, the user must enter the password again. The password is stored in PWTXT and may be any length, but must be terminated by a carriage return.

```
;Program PASSWORD.ASM: Wait until a secret password is entered
;                      Note: Control-C exits immediately
;
        .MODEL SMALL
        .DATA
PMSG    DB      0DH,0AH,'Password: $'
PWTXT   DB      'secret',0DH

        .CODE
        .STARTUP
        LEA    DX,PMSG      ;point to message string
        MOV    AH,9         ;display string function
        INT    21H          ;DOS call
GETPW:  LEA    SI,PWTXT     ;point to password string
GETCH:  MOV    AH,8         ;read keyboard without echo
        INT    21H          ;DOS call
        CMP    AL,[SI]      ;does character match?
        JNZ    GETPW        ;jump if no match found
        INC    SI           ;else, point to next password character
        CMP    AL,0DH       ;was last character a carriage return?
        JNZ    GETCH        ;continue reading password if not
        .EXIT

        END
```

The .STARTUP and .EXIT directives in the PASSWORD.ASM source file are used at the beginning and end of the executable code in the .CODE segment. These directives generate code themselves. .STARTUP produces the instructions that initialize the data and stack segment registers, set up the stack pointer, and clear the interrupt enable flag. .EXIT generates these two instructions:

```
MOV    AH,4CH
INT    21H
```

Function 4CH of DOS's INT   21 H is the function normally used by a program to return control to DOS.

All of the application programs that follow will utilize these directives.

Programming Exercise 7.1:   Write a subroutine that outputs the text string 'Enter a number:' to the display and then reads a two-digit integer, with echo. The value of the integer should be returned by the subroutine in register BL.

Programming Exercise 7.2:   Write a section of code that will loop 1 million times (via a nested loop). While the loop is executing, hit as many keys as you can. This should fill up the keyboard buffer. Use INT   16H, functions 00H and 01H to test the keyboard status and empty the keyboard buffer when the delay loop completes. *Note:* If 1 million loops do not provide at least 5 seconds of delay, use a bigger number.

Programming Exercise 7.3:   Write a routine that will output 'Num Lock On' or 'Num Lock Off' once each time the Num Lock button is pressed. Exit to DOS if 'Q' or 'q' is entered on the keyboard.

Programming Exercise 7.4:   Modify the PASSWORD program so that backspace editing of the password is allowed.

7.3   CONTROLLING THE VIDEO DISPLAY

The video display has two modes of operation: text and graphics. In this section, we will deal only with the text mode of operation.

When the PC was first introduced, it came with a monochrome monitor, which allowed only single-color text characters (white on black). A steady progression of improvements followed, beginning with the CGA (Color Graphics Adapter) display. This display allowed sixteen text colors on a 25 by 80 character screen, with the characters formed out of 8 by 8 blocks of *pixels*. A pixel is a single dot on the video screen. Characters are formed by turning individual pixels on and off. For example, the letter 'A' can be displayed in the following manner:

```
       1 2 3 4 5 6 7 8
  1        X X X
  2      X         X
  3      X         X
  4      X X X X X
  5      X         X
  6      X         X
  7      X         X
  8
```

where columns 1, 7, and 8, and row 8 are blank to allow for character spacing on the screen. The number of pixels that can be placed on a single screen is called the *resolution* of the video display. The CGA display had a maximum resolution of 640 (horizontal) by 200 (vertical) pixels.

After CGA came EGA (for Enhanced Graphics Adapter). This display standard still allowed sixteen-color text, but with higher resolution. Characters were formed out of 8 by 14 blocks of pixels, giving a maximum resolution of 640 by 350. The difference in displayed text was clearly visible, but the need for even higher resolution was still there; and thus, the EGA standard was soon followed by the VGA (Video Graphics Array) standard. Here, the characters are formed out of 8 by 16 blocks of pixels, giving a maximum text resolution of 720 by 400. The quality of text found with the VGA display is far superior to the original monochrome display. Newer standards have been introduced: the SVGA (Super VGA) and XGA (Enhanced Graphics Array). At any rate, all of the display standards mentioned support a text mode containing 25 lines of 80 characters per line. A block of memory is reserved by the display adapter and is referred to as **video RAM.** The ASCII codes for all characters displayed on the screen are stored within the video RAM, along with information that controls the way the characters are displayed. Circuitry on the display card plugged into your computer's motherboard constantly scans the video RAM and generates the appropriate sequence of pixels needed to obtain an intelligent display of characters.

Naturally, controlling the video display directly by manipulating the video RAM would require complex software. For this reason, a number of interrupt functions have been provided to make the programmer's job easier.

DOS INT   21H, Function 02H: Display Output

This interrupt writes a single character to the display screen. The ASCII code for the character must be placed into register DL and AH must be loaded with 02H. Control characters perform their associated function when output. For instance, if DL is loaded with 0DH, a carriage return is output and the entire display screen scrolls up one line. If 08H (backspace) is output the cursor will move backward one position.

DOS INT   21H, Function 09H: Display String

This interrupt has already been used many times in previous chapters. The string to be displayed must have a '$' as its last character. The starting address of the display string must be loaded into register DX and the string must reside within the current data segment. Register AH must be loaded with 09H to use this function.

The remaining interrupts provide a greater amount of control over the video display than DOS INT   21H, Functions 2 and 9. As you will see, many useful and interesting features can be manipulated through BIOS INT   10H.

BIOS INT   10H, Function 00H: Set Video Mode

This interrupt is used to set the video mode of the display. Register AH must be loaded with 00H. Register AL must be loaded with the number associated with the desired video mode. For purposes of this discussion, we will consider only video modes 1 and 3. Video mode 1 allows for 25 lines with 40 characters per line. Video mode 3 allows for 25 lines with 80 characters per line. When this interrupt is issued, the video mode is updated, the screen is cleared, and the cursor is placed in the upper left corner of the video screen.

Example 7.5: Trace the following instructions using DEBUG's **p** command:

```
MOV   AH,0
MOV   AL,1
INT   10
MOV   AH,0
MOV   AL,3
INT   10
```

When you execute the first INT 10 instruction, you should see the screen clear and then fill with a register display containing large letters, because the screen has been placed into 25 by 40 character mode. The second INT 10 instruction places the screen into 25 by 80 character mode. Once again, the screen should clear and fill with a register display. The letters should be much smaller this time.

BIOS INT 10H, Function 0FH: Read Current Video Mode

An interrupt related to Function 00H is Function 0FH. This interrupt is used to read the current video mode selected. This interrupt is useful in application programs to restore the video mode to its original state. Register AH must be loaded with 0FH. Upon return, AL will contain the number of the current video mode.

Example 7.6: Trace the following instructions using DEBUG's **p** command:

```
MOV   AH,0F
INT   10
PUSH  AX
XOR   AL,2
MOV   AH,0
INT   10
POP   AX
MOV   AH,0
INT   10
```

The XOR instruction is used to toggle the video mode from 1 to 3 (or 3 to 1), whichever is selected. The original video mode is restored by the last INT 10 instruction.

BIOS INT 10H, Function 02H: Set Cursor Position

This interrupt is used to place the cursor at any desired position on the text screen. When used with DOS INT 21H, Function 09H: display string, a text string can be output anywhere on the screen.

Register AH must be loaded with 02H for this function. The cursor's new row (0 to 24) and column (0 to 79) numbers must be loaded into registers DH and DL, respectively. In addition, the display page number (0 for all examples presented here) must be placed in register BH.

Example 7.7: The execution of the following instructions will create a strange-looking DEBUG display screen, but will illustrate the point of how the cursor can be positioned. Trace the instructions using DEBUG's **p** command.

```
MOV   AH,2
MOV   DH,1
MOV   DL,1
MOV   BH,0
INT   10
```

You should notice that the cursor jumps up to the top of the screen, causing the new register display to overwrite a previous one.

BIOS INT   10H, Function 03H: Read Current Cursor Position

This interrupt returns the row and column position of the current text cursor. Register AH must be loaded with 03H and register BL with the display page number. The row and column values are returned in DH and DL, respectively.

Example 7.8: Use the following instructions to determine the cursor's position at the time of the interrupt:

```
MOV   AH,3
MOV   BL,0
INT   10
```

Remember that a column value equal to 0 indicates that the cursor is on the left side of the screen.

BIOS INT   10H, Function 0AH: Write Character to Screen

This interrupt allows us to write individual characters (or groups of them) to the screen at the current cursor position. Register AH must be loaded with 0AH. The ASCII code for the screen character must be placed into AL. Register BH contains the display page number. The number of characters written to the screen must be placed in register CX.

Example 7.9: The following instructions write 32 dashes ('–') to the display screen at the current cursor position:

```
MOV  AH,0A
MOV  AL,2D
MOV  BH,0
MOV  CX,20
INT  10
```

This interrupt is useful for doing simple text-based graphics.

BIOS INT 10H, Function 09H: Write Character/Attribute to Screen

Like the previous interrupt, this interrupt may be used to write a character to the display screen. The added feature here is the ability to control the *attribute* of the character. A character's attribute indicates what its foreground and background colors will be, and whether it will blink. The 8 bits of the attribute bytes are assigned as follows:

7	6	5	4	3	2	1	0
Blink	Red	Green	Blue	Intensity	Red	Green	Blue
	└─ Background ─┘				└─ Foreground ─┘		

Thus, an attribute byte of 07H indicates a white character on a black background. An attribute byte of 24H gives a red character on a green background. There are 256 different attribute bytes available. It is interesting to experiment to find ones that are especially noticeable.

This interrupt requires the following register setup:

AH = 09H

AL = ASCII code of character

BH = Display page number

BL = Character's attribute

CX = Number of characters to write

Recall from the beginning of this section that the display card uses video RAM to store a copy of the ASCII codes representing the text shown on the video screen. Each character on the screen actually occupies 2 bytes of video RAM. The first byte is the character's ASCII code. The second byte is the character's attribute. Normally, the 25 by 80 color text mode has its video RAM located at segment B800 (physical address B8000). As an experiment, first fill up the video screen by doing a directory of your hard or floppy disk. Then use DEBUG to view the first few locations of video RAM. Try the following DEBUG command:

```
D B800:0
```

You should notice some correspondence between the characters at the top of the screen and the bytes displayed by the **d** command. If your display screen contains only white on black characters, every other byte should be a 07. Now use DEBUG's **e** command to enter new codes in place of the 07s. Are the characters changing color? If not, your video RAM might be located at a different base address. There are only 4000 bytes associated with a video page ($25 \times 80 \times 2$ bytes for each character), so it may not be easy to locate your video RAM if it is not at the standard address of B800:0. Do not worry, Example 7.10 will allow you to experiment with different attribute bytes and get guaranteed results.

Example 7.10:   The following instructions write three groups of characters to the screen. Can you determine what they will look like before execution?

```
MOV   AH,9
MOV   AL,41
MOV   BH,0
MOV   BL,2
MOV   CX,8
INT   10
MOV   AH,9
MOV   AL,42
MOV   BH,0
MOV   BL,47
MOV   CX,8
INT   10
MOV   AH,9
MOV   AL,43
MOV   BH,0
MOV   BL,81
MOV   CX,8
INT   10
```

Note that you may have to enter CLS at the DOS prompt to get the screen back to normal. Can you imagine what happens when the same color is used for foreground *and* background?

BIOS INT   10H, Function 08H: Read Character/Attribute from Screen

The function this interrupt performs is the opposite of the previous interrupt. The character code and attribute byte at the current cursor position are returned in registers AL and AH, respectively. Register AH must be loaded with 08H and register BH with the display page number to use this function.

BIOS INT   10H, Function 06H: Scroll Current Page Up

You are familiar with the contents of the display screen moving up one line each time you press enter (assuming that the current DOS or DEBUG prompt is at the bottom of the screen). This is accomplished by copying data within the video RAM from one group of locations to another. This process is performed automatically for us by Function 06H. Seven registers are used to control the operation of this interrupt function. Examine the following register requirements:

AH = 06H

AL = Number of rows to scroll up (0 for entire region)

BH = Attribute used (see following explanation)

CH = Row number at top of region

CL = Column number at top left of region

DH = Row number at bottom of region

DL = Column number at bottom right of region

The attribute byte loaded into register BH controls what foreground and background colors will be used in the scroll region when scrolling is completed.

Example 7.11:   The following instructions should create a huge blank square in the middle of the video display. Use DEBUG's **p** command to execute them.

```
MOV    AH,6
MOV    AL,0
MOV    BH,7
MOV    CH,4
MOV    CL,8
MOV    DH,14
MOV    DL,48
INT    10
```

Experiment with other coordinates and scroll values. How can the entire screen be cleared with this interrupt?

SHOWNAME: See Your Name in Lights

The SHOWNAME program displays a text string containing your name (or mine in this case) on the display and assigns different attributes to each letter for a colorful result. A number of interesting addressing modes are used to retrieve the character/attribute data for the video interrupts.

```
;Program SHOWNAME.ASM: Display name on color video screen.
;
        .MODEL SMALL
        .DATA
TEXT    DB    'James L. Antonakos',0
ATT1    DB    1,2,3,4,5,5,6,7,7,8,9,10,11,12,13,14,15,1
ATT2    DB    0EH,1EH,2EH,3EH,4EH,5EH,6EH,7EH,8EH,9EH,0AEH
        DB    0BEH,0CEH,0DEH,0EEH,0FH,1FH,2FH
ROW1    DB    12
ROW2    DB    18
COL     DB    20

        .CODE
        .STARTUP
        MOV    AH,0            ;set video mode
        MOV    AL,3            ;80 x 25 color
        INT    10H             ;video BIOS call
        MOV    AH,2            ;set cursor position
        MOV    BH,0            ;display page number
        MOV    DH,ROW1         ;row number
        MOV    DL,COL          ;column number
        INT    10H             ;video BIOS call
        LEA    BP,ATT1         ;point to first attribute array
        CALL   FAR PTR DISP    ;display first line of video text
        MOV    AH,2            ;set cursor position
        MOV    BH,0            ;display page number
        MOV    DH,ROW2         ;row number
```

```
        MOV    DL,COL           ;column number
        INT    10H              ;video BIOS call
        LEA    BP,ATT2          ;point to second attribute array
        CALL   FAR PTR DISP     ;display second line of text
        .EXIT

DISP   PROC   FAR
        MOV    SI,0             ;set up array pointer
NEXT:  MOV    AL,TEXT[SI]      ;get name character
        CMP    AL,0             ;exit if character is zero
        JZ     EXIT
        MOV    BH,0             ;display page number
        MOV    BL,[BP+SI]       ;get attribute
        MOV    CX,1             ;do 1 character
        MOV    AH,9             ;write character/attribute to screen
        INT    10H              ;video BIOS call
        INC    SI               ;point to next character/attribute
        ADD    DL,2             ;move two columns to the right
        MOV    AH,2             ;set cursor position
        INT    10H              ;video BIOS call
        JMP    NEXT             ;and continue
EXIT:  RET
DISP   ENDP

        END
```

ALLCHAR: Output Entire Video Display Character Set

The ASCII codes we normally use with the video interrupts number from 20H to 7FH.
What happens when a code from 0 to 1FH, or from 80H to 0FFH is used? In this event,
the display card substitutes a special *graphic* character. You have probably seen many of
them before. They are used to draw lines, make boxes, and describe mathematical equations (because many of them contain Greek symbols). The ALLCHAR program displays
a 16 by 16 matrix of characters, showing the entire character set of the video display.

The graphic characters begin halfway down the display.

```
;Program ALLCHAR.ASM: Display ASCII character set on screen.
;
        .MODEL SMALL
        .CODE
        .STARTUP
        MOV    AX,3             ;set video mode to 80x25 color
        INT    10H              ;BIOS call
        MOV    AL,0             ;select char
        MOV    BL,2             ;select green color attribute
        MOV    BH,0             ;page number 0
        MOV    CX,1             ;display one char
        MOV    DH,3             ;select row 3
NEXT:  MOV    DL,13            ;start in column 13
        CALL   VL16             ;write 16 characters to screen
        INC    DH               ;move to new line
        CMP    DH,19            ;have we reached the last line?
        JNZ    NEXT
        .EXIT

VL16   PROC   NEAR
        MOV    SI,16            ;prepare for 16 writes to screen
VOUT:  MOV    AH,2             ;set cursor position
        INT    10H              ;video BIOS call
```

```
             MOV     AH,9            ;write character and attribute
             INT     10H             ;video BIOS call
             ADD     DL,3            ;move right three columns
             INC     AL              ;select next character
             DEC     SI              ;have we done this 16 times yet?
             JNZ     VOUT
             RET
VL16         ENDP

             END
```

XYDRAW: An Interactive Screen Drawing Application

XYDRAW is an electronic sketch pad. The user selects the color (red, green, or blue) of a
star ('*') and draws objects with the star by moving it around with the arrow keys. Single-
letter keyboard commands are used to draw, skip (move the star without writing it to the
screen), change colors, and quit the program. XYDRAW uses many of the keyboard and
video routines previously discussed. Built-in software prevents the user from moving the
star off the display screen.

```
;Program XYDRAW.ASM: Draw simple figures on a color 25 by 80 screen.
;
             .MODEL SMALL
             .DATA
ROW          DB      12      ;initial cursor position
COL          DB      40
COLOR        DB      2       ;start with green
SKIP         DB      1       ;start with skip true (no drawing)
HELP         DB      'D: Draw, S: Skip, Move: Arrow Keys, '
             DB      'R/G/B: Set Color, Q: Quit $'

             .CODE
             .STARTUP
             MOV     AH,0            ;set video mode
             MOV     AL,3            ;80 x 25 color
             INT     10H             ;video BIOS call
             MOV     AH,2            ;set cursor position
             MOV     BH,0            ;display page number
             MOV     DH,24           ;bottom row
             MOV     DL,7            ;column number
             INT     10H             ;video BIOS call
             LEA     DX,HELP         ;set up pointer to help message
             MOV     AH,9            ;display string
             INT     21H             ;DOS call
             MOV     AH,2            ;set cursor position
             MOV     BH,0            ;display page number
             MOV     DH,ROW          ;row number
             MOV     DL,COL          ;column number
             INT     10H             ;video BIOS call
READKEY:     MOV     AH,0            ;read keyboard
             INT     16H             ;BIOS call
             CMP     AL,0            ;check scan code?
             JZ      CSC
             CMP     AL,'a'          ;test for lowercase
             JC      NOLC
             CMP     AL,'z'+1
             JNC     NOLC
             AND     AL,0DFH         ;convert to uppercase
NOLC:        CMP     AL,'Q'          ;quit?
             JNZ     X1
```

```
            JMP     EXIT
X1:         CMP     AL,'R'          ;switch to red?
            JZ      SETRED
            CMP     AL,'G'          ;switch to green?
            JZ      SETGREEN
            CMP     AL,'B'          ;switch to blue?
            JZ      SETBLUE
            CMP     AL,'S'          ;skip?
            JZ      SETSKIP
            CMP     AL,'D'          ;draw?
            JZ      SETDRAW
            JMP     READKEY
SETRED:     MOV     COLOR,4         ;set color to red
            JMP     READKEY
SETGREEN:   MOV     COLOR,2         ;set color to green
            JMP     READKEY
SETBLUE:    MOV     COLOR,1         ;set color to blue
            JMP     READKEY
SETSKIP:    MOV     SKIP,1          ;skip is true
            JMP     READKEY
SETDRAW:    MOV     SKIP,0          ;skip is false
            JMP     READKEY
CSC:        CMP     AH,48H          ;up arrow?
            JZ      GOUP
            CMP     AH,50H          ;down arrow?
            JZ      GODOWN
            CMP     AH,4BH          ;left arrow?
            JZ      GOLEFT
            CMP     AH,4DH          ;right arrow?
            JZ      GORIGHT
            JMP     READKEY         ;did not get a valid input
GOUP:       CMP     ROW,0           ;ignore if first row
            JNZ     GOUP2
            JMP     READKEY
GOUP2:      SUB     ROW,1
            JMP     SETCUR          ;reposition cursor
GODOWN:     CMP     ROW,23          ;ignore if last row
            JNZ     GODOWN2
            JMP     READKEY
GODOWN2:    ADD     ROW,1
            JMP     SETCUR
GOLEFT:     CMP     COL,0           ;ignore if first column
            JNZ     GOLEFT2
            JMP     READKEY
GOLEFT2:    SUB     COL,1
            JMP     SETCUR
GORIGHT:    CMP     COL,79          ;ignore if last column
            JNZ     GORIGHT2
            JMP     READKEY
GORIGHT2:   ADD     COL,1
SETCUR:     MOV     AH,2            ;set cursor position
            MOV     BH,0            ;display page number
            MOV     DH,ROW          ;row number
            MOV     DL,COL          ;column number
            INT     10H             ;video BIOS call
            CMP     SKIP,1          ;skip if true
            JZ      NOC
            MOV     AL,'*'          ;write colored * to screen
            MOV     BL,COLOR
            MOV     CX,1
            MOV     AH,9
```

```
        INT    10H              ;video BIOS call
NOC:    JMP    READKEY
EXIT:   .EXIT

        END
```

Programming Exercise 7.5: Write a subroutine that will write a red X into the center of the display screen.

Programming Exercise 7.6: Write a subroutine that will make the background blue for all characters on the first line of the display.

Programming Exercise 7.7: Write a routine that will draw the following diagram on the display screen:

You must make use of the special graphic characters to do this. Examine the output of the ALLCHAR program for hints.

Programming Exercise 7.8: Write a routine that outputs the following sequence of characters in the *same* screen position (with a short delay between each write to the screen):

$$- \quad \backslash \quad | \quad /$$

The sequence of characters should simulate a spinning bar.

Programming Exercise 7.9: Modify the SHOWNAME program so that the user is able to enter his or her name before it is displayed in color.

Programming Exercise 7.10: Modify the ALLCHAR program so that hexadecimal row and column numbers are displayed with the character set matrix.

Programming Exercise 7.11: Modify the XYDRAW program to allow for the following: (1) additional colors, (2) user selectable symbol for drawing, (3) previous move erasure, and (4) screen erase option.

Programming Exercise 7.12: Modify the XYDRAW program so that an *instant replay* option is available. In other words, after a screen drawing has been completed, pressing the instant replay key will cause the screen to blank and be redrawn automatically.

7.4   CONTROLLING THE SPEAKER

The motherboard contains circuitry capable of turning a small speaker on and off. By turning the speaker on and off at different rates, we can generate audio tones of various frequencies. DOS uses this technique to generate a short beep when the keyboard buffer is full. The beep is a warning to stop typing. Some application programs use the speaker in this fashion to indicate that an incorrect key has been pressed. Example 7.12 gives a simple way to beep the speaker.

Example 7.12:   Trace the following instructions using DEBUG's **p** command:

```
MOV   DL,7
MOV   AH,2
INT   21
```

ASCII code 07H is the code for BEL (bell), and will cause the speaker to beep when it is output.

Making the Speaker Beep

The duration and tone of the speaker beep generated in Example 7.12 is fixed. To change these parameters, it is necessary to manipulate a special timer circuit on the motherboard. Figure 7.1 shows a simplified schematic diagram of the motherboard speaker circuitry. The speaker is controlled by the signals output by two integrated circuits. The first chip

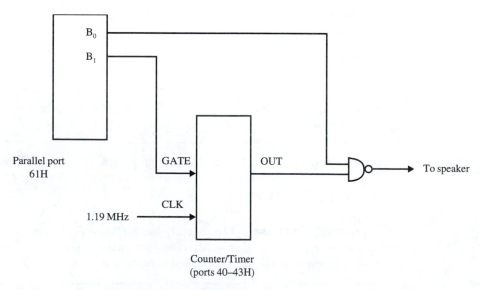

FIGURE 7.1   Simplified speaker control circuitry

operates as a parallel output port, with two outputs, B_0 and B_1, used to enable the speaker and the second chip, a counter/timer. As you can see, the B_1 output is connected to the GATE input of the counter/timer. The GATE input must be high for the counter/timer to operate. The B_0 output feeds one input of a two-input NAND gate. The other input of the NAND gate is connected to the counter/timer's OUT signal. The counter/timer will divide the 1.19-MHz CLK signal by a preloaded value, and output the resulting signal on OUT. When B_0 and B_1 are both high, the signal on OUT will generate a specific frequency tone on the speaker. For example, to generate a 1,000-Hz tone on the speaker, the counter/timer must divide the 1.19-MHz CLK input by 1,190. In order for us to generate tones of any frequency and duration, we must be able to load numbers into the counter/timer (for frequency division) and control the state of the B_0 and B_1 outputs.

Loading the counter/timer is a two-step process. Because the counter/timer actually contains three internal counters (Counter 0 through Counter 2), we must first select the specific counter that controls the speaker. This is Counter 2. Each counter is also capable of operating in either binary or BCD mode. There are also six different ways to use each counter.

To select the appropriate counter and set its various modes of operation, it is necessary to output a *control word* to the counter/timer chip. The control word needed for speaker operation is 0B6H. In addition, it is also necessary to output a counter value, so that the 1.19-MHz CLK signal can be divided. The following instructions show how Counter 2 is selected and loaded with the number 1,190:

```
MOV   AL,0B6H     ;select Counter 2, mode 3 binary counting
OUT   43H,AL      ;output control word to counter/timer
MOV   CX,1190     ;load counter value into CX
MOV   AL,CL       ;output lower byte of counter value
OUT   42H,AL
MOV   AL,CH       ;output upper byte of counter value
OUT   42H,AL
```

Once this is done, the counter/timer will wait for the GATE input to go high. When this happens, the signal at OUT will have a frequency of 1000 Hz. To make the GATE input high, we must output logic 1s on the B_0 and B_1 outputs. This is accomplished by the following three instructions:

```
IN    AL,61H      ;read current state of parallel port
OR    AL,3        ;set B0 and B1
OUT   61H,AL      ;output new state to parallel port
```

The duration of the generated tone is a function of how long we wait before resetting the bits on the parallel port. A delay subroutine can be used to produce the desired delay. When completed, the speaker is shut off like so:

```
IN    AL,61H      ;read current state of parallel port
AND   AL,0FCH     ;clear B0 and B1
OUT   61H,AL      ;output new state to parallel port
```

Now that we have the instructions necessary to turn the speaker on and off and generate a specific frequency, we can build an application to do something useful with the speaker.

TONES: A Simple Music Machine

The TONES program uses a data table to hold divisor/duration number pairs. A duration value of 1 has been predefined as a *pause* indicator to place gaps between the tones generated at the speaker. The length of the data table is not fixed, so a music sequence can be as long as desired. It is necessary to place a 0 only at the end of the data table. TONES plays a sequence of notes by outputting divisor values to the counter/timer one at a time.

```
;Program TONES.ASM: Play a simple sequence of tones on PC's speaker.
;
        .MODEL SMALL
        .DATA
;                               Note: Output frequency equals
;                               1.19 MHz / counter-value (divisor)
;       divisor  duration
TONES DW 1190,    2500      ;1st tone, freq: 1 KHz, duration: 2500
      DW 20,      1         ;pause
      DW 10000,   5000      ;2nd tone, freq: 119 Hz, longer duration (5000)
      DW 20,      1         ;pause
      DW 19833,   7500      ;3rd tone, freq: 60 Hz, 3rd longest duration
      DW 20,      1         ;pause
      DW 40000,   10000     ;4th tone, freq: 29 Hz, longest duration
      DW 20,      1         ;pause
      DW 238,     5000      ;5th tone, freq: 5 KHz, medium duration
      DW 0                  ;end-of-tones word

        .CODE
        .STARTUP
        MOV    SI,0             ;set up pointer to tones table
NEXT:   MOV    CX,TONES[SI]     ;read frequency division value
        CMP    CX,0             ;if frequency divisor = 0 we exit
        JZ     EXIT
        MOV    DX,TONES[SI+2]   ;else, read duration value
        ADD    SI,4             ;point to next divisor/duration pair
        CMP    DX,1             ;if duration = 1 we pause
        JZ     PAUSE
        CALL   SPKRON           ;turn speaker on
        CALL   LDTIMER          ;set speaker frequency
        CALL   DELAY            ;wait for chosen duration
        CALL   SPKROFF          ;turn speaker off
        JMP    NEXT             ;go load next divisor/duration pair
PAUSE:  MOV    DX,CX
PAUSE2: MOV    CX,50000         ;pause is a multiple of 50000 ticks
TICK:   LOOP   TICK
        DEC    DX               ;decrement pause counter
        JNZ    PAUSE2           ;until it is 0
        JMP    NEXT             ;go load next divisor/duration pair
EXIT:   .EXIT

SPKRON  PROC   NEAR
        IN     AL,61H           ;read current state of port 61h
        OR     AL,3             ;set speaker control bits
        OUT    61H,AL           ;output new state
        RET
SPKRON  ENDP

SPKROFF PROC   NEAR
        IN     AL,61H           ;read current state of port 61h
        AND    AL,0FCH          ;clear speaker control bits
        OUT    61H,AL           ;output new state
        RET
SPKROFF ENDP

DELAY   PROC   NEAR
WAIT1:  MOV    CX,500           ;a simple 500-loop delay
WAIT2:  LOOP   WAIT2
        DEC    DX               ;give another 500-loop delay
```

```
           JNZ    WAIT1              ;unless dx went to zero
           RET
DELAY      ENDP

LDTIMER PROC   NEAR
           MOV    AL,0B6H            ;timer 2 control word
           OUT    43H,AL
           MOV    AL,CL              ;output lower byte of count
           OUT    42H,AL
           MOV    AL,CH              ;output upper byte of count
           OUT    42H,AL
           RET
LDTIMER ENDP

           END
```

Programming Exercise 7.13:   Write a routine that will ramp the frequency of the speaker up from 1,000 Hz to 5,000 Hz in 100-Hz steps.

Programming Exercise 7.14:   Write a program that simulates a musical keyboard. The number keys (0 through 9) should each generate a different tone when pressed.

Programming Exercise 7.15:   Modify the TONES program so that the speed at which a melody is played is controlled by the F (for faster) and S (for slower) keys.

7.5   CONTROLLING THE PRINTER

The ability to send data directly to the printer is provided by DOS through its PRINT utility. PRINT is not a built-in component contained within COMMAND.COM. Instead, a program called PRINT.EXE is supplied with DOS that contains the ability to manage and print files.

Unfortunately, there is no way to access the PRINT command from within a custom application. For this reason, a number of interrupt functions are provided to allow the programmer access to the printer. With their use, it is possible to print information that may not be acceptable to DOS's PRINT utility. Or, if a desired printing format is not available with PRINT, it may be possible to implement it via the DOS and BIOS printer interrupts.

DOS INT   21H, Function 05H: Printer Output

This interrupt function outputs the ASCII code contained within register DL to the printer. Register AH must be loaded with 05H to use it.

Example 7.13:   It is often necessary to output a form-feed command to the printer to advance the printed page to the beginning of the next page. This is useful when a **Print-Screen** has just been performed. Because Print-Screen outputs only 25 lines of text to the printer, there is no possibility of the printer advancing to the next page by itself. The program shown below is a short and simple way of sending a form-feed code to the printer.

```
;Program FORMFEED.ASM: Output form-feed code to the printer.
;
        .MODEL SMALL
        .CODE
        .STARTUP
        MOV   AH,5          ;printer output function
        MOV   DL,0CH        ;ASCII form-feed code
        INT   21H          ;DOS call
        .EXIT

        END
```

Now, after doing a Print-Screen (or other printer operation), you can simply enter FORMFEED at the DOS prompt to advance to the next page. There is no need to touch the printer at all.

BIOS INT   17H, Function 00H: Print Character

This interrupt is similar to DOS INT   21H, Function 05H with the exception that it returns a status code (possibly indicating that something has gone wrong with the printer). For example, the FORMFEED program described in Example 7.13 will generate the following error if it is used when the printer's power is off:

```
Write fault error writing device PRN
Abort, Retry, Ignore, Fail?
```

This new interrupt will detect a printer error (if there is one) and return a status code in register AH. To use the interrupt, AH must first be loaded with 0. The ASCII character code sent to the printer must be placed in register AL. Also, the DX register must be loaded with the printer number (0 through 2), because DOS supports more than one printing device.

The error code returned in AH has the following format:

Bit 7: Printer not busy (when high)

Bit 6: Acknowledgment from printer (when high)

Bit 5: Out of paper (when high)

Bit 4: Printer selected (when high)

Bit 3: I/O error (when high)

Bit 2: Reserved

Bit 1: Reserved

Bit 0: Time-out error (when high)

Example 7.14: The following section of code shows how the status code may be checked for a paper-out condition during printing:

```
MOV   AL,[SI]      ;get next character to print
MOV   AH,0         ;print character function
INT   17H          ;BIOS call
TEST  AH,20H       ;test paper-out status bit
JNZ   PAPEROUT     ;jump if bit was high
```

Other status bits can be checked by changing the mask used in the TEST instruction.

BIOS INT   17H, Function 01H: Initialize Printer

This function requires AH to be loaded with 01H and DX with the printer number. Upon return, AH contains status information identical to that described in Function 00H.

Example 7.15: With the printer turned off, a status code of C8H is returned by Function 01H. This indicates the following conditions:

Printer not busy

Acknowledgment from printer

I/O error

With the printer turned on, a status code of 90H is returned. This indicates the following conditions:

Printer not busy
Printer selected

BIOS INT   17H, Function 02H: Read Printer Status

This function returns the status of the printer without affecting it in any way (like Initialize Printer does). Register AH must be loaded with 02H and DX with the printer number. Again, the status code is defined as it was in Function 00H.

Example 7.16: With the printer turned on but not selected (off-line), the following status code is returned: 00H. Because no bits in the status byte are high, we conclude that the printer is *not* selected (if it were, then bit 4 would be high).

PSTAT: Display Printer Status

The PSTAT program uses INT   17H, Function 02H to read the status of printer 0 and reports what it finds with text messages. The status bits are tested in order from LSB to MSB.

```
;Program PSTAT.ASM: Display printer status.
;
        .MODEL SMALL
        .DATA
```

```
PMSG    DB      'Status of Printer 0:',0DH,0AH,'$'
M1      DB      0DH,0AH,'Time-out Error$'
M2      DB      0DH,0AH,'Reserved$'
M3      DB      0DH,0AH,'Reserved$'
M4      DB      0DH,0AH,'I/O Error$'
M5      DB      0DH,0AH,'Printer Selected$'
M6      DB      0DH,0AH,'Out of Paper$'
M7      DB      0DH,0AH,'Acknowledge$'
M8      DB      0DH,0AH,'Printer Not Busy$'
MTAB    DW      M1,M2,M3,M4,M5,M6,M7,M8

        .CODE
        .STARTUP
        LEA     DX,PMSG         ;set up pointer to initial message
        MOV     AH,9            ;display string
        INT     21H             ;DOS call
        MOV     AH,2            ;get printer status
        MOV     DX,0            ;of printer 0
        INT     17H             ;BIOS call
        MOV     CX,8            ;set up loop counter
        LEA     SI,MTAB         ;set up pointer to message address table
AGAIN:  SHR     AH,1            ;put LSB into Carry flag
        PUSH    AX              ;save status
        JNC     NEXT            ;skip message if LSB was clear
        MOV     DX,[SI]         ;otherwise, get message address
        MOV     AH,9            ;display string
        INT     21H             ;DOS call
NEXT:   POP     AX              ;restore status
        ADD     SI,2            ;advance to next message address
        LOOP    AGAIN           ;repeat until done
        .EXIT

        END
```

PTEST: Output a Page of Text to the Printer

This program outputs 51 lines of text to the printer and finishes with a form-feed. Each line of text is as follows:

```
!"#$%&'()*+,-./0123456789:;<=>?@A  . . .  Z[\]^_`abcdefghijklmno
```

which is simply a string of successive, printable ASCII characters.

```
Program PTEST.ASM: Output a page of ASCII text to the printer.
;
        .MODEL SMALL
        .CODE
        .STARTUP
        MOV     CX,51           ;prepare loop counter for 51 lines of text
        JMP     CRLF            ;start with a carriage return, line feed
AGAIN:  MOV     DL,20H          ;first character to print is a blank
TOPTR:  MOV     AH,5            ;output character to printer
        INT     21H             ;DOS call
        INC     DL              ;advance to next ASCII character
        CMP     DL,70H          ;are we at the last character?
        JNZ     TOPTR           ;jump if we are not
CRLF:   MOV     DL,0DH          ;otherwise, load a carriage return
        MOV     AH,5            ;output character to printer
        INT     21H             ;DOS call
        MOV     DL,0AH          ;load a line feed
```

```
          MOV    AH,5            ;output character to printer
          INT    21H             ;DOS call
          LOOP   AGAIN           ;and repeat for a full page of text
          MOV    DL,0CH          ;load a form feed character
          MOV    AH,5            ;output character to printer
          INT    21H             ;DOS call
          .EXIT

          END
```

PRINTMSG: A Text-String Printing Application

PRINTMSG allows the user to enter a text string of up to 80 characters. The text string is then output to the printer. This application program is useful for adding titles to the top of pages where a screen dump is being performed. For example, in a tutorial on using the DIR command, PRINTMSG will allow you to put a title such as:

```
          Example Hard-disk Directory Listing
```

on the first line of the page. Then, after using DIR, the screen can be dumped to the printer with **Print-Screen.** It is also useful for adding comments to the printed page.

```
;Program PRINTMSG.ASM: Print a user-supplied text string.
;
          .MODEL SMALL
          .DATA
QMSG      DB     'Enter your text string below...',0DH,0AH,'$'
TMSG      DB     80 DUP(?)

          .CODE
          .STARTUP
          LEA    DX,QMSG         ;point to instruction string
          MOV    AH,9            ;display string
          INT    21H             ;DOS call
          LEA    DI,TMSG         ;point to text string buffer
GETCH:    MOV    AH,1            ;get character from standard input
          INT    21H             ;DOS call
          MOV    [DI],AL         ;save character in buffer
          INC    DI              ;point to next buffer position
          CMP    AL,0DH          ;was last character a carriage return?
          JNZ    GETCH           ;jump if no match
          LEA    SI,TMSG         ;point to text string buffer again
TOPTR:    MOV    DL,[SI]         ;get a buffer character
          MOV    AH,5            ;output character to printer
          INT    21H             ;DOS call
          INC    SI              ;point to next character
          CMP    DL,0DH          ;was last character a carriage return?
          JNZ    TOPTR           ;jump if no match
          MOV    DL,0AH          ;send a line-feed too
          MOV    AH,5            ;output character to printer
          INT    21H             ;DOS call
          .EXIT

          END
```

Programming Exercise 7.16: Write a program that automatically outputs your name to the printer.

Programming Exercise 7.17:   Write a program that will print the words GOOD JOB in large letters. For example, the word GOOD should look something like this:

```
    G G G           O O O           O O O         D D D D
  G       G       O       O       O       O       D       D
  G               O       O       O       O       D       D
  G     G G G     O       O       O       O       D       D
  G         G     O       O       O       O       D       D
  G         G     O       O       O       O       D       D
    G G G           O O O           O O O         D D D D
```

Programming Exercise 7.18:   Write a program that will accept a text string of 80 characters or less, and print out each character 40 times on each line, with the characters centered in the middle of the page. For example, if the user enters 'Hello!', the printer output should look like this:

```
HHHHHHHHHHHHHHHHHHHHHHHHHHHHHHHHHHHHHHHH
eeeeeeeeeeeeeeeeeeeeeeeeeeeeeeeeeeeeeeee
llllllllllllllllllllllllllllllllllllllll
llllllllllllllllllllllllllllllllllllllll
oooooooooooooooooooooooooooooooooooooooo
!!!!!!!!!!!!!!!!!!!!!!!!!!!!!!!!!!!!!!!!
```

7.6   USING THE COMMAND-LINE INTERFACE

You probably are familiar with DOS applications that allow file names or other information to be entered immediately after the program's name on the command line. For example, the screen editor EDIT that comes with DOS versions 5.0 and up allows a file name to be supplied, as in:

```
C> EDIT REPORT.TXT
```

DOS has built-in software that scans the command line with the intention of finding any of the following:

*   A DOS command, such as DIR, TYPE, or CD
*   An executable file name (with .EXE and .COM optional)
*   A batch file name (with .BAT optional)
*   A new drive letter (A: or C: to switch drives)

In the case of the EDIT REPORT.TXT command line, DOS will determine that EDIT refers to the EDIT.COM program, and load that program into memory for execution. What happens to the REPORT.TXT characters? DOS will recognize that the letters in REPORT.TXT are required by the EDIT program, and will use them during the program load process when it builds the *Program Segment Prefix* for EDIT.

The Program Segment Prefix

Any executable program (.EXE or .COM file) will have an associated Program Segment Prefix constructed for it before it loads into memory for execution. The Program Segment Prefix contains 256 bytes of data and is created automatically by DOS. The *prefix* term refers to the fact that the 256-byte block of memory reserved for the Program Segment Prefix is loaded into memory immediately before the code for the executing program.

The Program Segment Prefix contains all the information the executing program might need to know while it is in memory. For example, the REPORT.TXT characters that followed EDIT are included in the Program Segment Prefix in a number of specific places. This allows the EDIT program access to the entire DOS command line structure. The structure of the Program Segment Prefix is included here for reference and can be viewed with DEBUG's **d** command.

Program Segment Prefix Structure

Hex Offset	Description
0–1	INT 20H
2–3	Top of memory
4–9	Reserved
A–B	Terminate address IP
C–D	Terminate address CS
E–F	Crtl-break exit address IP
10–11	Crtl-break exit address CS
12–13	Critical error exit address IP
14–15	Critical error exit address CS
16–4F	Reserved
50–51	INT 21H
52–5B	Reserved
5C–6B	Unopened standard FCB1
6C–7F	Unopened standard FCB2
80	Parameter length
81–FF	Command tail parameters

The designers of DOS provided alternate techniques for issuing DOS calls and for terminating a program through the use of information contained within the Program Segment Prefix. For example, instead of using INT   21H in your program, you could use a long call to address 50H in the Program Segment Prefix (CALL FAR 50H). The FCB areas are *File Control Blocks* and will be explained in Chapter 9. The area that we will concentrate on in this discussion is the second half of the Program Segment Prefix.

Scanning the Command Tail

In this section we will concentrate on one portion of the Program Segment Prefix referred to as the *parameter area*. The parameter area can contain up to 127 characters of text and begins at offset 81H in the Program Segment Prefix. Example 7.17 will illustrate this point.

Example 7.17:   Examine the ECHO.ASM program that follows.

```
;Program ECHO.ASM: Echo command-line text to screen.
;
        .MODEL SMALL
        .CODE
        .STARTUP
START:  MOV   BL,ES:[80H]  ;get character count from PSP
        CMP   BL,0          ;are there any characters?
        JZ    EXIT
        MOV   SI,81H        ;set up pointer to command tail
EKHO:   MOV   DL,ES:[SI]    ;get command-tail character
        MOV   AH,2          ;display output
        INT   21H           ;DOS call
        INC   SI            ;advance to next character
        DEC   BL            ;finished?
        JNZ   EKHO          ;repeat until done
EXIT:   .EXIT

        END
```

The ECHO program echos all text on the command line following the name of the program. For instance, entering this command line:

```
ECHO Hello!
```

will cause

```
Hello!
```

to be output to the display screen. Any text entered after ECHO will be sent to the display, so even a nonsense command like:

```
ECHO My       desk is cluttered                with papers...
```

will echo the text

```
My       desk is cluttered             with papers...
```

DOS does not attempt to format the remaining text (referred to from now on as the *command tail*) by removing multiple blanks. What you see is what you get in the parameter area.

The ECHO program works by reading the command-tail characters out of the parameter area one by one and outputting them to the screen. The number of characters in the command tail, excluding the terminating carriage return, is stored as a byte at offset 80H in the Program Segment Prefix. We can examine the entire Program Segment Prefix by executing ECHO with DEBUG. The following printout shows how DEBUG is used to do this.

```
D:\8088-3E\FILES\CH7>debug echo.exe Hello!
-r
AX=0000  BX=0000  CX=0034  DX=0000  SP=0000  BP=0000  SI=0000  DI=0000
DS=2421  ES=2421  SS=2431  CS=2431  IP=0000   NV UP EI PL NZ NA PO NC
2431:0000 BA3424        MOV     DX,2434
-d 2421:0 ff
2421:0000  CD 20 00 A0 00 9A F0 FE-1D F0 4F 03 6A 1E 8A 03   . ........O.j...
2421:0010  6A 1E 17 03 6A 1E 59 1E-01 03 01 00 02 FF FF FF   j...j.Y.........
2421:0020  FF FF FF FF FF FF FF FF-FF FF FF 17 24 4C 01       ............$L.
2421:0030  2A 23 14 00 18 00 21 24-FF FF FF FF 00 00 00 00   *#....!$........
2421:0040  06 16 00 00 00 00 00 00-00 00 00 00 00 00 00 00   ................
2421:0050  CD 21 CB 00 00 00 00 00-00 00 00 00 00 48 45 4C   .!...........HEL
2421:0060  4C 4F 21 20 20 20 20 20-00 00 00 00 00 20 20 20   LO!      .....
2421:0070  20 20 20 20 20 20 20 20-00 00 00 00 74 6F 20 6E           ....to n
2421:0080  07 20 48 65 6C 6C 6F 21-0D 20 20 48 65 6C 6C 6F   . Hello! . Hello
2421:0090  21 20 0D 20 20 20 20 20-20 20 20 20 20 20 20 20   ! .
2421:00A0  20 20 20 20 20 20 20 20-20 20 20 20 20 20 20 20
2421:00B0  20 20 20 20 20 20 20 20-20 20 20 20 20 20 20 0D                  .
2421:00C0  3F 0D 0A 20 30 30 32 45-20 20 37 35 20 46 34 09   ?.. 002E  75 F4.
2421:00D0  09 09 20 20 20 20 20 20-20 20 4A 4E 5A 20 20 20   ..        JNZ
2421:00E0  45 4B 48 4F 20 20 20 20-20 20 20 20 3B 72 65 70   EKHO        ;rep
2421:00F0  65 61 74 20 75 6E 74 69-6C 20 64 6F 6E 65 0D 0A   eat until done..
-q
```

First notice how DEBUG is invoked. The DOS command line:

```
debug echo.exe Hello!
```

causes DEBUG to build a Program Segment Prefix for ECHO.EXE, place it into memory, then load ECHO.EXE. With an .EXE file, the DS and ES segment registers will be loaded with the starting address of the Program Segment Prefix. This address is 2421 in the example. Because the Program Segment Prefix is 256 bytes long (100H), the actual code of the ECHO program gets loaded into memory beginning at address 2431. This is indicated by the value of the CS register in the register display.

By displaying memory in the range 2421:0 to 2421:FF we can view the entire Program Segment Prefix. The information in locations 0 through 7FH are used for special functions by DOS and may not be altered. This portion of the Program Segment Prefix is useful during file operations.

Note the byte value 7 stored at address 80H in the Program Segment Prefix. This indicates the length of text contained in the parameter area. The command tail Hello! consists of 6 characters and we must also count the blank that separates the program name from the command tail. These characters are shown in boldface. The additional hellos found in the printout are a result of DOS trying to interpret Hello! as a file name.

Examine the ECHO program again. Notice how the length of the command tail is loaded into register BL by the MOV   BL,ES:[80H] instruction. Because the ES register automatically contains the address of the Program Segment Prefix, we need only supply the offset (80H in this case) to get at the desired information. The text of the command tail is accessed in a similar fashion by the MOV   DL,ES:[SI] instruction. Because SI is initialized to 81H, it begins with the first character in the command tail. Note that location 80H will contain a 0 if nothing is entered after the program name.

We can use the command tail to perform many different types of useful work. The programs that follow give just a few examples of how useful the parameter area can be.

DECTOHEX: A Decimal to Hexadecimal Conversion Application

This program converts a decimal number supplied on the command line into its hexadecimal equivalent. The input number must be in the range 0 to 65,535. An sample execution is as follows:

```
C> DECTOHEX  1000
03E8
```

Note that only the first number in the command tail will be converted.

```
;Program DECTOHEX.ASM: Convert decimal command-line input into hexadecimal.
;
          .MODEL SMALL
          .DATA
IDMSG    DB    'Invalid character found.$'
TLMSG    DB    'Input exceeds 65535.$'

          .CODE
          .STARTUP
          MOV   SI,81H        ;set up pointer to command tail
SKB:      MOV   BL,ES:[SI]    ;get a character
          INC   SI            ;and point to the next one
          CMP   BL,20H        ;is it a blank?
          JZ    SKB           ;jump if so
          CMP   BL,0DH        ;is it a carriage return?
          JZ    EXIT          ;jump if so
          MOV   AX,0          ;clear result
NEXT:     CMP   BL,'0'        ;let's make sure we have a digit
          JC    NOT09         ;jump if less than '0'
          CMP   BL,'9'+1
          JNC   NOT09         ;jump if greater than '9'
          SUB   BL,30H        ;remove ASCII bias
          MOV   CX,10         ;prepare for decimal multiply
          MUL   CX            ;adjust temp result
          CMP   DX,0          ;did we exceed 65535?
          JNZ   TOOBIG        ;jump if so
          MOV   BH,0
          ADD   AX,BX         ;add new digit to result
          JC    TOOBIG        ;jump if result needs 17 bits
          MOV   BL,ES:[SI]    ;get new character
          INC   SI            ;point to next character
          CMP   BL,20H        ;found end of number?
          JZ    SHOW          ;jump if so
          CMP   BL,0DH        ;found end of number?
          JZ    SHOW          ;jump if so
          JMP   NEXT          ;else, continue processing
NOT09:    LEA   DX,IDMSG      ;output invalid digit message
EROR:     MOV   AH,9          ;display string
          INT   21H           ;DOS call
          JMP   EXIT          ;and quit
TOOBIG:   LEA   DX,TLMSG      ;output number too large message
          JMP   EROR
SHOW:     MOV   BL,4          ;set up loop counter
AGAIN:    MOV   CL,4          ;set rotate count
          ROL   AX,CL         ;rotate one nybble
          PUSH  AX            ;save result
          AND   AL,0FH        ;mask off upper nybble
          ADD   AL,30H        ;add ASCII bias
          CMP   AL,'9'+1      ;test for alpha fixup
          JC    NOADD
```

```
            ADD    AL,7            ;add alpha bias
NOADD:      MOV    DL,AL           ;prepare for character output
            MOV    AH,2            ;output hexadecimal digit
            INT    21H             ;DOS call
            POP    AX              ;get result back
            DEC    BL              ;have we output 4 digits?
            JNZ    AGAIN           ;jump if we have not
EXIT:       .EXIT

            END
```

CRYPT: Encoding/Decoding Secret Messages

UWTLWFRRNSL NX KZS.

This discussion begins with an encrypted message. Can you determine what the secret message is? Through trial and error you will probably figure out the message in a few minutes. The technique used to encrypt the message is referred to as the *Caesar shift,* which is one of the simplest encryption techniques available. The letters in a message are shifted through the alphabet one or more positions. The sample message presented here has been shifted five letters forward in the alphabet. Thus, an A becomes an encrypted F, a B becomes a G, and so on. Can you figure out the message now? The CRYPT program presented here requires a specific syntax for its command line. The program will tell you what it is if you run it with a blank command tail, as follows:

```
C> CRYPT
Usage: CRYPT -<D or E><text>
```

The <D or E> portion selects decode or encode. The <text> portion is what will be encrypted or decrypted. The – is merely to enforce syntax and illustrate within the program how specific command-line parameters are scanned. CRYPT always uses a shift value of 5 to encode or decode a message.

Running the original secret message with the D option gives the following result:

```
C> CRYPT -D UWTLWFRRNSL NX KZS
PROGRAMMING IS FUN
```

This is, of course, the author's biased opinion.

```
;Program CRYPT.ASM: Encode and decode text strings.
;
            .MODEL SMALL
            .DATA
EMSG      DB    0DH,0AH,'Usage: CRYPT -<D or E><text>',0DH,0AH,'$'

            .CODE
            .STARTUP
            MOV    SI,81H          ;set up pointer to command tail
SKPBL:  MOV    AL,ES:[SI]      ;get a character
            INC    SI              ;advance to next character
            CMP    AL,20H          ;test for blank
            JZ     SKPBL
            CMP    AL,0DH          ;test for carriage return
            JZ     ERR
            CMP    AL,'-'          ;we need a dash
            JZ     NEXT
ERR:    LEA    DX,EMSG         ;set up pointer to error message
            MOV    AH,9            ;display string
            INT    21H             ;DOS call
```

```
              JMP    EXIT
      NEXT:   MOV    AL,ES:[SI]       ;get character
              INC    SI
              CMP    AL,'E'           ;encode?
              JZ     ENCO
              CMP    AL,'D'           ;decode?
              JZ     DECO
              JMP    ERR              ;neither, go output error message
      ENCO:   MOV    DH,5             ;set encryption value
      CRYPT:  MOV    DL,ES:[SI]       ;get character to encode
              INC    SI               ;advance to next character
              CMP    DL,20H           ;just echo blanks
              JZ     EKHO
              CMP    DL,0DH           ;test for carriage return
              JZ     EXIT
              SUB    DL,41H           ;remove alpha bias
              ADD    DL,DH            ;add crypting bias
              CMP    DL,26
              JL     NOCO1
              SUB    DL,26            ;correct alpha overflow
      NOCO1:  CMP    DL,0
              JGE    NOCO2
              ADD    DL,26
      NOCO2:  ADD    DL,41H           ;add alpha bias back
      EKHO:   MOV    AH,2             ;output character to screen
              INT    21H              ;DOS call
              JMP    CRYPT
      DECO:   MOV    DH,5             ;set decryption value
              NEG    DH               ;make it negative
              JMP    CRYPT
      EXIT:   .EXIT

              END
```

BSORT: An Integer Sorting Application

BSORT uses a modified version of the bubble-sort routine from Chapter 6 to sort single-digit integers supplied on the command line. A sample execution is as follows:

```
C> BSORT 9 2 8 3 5 7 4 1
Sorted array:  1 2 3 4 5 7 8 9
```

The program contains storage for up to 10 integers. Illegal characters are detected if present and execution aborted. BSORT contains three procedures: one to read numbers from the command tail, one to sort the numbers, and one to print them out.

```
;Program BSORT.ASM: Bubble sort up to 10 single-digit integers.
;
         .MODEL SMALL
         .DATA
ARRAY   DB   10 DUP(?)          ;storage for input array
N       DB   0                  ;number of data items in array
EMSG    DB   'Illegal character in input',0DH,0AH,'$'
SMSG    DB   'Sorted array: $'

         .CODE
         .STARTUP
         CALL  READEM           ;read in data from command line
         CMP   N,0              ;was there any data at all?
         JZ    BYE              ;jump if no data entered
```

```
                CALL   SORTEM            ;else, sort the data
                CALL   WRITEM            ;and display it
BYE:    .EXIT

READEM  PROC NEAR
        MOV   SI,81H              ;set up pointer to beginning of command tail
        MOV   DI,0               ;set up pointer to data storage
GET:    MOV   AL,ES:[SI]         ;get a command tail character
        INC   SI                 ;point to next character
        CMP   AL,0DH             ;is new character a carriage return?
        JZ    EXITR              ;jump if it is (all done now)
        CMP   AL,20H             ;else, is it a blank?
        JZ    GET                ;jump if it is (skip blanks)
        CMP   AL,'0'             ;make sure it is a valid integer
        JB    BAD
        CMP   AL,'9'
        JA    BAD
        MOV   ARRAY[DI],AL       ;save the integer in the data area
        INC   DI                 ;point to next data storage location
        INC   N                  ;increment number of data items
        JMP   GET                ;and continue
BAD:    LEA   DX,EMSG            ;point to error message string
        MOV   AH,9               ;display string
        INT   21H                ;DOS call
        MOV   N,0                ;set data item counter to zero
EXITR:  RET
READEM  ENDP

SORTEM  PROC NEAR
        MOV   DL,N               ;set up outer loop counter
        CMP   DL,1               ;check for a single data item
        JZ    EXITS              ;exit if there is only one item
        DEC   DL                 ;else, adjust outer loop counter
DOPASS: MOV   CL,DL              ;set up inner loop counter
        MOV   SI,0               ;set up pointer to data storage
CHECK:  MOV   AL,ARRAY[SI]       ;get first data item
        CMP   AL,ARRAY[SI+1]     ;and compare it with the second item
        JB    NOSWAP             ;jump if no need to swap them
        MOV   BL,ARRAY[SI+1]     ;else, perform the swap
        MOV   ARRAY[SI+1],AL
        MOV   ARRAY[SI],BL
NOSWAP: INC   SI                 ;point to next data item
        DEC   CL                 ;repeat until inner loop counter is zero
        JNZ   CHECK
        DEC   DL                 ;repeat until outer loop counter is zero
        JNZ   DOPASS
EXITS:  RET
SORTEM  ENDP

WRITEM  PROC NEAR
        LEA   DX,SMSG            ;set up pointer to result message
        MOV   AH,9               ;display string
        INT   21H                ;DOS call
        MOV   SI,0               ;set up pointer to data storage
        MOV   CL,N               ;set counter to number of data items
SHOW:   MOV   DL,ARRAY[SI]       ;get a data item
        INC   SI                 ;advance pointer to next item
        MOV   AH,2               ;output character to screen
        INT   21H                ;DOS call
        MOV   DL,20H             ;output a blank for spacing
        MOV   AH,2               ;output character to screen
        INT   21H                ;DOS call
        DEC   CL                 ;repeat until counter is zero
```

```
            JNZ   SHOW
            RET
WRITEM   ENDP

         END
```

Programming Exercise 7.19:   Use DEBUG to load the .EXE or program of your choice into memory and examine its Program Segment Prefix.

Programming Exercise 7.20:   Modify the DECTOHEX program so that multiple integers (separated by blanks) are allowed on the command line.

Programming Exercise 7.21:   Modify the DECTOHEX program so that integers in the range 0 to 100,000 are allowed.

Programming Exercise 7.22:   Modify the CRYPT program to allow specification of a two-digit shift value on the command line. A suggested syntax is:

```
CRYPT -E 14 ENCODE THIS MESSAGE
```

Programming Exercise 7.23:   Modify BSORT so that integers in the range 0 to 255 are allowed on the command line.

7.7   ADDITIONAL APPLICATIONS

In this section we will examine a number of additional programming applications that perform useful tasks. Some new DOS interrupt functions are introduced in the programs, and you are encouraged to study each application to understand how it works.

PCADD: An Integer Addition Program

This program asks the user for two integers (from 0 to 255) and adds them together. An overflow message is output if the sum exceeds 255. A sample execution is as follows:

```
C> PCADD
Enter number # 1 : 23
Enter number # 2 : 17

The sum is 040
```

Notice that there is no leading zero blanking in the result. The program uses a procedure called GETNUM to obtain an integer number from the user. A second procedure called SHOWNUM is used to display the integer sum.

```
;Program PCADD.ASM: A simple integer addition program.
;
        .MODEL SMALL
        .DATA
```

```
N1      DB      ?
N2      DB      ?
SUM     DB      ?
N1MSG   DB      0DH,0AH,'Enter number # 1 : $'
N2MSG   DB      0DH,0AH,'Enter number # 2 : $'
OMSG    DB      0DH,0AH,0DH,0AH,'Error! Number is larger than 255.',0DH,0AH,'$'
EMSG    DB      0DH,0AH,0DH,0AH,'Error! Illegal character.',0DH,0AH,'$'
SMSG    DB      0DH,0AH,0DH,0AH,'Overflow! Sum is larger than 255.',0DH,0AH,'$'
RMSG    DB      0DH,0AH,0DH,0AH,'The sum is $'

        .CODE
        .STARTUP
GN1:    LEA     DX,N1MSG        ;set up pointer to request message
        MOV     AH,9            ;display string
        INT     21H             ;DOS call
        CALL    GETNUM          ;get first integer
        CMP     AH,1            ;test error flag
        JZ      GN1             ;jump if error found
        MOV     N1,AL           ;save first number
GN2:    LEA     DX,N2MSG        ;set up pointer to second request message
        MOV     AH,9            ;display string
        INT     21H             ;DOS call
        CALL    GETNUM          ;get second integer
        CMP     AH,1            ;test error flag
        JZ      GN2             ;jump if error found
        MOV     N2,AL           ;save second number
        ADD     AL,N1           ;add numbers
        JNC     SSUM            ;jump if sum is less than 256
        LEA     DX,SMSG         ;set up pointer to overflow message
        MOV     AH,9            ;display string
        INT     21H             ;DOS call
        JMP     BBYE            ;exit to DOS
SSUM:   MOV     SUM,AL          ;save sum
        LEA     DX,RMSG         ;set up pointer to result message
        MOV     AH,9            ;display string
        INT     21H             ;DOS call
        MOV     AL,SUM          ;get sum back
        CALL    SHONUM          ;display result
BBYE:   .EXIT

GETNUM  PROC    NEAR
        SUB     AX,AX           ;initialize result
NEXT:   PUSH    AX              ;save temp result
        MOV     AH,1            ;read keyboard
        INT     21H             ;DOS call
        CMP     AL,13           ;is it a carriage return?
        JZ      DONE
        CMP     AL,'0'          ;test for valid digit
        JC      ERR
        CMP     AL,'9'+1
        JNC     ERR
        SUB     AL,30H          ;remove ASCII bias (0 <= AL <= 9)
        MOV     CL,AL           ;put new value into CX
        MOV     CH,0
        POP     AX              ;get temp result back
        MOV     BX,10           ;multiply temp result by 10
        MUL     BX
        ADD     AX,CX           ;add new value to temp result
        CMP     AX,256          ;are we > 255?
        JNC     OV
        JMP     NEXT            ;get next digit
ERR:    POP     AX              ;clean up stack
        LEA     DX,EMSG         ;set up pointer to error message
```

```
           JMP     EXIT              ;go display string
DONE:      POP     AX                ;get result from stack
           RET
OV:        LEA     DX,OMSG           ;set up pointer to error message
EXIT:      MOV     AH,9              ;display string
           INT     21H               ;DOS call
           MOV     AX,100H           ;set error flag
OK:        RET
GETNUM ENDP


SHONUM PROC    NEAR
           MOV     AH,0
           MOV     BL,100
           DIV     BL                ;divide value by 100
           PUSH    AX                ;save remainder
           MOV     DL,AL             ;copy quotient
           ADD     DL,30H            ;add ASCII bias
           MOV     AH,2              ;output 100's digit to display
           INT     21H               ;DOS call
           POP     AX                ;get remainder
           MOV     AL,AH             ;back into AL
           MOV     AH,0
           MOV     BL,10             ;divide value by 10
           DIV     BL
           PUSH    AX                ;save remainder
           MOV     DL,AL             ;copy quotient
           ADD     DL,30H            ;add bias
           MOV     AH,2              ;output 10's digit
           INT     21H               ;DOS call
           POP     AX                ;get remainder back
           ADD     AH,30H            ;add bias
           MOV     DL,AH
           MOV     AH,2              ;output 1's digit
           INT     21H               ;DOS call
           RET
SHONUM ENDP

           END
```

SHOWTIME: Display System Time (Using DOS INT   21H, Function 2CH)

This program reads the computer's system time clock and reports the time in military format. A sample execution is as follows:

```
C> SHOWTIME
The time is 10:53:24
```

which indicates it is almost 11:00 A.M. In the afternoon, the time would read something like this:

```
The time is 15:07:02
```

Military time reads from 0 hours to 23 hours. This is supported by the use of Function 2CH, which is called by loading register AH with 2CH. The results passed back are as follows:

CH = Hour (0 to 23)

CL = Minutes (0 to 59)

DH = Seconds (0 to 59)

DL = Hundredths of Seconds (0 to 99)

The SHOWTIME program contains a procedure called DIGITS that converts the byte value in AL into a two-digit ASCII integer.

Function 2CH is very useful in interactive applications where time must be determined or measured.

```
;Program SHOWTIME.ASM: Display system time (in military format).
;
            .MODEL SMALL
            .DATA
TMSG        DB          'The time is $'

            .CODE
            .STARTUP
            LEA         DX,TMSG         ;set up pointer to time message
            MOV         AH,9            ;display string
            INT         21H             ;DOS call
            MOV         AH,2CH          ;read system time
            INT         21H             ;DOS call
            MOV         AL,CH           ;load hours
            CALL        DIGITS          ;and display them
            MOV         DL,':'          ;load a colon
            MOV         AH,2            ;output character to screen
            INT         21H             ;DOS call
            MOV         AL,CL           ;load minutes
            CALL        DIGITS          ;and display them
            MOV         DL,':'          ;output another colon
            MOV         AH,2
            INT         21H
            MOV         AL,DH           ;load seconds
            CALL        DIGITS          ;and display them
            .EXIT

DIGITS      PROC        NEAR
            SUB         AH,AH           ;prepare for division by 10
            MOV         BL,10
            DIV         BL              ;AL will contain 10's digit, AH the 1's
            ADD         AL,30H          ;add ASCII bias
            MOV         DL,AL           ;prepare for output
            PUSH        AX              ;save the 1's digit
            MOV         AH,2            ;output character to display
            INT         21H             ;DOS call
            POP         AX              ;get 1's digit back
            ADD         AH,30H          ;add ASCII bias
            MOV         DL,AH           ;and output character
            MOV         AH,2
            INT         21H
            RET
DIGITS      ENDP

            END
```

SHOWDAY: Display Day of Week (Using DOS INT 21H, Function 2AH)

This function reads the date maintained by the system and returns the following information:

DH = Month (January = 1, December = 12)

DL = Day of Month (0 to 30)

CX = Year

AL = Day of the Week (Sunday = 0, Monday = 1, Saturday = 6)

To use this function, register AH must be loaded with 2AH.

The SHOWDAY program uses the value returned in AL to display the day of the week. A sample execution is as follows:

```
C> SHOWDAY
Today is Thursday
```

This is much more readable than 'Today is day 4'. A data table containing seven display strings (one for each day) is accessed through the use of a second data table containing the starting address of each day's display string. A similar technique may be used to display the current month.

```
;Program SHOWDAY.ASM: Display day of week.
;
            .MODEL SMALL
            .DATA
HMSG    DB          'Today is $'
SUN     DB          'Sunday$'
MON     DB          'Monday$'
TUE     DB          'Tuesday$'
WED     DB          'Wednesday$'
THU     DB          'Thursday$'
FRI     DB          'Friday$'
SAT     DB          'Saturday$'
DAYTAB  DW          SUN,MON,TUE,WED,THU,FRI,SAT

            .CODE
            .STARTUP
            LEA     DX,HMSG         ;set up pointer to 'today' message
            MOV     AH,9            ;display string
            INT     21H             ;DOS call
            MOV     AH,2AH          ;read system date
            INT     21H             ;DOS call
            ADD     AL,AL           ;double day of week value
            CBW                     ;convert into 16-bit value
            MOV     SI,AX           ;this becomes the table index value
            LEA     BX,DAYTAB       ;set up pointer to day address table
            MOV     DX,[BX+SI]      ;load address of day-of-week string
            MOV     AH,9            ;display string
            INT     21H             ;DOS call
            .EXIT

            END
```

DOSVER: Show Current DOS Version Number (Using DOS INT   21H, Function 30H)

Entering the VER command at the DOS prompt should give you a response similar to this:

```
C> VER
MS-DOS Version 6.20
```

The same thing is accomplished using the DOSVER program, which gets the DOS version number through the use of Function 30H. Register AH must be loaded with 30H to use the function, which returns the major version number ('6' in this case) in register AL and the minor version number ('2' in this case) in AH.

A sample execution is as follows:

```
C> DOSVER
DOS Version 6.2
```

Note that the minor version number is displayed slightly differently with DOSVER than with DOS.

```
;Program DOSVER.ASM: Show current DOS version number.
;
        .MODEL SMALL
        .DATA
VNS     DB    'DOS Version $'

        .CODE
        .STARTUP
        LEA   DX,VNS        ;set up pointer to output string
        MOV   AH,9          ;display string
        INT   21H           ;DOS call
        MOV   AH,30H        ;get DOS version number
        INT   21H           ;DOS call
        PUSH  AX            ;save copy of version number
        ADD   AL,30H        ;add ASCII bias to major version number
        MOV   DL,AL         ;prepare for output
        MOV   AH,2          ;output character to screen
        INT   21H           ;DOS call
        MOV   DL,'.'        ;load a period
        MOV   AH,2          ;output character to screen
        INT   21H           ;DOS call
        POP   AX            ;get version number back
        MOV   AL,AH         ;load minor version number
        MOV   BL,10         ;divide minor version number by 10
        SUB   AH,AH
        DIV   BL
        ADD   AL,30H        ;add ASCII bias
        MOV   DL,AL         ;prepare for output
        MOV   AH,2          ;output character to screen
        INT   21H           ;DOS call
        .EXIT

        END
```

HEXDUMP: An Application for Viewing the Contents of Memory

This program is very useful for snooping around inside the memory of your computer. HEXDUMP's output is similar to that of DEBUG's **d** command. A sample execution is as follows:

```
C> HEXDUMP
Segment address: 0
Starting address: 0
Ending address: 1FF
0000:0000   8A 10 1C 01 F4 06 70 00 16 00 03 05 F4 06 70 00
0000:0010   F4 06 70 00 54 FF 00 F0 4C E1 00 F0 6F EF 00 F0
0000:0020   75 18 43 07 AE 01 F4 09 6F EF 00 F0 6F EF 00 F0
0000:0030   6F EF 00 F0 6F EF 00 F0 B7 00 03 05 F4 06 70 00
0000:0040   B3 18 43 07 4D F8 00 F0 41 F8 00 F0 C5 18 43 07
0000:0050   39 E7 00 F0 A0 19 43 07 2E E8 00 F0 D2 EF 00 F0
0000:0060   D0 E3 00 F0 90 19 43 07 6E FE 00 F0 EE 06 70 00
0000:0070   53 FF 00 F0 A4 F0 00 F0 22 05 00 00 5F 5B 00 C0
0000:0080   94 10 1C 01 43 02 F4 09 04 03 F4 09 4A 01 C5 05
0000:0090   55 01 C5 05 DE 19 43 07 27 1A 43 07 BC 10 1C 01
0000:00A0   68 16 43 07 62 07 70 00 DA 10 1C 01 DA 10 1C 01
0000:00B0   DA 10 1C 01 DA 10 1C 01 3F 01 C5 05 85 02 DE E9
0000:00C0   EA D0 10 1C 01 FF 00 F0 DA 10 1C 01 DA 10 1C 01
0000:00D0   DA 10 1C 01 DA 10 1C 01 DA 10 1C 01 DA 10 1C 01
0000:00E0   DA 10 1C 01 DA 10 1C 01 DA 10 1C 01 DA 10 1C 01
```

```
0000:00F0   DA 10 1C 01 DA 10 1C 01 DA 10 1C 01 DA 10 1C 01
0000:0100   59 EC 00 F0 3D 05 00 F0 65 F0 00 F0 5F 57 00 C0
0000:0110   53 FF 00 F0 53 FF 00 F0 4D 05 00 F0 53 FF 00 F0
0000:0120   53 FF 00 F0 53 FF 00 F0 53 FF 00 F0 DB 11 01 C8
0000:0130   53 FF 00 F0 53 FF 00 F0 53 FF 00 F0 53 FF 00 F0
0000:0140   53 FF 00 F0 53 FF 00 F0 53 FF 00 F0 53 FF 00 F0
0000:0150   53 FF 00 F0 53 FF 00 F0 53 FF 00 F0 53 FF 00 F0
0000:0160   53 FF 00 F0 53 FF 00 F0 53 FF 00 F0 53 FF 00 F0
0000:0170   53 FF 00 F0 53 FF 00 F0 53 FF 00 F0 53 FF 00 F0
0000:0180   75 01 F4 09 00 00 00 00 00 00 00 00 00 00 00 00
0000:0190   00 00 00 00 00 00 00 00 00 00 00 00 B0 02 DA 02
0000:01A0   53 FF 00 F0 53 FF 00 F0 53 FF 00 F0 53 FF 00 F0
0000:01B0   53 FF 00 F0 E7 28 00 C0 53 FF 00 F0 53 FF 00 F0
0000:01C0   52 00 03 05 F5 EC 00 F0 6F EF 00 F0 6F EF 00 F0
0000:01D0   6F EF 00 F0 FC F0 00 F0 17 01 03 05 6F EF 00 F0
0000:01E0   00 00 00 00 00 00 00 00 00 00 00 00 00 00 00 00
0000:01F0   00 00 00 00 00 00 00 00 00 00 00 00 00 00 00 00
C>
```

In this example, HEXDUMP is used to display the first half of the computer's interrupt vector table. Through trial and error, other interesting memory areas can be located and examined. It is important to note that HEXDUMP requires uppercase letters in its input addresses. Address F0A0 and address f0a0 will not be treated the same, due to the way HEXDUMP's GETHEX routine operates. The solution is to convert all input text into uppercase. See Programming Exercise 7.29 for details.

```
;Program HEXDUMP.ASM: Display contents of any range of memory locations.
;
          .MODEL SMALL
          .DATA
SEGMSG DB    'Segment address:   $'
BMSG   DB    0DH,0AH,'Starting address: $'
EMSG   DB    0DH,0AH,'Ending address:    $'

          .CODE
          .STARTUP
          LEA    DX,SEGMSG    ;set up pointer to first message
          MOV    AH,9         ;display string
          INT    21H          ;DOS call
          CALL   GETHEX       ;get segment address
          MOV    ES,AX        ;use extra segment for viewing memory
          LEA    DX,BMSG      ;set up pointer to second message
          MOV    AH,9         ;display string
          INT    21H          ;DOS call
          CALL   GETHEX       ;get starting address
          MOV    SI,AX        ;use source index to access memory
          LEA    DX,EMSG      ;set up pointer to third message
          MOV    AH,9         ;display string
          INT    21H          ;DOS call
          CALL   GETHEX       ;get ending address
          MOV    DI,AX        ;use destination index to stop display
NEWLYN:   MOV    DL,0DH       ;load a carriage return
          MOV    AH,2         ;output character to screen
          INT    21H          ;DOS call
          MOV    DL,0AH       ;load a line feed
          MOV    AH,2         ;output character to screen
          INT    21H          ;DOS call
          MOV    AX,ES        ;get segment address
          CALL   HEXOUTW      ;display word in hexadecimal
          MOV    DL,':'       ;load a colon
          MOV    AH,2         ;output character to screen
```

```
            INT    21H              ;DOS call
            MOV    AX,SI            ;get current address
            CALL   HEXOUTW          ;display word in hexadecimal
            MOV    DL,20H           ;load a blank
            MOV    AH,2             ;output character to screen
            INT    21H              ;DOS call
            MOV    AH,2             ;output a second blank
            INT    21H              ;DOS call
            MOV    CX,16            ;prepare for 16 passes
NEWBYT:     MOV    AL,ES:[SI]       ;read memory
            INC    SI               ;advance to next location
            CALL   HEXOUTB          ;display byte in hexadecimal
            MOV    DL,20H           ;load a blank
            MOV    AH,2             ;output character to screen
            INT    21H              ;DOS call
            LOOP   NEWBYT           ;repeat loop if necessary
            CMP    SI,DI            ;time to quit?
            JL     NEWLYN
            .EXIT

HEXOUTW PROC NEAR
            PUSH AX                 ;save copy of 4-digit hex number
            MOV  AL,AH              ;output upper two digits
            CALL HEXOUTB
            POP  AX                 ;get copy of original number back
            CALL HEXOUTB            ;output lower two digits
            RET
HEXOUTW ENDP

HEXOUTB PROC NEAR
            PUSH AX                 ;save copy of 2-digit hex number
            ROL  AL,1               ;move upper 4 bits into lower 4 bit positions
            ROL  AL,1
            ROL  AL,1
            ROL  AL,1
            CALL DODIG              ;output hexadecimal equivalent character
            POP  AX                 ;get original number back
            CALL DODIG              ;output hex character for lower 4 bits
            RET
HEXOUTB ENDP

DODIG    PROC NEAR
            AND    AL,0FH           ;clear upper 4 bits
            ADD    AL,30H           ;add ASCII bias
            CMP    AL,'9'+1         ;is digit greater than 9?
            JC     NOADD            ;jump if it is not
            ADD    AL,7             ;otherwise, add 7 to correct alpha character
NOADD:      MOV    DL,AL            ;load character for output
            MOV    AH,2             ;output character to screen
            INT    21H              ;DOS call
            RET
DODIG    ENDP

GETHEX   PROC NEAR
            SUB    DX,DX            ;clear result
GH:         MOV    AH,1             ;read keyboard
            INT    21H              ;DOS call
            CMP    AL,0DH           ;did we get a carriage return?
            JZ     GHQ              ;jump if we did
            SUB    AL,30H           ;otherwise, subtract ASCII bias
            CMP    AL,0AH           ;is number still greater than 9?
            JC     RAA              ;jump if it is not
            SUB    AL,7             ;correct bits for A-F value
```

```
RAA:      SHL  DX,1              ;shift result 4 bits to left
          SHL  DX,1
          SHL  DX,1
          SHL  DX,1
          ADD  DL,AL             ;add new digit to result
          JMP  GH                ;and repeat
GHQ:      MOV  AX,DX             ;return result
          RET
GETHEX    ENDP

          END
```

Programming Exercise 7.24:   Modify the PCADD program so that numbers in the range 0 to 65,535 are allowed.

Programming Exercise 7.25:   Modify the PCADD program so that the user may select any of four mathematical operations (+, –, *, or /).

Programming Exercise 7.26:   Modify SHOWTIME so that the time is displayed with an A.M. or P.M. indicator.

Programming Exercise 7.27:   Modify SHOWDAY to output the month, day, and year, as well as the day of the week, as in:

```
Friday, November 7, 1997
```

Programming Exercise 7.28:   Use DOS INT   21H, Function 30H to find the actual minor version number value and modify DOSVER to display its two-digit equivalent.

Programming Exercise 7.29:   Modify HEXDUMP so that it accepts lowercase hexadecimal numbers and converts them into uppercase so that GETHEX returns the correct address.

Programming Exercise 7.30:   Modify HEXDUMP so that it also outputs the ASCII equivalent characters for each memory location displayed. Unprintable ASCII codes from 0 to 1FH, and any byte values from 80H to FFH, should cause a '.' to be output in their place. The output should look similar to this:

```
9000:FF00   00 00 00 00 00 00 07 00 01 00 07 00 01 00 07 00   ................
9000:FF10   61 9F A4 A7 61 9F 07 00 07 95 07 95 04 03 F4 09   a...a...........
9000:FF20   07 00 CE 5B 02 00 01 00 02 00 A6 71 F0 98 A6 71   ...[.......q...q
9000:FF30   02 00 07 95 07 95 04 03 F4 09 02 00 CE 5B 00 01   .............[..
9000:FF40   12 00 80 00 13 00 13 A2 B0 AD 00 01 80 00 00 00   ................
9000:FF50   04 A2 16 A2 16 A2 04 00 07 95 07 95 4E 11 4A A1   ............N.J.
```

```
9000:FF60   00 00 00 FF A5 A4 13 A2 81 00 07 95 4F 11 04 03    ............O...
9000:FF70   F4 09 4E 11 8F 2F 07 95 C5 05 00 00 24 00 00 00    ..N../......$...
9000:FF80   00 2C 00 2E 00 2D 00 3A 00 00 02 00 F5 0C 1C 01    .,...-.:........
9000:FF90   2C 00 00 00 00 00 00 00 00 00 00 00 00 00 00 4A    ,..............J
9000:FFA0   01 C5 00 FF FF 1A 6D 61 73 6D 5C 6C 69 6E 6B 20    ......masm\link
9000:FFB0   25 31 2C 2C 3B 0D 0A 1A F4 FF 1D 1C A4 32 00 18    %1,,;........2..
9000:FFC0   00 00 7F 00 01 01 00 18 43 07 46 70 10 3E 04 03    ..▮....C.Fp.>..
9000:FFD0   1D 14 02 08 4F 32 1D 1C 80 00 0D 1B F4 09 04 03    ....O2..........
9000:FFE0   5A 11 86 7A 51 09 5A 11 87 7A 80 00 C8 2E 00 00    Z..zQ.Z..z......
9000:FFF0   4D 08 00 4A 29 00 00 00 53 43 00 00 00 00 00 00    M..J)...SC......
```

7.8 TROUBLESHOOTING TECHNIQUES

The BIOS and DOS function calls rely on the contents of one or more processor registers, as well as the stack. Keep the following points in mind when using function calls.

- DOS automatically uses its own stack if one is not defined in your program (via the .STACK directive). There is generally room for a small number of items to be pushed, but keep in mind that the stack will also be used by the function call (for the return address and register storage).
- When tracing the execution of a function call, use DEBUG's **p** command, rather than the **t** command, to treat the function call (INT 21H for example) as a single instruction.
- Make sure the registers are properly initialized prior to the function call. Forgetting to put an offset into SI, or a function number in AH, will lead to unpredictable and possibly harmful results.
- Many function calls use the carry or zero flags to indicate the success or failure of the operation performed. Use JZ/JNZ or JC/JNC to control the action upon return from the function call.

Function calls greatly simplify the programmer's job, allowing control over many complex operations through the use of a handful of instructions. Use the function calls carefully and always double-check your register values.

SUMMARY

In this chapter, we examined many DOS and BIOS interrupt functions. These functions allow us to read the keyboard, output characters to the video display and printer, and access the computer's date and time. We also saw how the computer's speaker can be controlled and how the parameter area of a program's Program Segment Prefix can be used to read the contents of the command tail. Sample programs were included to demonstrate the use of each interrupt function. Each BIOS or DOS interrupt covered is included in Appendix E or F, respectively, and you are encouraged to use the appendixes for reference when writing your own programs.

Many more BIOS and DOS interrupts are available to the programmer, and a good portion of them will be discussed in Chapters 8 and 9.

STUDY QUESTIONS

1. What is the difference between reading the keyboard with DOS INT 21H, Function 01H, and with BIOS INT 16H, Function 00H?
2. What are some possible uses for the shift flag status returned by BIOS INT 16H, Function 02H?
3. How does PASSWORD detect the end of the secret password?
4. Explain how the current video mode can be restored after it has been changed to some other mode.
5. Why does a 25 by 80 character color display require 4,000 bytes of storage?
6. If B8000 is the first address of the 25 by 80 color display video RAM, determine the following:
   a) The address of the last character on the first line.
   b) The address of the first character on line 10.
   c) The address of the last character on the last line.
7. What are the attribute bytes for each of the following?
   a) White character on a green background.
   b) Green character on a red background.
   c) Blinking blue character on a black background.
   d) Red character on a yellow background.
8. How are the arrow keys recognized in the XYDRAW program?
9. What frequency is generated by the counter/timer chip if its counter has been loaded with 2380?
10. How does the TONES program turn the speaker on and off?
11. Why is it necessary to read the printer status?
12. How does PSTAT advance through the list of display strings?
13. Why is the ES register used in the ECHO program without being initialized by the program itself?
14. How does the CRYPT program deal with wrap-around at the end of the alphabet? For example, if the shift value is 5 and the letter to be encrypted is X, how does the program determine the resulting letter C?
15. Can you think of a way to allow BSORT to work with negative integers as well?
16. What is stored in the program segment prefix?

ADDITIONAL PROGRAMMING EXERCISES

1. Modify the PASSWORD program so that the password entered by the user is also stored in a data buffer (similar to PWTXT). Use the string instruction CMPSB to verify the password.
2. Write a program that will display your name, in color, on the screen, so that your name continually slides from right to left on the top line of the display.

3. Write a program that draws lines of stars ('*') connecting all four corners of the video display.

4. Add a MOVE command to XYDRAW that allows the row and column of the desired cursor position to be entered.

5. Write a program that allows the user to enter the desired frequency of the tone to generate on the speaker. Allow a frequency range of only 250 to 2,000 Hz.

6. Write a program that echoes characters entered on the keyboard to the printer as they are typed. If a carriage return is entered, send a carriage return and a line feed to the printer.

7. Modify the PRINTMSG program so that the 80-character user input limitation is enforced.

8. Write a HEXTODEC program that converts the hexadecimal number in the command tail to an integer and outputs it to the display.

9. Another simple encryption technique is called **transposition.** Consider this sample text:

```
PROGRAMS-ARE-FUN
```

The 16 letters in the sample text are rewritten the following way:

```
P R O G
R A M S
- A R E
- F U N
```

They are then read from top to bottom beginning with the first column to obtain the encrypted message:

```
PR--RAAFOMRUGSEN
```

Write a program that will implement transposition encryption.

10. Write a program that will find the integer average of all numbers supplied on the command line. For example,

```
AVE 1 2 3 4 5 6 7
```

should give a result of 4.

11. Write a program that generates a "random" number from 0 to 100 based on the value of the computer's time and/or date.

12. Write a new password program that allows the user only 10 seconds to type in the correct password.

13. Write a simple calculator program that evaluates single-digit integer expressions in the command tail. A few sample executions might look like this:

```
C> CALC 2 + 5
7
C> CALC 6 - 2
4
C> CALC 5 * 3 - 6
9
```

Parentheses are not allowed, but the order of operations must be enforced (* and / before + and −).

14. Write a program that displays the computer's time in the center of the display screen. The time should update automatically every second.

15. Write a program that displays the contents of its own Program Segment Prefix in a HEXDUMP-like fashion.

16. Write a set of timing functions that can be used to determine how long (in seconds) a block of code takes to execute.

17. Write a program that beeps the speaker once per second and displays the time on the display. Terminate when any key is pressed.

18. Write a screen saver program that, when executed, does the following:
    a) stores a copy of the current text screen.
    b) blanks the screen.
    c) once every 30 seconds, displays a user-supplied text string at a random location on the screen.

    Repeat (b) and (c) until any key is pressed. Then restore the original text screen.

CHAPTER 8

Advanced Programming Applications

OBJECTIVES

In this chapter you will learn about:

* Linking separate object files together
* Creating and using an object code library
* Creating and using source code macros
* Predicting the execution time of a block of code
* How interrupts are supported and used by DOS
* How two programs can execute "simultaneously"
* How memory is managed by DOS
* Using information from the PC's mouse
* Writing a memory-resident program
* Beating your computer at tic-tac-toe
* Testing for protected-mode operation
* How assembly language interfaces with C

8.1   INTRODUCTION

The material presented in this chapter is not a complete treatment of the many truly advanced things that can be accomplished by a good programmer and a PC. Instead, a good subset of the more interesting and useful advanced topics has been chosen. For some, memory-resident programming might be very useful; for others, it could be instruction execution time. You are encouraged to cover as many of the sections as you can to round out your programming experience.

Section 8.2 shows how separate source files can be assembled and linked. Section 8.3 discusses the use of macros to simplify source code. Section 8.4 details the process used to determine execution time of a group of instructions. This is followed by a second treatment of interrupts in Section 8.5. Section 8.6 shows how a specific DOS interrupt can be used to run more than one program at a time. Memory management is the subject of Section 8.7. Section 8.8 shows how mouse movements can be tracked with a simple program. In Section 8.9 a memory-resident program is discussed to illustrate how a **hot key** is created. An easy-to-beat tic-tac-toe game is presented in Section 8.10. The last three sections, 8.11 through 8.13, cover protected-mode detection, C language interfacing, and troubleshooting techniques, respectively.

8.2   USING THE EXTRN AND PUBLIC ASSEMBLER DIRECTIVES

Recall that the usual process of creating an executable program (.EXE or .COM) requires a number of steps. These steps are as follows:

1. Write source program (.ASM file)
2. Assemble program (via ML)
3. Link program (via LINK)

In a large organization, such as a computer software company, the programs being created are the work of many individual programmers, each working on a different portion of the main program. Their individual procedures can communicate through the use of memory locations and registers while the program is running, but how is the main program created from the many individual pieces written by each programmer? In this section, we will see how separate source files can be assembled and then linked into a single executable program.

Creating Separate Source Modules

Let us examine two source modules. The first module, from the source file DISPBIN.ASM, looks like this:

```
;Procedure DISPBIN.ASM: Display value of AL in binary on the screen.
;
        .MODEL SMALL
        .CODE

        PUBLIC DISPBIN          ;for linking

DISPBIN PROC FAR
        MOV    CX,8             ;set up loop counter
NEXT:   SHL    AL,1             ;move bit into carry flag
        PUSH   AX               ;save number
        JC     ITIS1            ;was the LSB a 1?
        MOV    DL,30H           ;load '0' character
        JMP    SAY01            ;go display it
ITIS1:  MOV    DL,31H           ;load '1' character
SAY01:  MOV    AH,2             ;display character function
```

```
        INT   21H                ;DOS call
        POP   AX                 ;get number back
        LOOP  NEXT               ;and repeat
        RET
DISPBIN ENDP

        END
```

Notice the use of the statement:

```
PUBLIC DISPBIN
```

This assembler directive informs the linker that the value of DISPBIN must be available at link time.

The second source module, from the source file TESTBIN.ASM, looks like this:

```
;Program TESTBIN.ASM: Test the DISPBIN display procedure.
;
        .MODEL SMALL
        .CODE

        EXTRN DISPBIN:FAR    ;for linking

        .STARTUP
        SUB   AL,AL          ;clear counter
AGAIN:  PUSH AX              ;save counter
        CALL DISPBIN         ;display counter in binary
        MOV   DL,20H         ;load blank character
        MOV   AH,2           ;display character function
        INT   21H            ;DOS call
        MOV   AH,2           ;output a second blank
        INT   21H
        POP   AX             ;get counter back
        INC   AL             ;increment it
        JNZ   AGAIN          ;and repeat until counter equals zero
        .EXIT

        END
```

The corresponding link statement in this source module is:

```
EXTRN DISPBIN:FAR
```

which informs the linker that DISPBIN is an external *far* label. The value of DISPBIN will be used by the assembler and the linker to create the correct code for the CALL DISPBIN instruction.

TESTBIN is sometimes referred to as a **driver** module, because it is being used to test the operation of the DISPBIN procedure.

When the first source module is assembled (via ML /c DISPBIN.ASM), the object file DISPBIN.OBJ is created. TESTBIN.OBJ is created by assembling the second source module (via ML /c TESTBIN.ASM).

The directory information for each file is as follows:

```
DISPBIN  OBJ           135 03-20-97   9:51a
TESTBIN  OBJ           168 03-20-97   9:55a
```

You might agree that the size of both object files is larger than that required by the machine code for each module's instructions. This is due to the fact that ML has placed additional information into each object file to indicate what public or external variables have been used.

Note that neither program can exist by itself. The only way to make an executable program out of them is to *combine* them. This is done with a simple linker statement:

```
LINK TESTBIN + DISPBIN,,;
```

which creates TESTBIN.EXE.

If TESTBIN is linked by itself (via LINK TESTBIN,,;), we get the following error message:

```
TESTBIN.OBJ(A) : error L2029 : 'DISPBIN' : unresolved external
```

So, the linker is able to determine if all necessary variable requirements have been met by matching every EXTRN reference it encounters with a PUBLIC directive for the same variable.

Building and Using an Object Code Library

An object code library is a file that contains a collection of many individually linked code modules. For example, an object code library called NUMOUT.LIB contains the code for each of the following routines:

Routine	Function
DISPBIN	Display AL in binary on screen
DISPHEX	Display AL in hex on screen
DISPHEX_16	Display AX in hex on screen
DISPBCD	Display AL in BCD on screen
DISPINT	Display AL as unsigned integer on screen
DISPINT_16	Display AX as unsigned integer on screen

These six routines provide many different ways for outputting numerical values. Having them already written and contained within a code library is convenient, because they need only be linked with another module (a driver or some other main program).

To create a code library, use the LIB utility supplied with DOS. The command:

```
C\> LIB NUMOUT.LIB
```

will cause the following action by LIB:

```
Library does not exist.   Create?
```

Answer **y** for yes to create the new library. LIB will then prompt for other inputs, but these can be ignored by hitting return:

```
Operations: <cr>
List file: <cr>
```

To verify that NUMOUT.LIB was created we can use the DIR command. The result is as follows:

```
NUMOUT    LIB        1,033 03-20-97  11:29a
```

The reason NUMOUT.LIB is not empty is that LIB requires library files to have a certain internal format that makes it possible to support multiple object code modules.

To add an object code module to an existing library, we use the following command:

```
C\> LIB NUMOUT + DISPBIN;
```

The LIB utility will add the code for DISPBIN to the NUMOUT.LIB library. This command is used for each of the six DISP modules.

To use a module contained within a library file, we use the linker in a different way. For example, to create TESTBIN.EXE using the NUMOUT library, we use this LINK command:

```
C\> LINK TESTBIN,,,NUMOUT
```

which causes the linker to search the NUMOUT library file for external references.

What happens when you decide to rewrite a procedure contained within a code library? The new procedure must be assembled to create an object file. Then the existing object file must be replaced with the new one. For example, to replace NUMOUT's DISPINT procedure with a new version of DISPINT, use the command:

```
C\> LIB -DISPINT +DISPINT
```

This allows the library file to be updated as the need arises.

There are many other useful output routines that could be added to NUMOUT.LIB. What kind of procedures might be placed into a NUMIN library?

Programming Exercise 8.1:   Write a DISPOCT procedure to display the value of AL in octal on the screen.

Programming Exercise 8.2:   Write a DISPBCD_16 procedure that outputs the BCD number in register AX on the screen.

Programming Exercise 8.3:   Write a driver program for each of the other five display routines. What do they all have in common?

8.3   USING MACROS

Macros provide the programmer with a powerful new way to write programs. In this section we will examine a number of standard macro forms and uses.

A Simple Macro

The following group of instructions are used to define a simple macro:

```
DISP_MSG    MACRO
            MOV    AH,9        ;display string function
            INT    21H         ;DOS call
            ENDM
```

The name of the macro is DISP_MSG, and it is defined on the first line of the macro definition (much like a procedure name is defined when used with PROC). The instructions enclosed between MACRO and ENDM (end macro) are the contents of the macro.

The operation of the macro during assembly is to replace all occurrences of DISP_MSG within the source file with the MOV and INT instructions. For example, the source statements:

```
LEA    DX,ABC
DISP_MSG
LEA    DX,DEF
DISP_MSG
```

specify two *calls* to the DISP_MSG macro. A macro call triggers the assembler into *expanding* a macro. The resulting source code is:

```
LEA    DX,ABC
MOV    AH,9      ;display string function
INT    21H       ;DOS call
LEA    DX,DEF
MOV    AH,9      ;display string function
INT    21H       ;DOS call
```

prior to assembly. The macro provides a convenient way for the programmer to replace groups of instructions with a single macro name. This saves the programmer time typing in the source file and leads to smaller source file sizes (although not in a smaller executable file).

Example 8.1:   Suppose that a programmer is busy with a program requiring frequent SHL operations on register AX. The statements:

```
SHL    AX,1
SHL    AX,1
SHL    AX,1
SHL    AX,1
```

can be replaced by a single macro named SHL_AX everywhere they occur. The macro definition will be as follows:

```
SHL_AX    MACRO
          SHL    AX,1
          SHL    AX,1
          SHL    AX,1
          SHL    AX,1
          ENDM
```

A different macro definition uses the special **REPT** directive to repeat the four SHL statements:

```
SHL_AX    MACRO
          REPT   4
              SHL    AX,1
              ENDM
          ENDM
```

Note that the two instructions:

```
MOV   CL,4
SHL   AX,CL
```

will work just as well.

Adding a Parameter to a Macro

A macro becomes more useful when we are able to use it for more than one thing. The DISP_MSG macro is useful from the standpoint that we are freed from typing:

```
MOV   AH,9
INT   21H
```

over and over again. This is improved by allowing the macro to expand into all three instructions required to display a string. Consider the SEND_MSG macro shown here:

```
SEND_MSG   MACRO   ADDR
           LEA     DX,ADDR     ;set up pointer to string
           MOV     AH,9        ;display string function
           INT     21H         ;DOS call
           ENDM
```

In this macro definition, the symbol ADDR is a **parameter** passed to the macro for expansion purposes. The source statement:

```
SEND_MSG ABC
```

causes the ADDR symbol to become ABC. The resulting expansion looks like this:

```
LEA   DX,ABC     ;set up pointer to string
MOV   AH,9       ;display string function
INT   21H        ;DOS call
```

The single source statement SEND_MSG ABC is certainly preferable to typing in all three instructions.

Because I/O is a big part of many programs, other macros designed to assist with I/O would be very useful. Two more character-based macros are as follows:

```
DISP_CHAR   MACRO   DATA
            MOV     DL,DATA     ;load ASCII character
            MOV     AH,2        ;display output function
            INT     21H         ;DOS call
            ENDM

CRLF        MACRO
            DISP_CHAR 0DH       ;output a cr
            DISP_CHAR 0AH       ;output a lf
            ENDM
```

The second macro, CRLF, actually is an example of a *nested* macro. A nested macro is a macro that contains a macro call within its definition. So, macros can make it easier to write other macros.

More than one parameter may be defined within a macro. In this case, parameter names must be separated by a comma in the macro definition, as in:

```
<Name>   MACRO   <parameter 1>,  <parameter 2>
```

Example 8.2:   The macro SAXPY defined here multiplies the AX register by XVAL and adds YVAL to the product:

```
SAXPY   MACRO   XVAL, YVAL
        MUL     XVAL
        ADD     AX,YVAL
        ENDM
```

When called with the following parameters:

```
SAXPY   BX, 25
```

we get:

```
MUL  BX
ADD  AX,25
```

Using Labels Within Macros

Suppose that we need to replace the following group of instructions with a macro:

```
        CMP  AL,10
        JNC  NOADD
        ADD  AL,7
NOADD:  ADD  AL,30H
        MOV  DL,AL
        MOV  AH,2
        INT  21H
```

The NOADD label will cause an assembly error if the macro is called more than once because the assembler will try to define a second NOADD symbol.

The solution to this problem is to define the label NOADD as a **local** variable, which means a variable that has meaning only *inside* the current macro expansion. The macro definition then becomes:

```
HEXOUT  MACRO
        LOCAL   NOADD
        CMP  AL,10
        JNC  NOADD
        ADD  AL,7
NOADD:  ADD  AL,30H
        MOV  DL,AL
        MOV  AH,2
        INT  21H
        ENDM
```

Every time HEXOUT is called, the assembler will define a new name for the NOADD label.

Conditional Macro Expansion

There are times when a program is written in such a way that it can be assembled in more than one way. For example, the program might be a numeric display application. One assembly might create an executable module for decimal numbers, another for hexadecimal numbers.

To allow a choice during assembly, a **conditional** statement must be used. Consider the following macro:

```
SHIFT   MACRO  SIZE
        IF SIZE EQ 8
           SHR    AL,1
        ELSE
           SHR    AX,1
        ENDIF
        ENDM
```

This macro uses the value of the SIZE parameter to determine which SHR instruction to generate. The SHR   AL,1 instruction is produced by SHIFT 8. SHR   AX,1 is generated when SHIFT 16 is used. The programmer may place as many statements as needed between IF and ELSE, and between ELSE and ENDIF.

The types of conditions that may be tested within the IF . . . ELSE statement are as follows:

Test	Meaning
EQ	Equal
NE	Not equal
LT	Less than
GT	Greater than
LE	Less than or equal
GE	Greater than or equal

Example 8.3:   The MULTIPLY macro shown here moves the NUMBER value into either register BL or BX, depending on the size of NUMBER. The appropriate MUL instruction is also generated.

```
MULTIPLY   MACRO  NUMBER
           IF NUMBER  GT 255
                MOV    BX,NUMBER
                MUL    BX
           ELSE
                MOV    BL,NUMBER
                MUL    BL
           ENDIF
           ENDM
```

Macro Operators

A number of operators are defined for use within a macro definition. Four of these operators, and their meanings, are:

%	Expression
<...>	Literal character string
;;	Macro comment
&	Substitute

The Expression operator (%) is used to interpret the *value* of a parameter, and use the value in the expansion. For example, the macro:

```
MOV_AL   MACRO    VAL
         MOV  AL,VAL
         ENDM
```

can be called with MOV_AL 2*30, or with MOV_AL %2*30, resulting in two different expansions, as shown here:

```
MOV_AL    2*30     -->    MOV  AL,2*30
MOV_AL    %2*30    -->    MOV  AL,60
```

It may be necessary in some cases to work with the value of a parameter instead of the actual parameter text.

The literal character string operators (<...>) are used in conjunction with the substitute operator (&) to place a specific character string into the expanded macro instruction. Consider the following string definitions:

```
STR1    DB    'This is a message',0DH,0AH,'$'
STR2    DB    'This is another message',0DH,0AH,'$'
```

If many messages need to be defined in the indicated format, the literal and substitute operators can be used in the following macro to allow easier string generation:

```
S_MAKE   MACRO   NUM, TXT
         STR&NUM     DB     '&TXT',0DH,0AH,'$'
         ENDM
```

The S_MAKE macro must be called like this:

```
S_MAKE    1, <This is a message>
S_MAKE    2, <This is another message>
```

You can see that the value of NUM was substituted into STR&NUM to generate STR1 and STR2. All characters between the literal operators < and > were substituted for &TXT.

Another example of this type of substitution expansion is as follows:

```
SHIFT   MACRO    DIR, REG
        SH&DIR   REG,1
        ENDM
```

This macro creates SHL or SHR instructions for *any* register, as in the following types of calls:

```
SHIFT    R, AX   -->    SHR   AX,1
SHIFT    L, BL   -->    SHL   BL,1
```

The macro comment operator (;;) prevents the comments included inside a macro definition from being repeated each time the macro is expanded. This avoids cluttering up the source file with a collection of identical comments.

TESTMAC: A Macro Expansion Example

The TESTMAC program shown here uses the macro structures previously covered to perform useful I/O with the user.

```
;Program TESTMAC.ASM: Test macro expansion.
;
        .MODEL SMALL
        .DATA
SMSG1   DB      'This message was output by the DISP_MSG macro.',0DH,0AH,'$'
SMSG2   DB      'This message was output by the SEND_MSG macro.',0DH,0AH,'$'
SMSG3   DB      ' decimal equals $'
SMSG4   DB      ' hexadecimal.',0DH,0AH,'$'
WMSG    DB      'Hit <C> to continue...$'

DISP_MSG   MACRO
           MOV    AH,9          ;;display string function
           INT    21H           ;;DOS call
           ENDM

SEND_MSG   MACRO  ADDR
           LEA    DX,ADDR        ;;set up pointer to string
           MOV    AH,9           ;;display string function
           INT    21H            ;;DOS call
           ENDM

DISP_CHAR  MACRO  DATA
           MOV    DL,DATA        ;;load ASCII character
           MOV    AH,2           ;;display output function
           INT    21H            ;;DOS call
           ENDM

CRLF       MACRO
           DISP_CHAR 0DH
           DISP_CHAR 0AH
           ENDM

NUM_OUT    MACRO  DATA, BASE
           IF BASE EQ 10
               MOV    AL,DATA     ;;load number
               MOV    BL,100      ;;load divisor
               CALL   DEC_OUT     ;;find and display 100's digit
               MOV    BL,10       ;;load divisor
               CALL   DEC_OUT     ;;find and display 10's digit
               ADD    AL,30H      ;;add ASCII bias to 1's digit
               DISP_CHAR AH       ;;display digit
           ELSE
               MOV    AL,DATA     ;;load number
               PUSH   AX          ;;save it
               MOV    CL,4        ;;load shift counter
               SHR    AL,CL       ;;shift upper nybble down
               CALL   HEXOUT      ;;displays MS hex digit
               POP    AX          ;;get number back
               CALL   HEX_OUT     ;;display LS hex digit
           ENDIF
           ENDM
```

```
HIT_C     MACRO
          LOCAL  WAIT_C, EXIT
          SEND_MSG WMSG
WAIT_C:   MOV    AH,1          ;;read keyboard function
          INT    21H           ;;DOS call
          CMP    AL,'c'        ;;is character a 'c'?
          JZ     EXIT
          CMP    AL,'C'        ;;is character a 'C'?
          JNZ    WAIT_C        ;;no, repeat until it is
EXIT:     CRLF                 ;;output cr and lf
          ENDM


          .CODE
          .STARTUP
          LEA    DX,SMSG1      ;set up pointer to message
          DISP_MSG             ;display message
          SEND_MSG SMSG2       ;display second message
          DISP_CHAR 'H'        ;output 'Hi.'
          DISP_CHAR 'i'
          DISP_CHAR '.'
          CRLF                 ;and cr and lf
          NUM_OUT 100, 10      ;display 100 in decimal
          SEND_MSG SMSG3       ;display decimal message
          NUM_OUT 100, 16      ;display 100 in hexadecimal
          SEND_MSG SMSG4       ;display hex message
          HIT_C                ;wait for a c/C
          HIT_C                ;wait for another c/C
          .EXIT


HEX_OUT   PROC   NEAR
          AND    AL,0FH        ;clear upper nybble
          ADD    AL,30H        ;add ASCII bias
          CMP    AL,'9'+1      ;do we need alpha fix?
          JC     NOADD7
          ADD    AL,7          ;add alpha bias
NOADD7:   DISP_CHAR AL         ;display hex digit
          RET
HEX_OUT   ENDP


DEC_OUT   PROC   NEAR
          SUB    AH,AH         ;prepare for division
          DIV    BL            ;find digit
          ADD    AL,30H        ;add ASCII bias
          PUSH   AX            ;save remainder
          DISP_CHAR AL         ;display digit
          POP    AX            ;get remainder back
          XCHG   AH,AL         ;save new number
          RET
DEC_OUT   ENDP

          END
```

Note the use of ;; in the macro definitions. This is done to keep the look of the resulting list file uncluttered.

Let us examine a few sections of output from the list file.

```
0017    8D 16 0000 R              LEA    DX,SMSG1      ;set up pointer to message
                                  DISP_MSG             ;display message
001B    B4 09            1        MOV    AH,9
001D    CD 21            1        INT    21H
```

```
                                    SEND_MSG SMSG2      ;display second message
001F    8D 16 0031 R    1              LEA     DX,SMSG2
0023    B4 09           1              MOV     AH,9
0025    CD 21           1              INT     21H
```

Note that when a macro name is encountered, as in DISP_MSG and SEND_MSG, the instructions generated by macro expansion are flagged with a 1. This indicates the nesting level for the current macro.

This is further illustrated during expansion of the CRLF macro, which calls the DISP_CHAR macro:

```
                                    CRLF                ;and cr and lf
0039    B2 0D           2              MOV     DL,0DH
003B    B4 02           2              MOV     AH,2
003D    CD 21           2              INT     21H
003F    B2 0A           2              MOV     DL,0AH
0041    B4 02           2              MOV     AH,2
0043    CD 21           2              INT     21H
```

You are encouraged to examine the list file for TESTMAC for other examples of how the macros were expanded.

The resulting output from TESTMAC's execution is as follows:

```
This message was output by the DISP_MSG macro.
This message was output by the SEND_MSG macro.
Hi.
100 decimal equals 64 hexadecimal.
Hit <C> to continue...noC
Hit <C> to continue...c
```

Recall that the HIT_C macro waits for c or C to be entered on the keyboard.

Programming Exercise 8.4: Write a new macro called SENDER that has the following calling conventions:

```
SENDER  C, X        --> output character 'X' via Function 2
SENDER  S, <Hello>  --> output Hello string via Function 9
```

Conditional macro expansion should be used to generate the appropriate instructions.

Programming Exercise 8.5: Modify the NUM_OUT macro to include binary and octal (base 8) output numbers.

8.4   INSTRUCTION EXECUTION TIMES

An important topic in the study of any microprocessor involves analysis of the execution time of programs, subroutines, or short sections of code. The most direct application of this study is in the design of programs that function under a time constraint. For example, high-resolution graphics operations, such as image rotation, filtering, and motion simulation, require all processing to be completed within a very short period of time (usually a few milliseconds or less). If analysis of the total instruction execution time for the graphics routine exceeds the allowed time of the system, a loss in image quality will most likely result. We will not get quite so involved with our analysis of the instruction times. Instead, we will look at one example subroutine and how its total execution time may be determined. We will use the 8088 microprocessor instruction times, and then examine the differences in the Pentium's execution of the same code.

Instruction Cycle Analysis

TOBIN is a subroutine that will convert a 4-digit BCD number in register BX into a binary number. The result is returned by TOBIN in AX. TOBIN is a good example to use for execution time determination because it contains two nested loops. The number of clock cycles for each instruction can be determined by referring to Appendix B. Table 8.1 is an example of how clock cycles are determined. The number of clock cycles an instruction takes to execute depends on a number of factors. Operand size is the first variable. Look at the clock cycles required by the MOV   CX instructions. The instruction itself takes four cycles, plus another four required by the 8088 to fetch a word operand from memo-

TABLE 8.1   Required instruction execution clock cycles in a simple programming loop

				Clock Cycles	
	Instructions		Overhead Cycles	Outer-Loop Cycles	Inner-Loop Cycles
TOBIN	PROC	FAR			
	SUB	AX,AX	3		
	MOV	DX,AX	2		
	MOV	CX,4	4 + 4		
NEXTDIGIT:	PUSH	CX		11 + 4	
	SUB	BP,BP	3		
	MOV	CX,4		4 + 4	
GETNUM:	RCL	BX,1			2
	RCL	BP,1			2
	LOOP	GETNUM			17/5
	MOV	CX,10		4 + 4	
	MUL	CX		118–133	
	ADD	AX,BP		3	
	POP	CX		8 + 4	
	LOOP	NEXTDIGIT		17/5	
	RET		18 + 8		
TOBIN	ENDP				

ry. The processor's 8-bit data bus requires two memory read cycles to obtain the word operand, resulting in the additional four clock cycles.

The addressing mode used by an instruction also affects the number of clock cycles required. Register addressing, as in the SUB   AX,AX and ADD   AX,BP instructions, requires fewer clock cycles than an instruction that must access memory during its execution, such as PUSH and POP. The clock cycles required for each addressing mode are included in Appendix B. You must note that the clock cycles shown for each instruction assume that the instructions have already been fetched and placed in the instruction queue. This, in many cases, results in very few clock cycles for some instructions. An exception is the LOOP instruction, which has two times listed. The smaller time is used when the jump does *not* take place. This makes sense, because the instruction following LOOP can simply be fetched from the instruction queue. Greater time is required when the LOOP does take place, owing to the fact that the queue must be flushed and reloaded. Other instructions have a variable number of cycles as a function of the data on which they operate. MUL is a good example of this, requiring anywhere from 118 to 133 cycles depending on the number of 1s in the operands forming the product. It is best to use the *worst case* execution time. Then you will avoid nasty situations such as having to explain why the routine does not always execute at the same speed.

The RET instruction requires an ample number of clock cycles to execute because it must pop the CS and IP return address off the stack. These operations require accesses to memory, which always increase the execution time.

Table 8.1 has three columns of clock cycles. The **overhead** column is for instructions that execute only once in the subroutine. They do not contribute significantly to the overall time, but cannot be ignored. The outer-loop and inner-loop cycles are repeated a number of times, and thus grow into a large number of clock cycles before the routine completes. Keep in mind that the overall execution time will be only an *estimate,* because other factors are normally present that affect execution time. These factors include the time required to initially load the instruction queue, or the time to reload it after a jump or LOOP instruction; however, our estimate will be within 10 percent of the actual execution time.

Execution Time Calculation

To compute the execution time, we must first determine the total number of clock cycles needed. The inner loop requires 21 cycles (worst case) for one pass. Because the inner loop is designed to execute four times, this gives 84 cycles for one completion of the inner-loop instructions. But these instructions are contained within the outer loop, which itself requires an additional 199 clock cycles (worst case). So, executing the outer loop once uses 283 clock cycles. The outer loop executes four times also, giving a total of 1,132 cycles so far. Adding in 39 overhead cycles (do you see why overhead is not significant?) results in a grand total of 1,171 clock cycles for the TOBIN subroutine.

How does this number translate into an execution time? Because each clock cycle has a period determined by its frequency, we need to know the clock frequency at which the processor is running. Suppose this frequency is 5 MHz. Each cycle will then have a period of 200 ns. Multiplying 200 ns by 1,171 gives 234.2 µs! This is the execution time of TOBIN.

In conclusion, it is interesting to note that TOBIN can convert more than 4,200 BCD numbers to binary in 1 second.

How the Pentium Does It

Table 8.2 lists the clock cycles required for each of TOBIN's instructions when executed on the Pentium. Note that most instructions in the Pentium require a *single clock cycle* to execute. This is a cornerstone of the RISC design philosophy, and is possible on the Pentium, thanks to a number of architectural features, such as instruction and data cache, instruction and address pipelining, and branch prediction.

MUL and LOOP are examples of instructions that take more than one instruction to execute. The task of multiplying large binary numbers cannot be performed in a single cycle without placing an enormous amount of logic gates in the instruction pipeline. This is impractical, and may even require a slower clock on the processor to ensure that the logic gates function properly. So, instead, the multiplication is done in stages over eleven clock cycles.

As before, LOOP instructions have a *pair* of cycle times associated with their execution. The 5/6 cycle count for LOOP means that five clock cycles are used when the LOOP takes place (CX is not zero) and six when the LOOP does not take place (CX is zero). Other instructions share this feature, most notably the conditional jumps. In general, instructions that change the flow of execution in a program cause the instruction pipeline to

TABLE 8.2   Required instruction execution clock cycles on the Pentium

				Clock Cycles	
		Instructions	Overhead Cycles	Outer-Loop Cycles	Inner-Loop Cycles
TOBIN	PROC	FAR			
	SUB	AX,AX	1		
	MOV	DX,AX	1		
	MOV	CX,4	1		
NEXTDIGIT:	PUSH	CX		1	
	SUB	BP,BP		1	
	MOV	CX,4		1	
GETNUM:	RCL	BX,1			1
	RCL	BP,1			1
	LOOP	GETNUM			5/6
	MOV	CX,10		1	
	MUL	CX		11	
	ADD	AX,BP		1	
	POP	CX		1	
	LOOP	NEXTDIGIT		5/6	
	RET		2		
TOBIN	ENDP				

stall for at least one additional cycle, as the pipeline changes its execution path. The RET instruction at the end of TOBIN falls into this category.

Several other factors contribute to a small percentage of uncertainty about the exact number of clock cycles required on a Pentium architecture, such as cache performance, operating system overhead (page faults, task switching), and the sequence of instructions being executed. It is still possible to get a good estimate of the number of cycles using the techniques developed for the 8088's analysis.

8.5   WORKING WITH INTERRUPT VECTORS

Recall from Chapter 5 that an **interrupt** is an event that causes the processor to stop its current program execution and perform a specific task to service the interrupt. In a PC, interrupts are used to keep accurate time, read the keyboard, operate the disk drives, and access the power of the disk operating system. DOS assigns a number of these interrupts (INT 21H, for example) for its own use, so we must be careful when accessing the interrupt vector table. When it is necessary to reassign an interrupt vector, the following functions should be used.

DOS INT   21H, Function 35H: Get Interrupt Vector

This function is selected when the AH register equals 35H. Upon entry, the interrupt number (0 to 255) must be in register AL. Upon exit, register BX contains the interrupt handler IP address, and the ES register contains the interrupt handler CS address.

Example 8.4:   The following instructions can be used in DEBUG to obtain the vector address of INT   21H:

```
MOV   AL,21
MOV   AH,35
INT   21
```

When these instructions are traced, we get:

```
-p

AX=0021  BX=0000  CX=0000  DX=0000  SP=FFEE  BP=0000  SI=0000  DI=0000
DS=1C46  ES=1C46  SS=1C46  CS=1C46  IP=0102   NV UP EI PL NZ NA PO NC
1C46:0102 B435          MOV     AH,35
-p

AX=3521  BX=0000  CX=0000  DX=0000  SP=FFEE  BP=0000  SI=0000  DI=0000
DS=1C46  ES=1C46  SS=1C46  CS=1C46  IP=0104   NV UP EI PL NZ NA PO NC
1C46:0104 CD21          INT     21
-p

AX=3521  BX=16B4  CX=0000  DX=0000  SP=FFEE  BP=0000  SI=0000  DI=0000
DS=1C46  ES=0BF4  SS=1C46  CS=1C46  IP=0106   NV UP EI PL NZ NA PO NC
1C46:0106 AE            SCASB
-
```

The resulting addresses in registers BX and ES give a vector address of 0BF4:16B4.

DOS INT   21H, Function 25H: Set Interrupt Vector

This function is selected when the AH register is loaded with 25H. Upon entry, the interrupt number (0 to 255) must be loaded into register AL. The CS:IP vector address of the new interrupt handler must be loaded into the DS and DX registers, respectively.

Attention! Even experienced programmers may get unexpected results when an interrupt vector is replaced. Usually, if an interrupt vector in a DOS machine is set improperly or does not have a valid handler at the new vector address, the machine is in danger of crashing. Be careful when using this function.

In Sections 8.6 and 8.9 you will examine applications designed to take over certain DOS/BIOS interrupt functions.

VECTORS: View Entire Interrupt Vector Table

The VECTORS program shown here displays all 256 vector addresses, as initialized by BIOS, DOS, and any other programs using them. A total of 64 vector addresses are displayed at a time, in the familiar CS:IP address format.

```
;Program VECTORS.ASM: View entire interrupt vector table.
;
        .MODEL SMALL
        .DATA
WMSG    DB      0DH,0AH,'Hit any key to continue...',0DH,0AH,0DH,0AH,'$'

        .CODE
        .STARTUP
        SUB     AL,AL           ;begin with vector 0
NEWV:   PUSH    AX              ;save vector number
        CALL    DISPHEX         ;display it
        MOV     DL,' '          ;load blank character
        MOV     AH,2            ;display character function
        INT     21H             ;DOS call
        MOV     DL,'='          ;load equal sign
        INT     21H             ;and output
        MOV     DL,' '          ;load another blank
        INT     21H             ;and output
        POP     AX              ;get vector number back
        PUSH    AX              ;and save it again
        MOV     AH,35H          ;get interrupt vector function
        INT     21H             ;DOS call
        MOV     AX,ES           ;get interrupt segment
        CALL    DISPHEX_16      ;display it
        MOV     DL,':'          ;load colon character
        MOV     AH,2            ;display character function
        INT     21H             ;DOS call
        MOV     AX,BX           ;get interrupt offset
        CALL    DISPHEX_16      ;display it
        MOV     DL,' '          ;load blank character
        MOV     AH,2            ;display character function
        INT     21H             ;DOS call
        INT     21H             ;output a second blank
```

```
                 POP     AX              ;get vector number back
                 MOV     DL,AL           ;make a copy of it
                 AND     DL,3            ;should we output a new line?
                 CMP     DL,3
                 JNZ     NNL
                 PUSH    AX              ;save vector number
                 MOV     DL,0DH          ;load carriage return
                 MOV     AH,2            ;display character function
                 INT     21H             ;DOS call
                 MOV     DL,0AH          ;also output a line feed
                 INT     21H
                 POP     AX              ;get vector number back
        NNL:     MOV     DL,AL           ;make copy of vector number
                 CMP     DL,255          ;finished?
                 JZ      NNP
                 AND     DL,63           ;should we pause here?
                 CMP     DL,63
                 JNZ     NNP
                 PUSH    AX              ;save vector number
                 LEA     DX,WMSG         ;set up pointer to pause message
                 MOV     AH,9            ;display string function
                 INT     21H             ;DOS call
                 MOV     AH,8            ;character input function
                 INT     21H             ;DOS call
                 POP     AX              ;get vector number back
        NNP:     INC     AL              ;advance to next vector
                 CMP     AL,0            ;did we wrap from 255 to 0?
                 JNZ     NEWV
                 .EXIT

        DISPHEX_16 PROC NEAR
                 PUSH  AX                ;save number
                 MOV   AL,AH             ;load upper byte
                 CALL  DISPHEX           ;go display in hex
                 POP   AX                ;get number back
                 CALL  DISPHEX           ;go display lower byte
                 RET
        DISPHEX_16 ENDP

        DISPHEX PROC NEAR
                 PUSH AX                 ;save number
                 SHR  AL,1               ;get upper nybble
                 SHR  AL,1
                 SHR  AL,1
                 SHR  AL,1
                 CALL HEXOUT             ;display hex character
                 POP  AX                 ;get number back
                 AND  AL,0FH             ;preserve lower nybble
                 CALL HEXOUT             ;display hex character
                 RET
        DISPHEX ENDP

        HEXOUT  PROC NEAR
                 CMP   AL,10             ;is AL greater than 10?
                 JC    NHA1              ;yes
                 ADD   AL,7              ;no, add alpha bias
        NHA1:    ADD   AL,30H            ;add ASCII bias
                 MOV   DL,AL             ;load output character
                 MOV   AH,2              ;display character function
                 INT   21H               ;DOS call
```

```
           RET
HEXOUT   ENDP

           END
```

The program uses DOS INT   21H, Function 35H to read each vector, one at a time, from 0 to 255.

A sample execution showing the first 64 vectors is as follows:

```
00 = 011C:108A   01 = 0070:06F4   02 = 09B4:0016   03 = 0070:06F4
04 = 0070:06F4   05 = 1587:04E4   06 = F000:E14C   07 = F000:EF6F
08 = 0BF4:1875   09 = 0BF4:1923   0A = F000:EF6F   0B = 09B4:006F
0C = F000:EF6F   0D = F000:EF6F   0E = 09B4:00B7   0F = 0070:06F4
10 = 0BF4:18B3   11 = F000:F84D   12 = F000:F841   13 = 0BF4:18C5
14 = F000:E739   15 = 0BF4:19A0   16 = F000:E82E   17 = F000:EFD2
18 = F000:E3D0   19 = 0BF4:1990   1A = F000:FE6E   1B = 0070:06EE
1C = F000:FF53   1D = F000:F0A4   1E = 0000:0522   1F = C000:5B5F
20 = 011C:1094   21 = 0BF4:16B4   22 = 0A76:02B1   23 = 0A76:014A
24 = 0A76:0155   25 = 0BF4:19DE   26 = 0BF4:1A27   27 = 011C:10BC
28 = 1587:02E8   29 = 0070:0762   2A = 011C:10DA   2B = 011C:10DA
2C = 011C:10DA   2D = 011C:10DA   2E = 0A76:013F   2F = 0F3A:306F
30 = 1C10:D0EA   31 = F000:FF01   32 = 011C:10DA   33 = 011C:10DA
34 = 011C:10DA   35 = 011C:10DA   36 = 011C:10DA   37 = 011C:10DA
38 = 011C:10DA   39 = 011C:10DA   3A = 011C:10DA   3B = 011C:10DA
3C = 011C:10DA   3D = 011C:10DA   3E = 011C:10DA   3F = 011C:10DA
Hit any key to continue...
```

The vector address for INT   21H is highlighted in bold. Note that the actual addresses on *your* computer probably will be different. The vector addresses change with each new version of DOS, and with the configuration specified at boot time via CONFIG.SYS and AUTOEXEC.BAT.

Examine the entire output of the VECTORS program. Are there any interrupts that contain vector address 0000:0000? There should be many of them, all indicating unused interrupts in the BIOS/DOS environment.

Programming Exercise 8.6:   Modify VECTORS so that the effective address of each interrupt routine is displayed. For example, a vector address of 011C:10DA has an effective address of 0229A.

8.6   MULTITASKING

In an ever-increasing effort to squeeze the most processing power out of the basic microprocessor, individuals have come up with ingenious techniques for getting a single CPU to do many wonderful things. Consider a standard, single-CPU microcomputing system. One user sits at a terminal entering commands, thinking, entering more commands, waiting for I/O from the system (as a file is loaded in from disk or tape), and so on and on. When computer specialists discovered that in this situation the CPU was wasting a great deal of time doing I/O, they thought of a way of getting more use out of the CPU. Suppose that circuitry could be added to the system to perform the I/O operations under the CPU's control. All the CPU would have to do is issue a command, such as "read the disk," and the disk controller would do the rest of the work. This would free up the processor for other things while the disk controller was busy reading the disk. What other things can the CPU do? The most obvious answer that came to mind was *service another user!* Thus, the age of multitasking was born. A microprocessor system capable of performing multitasking is able to communicate with several users, seemingly at once. Each user believes he or she is the only user on the system. What is actually happening is that one user gets a small slice of the processor's time, then another user gets another time slice, and the same goes for all other users. Figure 8.1 shows a simple diagram of this operation. In this figure we see that up to four users are executing their programs, seemingly at the same time. A single CPU can support more than one person at a time because of the high processing speed of the processor versus the slow thinking speed of the users. It is not difficult to see that the user will spend long periods of time thinking about what to do next or waiting for I/O to appear on the terminal. The time slice allocated to each user is designed so that it is long enough to perform a significant number of instructions without being so long as to be noticed by the user. For example, suppose that as many as sixteen users may be on the system at once. If the CPU must service each user once every 1/10th of a second, the time slice required is 6.25 ms, possibly enough to complete a user's current program. So, even though all users are forced to wait while the CPU services each of them (as shown in Figure 8.1), they probably do not even notice.

The software involved in supporting a multitasking system can become very involved, so we will examine only the basic details here. What exactly happens when one user's program is suspended so that the CPU can service a different user? Suppose that both users have written programs that use some of the same processor registers (for example, AX, BX, and SI are used by both programs). It becomes necessary to save one user's registers before letting the other user's program take over. This is referred to as a context switch. A context switch is used to save all registers for one user and load all registers used by the next user. Thus, a context switch is needed every time the CPU switches users. The routine presented here, TSLICE, will handle context switches for four different users in a round-robin fashion. This means that user 1 will not execute again until all other

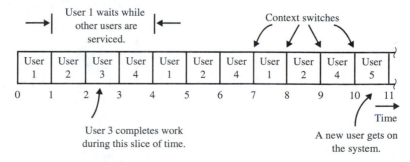

FIGURE 8.1 Multitasking with a single CPU

users have had their chance. To get the processor to do a context switch, we have to inform it that the user's time slice has expired. The easiest way to do this is to periodically interrupt the processor (by connecting a timer circuit to the processor's interrupt input). TSLICE is then actually an interrupt handler that is executed when a time slice is used up. TSLICE first saves all processor registers on the stack (the program counter and flags are already there, thanks to the processor's interrupt handling scheme), and then finds the next user who should execute. The registers for this user are then loaded from its stack area and execution resumes. This discussion assumes that all user programs remain in memory when not executing.

```
USER1       DW      64 DUP (?)      ;user stack areas
USER2       DW      64 DUP (?)
USER3       DW      64 DUP (?)
USER4       DW      64 DUP (?)
WHO         DB      ?               ;current user number (1-4)
STACK1      DW      ?               ;storage for user stack pointers
STACK2      DW      ?
STACK3      DW      ?
STACK4      DW      ?
  .
  .
  .
TSLICE:     PUSH    AX              ;save all registers
            PUSH    BX
            PUSH    CX
            PUSH    DX
            PUSH    SI
            PUSH    DI
            PUSH    BP
            PUSH    DS
            PUSH    ES
            MOV     AX,DATA         ;load system data segment
            MOV     DS,AX
            MOV     AL,WHO          ;get user number
            DEC     AL              ;need 0-3 user number for indexing
            CBW                     ;extend into 16 bits
            ADD     AX,AX           ;double accumulator
            MOV     DI,AX           ;load index
            LEA     BP,STACK1       ;point to stack pointer table
            MOV     [BP + DI],SP    ;save user stack pointer
            INC     WHO             ;increment user number
            ADD     DI,2            ;point to next stack pointer
            CMP     DI,8            ;wrap around to user 1?
            JNZ     GETSTACK
            MOV     WHO,1           ;reset user number
            MOV     DI,0            ;and index register
GETSTACK:   MOV     SP,[BP + DI]    ;load new stack pointer
            POP     ES              ;restore new user's registers
            POP     DS
            POP     BP
            POP     DI
            POP     SI
            POP     DX
            POP     CX
            POP     BX
            POP     AX
            IRET                    ;go service new user
```

One requirement of TSLICE is that all users use the same stack segment. This guarantees that all PUSHes, CALLs, and interrupts generated by any user will place data within memory governed by the system software. The complexity of this code results from the need to wrap around the buffer containing the stack pointers, when the context switch is from user 4 to user 1. In Chapter 16, we will see how multitasking is built into protected mode, and how multiple tasks are managed.

BIOS INT   1CH: Timer Tick

This interrupt is called by BIOS 18.2 times each second, the rate at which the computer's time-of-day clock is serviced. To run a program in *background* (such as PRINT or DOSKEY) or to switch between programs in an orderly fashion, this interrupt must be *hooked*. DOS interrupts 25H and 35H can be used to hook an interrupt. Any application that wants to run in background may use INT   1CH to install itself in the interrupt chain for Timer Tick. One note of caution: because INT   1CH is issued 18.2 times per second, only 1/18.2, or 54.94 milliseconds, is allowed for **all** programs sharing INT   1CH. Thus, the tasks performed by each program hooked into the Timer Tick interrupt should execute quickly. The specific details of hooking an interrupt are covered in Section 8.9.

It is also very important to preserve the state of all processor registers to ensure that the program interrupted by INT   1CH resumes normally.

SWITCHER: A Simple Task-Switching Application

The SWITCHER program presented here allows two individual procedures (tasks) to run at different rates of execution. There is a significant difference between the SWITCHER program and the TSLICE procedure. TSLICE assumes that programs have been interrupted *during* execution, and thus saves all information on the current stack and then switches stack areas to restart a suspended process. SWITCHER operates differently by assuming that the individual tasks *complete* execution between successive interrupts. This allows the complicated stack switching to be eliminated.

```
;Program SWITCHER.ASM: Automatically switch between two simple tasks.
;
            .MODEL SMALL
            .CODE
TASK    DB    0           ;task number
COUNT   DB    '0'         ;initial count
T1KNT   DB    4           ;count rate value
ALPHA   DB    'A'         ;initial alpha
T2KNT   DB    12          ;alpha rate value
OLDIPCS DD    ?           ;old interrupt address

            .STARTUP
            SUB   AH,AH           ;set video mode function
            MOV   AL,3            ;25 by 80 color
            INT   10H             ;BIOS call
            MOV   AH,35H          ;get interrupt vector function
            MOV   AL,1CH          ;timer interrupt number
            INT   21H             ;DOS call
            MOV   WORD PTR CS:OLDIPCS,BX      ;save old IP
            MOV   WORD PTR CS:OLDIPCS[2],ES   ;save old CS
```

```
               MOV   AX,CS          ;load current CS value
               MOV   DS,AX
               LEA   DX,CS:SWAP     ;load address of new ISR
               MOV   AH,25H         ;set interrupt vector function
               MOV   AL,1CH         ;timer interrupt number
               INT   21H            ;DOS call
      WAIT4KY: MOV   AH,1           ;read keyboard status function
               INT   16H            ;BIOS call
               JZ    WAIT4KY        ;loop until any key is pressed
               MOV   DX,WORD PTR CS:OLDIPCS      ;load old interrupt IP
               MOV   DS,WORD PTR CS:OLDIPCS[2]   ;load old interrupt CS
               MOV   AL,1CH         ;timer interrupt number
               MOV   AH,25H         ;set interrupt vector function
               INT   21H            ;DOS call
               MOV   AH,1           ;read keyboard function
               INT   21H            ;DOS call
               .EXIT

      SWAP:    PUSHF                ;save registers used here
               PUSH AX
               PUSH BX
               PUSH CX
               PUSH DX
               NOT   CS:TASK        ;switch tasks
               CMP   CS:TASK,0      ;task 2's turn now?
               JNZ   ET2
               CALL  TASK1          ;perform task 1
               JMP   BYE            ;go resume interrupt chain
      ET2:     CALL  TASK2          ;perform task 2
      BYE:     POP   DX             ;restore registers
               POP   CX
               POP   BX
               POP   AX
               POPF
               JMP   CS:OLDIPCS     ;go execute old ISR

      TASK1    PROC NEAR
               DEC   CS:T1KNT            ;decrement rate value
               JNZ   ER1                 ;exit if task is still asleep
               MOV   CS:T1KNT,4          ;else, reload rate value
               MOV   AH,2                ;set cursor position function
               MOV   BH,0                ;display page 0
               MOV   DH,12               ;row 12
               MOV   DL,37               ;column 37
               INT   10H                 ;BIOS call
               MOV   AL,CS:COUNT         ;load count value
               MOV   BL,2                ;color is green
               MOV   CX,1                ;write one character
               MOV   AH,9                ;write character/attribute function
               INT   10H                 ;BIOS call
               INC   CS:COUNT            ;increment count value
               CMP   CS:COUNT,'9'+1      ;check for wrap around
               JNZ   ER1
               MOV   CS:COUNT,'0'        ;set initial count value
      ER1:     RET
      TASK1    ENDP

      TASK2    PROC NEAR
               DEC   CS:T2KNT            ;decrement rate value
               JNZ   ER2                 ;exit if task is still asleep
               MOV   CS:T2KNT,12         ;else, reload rate value
```

```
              MOV   AH,2              ;set cursor position function
              MOV   BH,0              ;display page 0
              MOV   DH,12             ;row 12
              MOV   DL,43             ;column 43
              INT   10H               ;BIOS call
              MOV   AL,CS:ALPHA       ;load alpha value
              MOV   BL,1              ;color is blue
              MOV   CX,1              ;write one character
              MOV   AH,9              ;write character/attribute function
              INT   10H               ;BIOS call
              INC   CS:ALPHA          ;increment alpha value
              CMP   CS:ALPHA,'Z'+1    ;check for wrap around
              JNZ   ER2
              MOV   CS:ALPHA,'A'      ;set initial alpha value
ER2:          RET
TASK2         ENDP

              END
```

The first task in SWITCHER displays a running count from 0 to 9, in green, on the screen. A short delay is added between outputs.

The second task displays the alphabet, from A to Z, in blue on the screen also, with a much longer time delay between outputs. The time between outputs in both tasks is controlled by counting the number of Timer Tick interrupts that have occurred. Task 1 executes its display code every 4 ticks. Task 2 executes its display code every 12 ticks. The interrupt service routine SWAP uses the TASK byte to determine which task should execute. SWAP complements the TASK byte with each Timer Tick, causing alternate calls to TASK1 and TASK2. SWAP preserves the state of all registers used in the TASK procedures and jumps to the address of the old interrupt service routine for INT   1CH when it completes execution.

Any keystroke will cause the program to exit to DOS. SWITCHER will restore the old interrupt vector and return to DOS.

Programming Exercise 8.7:   Modify the SWITCHER program so that four tasks are switched. The two new tasks should count *down* from 9 to 0 and from Z to A.

Programming Exercise 8.8:   Modify the SWITCHER program so that both TASK procedures are installed in the chain for INT   1CH. That is, eliminate the SWAP code and have INT   1CH process both TASK procedures for every Timer Tick. It will be necessary to set the interrupt vector twice to accomplish this.

8.7   MEMORY MANAGEMENT

In this section, we will see how entire blocks of memory can be assigned through the use of a **memory management** routine. This is a required feature in all operating systems that load multiple programs into memory for shared execution (as in multitasking). We can easily see why memory management is needed by examining Figure 8.2. In this figure, the memory space of a typical system is examined. Initially, three jobs (A, B, and C) are running (1).

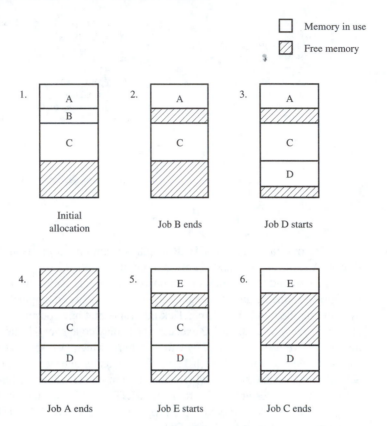

FIGURE 8.2   Memory allocation in an operating system

Because they were the first three to begin execution, all were assigned consecutive blocks of memory. In (2), we see that job B (or program B) terminates. The memory allocated to job B is returned to the storage pool. In (3), job D begins execution. Because the memory required by this job exceeded what was available in the area vacated by job B, this job is loaded into the first available space that is big enough. When job A ends in (4), its space is also returned to the storage pool. Job E quickly takes this space over in (5). Finally, job C terminates in (6).

The purpose of the memory manager is to keep track of all free blocks of memory. When the operating system needs to load a new job into memory, it will inform the memory manager of how much memory is needed by the new job. The memory manager will either find a big enough space and return the starting address of the block, or indicate that not enough memory is available for the job to execute at the present time. In this case, the job will have to wait until more memory becomes available.

There are three built-in DOS functions explicitly designed to help programs manage memory in the ways just described.

DOS INT   21H, Function 48H: Allocate Memory

This function is selected when register AH equals 48H. Upon entry to the interrupt, the BX register must contain the number of *paragraphs* requested for allocation. DOS allocates memory in chunks of sixteen locations called paragraphs, so a 1,024-byte storage block is composed of 1,024/16, or 64 paragraphs.

If DOS is able to allocate the requested number of paragraphs, it will return the starting segment address of the memory block in register AX. If DOS cannot allocate enough memory, the carry flag will be set and register AX will contain error code information. The BX register will contain the number of paragraphs in the largest block of available memory.

Example 8.5: Show the code required to attempt allocation of a 32KB block of storage. If successful, load the ES register with the allocation block segment address. If unsuccessful, jump to NO_MEM.

A 32KB block of memory requires 2,048 paragraphs. Thus, we have the following request code:

```
MOV   AH,48H
MOV   BX,2048
INT   21H
JC    NO_MEM
MOV   ES,AX
```

Offset 0 in the Extra Segment is the first location in the newly allocated 32KB block.

DOS INT   21H, Function 49H: Free Allocated Memory

This function is selected when the AH register is loaded with 49H. Upon entry, the ES register must contain the segment address of the block being freed. If the carry flag is cleared upon return, the memory was freed successfully. If the carry flag is set, register AX will contain the error code.

This function *must* be used with a segment address previously obtained by a call to function 48H.

DOS INT   21H, Function 4AH: Modify Allocated Memory Blocks

This function is selected when the AH register equals 4AH. Upon entry, the ES register must contain the segment address of the block being modified. The BX register must contain the number of paragraphs that will *remain* after reallocation.

As with the other functions, the operation is successful if the carry flag is clear upon return. Otherwise, register AX contains an error code, and register BX contains the maximum number of paragraphs available for an increase in allocation.

Example 8.6: A modify request is generated as follows:

```
MOV   AH,4AH
MOV   BX,200H
INT   21H
```

Upon return, the carry flag is set and the BX register equals 118H. What does this mean?

Because the carry flag is set, we know that DOS was not able to reallocate memory. The value in register BX indicates that more memory was requested (200H paragraphs) than is available (118H paragraphs).

MEMMAN: A Memory Usage Application

The MEMMAN program presented here is structured as a .COM file for allocation purposes. When DOS executes a .COM program, it loads the program into memory (at some initial allocation segment address), and then allocates *all available* memory to the program. If it is necessary for the .COM program to allocate and free memory blocks, it must first modify its own memory allocation to return memory blocks to DOS. Notice the EOP label at the end of the MEMMAN program. The value of this label is used to calculate the length of the program's required memory space.

```
;Program MEMMAN.ASM: A memory-management demonstrator.
;Note: Program is written in .COM format (ORG at 100H) and
;must use the .TINY model.
;
        .MODEL  TINY
        .CODE
        ORG     100H            ;code must be relative to 100H
START:  JMP     GO              ;jump over data area

ERR0    DB      'Error! Could not allocate memory.',0DH,0AH,'$'
OK0     DB      '50K-bytes of memory has been allocated at segment $'
ERR1    DB      'Error! Could not modify allocated memory.',0DH,0AH,'$'
OK1     DB      'Allocated memory has been modified.',0DH,0AH,'$'
CRLF    DB      0DH,0AH,'$'
AMB     DW      64 DUP(?)

GO:     LEA     BX,EOP          ;get program size in bytes
        MOV     CL,4            ;shift count
        SHR     BX,CL           ;compute paragraph size
        INC     BX              ;round up to next paragraph
        MOV     AH,4AH          ;modify allocated memory function
        INT     21H             ;DOS call
        JNC     MM1             ;jump if modification successful
        LEA     DX,ERR1         ;set up pointer to not modified message
        MOV     AH,9            ;display string function
        INT     21H             ;DOS call
        JMP     EXIT
MM1:    LEA     DX,OK1          ;set up pointer to memory modified message
        MOV     AH,9            ;display string function
        INT     21H             ;DOS call
        SUB     DI,DI           ;clear index
MM3:    MOV     BX,3200         ;request 3200 paragraphs (50K bytes) of memory
        MOV     AH,48H          ;allocate memory function
        INT     21H             ;DOS call
        JNC     MM2             ;jump if allocated
        LEA     DX,ERR0         ;set up pointer to not allocated message
        MOV     AH,9            ;display string function
        INT     21H             ;DOS call
        JMP     EXIT
MM2:    MOV     AMB[DI],AX      ;save segment address of new memory block
        LEA     DX,OK0          ;set up pointer to segment address message
```

```
            MOV    AH,9          ;display string function
            INT    21H           ;DOS call
            MOV    AX,AMB[DI]     ;load allocated segment address
            ADD    DI,2          ;advance pointer to allocation table
            CALL   DHEX16        ;display segment address of new block
            LEA    DX,CRLF       ;set up pointer to crlf string
            MOV    AH,9          ;display string function
            INT    21H           ;DOS call
            JMP    MM3           ;go try for another 50K-byte block
EXIT:    .EXIT

DHEX16 PROC NEAR
            PUSH   AX            ;save number
            MOV    AL,AH         ;get upper byte
            CALL   DHEX          ;convert and display
            POP    AX            ;get number back
            CALL   DHEX          ;convert and display lower byte
            RET
DHEX16 ENDP

DHEX PROC NEAR
            PUSH AX              ;save value
            SHR    AL,1          ;shift down upper nybble
            SHR    AL,1
            SHR    AL,1
            SHR    AL,1
            CALL DDIG            ;display hex digit
            POP    AX            ;get value back
            AND    AL,0FH        ;set lower nybble value
            CALL DDIG            ;display hex digit
            RET
DHEX ENDP

DDIG     PROC NEAR
            CMP    AL,10         ;is AL less than 10?
            JC     NHA1          ;jump if yes
            ADD    AL,7          ;otherwise, add alpha bias
NHA1:       ADD    AL,30H        ;add ASCII bias
            MOV    DL,AL         ;load character for output
            MOV    AH,2          ;display character function
            INT    21H           ;DOS call
            RET
DDIG     ENDP

EOP:     NOP                     ;end of program code, used
                                 ;for length calculations
            END    START        ;must specify starting address
```

When DOS launches a .COM file, the values of all four segment registers are identical and equal to the load address of the program segment prefix. An example DEBUG session shows MEMMAN's initial segment register addresses:

```
C:\>DEBUG MEMMAN.COM
-r
AX=0000  BX=0000  CX=01B9  DX=0000  SP=FFFE  BP=0000  SI=0000  DI=0000
DS=1C64  ES=1C64  SS=1C64  CS=1C64  IP=0100    NV UP EI PL NZ NA PO NC
1C64:0100 E92C01        JMP      022F
-q
```

Notice that all segment registers contain 1C64. This is the segment of the initial allocation block for the MEMMAN.COM program. This segment address must be used when MEMMAN attempts to change its allocation through function 4AH. When making the current allocation smaller, MEMMAN must be careful to save enough memory (in paragraphs) for its code requirements. By using the address value of the EOP label, MEMMAN is able to determine the correct number of required paragraphs.

Assuming that MEMMAN is able to return memory, it then attempts to request as many 50KB blocks as it can. The segment address of each successful allocation is displayed, as shown in the following sample execution:

```
C:\>MEMMAN
Allocated memory has been modified.
50K-bytes of memory has been allocated at segment 16CB
50K-bytes of memory has been allocated at segment 234C
50K-bytes of memory has been allocated at segment 2FCD
50K-bytes of memory has been allocated at segment 3C4E
50K-bytes of memory has been allocated at segment 48CF
50K-bytes of memory has been allocated at segment 5550
50K-bytes of memory has been allocated at segment 61D1
50K-bytes of memory has been allocated at segment 6E52
50K-bytes of memory has been allocated at segment 7AD3
50K-bytes of memory has been allocated at segment 8754
Error! Could not allocate memory.
```

The program was able to allocate ten 50KB blocks before DOS ran out of memory. To get a handle on the segment addresses returned, notice that they increase in value as the allocations progress. If we subtract any two consecutive segment addresses, we will get the block size in paragraphs. To check, 6E52 – 61D1 = 0C81, which corresponds to 3,201 paragraphs. A 50KB block requires 3,200 paragraphs. The extra paragraph allocated is used by DOS to maintain allocation information.

It may be interesting to watch what happens when MEMMAN is executed when you have *shelled to DOS* from another application. For instance, after starting up WordPerfect® 5.1 and shelling to DOS via Control-F1, MEMMAN gives the following result:

```
C:\WP51\>MEMMAN
Allocated memory has been modified.
50K-bytes of memory has been allocated at segment 79A6
50K-bytes of memory has been allocated at segment 8627
50K-bytes of memory has been allocated at segment 92A8
Error! Could not allocate memory.
```

The program is now only able to allocate three 50KB blocks, because the rest of memory is already allocated to the suspended WordPerfect program.

Programming Exercise 8.9:   Modify the MEMMAN program to allocate blocks of 4,095 paragraphs. What do you notice about the segment addresses displayed?

Programming Exercise 8.10:   Write a memory allocation program that alternately requests 16KB, 32KB, and 64KB blocks and saves their segment addresses. When no more blocks can be allocated, return all 32KB blocks allocated, then request as many 8KB blocks as possible.

8.8 USING THE MOUSE

Anyone who has used a mouse to point and click has probably discovered that it is easy to handle. With a software mouse driver installed, the PC keyboard has some electronic competition for input services. It is often more convenient to use a mouse to navigate through a program than a keyboard. In this section, we will see how to incorporate mouse functions in our own programs.

Assuming that a MOUSE.SYS driver was loaded during boot time (or that MOUSE.COM has been executed), the mouse functions should be available through INT 33H. More than 50 individual mouse functions are provided through INT 33H. We will examine just a few of them in this section.

MOUSE INT 33H, Function 00H: Mouse Reset and Status

This function is selected by loading 00H into register AH. The interrupt returns status information in registers AX and BX. If AX contains 0FFH upon return, the mouse was initialized. If AX contains 0, no mouse functions are available. Register BX contains the status of the mouse buttons. If the left button is pressed, bit-0 of BX will be set. If the right button is pressed, bit-1 will be set.

When the mouse is reset, a number of parameters are initialized. A few of these mouse parameters are as follows:

```
Cursor position:      Center of the screen
Cursor:               Hidden
Text Cursor:          Inverse video box
Graphics Cursor:      An arrow
Display page number:  0
```

If no mouse driver has been loaded, INT 33H usually will have a vector value of 0000:0000 or, if a vector exists, the only instruction in the interrupt service routine might be IRET. So, it is necessary to examine register AX after a mouse reset attempt.

Example 8.7: How can the presence of a mouse be tested? After performing a mouse reset function, register AX can be tested for 0 like so:

```
OR    AX,AX      ;set zero flag if AX = 0
JZ    NO_MOUSE   ;jump if no mouse found
```

MOUSE INT 33H, Function 01H: Show Mouse Cursor

This function is selected when AH equals 01H. The mouse cursor is turned on and will move when the mouse is used. In text mode, the mouse cursor is an inverse video box the size of a single character. In graphics mode, the mouse cursor is a white arrow with a black outline.

MOUSE INT   33H, Function 02H: Hide Mouse Cursor

This function is selected when AH is loaded with 02H. The mouse cursor is turned off. Because DOS does not automatically turn the mouse cursor off when a .COM or .EXE file terminates, this function must be used prior to exit.

MOUSE INT   33H, Function 03H: Get Button Status and Mouse Position

This function is selected when AH equals 03H. Upon return, the button status of the mouse is saved in register BX. The horizontal and vertical positions of the mouse are saved in registers CX and DX, respectively.

The lower 2 bits of register BX are used to represent the state of the left and right mouse buttons. If either button is pressed, its associated bit in register BX will be set. Bit-0 indicates left-button status and bit-1 indicates right-button status. Register CX will equal 0 when the mouse is on the left side of the screen. Register DX will equal 0 when the mouse is at the top of the screen. The maximum values of registers CX and DX depend on the video mode selected.

Example 8.8:   In this example, we see one way the mouse status can be read and analyzed. The instructions place a 1 in registers AL and AH, depending on the button status returned by INT   33H.

```
MOV    AH,3         ;get mouse status/position function
INT    33H          ;MOUSE call
SUB    AX,AX        ;clear result
SHR    BX,1         ;shift left button bit into carry flag
ADC    AL,0         ;adjust AL for left-button status
SHR    BX,1         ;repeat for right-button status
ADC    AH,0
```

The button status registers can be tested elsewhere with code like this:

```
OR     AL,AL        ;is button pushed?
JZ     NOT_PUSHED
```

MOUSETST: A Mouse Application

There are many more mouse functions available, but the four presented here should enable us to write a simple mouse-interactive program. The MOUSETST program shown here uses all four functions to control and access the mouse.

```
;Program MOUSETST.ASM: Display mouse's horizontal/vertical position.
;
        .MODEL SMALL
        .DATA
NODRV   DB      'No mouse driver installed.',0DH,0AH,'$'
NOMSE   DB      'Mouse not responding.',0DH,0AH,'$'
MEXIT   DB      0DH,0AH,'Press left mouse button to exit...',0DH,0AH,'$'
SPOS    EQU     12*160

        .CODE
```

```
            .STARTUP
            MOV     AH,35H          ;get interrupt vector function
            MOV     AL,33H          ;INT 33H is for the mouse
            INT     21H             ;DOS call
            MOV     AX,ES           ;is vector 0000:0000?
            OR      AX,BX
            JNZ     GO
            LEA     DX,NODRV        ;set up pointer to no-driver message
ERRD:       MOV     AH,9            ;display string function
            INT     21H             ;DOS call
            JMP     EXIT            ;exit to DOS
GO:         SUB     AX,AX           ;initialize mouse function
            INT     33H             ;MOUSE call
            OR      AX,AX           ;does mouse exist?
            JNZ     NEXT
            LEA     DX,NOMSE        ;set up pointer to no-mouse message
            JMP     ERRD            ;go process error
NEXT:       MOV     CX,25           ;set up loop counter
CLRSC:      MOV     DL,0AH          ;load line-feed code
            MOV     AH,2            ;display character function
            INT     21H             ;DOS call
            LOOP    CLRSC
            LEA     DX,MEXIT        ;set up pointer to exit message
            MOV     AH,9            ;display string function
            INT     21H             ;DOS call
            MOV     AX,1            ;show mouse cursor function
            INT     33H             ;MOUSE call
            MOV     AX,0B800H       ;load address of video RAM segment
            MOV     DS,AX
            MOV     SI,SPOS         ;set up pointer to screen position
            MOV     BYTE PTR [SI+54],'X'    ;write X: to screen
            MOV     BYTE PTR [SI+56],':'
            MOV     BYTE PTR [SI+94],'Y'    ;write Y: to screen
            MOV     BYTE PTR [SI+96],':'
RDMSE:      MOV     AX,3            ;get status/position function
            INT     33H             ;MOUSE call
            MOV     AX,CX           ;read X position
            MOV     DI,SPOS+60      ;set up pointer to X's screen position
            CALL    DISP            ;display number on screen
            MOV     AX,DX           ;read Y position
            MOV     DI,SPOS+100     ;set up pointer to Y's screen position
            CALL    DISP            ;display number on screen
            CMP     BL,1            ;was left button pressed?
            JNZ     RDMSE           ;continue if not
            MOV     AX,2            ;hide mouse cursor function
            INT     33H             ;MOUSE call
EXIT:       .EXIT

DISP PROC NEAR
            PUSH    CX              ;save CX
            MOV     CX,4            ;set up loop counter
NYBL:       ROL     AX,1            ;get upper nybble of AX
            ROL     AX,1
            ROL     AX,1
            ROL     AX,1
            PUSH    AX              ;save AX
            AND     AL,0FH          ;mask out upper nybble
            ADD     AL,30H          ;add ASCII bias
            CMP     AL,3AH          ;are we greater than 9?
            JC      NO7             ;no, go save digit
            ADD     AL,7            ;correct to hex-alpha
```

```
NO7:    MOV     [DI],AL         ;save digit in memory
        POP     AX              ;get AX back
        ADD     DI,2            ;advance memory pointer
        LOOP    NYBL            ;repeat until done
        POP     CX              ;get CX back
        RET
DISP    ENDP

        END
```

The initial group of instructions attempts to determine if a mouse driver is present. If so, the screen is cleared and a help message displayed, along with the mouse's horizontal (X) and vertical (Y) position, as in:

```
X:   0140    Y: 0060
```

The X and Y numbers will automatically be updated as the mouse is moved around. This is accomplished by converting the horizontal and vertical positions (in CX and DX) into four-digit integers and writing them directly to the screen's video RAM. The program exits to DOS when the left button is pressed.

Programming Exercise 8.11: Modify the MOUSETST program so that the keyboard can be used to show and hide the mouse. Hide the mouse if 'H' is hit, and show the mouse if 'S' is pressed.

Programming Exercise 8.12: Modify the MOUSETST program so that the program may also exit if 'Q' is pressed on the keyboard.

Programming Exercise 8.13: Modify the MOUSETST program so that a *target* is displayed on the screen, such as [*]. Exit to DOS only if the mouse button is pushed while the cursor is over the *.

Programming Exercise 8.14: Write a program that will make the speaker beep whenever the mouse gets near the center of the screen.

8.9   WRITING A MEMORY-RESIDENT PROGRAM

A memory-resident program is a program that remains in memory after it returns control to DOS. Normally, when an .EXE or a .COM program completes execution, control is returned to DOS with the instructions:

```
MOV  AH,4CH    ;terminate program function
INT  21H       ;DOS call
```

which are generated by the .EXIT directive. This is, in fact, a DOS call that never returns. When this interrupt function is issued, DOS reclaims all memory allocated to the terminating program and starts up the command processor (COMMAND.COM) again. For a program to remain resident in memory (referred to as a **TSR** for Terminate and Stay Resident program), a different function is needed.

DOS INT   21H, Function 31H: Terminate and Stay Resident

This function is selected when register AH equals 31H. Upon entry, the DX register must contain the number of *paragraphs* that will remain resident. The number of paragraphs must be large enough to hold the program segment prefix and all necessary code and data segments of the TSR.

Example 8.9:   What is the memory size of the resident program after these instructions execute?

```
MOV   DX,60
MOV   AH,31H
INT   21H
```

The DX register specifies 60 paragraphs, which equals 60 * 16 = 960 bytes.

A second method used to terminate and stay resident, which should be used only with .COM files, is INT   27H. This interrupt requires that the DX register be loaded with the size of the program in *bytes*. INT   27H uses the current CS address to preserve the bytes indicated by DX.

Writing a terminate and stay resident program requires more than simply issuing Function 31H (or INT   27H). Some method must be used to give the program control after it terminates. Let us examine one common approach.

Hooking an Interrupt

When a program terminates and stays resident, it must provide for a method of getting control so that it can execute. Otherwise, it will simply remain in memory, not executing, until the computer is rebooted.

One common method used to give control to memory-resident programs is through the use of an interrupt *hook*. An interrupt hook is accomplished through the use of DOS INT   21H, Functions 25H and 35H. A TSR program may use the following steps to hook an interrupt for itself:

1. Get vector of interrupt to hook (with INT   21H, Function 35H).
2. Save old interrupt vector.
3. Make a new vector for the interrupt.
4. Set new vector of hooked interrupt (with DOS INT   21H, Function 25H).

The new interrupt vector usually points to the address of a procedure contained within the memory-resident program. Figure 8.3 shows the result of an interrupt hook for INT 16H. The original vector for INT   16H (F000:2EE8) is read with Function 35H and saved within the data area of the TSR. The CS:IP address (1587:01CD) of the procedure within the TSR that will handle the interrupt replaces the old vector for INT   16H (using Function 25H). Thus, an interrupt *chain* has been created. An INT   16H will now cause the TSR to get control. Because the old vector of INT   16H is saved in the TSR's data area, the TSR can JMP or CALL the original INT   16H handler as required.

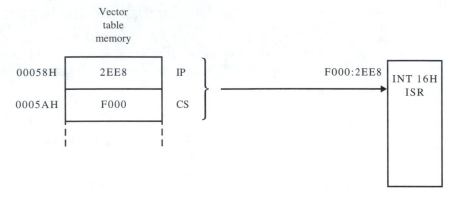

(a) Original interrupt vector

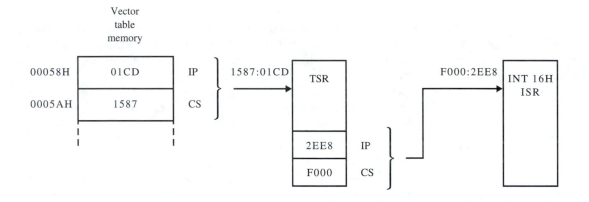

(b) Modified interrupt chain

FIGURE 8.3   Hooking interrupt 16H: (a) original interrupt vector; (b) modified interrupt chain

Example 8.10:   Specifically, INT 16H may be hooked through the following instructions:

```
;storage for old interrupt vector
OLDIPCS    DD    ?    ;need a double word for CS:IP
;save current vector address
MOV    AL,16H    ;interrupt number
MOV    AH,35H    ;get interrupt vector function
INT    21H       ;DOS call
MOV    WORD PTR OLDIPCS,BX      ;save INT  16H instruction pointer
MOV    WORD PTR OLDIPCS[2],ES   ;save INT  16H code segment
;set new vector address
MOV    AL,16H    ;interrupt number
MOV    AH,25H    ;set interrupt vector function
LEA    DX,NEWISR    ;load IP of new interrupt service routine
```

```
PUSH   CS          ;put copy of CS into DS
POP    DS
INT    21H         ;DOS call
```

The code beginning at the address specified by NEWISR will get control whenever INT 16H is issued.

The TSR may give control to the old interrupt service routine in two ways: through the use of a FAR JMP or a FAR CALL. Considering the code from Example 8.10, a FAR JMP is performed by the instruction:

```
JMP    OLDIPCS
```

In this case, the IRET at the end of the original ISR will cause a return to the program that issued the interrupt.

To use a FAR CALL it is necessary to adjust the processor's stack pointer. Remember that a FAR CALL pushes two words onto the stack, the CS:IP return address. The original ISR's IRET instruction will pop *three* words from the stack, the CS:IP return address *and* the flags. So, to use a FAR CALL within the TSR, we need the following code:

```
PUSHF              ;adjust stack
CALL   OLDIPCS     ;execute original ISR
```

Either way, it is necessary for the TSR to pass control to the original ISR at some point, or risk interfering with the normal operation of the computer.

PHONES: A Pop-Up Telephone Pad

The PHONES program presented here installs itself as a memory-resident program and hooks INT 16H so that it can watch the keystrokes entered on the keyboard. When a special key is pressed (Alt-P), the TSR performs its intended task and writes a small telephone directory to the display. All other keyboard keys are ignored.

The telephone directory is output as a multiline white character on blue background message. The screen information overwritten by the telephone directory is lost.

When Alt-P is entered (after PHONES has been installed as a TSR), the screen should look something like this:

```
Alan .................... 555 - 5017
Charlie ................. 555 - 5325
Dave .................... 555 - 7854
Michele ................. 555 - 7514
Mike .................... 555 - 7015
```

The telephone information will scroll off the screen line by line as Enter is used.

```
;Program PHONES.ASM: Memory-resident telephone pad.
;Using .COM format.
;
       .MODEL TINY
       .CODE
       ORG    100H      ;needed for .COM structure
START: JMP    INIT
```

```
OLDIPCS  DD      ?
PMSG     DB      ' Alan ................... 555 - 5017 ',1
         DB      ' Charlie ................ 555 - 5325 ',1
         DB      ' Dave ................... 555 - 7854 ',1
         DB      ' Michele ................ 555 - 7514 ',1
         DB      ' Mike ................... 555 - 7015 ',1
         DB      0
IPOS     EQU     20*2                    ;initial screen position
SPOS     DW      IPOS
HOTKEY   EQU     1900H                   ;19H for ALT-P, 00H for extended key

POPUP    PROC    FAR
         PUSHF                           ;save flags
         CMP     AH,0                    ;is function 00H requested?
         JNZ     EXTKEY                  ;no, go look for other function
         JMP     RDKEY                   ;go read keyboard
EXTKEY:  CMP     AH,10H                  ;is function 10H requested?
         JNZ     EXIT                    ;no, go process old ISR
RDKEY:   PUSHF                           ;fix stack for CALL
         CALL    CS:OLDIPCS              ;process old ISR (read keyboard)
         CMP     AX,HOTKEY               ;does key match popup code?
         JNZ     BYE                     ;no, just exit now
         PUSH    DI                      ;save registers
         PUSH    SI
         PUSH    DS
         MOV     AX,0B800H               ;load address of video RAM
         MOV     DS,AX
         LEA     SI,CS:PMSG              ;set up pointer to popup string
N1:      MOV     DI,CS:SPOS              ;set up pointer to screen position
N2:      MOV     AL,CS:[SI]              ;read popup character
         INC     SI                      ;advance to next character
         CMP     AL,1                    ;was it end-of-line code?
         JZ      N3                      ;yes, go update position
         OR      AL,AL                   ;was it end-of-string character?
         JZ      N4                      ;yes, go restore registers
         MOV     [DI],AL                 ;otherwise, write character to screen
         MOV     BYTE PTR [DI+1],17H     ;force white on blue attribute
         ADD     DI,2                    ;advance to next screen position
         JMP     N2                      ;and repeat
N3:      ADD     CS:SPOS,160             ;advance screen position to next line
         JMP     N1                      ;and continue
N4:      MOV     CS:SPOS,IPOS            ;reset screen position
         POP     DS                      ;restore registers
         POP     SI
         POP     DI
         MOV     AX,3920H                ;return <space> for <Alt-P>
BYE:     POPF                            ;restore flags
         IRET                            ;return from interrupt
EXIT:    POPF                            ;restore flags
         JMP     CS:OLDIPCS              ;go process old interrupt
POPUP    ENDP

INIT:    MOV     AL,16H                  ;keyboard interrupt number
         MOV     AH,35H                  ;get interrupt vector function
         INT     21H                     ;DOS call
         MOV     WORD PTR CS:OLDIPCS,BX          ;save old interrupt IP
         MOV     WORD PTR CS:OLDIPCS[2],ES       ;save old interrupt CS
         LEA     DX,CS:POPUP             ;load address of popup ISR
         MOV     AL,16H                  ;keyboard interrupt number
         MOV     AH,25H                  ;set interrupt vector function
```

```
        INT     21H                 ;DOS call
        LEA     DX,CS:INIT          ;load required memory size in bytes
        MOV     CL,4                ;set up shift count
        SHR     DX,CL               ;convert size into paragraphs
        INC     DX                  ;round up for good measure
        MOV     AH,31H              ;keep program resident function
        INT     21H                 ;DOS call

        END     START               ;must specify starting address
```

The PHONES program is designed as a .COM program (via the .TINY directive) with all code *and data* contained within the code segment. This accounts for the operand references containing the segment override prefix CS, as in CS:[SI] and CS:OLDIPCS.

The INIT portion of the program is used to install the TSR and is not required after the interrupt chain has been modified. The instructions:

```
LEA     DX,CS:INIT
MOV     CL,4
SHR     DX,CL
INC     DX
MOV     AH,31H
INT     21H
```

are used to compute the length of the TSR in paragraphs, *up to* the INIT code.

We can examine the effects on allocated memory before and after execution of PHONES with DOS's MEM/C command. For example, before PHONES is executed, MEM/C shows the following allocation:

```
Modules using memory below 1 MB:
  Name           Total        =    Conventional   +   Upper Memory
  -----       ----------------     ----------------     --------------
  MSDOS        17,565   (17K)       17,565   (17K)           0    (0K)
  HIMEM         1,168    (1K)        1,168    (1K)           0    (0K)
  EMM386        3,120    (3K)        3,120    (3K)           0    (0K)
  COMMAND       2,928    (3K)        2,928    (3K)           0    (0K)
  SMARTDRV     27,504   (27K)       11,104   (11K)      16,400   (16K)
  MSCDEX       36,224   (35K)       36,224   (35K)           0    (0K)
  SETVER          800    (1K)            0    (0K)         800    (1K)
  MOUSE        15,808   (15K)            0    (0K)      15,808   (15K)
  MTMCDE       57,664   (56K)            0    (0K)      57,664   (56K)
  DBLSPACE     39,648   (39K)            0    (0K)      39,648   (39K)
  DOSKEY        4,144    (4K)            0    (0K)       4,144    (4K)
  Free        607,168  (593K)      583,008  (569K)      24,160   (24K)
```

When PHONES.COM is executed, we see the following change in memory allocation:

```
Modules using memory below 1 MB:
  Name           Total        =    Conventional   +   Upper Memory
  --------    ----------------     ----------------     --------------
  MSDOS        17,565   (17K)       17,565   (17K)           0    (0K)
  HIMEM         1,168    (1K)        1,168    (1K)           0    (0K)
  EMM386        3,120    (3K)        3,120    (3K)           0    (0K)
  COMMAND       2,928    (3K)        2,928    (3K)           0    (0K)
  SMARTDRV     27,504   (27K)       11,104   (11K)      16,400   (16K)
  PHONES          832    (1K)          832    (1K)           0    (0K)
  MSCDEX       36,224   (35K)       36,224   (35K)           0    (0K)
  SETVER          800    (1K)            0    (0K)         800    (1K)
```

```
MOUSE      15,808   (15K)         0   (0K)     15,808   (15K)
MTMCDE     57,664   (56K)         0   (0K)     57,664   (56K)
DBLSPACE   39,648   (39K)         0   (0K)     39,648   (39K)
DOSKEY      4,144    (4K)         0   (0K)      4,144    (4K)
Free      606,336  (592K)   582,176 (569K)     24,160   (24K)
```

In addition to the change in allocated memory, we should examine the differences in the interrupt vector table for INT 16H. The addresses for INT 16H vector are 00058H through 0005BH. These addresses can be examined using DEBUG, with the command:

```
-D 0:58 L 4
```

When this is done *before* PHONES is executed, we get:

```
E8 2E 00 F0
```

which indicates the address F000:2EE8. This is the original address of the INT 16H handler (somewhere in the system BIOS ROM).

After PHONES is executed, DEBUG will show the vector:

```
CD 01 87 15
```

which indicates the address 1587:01CD, the address of the POPUP code in PHONES.

So, we have proof that PHONES has been installed as a TSR program. Because it hooked into INT 16H, it will gain control frequently (every time a key is hit). Whenever Alt-P is pressed, the telephone information in the PMSG string will be copied to screen RAM. Notice that PMSG is not structured as a character string terminated by a '$' character (for use with Function 9: display string), but instead uses the number 1 to represent the end of a line of text, and a 0 to indicate the end of the list. A small group of instructions is used to copy the PMSG data to screen RAM and adjust the screen attributes.

Attention! It is extremely important to preserve the state of the processor registers used within the TSR! Only a small amount of stack space is provided by DOS for interrupt handling, so use and save registers sparingly.

If the TSR issues an INT of its own (possibly to an INT 21H function), there may not be enough stack space available to save everything. It may be necessary for the TSR to create its own stack environment while it has control. Then, before exiting, the original stack area must be reassigned.

Programming Exercise 8.15:   Modify the PHONES program so that the screen text overwritten by the telephone directory is saved in memory and then restored when any key is hit after Alt-P.

Programming Exercise 8.16:   Try to rewrite PHONES with a call to INT 21H, Function 09H to display the telephone directory. What problems, if any, do you encounter?

8.10   TICTAC: A GAME FOR A CHANGE

Writing a game program may not be considered an advanced programming assignment, but the effort still has its benefits. Many games are simple and fun to play, and do not require a tremendous amount of programming experience to develop. Also, it is often easy to spot incorrect program operation when you have a good idea of what the program should be doing next. Debugging a game program usually offers many valuable programming lessons.

In this section we will examine a simple tic-tac-toe program that does not take much effort to beat. In the human versus computer TICTAC.ASM program presented here, the strategy of the computer is straightforward: put an O in the first empty position encountered. It is difficult to lose a game against the computer due to this simple strategy.

Even though the computer's strategy lacks skill, writing the code for the entire program is still a challenge. The game can be broken down into a number of individual procedures. These procedures do the following:

*   Display the game board
*   Get a valid move from the user
*   Determine a valid move for the computer
*   Check to see who won

You may notice that other games, including checkers and chess, perform the same operations. It is very satisfying to write new game procedures one by one and watch the game develop into a playable program.

Examine the TICTAC.ASM program.

```
;Program TICTAC.ASM: Play Tic Tac Toe with the computer.
;
        .MODEL SMALL
        .DATA
GMSG    DB   'Computer TIC TAC TOE.',0DH,0AH
        DB   'User is X, computer is O',0DH,0AH,0DH,0AH,'$'
BOARD   DB   '123456789'
BTXT    DB   0DH,0AH
        DB   '   |   |   ',0DH,0AH
        DB   '-----------',0DH,0AH
        DB   '   |   |   ',0DH,0AH
        DB   '-----------',0DH,0AH
        DB   '   |   |   ',0DH,0AH,0DH,0AH,'$'
BPOS    DB   2,6,10,24,28,32,46,50,54
PMSG    DB   'Enter your move (0 to 9): $'
PIM1    DB   0DH,0AH,'That move does not make sense, try again.',0DH,0AH,'$'
PIM2    DB   0DH,0AH,'That square is occupied, try again.',0DH,0AH,'$'
CMSG    DB   'I choose square $'
CRLF    DB   0DH,0AH,'$'
WINS    DW   1,2,3, 4,5,6, 7,8,9          ;any row
        DW   1,4,7, 2,5,8, 3,6,9          ;any column
        DW   1,5,9, 3,5,7                 ;either diagonal
XWIN    DB   'X wins the game!',0DH,0AH,'$'
OWIN    DB   'O wins the game!',0DH,0AH,'$'
MTIE    DB   'The game is a tie.',0DH,0AH,'$'
```

```
              .CODE
              .STARTUP
              LEA    DX,GMSG    ;set up pointer to greeting
              MOV    AH,9       ;display string function
              INT    21H
              CALL   SHOWBRD    ;display board
NEXT:  CALL   PMOVE      ;get player move
              CALL   SHOWBRD    ;display board
              CALL   CHECK      ;did player win or tie?
              JZ     EXIT
              CALL   CMOVE      ;let computer move
              CALL   SHOWBRD    ;display board
              CALL   CHECK      ;did computer win or tie?
              JZ     EXIT
              JMP    NEXT       ;continue with game
EXIT:  .EXIT


SHOWBRD PROC  NEAR
              MOV    CX,9                  ;set up loop counter
              SUB    SI,SI                 ;set up index pointer
LBC:   MOV    AL,BPOS[SI]           ;get a board position
              CBW                          ;convert to word
              MOV    DI,AX                 ;set up pointer to board string
              MOV    AL,BOARD[SI]          ;get player symbol
              MOV    BTXT[DI],AL           ;write into board string
              INC    SI                    ;advance index pointer
              LOOP   LBC                   ;repeat for all nine positions
              LEA    DX,BTXT               ;set up pointer to board string
              MOV    AH,9                  ;display string function
              INT    21H                   ;DOS call
              RET
SHOWBRD ENDP


PMOVE   PROC NEAR
              LEA    DX,PMSG               ;set up pointer to player string
              MOV    AH,9                  ;display string function
              INT    21H                   ;DOS call
              MOV    AH,1                  ;read keyboard function
              INT    21H                   ;DOS call
              CMP    AL,'1'                ;ensure user input is a digit
              JC     BPM
              CMP    AL,'9'+1
              JNC    BPM
              SUB    AL,31H                ;remove ASCII bias
              CBW                          ;convert to word
              MOV    SI,AX                 ;set up index pointer
              MOV    AL,BOARD[SI]          ;get board symbol
              CMP    AL,'X'                ;is position occupied?
              JZ     PSO
              CMP    AL,'O'
              JZ     PSO
              MOV    BOARD[SI],'X'         ;save player move
              LEA    DX,CRLF               ;set up pointer to newline string
              MOV    AH,9                  ;display string function
              INT    21H                   ;DOS call
              RET
BPM:   LEA    DX,PIM1               ;set up pointer to illegal string
STP:   MOV    AH,9                  ;display string function
              INT    21H                   ;DOS call
              JMP    PMOVE                 ;go give user a second chance
```

```
PSO:    LEA  DX,PIM2           ;set up pointer to occupied string
        JMP  STP               ;go process error message
        RET
PMOVE   ENDP


CMOVE   PROC NEAR
        SUB  SI,SI             ;clear index pointer
NCM:    MOV  AL,BOARD[SI]      ;get board symbol
        CMP  AL,'X'            ;is position occupied?
        JZ   STN
        CMP  AL,'O'
        JZ   STN
        MOV  BOARD[SI],'O'     ;save computer move (not very tough, is it?)
        MOV  AX,SI             ;save move value
        PUSH AX
        LEA  DX,CMSG           ;set up pointer to choice string
        MOV  AH,9              ;display string function
        INT  21H              ;DOS call
        POP  DX               ;get move value back
        ADD  DL,31H            ;add ASCII bias
        MOV  AH,2              ;display character function
        INT  21H              ;DOS call
        LEA  DX,CRLF           ;set up pointer to newline string
        MOV  AH,9              ;display string function
        INT  21H              ;DOS call
        RET
STN:    INC  SI               ;advance to next position
        JMP  NCM              ;go check next position
CMOVE   ENDP


CHECK   PROC NEAR
        SUB  SI,SI             ;clear index pointer
        MOV  CX,8             ;set up loop counter
CAT:    MOV  DI,WINS[SI]      ;get first board position
        MOV  AH,BOARD[DI-1]   ;get board symbol
        MOV  DI,WINS[SI+2]    ;get second board position
        MOV  BL,BOARD[DI-1]   ;get board symbol
        MOV  DI,WINS[SI+4]    ;get third board position
        MOV  BH,BOARD[DI-1]   ;get board symbol
        ADD  SI,6            ;advance to next set of positions
        CMP  AH,BL            ;do all three symbols match?
        JNZ  NMA
        CMP  AH,BH
        JNZ  NMA
        CMP  AH,'X'          ;does match contain X?
        JNZ  WIO
        LEA  DX,XWIN          ;set up pointer to x-wins string
        JMP  EXC             ;go process string
WIO:    LEA  DX,OWIN          ;set up pointer to o-wins string
        JMP  EXC             ;go process string
NMA:    LOOP CAT             ;no match, try another group
        SUB  SI,SI             ;clear index pointer
CFB:    MOV  AL,BOARD[SI]      ;get board symbol
        CMP  AL,'X'          ;is symbol X?
        JZ   IAH
        CMP  AL,'O'          ;is symbol O?
        JZ   IAH
        RET                  ;no tie yet
IAH:    INC  SI              ;advance to next position
        LOOP CFB             ;go check another board symbol
```

```
             LEA    DX,MTIE          ;set up pointer to tie message
   EXC:      MOV    AH,9             ;display string function
             INT    21H              ;DOS call
             SUB    AL,AL            ;set zero flag
             RET
   CHECK     ENDP

             END
```

The tic-tac-toe board is contained within the text string BTXT. The positions in the BTXT array that will contain board symbols are stored in the BPOS array. For example, the position of the first square in the first row is 2. The position of the first square in the second row is 24. The position of the last square in the third row is 54. These position values can be verified by carefully counting character places in BTXT. Together with the symbols stored in the BOARD array, the entire board can be analyzed. A sample tic-tac-toe board (generated during a game) might look like this:

```
         X | 0 | 3
         --------------
         X | 5 | 6
         --------------
         7 | 8 | 9
```

where each unoccupied square contains a digit. The board is displayed by the SHOWBRD procedure.

The user enters a move through the PMOVE procedure. This procedure gets a user move from the keyboard and checks to ensure that the move is valid (a legal digit representing an unoccupied board position). Then the BOARD array is updated. The BOARD array for the sample tic tac toe board previously shown is:

```
XO3X56789
```

If the computer puts an O into position 3, BOARD will become:

```
XOOX56789
```

Once the BOARD array is updated, it is checked for a win. This is done by the CHECK procedure, which tests all rows, columns, and diagonals for three Xs or three Os. It indicates either win (or a tie) by setting the zero flag upon return. CHECK uses the WINS array to examine eight groups of three board symbols, which represent all three rows and columns, and both diagonals.

Programming Exercise 8.17: Modify the TICTAC program so that the computer's first move is the center square. If the center square is already occupied, one of the four corner squares should be chosen.

Programming Exercise 8.18: Modify the TICTAC program so that the user or computer may go first.

Programming Exercise 8.19:   Modify the TICTAC program so that the user may choose X or O.

Programming Exercise 8.20:   Write a BLOCK procedure that examines the tic-tac-toe board and determines if any row, column, or diagonal contains exactly two Xs and a blank. Return the position of the blank (1–9) or 0 if no blocking move is found.

8.11   PROTECTED-MODE DETECTION

The 80386, 80486, and Pentium microprocessors are capable of operating in two basic modes of operation: *real* mode and *protected* mode. In real mode, they act like really fast 8086s. Registers are 16 bits wide. Memory space is limited to 1MB. No special instructions are allowed.

In protected mode, the full power of the processor is available. Registers are 32 bits wide, and the memory space can be as large as 4 **gigabytes** (4 billion bytes).

DOS is an example of a program that runs in real mode, because DOS was created for an 8086-based machine. Therefore, DOS has access to only 1MB of memory. An example of a program that runs in protected mode is Windows. As a matter of fact, most Windows application programs also run in protected mode. This can be verified easily. Find an .EXE program in the WINDOWS directory, such as CALC.EXE. If you execute CALC from DOS, you will get the following result:

```
This program requires Microsoft Windows.
```

CALC.EXE was able to determine that the computer did not have protected-mode capability. However, when Windows loads and executes, a DOS Protected-Mode Interface (DPMI) is established through Windows code. This DPMI is accessible through the use of *multiplex* interrupt 2FH. The multiplex interrupt is provided by DOS to support background programs (such as PRINT or DOSVER). The multiplex interrupt requires AX to be equal to 1687H to check DPMI status. If a DPMI program is active, we get the following:

AX: 0

BX: 32-bit support flag

CL: Processor type

DX: DPMI version number

SI: Private data paragraph count

DI: Entry point offset

ES: Entry point segment

The DPMISTAT program shown here uses this information to display the DPMI status.

```
;Program DPMISTAT.ASM: Check for DOS Protected-Mode Interface.
;
            .MODEL SMALL
            .DATA
NODPMI    DB    'No DOS Protected-Mode Interface detected.',0DH,0AH,'$'
NOTSUP    DB    '32-bit programs are NOT supported.',0DH,0AH,'$'
ARESUP    DB    '32-bit programs ARE supported.',0DH,0AH,'$'
P286MSG   DB    'Processor is an 80286.',0DH,0AH,'$'
P386MSG   DB    'Processor is an 80386.',0DH,0AH,'$'
P486MSG   DB    'Processor is an 80486 or Pentium.',0DH,0AH,'$'
PUNMSG    DB    'Unidentified processor.',0DH,0AH,'$'

            .CODE
            .STARTUP
            MOV    AX,1687H      ;get mode switch entry point
            INT    2FH           ;MULTIPLEX interrupt
            OR     AX,AX         ;DMPI present if AX = 0
            JZ     SHOWB         ;go show 32-bit status
            LEA    DX,NODPMI     ;set up pointer to nodpmi message
            MOV    AH,9          ;display string function
            INT    21H           ;DOS call
            JMP    EXIT
SHOWB:      OR     BX,BX         ;are 32-bit programs supported?
            JNZ    YES32         ;jump if yes
            LEA    DX,NOTSUP     ;set up pointer to not supported message
            MOV    AH,9          ;display string function
            INT    21H           ;DOS call
            JMP    SHOWP         ;go show processor type
YES32:      LEA    DX,ARESUP     ;set up pointer to supported message
            MOV    AH,9          ;display string function
            INT    21H           ;DOS call
SHOWP:      CMP    CL,2          ;is processor 80286?
            JNZ    P2
            LEA    DX,P286MSG    ;set up pointer to 80286 message
            JMP    SCPU          ;go display string
P2:         CMP    CL,3          ;is processor 80386?
            JNZ    P3
            LEA    DX,P386MSG    ;set up pointer to 80386 message
            JMP    SCPU          ;go display string
P3:         CMP    CL,4          ;is processor 80486?
            JNZ    P4
            LEA    DX,P486MSG    ;set up pointer to 80486 message
            JMP    SCPU          ;go display string
P4:         LEA    DX,PUNMSG     ;set up pointer to unidentified message
SCPU:       MOV    AH,9          ;display string function
            INT    21H           ;DOS call
EXIT:       .EXIT

            END
```

When the program is executed from the DOS environment, the result is:

```
No DOS Protected-Mode Interface detected.
```

But, if we start up Windows (via WIN /3) and then choose MS-DOS Prompt from the Main window, we will get access to DOS from *inside* Windows.

When DPMISTAT is executed now, we get:

```
32-bit programs ARE supported.
Processor is an 80486 or Pentium.
```

So, checking for the presence of a DPMI program is important for programs that desire access to the 32-bit processing power of protected mode.

8.12   INTERFACING C WITH ASSEMBLY LANGUAGE

The C programming language has become very popular over the last few years, primarily because it is extremely *portable*. A portable programming language has the property that its programs are easily transferred to other machines. For example, a C program written on a VAX™ mainframe will usually compile without errors when ported to an MS-DOS (80x86-based) or Macintosh® (680x0-based) machine. The purpose of the C compiler is to convert the statements in a C program into the correct sequence of machine language instructions for the host processor.

In this section, we will examine the machine code generated by a C compiler and the standard way the C compiler uses memory to access information.

One way to interface C with assembly language is to put the instructions directly into the C program. Examine the following C program:

```
main ()
{
        asm     mov     dl,0x30     /* begin counter at '0' */
        asm     mov     ah,2        /* display character function */
        next:
        asm     int     0x21        /* DOS call */
        asm     inc     dl          /* increment counter */
        asm     cmp     dl,0x3a     /* is counter > '9' ? */
        asm     jnz     next
}
```

Notice that the program is written almost entirely in assembly language, with the exception of the **main( )** keyword, and the braces {} surrounding the statements. The C keyword **asm** is used to perform *in-line assembly* of the assembly language statements in the C program. The six assembly language statements form a loop that outputs the digits 0 through 9 to the display.

Normally, this C program would be compiled to create an executable program. This process, however, will not allow us to see the final assembly language generated by the C compiler. So, a different technique will be used to create an assembly language *source file* from the C program. A program available with many C compiler packages is used to convert the .C program into an .ASM program. This program is called TCC.EXE and is used like so:

```
TCC -S -EFILENAME.ASM FILENAME.C
```

This DOS command tells TCC to create FILENAME.ASM using FILENAME.C. The result of TCC's execution is as follows:

```
_TEXT   segment byte public 'CODE'
    ;
    ;     main()
    ;
```

```
            assume  cs:_TEXT
_main       proc    near
            push    bp
            mov     bp,sp
    ;
    ;       {
    ;               asm     mov     dl,0x30     /* begin counter at '0' */
    ;
                    mov     dl,030H
    ;
    ;               asm     mov     ah,2        /* display character function */
    ;
                    mov     ah,2
@1@86:
    ;
    ;               next:
    ;               asm     int     0x21        /* DOS call */
    ;
                    int     021H
    ;
    ;               asm     inc     dl          /* increment counter */
    ;
                    inc     dl
    ;
    ;               asm     cmp     dl,0x3a     /* is counter > '9' ? */
    ;
                    cmp     dl,03aH
    ;
    ;               asm     jnz     next
    ;
                    jne     short @1@86
    ;
    ;       }
    ;
            pop     bp
            ret
_main       endp
_TEXT       ends
            end
```

The assembly language generated by the asm keyword is almost identical to that used in the asm statement, except when numbers or labels are used in the operand field. For example, the 0x30 operand in the first statement is used to represent the value 30 hexadecimal. This is the standard way a hexadecimal number is represented in C. Recall that the assembly language required by ML requires 30H to specify 30 hexadecimal. This is illustrated in the instruction mov   dl,30H.

When a label is used, TCC must generate a unique name for it. The label next has the name @1@86 associated with it. Notice that the asm   jnz   next statement created the instruction jne   short   @1@86. Both are functionally equivalent.

Finally, TCC supplies entry and exit code to allow execution of the program in the DOS environment. The instructions for program entry are:

```
push   bp
mov    bp,sp
```

and for program exit, we have:

```
pop    bp
ret
```

These two groups of instructions indicate that the BP and SP registers play an important role in C's interface with assembly language. This role is well defined and can best be explained through the examination of another C program and its associated assembly language code (generated by TCC).

The following C program contains a function called **addem** that uses two input parameters, x and y, both of which are defined as integers.

```
#include <stdio.h>

void addem(int x, int y);

main()
{
        int a, b;

        a = 5;
        b = 7;
        addem(a, b);
}

void addem(int x, int y)
}
        int c;

        c = x + y;
        printf("The sum is %3d\n" , c);
}
```

This can be seen in the function's definition statement:

```
void addem(int x, int y);
```

The function definition statement provides important information to the C compiler, which must ensure that the parameter values are accessible when the function is called in the main program. For example, when the C compiler encounters the statement:

```
addem(a, b);
```

it creates a working environment for the function according to a standard known as the *C calling convention*. This calling convention dictates that parameter values used in a function call be pushed onto a run-time stack in right-to-left order. Thus, the value of variable b (parameter y) is pushed onto the run-time stack, followed by the value of variable a (parameter x). This is illustrated in Figure 8.4. Notice that the integer values of both variables (7 and 5) each occupy one word of stack space. This is due to the C compiler's use of a word to represent an integer.

After the function parameters are pushed, the program's return address is pushed. This is the address that will get control when the function completes execution. Thus, a portion of the run-time stack has been constructed to allow execution of the addem( ) function. A second portion, which was constructed for the main( ) routine, has been previously loaded onto the stack. The first value pushed onto the stack in this portion was the original value of the BP register when main( ) got control. The BP register is then updated to point to the bottom of main( )'s run-time stack. For purposes of this discussion, a set of addresses has been included in Figure 8.4 to help illustrate exactly where information is being saved. Thus, the bottom of the run-time stack is the word located at address 1012.

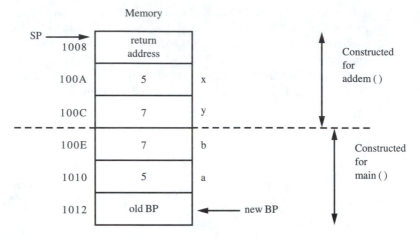

FIGURE 8.4   Run-time stack during a C function call

The new value of the BP register is used by main( ) to access all information on the run-time stack. When the compiler encounters:

```
int a, b;
```

it reserves two words of run-time stack space for the values of variables a and b. Figure 8.4 shows that these two words immediately follow the original BP register value. The compiler may now refer the variable a using the operand [bp-2]. Variable b is accessed through the operand [bp-4]. Note that the order of the variables stored in the run-time stack is identical to the order in which they were declared. This is exactly opposite the order used to push parameter values for a function.

To better understand the run-time stack usage, let us examine the assembly language file created by TCC for our current C program.

```
_TEXT   segment byte public 'CODE'
    ;
    ;    main()
    ;
        assume  cs:_TEXT
_main   proc    near
        push    bp
        mov     bp,sp
        sub     sp,4
    ;
    ;    {
    ;
    ;            int a, b;
    ;
    ;            a = 5;
    ;
        mov     word ptr [bp-2],5
    ;
    ;            b = 7;
    ;
        mov     word ptr [bp-4],7
    ;
    ;            addem(a, b);
    ;
        push    word ptr [bp-4]
```

```
            push    word ptr [bp-2]
            call    near ptr _addem
            pop     cx
            pop     cx
    ;
    ;       }
    ;
            mov     sp,bp
            pop     bp
            ret
_main       endp
    ;
    ;       void addem(int x, int y)
    ;
            assume  cs:_TEXT
_addem      proc    near
            push    bp
            mov     bp,sp
            sub     sp,2
    ;
    ;       {
    ;              int c;
    ;
    ;              c = x + y;
    ;
            mov     ax,word ptr [bp+4]
            add     ax,word ptr [bp+6]
            mov     word ptr [bp-2],ax
    ;
    ;              printf("The sum is %3d\n", c);
    ;
            push    word ptr [bp-2]
            mov     ax,offset DGROUP:s@
            push    ax
            call    near ptr _printf
            pop     cx
            pop     cx
    ;
    ;       }
    ;
            mov     sp,bp
            pop     bp
            ret
_addem      endp
_TEXT       ends
_DATA       segment word public 'DATA'
s@          label   byte
            db      'The sum is %3d'
            db      10
            db      0
_DATA       ends
            end
```

The first group of instructions in _main are:

```
push    bp
mov     bp,sp
sub     sp,4
```

These instructions save the original value of the BP register, adjust the BP register so that it points to the bottom of the run-time stack, and adjust the SP register so that space is reserved for the integer variables a and b (as shown in Figure 8.4). The statement:

```
a = 5;
```

is coded in assembly language as:

```
mov    word ptr [bp-2],5
```

This translates to the address 1010 in the run-time stack of Figure 8.4. A similar method is used to initialize the value of variable b, which is stored at address 100E in the run-time stack.

When the addem(a, b); statement is encountered, the compiler generates instructions to place addem's parameter values onto the run-time stack. These instructions are as follows:

```
push   word ptr [bp-4]
push   word ptr [bp-2]
```

Remember that the [bp-4] value is the value of variable b and that [bp-2] refers to the value of variable a. Thus, the parameter values for the addem( ) function have been pushed onto the run-time stack in right-to-left order (y then x). The return address is automatically pushed by the:

```
call   near ptr _addem
```

instruction.

When the _addem procedure gets control, it performs the same initial run-time stack allocation that main( ) did. The original value of the BP register is saved and then reloaded, and the SP register is updated to reserve room for the variable c used by addem( ). Figure 8.5 shows the new run-time stack.

Now that the _addem procedure is executing, how does it access its x and y parameter values? By the nature of the C calling convention, the parameter's values are accessed by use of the BP register. However, because a return address was pushed onto the

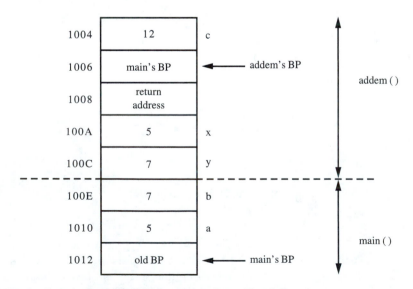

FIGURE 8.5   Run-time stack while executing the addem() function

run-time stack *after* the two parameter values, the offsets used with the BP register must be changed. For example, in the statement:

$$c = x + y$$

the value of variable x is accessed through the operand [bp+4] and variable y through [bp+6]. This is necessary to skip over the return address in the run-time stack. Use the address values in Figure 8.5 to prove the correctness of these operand addresses.

You may agree that the run-time stack gets very cluttered with information during a program's execution. Having to skip over return addresses and reserve room for variable storage makes it hard to keep track of where the actual return address for any given procedure is. The C calling convention offers an easy solution to this problem. Because the SP and BP registers are updated upon entry to any routine, upon exit we need only:

```
mov   sp,bp
pop   bp
ret
```

to always return correctly.

Example 8.11:   The following C program calls an external integer function xplusy():

```
#include <stdio.h>

extern int xplusy(int x, int y);

main()
{
    int a,b;
    a = 5;
    b = 7;
    printf("The sum is %d\n",xplusy(a,b));
}
```

The xplusy() function is an assembly language routine. A number of assembler directives may be used when writing xplusy() in order to implement the C calling convention and correctly manipulate the run-time stack. Here is what xplusy() looks like:

```
        .MODEL  SMALL,C
        .CODE

        PUBLIC xplusy

xplusy  PROC    NEAR C, x:WORD, y:WORD
        LOCAL   z:WORD

        MOV     AX,x       ;load value of x into AX
        ADD     AX,y       ;add value of y to AX
        MOV     z,AX       ;store AX in z
        RET

xplusy  ENDP

        END
```

The PUBLIC keyword must be used because xplusy() is an external reference to the ADDER.C program. The PROC NEAR C directive (along with .MODEL SMALL,C) informs the assembler that procedures must have the standard entry and exit code required by the run-time stack. This code is automatically generated when the assembler encounters the first instruction of the procedure (MOV AX,x) and the last (RET). The PROC directive also allows us to specify the formal parameters used by the procedure (x:WORD and y:WORD) and their data type. A WORD corresponds to an integer in C.

The LOCAL directive is used to reserve room on the run-time stack for variables that are local to the procedure (z:WORD). All variables are referenced by the standard [BP +/– offset] notation.

Since xplusy() is an integer function, it must return an integer value. This value is returned in AX. Simple data types, such as chars and ints, are returned in the accumulator.

Figure 8.6 details the resulting machine code of the xplusy() function. Note the offsets associated with x [BP+4], y [BP+6], and z [BP-2]. These are the expected locations within the run-time stack for these variables.

When the ADDER program is compiled and linked with xplusy(), the resulting execution is as follows:

```
The sum is 12
```

This simple example should help you begin writing your own assembly language functions for C.

Understanding how the C calling convention operates allows you to write your own interface code for existing C routines. Through correct use of the run-time stack (via the BP register), you should be able to access and use compiler-generated C code for your own purposes.

Programming Exercise 8.21: Write an in-line assembly language C program that displays your initials on the screen.

```
55           PUSH   BP          ⎫  Generated by
8B EC        MOV    BP,SP       ⎬  PROC NEAR C

83 C4 FE     ADD    SP,-02         Generated by LOCAL z:WORD

8B 46 04     MOV    AX,[BP+04]     x:WORD

03 46 06     ADD    AX,[BP+06]     y:WORD

89 46 FE     MOV    [BP+02],AX     z:WORD

8B E5        MOV    SP,BP       ⎫  Automatically generated
5D           POP    BP          ⎬  when RET is seen in source file

C3           RET
```

FIGURE 8.6   xplusy() machine code

Programming Exercise 8.22: Change the variable declarations for a, b, and c from int to float in the second C program. Make any other necessary changes to ensure correct compilation and then create an assembly language file using TCC. What is pushed onto the run-time stack now?

Programming Exercise 8.23: Explain what is done to the run-time stack in the second C program to support the printf( ) function.

Programming Exercise 8.24: Write an assembly language C function called max() that compares two integers x and y and returns the value of the larger integer.

8.13 TROUBLESHOOTING TECHNIQUES

Many different types of programming applications were covered in this chapter. Let us take a look at some of the techniques presented, as they may be useful when beginning a new application.

- Create source files that contain code for all of the things you normally do in a program, such as display strings, input and output numbers, or scan the command tail. Assemble the source files and combine them into a library of code modules that you can link to. This will allow you to reuse your code when writing new applications.
- The execution time of a block of code depends on many factors, such as clock speed, processor type (8086 vs. Pentium), and the number of branch instructions. The calculated execution time should be thought of as an estimate and used as a guide to control further code development.
- Use memory-resident programs with caution. It is easy to crash the system with an improperly written TSR. If you intercept an interrupt to gain access to the TSR program, remember to restore the original interrupt vector before exiting.
- Add memory management and mouse support to your list of PC software/hardware that you can control. You may wish to create a subdirectory that contains working examples of all the different control applications (speaker and keyboard programs from Chapter 7, MEMMAN and MOUSE from this chapter). The example programs can serve as templates for further development.
- Do not attempt any protected-mode programming without further research into protected-mode operation. Chapter 16 provides a detailed look at protected mode.
- Review the way the stack is used when interfacing a C program to an assembly language routine. Stack-based parameter passing is very useful and should be well understood.

Your own special programing techniques will develop over time. You may see an example program in a magazine that does something neat and adopt it (start using it in your code).

Programs written by others are a good source of information for you. Search the Web for assembly language examples; you will find plenty of archives. Go to a local computer show and buy a used assembly language book. It pays to have lots of references.

SUMMARY

In this chapter we examined many advanced topics that have direct application to real-world needs. Instruction execution time, interrupts, multitasking, and memory management are all important areas for the serious programmer. Interfacing with the mouse and working with memory-resident code add up to more powerful programming applications. Specific programming techniques, such as assembling and linking separate source modules, and macro usage were also covered. Some technical details of protected mode and C assembly language structure were also explained.

STUDY QUESTIONS

1. Explain the meaning and use of the PUBLIC and EXTRN assembler directives.
2. Show the linker command needed to create the executable XY.EXE from the object files XY.OBJ and FUNCT.OBJ.
3. Explain what is meant by a nested macro.
4. Explain why a label must be declared LOCAL if a macro is used more than once.
5. Compute the execution time of this section of code; assume a 100-MHz clock frequency.

```
            MOV    CX,1000H
NEXT:       ADD    AL,2
            MOV    [SI],AL
            LOOP   NEXT
```

6. Repeat Question 5, assuming that all memory references take an additional two cycles.
7. What are the addresses in the interrupt vector table for INT 21H? How can DEBUG be used to display the interrupt vector?
8. How can DOS INT 21H, Function 35H be used to get the interrupt vector for INT 10H?
9. Use DOS INT 21H, Function 25H to store the interrupt vector 0AE3:09BF.
10. Why is it necessary for the SWITCHER program to restore the original INT 1CH interrupt vector before exiting?
11. Explain how memory is managed by DOS.
12. What instructions are needed to allocate 8,192 bytes of memory?
13. How is mouse data accessed? When is mouse information available?
14. How does MOUSETST check for the presence of a mouse?
15. What must be done to make a program memory resident?
16. Why do you have to save the original interrupt vector when hooking an interrupt?

17. Explain how the tic-tac-toe board is managed in TICTAC.
18. What are the differences between real mode and protected mode?
19. How is the stack used in a C program?
20. What is *in-line* code in a C program?

ADDITIONAL PROGRAMMING EXERCISES

1. Write a series of string procedures, such as STRLEN (string length), STRCAT (string catenation), STRCPY (string copy), and STRCHR (string has character). Create an object code library for the string procedures.
2. Write a macro that will set or clear a specific bit in register AX. For example, BITOP S,4 will set bit 4 of AX. BITOP   C,11 will clear bit 11.
3. Write a macro called MAX that returns the maximum value contained in parameters A and B. For example, MAX   50,36 returns the value 50. MAX   DX, 2000 returns either the value of DX or 2000, whichever is higher. The maximum value must always be returned in register AX.
4. Write an interrupt service routine that handles these four functions:

```
AH = 05H: NOT   AL
AH = 10H: AND   AL,BL
AH = 20H: OR    AL,BL
AH = 80H: XOR   AL,BL
```

Install the routine using DOS INT   21H, Functions 35H and 25H.
5. Write a program that will hook INT   16H and keep track of the number of keys pressed on the keyboard. Save the key count as a word value and show how DEBUG can display the count.
6. Modify the MEMMAN program to reserve memory blocks that double in size with each request. The first block size should be 64 bytes. How many blocks are assigned? What are their segment addresses?
7. Write a program that will request three 1KB blocks of memory. Add the bytes in the first block to the bytes in the second block, and store the sum in the third block. Use DEBUG to demonstrate that the third block contains the correct results.
8. Write a mouse application that places a red * on the screen at the current mouse coordinates whenever the left button is pushed.
9. Write a memory-resident program that beeps the speaker three times at the beginning of each hour, and once on the half-hour.
10. Write a blackjack program that allows the user to play one hand of blackjack against the computer. The computer should deal two random cards to the user and itself, evaluate the points in both hands, and ask the user for hits, if necessary.
11. Generate assembly language for the following C program:

```
#include <stdio.h>
int adder(int x, int y);
main()
{
     int  a = 10, b = 20;
```

```
        printf("The sum is %d\n",adder(a,b));
}
int adder (int x, int y)
{
        return (x + y);
}
```

12. Write a program that disassembles the machine code beginning at a specific address, similar to the unassemble command in DEBUG.

CHAPTER 9

Using Disks and Files

OBJECTIVES

In this chapter you will learn about:

- The differences between floppy and hard disks
- What the boot sector is for
- What the file allocation table is for
- How to access disk sectors through BIOS INT 13H
- How to navigate through DOS's directory structure
- Using DOS INT 21H functions to access disk files

9.1 INTRODUCTION

The use of a disk adds a new dimension to the type of applications possible on the PC. In this chapter we will examine the operation of the PC's disk subsystem in detail.

Section 9.2 discusses the organization of floppy and hard disks. Section 9.3 explains how individual sectors are accessed through BIOS INT 13H. Section 9.4 outlines the directory structure used by DOS and covers DOS INT 21H directory functions. Text files are covered in Section 9.5, followed by the creation of a text file in Section 9.6. Random access of a file is explained in Section 9.7. A number of useful file functions are shown in Section 9.8. Finally, file-based troubleshooting tips are provided in Section 9.9.

9.2   ORGANIZATION OF FLOPPY AND HARD DISKS

In this section we will examine how information on a floppy disk or hard drive is organized. The material presented here lays the groundwork for the disk and file interrupt functions that follow in the remaining sections.

3.5- and 5.25-Inch Disks

Both of these types of disks are known as *floppy* disks due to the nature of their construction. Both disks contain a round, flexible piece of plastic (similar to a very thin phonograph record) whose surface is coated with a magnetic oxide. The disk is mounted inside a protective package to allow for easy insertion into the floppy drive.

Both disks are capable of storing different amounts of information. This information is stored on the disk in data areas called **sectors.** Normally, a sector contains 512 bytes of information. Sectors are arranged in groups of nine or more on a single **track,** as shown in Figure 9.1.

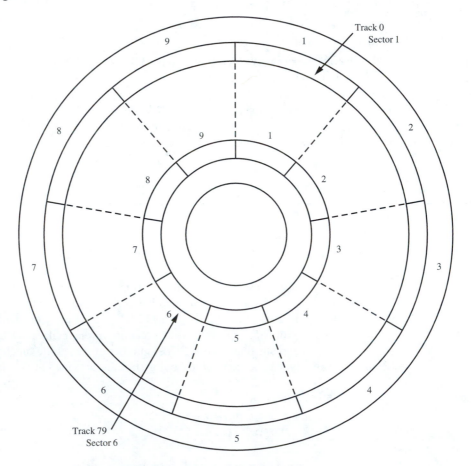

FIGURE 9.1   Sectors and tracks on a 3.5-inch floppy disk

Note that the 3.5-inch disk shown in Figure 9.1 contains 80 tracks. These 80 tracks are the circular portions of the disk surface scanned by the read/write head in the floppy drive. Thus, with the read/write head in a particular position, say over track 34, all sectors in track 34 will pass under the read/write head as the disk rotates.

Most floppy drives today contain two read/write heads so that both sides of the disk can be used. This *doubles* the number of sectors associated with any given track, and leads to the term *double-sided* used to describe a particular disk.

When a disk is first manufactured, its surface contains a random jumble of magnetic polarities and cannot be used for storage until it is **formatted.** You may be familiar with DOS's FORMAT command. This command prepares a floppy disk for use by laying down all of the necessary tracks, organizing the raw disk surface into separate sectors, and initializing some special sectors with information (which we will discuss shortly).

Double-sided 5.25- and 3.5-inch disks may be formatted as follows:

Disk	Tracks	Sectors/Track	Capacity
5.25	40	9	360KB
5.25	80	15	1.2MB
3.5	80	9	720KB
3.5	80	18	1.44MB

The capacity of a disk is found by the following formula:

$$\text{Capacity} = \text{Sides} \times \text{Tracks} \times \text{Sectors/Track} \times \text{Bytes/Sector}$$

So, for a 3.5-inch disk with nine sectors/track, we have:

$$\text{Capacity} = 2 \times 80 \times 9 \times 512 = 737{,}280 \text{ bytes}$$

To convert 737,280 bytes into kilobytes, divide by 1,024. This gives 737,280/1,024 = 720KB.

The 15 sector/track 5.25-inch disk is referred to as a *high-density* disk. The 18 sector/track 3.5-inch disk is also a high-density disk. Disk manufacturers indicate the storage capabilities of a floppy through the **2DD** (two-sided, double-density) and **2HD** (two-sided, high-density) part numbers.

Hard Disks

Hard disks are popular for two major reasons. First, their storage capacity is rated in *megabytes*. A 400MB hard drive stores the equivalent of 284 1.44MB 3.5-inch disks. Second, hard drives transfer data much faster than floppy drives. A 100KB program stored on a 3.5-inch 720KB floppy requires almost *4 seconds* to load into memory. The same 100KB program loads in less than *50 milliseconds* from a typical hard drive. This is 1/80th the floppy transfer time. The reduction in transfer time is largely due to the increased rotational speed of the hard disk.

How do hard drives store so much information? Consider the following parameters for two different hard drives:

Heads	Cylinders	Sectors/Track	Capacity
16	683	38	202MB
16	936	17	124MB

Note that there are many more read/write heads in a hard disk. Typically, a hard drive will contain a number of recording surfaces called **platters,** stacked one on top of each other. Each platter is similar to a floppy disk, except it is made of a more rigid material (so that it can spin faster), and can fit more tracks per inch than a floppy disk. Figure 9.2 shows an example of a four-platter, eight-head hard drive.

Because the read/write heads are mounted vertically, they all access the same portion of each platter as the platters spin. Thus, the eight heads define a three-dimensional *cylinder* of tracks. So, cylinder 0 refers to track 0 on all sides of each platter.

To get the hard drive's storage capacity, use the following formula:

$$Capacity = Heads \times Cylinders \times Sectors/Track$$

Example 9.1:   An 80MB hard drive contains 16 heads and 17 sectors/track. How many cylinders does the drive use?

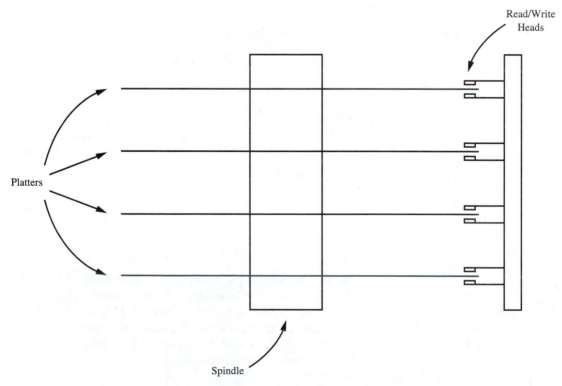

FIGURE 9.2   Typical hard-drive construction

First, find the total capacity in bytes: $80 \times 1{,}024 \times 1{,}024 = 83{,}886{,}080$ bytes. Divide this number by 512 to get the number of sectors. This gives 163,840 sectors. With 17 sectors/track, we get 9,637 tracks. Dividing this number by 16 heads gives the number of cylinders required, which is 603 (the actual answer of 602.35 must be rounded up).

The Boot Sector

The first sector on any disk is sector 1 of track 0. This sector is called the **boot** sector and has a special function. When a disk is formatted, it is formatted either as a **non-system** disk or as a **system** disk. A system disk is capable of starting DOS when it is used to boot up the computer. A non-system disk gives an error message similar to:

```
Non-System disk or disk error.
Replace and hit any key to continue.
```

when it is accidentally used to boot the computer. The boot sector controls what happens in both cases. Suppose that the floppy disk in drive A is formatted with the DOS command:

```
FORMAT A:
```

In this case, the boot sector will contain code whose only function is to output the non-system disk error message.

When a disk is formatted with the DOS command:

```
FORMAT A: /S
```

the /S refers to *system* and instructs FORMAT to place a copy of BIOS, DOS, and COMMAND.COM on the disk after it has been formatted. In this case, the boot sector is responsible for beginning the load of BIOS (which then loads DOS and COMMAND.COM).

Example 9.2: We can use DEBUG to read and display the boot sector of any drive with the L command. The format of the L command is:

```
L [Address] [Drive] [Starting Sector] [Number of Sectors]
```

The **Drive** parameter must be 0 for drive A, 1 for drive B, 2 for drive C, and so on. DEBUG numbers disk sectors beginning with 0. This means that loading sector 0 causes a copy of the boot sector to be placed in RAM beginning at **Address.** The DEBUG command:

```
L 100 0 0 1
```

loads a copy of the boot sector from drive A into memory beginning at address 100H. Because the boot sector contains 512 bytes of information, the memory from address 100H to 2FFH may be examined with the **d** command to view the boot sector contents.

The structure of the boot sector is fixed and contains the following parts:

- Initial jump instruction (3 bytes)
- Vendor identification (8 bytes)
- BIOS parameter block and drive ID (51 bytes)
- Boot code (450 bytes)

The initial jump instruction is used to jump over the vendor and BIOS parameter areas when the boot sector information is loaded and executed at boot time. The vendor identification area describes the operating system for which the disk was formatted, and usually is a string of characters (like MSDOS5.0).

The BIOS parameter block provides information required by BIOS to know what type of disk is being used and the characteristics of the disk. The BIOS parameter block has the following components:

- Bytes/sector (2 bytes)
- Sectors/cluster (1 byte)
- Number of reserved sectors (2 bytes)
- Number of FAT copies (1 byte)
- Maximum number of root directory entries (2 bytes)
- Number of disk sectors (2 bytes)
- Media descriptor byte (1 byte)
- Size of FAT in sectors (2 bytes)
- Sectors/track (2 bytes)
- Number of drive heads (2 bytes)
- Number of hidden sectors (4 bytes)
- Number of sectors for drives larger than 32MB (4 bytes)

The BIOS parameter block is followed by specific drive identification information, which is:

- Drive number (1 byte)
- Reserved (1 byte)
- Boot signature 29H (1 byte)
- Volume ID number (4 bytes)
- Volume label (11 bytes)
- File system type (8 bytes)

Let us take a look at what value these parameters have on a typical disk.

Example 9.3:   The initial output from the **d** command of Example 9.2 is as follows:

```
-L 100 0 0 1
-D
1C3F:0100   EB 3C 90 4D 53 44 4F 53-35 2E 30 00 02 02 01 00    .<.MSDOS5.0.....
1C3F:0110   02 70 00 A0 05 F9 03 00-09 00 02 00 00 00 00 00    .p..............
1C3F:0120   00 00 00 00 00 00 29 FD-15 3F 15 4A 4C 41 57 4F    ......)..?.JLAWO
1C3F:0130   52 4B 20 20 20 20 46 41-54 31 32 20 20 20 FA 33    RK    FAT12   .3
1C3F:0140   C0 8E D0 BC 00 7C 16 07-BB 78 00 36 C5 37 1E 56    .....|...x.6.7.V
1C3F:0150   16 53 BF 3E 7C B9 0B 00-FC F3 A4 06 1F C6 45 FE    .S.>|.........E.
1C3F:0160   0F 8B 0E 18 7C 88 4D F9-89 47 02 C7 07 3E 7C FB    ....|.M..G...>|.
1C3F:0170   CD 13 72 79 33 C0 39 06-13 7C 74 08 8B 0E 13 7C    ..ry3.9..|t....|
-
```

The BIOS parameter block and drive ID area are shown in bold. The first two bytes, 00 02, indicate there are 200H, or 512 bytes/sector. Remember to swap the upper and lower bytes read out of memory to obtain the actual word value.

The next byte, 02, indicates that there are two sectors/cluster. A *cluster* is simply a group of sectors. There are fewer cluster numbers than sector numbers. DOS uses cluster numbers when accessing information on a disk for reasons of efficiency.

The remaining BIOS parameters are as follows:

Number of reserved sectors: 01 00 (1)

Number of FAT copies: 02

Maximum number of root directory entries: 70 00 (112)

Number of disk sectors: A0 05 (1440)

Media descriptor byte: F9

Size of FAT in sectors: 03 00 (3)

Sectors/track: 09 00 (9)

Number of drive heads: 02 00 (2)

Number of hidden sectors: 00 00 00 00 (0)

Number of sectors for large drives: 00 00 00 00 (0)

The drive information looks like this:

Drive number: 00

Reserved: 00

Boot signature: 29

Volume ID number: FD 15 3F 15 (153F–15FD)

Volume label: 4A 4C 41 57 4F 52 4B 20 20 20 20 (JLAWORK)

File system type: 46 41 54 31 32 20 20 (FAT12)

The data area of the boot sector follows the BIOS parameter block and drive ID area. This is where the boot code (accessed by the initial jump instruction) resides. For the disk used in Examples 9.2 and 9.3, the last 128 bytes of the boot sector look like this:

```
-d
1C3F:0280   C3 B4 02 8B 16 4D 7C B1-06 D2 E6 0A 36 4F 7C 8B    .....M|.....60|.
1C3F:0290   CA 86 E9 8A 16 24 7C 8A-36 25 7C CD 13 C3 0D 0A    .....$|.6%|.....
1C3F:02A0   4E 6F 6E 2D 53 79 73 74-65 6D 20 64 69 73 6B 20    Non-System disk
1C3F:02B0   6F 72 20 64 69 73 6B 20-65 72 72 6F 72 0D 0A 52    or disk error..R
1C3F:02C0   65 70 6C 61 63 65 20 61-6E 64 20 70 72 65 73 73    eplace and press
1C3F:02D0   20 61 6E 79 20 6B 65 79-20 77 68 65 6E 20 72 65     any key when re
1C3F:02E0   61 64 79 0D 0A 00 49 4F-20 20 20 20 20 20 53 59    ady...IO      SY
1C3F:02F0   53 4D 53 44 4F 53 20 20-20 53 59 53 00 00 55 AA    SMSDOS   SYS..U.
-
```

The text message at the end of the boot code indicates that the disk is not a system disk.

The two file names IO.SYS and MSDOS.SYS are included in the boot code to help determine if the disk is bootable. The boot code may search for these two files in an attempt to bring up DOS. These two files *must* be present to bring up the operating system. They are present only on a system disk.

The File Allocation Table

Recall from the previous discussion that a number of items in the BIOS parameter block and drive ID area referred to the FAT, or **File Allocation Table,** which is the master index of where *all* information is stored on a disk. The FAT stores information about every cluster of the disk residing in the **data area.** The data area of the disk refers to all sectors that follow the reserved disk sectors. Reserved sectors are used for boot, FAT, and root directory information.

The number of reserved sectors and their assignments are different for each type of disk. Consider the following information:

Disk	FAT Sectors	DIR Sectors	DATA Sectors
5.25, 360KB	4	7	708
5.25, 1.2M	14	14	2,371
3.5, 720KB	6	7	1,426
3.5, 1.44M	18	14	2,847

Each type of disk has a different, and plentiful, number of free data sectors.

The free data sectors are grouped into clusters, with the first free cluster numbered 2. Thus, the first file stored on a disk begins at cluster 2.

The FAT stores a special number for each cluster on the disk. This number, called a FAT *entry,* is 12 bits long for floppy disks. Hard drives usually require FAT entries of 16 bits, because they contain a significantly larger number of clusters. Let us examine the operation of a 12-bit FAT.

The first two entries of the FAT are reserved by DOS (to store the media descriptor for the disk). So, no FAT entries may have the values 0 or 1. This is why the first free cluster is number 2.

The cluster entry in a FAT may contain any of the following values:

- 000: Cluster is available
- 002–FEF: Number of next cluster in file or directory
- FF0–FF6: Reserved
- FF7: Bad sector in cluster
- FF8–FFF: Last cluster in file or directory

Because FAT cluster entries are 12 bits long, they do not completely fill a 16-bit word. DOS places two FAT entries together, resulting in a 24-bit pattern that fits exactly into 3 bytes. The structure of this 24-bit FAT entry pair is shown in Figure 9.3. An example should serve to explain how this structure is used.

FIGURE 9.3 Structure of a FAT entry pair

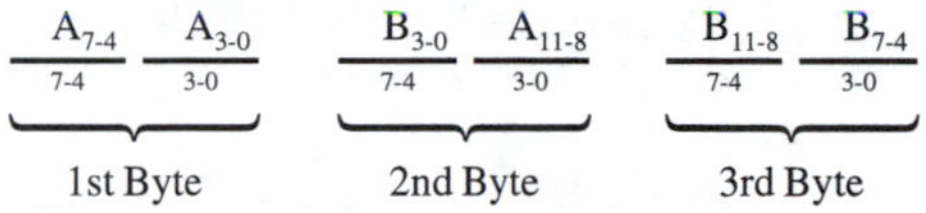

A_{7-4}	A_{3-0}		B_{3-0}	A_{11-8}		B_{11-8}	B_{7-4}
7-4	3-0		7-4	3-0		7-4	3-0

1st Byte   2nd Byte   3rd Byte

Example 9.4: The 3 bytes of a FAT entry pair are 03 40 00. What are the two FAT entries?

Solution: If the first 2 bytes are read as a word, we get 4003H. Keeping only the lower 12 bits of this word gives us 003H. This is the first FAT entry. If the second 2 bytes are read as a word, we get 0040H. Shifting this number to the right 4 bit positions, we get 004H, the second FAT entry.

The FAT entries for a particular file form a *chain*. Remember that when a FAT entry is between 002H and FEFH, it represents the *next* cluster in a chain of clusters. A simplified FAT is shown in Figure 9.4. The FAT contains entries for two files.

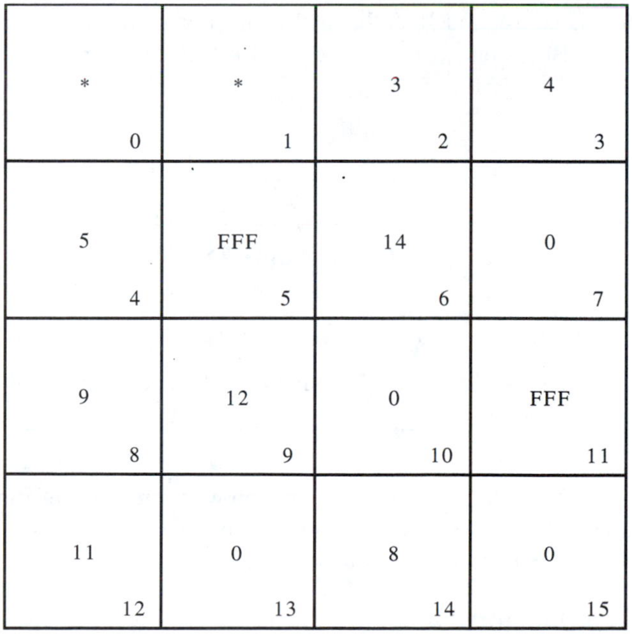

*	0	*	1	3	2	4	3
5	4	FFF	5	14	6	0	7
9	8	12	9	0	10	FFF	11
11	12	0	13	8	14	0	15

* Reserved

FIGURE 9.4 Simplified FAT with cluster chains

If the first file begins with cluster 2, we get the cluster chain:

$$2 \quad 3 \quad 4 \quad 5$$

The chain ends at cluster 5 due to the FFFH entry. If the starting cluster for the second file is 6, the resulting cluster chain is:

$$6 \quad 14 \quad 8 \quad 9 \quad 12 \quad 11$$

This illustrates an important point: files may be *physically* stored all over a disk, and still be *logically* connected, through their associated cluster chain.

Clearly, any damage to a disk's FAT may result in a catastrophic loss of data.

Test your understanding of cluster chains with Example 9.5.

Example 9.5:   The first sector of the FAT for the disk in drive A is loaded into memory at 100H, and displayed. The information is as follows:

```
-L 100 0 1 1
-D
1C3F:0100   F9 FF FF 03 40 00 05 60-00 07 80 00 FF FF FF FF
1C3F:0110   CF 00 0D F0 FF 0F 00 01-11 F0 FF 13 40 01 15 60
1C3F:0120   01 17 80 01 19 A0 01 1B-C0 01 1D E0 01 1F 00 02
1C3F:0130   FF 0F 00 00 00 00 00-00 00 00 00 00 00 00 00
```

Assume that there are six files, with cluster chains beginning at the following positions: 2, 9, 10, 11, 14, and 18. What are the resulting cluster chains?

Based on examination of the FAT information, the cluster chains are as follows:

File 1:	2	3	4	5	6	7	8
File 2:	9						
File 3:	10						
File 4:	11	12	13				
File 5:	14	15	16	17			
File 6:	18	19	20	21	22	23	24	25
	26	27	28	29	30	31	32	

The first cluster chain is highlighted in bold, as is the chain for file 5.

The FAT is used in conjunction with directory information to access files. As in Example 9.5, it is necessary to know what the starting cluster of a file is. We will now see how DOS maintains directory information.

The Directory

The directory structure used by DOS resembles an upside-down tree. The top of the directory is referred to as *root*, or the root directory. The root directory may contain files and/or other directories (called *subdirectories*). As Figure 9.5 shows, the root directory contains two files (HELLO.TXT and RUNME.EXE) and a subdirectory called WORK.

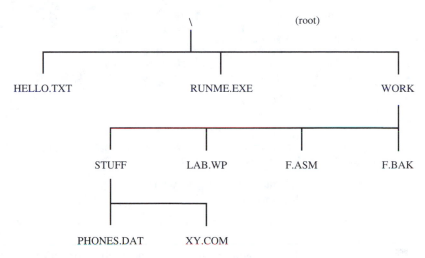

FIGURE 9.5 Sample directory tree

The WORK subdirectory contains three files and another subdirectory called STUFF, which contains two files. Thus, we see that files are associated with the directories within which they exist. This requires us to specify a *path* to the file when we wish to access it. The path for each file in the sample directory of Figure 9.5 is as follows:

```
\HELLO.TXT
\RUNME.EXE
\WORK
\WORK\LAB.WP
\WORK\F.ASM
\WORK\F.BAK
\WORK\STUFF
\WORK\STUFF\PHONES.DAT
\WORK\STUFF\XY.COM
```

DOS uses reserved sectors to store the necessary directory information for all files on a disk. The directory sectors immediately follow the second copy of the FAT. Let us see what a sample directory looks like.

Example 9.6:   The DEBUG command:

```
-L 100 0 7 1
```

is used to read the first directory sector on a 720K 3.5-inch disk in drive A. Sector 7 contains the first directory sector, because sectors 0 through 6 have been used for boot and FAT information.

The resulting display of sector data is as follows:

```
1C3F:0100   4D 42 54 20 20 20 20 20-20 20 20 20 00 00 00 00   MBT           ...
1C3F:0110   00 00 00 00 00 00 1A 9A-7B 1E 02 00 5D 18 00 00   ........{...]...
1C3F:0120   41 4C 54 54 20 20 20 20-57 50 4D 20 00 00 00 00   ALTT    WPM ....
```

```
1C3F:0130   00 00 00 00 00 00 77 96-7C 1E 09 00 A8 00 00 00   ......w.|.......
1C3F:0140   49 4E 54 31 37 20 20 20-57 50 20 20 00 00 00 00   INT17   WP  ....
1C3F:0150   00 00 00 00 00 00 CD A0-7B 1E 0A 00 F0 03 00 00   ........{.......
1C3F:0160   49 4E 54 31 33 20 20 20-57 50 20 20 00 00 00 00   INT13   WP  ....
1C3F:0170   00 00 00 00 00 00 1D A5-7B 1E 0B 00 51 09 00 00   ........{...Q...
1C3F:0180   49 4E 54 31 30 20 20 20-57 50 20 20 00 00 00 00   INT10   WP  ....
1C3F:0190   00 00 00 00 00 00 84 99-7C 1E 0E 00 ED 0F 00 00   ........|.......
1C3F:01A0   41 20 20 20 20 20 20 20-57 50 20 20 00 00 00 00   A       WP  ....
1C3F:01B0   00 00 00 00 00 00 06 43-7D 1E 12 00 81 39 00 00   ......C}.....9..
1C3F:01C0   00 00 00 00 00 00 00 00-00 00 00 00 00 00 00 00   ................
1C3F:01D0   00 00 00 00 00 00 00 00-00 00 00 00 00 00 00 00   ................
1C3F:01E0   00 00 00 00 00 00 00 00-00 00 00 00 00 00 00 00   ................
1C3F:01F0   00 00 00 00 00 00 00 00-00 00 00 00 00 00 00 00   ................
```

The display indicates that a directory entry contains 32 bytes of information. Six files are listed in the directory.

The 32-byte directory entry stores all necessary file information required by DOS. The format of the directory entry looks like this:

- File name (8 bytes)
- Extension (3 bytes)
- Attributes (1 byte)
- Reserved (10 bytes)
- Time (2 bytes)
- Date (2 bytes)
- First cluster (2 bytes)
- File size (4 bytes)

These parameters have the following values for the file MBT in Example 9.6:

- File name: 4D 42 54 20 20 20 20 20
- Extension: 20 20 20
- Attributes: 20
- Reserved: 00 00 00 00 00 00 00 00 00 00
- Time: 1A 9A
- Date: 7B 1E
- First cluster: 02 00
- File size: 5D 18 00 00

The file name (MBT) is left justified and padded with blanks, as is the extension (which is blank). The file attribute byte (20H) indicates that the file has the **Arc** (for Archive) bit set. The format of the attribute byte is this:

7	6	5	4	3	2	1	0
0	0	Arc	Dir	Vol	Sys	Hid	RO

where Dir stands for Directory, Vol for Volume label, Sys for System, Hid for Hidden, and RO for Read Only. A file may possess more than one of these attributes at a time.

Example 9.7:   The attribute byte of a file is 23H. What are the file attributes? In binary, the attribute byte is 00100011. This indicates that Arc, Hid, and RO are the active attributes for the file.

The Time (1A 9A) indicates when the file (or directory) was created or last updated. The format of the 16-bit Time word is as follows:

15 through 11	10 through 5	4 through 0
Hour (0–23)	Minutes (0–59)	Seconds (0–29)

where "Seconds" must be doubled to get the actual time in seconds. The Time word for the MBT file is 9A1AH. In binary, we have:

$$1 \ 0 \ 0 \ 1 \ 1 \quad 0 \ 1 \ 0 \ 0 \ 0 \ 0 \quad 1 \ 1 \ 0 \ 1 \ 0$$

Hour	Minutes	Seconds
19	16	26(52)

which indicates 7:16 P.M.

The Date word (7B 1E) specifies the date the file (or directory) was created or last updated. The format of the 16-bit Date word is as follows:

15 through 9	8 through 5	4 through 0
Year	Month	Day

where Year is relative to 1980 (when DOS first appeared). The Date word in our example file is 1E7B. In binary, we have:

$$0 \ 0 \ 0 \ 1 \ 1 \ 1 \ 1 \ 0 \ 0 \ 1 \ 1 \ 1 \ 1 \ 0 \ 1 \ 1$$

15	3	27

which indicates a date of March 27, 1995.

The starting cluster of the MBT file is 002 (as shown in Example 9.5). The file size is 0000185DH, or 6,237 bytes. This information is verified through the DIR command, which gives:

```
MBT    6,237 03-27-95    7:16p
```

Because each directory entry requires 32 bytes, a total of 16 directory entries can be stored in a 512-byte sector. For disks that use seven directory sectors, a total of 112 files can be stored in the root directory. If more files need to be stored, it is necessary to create one or more subdirectories.

One last bit of information is needed to complete this discussion. When DOS is searching the directory information, it may encounter a file (or directory) that has been deleted. The directory entry for a deleted file (or directory) is *not* erased. Instead, the byte value E5H is written over the first character of the file or directory name.

The end of the directory information is indicated by a 0 in the first character of a file or directory name.

The remaining sections of the chapter will explore the various BIOS and DOS interrupts that handle disk and file I/O.

Programming Exercise 9.1:   Use DEBUG to load the boot sector of your own floppy disk into memory. Analyze the information contained within the BIOS parameter block.

Programming Exercise 9.2:   Use DEBUG to load the boot sector for hard drive C. What are the some of the differences between drive C's boot sector and drive A's (from Example 9.3)?

Programming Exercise 9.3:   Use DEBUG to load the first FAT for a disk in drive A. What are the first ten FAT entries?

Programming Exercise 9.4:   Use the directory information from Example 9.6 to verify that it represents the following additional file information:

```
ALTT      WPM          168 03-28-95    6:51p
INT17     WP         1,008 03-27-95    8:06p
INT13     WP         2,385 03-27-95    8:40p
INT10     WP         4,077 03-28-95    7:12p
A         WP        14,721 03-29-95    8:24a
```

9.3   READING AND WRITING DISK SECTORS

BIOS interrupt 13H provides a number of functions related to operation of the disk system. Four of these functions will be covered in this section. These functions will allow us to access any (and all) sectors that we desire.

BIOS INT   13H, Function 00H: Reset Disk System

This function is selected when AH equals 00H. The drive to reset must be loaded into register DL prior to the function call. DL must equal 0 for drive A or 1 for drive B. Upon return, the carry flag indicates success (when cleared) or failure (when set). If the reset operation failed, register AH will contain an error code. The error code will be one of the following:

Error Code	Meaning
00H	No error
01H	Invalid function request
02H	Address mark not found
03H	Write protect error
04H	Sector not found
06H	Disk change line active

08H	DMA overrun on operation
09H	Data boundary error
0CH	Media type not found
10H	Uncorrectable ECC or CRC error
20H	General controller failure
40H	Seek operation failed
80H	Timeout

The disk head will be relocated to track 0 by this function.

BIOS INT   13H, Function 01H: Read Disk Status

This function requires AH to be loaded with 01H. The drive number must be in register DL prior to the function call. The carry flag and contents of AH, upon return, are the same as those found in function 00H.

BIOS INT   13H, Function 02H: Read Disk Sector

Register AH must be loaded with 02H to use this function. A number of other registers must be set up as follows:

AL = Number of sectors (1 to 18 depending on disk type)
CH = Track number (0 to 79 depending on disk type)
CL = Sector number (1 to 18 depending on disk type)
DH = Head number (0 or 1)
DL = Drive number (0 for A, 1 for B)
BX = Pointer to buffer
ES = Buffer segment

The buffer is a data area supplied by the user where the disk information is stored once it is read from the disk. This data area is commonly called the **Disk Transfer Address (DTA)**.

Upon return, the carry flag indicates success or failure (cleared or set, respectively). Register AH contains the error code, if any, and the AL register contains the number of sectors read.

BIOS INT   13H, Function 03H: Write Disk Sector

This function is selected when the AH register has been loaded with 03H. The other registers must be set up as they are in Function 02H. The carry flag is affected in the usual way. Note that writing a sector to disk does *not* update the FAT. The higher level DOS interrupts perform FAT housekeeping.

BOOTSEC: A Boot-Sector Display Program

The BOOTSEC program shown here reads the 512-byte boot sector from the disk in drive A and displays it in hexadecimal and ASCII format. The ASCII portion is included to help identify text-based data areas.

```
;Program BOOTSEC.ASM: Display contents of disk A's boot sector (with ASCII).
;
        .MODEL SMALL
        .DATA
BMSG    DB      'Boot sector for disk in drive A:',0DH,0AH,0DH,0AH,'$'
EMSG1   DB      'Error! Can not reset disk system.',0DH,0AH,'$'
EMSG2   DB      'Error! Can not read boot sector.',0DH,0AH,'$'
BLANK   DB      ' $'
PRESS   DB      0DH,0AH,'Press any key for next 256 bytes...'
CRLF    DB      0DH,0AH,'$'
DBUF    DB      512 DUP(0)

        .CODE
        .STARTUP
        MOV     AX,DS
        MOV     ES,AX           ;set up extra segment
        MOV     AH,0            ;reset disk function
        MOV     DL,0            ;drive A
        INT     13H             ;BIOS call
        JNC     OK
        LEA     DX,EMSG1        ;set up pointer to error message
PRE:    MOV     AH,9            ;display string function
        INT     21H             ;DOS call
        JMP     EXIT
OK:     CALL    READ            ;read boot sector, first try
        JNC     OBM
        CALL    READ            ;read boot sector, second try
        JNC     OBM
        CALL    READ            ;read boot sector, last try
        JNC     OBM
        LEA     DX,EMSG2        ;set up pointer to error message
        JMP     PRE             ;go process error
OBM:    LEA     DX,BMSG         ;set up pointer to boot message
        MOV     AH,9            ;display string function
        INT     21H             ;DOS call
        SUB     SI,SI           ;clear index pointer
        CALL    SHOWDTA         ;show first half of DTA
        LEA     DX,PRESS        ;set up pointer to press message
        MOV     AH,9            ;display string function
        INT     21H             ;DOS call
        MOV     AH,1            ;read keyboard function
        INT     21H             ;DOS call
        LEA     DX,CRLF         ;set up pointer to crlf string
        MOV     AH,9            ;display string function
        INT     21H             ;DOS call
        CALL    SHOWDTA         ;show second half of DTA
EXIT:   .EXIT

READ    PROC    NEAR
        LEA     BX,DBUF         ;set up pointer to DTA
        MOV     AH,2            ;read disk sectors function
        MOV     AL,1            ;read one sector
        MOV     DL,0            ;from drive A
        MOV     DH,0            ;head 0
        MOV     CX,1            ;sector 1
        INT     13H             ;BIOS call
        RET
READ    ENDP

SHOWDTA PROC    NEAR
        MOV     CX,16           ;load line counter
```

```
NLYN:    PUSH  CX              ;save line counter
         MOV   AX,SI           ;load pointer address
         CALL  DISPHEX2        ;output hex address
         MOV   DL,':'          ;load colon
         MOV   AH,2            ;display character function
         INT   21H             ;DOS call
         MOV   CX,16           ;load byte counter
NBYT:    LEA   DX,BLANK        ;set up pointer to blank string
         MOV   AH,9            ;display string function
         INT   21H             ;DOS call
         MOV   AL,ES:DBUF[SI]  ;get next byte
         CALL  DISPHEX         ;output it
         INC   SI              ;advance pointer
         LOOP  NBYT            ;repeat for next byte
         CALL  SHOWASC         ;output ASCII equivalents
         LEA   DX,CRLF         ;set up pointer to crlf string
         MOV   AH,9            ;display string function
         INT   21H             ;DOS call
         POP   CX              ;get line counter back
         LOOP  NLYN            ;and repeat until done
         RET
SHOWDTA  ENDP

SHOWASC  PROC  NEAR
         LEA   DX,BLANK        ;set up pointer to blank string
         MOV   AH,9            ;display string function
         INT   21H             ;DOS call
         LEA   DX,BLANK        ;set up pointer to blank string
         MOV   AH,9            ;display string function
         INT   21H             ;DOS call
         SUB   SI,16           ;adjust pointer
         PUSH  CX              ;save current counter
         MOV   CX,16           ;load character counter
GCHR:    MOV   DL,ES:DBUF[SI]  ;get next byte
         INC   SI              ;advance pointer
         AND   DL,7FH          ;convert to 7-bit ASCII
         CMP   DL,20H          ;is it printable?
         JC    BLNK            ;no, go output a blank
         CMP   AL,80H          ;is it still printable?
         JNC   BLNK
AOUT:    MOV   AH,2            ;display character function
         INT   21H             ;DOS call
         LOOP  GCHR            ;repeat until done
         POP   CX              ;get counter back
         RET
BLNK:    MOV   DL,20H          ;load blank character
         JMP   AOUT
SHOWASC  ENDP


DISPHEX PROC NEAR
         PUSH AX               ;save number
         SHR  AL,1             ;shift upper nybble down
         SHR  AL,1
         SHR  AL,1
         SHR  AL,1
         CMP  AL,10            ;is nybble value less than 10?
         JC   NHA1             ;yes, go convert and display
         ADD  AL,7             ;add alpha bias
NHA1:    ADD  AL,30H           ;add ASCII bias
         MOV  DL,AL            ;load character
```

```
              MOV    AH,2            ;display character function
              INT    21H             ;DOS call
              POP    AX              ;get number back
              AND    AL,0FH          ;work with lower nybble
              CMP    AL,10           ;is it less than 10?
              JC     NHA2            ;yes, go convert and display
              ADD    AL,7            ;add alpha bias
NHA2:         ADD    AL,30H          ;add ASCII bias
              MOV    DL,AL           ;load character
              MOV    AH,2            ;display character function
              INT    21H             ;DOS call
              RET
DISPHEX ENDP

DISPHEX2 PROC   NEAR
              PUSH   AX              ;save number
              XCHG   AH,AL           ;get upper byte
              CALL   DISPHEX         ;output it
              POP    AX              ;get number back
              CALL   DISPHEX         ;output lower byte
              RET
DISPHEX2 ENDP

              END
```

The program will attempt to read the boot sector *three* times before reporting failure. This is a common practice when working with BIOS INT   13H and is necessary to allow the disk motor to get up to speed (if it was off). The DBUF source statement defines the DTA for Function 02H. The 512 bytes written into DBUF by the READ procedure are displayed with the SHOWDTA procedure.

If BOOTSEC is executed without a floppy in drive A, the following error message is output:

```
Error! Cannot read boot sector.
```

With a disk available, BOOTSEC displays the first 256 bytes of the boot sector, and then pauses to let the user examine the information. The user may press any key on the keyboard to see the second 256 bytes. A sample execution is as follows:

```
Boot sector for disk in drive A:

0000: EB 3C 90 4D 53 44 4F 53 35 2E 30 00 02 02 01 00   k< MSDOS5.0
0010: 02 70 00 A0 05 F9 03 00 09 00 02 00 00 00 00 00    p   y
0020: 00 00 00 00 00 00 29 F3 14 77 36 4E 4F 20 4E 41         )s w6NO NA
0030: 4D 45 20 20 20 20 46 41 54 31 32 20 20 20 FA 33   ME    FAT12   z3
  .
  .
  .
00E0: A6 75 0A 8D 7F 20 B9 0B 00 F3 A6 74 18 BE 9E 7D   &u   9  s&t > }
00F0: E8 5F 00 33 C0 CD 16 5E 1F 8F 04 8F 44 02 CD 19   h_ 3@M ^   D M

Press any key for next 256 bytes...

0100: 58 58 58 EB E8 8B 47 1A 48 48 8A 1E 0D 7C 32 FF   XXXkh G HH   |2
0110: F7 E3 03 06 49 7C 13 16 4B 7C BB 00 07 B9 03 00   wc  I|  K|;  9
  .
  .
  .
0190: CA 86 E9 8A 16 24 7C 8A 36 25 7C CD 13 C3 0D 0A   J i  $| 6%|M C
01A0: 4E 6F 6E 2D 53 79 73 74 65 6D 20 64 69 73 6B 20   Non-System disk
```

```
01B0: 6F 72 20 64 69 73 6B 20 65 72 72 6F 72 0D 0A 52   or disk error  R
01C0: 65 70 6C 61 63 65 20 61 6E 64 20 70 72 65 73 73   eplace and press
01D0: 20 61 6E 79 20 6B 65 79 20 77 68 65 6E 20 72 65    any key when re
01E0: 61 64 79 0D 0A 00 49 4F 20 20 20 20 20 20 53 59   ady  IO       SY
01F0: 53 4D 53 44 4F 53 20 20 20 53 59 53 00 00 55 AA   SMSDOS   SYS  U*
```

From the display it is possible to determine the type of FAT used (FAT12), the disk's volume label (NO NAME), the DOS version the disk was formatted under (MSDOS5.0), and that the disk is not a system disk (see the message beginning at display address 019EH).

Programming Exercise 9.5: Modify the BOOTSEC program to emphasize the information contained in the BIOS parameter area. For example, output the message "BIOS parameter block:" before the first byte is displayed.

Programming Exercise 9.6: The boot sector normally loads at address 700H. Use DEBUG to load the boot sector from a non-system disk into memory at 700H. Unassemble the initial JMP instruction to find out where the boot code begins. Then unassemble the boot code. Does it make use of INT   13H?

Programming Exercise 9.7: Write a procedure that will output the correct error message to the display based on the error code found in register AH.

9.4   DIRECTORY FUNCTIONS

This section takes us from the low-level BIOS functions of Section 9.3 to the high-level DOS functions that support disk I/O. We will concentrate on directory functions here, and cover file I/O in the remaining sections.

DOS INT   21H, Function 47H: Get Current Directory

This function requires register AH to be 47H prior to the function call. Register DL contains the drive ID (identification). Unlike BIOS INT   13H functions, some of the DOS-based file functions refer to drive A as 1 (instead of 0). Drive B is 2, drive C is 3, and so on. This is one of those functions. A data area must be available, pointed to by DS:SI, for Function 47H to return path information (in ASCII).

If successful, the carry flag will be cleared. Otherwise, register AX will contain an error code from the following list:

Error Code	Meaning
00H	Successful
01H	Invalid function number
02H	File not found
03H	Path not found

(continued on next page)

Error Code	Meaning
04H	No more handles available
05H	Access denied
06H	Invalid handle
07H	Bad memory control blocks
08H	Insufficient memory
09H	Invalid memory block address
0AH	Invalid environment
0BH	Invalid format
0CH	Invalid access code
0DH	Invalid data
0EH	Reserved
0FH	Invalid drive specification
10H	Removing current directory
11H	Not same device
12H	No more files to be found
13H	Disk is write protected
14H	Unknown disk
15H	Drive is not ready
16H	Unknown command
17H	Data error (CRC)
18H	Bad request length
19H	Seek error
1AH	Unknown media type
1BH	Sector not found
1CH	Printer out of paper
1DH	Write fault
1EH	Read fault
1FH	General failure

Example 9.8: The DEBUG session shown here determines and displays the current directory information:

```
-A
1C3F:0100 MOV AH,47
1C3F:0102 MOV DL,0
1C3F:0104 MOV SI,200
1C3F:0107 INT 21
1C3F:0109
-T

AX=4700  BX=0000  CX=0000  DX=0000  SP=FFEE  BP=0000  SI=0000  DI=0000
DS=1C3F  ES=1C3F  SS=1C3F  CS=1C3F  IP=0102    NV UP EI PL NZ NA PO NC
```

```
1C3F:0102 B200            MOV     DL,00
-T

AX=4700  BX=0000  CX=0000  DX=0000  SP=FFEE  BP=0000  SI=0000  DI=0000
DS=1C3F  ES=1C3F  SS=1C3F  CS=1C3F  IP=0104    NV UP EI PL NZ NA PO NC
1C3F:0104 BE0002          MOV     SI,0200
-T

AX=4700  BX=0000  CX=0000  DX=0000  SP=FFEE  BP=0000  SI=0200  DI=0000
DS=1C3F  ES=1C3F  SS=1C3F  CS=1C3F  IP=0107    NV UP EI PL NZ NA PO NC
1C3F:0107 CD21            INT     21
-P

AX=0100  BX=0000  CX=0000  DX=0000  SP=FFEE  BP=0000  SI=0200  DI=0000
DS=1C3F  ES=1C3F  SS=1C3F  CS=1C3F  IP=0109    NV UP EI PL NZ NA PO NC
1C3F:0109 031F            ADD     BX,[BX]              DS:0000=20CD
-D 200 L 10
1C3F:0200  41 4C 50 5C 43 48 39 00-4E 0E 4E 19 4E 24 4E 2F    ALP\CH9.N.N.N$N/
```

The path name returned by Function 47H is shown in bold in the DEBUG display. The length of the path name depends on what the current directory is, so DOS indicates the end of the path name string with a byte equal to 0. This is commonly referred to as an **ASCIIZ** string, meaning an ASCII string followed by 0. ASCIIZ strings are used to communicate with the high-level DOS disk functions.

DOS INT 21H, Function 3BH: Set Current Directory

This function requires register AH to be 3BH prior to its call. Upon entry DS:DX must point to an ASCIIZ string specifying the new directory path.

If successful, the carry flag is cleared upon return, otherwise register AX contains the error code.

Example 9.9: To make the current directory the root directory, the DX register must point to the following ASCIIZ string:

```
ROOT    DB      '\',0
```

To change the current directory to root, use these instructions:

```
LEA   DX,ROOT
MOV   AH,3BH
INT   21H
JC    DISK_ERROR
```

DOS INT 21H, Function 39H: Create Subdirectory

This function requires register AH to be 39H prior to its call. Upon entry DS:DX must point to an ASCIIZ string specifying the new subdirectory.

If successful, the carry flag is cleared upon return; otherwise, register AX contains the error code.

Example 9.10:   What is the difference between these two ASCIIZ strings?

```
FILE1    DB    '\WORK\ABC.TXT',0
FILE2    DB    'WORK\ABC.TXT',0
```

The difference is that FILE1 begins its path at the root directory and FILE2 begins at the current directory (which may or may not be root).

DOS INT   21H, Function 3AH: Delete Subdirectory

This function requires register AH to be 3AH prior to its call. Upon entry DS:DX must point to an ASCIIZ string specifying the directory to delete. The directory *must be empty* for this function to complete successfully.

If successful, the carry flag is cleared upon return; otherwise, register AX contains the error code.

DOS INT   21H, Function 19H: Get Current Drive

It is often necessary to know the current drive letter when working with directory paths. This function returns the current drive number in register AL and must be called with the AH register equal to 19H.

This DOS function numbers drives beginning with 0 (like INT   13H did), so drive A = 0, drive B = 1, and so on.

PATHNAME: Display Current Directory Path

The PATHNAME program presented here uses Functions 19H and 47H to get the current drive number and directory path. Because the ASCIIZ string returned by Function 47H is terminated with a 0, Function 09H (Display String) cannot be used to output the path name. Instead, the path name is output character by character until the 00H terminator is found.

```
;Program PATHNAME.ASM: Display current directory path.
;
        .MODEL SMALL
        .DATA
PMSG    DB    'Path for current directory: $'
DRIVE   DB    ?
        DB    ':\$'
PATH    DB    128 DUP(?)
CRLF    DB    0DH,0AH,'$'
EMSG    DB    'Error getting pathname.',0DH,0AH,'$'

        .CODE
        .STARTUP
        MOV    AH,19H        ;get current drive function
        INT    21H           ;DOS call
        ADD    AL,41H        ;add alpha bias
        MOV    DRIVE,AL      ;save for later
        LEA    SI,PATH       ;set up pointer to pathname buffer
```

```
                SUB     DL,DL         ;current drive
                MOV     AH,47H        ;get current directory
                INT     21H           ;DOS call
                JC      EROR          ;error if carry set
                LEA     DX,PMSG       ;set up pointer to dir message
                MOV     AH,9          ;display string function
                INT     21H           ;DOS call
                LEA     DX,DRIVE      ;set up pointer to drive string
                MOV     AH,9          ;display string function
                INT     21H           ;DOS call
                LEA     SI,PATH       ;set up pointer to path
SHPA:           MOV     DL,[SI]       ;get path character
                OR      DL,DL         ;is it zero?
                JZ      FIN
                MOV     AH,2          ;display character function
                INT     21H           ;DOS call
                INC     SI            ;advance to next position
                JMP     SHPA
FIN:            LEA     DX,CRLF       ;set up pointer to newline string
                MOV     AH,9          ;display string function
                INT     21H           ;DOS call
                JMP     EXIT
EROR:           LEA     DX,EMSG       ;set up pointer to error message
                MOV     AH,9          ;display string function
                INT     21H           ;DOS call
EXIT:           .EXIT

                END
```

A sample execution is as follows:

```
Path for current directory: D:\ALP\CH9
```

The value 41H (code for an ASCII 'A') is added to the drive number returned by Function 19H to determine the correct drive letter.

Programming Exercise 9.8: Write a program that will read the current directory path and use it to move up one directory (one directory level closer to root). For example, if the current directory path is \ALP\CH9, the program must change the directory to \ALP.

Programming Exercise 9.9: Write a program to create the directory \DIRTEST and another to delete the directory \DIRTEST. Run the first program. Use the DIR command to verify that the \DIRTEST directory was created. Then run the second program and use DIR to verify that \DIRTEST was deleted.

9.5   READING TEXT FILES

In this section we will examine three useful file functions. They will be used to open, read, and close a sample text file.

A Sample Text File

The text file used in this section is called BUFF.ASM, and looks like this:

```
;Procedure KEYBUFF: Place keystrokes into buffer until CR is entered.
;
            .MODEL SMALL

            PUBLIC  KEYBUFF      ;for linking

            .CODE
KEYBUFF     PROC    FAR
            MOV     CL,0         ;clear character counter
NEXTKEY:    MOV     AH,1         ;read keyboard function
            INT     21H          ;DOS call
            MOV     [DI],AL      ;save character in buffer
            INC     DI           ;advance buffer pointer
            INC     CL           ;increment character counter
            CMP     AL,0DH       ;was last keystroke a CR?
            JNZ     NEXTKEY      ;jump if not
            RET
KEYBUFF     ENDP

            END
```

The text file was created with a word processor and saved in ASCII format. This means that all of the bytes in the file represent ASCII characters. Remember that ASCII codes represent numbers, letters, punctuation, *and control characters,* such as carriage return (0DH), line feed (0AH), and Control-Z (1AH).

DOS INT   21H, Function 3DH: Open File with Handle

This function must be used before a file can be read (or written). Register AH must equal 3DH when this function is called. The ASCIIZ string of the file to be opened must be pointed to by DS:DX. The AL register must contain the access code. The following access codes are available:

Access Code	Meaning
-----000	Read access
-----001	Write access
-----010	Read/write access
----0---	Reserved
-000----	Compatibility mode
-001----	Read/write access denied
-010----	Write access denied
-011----	Read access denied
-100----	Full access denied
0-------	File inherited by child process
1-------	File private to current process

We will concentrate only on the first three file-access codes for read, write, and read/write.

If the open function is unsuccessful, the carry flag is set and register AX contains the error code. If the open function is successful, the carry flag is cleared and register AX contains the file's *handle*. A handle is a 16-bit number assigned by DOS when a file is opened (or created) that uniquely identifies the file. The file handle is used with other disk functions to access the opened file. This makes our programming job easy, because we need to use only a 16-bit file handle. Also, many different files (all with separate file handles) may be open at the same time. Each is accessed by the DOS functions through the use of their respective file handles.

An older, outdated method for accessing files is through the use of **File Control Blocks** (FCBs). The program segment prefix created by DOS for every executable program contains two FCBs, located at offsets 005C and 006C, respectively. Each FCB stores information about one file. The file name, time and date of creation or last update, and file length are some of the items contained in the FCB. Unfortunately, there is no way to specify a path name within an FCB. This restriction requires us to use a handle instead, but this is to our benefit, since it is easy to pass a 16-bit handle during a function call.

Example 9.11:   The source statements below are required to open a specific file and save its handle:

```
FILENAME    DB      'AFILE.TXT',0   ;ASCIIZ string
HANDLE      DW      ?
 .
 .
 .
            LEA     DX,FILENAME     ;set up pointer to filename
            MOV     AL,0            ;read access
            MOV     AH,3DH          ;open file function
            INT     21H             ;DOS call
            JC      DISK_ERROR
            MOV     HANDLE,AX       ;save file handle if no error
```

DOS INT   21H, Function 3FH: Read from File

This function is selected when register AH equals 3FH. The BX register must contain the handle of the file to open. The number of bytes to read must be in register CX. The DTA must be pointed to by DS:DX.

If successful, the carry flag will be cleared and the number of bytes read from the file will be in the AX register. If unsuccessful, the carry flag will be set and register AX will contain the error code.

Example 9.12:   The source statements shown here read 80 bytes at a time from an opened file:

```
FBUFF       DB      80 DUP(?)       ;file DTA
HANDLE      DW      ?
 .
```

```
        .
        .
        LEA     DX,FBUFF        ;set up pointer to DTA
        MOV     CX,80           ;read 80 bytes
        MOV     BX,HANDLE       ;load file handle
        MOV     AH,3FH          ;read from file function
        INT     21H             ;DOS call
        JC      DISK_ERROR
```

DOS INT   21H, Function 3EH: Close File with Handle

This function must be used when a file is finished being read or written, and is selected when
the AH register equals 3EH. Upon entry, the BX register must contain the file handle.

If successful, the carry flag will be cleared and the AX register will contain a copy of
the file handle. If unsuccessful, the carry flag will be set and register AX will contain the
error code.

READFILE: A Text-File Viewer

The sample text file BUFF.ASM (covered in the first part of this section) is opened, read
(and displayed on the screen), and then closed by the READFILE program shown here.

```
;Program READFILE.ASM: Read and display a text file.
;
            .MODEL SMALL
            .DATA
FILENAME    DB      'buff.asm',0
HANDLE      DW      ?
FBUFF       DB      ?       ;file data buffer
OEMSG       DB      'Cannot open BUFF.ASM.$'
RFMSG       DB      'Cannot read BUFF.ASM.$'
CFMSG       DB      'Cannot close BUFF.ASM.$'

        .CODE
        .STARTUP
        CALL OPENFILE       ;open BUFF.ASM
        JC   EXIT           ;jump if error
        CALL READFILE       ;read BUFF.ASM
        CALL CLOSEFILE      ;close BUFF.ASM
EXIT:   .EXIT

OPENFILE PROC NEAR
        MOV     AH,3DH          ;open file with handle function
        LEA     DX,FILENAME     ;set up pointer to ASCIIZ string
        MOV     AL,0            ;read access
        INT     21H             ;DOS call
        JC      OPENERR         ;jump if error
        MOV     HANDLE,AX       ;save file handle
        RET
OPENERR: LEA    DX,OEMSG        ;set up pointer to error message
        MOV     AH,9            ;display string function
        INT     21H             ;DOS call
        STC                     ;set error flag
        RET
OPENFILE ENDP
```

```
READFILE PROC NEAR
         MOV  AH,3FH        ;read from file function
         MOV  BX,HANDLE     ;load file handle
         LEA  DX,FBUFF      ;set up pointer to data buffer
         MOV  CX,1          ;read one byte
         INT  21H           ;DOS call
         JC   READERR       ;jump if error
         CMP  AX,0          ;were 0 bytes read?
         JZ   EOFF          ;yes, end of file found
         MOV  DL,FBUFF      ;no, load file character
         CMP  DL,1AH        ;is it Control-Z <EOF>?
         JZ   EOFF          ;jump if yes
         MOV  AH,2          ;display character function
         INT  21H           ;DOS call
         JMP  READFILE      ;and repeat
READERR: LEA  DX,RFMSG      ;set up pointer to error message
         MOV  AH,9          ;display string function
         INT  21H           ;DOS call
         STC                ;set error flag
EOFF:    RET
READFILE ENDP

CLOSEFILE PROC NEAR
         MOV  AH,3EH        ;close file with handle function
         MOV  BX,HANDLE     ;load file handle
         INT  21H           ;DOS call
         JC   CLOSERR       ;jump if error
         RET
CLOSERR: LEA  DX,CFMSG      ;set up pointer to error message
         MOV  AH,9          ;display string function
         INT  21H           ;DOS call
         STC                ;set error flag
         RET
CLOSEFILE ENDP

         END
```

The program contains an ASCIIZ string that specifies the BUFF.ASM file. This means that READFILE will access only one specific file.

READFILE reads 1 byte at a time from the opened file until one of two events happens:

1. Zero bytes are read by function 3FH
2. The byte 1AH (end-of-file character) is read

The reason READFILE terminates reading when it encounters 1AH is that some word processors use 1AH as an end-of-file character.

If the BUFF.ASM file is in the same directory as the READFILE program, executing READFILE will cause the BUFF.ASM file to be displayed on the screen. If READFILE does not find BUFF.ASM in the current directory, the following error message is output:

```
Cannot open BUFF.ASM.
```

Similar error messages are output if READFILE cannot read or close BUFF.ASM.

Programming Exercise 9.10:   How can the BUFF.ASM procedure be included within the READFILE program to allow *any* file to be opened and read? Modify READFILE accordingly.

Programming Exercise 9.11: Modify READFILE so that line numbers are displayed as the file is output. For example, the output might look like this

```
1: ;Procedure KEYBUFF: Place keystrokes into buffer until CR is entered.
2: ;
3:         .MODEL SMALL
4:
5:         PUBLIC  KEYBUFF     ;for linking
```

Programming Exercise 9.12: Modify READFILE so that lowercase characters are converted into uppercase before being displayed.

9.6   CREATING A TEXT FILE

The READFILE program covered in the previous section assumed that its text file BUFF.ASM already existed in the current directory. What happens when a file does *not* exist? How is a file created? How is new information stored in a file? These are the issues addressed in this section.

DOS INT   21H, Function 3CH: Create File

This function is selected when the AH register equals 3CH. Upon entry, the ASCIIZ string of the file to be created must be pointed to by DS:DX. The file attribute must be in the CX register. Only four file attributes are allowed:

File Attribute	Meaning
00H	Normal
01H	Read-only
02H	Hidden
03H	System

If successful, a directory entry for the file is created if the file does not exist, and the file handle is returned in register AX (the carry flag is cleared).

If the function is unsuccessful, the carry flag is set and the AX register contains the error code.

If the ASCIIZ string indicates a file that already exists, the file length is *reset to 0!* This means that all information will be lost if an existing file is "created." It might be a good idea to try to open the file first (using the ASCIIZ string). If the file does not exist, then it may be safely created.

DOS INT   21H, Function 40H: Write to File

This function is selected when register AH equals 40H. The handle of the file to be written to must be in the BX register prior to the function call. Also, register CX must contain the number of bytes to write, and the DTA must be pointed to by DS:DX.

If the function is successful, the carry flag will be cleared and the AX register will contain the number of bytes written. If the function is unsuccessful, the carry flag will be set and register AX will contain the error code.

MAKETEXT: Create a Text File

The MAKETEXT program shown here creates a text file called LETTER.TXT. The user is allowed to enter as many lines of text as desired. Hitting the carriage return at the beginning of a line terminates the input sequence and closes the text file.

```
;Program MAKETEXT.ASM: Make a text file.
;
            .MODEL SMALL
            .DATA
FILENAME    DB    'letter.txt',0
HANDLE      DW    ?
FBUFF       DB    80 DUP(?)        ;file data buffer
INPSTR      DB    'Enter your text line by line. A blank line exits.'
CRLF        DB    0DH,0AH,'$'
OEMSG       DB    'Cannot create LETTER.TXT.$'
WFMSG       DB    'Cannot write to LETTER.TXT.$'
CFMSG       DB    'Cannot close LETTER.TXT.$'

            .CODE
            .STARTUP
            CALL  CREATE           ;create LETTER.TXT
            JC    EXIT             ;jump if error
            LEA   DX,INPSTR        ;set up pointer to instruction message
            MOV   AH,9             ;display string function
            INT   21H              ;DOS call
            CALL  WRITEFILE        ;write to LETTER.TXT
            CALL  CLOSEFILE        ;close LETTER.TXT
EXIT:   .EXIT

CREATE      PROC NEAR
            MOV   AH,3CH           ;create file with handle function
            LEA   DX,FILENAME      ;set up pointer to ASCIIZ string
            MOV   CX,0             ;normal attribute
            INT   21H              ;DOS call
            JC    CREATERR         ;jump if error
            MOV   HANDLE,AX        ;save file handle
            RET
CREATERR: LEA   DX,OEMSG          ;set up pointer to error message
            MOV   AH,9             ;display string function
            INT   21H              ;DOS call
            STC                    ;set error flag
            RET
CREATE      ENDP

WRITEFILE PROC  NEAR
            LEA   DI,FBUFF         ;set up pointer to text buffer
            CALL  KEYBUFF          ;get a line of text
            CMP   CL,1             ;just CR entered?
            JZ    EOFF
            MOV   BYTE PTR [DI],0AH ;fix end of text buffer
            INC   CL               ;advance character counter
            LEA   DX,CRLF          ;set up pointer to newline string
            MOV   AH,9             ;display string function
            INT   21H              ;DOS call
            MOV   AH,40H           ;write to file function
```

```
                    MOV     BX,HANDLE               ;load file handle
                    LEA     DX,FBUFF                ;set up pointer to data buffer
                    SUB     CH,CH                   ;fix bytes-to-write value
                    INT     21H                     ;DOS call
                    JC      WRITERR                 ;jump if error
                    JMP     WRITEFILE               ;and repeat
          WRITERR:  LEA     DX,WFMSG                ;set up pointer to error message
                    MOV     AH,9                    ;display string function
                    INT     21H                     ;DOS call
                    STC                             ;set error flag
          EOFF:     RET
          WRITEFILE ENDP

          KEYBUFF   PROC    NEAR
                    MOV     CL,0            ;clear character counter
          NEXTKEY:  MOV     AH,1            ;read keyboard function
                    INT     21H             ;DOS call
                    MOV     [DI],AL         ;save character in buffer
                    INC     DI              ;advance buffer pointer
                    INC     CL              ;increment character counter
                    CMP     AL,0DH          ;was last keystroke a CR?
                    JNZ     NEXTKEY         ;jump if not
                    RET
          KEYBUFF   ENDP

          CLOSEFILE PROC NEAR
                    MOV     AH,3EH          ;close file with handle function
                    MOV     BX,HANDLE       ;load file handle
                    INT     21H             ;DOS call
                    JC      CLOSERR         ;jump if error
                    RET
          CLOSERR:  LEA     DX,CFMSG        ;set up pointer to error message
                    MOV     AH,9            ;display string function
                    INT     21H             ;DOS call
                    STC                     ;set error flag
                    RET
          CLOSEFILE ENDP

                    END
```

Note the use of the KEYBUFF procedure to get a line of text from the user. No editing of the input lines is possible, because the backspace and delete key codes will simply be stored in the buffer as characters.

A sample execution is as follows:

```
Enter your text line by line. A blank line exits.
Ken,<CR>
The new assembler works great!<CR>
Thanks.<CR>
                Jim<CR>
<CR>
```

The <CR> is used to show you where the user pressed enter.

Because no error messages were output, the file was successfully created and written to. This can be verified with the DOS command:

```
TYPE LETTER.TXT
```

Programming Exercise 9.13: Modify MAKETEXT so that blank lines are allowed to be input and written to the LETTER.TXT file. Use some other technique to exit (such as a single Q on an input line).

Programming Exercise 9.14: Write a program that will create a text file called ALPHA.BET. Write the alphabet to the file like this:

```
aAa
bbBBbb
cccCCCccc
etc.
```

9.7   ACCESSING AN EXISTING FILE

Up to now we have been able only to read existing file information or create a brand-new file. How can information from a file be accessed *randomly* (from any position)? How can new data be written (or **appended**) to the end of a file? Both actions require control over the file *pointer*. The file pointer is a number maintained by DOS that indicates where the next read or write operation will take place in a file. To randomly read a file, we need only to position the file pointer where we want it and then execute the read function. The same is possible for the write operation. Let us see how the file pointer can be manipulated.

DOS INT   21H, Function 42H: Position File Pointer

This function is selected when register AH equals 42H. The file handle must be in the BX register prior to the function call. The code representing the positioning method must be in register AL. There are three ways to position the file pointer. They are:

Method Code	Meaning
0	Absolute byte offset from beginning of file
1	Relative byte offset from current position
2	Absolute byte offset from end of file

The byte offset value can be very large (for large files) and thus is specified in *two* registers, CX and DX. The CX register contains the upper 16 bits of the offset value. Note that offsets of 0 have value, because they can be used to get to the beginning or the end of a file.

If the function is successful, the carry flag will be cleared and the new file pointer position will be returned in registers DX and AX (with DX most significant). If the function is unsuccessful, the carry flag is set and register AX contains the error code.

APPENDER: Add New Text to an Existing File

The APPENDER program shown here uses function 42H to position the file pointer to
the end of the LETTER.TXT file. Once the pointer is there, a new line of text is written to the file. Method 2 is used with an offset of 0 to move the pointer to the end of
the file.

```
;Program APPENDER.ASM: Append a short message line to a text file.
;
            .MODEL SMALL
            .DATA
FILENAME    DB    'letter.txt',0
HANDLE      DW    ?
FBUFF       DB    0DH,0AH,'This line was not here before...',0DH,0AH
STAT        DB    'The file LETTER.TXT has been appended',0DH,0AH,'$'
FLEN        EQU   STAT - FBUFF      ;calculate length of FBUFF
OEMSG       DB    'Cannot open LETTER.TXT.$'
MEMSG       DB    'Cannot position file pointer.$'
WFMSG       DB    'Cannot write to LETTER.TXT.$'
CFMSG       DB    'Cannot close LETTER.TXT.$'

            .CODE
            .STARTUP
            CALL OPENFILE            ;open LETTER.TXT
            JC   EXIT                ;jump if error
            CALL MOVEPTR             ;move file pointer to end of file
            CALL WRITEFILE           ;write to LETTER.TXT
            CALL CLOSEFILE           ;close LETTER.TXT
EXIT:   .EXIT

OPENFILE    PROC NEAR
            MOV  AH,3DH              ;open file with handle function
            LEA  DX,FILENAME         ;set up pointer to ASCIIZ string
            MOV  AL,2                ;read/write access
            INT  21H                 ;DOS call
            JC   OPENERR             ;jump if error
            MOV  HANDLE,AX           ;save file handle
            RET
OPENERR:    LEA  DX,OEMSG            ;set up pointer to error message
            MOV  AH,9                ;display string function
            INT  21H                 ;DOS call
            STC                      ;set error flag
            RET
OPENFILE    ENDP

MOVEPTR     PROC NEAR
            MOV  AH,42H              ;position file pointer function
            MOV  AL,2                ;offset from end method
            MOV  BX,HANDLE           ;load file handle
            MOV  CX,0                ;offset = 0 from end
            MOV  DX,0
            INT  21H                 ;DOS call
            JC   MOVERR
            RET
MOVERR:     LEA  DX,MEMSG            ;set up pointer to error message
            MOV  AH,9                ;display string function
            INT  21H                 ;DOS call
            STC                      ;set error flag
            RET
MOVEPTR     ENDP
```

```
WRITEFILE PROC   NEAR
          MOV    AH,40H          ;write to file function
          MOV    BX,HANDLE       ;load file handle
          LEA    DX,FBUFF        ;set up pointer to data buffer
          MOV    CL,FLEN         ;load buffer length
          SUB    CH,CH           ;fix bytes-to-write value
          INT    21H             ;DOS call
          JC     WRITERR         ;jump if error
          RET
WRITERR:  LEA    DX,WFMSG        ;set up pointer to error message
          MOV    AH,9            ;display string function
          INT    21H             ;DOS call
          STC                    ;set error flag
EOFF:     RET
WRITEFILE ENDP

CLOSEFILE PROC   NEAR
          MOV    AH,3EH          ;close file with handle function
          MOV    BX,HANDLE       ;load file handle
          INT    21H             ;DOS call
          JC     CLOSERR         ;jump if error
          RET
CLOSERR:  LEA    DX,CFMSG        ;set up pointer to error message
          MOV    AH,9            ;display string function
          INT    21H             ;DOS call
          STC                    ;set error flag
          RET
CLOSEFILE ENDP

          END
```

There is no output from the program unless an error occurs. To test APPENDER's operation, first find the size of the LETTER.TXT file with the DIR command, like this:

```
DIR LETTER.TXT
```

Then run APPENDER and use the DOS DIR command again. You should see that the file size has gotten larger.

Programming Exercise 9.15: Suppose that a certain text file has a size that is smaller than 1,024 bytes. Write a program that will write a new line of text (your name, for instance) to the beginning of the file, *without* overwriting any current file information.

Programming Exercise 9.16: Write a program that will display every tenth character of a text file. Use method 1 with Function 42H to move the file pointer.

9.8 MISCELLANEOUS FILE AND DISK FUNCTIONS

In this section we will examine a number of additional file and disk functions that might be useful when writing applications.

DOS INT   21H, Function 43H: Get or Set File Attributes

This function is selected when register AH equals 43H. Upon entry, the value of the AL register selects one of the following:

$$AL = 0: \text{Get attribute}$$

$$AL = 1: \text{Set attribute}$$

The CX register must contain the new attribute value. The ASCIIZ string for the file must be pointed to by DS:DX.

If the function is successful, the carry flag will be cleared and CX will contain the old attribute (if setting). If the function is unsuccessful, the carry flag will be cleared and register AX will contain the error code.

Example 9.13:   The source statements shown here set the attributes of the OUTPUT. LST file to hidden and read-only:

```
FILENAME    DB      'OUTPUT.LST',0
.
.
.
            MOV     AL,1            ;set attribute
            MOV     CX,3            ;hidden and read-only
            LEA     DX,FILENAME
            MOV     AH,43H          ;get/set attribute function
            INT     21H            ;DOS call
            JC      DISK_ERROR
```

DOS INT   21H, Function 56H: Rename File

This function is selected when the AH register equals 56H. The old and new names of the file must be stored as ASCIIZ strings and accessed as followed:

```
DS:DX points to old ASCIIZ string (current file name)
ES:DI points to new ASCIIZ string (new file name)
```

Upon the function's return, the carry flag will be cleared if successful and set otherwise. The AX register contains the error code when the function is unsuccessful.

DOS INT   21H, Function 57H: Get or Set File Date and Time

This function is selected when register AH equals 57H. The operation (get or set) is selected by the value in the AL register, as follows:

```
AL = 0: Get time and date
AL = 1: Set time and date
```

If setting, the new time and date values must be in registers CX and DX, respectively.

Upon the function's return, the CX and DX registers will contain the file time and date (if getting). If unsuccessful, the error code is returned in register AX.

The format for the time and date is as follows:

CX: HHHHH MMMMMM SSSSS

where H = Hours, M = Minutes, and S = double Seconds

DX: YYYYYYY MMMM DDDDD

where Y = Year, M = Month, and D = Day

DOS INT 21H, Function 1AH: Set Disk Transfer Area (DTA)

This function is selected when the AH register equals 1AH. The new DTA must be pointed to by DS:DX. Normally, the DTA is located at address 80H within the program segment prefix, and is 128 bytes long. This interrupt allows a larger DTA to be defined.

DOS INT 21H, Function 1BH: Get Current Drive Information

This function is selected when the AH register equals 1BH. Upon return, these registers have been updated:

AL = Sectors per allocation unit

CX = Bytes per sector

DX = Number of allocation units

DS:BX = Pointer to FAT ID byte

This information is useful for a program performing extensive disk work.

DOS INT 21H, Function 1CH: Get Drive Information

This function is selected when the AH register equals 1CH and is identical in operation to Function 1BH with one exception: The drive number must be specified in register DL upon entry. Recall that drive A = 1 in regard to the drive number.

DOS INT 21H, Function 4EH: Find File

This function is selected when the AH register equals 4EH. Upon entry, DS:DX must point to the ASCIIZ string of the file to be searched for.

Upon return, the AX register will contain a return code. If the return code is 0, the search was successful. Any other value indicates an unsuccessful search.

DOS INT 21H, Function 41H: Delete File

This function is selected when the AH register equals 41H. Upon entry, DS:DX must point to the ASCIIZ string of the file to be deleted.

If the function is successful, the carry flag will be cleared. If the function is unsuccessful, the carry flag will be set and register AX will contain the error code.

DISKCAP: Display Disk Capacity

The DISKCAP program uses Function 36H to obtain the capacity of the current drive. A binary-to-integer display procedure (DISPINT) is used to show the resulting values.

```
;Program DISKCAP.ASM: Show capacity of current drive.
;
            .MODEL SMALL
            .DATA
MSG1    DB      'Allocation units on drive:   $'
MSG2    DB      0DH,0AH,'Bytes per sector:            $'
MSG3    DB      0DH,0AH,'Available allocation units: $'
MSG4    DB      0DH,0AH,'Sectors per cluster:         $'
EMSG    DB      'Drive is invalid.$'
AU      DW      ?
BPS     DW      ?
AAU     DW      ?
SPC     DW      ?

            .CODE
            .STARTUP
            MOV     AH,36H      ;get disk free space
            MOV     DL,0        ;use current drive
            INT     21H         ;DOS call
            CMP     AX,0FFFFH   ;was the drive valid?
            JZ      ERR
            MOV     SPC,AX      ;sectors per cluster
            MOV     AU,DX       ;allocation units on drive
            MOV     BPS,CX      ;bytes per sector
            MOV     AAU,BX      ;available allocation units
            LEA     DX,MSG1     ;set up pointer to 1st message
            MOV     AH,9        ;display string function
            INT     21H         ;DOS call
            MOV     AX,AU       ;get allocation units
            CALL    DISPINT     ;display value
            LEA     DX,MSG2     ;set up pointer to 2nd message
            MOV     AH,9        ;display string function
            INT     21H         ;DOS call
            MOV     AX,BPS      ;get bytes per sector
            CALL    DISPINT     ;display value
            LEA     DX,MSG3     ;set up pointer to 3rd message
            MOV     AH,9        ;display string function
            INT     21H         ;DOS call
            MOV     AX,AAU      ;get available allocation units
            CALL    DISPINT     ;display value
            LEA     DX,MSG4     ;set up pointer to 4th message
            MOV     AH,9        ;display string function
            INT     21H         ;DOS call
            MOV     AX,SPC      ;get sectors per cluster
            CALL    DISPINT     ;display value
            JMP     EXIT        ;and return to DOS
ERR:        LEA     DX,EMSG     ;set up pointer to error message
            MOV     AH,9        ;display string function
            INT     21H         ;DOS call
EXIT:       .EXIT

DISPINT PROC    NEAR
            MOV     BX,10000    ;load divisor
            CALL    TODECI      ;find and display 10,000's digit
            MOV     BX,1000     ;load divisor
            CALL    TODECI      ;find and display 1,000's digit
            MOV     BX,100      ;load divisor
            CALL    TODECI      ;find and display 100's digit
            MOV     BX,10       ;load divisor
            CALL    TODECI      ;find and display 10's digit
            ADD     AL,30H      ;add ASCII bias to 1's digit
```

```
                MOV    DL,AL        ;load character for output
                MOV    AH,2         ;display character function
                INT    21H          ;DOS call
                RET

TODECI   PROC   NEAR
         SUB    DX,DX        ;clear DX
         DIV    BX           ;find digit
         ADD    AL,30H       ;add alpha bias
         PUSH   DX           ;save remainder
         MOV    DL,AL        ;load character for output
         MOV    AH,2         ;display character function
         INT    21H          ;DOS call
         POP    AX           ;get remainder back
         RET
TODECI   ENDP
DISPINT  ENDP

         END
```

If the current drive is a hard drive, DISKCAP's execution might output information similar to this:

```
Allocation units on drive:   28179
Bytes per sector:            00512
Available allocation units:  08325
Sectors per cluster:         00016
```

On a 720K floppy drive, the results are significantly different:

```
Allocation units on drive:   00713
Bytes per sector:            00512
Available allocation units:  00479
Sectors per cluster:         00002
```

The data presented can be used to develop other important information, such as how many bytes are available.

Programming Exercise 9.17: Modify the DISKCAP program so that it also outputs the number of bytes free on the current drive.

Programming Exercise 9.18: Write a program that will allow you to enter a file name and then change the date and time of the file to the current date and time.

Programming Exercise 9.19: Write a program called HIDE that will set the Hidden attribute of the file specified by the user.

9.9   TROUBLESHOOTING TECHNIQUES

The use of files from a floppy or hard disk greatly improves the capabilities of a program. A few tips to keep in mind when working with files:

- Do not fool around with file operations using DEBUG, especially if you are writing sectors. It is easy to enter an incorrect sector number and damage important information, such as the FAT or root directory.
- Build a proper ASCIIZ string, such as

```
RESULTS DB '\C:\DATA\RESULTS.DAT',0
```

- Use the DOS file functions carefully, particularly when deleting a file or subdirectory.
- Use file handles to access open files.
- Close a file when finished with it. Do not rely on DOS to close it for you.
- It is important to perform error checking upon the return of a file function, to guarantee that the operation was successful. If the disk is full during a write operation, we must be told and action taken (discard the file or save it somewhere else).

Many interesting file applications are possible, from image processing to virtual reality game graphics, through the use of data stored in files. Begin developing your own file applications—it will be worth the experience.

SUMMARY

In this chapter we examined the operation of disks and files. The basic structure of a floppy disk, from boot sector to FAT and directory sectors, was covered. A number of BIOS and DOS interrupt functions were explained and used to perform useful work on disks and disk files. These various interrupt functions should be used carefully, because improper use may lead to file corruption and even erasure.

STUDY QUESTIONS

1. What are the differences between 5.25- and 3.5-inch disks?
2. How do hard drives store so much information?
3. Why are hard drives faster than floppy drives?
4. What is the purpose of the boot sector?
5. What is the BIOS parameter block? Where is it stored?
6. A FAT entry pair has the following data: 12 10 08. What two cluster numbers are stored?
7. Describe the operation and purpose of a cluster chain.
8. What is a cluster? Why are cluster numbers used in the FAT instead of sector numbers?
9. How many bytes does a directory entry use? How many directory entries are possible with fourteen directory sectors?
10. How does DOS represent a deleted file or directory?

11. The attribute byte in a directory entry is 12H. What does this mean?
12. What is the difference in disk access capability between BIOS INT   13H and DOS INT   21H?
13. What is an ASCIIZ string? Show the necessary source statement for an ASCIIZ string describing the file ABC.DAT.
14. What is a file handle? Where does it come from? What is it used for?
15. Show the instructions needed to open the file SPIN.ASM for read/write access.
16. What happens when an existing file is *created* with DOS INT   21H, Function 3CH?
17. What are the three ways a file pointer may be moved?
18. Where is track zero?
19. What is stored on track zero?
20. What does the boot code do?

ADDITIONAL PROGRAMMING EXERCISES

1. Write a program that reads the FAT of drive A and displays the decoded FAT entry values.
2. Write a program that displays all of the subdirectories of root.
3. Write a program that simulates the DIR command. Differentiate files from directories by using square brackets [ ] around directory names.
4. Write a program that prints out the file name and first line of text from every .ASM file in the current directory.
5. Write a program that opens *any* file and displays its contents in hexadecimal.
6. Write a program that creates a text file called HELLO.TXT only if the file does not yet exist.
7. Write a program that counts the number of lines in a text file. Assume all lines are terminated by a carriage return (0DH). Display the count as follows:

```
The file contains 143 lines of text.
```

8. Write a program that *double spaces* a text file.
9. Write a program that counts the number of occurrences for each letter found in a text file. Display the results as follows:

```
Number of As: 29
Number of Bs: 7
    .
    .
    .
Number of Zs: 2
```

10. Write a program that writes fifty blocks of 11 bytes to an output file. Each block of 11 bytes should contain a valid phone number, as in:

```
DB   '16075551234'
```

11. Write a program that searches the output file created in Question 10 for a number supplied by the user.

12. Write a program that prints out the cluster chain of the file specified by the user.

13. Use a data file to store a 16 by 16 group of bytes that represent a maze. Byte values are 00H or FFH, meaning floor and wall, respectively.

14. Read the maze data file from Question 13 and display it on the screen as a grid of blank (floor byte) or '*' (wall byte) characters.

15. Write a program that searches a file (name supplied by the user) for a text string (also supplied by the user). Display the offset into the file of the start of the string, as in:

```
string found at offset 3E86
```

PART 4

Hardware Architecture

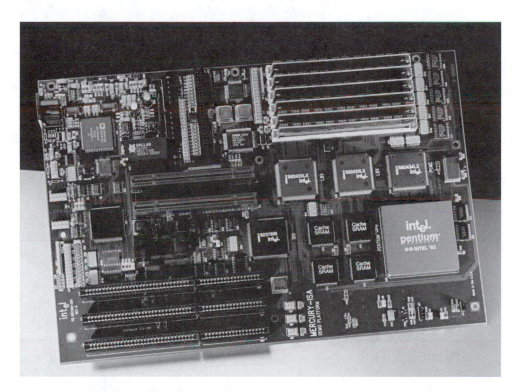

A typical Pentium motherboard for a personal computer.

CHAPTER 10

Hardware Details of the 8088

OBJECTIVES

In this chapter you will learn about:

- The general specifications of the 8088 microprocessor
- The processor's control signal names and functions
- General signal relationships and timing
- Methods by which the 8088 can interface with external devices
- The external interrupt signals and their operations
- The 8088 bus controller
- The method used to access an 8085 peripheral

10.1   INTRODUCTION

Before using any microprocessor, it is necessary to understand both its hardware requirements and its software functions. In this chapter we will examine all 40 pins of the 8088's package and see what their uses are in a larger system employing the 8088 as its CPU. We will not concentrate on interfacing, because this important topic is covered in Chapters 11, 12, and 13. Upon completion of this chapter, we should, however, know about the various signals of the processor to begin interfacing it with support circuitry, which includes memories, I/O devices, and coprocessors. The hardware architecture of the Pentium, which is covered in Chapter 15, is radically different from that of the 8088. Even so, the experience gained working with the 8088's hardware is applicable to current microprocessor technology, especially since the motherboards of today's personal computers still support the original bus architecture and signals. The information presented in this chapter, and in Chapters 11 through 13, will prepare you for the design of a working 8088 system in Chapter 14.

Section 10.2 gives a quick overview of the capabilities of the 8088, its memory addresssing capabilities, available clock speeds, and various other functions. Section 10.3 covers all 40 pins of the 8088 in detail. The pins are separated into eight functional groups, such as interrupt control, system control, processor status. Block diagrams and timing waveforms are given where applicable, except where they might apply to interfacing. Section 10.4 describes the operation of the 8284 clock generator, an essential integrated circuit used in the 8088-based systems. A second essential component is the 8288 bus controller, which is covered in Section 10.5. Timing diagrams for certain processor operations are examined in Section 10.6. Section 10.7 covers the various types of bus connectors found on the motherboard in many personal computers. Hardware troubleshooting techniques are presented in Section 10.8.

10.2 CPU SPECIFICATIONS

Although we covered some of the 8088's specifications in Chapter 2, it will be useful to cover them again, this time adding more detail.

The 8088 is a 16-bit microprocessor that communicates with the outside world via an 8-bit bidirectional data bus. This requires the 8088 to perform *two* read cycles to capture 16-bit chunks of data. This has the effect of increasing memory access time and program execution time. Although the increase in time is not significant, it is important to remember this feature of the 8088. The 8086, having a 16-bit data bus, needs to perform only *one* read cycle to fetch the same data and thus executes programs slightly faster than the 8088.

The 8088's 20-bit address bus can access over *1 million* bytes of memory (1,048,576 bytes, to be exact). We commonly refer to this number as 1 **megabyte** of memory. Control signals are provided that enable external circuitry to take over the 8088's buses (a must for DMA operations), and two interrupt lines are included to provide maskable and nonmaskable interrupt capability. A number of status outputs are available, which may be used to decode any of eight internal CPU states, and other control signals are provided to allow interfacing with 8085 and 8088 peripherals.

The 8088 comes with maximum clock speeds of 5 or 8 MHz as of this writing, and has been on the market long enough to be purchased at a reasonable cost. But the power of the 8088 can be tapped only if we know how to use it. So let's begin examining the functional operation of the processor.

Minmode Operation

The 8088 has two functional modes of operation: minimum mode (minmode) and maximum mode (maxmode). Certain pins on the processor have been designed for dual purposes, one for minmode and the other for maxmode. Figure 10.1 shows a pinout diagram of the 8088. Notice how pins 24 through 31 and pin 34 have two sets of signal names. In minmode, these nine signals are $\overline{\text{INTA}}$, ALE, $\overline{\text{DEN}}$, DT/$\overline{\text{R}}$, IO/$\overline{\text{M}}$, $\overline{\text{WR}}$, HLDA, HOLD, and SS0. They correspond to control signals needed to operate memory and I/O devices connected to the 8088, and are compatible with signals used in older 8085-based systems.

FIGURE 10.1   8088 pin assignments

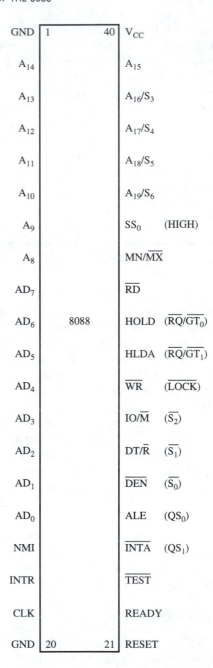

Note: ( ) denotes a maxmode signal

Because the 8088 generates these signals in minmode, fewer chips are needed in the overall system; however, some functions are unavailable when the 8088 operates in minmode, including bus request/grant operations and coprocessor capability.

Maxmode Operation

When the 8088 operates in maxmode, the nine signals we just examined change their functions. The new signals become QS_1, QS_0, $\overline{S}_0$, $\overline{S}_1$, $\overline{S}_2$, $\overline{LOCK}$, $\overline{RQ/GT}_1$, $\overline{RQ/GT}_0$, and HIGH. The lack of control signals (which exist only in minmode operation) now requires the use of the 8288 bus controller to generate memory and I/O read/write signals. Because an external chip is now generating these control signals, the processor is free to expand its functional capability. This allows the use of an 8087 coprocessor, and provides bus request/grant operation and queue status. All of these functions will be explained in detail in the next section.

10.3 CPU PIN DESCRIPTIONS

Refer to Figure 10.1 for another look at the 40 pins of the 8088's Dual In-line Package. There are eight groups of pins that we will examine in this section. Each group performs a specific function, necessary to the proper operation of the 8088.

V_{CC}, GND, and CLK

This group deals with the processor power and clock inputs. Note that there are two pins for ground (GND) and one for V_{CC}. Both grounds must be used for proper operation. The 8088 operates on a single, positive supply of 5 V ± 10 percent (with some versions having a 5 percent tolerance), and will dissipate 2.5 watts of power at this voltage. The specified supply current is 340 mA at room temperature.

The CLK input requires a digital waveform with a 33 percent duty cycle. This waveform is shown in Figure 10.2. A 33 percent duty cycle means that the digital level is high one-third of the time. The minimum clock period is 200 ns, corresponding to a frequency of 5 MHz. The maximum clock period is 500 ns, resulting in a minimum clock speed of 2 MHz. Rise and fall times should be kept under 10 ns. The TTL-compatible clock signal is generated by the 8284 clock generator covered in Section 10.4. Even though the clock input is internally buffered, the clock signal should be kept at a *constant* frequency (via the 8284 and a crystal oscillator) for best operation.

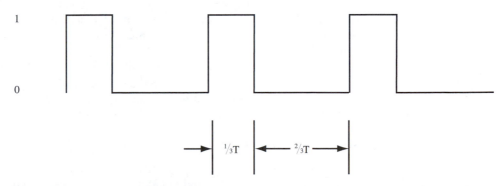

T = time of one cycle

FIGURE 10.2 Three cycles of the CLK waveform

TABLE 10.1   8088 status signals

$\overline{S_2}$	$\overline{S_1}$	$\overline{S_0}$	Indicated Operation
0	0	0	Interrupt acknowledge
0	0	1	I/O read
0	1	0	I/O write
0	1	1	Halt
1	0	0	Code access
1	0	1	Memory read
1	1	0	Memory write
1	1	1	Passive

MN/$\overline{\text{MX}}$

This pin is used to control the 8088's mode of operation. Remember from Section 10.2 that the 8088 may operate in minmode or maxmode. The specific mode is selected with the MN/$\overline{\text{MX}}$ pin. Maxmode is enabled when MN/$\overline{\text{MX}}$ is connected to ground. Minmode is enabled when MN/$\overline{\text{MX}}$ is pulled high (through an appropriate pullup resistor). As we saw before, nine of the 8088's signals have two functions, with each function depending on the mode of processor operation.

$\overline{S_0}$, $\overline{S_1}$, and $\overline{S_2}$

These three signals are the 8088's status outputs and are active only when the processor is in maxmode. They are used to indicate internal processor operations. The 8288 bus controller (covered in Section 10.5) uses the status outputs to generate memory and I/O read/write signals. Table 10.1 shows the eight different conditions that can be indicated by the status outputs. The *interrupt acknowledge* status can be used by external circuitry to manipulate the processor's interrupt mechanism. The *code access* status indicates when the processor is fetching instructions. When the 8088 has completed a bus cycle, the status outputs will indicate the passive state. An easy way to decode all eight processor states is to use a three- to eight-line decoder, such as the 74LS138; this is left for you to do as a homework problem. Normally in a small system we have no use for most of the decoded cycle states. In fact, an interrupt acknowledge signal ($\overline{\text{INTA}}$) is already provided for us when operating in minimum mode.

The status outputs are capable of tri-stating when the 8088 enters into hold acknowledge (see the description of HOLD and HLDA in the next section).

A tri-state signal may be in one of three states at any time: low (0) state, high (1) state, and high-impedance (open) state, as indicated in Figure 10.3.

RESET, READY, HOLD, and HLDA

This group of signals is used for system control. RESET and READY operate the same way in both minmode and maxmode. HOLD and HLDA (*hold acknowledge*) work only when the processor is in minmode.

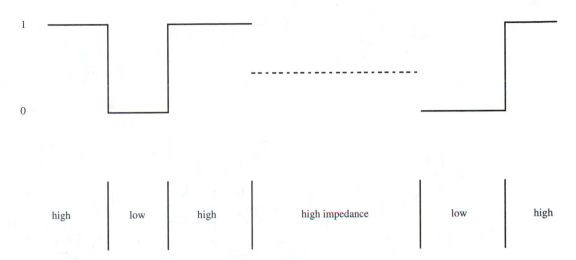

FIGURE 10.3 Tri-state signal levels

RESET is the signal that really gets the processor running after a power-up. It is very important to RESET the 8088 when power is first applied to guarantee that the 8088 begins doing intelligent things. A high logic level is needed to activate the RESET input, which must remain high for at least four clock cycles to ensure proper operation. RESET can be applied during program execution as well, as a sort of panic button used to start the program over from the beginning.

Issuing a RESET causes the 8088 to fetch the first instruction from memory, beginning at address FFFF0H. This requires some type of ROM data at that address to guarantee that the 8088 begins running correctly.

READY is an input that informs the processor that the selected memory or I/O device is ready to complete a data transfer. READY is often used to synchronize the fast processor with a slow memory or I/O device that may need an extended bus cycle to perform a read or write operation. A high logic level on READY indicates that the 8088 may go ahead and complete the bus cycle. A low logic level causes the processor to extend the bus cycle with all signals frozen at their current logic levels. This means that the processor keeps the address lines, data bus, and control signals in their current states so that the external device may use them for a longer period of time.

HOLD is a minmode input that is used to place the processor into a **suspended** execution state. While the 8088 is in a hold state, it does not continue program execution. In fact, many of the processor's outputs are automatically tri-stated to prevent conflict on the system bus. The most common use of HOLD is in a computer system having two or more processors. If the processors share memory (such as EPROM or RAM space), only one processor may access memory at a time. When a processor wishes to take control of the buses and access memory, it must first issue a HOLD request to the other processors in the system, which will suspend their processing and release the system bus. A high logic level is needed to activate HOLD. Furthermore, the 8088 will remain held only as long as the HOLD input remains high. The processor will resume program execution where it left off as soon as HOLD goes low.

HLDA (hold acknowledge) is a minmode output used to indicate to external devices that the 8088 has suspended execution (via HOLD). HLDA will go to a high logic level to show that the processor has actually stopped execution. Technically, another device should take over the system bus only *after* HLDA goes high. When execution resumes (i.e., HOLD has been taken low), HLDA will go low.

NMI, INTR, and $\overline{\text{INTA}}$

These three signals control the activity of external hardware interrupts. NMI and INTR are inputs and function identically in either processor mode. $\overline{\text{INTA}}$, an output, is available only in minmode.

External hardware interrupts are used to suspend current program execution and vector the processor to a special set of instructions called an **interrupt service routine (ISR).** Interrupts are used to perform high-priority tasks without affecting the main processor program (except for a loss of execution time). A very useful application of hardware interrupts is keeping track of time. It is not difficult to convert the 60-cycle powerline frequency into a digital signal (with the use of a Schmitt trigger). The resulting 60-Hz digital signal is then connected to NMI or INTR, generating 60 interrupts every second. The corresponding ISR that services the interrupt decrements a counter each time it runs. If the counter is initially set to 60, it is easy for the computer to know when 1 second has passed.

NMI (*nonmaskable interrupt*) requires a *rising edge* to be recognized, as indicated in Figure 10.4. It cannot be internally disabled (masked) by software, hence its name. NMI generates a type-2 interrupt. We have seen (in Chapter 5) that there are 255 different types of interrupts. NMI is recognized by the 8088 at the end of the currently executing instruction. The address of the type-2 ISR is then read from a table containing all ISR addresses.

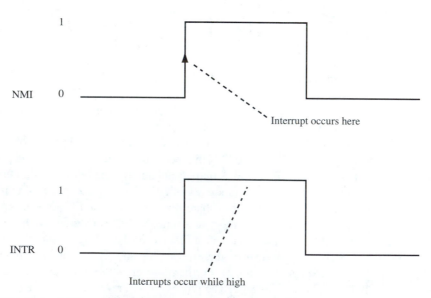

FIGURE 10.4   NMI and INTR signals

This table is stored in memory and is called the **interrupt vector table.** The 8088 then jumps to the ISR address to process the interrupt code.

INTR (interrupt request) requires a *high* logic level to be recognized. Leaving INTR in a high state could cause repeated interrupts, so caution is advised. A bit in the processor's status register that is used to enable/disable INTR is called the **interrupt enable flag (IF).** Using the IF to disable INTR can be done easily with software, and is referred to as **masking.** INTR operates in much the same way as NMI except for the way the interrupt type is generated. NMI automatically causes a type-2 interrupt. The interrupt type for INTR is actually read from the processor's data bus during an **interrupt acknowledge cycle.** The 8-bit interrupt vector read from the data bus is internally converted into the proper address for the interrupt vector table.

A priority is assigned to each interrupt, with NMI having the higher priority of the two. The priority is used to determine which interrupt is recognized first if both occur at the same time. The higher priority of NMI guarantees that it is recognized before INTR. A comparison of NMI and INTR is shown in Table 10.2.

$\overline{\text{INTA}}$ (interrupt acknowledge) is an active low output that operates in minmode and is used to indicate that the 8088 has received an INTR and is beginning interrupt processing. The external circuitry connected to INTR should use $\overline{\text{INTA}}$ to control when the 8-bit interrupt vector is placed onto the data bus. The exact timing of $\overline{\text{INTA}}$ will be covered in Section 10.6.

$\overline{\text{RQ/GT}}_0$, $\overline{\text{RQ/GT}}_1$, and $\overline{\text{LOCK}}$

These three maximum mode signals are used to interface the 8088 with other devices capable of taking over the system bus. In the description of the HOLD input we saw one way two 8088s could share a common system bus and memory. The three signals presented here offer a second technique. $\overline{\text{RQ/GT}}_0$ and $\overline{\text{RQ/GT}}_1$ are request/grant signals used by other devices called **bus masters** to take over the 8088's system bus. An example is the 8087 coprocessor, which periodically takes over the system bus to read data or write results into memory. Both $\overline{\text{RQ/GT}}$ signals are bidirectional, meaning that they act as inputs *and outputs.* $\overline{\text{RQ/GT}}_0$ has priority over $\overline{\text{RQ/GT}}_1$, and both have internal pullup resistors. The following discussion will concern $\overline{\text{RQ/GT}}_0$ only, but applies to $\overline{\text{RQ/GT}}_1$ as well.

When another bus master decides to take over the system bus, it will pull $\overline{\text{RQ/GT}}_0$ low for one clock period. When the 8088 is ready to release the system bus, it will use $\overline{\text{RQ/GT}}_0$ as an output to inform the new bus master. A second low-level pulse of one clock period does this. The 8088 then enters a hold acknowledge state until the new bus master is ready to give the system bus back. This is done by a third low-level, one-clock-period pulse on $\overline{\text{RQ/GT}}_0$ (from the new bus master back to the 8088). So, the normally high $\overline{\text{RQ/GT}}_0$ signal went from being an input to an output and back to an input again.

TABLE 10.2 Comparison of NMI and INTR

Interrupt	Logic Level Needed to Trigger	Disabled via Software	Priority
NMI	Rising edge	No	High
INTR	High	Yes	Low

$\overline{\text{LOCK}}$ is an active low output used to inform other possible bus masters that the 8088's system bus is not available for takeover. A special instruction called a **LOCK prefix** activates the $\overline{\text{LOCK}}$ signal, which will go back to its inactive state after execution of the instruction following the LOCK prefix instruction. If an $\overline{\text{RQ/GT}}$ sequence is requested while the bus is $\overline{\text{LOCK}}$ed, it will not be acted on until the bus becomes un$\overline{\text{LOCK}}$ed.

ALE, $\overline{\text{DEN}}$, DT/$\overline{\text{R}}$, $\overline{\text{WR}}$, $\overline{\text{RD}}$, and IO/$\overline{\text{M}}$

Five of these signals operate in minmode. Only $\overline{\text{RD}}$ operates in both minmode and maxmode. Some of these signals are used to interface the 8088 with 8085 peripherals. All six signals are outputs, and all but ALE will be tri-stated during hold acknowledge.

ALE (*address latch enable*) is an output signal used to demultiplex the 8088's address/data bus. ALE is usually used to control an external latch capable of storing the lower 8 bits of the processor's address bus. ALE goes high to indicate that the 8088 is outputting address information. Take another look at Figure 10.1 and note the signals AD_0 through AD_7. These eight signals perform two functions. They operate as A_0 through A_7 during the beginning of a bus cycle, and as D_0 through D_7 for the rest of the bus cycle. So, AD_0 through AD_7 are constantly switching back and forth between address bus mode and data bus mode. The processor uses ALE to indicate when AD_0 through AD_7 contain address information. Figure 10.5 shows one way the address/data lines may be demultiplexed. A 74LS373 octal D latch is used to capture A_0 through A_7 when enabled by ALE. This capture is illustrated by Figure 10.6. We will see another example of how the address/data lines are demultiplexed when we cover the 8288 bus controller in Section 10.5.

$\overline{\text{DEN}}$ (*data enable*) is an active low output used in a minmode system to control a bidirectional buffer (also called a **transceiver**) connected to the processor's data bus. The bidirectional buffer is used to buffer data going both ways on the data bus (into the 8088 and out of the 8088). Because there may be times when we want to disconnect the data bus from the processor (e.g., when we are sharing memory with another processor), $\overline{\text{DEN}}$ gives us a way to turn the transceiver off. In Section 10.5 we will see how $\overline{\text{DEN}}$ is generated in a maxmode system.

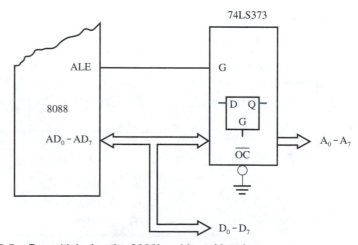

FIGURE 10.5   Demultiplexing the 8088's address/data bus

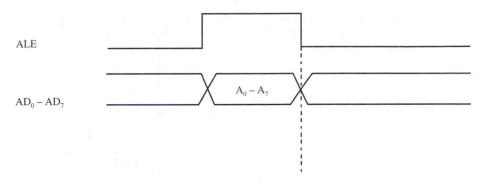

A$_0$ – A$_7$ latched when ALE goes low

FIGURE 10.6   Latching the lower address byte

DT/$\overline{R}$ (*data transmit/receive*) is an output used in a minmode system to control the direction of data flow in the bidirectional buffer used on the data bus. When DT/$\overline{R}$ is low, data should flow into the 8088. When DT/$\overline{R}$ is high, the 8088 is outputting data. We will see an example of how DT/$\overline{R}$ is used in Section 10.5.

$\overline{WR}$ (write) is an active low output used to indicate when the processor is writing to a memory or I/O location.

$\overline{RD}$ (read) is an active low output used to indicate when the processor is reading a memory or I/O location.

IO/$\overline{M}$ is an output that indicates whether the current bus cycle is a memory access or an I/O access. A memory access is indicated by a logic zero on IO/$\overline{M}$, and I/O access by a logic one. IO/$\overline{M}$ is used with $\overline{RD}$ and $\overline{WR}$ to generate separate read and write signals for memory and I/O devices. Figure 10.7 shows how OR gates can be used to decode different read/write operations. In Section 10.5 we will see that some of the signals generated by the 8288 bus controller are identical in operation to those generated in Figure 10.7.

A$_8$ Through A$_{19}$, AD$_0$ Through AD$_7$

These signals constitute the 8088's 20-bit address bus and 8-bit data bus. AD$_0$ through AD$_7$ are the processor's multiplexed address/data bus. These signals can behave like A$_0$ through A$_7$ or D$_0$ through D$_7$ (depending on the state of ALE). Usually an external latch is used to store address information when it is present. The other address lines, A$_8$ through A$_{19}$, make up the rest of the 8088's address bus. These address lines do not have to be latched because their outputs are always valid. Together, the 20 address lines can access 1MB of memory. The ability to directly address this many address locations provides the programmer with a very flexible programming environment. Older systems required that special software be used to manage memory in "pages" that were some portion of the system's main address space. This software is not even needed now in some applications, because of the great increase in address lines and, hence, memory locations. The 8088 has special segment registers that manage 64KB blocks of memory starting on any 16-byte boundary, so it is relatively easy to manage the system's addressing space.

A$_0$ through A$_{15}$ have a dual role: they are also used to access I/O devices (depending on the state of IO/$\overline{M}$). Sixteen address lines provide for up to 65,536 I/O locations. We will see more about memory and I/O devices in Chapters 11 and 12.

FIGURE 10.7 Decoding 8088 memory and I/O read/write signals

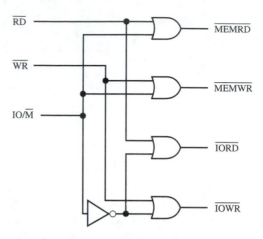

Signal Summary

Table 10.3 summarizes all of the signals we have just covered. It shows whether a signal is an input or an output (or both), if it has tri-state capability, and in which mode (min or

TABLE 10.3   8088 signal summary

Signal	Input	Output	Tri-state	Minmode	Maxmode
CLK	✓			✓	✓
MN/$\overline{\text{MX}}$	✓			✓	✓
$\overline{S}_0, \overline{S}_1, \overline{S}_2$		✓	✓		✓
RESET	✓			✓	✓
READY	✓			✓	✓
HOLD	✓			✓	
HLDA		✓		✓	
NMI	✓			✓	✓
INTR	✓			✓	✓
$\overline{\text{INTA}}$		✓		✓	
$\overline{\text{RQ}}/\overline{\text{GT0}}$	✓	✓			✓
$\overline{\text{RQ}}/\overline{\text{GT1}}$	✓	✓			✓
$\overline{\text{LOCK}}$		✓	✓		✓
ALE		✓		✓	
$\overline{\text{DEN}}$		✓	✓	✓	
DT/$\overline{\text{R}}$		✓	✓	✓	
$\overline{\text{WR}}$		✓	✓	✓	
$\overline{\text{RD}}$		✓	✓	✓	✓
IO/$\overline{\text{M}}$		✓	✓	✓	
AD_0–AD_7	✓	✓	✓	✓	✓
A_8–A_{19}		✓	✓	✓	✓

max) it is active. In the following sections we will see how many of these signals are used in an actual system.

10.4   THE 8284 CLOCK GENERATOR

As we saw in the previous section, CLK, RESET, and READY are three of the most important signals in the overall operation of the processor. A small number of logic gates would be needed to implement the correct signals on CLK and RESET. Most of this circuitry is already provided for us in an 18-pin DIP called the 8284 clock generator. Figure 10.8 shows a pin diagram of the 8284. Two of the pins, X_1 and X_2, are meant to be directly connected to a crystal. The internal clock circuitry of the 8284 then generates the proper CLK signal for the 8088. Because the frequency of CLK will be one-third the crystal frequency (due to an internal frequency divider), the 8284 provides the OSC output, whose frequency is the same as the crystal's. When using a crystal, the 8284's $F/\overline{C}$ input must be grounded. When $F/\overline{C}$ is high, the 8284's EFI pin must be connected to an external oscillator, or some type of timing circuit that generates a TTL signal at the proper frequency. In this case, the frequency of OSC will match the frequency of the EFI input. In addition to $F/\overline{C}$, a second signal, CSYNC, is used to provide clock synchronization when EFI is used as the frequency source. When a crystal is used, CSYNC must be taken low.

Another signal, $\overline{RES}$, is the 8284's reset input. This input is normally connected to a simple resistor-capacitor network. When power is first applied, the RC network allows a logic zero to remain on the $\overline{RES}$ input for a short period of time. The internal circuitry of the 8284 uses $\overline{RES}$ to generate the processor's RESET signal. Together, the RC network and the 8284 provide a *power-on reset* signal to the 8088.

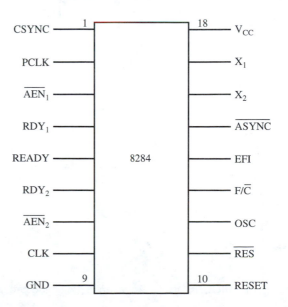

FIGURE 10.8   8284 clock generator

Two ready inputs are provided on the 8284, RDY_1 and RDY_2. Together with $\overline{AEN}_1$ and $\overline{AEN}_2$, the 8284 generates the processor's READY signal. In a system where memory and I/O devices are used, one RDY signal can be used with the memory circuitry and the other for the I/O circuitry. The $\overline{AEN}$ signals are used as *qualifiers* for the RDY inputs. For example, to use RDY_1, $\overline{AEN}_1$ must be low. It may be necessary to generate a **wait state** in the system, due to the use of slow memory or I/O devices. The RDY inputs are designed for this purpose. A fifth signal, $\overline{ASYNC}$, is used to select the number of stages of synchronization on READY. Only one stage is used when $\overline{ASYNC}$ is high.

Many of the devices in an 8088 system require a clock that is slower than the processor's (the UART is one example). The PCLK output of the 8284 has a frequency that is one-sixth that of the crystal (or EFI) frequency. Together with EFI, $F/\overline{C}$, and X_1 and X_2, we have a number of ways of controlling the timing of our system with the 8284. Figure 10.9 shows

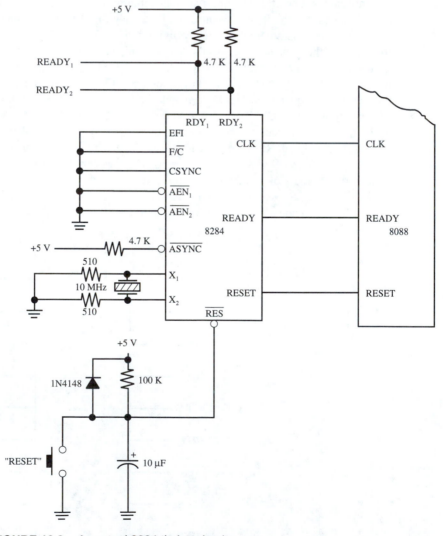

FIGURE 10.9   An actual 8284 timing circuit

one way the 8284 can be connected to the 8088. Notice that a 10-MHz crystal is connected to X_1 and X_2, and that F/$\overline{C}$ is grounded. The 8284 will thus generate a CLK frequency just over 3.3 MHz! How fast would the processor run if a 12-MHz crystal were used?

The RC network on $\overline{RES}$ is used to provide the power-on reset signal. A push button is connected across the capacitor to allow manual resets with power still applied. Figure 10.9 represents the timing circuitry used in the single board computer of Chapter 14.

10.5   THE 8288 BUS CONTROLLER

We have already seen a number of differences brought about by the 8088's minmode and maxmode operation. To get the maxmode signals, we sacrifice other signals that have important uses. For example, in maxmode we get $\overline{RQ}/\overline{GT}$ signals but have to give up HOLD and HLDA. Other signals are replaced as well, leaving us with no ability to decode read/write accesses in maxmode *unless we use the status outputs.* Because $\overline{S}_0$, $\overline{S}_1$, and $\overline{S}_2$ are available in maxmode, we use them as inputs to a special 8288 bus controller, which in turn decodes the missing signals. Figure 10.10 shows a pin diagram of the 8288. Notice the three status inputs and also the three *outputs* ALE, DEN, and DT/$\overline{R}$. Three of the signals that were lost when we switched to maxmode are now being generated by the 8288. Other signals that we require are $\overline{MRDC}$ (*memory-read command*), $\overline{MWTC}$ (*memory-write command*), $\overline{IORC}$ (*I/O-read command*), and $\overline{IOWC}$ (*I/O-write command*). These signals control the memory and I/O devices. Two additional signals, $\overline{AMWC}$ (*advanced memory-write command*) and $\overline{AIOWC}$ (*advanced I/O-write command*), are provided to give memory and I/O circuitry advance warning that they will be accessed (so that they can get a jump on the decoding process).

FIGURE 10.10   8288 bus controller

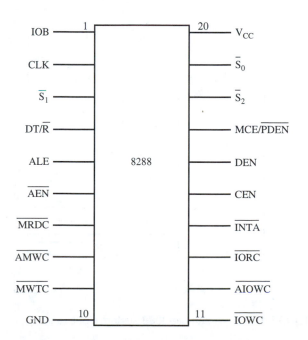

$\overline{\text{INTA}}$ is decoded and also available as an output. Remember that this signal indicates that the processor received an interrupt request on INTR.

The 8288 can be used to control the operation of the buffers and bidirectional latches used on the address and data buses. The ALE, DEN, and DT/$\overline{\text{R}}$ outputs of the 8288 operate like the actual minmode signals, with the exception of DEN being inverted. Figure 10.11 shows how the 8288 connects to the 8088 and its bus circuitry in a typical application. The inverter on the DEN output is used to enable the 8286 data bus transceiver. DT/$\overline{\text{R}}$ controls the direction of data in the 8286. The latching of the lower

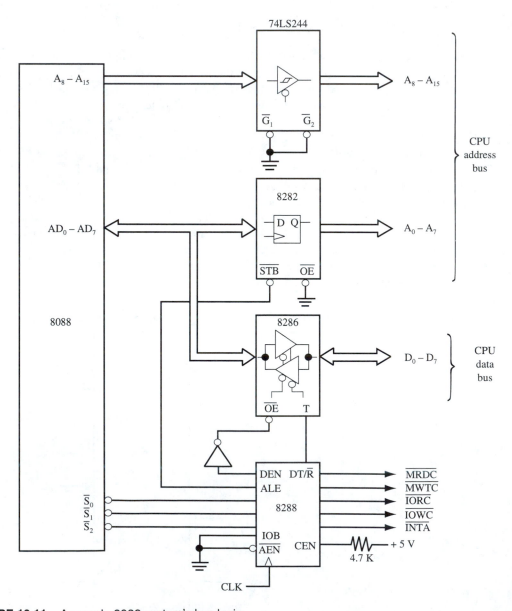

FIGURE 10.11   A sample 8088 system's bus logic

eight address lines is accomplished by ALE and an 8282 octal latch. Address lines A_8 through A_{15} are buffered with a 74LS244 octal buffer. Similar buffering is also needed on address lines A_{16} through A_{19}, which are not shown since they are unused in the application (not unusual, since many minimal systems do not require more than 64KB of memory).

The 8288 is capable of additional operations, one being I/O-bus mode. Here the IOB (*I/O bus mode*) input must be high, CEN (*command enable*) must be high, and $\overline{AEN}$ (*address enable*) should be low to enable the buses. These signals will also make MCE/$\overline{PDEN}$ operate like DEN, except only during I/O operations. The 8288 can thus be used to control a special I/O bus, if necessary. IOB, CEN, and $\overline{AEN}$ can also configure MCE/$\overline{PDEN}$ to operate as MCE (*master cascade enable*), a signal used during interrupt cycles. Other configurations allow multiple 8288s (with their own 8088s) to share a common system bus. In Figure 10.11, IOB and $\overline{AEN}$ are grounded, and CEN is pulled high. This selects system-bus mode and makes MCE/$\overline{PDEN}$ operate as MCE. The system bus cannot be disabled in this configuration.

10.6   SYSTEM TIMING DIAGRAMS

When timing is a critical issue in the design of a new 8088-based system, it pays to know how to interpret the CPU timing diagrams supplied by the manufacturer. This section will provide more details on the processor's control and timing signals by analyzing a number of timing scenarios.

We will start with the timing of a typical bus cycle.

CPU Bus Cycle

The 8088 routinely performs a number of different types of bus cycles. Memory read, memory write, I/O read, I/O write, and interrupt acknowledge are most of the important ones. A bus cycle is composed of four or more **T states,** numbered T_1, T_2, T_3, and T_4, where each T state represents the time of one CLK period. For example, if an 8088 has a CLK frequency of 4 MHz, the time of one T state is 250 ns, and a four-state bus cycle takes 1 μs. A T state is the smallest amount of time in which the processor can perform *any* function. Because all instructions are composed of a number of T states, instruction execution time depends on the CLK frequency. Figure 10.12 shows the simplified timing for a CPU bus cycle in a minmode system. Each new CLK pulse defines a new T state. Because the CLK must have a 33 percent duty cycle, you will notice that the CLK signal is low for two-thirds of its period and high for the last third. Note that ALE pulses high during the first T state. This means that the multiplexed address/data lines AD_0 through AD_7 are in address mode and should be latched by external circuitry. During T_1 we also see address lines A_8 through A_{19} changing to their next state. IO/$\overline{M}$ and DT/$\overline{R}$ may also need to change state during T_1, depending on the type of access (memory or I/O) and the direction of data on the bus (read or write). During T_2, the processor's $\overline{RD}$ and $\overline{WR}$ lines change state, in preparation for the data read or write that will occur in state T_3. What else happens in T_2? The upper four address lines change state to become four additional status outputs. Generally, these status bits indicate which type of segment register is being used during the bus cycle.

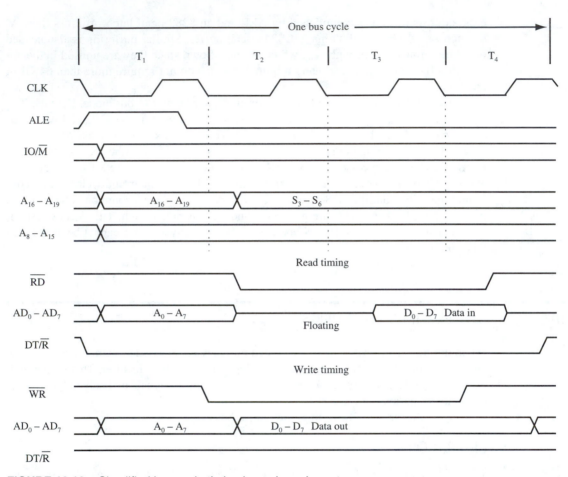

FIGURE 10.12   Simplified bus cycle timing in a minmode system

T_3 is the state in which the selected memory or I/O device should complete the data transfer. If the accessed memory or I/O logic cannot complete its job during T_3, it can extend the bus cycle via the processor's READY logic. The 8088 will insert **wait states** between T_3 and T_4 until READY indicates that the cycle may continue normally. Wait states have a duration of one CLK period also, so the time of any bus cycle will always be a multiple of CLK periods. During the end of T_4, all control signals and buses switch back to their inactive levels in preparation for the next cycle.

The processor may not automatically begin a new bus cycle after completing the current one. Suppose that the most recent bus cycle was the final fetch cycle for a multiply instruction. During the many CLK cycles it will take to internally perform the binary multiplication, the processor has no use for the bus, which may *idle* until completion of the multiply instruction. During this idle time, other devices are free to use the processor's bus, as long as they are done with it by the time the processor resumes its next bus cycle.

Interrupt Acknowledge Cycle

A minmode system uses one bus cycle to perform an interrupt acknowledge. Figure 10.13(a) shows the events that occur in this cycle. During T_1 the processor tri-states the address bus. During T_2 the $\overline{INTA}$ output is asserted, remaining low until it becomes inactive in state T_4. The interrupting device should respond to activity on $\overline{INTA}$ by placing an 8-bit **interrupt type** onto the data bus, which will be captured by the processor before it deactivates $\overline{INTA}$.

In a maxmode system, two bus cycles are used for interrupt acknowledge. Figure 10.13(b) shows the events that occur in each cycle. In the first cycle the address bus is again placed into a high impedance state. In addition, $\overline{LOCK}$ is asserted to inform other devices on the system bus that they cannot take over the bus until completion of the second cycle. $\overline{INTA}$ is asserted twice, once during each cycle. Special interrupt peripherals exist that have been designed to respond to the interrupt acknowledge cycles. One example is the 8259 programmable interrupt controller, which we will examine in Chapter 13.

HOLD/HLDA Timing

From Section 10.3 we know that it is possible to place the 8088 into a HOLD state (the processor idles) by asserting HOLD. The processor will respond by placing its buses in

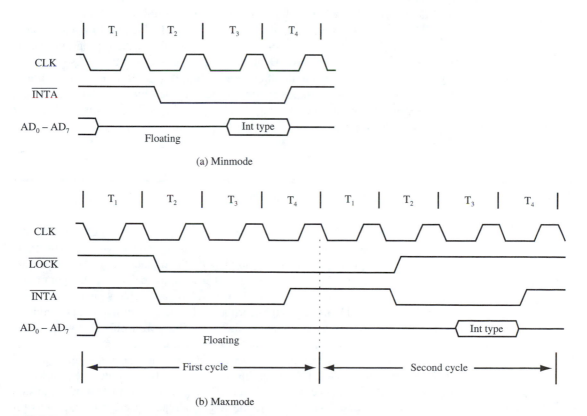

FIGURE 10.13   Interrupt acknowledge timing

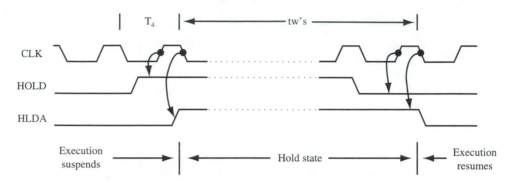

FIGURE 10.14   HOLD timing

the high impedance state and will acknowledge that it is in a HOLD state with HLDA. Figure 10.14 shows the corresponding timing relationship between HOLD and HLDA. The state of the HOLD input is sampled on every rising edge of CLK. If it is high, the processor will activate HLDA at the end of T_4 (or at the end of the current idle state). The processor may need a number of internal states to complete what it was doing (e.g., finishing a multiply instruction), but once HLDA is asserted it will no longer use its buses until it sees HOLD go low. While the 8088 is in the hold state, it will continue to sample HOLD on each rising edge of CLK. When HOLD does go low, the processor will reset HLDA at the beginning of the next T state and resume program execution. A technique that activates HOLD at the end of every instruction, so that the processor's buses can be examined, is called **single stepping.** So, a simple circuit could be used that would allow only one instruction to be executed by the 8088 each time a button was pressed. Think of how you might design such a single-step circuit (starting with a flip-flop might help).

10.7   PERSONAL COMPUTER BUS STANDARDS

The relatively low expense and high power of the PC has made it a popular teaching machine. Software can be written, tested, and executed easily, and hardware interfacing is made simple by the use of a standard connection bus on the motherboard of the machine. If you were to make a list of which processor signals you might need to connect your computer to a disk drive, or a serial data terminal, or a video card, what signals would you choose? Certainly we would need data lines for the exchange of information, and address lines to select the device we wish to communicate with. Then, we would also need all the necessary control signals, such as memory and I/O read/write, ALE.

The list we have been considering was made up many years ago and agreed upon by all people involved in the personal computer business. Figure 10.15 shows the pin assignments for a standard 62-pin connector found on all PC motherboards that allow expansion with the 8088 microprocessor. In addition to the signals already discussed, provisions for power and many levels of interrupts are also included. Do not confuse address lines A_0 through A_{19} with their respective connector pin names.

FIGURE 10.15 Personal computer standard pin assignments (ISA connector)

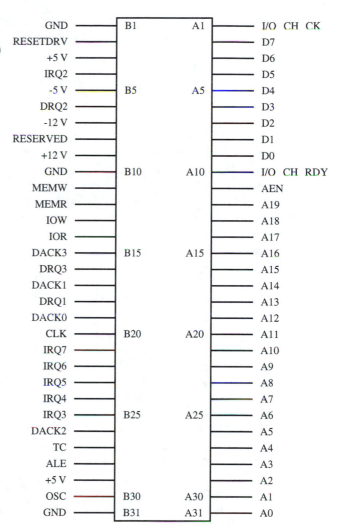

This connector is referred to as an **ISA (Industry Standard Architecture)** connector. It is also known as the PC-XT connector. Because it was designed for use with the original 8088 microprocessor, it contains only an 8-bit data path (D_0 through D_7).

When the 8086 and 80286 microprocessors found their way onto motherboards, it was necessary to provide a 16-bit data path to the connectors. A 36-pin extension connector was added to the original 62-pin connector, containing eight new data lines (D_8 through D_{15}) and additional interrupt and DMA signals. This 16-bit ISA connector is commonly called the PC-AT connector (the computer it was first used in). Figure 10.16 shows the structure of the PC-AT connector pair. Plug-in cards that require the full 16-bit data bus have two edges with connector traces. Older 8-bit cards have only one edge of connector traces, and do not use the additional 36-pin socket.

A recent improvement to the PC-AT connector is the **EISA (Extended ISA)** standard. This connector supports 80386, 80486, and Pentium microprocessors by providing a full 32-bit data bus. Additional address lines are provided to allow memory expansion as

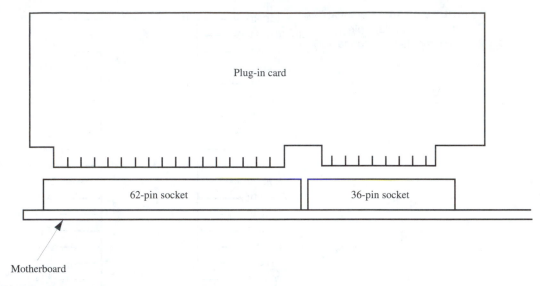

FIGURE 10.16   PC-AT connector sockets

well. The EISA connector is very similar to the PC-AT connector, with the original 62-pin and 38-pin definitions unchanged, and new connector pins added *between* the old connector pins, and at a lower depth! This prevents an old ISA board from making contact with the newer EISA signals. Three special bus-controlling chips are used to manage data transfers through the EISA connectors.

Two other bus standards are the **Local** bus and IBM's **MicroChannel**™ architecture. A Local bus connector provides the fastest communication possible between a plug-in card and the machine by bypassing the EISA chip-set and connecting directly to the CPU! Local bus video cards and hard-drive controllers are popular due to their high-speed data transfer capability.

IBM's MicroChannel architecture is entirely different from all of the other bus standards. Cards manufactured for the MicroChannel bus do not even plug into the ISA or EISA connector. This standard is not very popular outside of IBM's equipment, even though it provides 32-bit capability.

Overall, it is safe to say that connector technology has advanced to keep up with the improvements in microprocessors.

10.8   TROUBLESHOOTING TECHNIQUES

The hardware operation of the 8088 is complex and requires great patience during design or troubleshooting sessions. Your basic knowledge of 8088 hardware architecture should include operation of all its signals, and their functional groupings:

- System signals CLK, RESET, READY, HOLD, HLDA, and MN/$\overline{\text{MX}}$.
- The signals involved with memory and I/O accesses. These are data/address lines AD_0 through AD_7, address lines A_8 through A_{19}, ALE, $\overline{\text{RD}}$, $\overline{\text{WR}}$, DEN, DT/$\overline{\text{R}}$, and IO/$\overline{\text{M}}$.

- The signals involved with interrupts. These are NMI, INTR, and $\overline{\text{INTA}}$.
- The processor status signals $\overline{S}_0$ through $\overline{S}_2$.

The groupings are important because they point the way at the beginning of a design or while troubleshooting. Knowing the relationships between signals allows you to make informed decisions about what to do next, such as how to design a memory address decoder or determine why the READY input stays low in a faulty system.

SUMMARY

In this chapter we examined the operation of the 8088's hardware signals. We saw that there are actually two sets of processor signals, one for minmode operation and the other for maxmode operation. Minmode signals can be directly decoded by memory and I/O circuits, resulting in a system with minimal hardware requirements. Maxmode systems are generally more complicated, resulting from the use of the 8288 bus controller and the new maxmode signals that allow for bus grants.

We saw that the 8088 can access 1MB of memory, and that it contains two hardware interrupt mechanisms and uses a multiplexed address/data bus. Hardware examples showing how the bus is demultiplexed, how memory and I/O control signals are generated, and how the 8284 clock generator and 8288 bus controller are used were also given. The chapter finished with a short look at CPU timing diagrams and the interface connectors for PCs. In Chapters 11 through 14 we will draw on the information presented in this chapter, so use it as a handy reference.

STUDY QUESTIONS

1. What are some of the differences between minmode and maxmode operation?
2. How is the 8088 put into minmode operation?
3. Sketch four cycles of the 8088's CLK signal if its frequency is 3.125 MHz. Compute the time, in nanoseconds, of the low-portion and high-portion of each cycle.
4. If an instruction requires 20 states to complete, what is the instruction execution time if the CLK period is 250 ns?
5. What is the processor's CLK frequency if a 10-MHz crystal is used with the 8284?
6. Design a circuit that will turn an LED on when the status outputs $\overline{S}_0$ through $\overline{S}_2$ indicate the HALT state.
7. Show how a 3- to 8-line decoder (74LS138) can be used to decode the status assignments found in Table 10.1. The 74LS138 is shown in Figure 10.17.
8. What address is used first after a RESET?
9. Why do 8088-based systems need EPROM at the high end of memory?
10. How is the 8088 slowed down enough to communicate with a slow memory device?
11. What signal(s) might a second 8088 use to take over the buses of another 8088?
12. What are the differences between NMI and INTR?

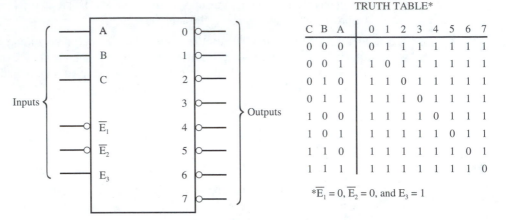

TRUTH TABLE*

C	B	A	0	1	2	3	4	5	6	7
0	0	0	0	1	1	1	1	1	1	1
0	0	1	1	0	1	1	1	1	1	1
0	1	0	1	1	0	1	1	1	1	1
0	1	1	1	1	1	0	1	1	1	1
1	0	0	1	1	1	1	0	1	1	1
1	0	1	1	1	1	1	1	0	1	1
1	1	0	1	1	1	1	1	1	0	1
1	1	1	1	1	1	1	1	1	1	0

$*\overline{E_1} = 0, \overline{E_2} = 0,$ and $E_3 = 1$

FIGURE 10.17   For Question 10.7

13. What would happen if INTR were stuck high?
14. What is so special about the operation of ALE?
15. What other types of data latches could be used in Figure 10.5 in place of the 74LS373?
16. If a minimal 8088 system uses only address lines A_0 through A_{17}, how many memory locations is it capable of accessing?
17. Show how the address 3A8C4 would be represented on the 20-bit address bus.
18. What is one advantage of a multiplexed address/data bus?
19. What is one disadvantage of a multiplexed address/data bus?
20. Show how NAND gates could be used to decode the signals in Figure 10.7.
21. Show the connections needed to feed a 12-MHz clock signal into the 8284 (note the lack of external crystal now).
22. In Figure 10.11, why does the data bus require a bidirectional driver?
23. During which state is ALE active?
24. Design a single-step circuit that allows only one instruction to execute each time a push button is pressed.
25. When two or more 8088s share a common memory system, what do you expect happens to the overall bus activity of the system? What about the bus activity of each processor?
26. What are the states of $IO/\overline{M}$, $\overline{RD}$, and $\overline{WR}$ when the 8088 is:
    a) writing to memory
    b) reading from an I/O device
27. What is placed on D_0 through D_7 during an interrupt acknowledge cycle?
28. Explain the differences between ISA, EISA, and Local bus connectors.
29. How can two 8088 CPUs be connected so that they *share* a 64KB RAM and have separate 16KB EPROMs?
30. Design a circuit that will reset the processor if an NMI occurs while the 8088 is being held (via HOLD).

CHAPTER 11

Memory System Design

OBJECTIVES

In this chapter you will learn about:

- The importance of bus buffering
- How the 8088 addresses (accesses) memory
- The design of custom memory address decoders
- The difference between full- and partial-address decoding
- How wait states may be inserted into memory read/write cycles
- The differences between static and dynamic RAMs
- How a dynamic RAM is addressed and what purpose refresh cycles serve
- DMA (direct memory access)

11.1 INTRODUCTION

The internal memory capacity of any microprocessor, with the exception of single-chip microprocessors, is severely limited. The 8088 itself has only a handful of 16-bit locations in which it can store numbers, and these locations are the actual data registers available to the programmer. The need for larger, external memories quickly becomes apparent, especially if an application involves number crunching or word processing. The purpose of this chapter is to explore ways of adding external memory to 8088-based systems. We will examine how the 8088's various control signals (ALE, DT/$\overline{R}$, $\overline{WR}$, $\overline{RD}$, and IO/$\overline{M}$) are used to supply memory read and write signals to read-only memories and both static and dynamic random access memories.

In addition we will see how an external device called a **bus master** takes over control of the 8088's memory system during a process called **direct memory access.**

The information provided in this chapter should enable you to design future memory systems from scratch. Also, the concepts presented in this chapter, such as designing an address decoder, are easily extended to more advanced architectures, such as 32- and 64-bit-wide memory systems.

Section 11.2 explains the 8088's address and data buses. The importance of bus buffering is discussed in Section 11.3. Section 11.4 shows how the 8088 accesses memory, Section 11.5 covers the design of a memory address decoder, and Section 11.6 introduces the partial-address decoder. Section 11.7 explores the use of a shift register to generate wait states. Section 11.8 contains a complete 8KB RAM/EPROM memory. In Section 11.9, we show how dynamic RAM can be used with the 8088. Section 11.10 explains how DMA works. Memory-mapped I/O is covered in Section 11.11. Some hardware troubleshooting techniques are given in Section 11.12.

11.2　THE 8088 ADDRESS AND DATA BUSES

As previously discussed, the 8088 microprocessor has an 8-bit data bus, and a 20-bit address bus that can access 1MB of external memory. The lower eight address lines are multiplexed together with the eight data lines, resulting in signals AD_0 through AD_7. In maximum mode, this multiplexed bus is decoded by external hardware, specifically by the 8288 bus controller covered in Chapter 10, with the aid of the processor's status outputs. The 8288 takes care of latching the lower eight address lines and controlling the direction of a bidirectional buffer on the data bus. In minimum mode, the processor outputs the necessary control signals directly (via ALE, $IO/\overline{M}$, $\overline{RD}$, $\overline{WR}$, and others). Thus, we end up with data lines D_0 through D_7 and address lines A_0 through A_{19}. We will see that all of these signals are needed to communicate with the RAM and EPROM devices contained in the memory system.

11.3　BUS BUFFERING

Every microprocessor-based memory system, whether EPROM or RAM, will have standard buses connecting it to the microprocessor, whose functions are to direct the flow of information to and from the memory system. These buses are generally called the **control** bus, the **data** bus, and the **address** bus. Figure 11.1 shows the relationship between the CPU, the buses, and the memory system. Note that the address bus is unidirectional, which means that data on the address bus goes one way, from the CPU to the memory system. The data and control buses, on the other hand, are bidirectional. Data may be written to or read from memory, hence the need for a bidirectional data bus. We will soon see why the control bus is also bidirectional.

FIGURE 11.1 Memory bus structure

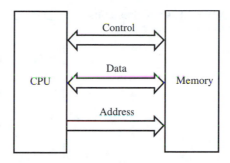

Whether they are bidirectional or not, some care must be taken when the buses are connected to the memory section. It is possible to overload an address or data line by forcing it to drive too many loads. As always, it is important *not* to exceed the fanout of a digital output. If, for example, a certain output is capable of sinking 2 mA, how many 0.4 mA inputs can it drive? The answer is five, which we get by dividing the output sink current by the required input current. If more than 5 inputs are connected, the output is overloaded and its ability to function properly is diminished. Clearly, the possibility of overloading the 8088's address or data buses exists when they are connected to external memory. For this reason, we will **buffer** the address and data buses. This concept is illustrated in Figure 11.2.

Figure 11.3 shows how address lines A_8 through A_{15} are buffered by connecting them to a standard high-current buffer, the 74LS244 octal line driver/receiver. An address line on the 8088 is capable of sinking 2 mA all by itself. When the output of the 74LS244 is used instead, the address line has an effective sink current of 24 mA. This means that twelve times as many gates can be driven. Buffering the address lines allows the CPU to

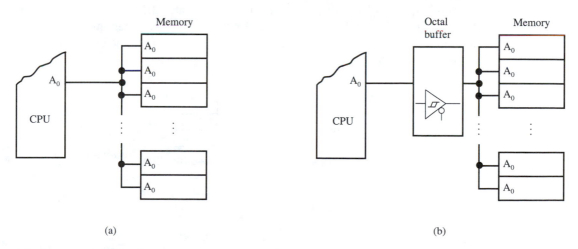

FIGURE 11.2 Microprocessor address line: (a) cannot drive required number of memory devices; (b) drives all memory devices via octal buffer

FIGURE 11.3   Address bus buffering

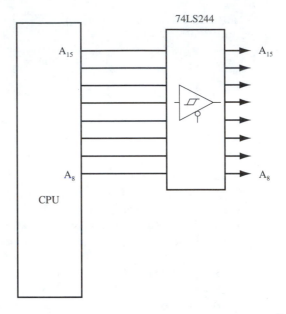

drive all the devices in our memory system, without the added worry of overloading the address line.

Buffering the data bus is a little trickier because the data bus is bidirectional. Data must now be buffered in both directions. Figure 11.4 shows how this bidirectional buffering is accomplished. The 8286 is an octal bus transceiver. Data flow through this device is controlled by the T (*transmit*) input, which tells the buffer to pass data from left to right, or from right to left. Left-to-right data is CPU output data. Right-to-left data is considered CPU input data. The natural choice for controlling the direction of the 8286 is the 8088's DT/$\overline{\text{R}}$ line, which always indicates the direction of data on the 8088 data bus. DT/$\overline{\text{R}}$ is directly generated by the 8088 in minimum mode, and by the 8288 in maximum mode. $\overline{\text{DEN}}$ is used to disable the 8286 when the bus is idle or contains address information.

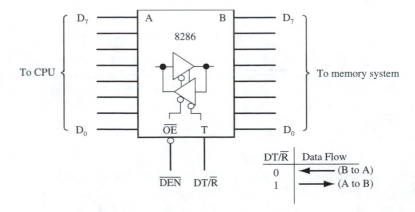

FIGURE 11.4   Data bus buffering

In conclusion, then, remember that address and data buses should be buffered so that many gates can be connected to them instead of the few that can be directly driven by the unbuffered address or data line. All designs presented in this chapter will assume that the buses are already well buffered.

11.4   ACCESSING MEMORY

In addition to well-buffered address and data buses, a control bus must also be used to control the operation of the memory circuitry. The three operations we have to consider are the following:

1. Read data from memory
2. Write data to memory
3. Do not access memory

The first two cases represent data that gets transferred between the 8088 and memory. The third case occurs when the 8088 is performing some other duty (internal instruction execution, perhaps) and has no need for the memory system. Thus, it appears that the 8088 either accesses memory or does not access it. Does a processor signal (or group of signals) exist that tells external circuitry that the 8088 needs to use its memory? Yes, a number of signals indicate this need. In minimum mode, the processor will output a 0 on $IO/\overline{M}$ to indicate that a memory reference is beginning. This signal is combined with $\overline{RD}$ and $\overline{WR}$ to form memory-read and memory-write signals for the memory system (as you can see in Figure 11.5). We use the active-low $\overline{MEMRD}$ and $\overline{MEMWR}$ signals to select and enable devices in the memory system.

In maximum mode, the 8288 bus controller decodes the processor status outputs and generates active-low $\overline{MRDC}$ (memory-read command) and $\overline{MWTC}$ (memory-write command) signals. The presence of a 0 on either signal indicates a memory access.

Figure 11.6 shows a simplified timing diagram for a memory-read cycle. The cycle is composed of four T states, with each T state equivalent to one clock cycle. In T_1 the processor outputs a full 20-bit address on address lines A_8 through A_{19} and AD_0 through AD_7. ALE has also gone high, indicating that the multiplexed address/data bus contains address information. Because this is a memory access, the processor also has output a 0 on $IO/\overline{M}$, which will remain low for the duration of the bus cycle.

In state T2 the processor tri-states the multiplexed address/data bus in preparation for the data read which will take place in T3. Address lines A16 through A19 switch over to status outputs S3 through S6, and a zero is output on $\overline{RD}$ to specify a memory-read cycle to external hardware. It is the responsibility of the memory circuitry to use $IO/\overline{M}$, $\overline{RD}$,

FIGURE 11.5   Decoding memory read and write signals in minimum mode

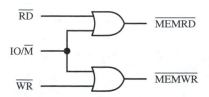

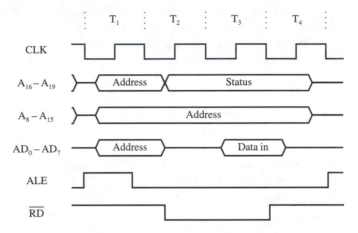

FIGURE 11.6   Memory read cycle timing

ALE, and the address lines in such a way that the data is placed onto the data bus only when IO/$\overline{\text{M}}$ and $\overline{\text{RD}}$ are both low.

Figure 11.7 shows the same basic timing for a memory-write cycle. The most noticeable difference (aside from the use of $\overline{\text{WR}}$ instead of $\overline{\text{RD}}$) is the activity on the data bus. Unlike the read cycle, the data bus switches from address-out information to data-out information and keeps a valid copy of the output data on the bus for the remainder of the cycle. This should eliminate any setup times required by the memory chips.

In the next section we will see how a memory address decoder uses the address bus and the read/write signals to enable RAM and EPROM memories.

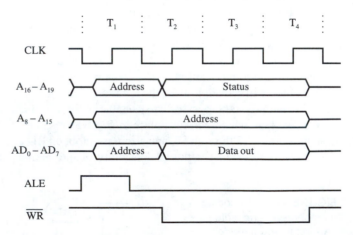

FIGURE 11.7   Memory write cycle timing

11.5   DESIGNING A MEMORY ADDRESS DECODER

The sole function of a memory address decoder is to monitor the state of the address bus and determine when the memory chips should be enabled. But what is meant by *memory chips?* These are the actual RAMs or EPROMs the designer wants to use in the computer. So, before the design begins, it must be decided how much memory is needed. If 8KB of EPROM is enough, then the designer knows that 13 address lines are needed to address a specific location inside the EPROM (because 2 raised to the 13th power is 8,192). How many address lines are needed to select a specific location in a 32K memory? The answer is 15, because 2 to the 15th is 32,768! The first step in designing a memory address decoder is determining how many address lines are needed just for the memory device itself. Any address lines remaining are used in the address decoder.

Figure 11.8(a) shows a block diagram of a memory address decoder connected to a memory chip. Figure 11.8(b) shows a simplified timing diagram representing the activity on the address bus and the $\text{IO}/\overline{\text{M}}$ output. The memory address decoder waits for a

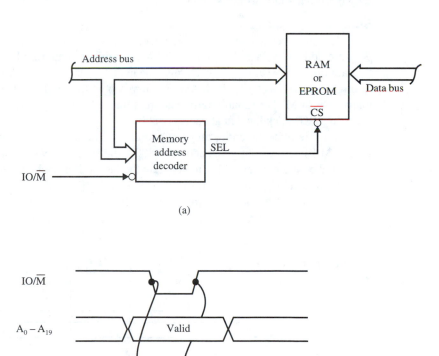

(a)

(b)

FIGURE 11.8   Simple memory address decoder: (a) block diagram; (b) timing

particular pattern on the address lines and a low on IO/$\overline{\text{M}}$ before making $\overline{\text{SEL}}$ low. When these conditions are satisfied, the low on $\overline{\text{SEL}}$ causes the $\overline{\text{CS}}$ (chip select) input on the memory chip to go low, which enables its internal circuitry, thus connecting the RAM or EPROM to the processor's data bus. When the address bus contains an address different from the one the address decoder expects to see, or if IO/$\overline{\text{M}}$ is high, the output of the decoder will remain high, disabling the memory chip and causing its internal buffers to tristate themselves. Thus, the RAM or EPROM is effectively disconnected from the data bus.

The challenge presented to us, the designers of the memory address decoder, is to chip-enable the memory device at the correct time. The following example illustrates the steps involved in the design of a memory address decoder.

Example 11.1:   A circuit containing 32KB of RAM is to be interfaced to an 8088-based system, so that the first address of the RAM (also called the **base** address) is at 48000H. What is the entire range of RAM addresses? How is the address bus used to enable the RAMs? What address lines should be used?

Solution: Figure 11.9 shows how the memory lines are assigned.

Because we are using a 32KB device, we need 15 address lines to select one of 32K possible addresses. We always use the lowest numbered address lines first (the least significant ones). We start with A_0 and use the next 14 just for the RAM. This means that A_0 through A_{14} go directly to the RAM circuitry, where they will be used to select a location inside the RAM. The remaining five address lines, A_{15} through A_{19}, are used to *select* the specific 32KB bank located at address 48000.

To determine the entire range of addresses, first make all the don't cares (the X's in Figure 11.9) zeros. That gives address 48000, the first address in the range of addresses. Next, make all don't cares high to generate the last address, which becomes 4FFFF.

Note that A_{18} and A_{15} are high when the 32KB RAM bank is being accessed, while the other three upper address bits are low. This particular pattern of 1s and 0s is one of 32 possible binary combinations that may occur on the upper five address bits.

FIGURE 11.9   Memory address range decoding

4	8 → F	0 → F	0 → F	0 → F
A_{19} A_{18} A_{17} A_{16}	A_{15} A_{14} A_{13} A_{12}	A_{11} A_{10} A_9 A_8	A_7 A_6 A_5 A_4	A_3 A_2 A_1 A_0
0 1 0 0	1 X X X	X X X X	X X X X	X X X X

X—Don't care (Use 0 or 1)

These 5 address lines set the base address of the memory.

These 15 address lines will select one of 2^{15} (or 32,768) locations inside the RAMs.

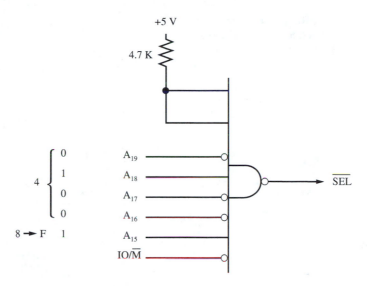

FIGURE 11.10   Memory address decoder for 48000 to 4FFFF range

We need to detect a *single* pattern, so that the RAM circuit responds only to the address range 48000H to 4FFFFH. The circuit of Figure 11.10 is one way to do this. The output of the 8-input NAND gate is the output of the memory address decoder, which in turn gets connected to the chip-enable inputs of the 32KB RAM bank. The only time the output of the NAND gate will go low is when *all* of its inputs are high. Because some of the upper address bits are low when the desired memory range is present on the address bus, they must be inverted before they reach the NAND gate. Even though A_{19}, A_{17}, and A_{16} are low, the NAND gate receives three 1s from them, via the inverters. Because A_{18} and A_{15} are already high, there are now five 1s present on the input of the NAND gate. When IO/$\overline{\text{M}}$ goes low, indicating a valid memory address, the last required logic 1 is presented to the NAND gate (via another inverter), and its output goes low, enabling the 32KB RAM bank.

In general, a memory address decoder is used to reduce many inputs to a single output. The inputs are address lines and control signals. The single output is usually an enable signal sent to the memory section. Various TTL gates are used depending on the addressing requirements. The following examples present only a few of the hundreds of ways we can design memory address decoders to suit our needs.

Example 11.2:   A 16KB EPROM section, with a starting address of 30000, is to be added to an existing memory system. The following circuitry will properly decode the entire address range, 30000 to 33FFF. The 8-input NAND gate in Figure 11.11 is used

FIGURE 11.11   Memory address decoder for 30000 to 33FFF range

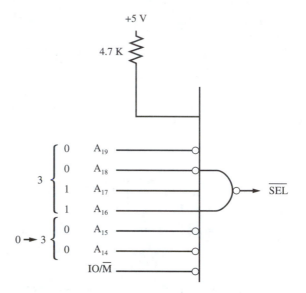

to detect the 3 pattern on the upper four address bits, and the 0s on A_{15} and A_{14}. Six address lines are used in this decoder, because the other 14 are needed to address one of 16K possible byte locations in the EPROM. The EPROM is directly addressed by A_0 through A_{13}.

Let us pause for a moment to consider a few points. In Figure 11.5 we saw how $IO/\overline{M}$ was combined with $\overline{RD}$ and $\overline{WR}$ to create the $\overline{MEMRD}$ and $\overline{MEMWR}$ signals used in a minmode system. In this case, we do not have to use $IO/\overline{M}$ again in the address decoder, because this would lead to redundancy in our design. We can, however, eliminate the use of the OR gates to decode the $\overline{MEMRD}$ and $\overline{MEMWR}$ signals and use $\overline{RD}$ and $\overline{WR}$ directly, but in this case we *must* use $IO/\overline{M}$ in the address decoder. The point is this: in a minmode system, $IO/\overline{M}$ must be used somewhere in the memory system. Without it, the memory cannot differentiate between memory addresses and I/O-port addresses.

In a maxmode system, the $IO/\overline{M}$ signal is not even generated, so it is not possible to include it in our designs. Instead, we use the 8-input NAND gate (or some other combination of gates) to decode only the address we wish to recognize. The $\overline{MRDC}$ and $\overline{MWTC}$ signals generated by the 8288 bus controller will be connected directly to the memory device being accessed.

Example 11.3:   Two 32KB memories, an EPROM with a starting address of 60000 and a RAM with a starting address of 70000, are needed for a new maxmode memory system. Figure 11.12(a) shows how the EPROM is enabled, and Figure 11.12(b) shows how the RAM is enabled. In this design, it is only necessary to use a 4-input NAND gate to do the decoding of the "6" or "7" part of the address range.

FIGURE 11.12 Memory address decoders for two different ranges: (a) EPROM bank at 60000; (b) RAM bank at 70000

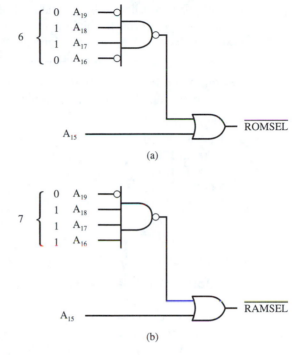

Experienced digital designers can detect binary patterns, and the reward in finding a pattern is generally a reduction in the digital circuitry needed to implement a desired function. Did you notice that the address ranges for the RAM and EPROM in the previous example are very similar? In fact, they are identical, except for the A_{16} address bit. Let us look at another example to see how we can use pattern detection to simplify the required hardware.

Example 11.4: The RAM and EPROM sections of Example 11.3 are enabled by the simplified decoder presented in Figure 11.13. Do you see how the NAND gate is used to detect

FIGURE 11.13 Combined RAM/EPROM decoder

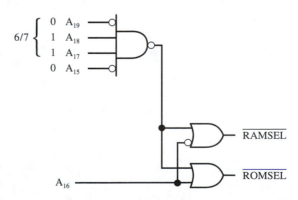

the "6/7" pattern, and how the A_{16} address line is used to enable the RAM *or* the EPROM? In this case, we are able to eliminate one of the 4-input NAND gates and one inverter. The next example shows how we can design a decoder to respond to *eight* different address ranges.

Example 11.5:   A 256KB RAM memory is composed of eight 32KB RAMs. The address ranges for the RAMs are as follows:

1. 00000 to 07FFF
2. 08000 to 0FFFF
3. 10000 to 17FFF
4. 18000 to 1FFFF
5. 20000 to 27FFF
6. 28000 to 2FFFF
7. 30000 to 37FFF
8. 38000 to 3FFFF

How might all eight RAMs be selectively enabled by one device?

Solution: Our first thought may be to use eight individual memory address decoders, one for each address range and 32KB RAM. But this would be an unnecessary waste of circuitry. If we instead look for a pattern, we see that address lines A_{19} and A_{18} are always low in the memory range 00000 to 3FFFF. In addition to this important piece of information, each RAM requires 15 address lines, A_0 through A_{14}, to

FIGURE 11.14   Multibank address decoder

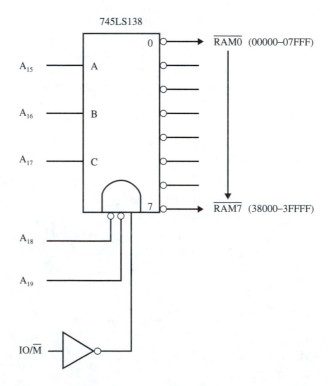

select one of 32KB locations within the RAM. This leaves us with address lines A_{15}, A_{16}, and A_{17} actually indicating a specific 32KB memory range. When these three address lines are all low, address range 00000 to 07FFF is selected. When A_{15} is high, and A_{16} and A_{17} low, address range 08000 to 0FFFF is selected. The last range, 38000 to 3FFFF, is selected when A_{15}, A_{16}, and A_{17} are all high. What we need then is a circuit that can decode these eight possible conditions by using only the three address lines. Figure 11.14 shows the required circuitry.

In this circuit, a 74LS138 three- to eight-line decoder is used to decode the different memory ranges. The 74LS138 has three select inputs and three control inputs. The select inputs are connected to address lines A_{15}, A_{16}, and A_{17}. The 3-bit binary number present on the select inputs will pull the selected output of the 74LS138 low (assuming that the 74LS138 is enabled), thus activating a specific RAM bank. To enable the 74LS138, two lows and a high must be placed on its control inputs. The two lows are generated by A_{19} and A_{18}. The IO/$\overline{\text{M}}$ signal is inverted to generate the last control input.

By using special integrated circuits like the 74LS138 and a simple pattern recognition technique, we are able to greatly simplify the hardware required to generate all of our memory enables.

The last four examples have shown how we can decode a specific range of memory addresses using the full address bus of the 8088. In the next section, we will see how to further simplify our decoder, by using a technique called **partial-address decoding.**

11.6   PARTIAL-ADDRESS DECODING

Although the 8088 is capable of addressing over 1 million bytes of memory, it would be safe to assume that most applications would require much smaller memories. A good example might be an educational 8088 single-board computer, much like the one presented in Chapter 14, using only 8K words of EPROM and 8K words of RAM. This type of system needs only 14 address lines. The first 13, A_0 through A_{12}, go directly to the EPROM and RAM, and the last address line, A_{13}, is used to select either the EPROM or the RAM. Figure 11.15 details this example system.

In this figure, A_{13} is connected directly to the $\overline{\text{CS}}$ input of the RAM and is *inverted* before it gets to the $\overline{\text{CS}}$ input of the EPROM. So, whenever A_{13} is low, the RAM is enabled, and whenever A_{13} is high, the EPROM is enabled. We use a single inverter to do all the decoding in our memory section!

But what about the other address lines, A_{14} through A_{19}? They are ignored, and here is why: when the 8088 is powered-up, a reset causes the processor to look first at memory location FFFF0. The 8088 is looking at memory location FFFF0 to get its initial instruction. We had better make sure good data is in that location at power-up. If we use an 8KB EPROM at FE000, we can be ensured that the correct information will be present.

FIGURE 11.15   Partial address decoding for RAM/EPROM

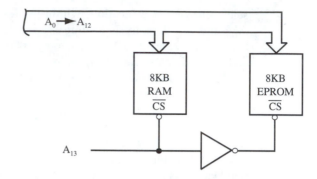

Going back to Figure 11.15, it is clear that the EPROM will be enabled at power-on, because A_{13} will be high when the processor tries to access location FFFF0.

But we still do not know why the other six upper address lines, A_{14} through A_{19} in this case, can be ignored. The answer lies in Figure 11.15. Do you see any address lines other than A_{13} being used to enable or disable the EPROM or RAM memories? No! Because we ignore these address lines, it does not matter if they are high or low. In this fashion we can read from memory locations FFFF0, 3BFF0, 07FF0, or C3FF0 and get the same data each time. The upper address bits have no effect on our memory circuitry, because we are using only the lower 14 address lines.

Partial-address decoding gives us a way to get the job done with a minimum of hardware. Because fewer address lines have to be decoded, less hardware is needed. This is its greatest advantage. A major disadvantage is that future expansion of memory is difficult, and usually requires a redesign of the memory address decoder. This may turn out to be a difficult, or even impossible, job. The difficulty lies in having to add hardware to the system. If a system manufactured by one company has been distributed to a number of users, making changes to all systems becomes a challenge. Furthermore, individuals wanting to make changes themselves may mistakenly place a new memory device into a partially decoded area. This will unfortunately result in two memories being accessed at the same time, probably resulting in invalid data during reads.

As long as these dangers and limitations are understood, partial-address decoding is a suitable compromise and acceptable in small systems.

Two more examples are presented to further show the simplicity of partial address decoding.

Example 11.6:   A 16KB block of memory, composed of two 8KB EPROMs, is to have a starting address of 4000H. What is the address range for each EPROM?

What circuitry is needed to implement a partial address decoder for a minmode system?

Solution: Figure 11.16(a) shows the address decoding table for the 16KB block of storage. The base address of 4000H is written down in binary, with the lower 13 bits associated with A_0 through A_{12}, the address lines needed to select locations within each

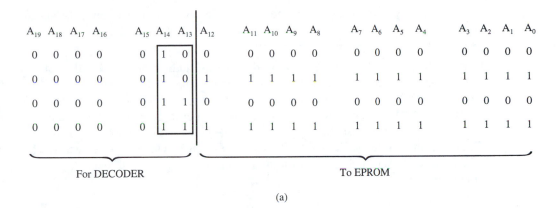

A_{19}	A_{18}	A_{17}	A_{16}		A_{15}	A_{14}	A_{13}	A_{12}		A_{11}	A_{10}	A_9	A_8		A_7	A_6	A_5	A_4		A_3	A_2	A_1	A_0
0	0	0	0		0	1	0	0		0	0	0	0		0	0	0	0		0	0	0	0
0	0	0	0		0	1	0	1		1	1	1	1		1	1	1	1		1	1	1	1
0	0	0	0		0	1	1	0		0	0	0	0		0	0	0	0		0	0	0	0
0	0	0	0		0	1	1	1		1	1	1	1		1	1	1	1		1	1	1	1

For DECODER   To EPROM

(a)

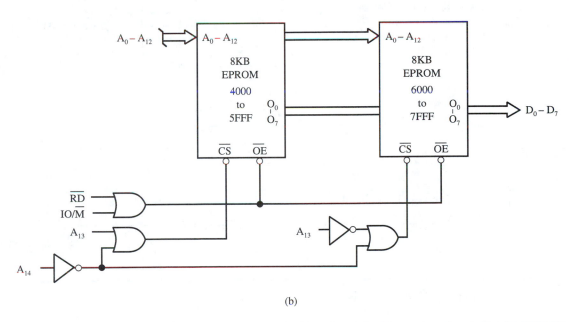

(b)

FIGURE 11.16   16KB EPROM storage using partial addressing: (a) address decoding table; (b) EPROM circuitry

8KB EPROM. Note that A_{14} is high and A_{13} is low for all possible binary patterns of 0s and 1s on A_0 through A_{12}. This sets the address range for the first EPROM, which is 4000 to 5FFF. An OR gate is used to recognize the 1 0 pattern on A_{14} and A_{13} (as shown in Figure 11.16(b)). Continuing with this technique, we see that A_{14} and A_{13} are both high when the second EPROM is being accessed. This translates into the address range 6000 to 7FFF, or a total address range of 4000 to 7FFF. Another OR gate is used to decode the 1 1 pattern on A_{14} and A_{13}. Since this memory is used in a minmode system, IO/$\overline{\text{M}}$ and $\overline{\text{RD}}$ are combined with a third OR gate and used to control the output-enable input of both EPROMs.

You may want to practice by redesigning this circuit with two-input NAND gates.

Example 11.7:   A 32KB EPROM needs a starting address of 30000, and a 32KB RAM needs a starting address of 20000. The circuitry in Figure 11.17 shows how these addresses are partially decoded. In this example, three-input NAND gates are used to do the decoding. All three inputs must be high for the output to go low (and enable the memories). $IO/\overline{M}$ is inverted, so that it presents a 1 when low. A_{17} is connected directly, because it is high in both the RAM and EPROM address ranges. Only A_{16} changes. It is low for the RAM range and high for the EPROM range. Address lines A_0 through A_{14} are used by the memories themselves.

FIGURE 11.17   Partial address decoder for 32KB EPROM at 30000, and 32KB RAM at 20000

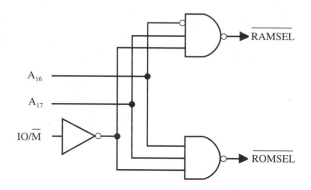

Can you determine the range of addresses for each memory?

11.7   GENERATING WAIT STATES

READY is a signal that tells the 8088 CPU that data may be read from or written into memory. We have ignored this signal so far, so that we could develop an understanding of how memory address decoders work. The first function of the memory address decoder is to monitor the address bus and activate the RAMs or EPROMs when a specific address, or range of addresses, is seen. The second function of the decoder is to tell the CPU to *wait* until the memories have been given enough time to become completely active. A typical RAM might require 200 ns to become active after it gets enabled. This is due to the time required by the internal RAM circuitry to correctly decode the supplied address and turn on its internal buffers. If the decoder did not tell the CPU to wait for 200 ns while this was happening, problems such as data loss might arise. The READY signal gives us a way to slow down the 8088 so that it can use slow memories.

In Figure 11.18(a) we see that the output of the address decoder is connected to the memory and to a delay circuit. The delay circuit is used to extend the bus cycle for a time equal to the access time of the memories. The 8088 samples READY during state T_2 of

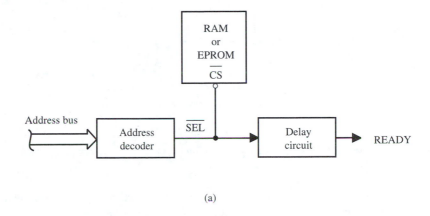

(a)

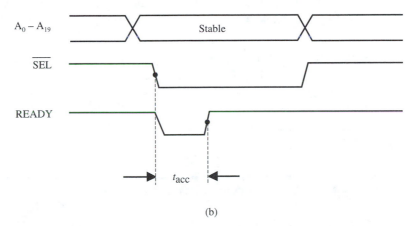

(b)

FIGURE 11.18 READY operation during memory access: (a) block diagram of delay circuit; (b) associated timing

each bus cycle, and will not proceed into T_3 until READY is at a high level. Pulling READY low with the delay circuit inserts a wait state into the bus cycle. The wait state time is equal to t_{acc}, the access time of the memory. This is shown in Figure 11.18(b).

How do we delay the generation of READY? A very simple solution is to use a shift register. Consider the 2-bit shift register of Figure 11.19. It will take two clock pulses for information at the first D input to get to the second Q output. So, if SEL goes low, we expect READY to go low for two clock pulses. If an 8-MHz clock is used, the flip-flops are clocked every 125 ns, which results in a wait state of 250 ns!

The delay time needed depends on the type of memory being used, the clock frequency, and the size (in stages) of the shift register. A one-shot (mono-stable multivibrator) could be used as well, but would not be as stable as the digital circuit due to the nature of the resistor/capacitor network needed.

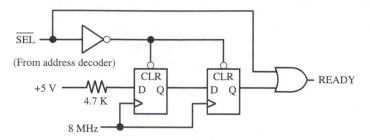

FIGURE 11.19   Using a 2-bit shift register to generate a wait state

We finish this section with an example of a delay circuit. In our next section we will see a complete schematic of an 8KB RAM/EPROM memory.

Example 11.8:   A delay circuit is composed of three D-type flip-flops connected as a 3-bit shift register driven by a 4-MHz clock. Compute the length of the delay generated by this circuit.

Solution: The length of delay is three times the period of the clock. A 4-MHz clock has a period of 250 ns; therefore, the delay time is 750 ns.

Can the delay time in this circuit be doubled (to 1.5 μs) by adding only one more flip-flop? The answer is yes, and is left for you to prove as a homework problem.

11.8   A COMPLETE RAM/EPROM MEMORY

Now that we have covered all the required basics, a complete memory design is presented. Figure 11.20(a) shows the required hardware necessary for 8KB of EPROM (located at base address FE000) and 8KB of RAM (located at 00000).

The control signals are associated with a minimum mode system. The addresses for each memory device are fully decoded. The 8-input NAND gate is used to enable the EPROM, and three 3-input NOR gates and a 3-input AND gate are used to enable the RAM. Different logic is required in each decoder, because the EPROM address requires recognition of seven 1s and the RAM decoder must recognize seven 0s. The 2764 is an 8KB EPROM, with an internal address range of 0000 to 1FFF, making its system address range FE000 to FFFFF. The 6264 is an 8KB static RAM with a system address range from 00000 to 01FFF. The memories were placed into the 8088's memory space in such a way that the EPROM is enabled upon reset, and the RAM is available for interrupt vector, program, and data storage.

Figure 11.20(b) shows an almost identical memory system, with a few changes made so that it can be placed into a maximum mode system. $\overline{RD}$ and $\overline{WR}$ now become $\overline{MRDC}$ and $\overline{MWTC}$ (generated by the 8288 bus controller). IO/$\overline{M}$ disappears, allowing us to eliminate one of the NOR gates in the RAM section and requiring the addition of an inverter (the one that was previously used for IO/$\overline{M}$ in the EPROM section).

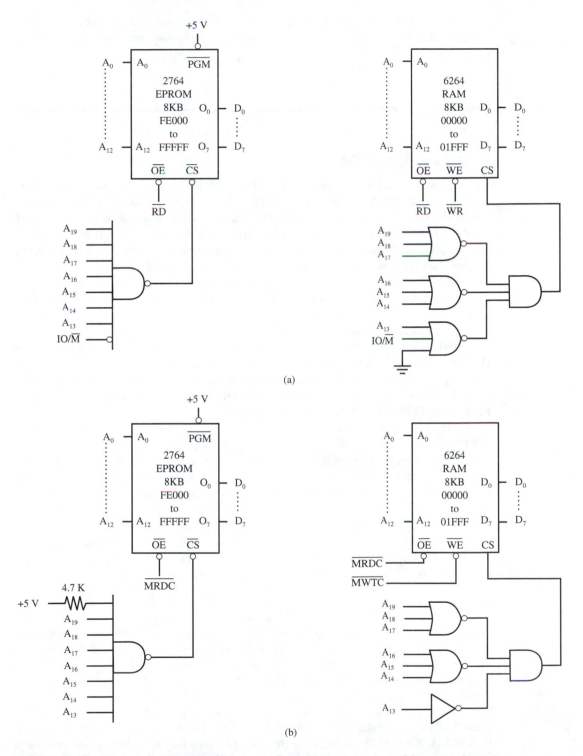

FIGURE 11.20   Complete RAM/EPROM memory: (a) minmode system; (b) maxmode system

While 8KB of RAM is enough for small educational systems, other systems may require much more memory. In the next section, we will see how dynamic RAM can be interfaced to the 8088.

11.9   DYNAMIC RAM INTERFACING

What Is Dynamic RAM?

Dynamic RAM is a special type of RAM memory that is currently the most popular form of memory used in large memory systems for microprocessors. It is important to discuss a few of the specific differences between static RAMs and dynamic RAMs. Static RAMs use digital flip-flops to store the required binary information, whereas dynamic RAMs use MOS capacitors. Because of the capacitive nature of the storage element, dynamic RAMs require less space per chip, per bit, and thus have larger densities.

In addition, static RAMs draw more power per bit. Dynamic RAMs employ MOS capacitors that retain their charges (stored information) for short periods of time, whereas static RAMs must saturate transistors within the flip-flop to retain the stored binary information, and saturated transistors dissipate maximum power.

A disadvantage of the dynamic RAM stems from the usage of the MOS capacitor as the storage element. Left alone, the capacitor will eventually discharge, thus losing the stored binary information. For this reason the dynamic RAM must be constantly **refreshed** to avoid data loss. During a refresh operation, all of the capacitors within the dynamic RAM (called DRAM from now on) are recharged.

This leads to a second disadvantage. The refresh operation takes time to complete, and the DRAM is unavailable for use by the processor during this time.

Older DRAMs required that all storage elements inside the chip be refreshed every 2 ms. Newer DRAMs have an extended 4-ms refresh time, but the overall refresh operation ties up an average of 3 percent of the total available DRAM time, which implies that the CPU has access to the DRAM only 97 percent of the time. Because static RAMs require no refresh, they are available to the CPU 100 percent of the time, a slight improvement over DRAMs.

In summary, we have static RAMs that are fast, require no refresh, and have low bit densities. DRAMs are slower and require extra logic for refresh and other timing controls, but are cheaper, consume less power, and have very large bit densities.

Accessing Dynamic RAM

A major difference in the usage of DRAMs lies in the way in which the DRAM is addressed. A 64K bit DRAM requires 16 address bits to select one of 65,536 possible bit locations, but its circuitry contains only 8 address lines. A study of Figure 11.21 will show how these 8 address lines are expanded into 16 address lines with the help of two additional control lines: $\overline{RAS}$ and $\overline{CAS}$.

The 8 address lines are presented to row and column address buffers, and latched accordingly by the application of the $\overline{RAS}$ and $\overline{CAS}$ signals. To load a 16-bit address into the DRAM, 8 bits of the address are first latched by pulling $\overline{RAS}$ low. Then the other 8 address bits are presented to A_0 through A_7, and $\overline{CAS}$ is pulled low. By adding just one

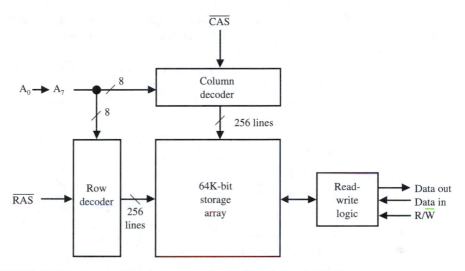

FIGURE 11.21   Internal block diagram of a 64K-bit dynamic RAM

more address line to the DRAM, the addressing capability is increased by a factor of 4, because one extra address line signifies an extra row and column address bit. This explains why DRAMs tend to quadruple in size with each new release.

The actual method for addressing the DRAM is presented in Figure 11.22. First, the 8 row address bits are applied to A_0 through A_7, and $\overline{RAS}$ is pulled low. Then A_0 through A_7 receive column address information, and $\overline{CAS}$ is pulled low. After a short delay, the circuitry inside the DRAM will have decoded the full 16-bit address, and reading or writing may commence.

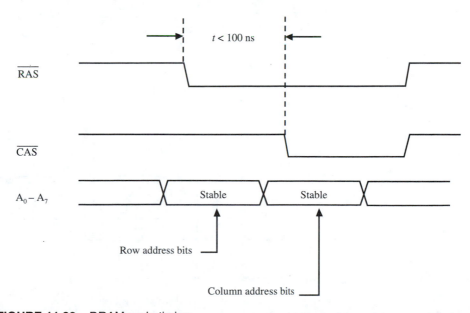

FIGURE 11.22   DRAM cycle timing

The row address strobe and column address strobe signals must be generated within 100 ns of each other to avoid data loss. The specific timing requirements for the DRAM depend on the manufacturer.

External logic is needed to generate the $\overline{RAS}$ and $\overline{CAS}$ signals, and also to take care of presenting the right address bits to the DRAMs. The circuit of Figure 11.23 shows an example of the required logic.

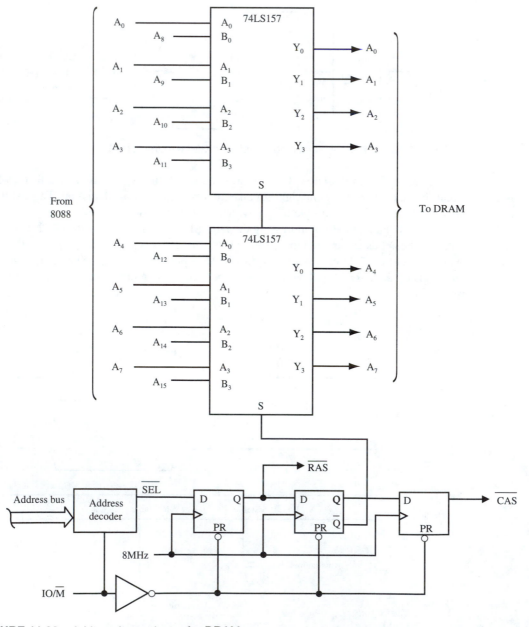

FIGURE 11.23   Address bus selector for DRAM

The operation of this circuit is as follows. The address decoder monitors the address bus for an address in the desired DRAM range, and outputs a logic 0 when it sees one. Normally the three Q outputs of the shift register are all high. The first clock pulse will shift the logic 0 from the address decoder to the output of the first flip-flop, causing $\overline{RAS}$ to go low. Because the output of the second flip-flop is still high, the 74LS157s (quad 2-line to 1-line multiplexers) are told to pass processor address lines A_0 through A_7. This is how we load the ROW address bits into the DRAM.

The second clock pulse will shift the logic 0 to the second Q output (the first is still low also), which causes the 74LS157s to select the processor address lines A_8 through A_{15}. These address bits are recognized and latched by the DRAM when the third clock pulse occurs, because the logic 0 has been shifted to the third Q output, which causes $\overline{CAS}$ to go low. The DRAM has been loaded with a full 16-bit address, and reading or writing may commence. At the end of the read or write cycle, $IO/\overline{M}$ will go high, presetting all three flip-flops via the preset line, and the shift register reverts back to its original state. This sequence will repeat every time the address decoder detects a valid address.

Figure 11.24 shows a complete DRAM addressing circuit with read-write logic. When the 74LS138 detects a valid memory address, one of its eight outputs will go low, removing the 74LS175 quad D flip-flop from its forced-clear state. All four Q outputs are high at this time. As a logic 1 shifts through the 74LS175 (connected as a 4-bit shift register), the $\overline{RAS}$, $\overline{WE}$, and $\overline{CAS}$ signals will be generated. The resistors in the address and control lines are called **damping** resistors, and are used to control the waveshape of the digital signals to the DRAMs. The damping resistors reduce ringing and other noise that would normally occur in a high-speed digital system. The only circuitry missing from Figure 11.24 is the required refresh logic, which we will study in the next section.

Refreshing Dynamic RAM

Previously we learned that DRAMs need to be refreshed, or the MOS capacitors that retain the binary information will discharge and data will be lost. Older DRAMs required that all cells (storage elements) be refreshed within 2 ms. Although the process of reading or writing a DRAM cell is a form of refresh, it is possible that entire banks of DRAM remain inactive while the CPU addresses other memories or I/O devices, so a safe designer will include a refresh circuit in the new DRAM system.

Newer DRAMs (such as the MCM6664) contain a single control line called $\overline{REF}$ that automatically refreshes the DRAM whenever it is pulled low. We will instead look at the process that is used to refresh a DRAM and the circuitry needed to control the process.

DRAMs are internally designed as a grid of memory cells arranged as a matrix, with an equal number of rows and columns (hence the $\overline{RAS}$ and $\overline{CAS}$ control signals). A 4K bit DRAM would need 12 address lines: 6 for the row decoder and 6 for the column decoder. Each decoder would pick one row and column out of a possible 64. During a refresh operation, all 64 column-cells would be refreshed by the application of a single $\overline{RAS}$ signal. This is called **RAS-only refresh.** To refresh all 4096 bits, it is necessary only to $\overline{RAS}$ select all 64 rows. A larger DRAM, a 64K bit one for example, would require $\overline{RAS}$ selecting more rows (256 in this case). The easiest way to ensure that all rows get selected during a refresh operation is to use a binary counter and connect the output of the counter to the DRAM address lines during a refresh. To ensure that the DRAMs get refreshed periodically, a timer is needed to generate a REFRESH signal. The REFRESH signal will suspend

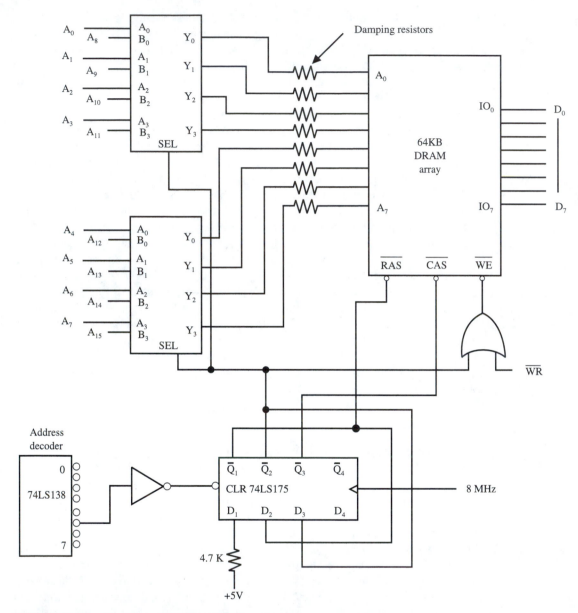

FIGURE 11.24   Complete DRAM addressing circuit

processor activity while the DRAM is refreshed. Figure 11.25 shows how a 555 timer can be used to generate a REFRESH signal every 100 μs. The 555 timer clocks a D-type flip-flop, whose output is REFRESH. When the refresh cycle is completed, $\overline{\text{DONE}}$ is used to preset the flip-flop and remove the REFRESH request, until the 555 times out again. Figure 11.26 shows how the refresh timer, together with the $\overline{\text{RAS}}$ refresh circuitry, is used to refresh the DRAMs. When the 555 timer initiates a refresh cycle, REFRESH will go high, issuing a HOLD request to the 8088. The processor will respond by asserting

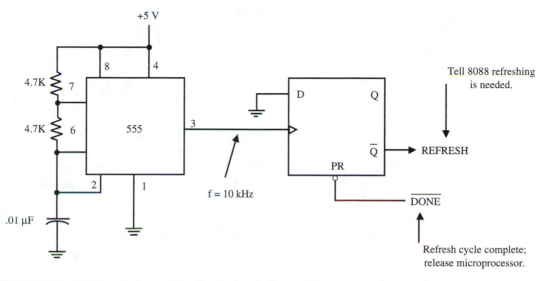

FIGURE 11.25   555 timer generates refresh signals every 100 µs

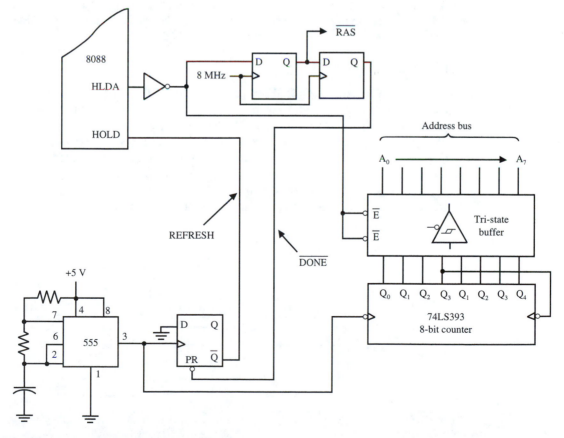

FIGURE 11.26   DRAM refresh generator

HLDA, which allows a 0 to be clocked into the 2-bit shift register used to control $\overline{\text{RAS}}$ and $\overline{\text{DONE}}$. When $\overline{\text{RAS}}$ is active, bits A_0 through A_7 of the address bus will contain the 8-bit counter value (the current state of the 74LS393). When $\overline{\text{DONE}}$ is active, the refresh flip-flop is preset, which removes the HOLD signal. This causes the processor to release HLDA, which in turn causes the 2-bit shift register to be loaded with 1s. At this point, the bus request is over, and the processor resumes execution. Because the 555 timer also clocks the 8-bit counter, a unique row address is generated each refresh cycle.

A Dynamic RAM Controller

We may conclude that the circuitry required to address, control, and refresh DRAMs is both complicated and extensive (which may translate into *expensive*). There must be a simpler way.

There is!

Various companies make DRAM controller devices that take care of all refreshing and timing requirements needed by the dynamic RAMs. All circuitry is contained in a single package in most cases. The DRAM controller does its work independently of the processor. This means that the DRAM controller will issue a wait to the processor when the processor tries to access memory during a refresh cycle. Figure 11.27 shows how a DRAM controller is used in an 8088-based system. Using a dedicated DRAM controller minimizes the time required to design, debug, and eventually troubleshoot DRAM memory systems. It may also be more cost effective in the long run.

Dynamic RAM Summary

Our study of DRAMs has shown that they are slow and require complicated circuitry to get them to work (unless a DRAM controller is used). On the other hand, DRAMs are cheaper, per bit, than static RAM, they consume less power, and have much larger bit densities. With the advance of the microcomputer into the word processing arena, where very large memories are needed to store and manipulate text, dynamic RAM becomes a very economical solution. Image processing, large informational databases, and virtually any large storage system make the use of dynamic RAMs an ideal choice. Furthermore, interfacing dynamic RAMs is made easier with the use of a DRAM controller.

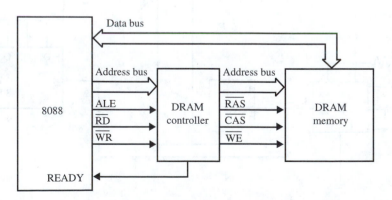

FIGURE 11.27   DRAM controller interfacing

11.10   DIRECT MEMORY ACCESS

Direct memory access, usually called DMA for short, is a process in which a device external to the CPU requests the CPU's buses (address bus, data bus, and control bus) for its own use. Examples of external circuits that might wish to perform DMA are video pattern generators, which share video RAM with the CPU, and high-speed data transfer circuits such as those used in hard disks.

In general, a DMA process consists of a **slave** device requesting the use of the **master's** buses. In a microprocessor-based system, the master is usually the CPU. Once the slave device has control of the bus, it can read or write to the system memory as necessary. When the slave device is finished, it releases control of the master's buses, and system operation returns to normal.

An example of why DMA is a useful technique can be illustrated in the following way. Suppose you wish to add a hard disk storage unit to your microcomputer. The hard disk boasts a data transfer rate of 5 million bytes per second. This comes to 1 byte transferred every 200 ns! Most microprocessors would be hard put to execute even *one* instruction in 200 ns, much less the multiple number of instructions that would be required to read the byte from the hard disk, place it in memory, increment a memory pointer, and then test for another byte to read. A DMA controller would be very handy in this example. It would merely take over the CPU's buses, write all the bytes into memory very quickly, and then return control to the CPU.

To perform DMA on the 8088, two signals may be used. They are $\overline{\text{RQ/GT}}_0$ (*request/grant*) and $\overline{\text{RQ/GT}}_1$. Both of these signals are bidirectional; they are both inputs and outputs. A logic zero must be placed on either signal to request the processor's bus, with $\overline{\text{RQ/GT}}_0$ having priority over $\overline{\text{RQ/GT}}_1$. The processor will acknowledge the takeover when the current bus cycle finishes and output a 0 on the appropriate signal line. The low-level request signal must be at least one CLK period long for the 8088 to recognize it. The processor will output a 0 for one CLK period during the next T_4 or T_1 state to acknowledge the takeover. It will then enter a "hold acknowledge" state until the new bus master sends another low-level pulse to $\overline{\text{RQ/GT}}$. All bus takeovers must consist of this three-pulse sequence. It is important to note that the device performing the DMA is responsible for maintaining the DRAM refresh requirements, either by performing them itself, or by allowing them to happen normally with existing circuitry.

Still another way to transfer data to an external device is through a technique called **memory-mapped I/O,** which is slower than DMA but just as useful. It is even possible for a system to employ both DMA and memory-mapped I/O, using DMA for the high-speed transfers and memory-mapped I/O for the slow ones.

11.11   MEMORY-MAPPED I/O

Normally, a memory location, or group of locations, is used to store program data and other important information. Data is written into a particular memory location and read later for use. Through a process called memory-mapped I/O, we remove the storage capability of the memory location and instead use it to communicate with the outside world. Imagine that you have a keyboard that supplies an 8-bit ASCII code (complete with parity) whenever you press a key. Your job is to somehow get this parallel information into your computer. By using memory-mapped I/O, a memory *location* may be set aside that, when read, will contain the 8-bit code generated by the keyboard. Conversely, data may be sent to the outside world by writing to a memory-mapped output location. The 8088 CPU is capable of performing memory-mapped I/O in either byte or word lengths. All that is required is a memory address decoder, coupled with the appropriate bus circuitry. For a memory-mapped output location, the memory address decoder provides a clock pulse to an octal flip-flop capable of storing the output data. A memory-mapped input location would use the memory address decoder to enable a tristate octal buffer, placing data from the outside world onto the CPU's data bus when active. Figure 11.28 shows the circuitry for an 8-bit memory-mapped I/O

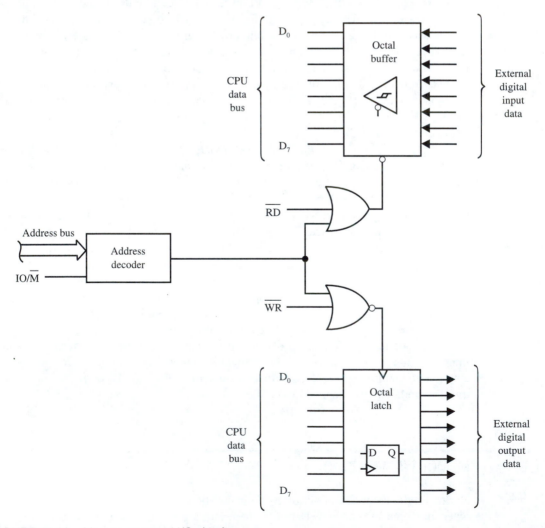

FIGURE 11.28   Memory-mapped I/O circuitry

location, sometimes referred to as a memory-mapped I/O port. The memory address decoder may be used for both input and output.

11.12   TROUBLESHOOTING TECHNIQUES

As we have seen before, a good knowledge of binary numbers is beneficial when working with microprocessors, and necessary to the efficient design of address decoders (and other interfacing circuitry). Once the decoders are designed, however, they must be tested to see whether they perform as required. Testing the operation of a memory address decoder can be accomplished any number of ways. The circuit can be set up on a breadboard, simulated via software, or just plain stared at on paper until it seems correct.

In situations like the troubleshooting phase of the single-board computer project in Chapter 14, testing the memory address decoder (similar to Figure 11.14) is often necessary. In addition to checking the wiring connections visually (or via a continuity tester or DMM), a logic analyzer is connected so that the waveforms can be examined. Even an old eight-channel logic analyzer can be used to diagnose difficult problems in a microprocessor-based system. Oscilloscopes typically fall short in showing the associated high-speed timing relationships (unless they are storage scopes).

The logic analyzer is connected so that all of the inputs to the address decoder are sampled, and as many of the outputs, as necessary. In addition, the logic analyzer is set up so that it triggers (and begins capturing data) at RESET, and the initial activity of the processor's address bus can be observed. If the single-board's EPROM is not enabled, the system will not function at all. The logic analyzer will show how the address decoder responds at power on.

The logic analyzer can also be used to examine the data coming out of the EPROM or RAM. Sampling the data and a few of the address lines should be enough to verify whether the data is correct. It is sometimes possible to spot switched data or address lines this way.

These techniques also apply to I/O circuitry, which we examine in the next chapter.

SUMMARY

In this chapter we studied some of the most common methods used in the design of memory circuitry for microprocessor-based systems. Full- and partial-address decoding, memory-mapped I/O, direct memory access, and the logical requirements for static and dynamic RAMs were all covered. A good designer will employ many of these techniques in an effort to construct a new system that is logically simple and elegant but also functional and easy to troubleshoot. The end-of-chapter questions are designed to further test your knowledge of these topics. You are encouraged to work *all* of them to increase your ability to design memory address decoders, partial-address decoders, and complete memory systems.

STUDY QUESTIONS

1. Explain the different functions associated with processor signals AD_0 through AD_7.
2. How does external circuitry know when address information is present on the multiplexed address/data bus?
3. List the different control bus signals used in minimum mode and maximum mode.
4. How can the $\overline{RD}$, $\overline{WR}$, and $IO/\overline{M}$ outputs be used to detect *any* kind of access to memory? Design a circuit that will output a 0 on $\overline{MEMORY}$ whenever a memory read or write occurs.
5. If a state time on an 8088-based system is 250 ns, what is the minimum time spent doing a memory read?
6. When (and why) are wait states inserted into memory accesses?
7. How many address lines are needed for a 128KB memory? For a 2MB memory?
8. For the state time of Question 5, what is the time spent doing a memory read with two wait states?
9. Two 2KB EPROMs are used to make a 4KB memory. How many address lines are needed for the EPROMs? What upper address lines must be used for the decoder?
10. For the memory of Question 9, what is the address of the last memory location, if the starting address of the EPROM is E4000?
11. Design a memory address decoder for the EPROM memory of Question 10, using a circuit similar to that in Figure 11.10.
12. Repeat Questions 9 through 11 for these memory sizes and starting addresses:
    a) 8KB, base address of CC000
    b) 32KB, base address of 80000
    c) 256KB, base address of 00000
13. What are the decoded address ranges for the circuits in Figure 11.29?
14. What signal (or signals) is missing from the address decoder in Figure 11.29? Modify the decoders to include the missing signal (or signals).
15. What are the address range groups for the decoder in Figure 11.30?
16. Use a circuit similar to that of Figure 11.30 to decode these address ranges:

    18000 to 187FF

    18800 to 18FFF

    19000 to 197FF

    19800 to 19FFF

    1A000 to 1A7FF

    1A800 to 1AFFF

    1B000 to 1B7FF

    1B800 to 1BFFF

17. What are two main advantages gained in using partial-address decoding? Two disadvantages?

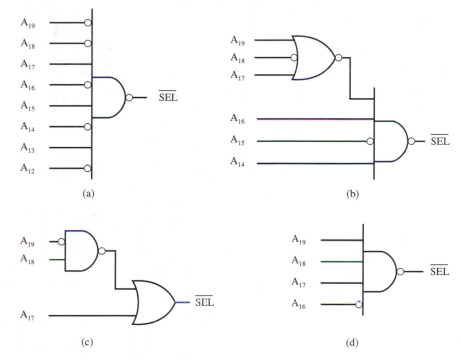

FIGURE 11.29   For Questions 13 and 14

18. Give three possible address ranges for each decoder in Figure 11.31. Address lines A_0 through A_{13} are used by the memories.
19. Suppose that three different memory decoders have output signals $\overline{RAMA}$, $\overline{RAMB}$, and $\overline{ROM}$. Design a circuit to generate a READY delay of 200 ns using a 100-ns-period clock and a circuit similar to that of Figure 11.19. Any of the three signals going low triggers the generator.
20. Use an 8-bit parallel-out shift register to design a variable wait-state circuit. Assume that the shift register is clocked once every 125 ns, and that 0 to 7 125 ns wait states are allowed.

FIGURE 11.30   For Questions 15 and 16

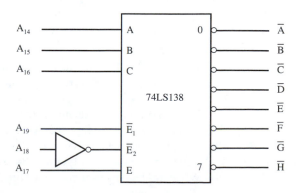

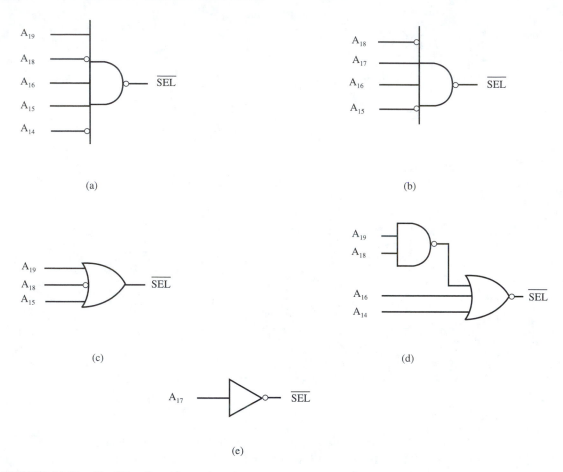

FIGURE 11.31   For Question 18

21. Design a 32KB memory using 8KB EPROMs. Show the address and data line connections to all EPROMs and the circuitry needed to switch between the four 8KB sections.
22. How do the $\overline{RAS}$ and $\overline{CAS}$ lines on a DRAM eliminate half of the required chip address lines?
23. Why does the size of a DRAM go up by a factor of 4 for each single address line that is added?
24. Why do DRAMs consume less power than static RAMs?
25. Explain how DRAM refreshing could be accomplished using an interrupt service routine.
26. How does program execution change on a system that supports DMA?
27. What is the 8088 doing while its external buses are involved in a DMA transfer?
28. Design a partial address decoder for a 64KB EPROM with a base address of 40000.
29. Redesign the circuit of Example 11.6 using NAND gates.
30. What are the ranges of addresses for the partial address decoders of Example 11.7?

CHAPTER 12

I/O System Design

OBJECTIVES

In this chapter you will learn about:

- I/O addressing space
- The design of full and partial I/O address decoders
- The operation of buffered input ports
- The operation of latched output ports
- Parallel I/O with the 8255 PPI
- Serial I/O with the 8251 UART

12.1   INTRODUCTION

In Chapter 11 we saw how the 8088 is connected to its memory system. Program data and instructions are stored in memory, and accessed over a system bus consisting of address, data, and control signals. Many times, a microprocessor is used to control a process that requires an exchange of data between the processor and the hardware used by the process. For example, an 8088 controlling an assembly line may receive input data from switches or photocells that give the position of an assembly making its way down the line. The 8088 will be able to test the state of each sensor by reading its status with an input port. Indicator lights, solenoids, and video display terminals associated with the assembly line may be driven by a few of the 8088's output ports. In this chapter we will examine the design and operation of input and output ports, and see how they are used to communicate with the outside world.

As usual, our discussion applies to the advanced 80x86 machines as well, since the original bus architecture of the 8088 is still supported.

Section 12.2 discusses the processor's I/O addressing space. Section 12.3 shows how a port address decoder is designed. The operation of input and output ports is covered in Sections 12.4 and 12.5, respectively followed by binary counter and D/A conversion applications in Section 12.6. The 8255 PPI and 8251 UART are explained in Sections 12.7 and 12.8, respectively, including examples of serial data transmission, A/D conversion, and parallel I/O. Troubleshooting techniques for I/O operations are covered in Section 12.9.

12.2   THE 8088 PORT ADDRESSING SPACE

Chapter 11 showed that the 8088 had a memory address space whose 1MB size is determined by the processor's 20-bit address bus. Unique binary patterns on A_0 through A_{19} select one of the 1,048,576 locations within the processor's memory space. The 8088's **I/O addressing space** is smaller, containing just 65,536 possible input/output ports, accessed by A_0 through A_{15}. Data transfer between ports and the processor is over the CPU data bus, and may contain 8 or 16 bits of data. The accumulator (AL for 8-bit transfers and AX for 16-bit transfers) is the only processor register involved in an I/O operation, with the single exception of the use of DX as the port address register. Figure 12.1 shows the organization of the 8088's I/O space. Ports 00 through FF may be addressed by one form of the IN and OUT instructions, and the entire I/O space (ports 0000 to FFFF) by another form of IN and OUT that uses register DX. In the first form, the port address may be used directly in the instruction, as in:

FIGURE 12.1   The 8088's
I/O addressing space

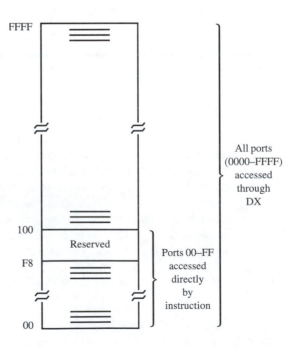

```
IN      AL,80H
IN      AX,6
OUT     3CH,AL
OUT     0A0H,AX
```

The port address in this form is limited to the range 00 to FF. Intel Corporation reserves the right to use ports F8 through FF for its own needs, and programmers are urged to consider other addresses in their designs. Port addresses appear on address lines A_0 through A_7 when these instructions are executed.

The second way a port address may be specified is by placing it into register DX and using one of these instructions:

```
IN      AL,DX
IN      AX,DX
OUT     DX,AL
OUT     DX,AX
```

A good practice to follow is to use even port addresses for word transfers. This ensures the fastest 2-byte transfer the processor can perform. Word read or writes to odd I/O addresses require additional clock cycles.

When using DX as the port address register, the I/O addressing space becomes 0000 to FFFF, with the port address showing up on A_0 through A_{15}. This represents 65,536 8-bit ports (or 32K 16-bit ports). In the next section we will see how a port address is recognized and decoded.

12.3   DESIGNING A PORT ADDRESS DECODER

The port address decoder is a circuit designed to recognize the execution of an I/O instruction. In minimum mode, the processor uses the IO/$\overline{\text{M}}$ signal to indicate an I/O access by placing a high logic level on it, along with the port address on the address bus. In maximum mode, the 8288 bus controller decodes an I/O access and generates active $\overline{\text{IORC}}$ (*I/O read command*) and $\overline{\text{IOWC}}$ (*I/O write command*) signals. These control signals must be incorporated within the design of the port address decoder to distinguish port addresses from memory addresses.

Full Port Address Decoding

In this type of port address decoder, we include as many address lines as possible in the decoder. For example, Figure 12.2(a) shows the logic needed to do a full decode of port address 4F in a minmode system. Because the port address is between 00 and FF, it can be directly included in the I/O instruction (as in IN   AL,4FH) and we need only examine A_0 through A_7 for the required port address. We then combine the valid port address with IO/$\overline{\text{M}}$ to generate the $\overline{\text{RD4F}}$ and $\overline{\text{WR4F}}$ signals. These two outputs can also be generated with $\overline{\text{IORC}}$ and $\overline{\text{IOWC}}$, as shown in Figure 12.2(b).

When groups of ports are needed, a technique similar to the one used to map multiple RAMs or EPROMs is used. In Figure 12.3 we see how a 3-line to 8-line decoder is

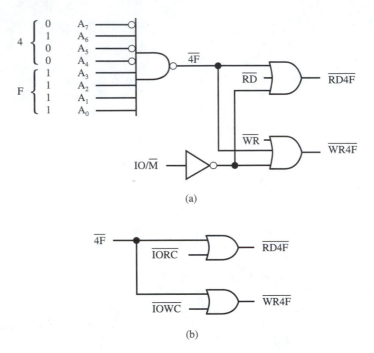

(a)

(b)

FIGURE 12.2   Port address decoder: (a) minmode system; (b) maxmode system

used to decode signals for eight different input ports. Address lines A_0 through A_2 are used by the 74LS138 to select one of its eight outputs (output 0 when all three are low and output 7 when all three are high). Address lines A_3 through A_7 enable the 138 when the correct address is present. The correct port address is any address that matches 10011---. The *base* port address, when A_0 through A_2 are low, is 98H. The last port address is 9FH. So IN   AL,98H will cause the $\overline{IN}_0$ signal to strobe low, IN   AL,99H will activate $\overline{IN}_1$, and so on. What needs to be changed to get this circuit to

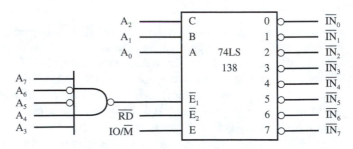

FIGURE 12.3   Decoding multiple input port addresses 98H through 9FH

work for output ports in the same range? What must be done to decode port addresses 80 through 88?

If the port address has been placed in register DX, address lines A_0 through A_{15} must be used in the decoder. Figure 12.4 shows how port address A4C0 is decoded in a max-mode system. One 8-input NAND gate is used for each half of the address bus. One NAND gate recognizes A4 and the other C0. The outputs of each NAND gate are combined with $\overline{\text{IORC}}$ and $\overline{\text{IOWC}}$ to generate the required I/O signals.

An important point to remember when working with these kinds of address decoders is that they decode a *fixed* port address, or range of addresses. Manufacturers who design I/O boards for consumer use (within their PCs) know that each user who buys a board may have a different I/O address in mind, depending on how each system is configured with other hardware items. It would be much more convenient to design a port address decoder that allows selection of a port address through a DIP switch. In this case, some type of binary **comparator** must be used to compare the port address on the address bus with the desired port address represented by a DIP switch. Figure 12.5 shows how two 74LS85 4-bit magnitude comparators can be used to recognize an 8-bit port address. Each magnitude comparator determines if the address signals on the four A inputs are equal to the DIP switch information on the four B inputs. Each comparator has an A = B input and A = B output. To perform an 8-bit comparison, the 85s are cascaded by connecting the first A = B output to the second A = B input. The A = B output of the second comparator will go high only when there is an 8-bit address match. $\overline{\text{SEL}}$ will go low if the match exists when IO/M is high. $\overline{\text{RD}}$ and $\overline{\text{WR}}$ can be combined with $\overline{\text{SEL}}$ to get the required I/O read and write signals.

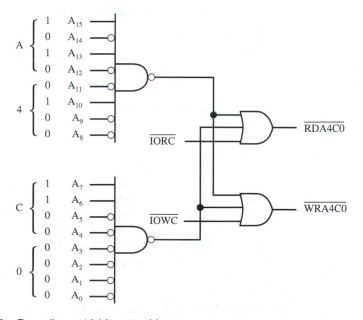

FIGURE 12.4   Decoding a 16-bit port address

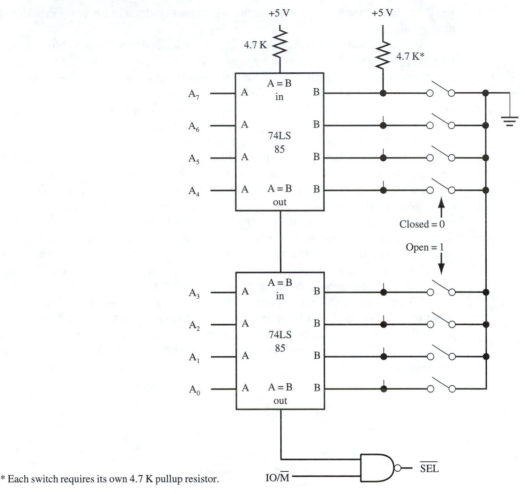

FIGURE 12.5   Using comparators in an address decoder

Partial Port Address Decoding

In a minimal system requiring only a handful of I/O ports, we can use **partial port address decoding** to reduce the hardware needed to produce I/O read and write signals. As we saw with partial address decoders for memory systems, many address lines are not included in the decoder, setting the stage for groups of addresses having the ability to enable the same I/O port. For example, in Figure 12.6(a) a two-input NAND gate is used to decode any port address with A_7 high. This would correspond to port addresses 80 through FF. Any reference to a port in this range will cause $\overline{\text{SEL}}$ to go low. Inverting A_7 before it gets to the NAND gate will decode ports in the range 00 to 7F. Figure 12.6(b) uses a four-input NAND gate to generate $\overline{\text{SEL}}$ for a 16-bit port address in the range 8000 to 9FFF. Any one of the 8192 port addresses in this range will cause $\overline{\text{SEL}}$ to go low. For practice, redesign the decoder of Figure 12.3 so that the eight port addresses are decoded whenever the decoder sees a port between 40 and 7F.

FIGURE 12.6   Partial
address decoders

(a) For 8-bit port addresses 80 to FF

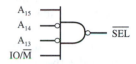

(b) For 16-bit port addresses 8000 to 9FFF

12.4   OPERATION OF A BUFFERED INPUT PORT

An input port is used to gate data from the outside world onto the CPU data bus, where
it is captured by the processor and stored in the accumulator. It is not possible to simply
place the digital data onto the data bus any time we please, for this will surely interfere
with the program instruction and data bytes constantly being read from memory. The
processor indicates the appropriate time to do the I/O read by activating IO/$\overline{\text{M}}$ and $\overline{\text{RD}}$
(or with $\overline{\text{IORC}}$). In addition to the port address decoder, we then need a device to gate
the input data onto the bus during the read operation, and to isolate the input data from
the bus when the processor is not reading the input port. A tristate buffer is used for this
purpose. Figure 12.7 shows a design for an 8-bit input port located at the fully decoded

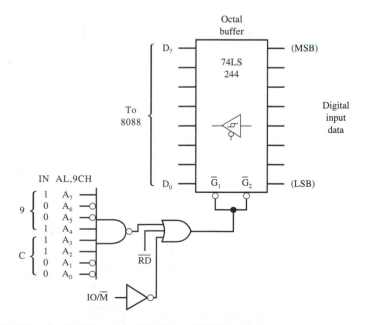

FIGURE 12.7   An 8-bit input port (address 9CH)

port address 9C. The 74LS244 octal buffer is a tristate device, meaning that its outputs are capable of going into a high-impedance state, effectively disconnecting the 244 from the processor's data bus. The high-impedance state can be thought of as an open switch. This is the normal state of the 244. When a valid port address appears on the address bus, the 244 will be enabled by the OR gate, allowing its outputs to resemble the digital data present on the eight inputs. It is important to remember that the 244 is only enabled when the processor is reading the input port! This happens when the instruction IN AL,9CH is executed.

A 16-bit input port requires an additional octal buffer and enabling logic. If port address 9C is to be used for the 16-bit input port, the second buffer must be enabled when port address 9D is seen on the address bus. The processor will use A_0 to enable each half of the 16-bit port.

12.5   OPERATION OF A LATCHED OUTPUT PORT

There is more to an output port than a simple reversal of the direction that data takes on the data bus. Data output to a port comes from the accumulator and is placed onto the data bus only for a short period of time! We have only a few hundred nanoseconds at the most to capture the output data before it is replaced by other data (an instruction

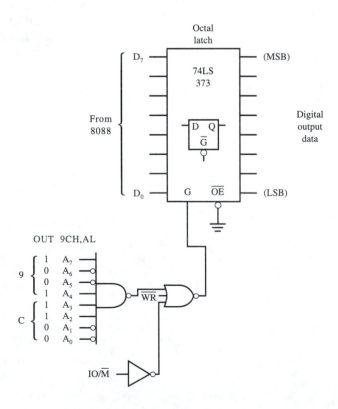

FIGURE 12.8   An 8-bit output port (address 9CH)

read from memory perhaps). It is necessary to *store* a copy of the output data when it appears. This is why output ports must contain storage elements. The processor's control signals are used to tell the output port circuitry when to store what is on the data bus. The output port in Figure 12.8 is designed to have the same I/O address as the input port of the previous section. You are encouraged to redesign both ports so that they are contained in a single circuit.

The NOR gate is used to place a high level on the latch input (G) of the 74LS373. Whatever data is present on the CPU data bus when G is high will be stored when G goes back low. Thus, the 373 is used to take a snapshot of the contents of the data bus, and store the data until the next I/O write to port 9C (which takes place when OUT 9CH,AL is executed). An edge-triggered storage device (such as the 74LS374) may be used in place of the 373.

12.6 SIMPLE I/O APPLICATIONS

In this section we will examine two applications of I/O ports. The first application involves a binary counter, whose 8-bit count is displayed with light-emitting diodes connected to an output port. An input port wired to a set of switches is used to control the speed of the binary counter (all switches down is the fastest, all up is the slowest).

The second application shows how a data table containing sampled data from a sine wave can be output sequentially to a digital-to-analog converter to recreate the sine wave.

The Binary Counter

Figure 12.9 shows the logic needed to implement an input/output port with a base address of DE. Because A_0 is not used in the port address decoder, accesses to port DF will also work. The input port uses the 74LS244 octal buffer to read a set of DIP switches. An open switch makes a 1. The information present on the DIP switches is read with IN AL,0DEH and then moved into register CX. Register CX is used in a delay subroutine that is called between updates of the binary counter. Thus, the DIP switches have control over how fast or slow the binary counter counts. The count is displayed on a set of light-emitting diodes by outputting the count to port DE with OUT 0DEH,AL, which stores the count in the 74LS373.

The software required for this application is as follows:

```
BINCNT:     MOV     AL,0                ;start count at 00
DISPCNT:    OUT     0DEH,AL             ;send count to LEDs
            CALL    FAR PTR DELAY       ;pause
            INC     AL                  ;increment counter
            JMP     DISPCNT             ;repeat forever

DELAY       PROC    FAR
            MOV     BL,AL               ;save copy of counter
            IN      AL,0DEH             ;read DIP switches
            MOV     CH,AL               ;save speed byte
            MOV     CL,1                ;ensure at least one loop
WAIT:       NOP
```

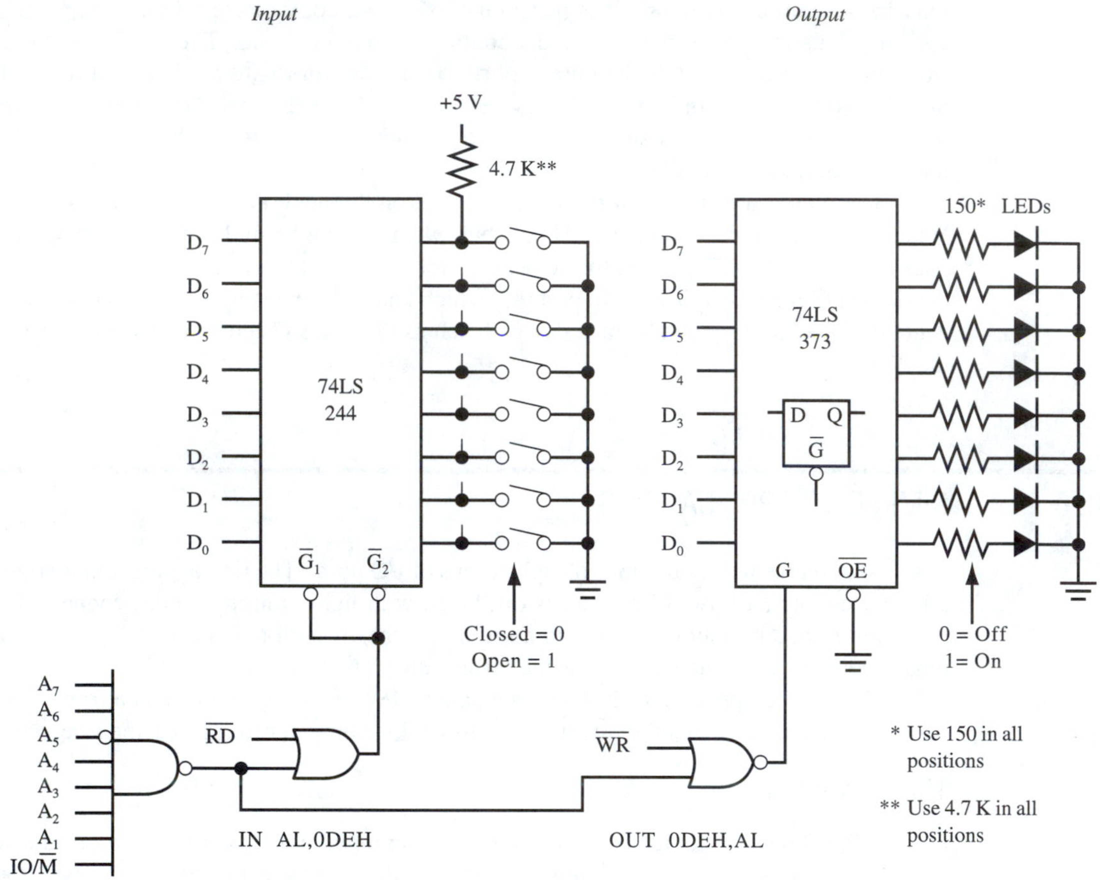

FIGURE 12.9   An input/output port

```
              NOP
              LOOP    WAIT                ;waste a little time
              MOV     AL,BL               ;get counter value back
              RET
DELAY         ENDP
```

With the DIP switches all closed, a 0 is read into the accumulator, giving CX an initial value of 0001. This will cause DELAY to execute the WAIT loop once before returning. The counter is updated at the fastest rate in this case. Opening all of the DIP switches causes FF to be read into AL, giving CX an initial value of FF01. This will cause over 65,000 WAIT loops to be executed, giving the longest possible delay between updates, and thus the slowest count.

A useful exercise to complete this application would be to determine the amount of time between outputs to the display port (the LEDs) for a particular setting of the input switches.

Sine Wave Generator

In this application, an 8-bit digital-to-analog converter is connected to an output port. The circuitry of Figure 12.10 shows how the 1408 DAC is wired to a 74LS373. The 1408 is designed to sink current at its output, the amount of current between 0 mA and some maximum value (not to exceed 2 mA), depending on the binary number present on its inputs. The output current is converted into voltage by the 741 op-amp so that a +/− range is generated. We will see that −2.5 V is created when 00 is at the DAC input, a value of 80 produces 0 V, and FF gives +2.5 V.

Square waves can be generated by sending out two alternating binary values. For example, outputting 00...FF...00...FF... produces a square wave with a peak-to-peak voltage of 5 V. Sending 20...60...20...60... also produces a square wave, but with a different overall voltage. The software shown here generates a sine wave by outputting each value from the SINE data table. One pass through the table produces one cycle of a sine wave at the DAC output. The information in the SINE data table was computed by breaking one cycle of a sine wave into 256 equal parts of 1.4 degrees. At each degree increment (0, 1.4, 2.8, etc.), the value of $\sin(x)$ is computed and multiplied by 128. The resulting value is then converted into binary. Try a few conversions for yourself and you should get the same pattern that appears in the table.

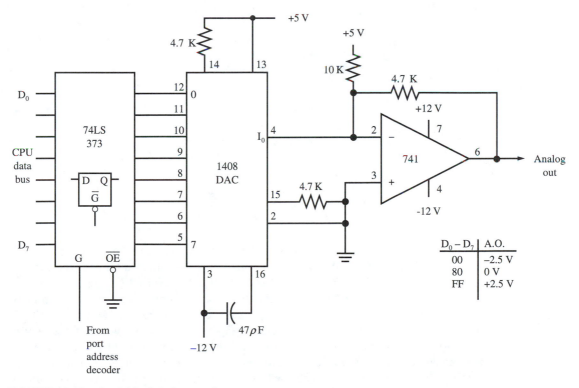

FIGURE 12.10   An 8-bit digital-to-analog converter

The frequency of the sine wave can be altered by playing with the DELAY subroutine.

```
In current data segment...
SINE     DB    82H,85H,88H,8BH,8EH,91H,94H,97H
         DB    9BH,9EH,0A1H,0A4H,0A7H,0AAH,0ADH,0AFH
         DB    0B2H,0B5H,0B8H,0BBH,0BEH,0C0H,0C3H,0C6H
         DB    0C8H,0CBH,0CDH,0D0H,0D2H,0D4H,0D7H,0D9H
         DB    0DBH,0DDH,0DFH,0E1H,0E3H,0E5H,0E7H,0E9H
         DB    0EBH,0ECH,0EEH,0EFH,0F1H,0F2H,0F4H,0F5H
         DB    0F6H,0F7H,0F8H,0F9H,0FAH,0FBH,0FBH,0FCH
         DB    0FDH,0FDH,0FEH,0FEH,0FEH,0FEH,0FEH,0FFH
         DB    0FEH,0FEH,0FEH,0FEH,0FEH,0FDH,0FDH,0FCH
         DB    0FBH,0FBH,0FAH,0F9H,0F8H,0F7H,0F6H,0F5H
         DB    0F4H,0F2H,0F1H,0EFH,0EEH,0ECH,0EBH,0E9H
         DB    0E7H,0E5H,0E3H,0E1H,0DFH,0DDH,0DBH,0D9H
         DB    0D7H,0D4H,0D2H,0D0H,0CDH,0CBH,0C8H,0C6H
         DB    0C3H,0C0H,0BEH,0BBH,0B8H,0B5H,0B2H,0AFH
         DB    0ADH,0AAH,0A7H,0A4H,0A1H,9EH,9BH,97H
         DB    94H,91H,8EH,8BH,88H,85H,82H,7EH
         DB    7BH,78H,75H,72H,6FH,6CH,69H,66H
         DB    62H,5FH,5CH,59H,56H,53H,50H,4EH
         DB    4BH,48H,45H,42H,3FH,3DH,3AH,37H
         DB    35H,32H,30H,2DH,2BH,29H,26H,24H
         DB    22H,20H,1EH,1CH,1AH,18H,16H,14H
         DB    12H,11H,0FH,0EH,0CH,0BH,9,8
         DB    7,6,5,4,3,2,2,1,0,0,0,0,0,0,0,0
         DB    0,0,0,0,0,0,0,1,2,2,3,4,5,6,7,8
         DB    9,0BH,0CH,0EH,0FH,11H,12H,14H
         DB    16H,18H,1AH,1CH,1EH,20H,22H,24H
         DB    26H,29H,2BH,2DH,30H,32H,35H,37H
         DB    3AH,3DH,3FH,42H,45H,48H,4BH,4EH
         DB    50H,53H,56H,59H,5CH,5FH,62H,66H
         DB    69H,6CH,6FH,72H,75H,78H,7BH,7FH
         .
         .
         .
WAVER:   MOV   CX,256            ;init loop counter
         MOV   SI,0              ;and pointer to data
SINOUT:  MOV   AL,SINE[SI]       ;read sine wave data
         OUT   0DEH,AL           ;send to 1408 DAC
         INC   SI                ;point to next item
         CALL  FAR PTR DELAY     ;pause between outputs
         LOOP  SINOUT            ;repeat forever
         JMP   WAVER
```

The OUT instruction uses the same port address as the one implemented in the binary counter application. An analysis of the instruction cycles required for one pass through the SINOUT loop, *not including* the CALL to DELAY, gives 17 cycles. Multiplying this by 256 gives 4,352 CLK cycles used in the creation of one sine wave! If the processor is running at 5 MHz, this gives a sine wave frequency of over 1,100 Hz. A DELAY subroutine requiring 50 additional cycles produces a sine wave with a frequency of only 290 Hz. These estimates should give you an indication as to the limits of the sine wave generator. High-frequency waveforms will have to be created by other methods. Even so, this simple circuit can be used to create interesting audio effects by connecting the output of the 1408 to an amplifier. It is only a matter of sending the right data to the 1408.

12.7   PARALLEL DATA TRANSFER: THE 8255 PPI

In the previous section we interfaced switches, lights, and a digital-to-analog converter to an I/O port. The number of applications is endless, with each one performing a different function. Even so, they all have at least one thing in common: they all use *parallel* I/O. In parallel I/O, all data bits are sent or received at the same time, as a group. This is very necessary in many applications! What would the binary counter look like if each LED were not updated at the same time as every other LED?

Many applications require more than one I/O port to get the job done. Peripheral designers realized this years ago and came up with a parallel I/O peripheral containing three separate I/O ports, all of which are programmable. This device is the 8255 programmable peripheral interface. In this section we will see how the 8255 is interfaced to the 8088 and programmed.

Interfacing the 8255

Figure 12.11 shows a diagram of the 8255 and its I/O and control signals. Twenty-four of the 8255's 40 pins are dedicated to the three programmable ports A, B, and C. These three ports, and a fourth one called a control port, are accessed via $\overline{RD}$, $\overline{WR}$, $\overline{CS}$, and address lines A_0 and A_1. A RESET input is included to initialize the 8255 when power is first applied. Figure 12.11 shows how an 8-input NAND gate is used to decode port addresses A0 through A3. When the address bus contains one of these four port addresses during an I/O access, $\overline{CS}$ will be pulled low. The 8255 will internally decode the states of A_0 and A_1 and determine which port to access. In this example, port A has port address A0. Ports B and C are accessed through ports A1 and A2, respectively, and the

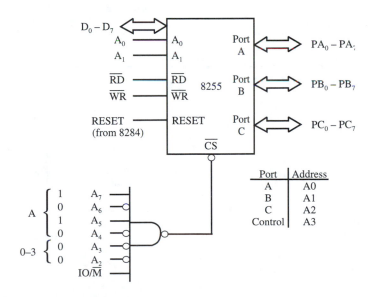

FIGURE 12.11   Interfacing the 8255 PPI

control port is at A3. It is very easy to determine the four port addresses by adding 0, 1, 2, and 3 to the base port address. The base port address is found by picking the upper six address lines (A_2 through A_7) to be what you need and assuming 0 for A_1 and A_0.

The nicest feature of the 8255 is that different hardware circuits can be connected to ports A, B, and C, with the direction (input or output) of each port configured with initial programming. This allows an 8088-based system with an 8255 in it to be used for many different purposes.

Programming the 8255

The 8255 has three modes of operation. The first is *mode 0: basic input/output.* In this mode, ports A, B, and C can be individually programmed as input or output ports. Port C is divided into two 4-bit halves, directionally independent from each other. So, there are sixteen combinations of input and output configurations available with this mode. A RESET automatically causes the 8255 to enter mode 0 with all ports programmed for input.

Input data is not latched. Data must be present when the port is being read by the processor. Output data is latched, as we would normally expect in an output port.

To program the 8255 for mode 0 operation and set the direction of each port, a mode word must be output to the control port. The definition of the mode word is shown in Figure 12.12. The MSB is the **mode-set flag,** which must be a 1 to program the 8255. Bits 5 and 6 are used to select the 8255's mode. 00 selects mode 0, 01 selects mode 1, and mode 2 is selected when bit 6 is high. Bit 2 is also used as a select bit for modes 0 and 1. The other 4 bits set the direction of ports A and B and both halves of C. A 0 indicates an output port and a 1 indicates an input port. To configure the 8255 for mode 0, all ports programmed for input, the mode word must be 10011011 (9BH). This byte must be output to

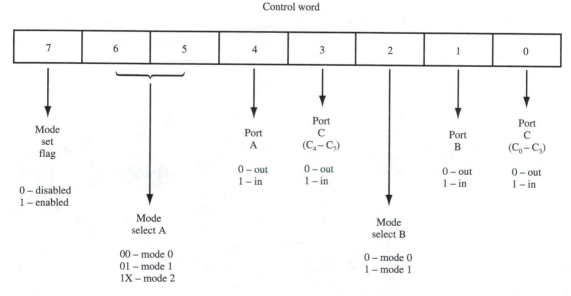

FIGURE 12.12   8255 mode word format

the control port to configure the 8255. The following two instructions will initialize the 8255 after a reset:

```
MOV     AL,9BH
OUT     0A3H,AL
```

Remember that the 8255 of Figure 12.11 has its control port at address A3.

Once the 8255 is programmed, the ports can be accessed with the appropriate IN instruction, such as IN   AL,0A0H (which reads port A). What mode word is needed to program port A for input, port B for output, and both halves of port C for input? You should get 99H when using the mode word format of Figure 12.12.

Assume that the 8255 has a DIP switch wired to port A and a set of LEDs wired to port B. The following code can be used to repeatedly read the switches and send their states to the LEDs.

```
READEM:  MOV    AL,99H      ;configure 8255 for Ain, Bout, mode 0
         OUT    0A3H,AL
GETSW:   IN     AL,0A0H     ;read switches
         OUT    0A1H,AL     ;send data to lights
         JMP    GETSW
```

In this case, a closed switch turns an LED off.

The next mode is *mode 1: strobed input/output.* In this mode, the 8255 uses port C as a *handshaking* port. Handshaking signals are commonly used in printers to sense the status of the paper-out sensor and the printer's readiness to accept new data. Ports A and B can be programmed for input or output. Data are latched in both directions. If port A is programmed for input, a strobe signal is needed on PC_4 to write data into port A. The 8255 will acknowledge the new input data by outputting a 1 on PC_5. These two signals on port C are defined as shown in Figure 12.13(a). PC_5 is IBF_a, input buffer A full. IBF is cleared when the processor reads port A. Port B operates in the same way, using PC_2 and PC_1 as handshaking signals. Both ports have the capability of causing an interrupt when data is strobed into them. The INTR output will go high when IBF goes high *and* the internal interrupt-enable bit is set. PC_4 and PC_2 make up the interrupt-enable bits for ports A and B. Setting PC_4 will cause $INTR_a$ to go high when data is strobed into port A. Reading the input port will clear the interrupt request. This interrupt mechanism is a useful alternative to using software to constantly poll the input port. Polling wastes a lot of time waiting for input data that may not be there. Interrupting the processor only when new data has arrived results in more efficient program execution. This is one of the advantages of mode 1.

As Figure 12.13(a) shows, PC_6 and PC_7 are available for general purpose I/O when port A is programmed for input. The mode word needed to program the 8255 for mode 1, Ain, Bin, and PC_6 and PC_7 out is B6.

Figure 12.13(b) shows how the 8255 is configured for output in mode 1. Here we have *output buffer full* ($\overline{OBF}$) and *acknowledge* ($\overline{ACK}$) signals used to handshake with the output circuitry. $\overline{OBF}$ will go low when the processor writes to port A or B. This signal will remain low until a low pulse arrives on $\overline{ACK}$. $\overline{ACK}$ is used to indicate that the new output data was received. $\overline{ACK}$, together with interrupt-enable, can be used to generate an interrupt with INTR. This would interrupt the processor when the new output data has been read, and avoid the need to poll the $\overline{ACK}$ signal. To program the 8255 for mode 1, Aout, and Bout, use mode word A4.

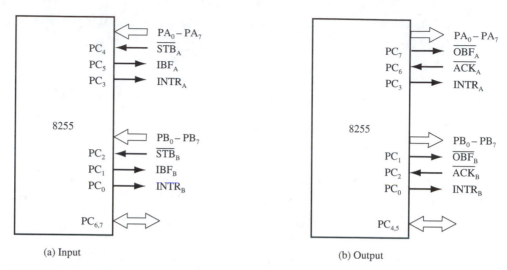

(a) Input   (b) Output

FIGURE 12.13   Mode 1 port definitions

You may notice that the port C bits are assigned differently in the output config-uration. For example, PC_4 and PC_5 are now used for general purpose I/O. The type of hardware configuration must be decided on, and then connected to the appropriate bits in port C.

The last mode is *mode 2: strobed bidirectional I/O*. This mode allows port A to op-erate as an 8-bit bidirectional bus. This is needed to allow the 8255 to be interfaced with 8-bit peripherals such as UARTS, which require a bidirectional data bus. Bits in port C are again used for handshaking and general purpose I/O, as indicated by Figure 12.14. Port B can operate as an input port or output port in mode 0 or mode 1. When operating port B in mode 0 (with port A in mode 2), PC_0 through PC_2 are available for general purpose I/O. The definitions for PC_0 through PC_2 in mode 1 apply when port B is operated in mode 1 with port A in mode 2.

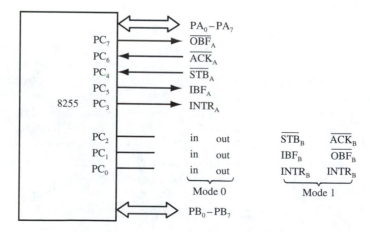

FIGURE 12.14   Mode 2 operation

As before, the INTR output can be used to interrupt the processor for input or output operations. When the CPU writes data to port A, $\overline{OBF}$ will go low. To enable the output buffer on port A and read the data, $\overline{ACK}$ must be pulled low. Data is written into port A by pulling $\overline{STB}$ low. This will cause IBF to go high until the data is read by the processor.

An 8255 Application: A/D Conversion

The 8255 in Figure 12.15 is configured for mode 0, port A in, port B out, PC_0 through PC_3 in, and PC_4 through PC_7 out. The 8255 has a base address of 40H. An 8-bit analog-to-digital converter (ADC0804) is connected to ports A and C. Port C is used to control the $\overline{RD}$

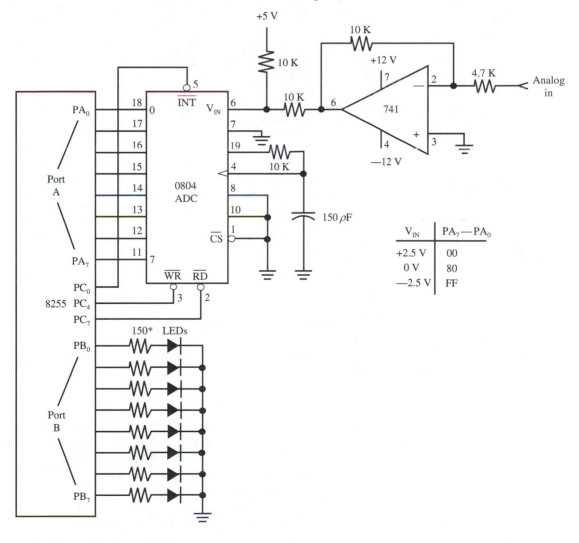

FIGURE 12.15 8-bit analog-to-digital converter

and $\overline{\text{WR}}$ inputs on the 0804, and to read the end-of-conversion status. The 741 op-amp converts a +/− input voltage swing into a 0–5 V signal that can be digitized by the 0804. To start a conversion, the 0804's $\overline{\text{WR}}$ line must be pulled low. This will force the $\overline{\text{INT}}$ output high until the conversion is complete. When $\overline{\text{INT}}$ goes low, the 0804 can be read by pulling the $\overline{\text{RD}}$ input low and reading port A.

The following subroutine is called to perform a single conversion and return the results in AH:

```
VCON    PROC    FAR
        MOV     AL,80H      ;start a conversion by pulling
        OUT     42H,AL      ;WR low
        MOV     AL,90H      ;now pull WR high
        OUT     42H,AL
EOC:    IN      AL,42H      ;read port C
        AND     AL,1        ;test bit-0 (INT)
        JNZ     EOC         ;wait for end of conversion
        MOV     AL,10H      ;enable RD
        OUT     42H,AL
        IN      AL,40H      ;read port A (the 0804)
        OUT     41H,AL      ;echo data to port B
        MOV     AH,AL       ;return result in AH
        MOV     AL,90H      ;get RD back to normal
        OUT     42H,AL
        RET
VCON    ENDP
```

The data read from the 0804 is sent out to the LEDs on port B. This gives a visual indication that everything is working properly. Being able to split up port C makes the interface with the 0804 easy to accomplish.

The 8255 is configured and initialized in this way:

```
MOV     AL,91H      ;mode 0, Ain, Bout, CLin, CHout
OUT     43H,AL      ;send to control port
MOV     AL,90H      ;RD and WR both high
OUT     42H,AL      ;send to port C
```

A routine to digitize a waveform presented to the analog input would require successive CALLs to VCON, storing AH in a data table each time VCON returns. Once the waveform has been digitized, the data bytes that represent it can be altered and then output to a digital-to-analog converter for playback.

The Centronics Parallel Printer Interface

Another application involving parallel data transfer is the use of a parallel printer. A parallel printer connection, such as the Centronics™ standard, provides for communication between the computer and the parallel printer. ASCII codes are output to the printer, and printer status is monitored by the computer through signals to a DB25 connector, as shown in Figure 12.16. On the PC, three ports are used to interface with the printer. A data port (address 378H) outputs 8-bit ASCII information to the printer. A control port (address 37AH) supplies a number of control signals (such as *strobe* and *initialize printer*), and a status port (address 379H) monitors printer status. In general, to print a character, the following sequence must take place:

•   Output ASCII code to data port
•   Output a low-going strobe pulse
•   Wait for a low-going acknowledge pulse

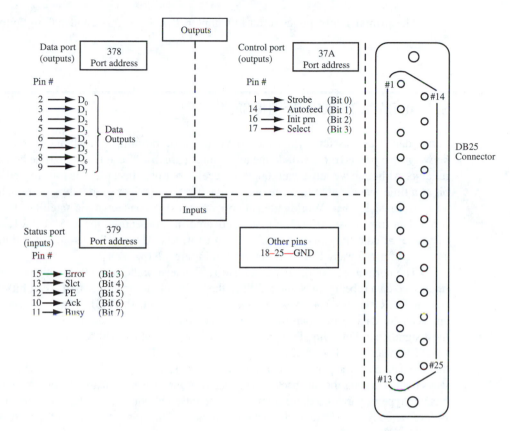

FIGURE 12.16   Centronics parallel printer connections

This sequence, illustrated in Figure 12.17, performs handshaking between the printer and the computer, and guarantees that the computer does not send data to the printer faster than the printer can accept it.

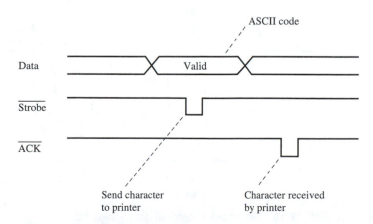

FIGURE 12.17   Printer/computer handshaking

The printer interrupts covered in Chapter 7 perform this handshaking automatically.

12.8   SERIAL DATA TRANSFER: THE 8251 UART

Serial data transmission offers the convenience of running a small number of wires between two points (three will do the job in most cases), while at the same time being very reliable. Although we must wait longer to receive our data because it is transmitted only 1 bit at a time, we are able to place our communication devices (computers, terminals), far away from each other. Worldwide networks now exist, connected via satellites, based on serial data transmission. The peripheral covered in this section, the 8251 UART, implements serial data transmission in a variety of formats. The standard serial data transmission waveform for any UART is depicted in Figure 12.18.

The normal state of the line is a logic 1. This level indicates that no activity is present (that is, no data is being transmitted). When the line level falls to a logic 0 (the start bit), the receiving UART knows that a new character is being transmitted. The data bits representing the character (or data) being transmitted are clocked out in the order shown, least significant to most significant. Following the data bits is the parity bit, which will be used by the receiving UART to determine the accuracy of the data it received. The parity bit in Figure 12.18 shows that the data has even parity. The last bits in any transmission are the stop bits, which are always high. This gets the line back into its inactive state. We are able to set the number of data bits, the type of parity used, the number of stop bits, and other parameters through software. Before we consider how to do this, let us examine the hardware operation of the 8251.

Interfacing the 8251

The 8251 was originally designed to be used with the 8080 and 8085 microprocessors, 8-bit machines that preceded the 8088. The 8088 interfaces with the 8251 easily, requiring the usual address decoder and a few control signals. Figure 12.19 shows a complete serial data circuit for the 8088. The 8251 is connected to the processor's data bus and $\overline{\text{IORC}}$ and $\overline{\text{IOWC}}$

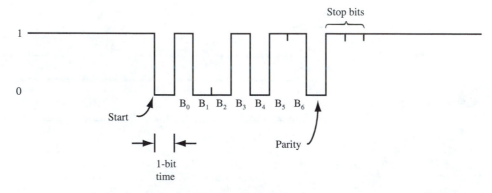

FIGURE 12.18   Standard TTL serial data waveform

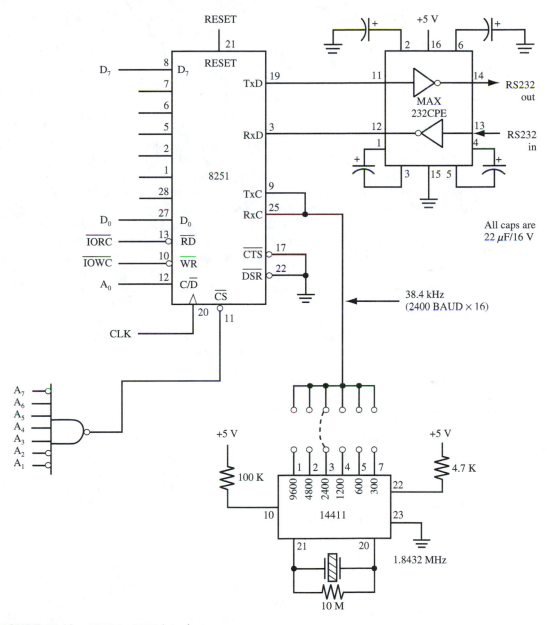

FIGURE 12.19   8251 to 8088 interface

signals. Because these signals are active only during I/O operations, the address decoder need only examine the state of the address bus. The NAND gate recognizes port addresses 78H and 79H. Address line A_0 is not used in the decoder. Instead, it is connected to the 8251's C/$\overline{D}$ input. This pin selects internal *control* or *data* registers. So, if A_0 is low (port 78H) and $\overline{IORC}$ is active, the UART's receive register will be read. If $\overline{IOWC}$ is active with A_0 low, the UART's transmitter register will be written to and a new transmission started.

The control functions are selected when A_0 is high (port 79H). Reading from port 79H gets 1 byte of status information; writing to it selects functions such as data size and parity. $\overline{\text{CTS}}$ (*clear to send*) and $\overline{\text{DSR}}$ (*data set ready*) are handshaking signals normally used when the 8251 is connected to a modem. They are grounded to keep them enabled. TxC and RxC are the transmitter and receiver clocks. The frequency of the TTL signal at these inputs determines the bit rate and time of the transmitter and receiver. It is common to run the UART at a clock speed sixteen times greater than the baud rate. So, a 2,400 baud transmission rate requires a 38.4 kHz clock (multiply 2,400 by 16). This frequency and other standard baud rate frequencies are generated automatically by the 14,411 baud-rate generator. All that is needed is a 1.8432 MHz crystal.

Attempting to transmit a digital (0–5 V) signal over a long length of wire causes distortion in the signal shape due to the line capacitance. It was discovered that making the signal switch from a positive voltage to a *negative* voltage helps to eliminate the distortion. Higher baud rates are possible using the +/– swinging signal. A standard was developed for this type of signal, called the RS232C standard. Take another look at the waveform in Figure 12.18. This is the TTL waveform that comes out of the UART's transmitter. The RS232 waveform that gets transmitted over the wires is inverted and swings plus and minus. So, a high level on the TTL waveform creates a low (negative) level on the RS232C waveform. An integrated circuit capable of performing the RS232C-to-TTL conversions is the MAX232CPE. This chip is especially useful because, by adding four 22 μF electrolytic capacitors, the MAX232 generates its own +/–10 V supply while needing only the standard 5 V. Older chips, such as the 1488 line driver and 1489 line receiver, required additional external power supplies. The MAX232 has two separate RS232C drivers/receivers for systems requiring two serial data channels.

Programming the 8251

Because the 8251 is connected to RESET, we are assured that the 8251 is functional after a power-on. It is still necessary to program the 8251 to ensure that the correct number of data bits will be used, that the parity will be generated as expected, and so on. To program the 8251, a series of bytes are output to the control port (79H from our example). The first byte is called the mode instruction. The format of this byte is shown in Figure 12.20. The 8251 can operate in **asynchronous mode** or **synchronous mode.** In asynchronous mode, the baud rate is determined by the lower 2 bits in the mode instruction. If these 2 bits are low, synchronous mode is selected.

The number of data bits used in a transmission is selected by bits 2 and 3. To enable generation of a parity bit, bit 4 must be set. Odd or even parity is chosen by the setting of bit 5. Finally, the number of stop bits is chosen by the upper 2 bits in the mode instruction. The waveform of Figure 12.18 contained 7 data bits, an even parity bit, and 2 stop bits. The required mode instruction byte is FA. To program the 8251, use:

```
MOV     AL,0FAH
OUT     79H,AL
```

Because an X16 clock was selected, the 8251 will operate in asynchronous mode. Synchronous mode is used for high-speed data transmission (not usually needed for communication with a serial display terminal). Synchronous mode is selected by making the lower two mode instruction bits 0. In this case, the upper two mode instruction bits do not

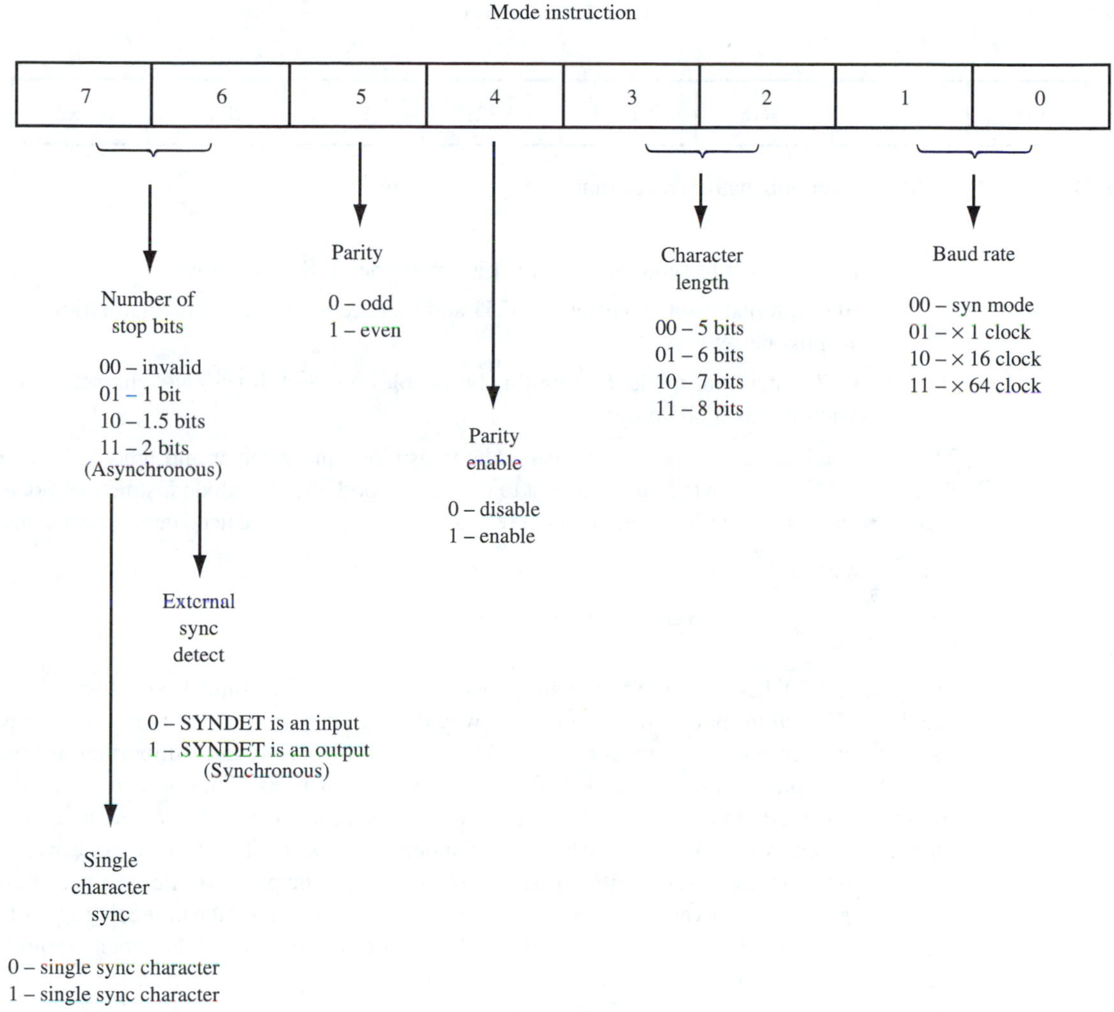

Mode instruction

7	6	5	4	3	2	1	0

Parity

0 – odd
1 – even

Number of
stop bits

00 – invalid
01 – 1 bit
10 – 1.5 bits
11 – 2 bits
(Asynchronous)

Parity
enable

0 – disable
1 – enable

Character
length

00 – 5 bits
01 – 6 bits
10 – 7 bits
11 – 8 bits

Baud rate

00 – syn mode
01 – × 1 clock
10 – × 16 clock
11 – × 64 clock

External
sync
detect

0 – SYNDET is an input
1 – SYNDET is an output
    (Synchronous)

Single
character
sync

0 – single sync character
1 – single sync character

FIGURE 12.20   8251 mode instruction format

set the number of stop bits, but rather the number of sync characters transmitted and the function of the SYNDET pin.

A second byte must be output to the control port to complete the initialization of the 8251. This byte is called the *command* instruction. The bits are assigned as shown in Figure 12.21; they have the following meanings:

Bit 0: transmit enable. Enable transmitter when this bit is set.

Bit 1: data terminal ready. Setting this bit will force the $\overline{\text{DTR}}$ output low.

Bit 2: receive enable. Enable receiver when this bit is set.

Bit 3: send break character. Setting this bit forces TxD low.

Bit 4: error reset. Setting this bit clears the PE, OE, and FE error flags.

Command instruction

7	6	5	4	3	2	1	0
EH	IR	RTS	ER	SBRK	RxE	DTR	TxE

FIGURE 12.21   8251 command instruction format

Bit 5: request to send. Setting this bit forces the $\overline{CTS}$ output low.

Bit 6: internal reset. To reset the 8251 and prepare for a new mode instruction, this bit must be set.

Bit 7: enter hunt mode. Setting this bit enables a search for SYNC characters (in synchronous mode only).

The command instruction needed to enable the transmitter and receiver and ignore all other functions is 05H. This byte must be output to the control port after the mode instruction. So, to totally initialize the 8251 for operation in the circuit of Figure 12.19, we need these instructions:

```
MOV    AL,0FAH      ;mode instruction
OUT    79H,AL
MOV    AL,5         ;command instruction
OUT    79H,AL
```

Once the UART has been programmed we have no need for the control port. Instead, we use the 8251's *status* port to help control the way data are transmitted and received. Figure 12.22 shows the bit assignments in the 8251's status port. Particularly important are the TxRDY (*transmitter ready*) and RxRDY (*receiver ready*) flags. They tell us when the transmitter is ready to transmit a new character and when the receiver has received a complete character. A number of error bits are included to show what may have gone wrong with the last reception. PE is parity error, and will go high if the parity of the received character is wrong. OE is overrun error, and will be set if a new character is received before the processor read the last one. FE stands for framing error, and goes high when stop bits are not detected. SYNDET (*sync character detected*) will go high when a sync byte is received in synchronous mode. DSR (*data set ready*) will go high whenever $\overline{DSR}$ is low.

The programmer must use the 8251's status bits to ensure proper serial data communication. Figure 12.23 shows how the first two bits are used to implement a simple serial input/output procedure.

Both flowcharts indicate that repeated testing of the RxRDY/TxRDY bits may be necessary. For example, to show the importance of this repeated testing, consider the following case. Suppose that an 8251 is configured to transmit and receive data at 1,200 baud, with 7 data bits, odd parity, and 1 stop bit. How long does it take to fully transmit or receive

Status byte

7	6	5	4	3	2	1	0
DSR	SYNDET	FE	OE	PE	TxEMPTY	RxRDY	TxRDY

FIGURE 12.22   8251 status byte

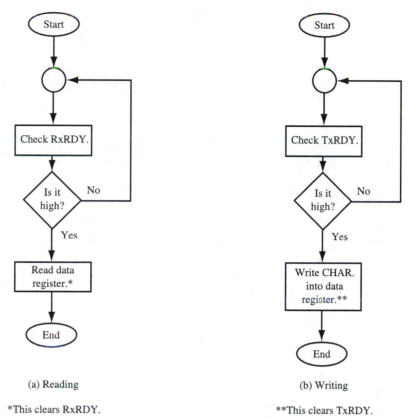

(a) Reading

(b) Writing

*This clears RxRDY.

**This clears TxRDY.

FIGURE 12.23 I/O flowcharts

a character? At 1,200 baud, the bit time is just over 833 µs, and the selected word length of 10 bits makes the total time to receive or transmit a single character roughly 8.3 ms.

It is not difficult to imagine how many instructions the 8088 might be able to execute in 8.3 ms. Would a few thousand be unreasonable? Probably not. Therefore, we use the status bits to actually slow down the 8088, so that it does not try to send or receive data from the 8251 faster than the 8251 can handle.

The two short routines that follow show how a character input and a character output routine might be written in 8088 code.

Character Input

The character input routine is used to read a character from the receiver, returning it in AL. It is necessary to check the state of RxRDY before reading the receiver.

```
CHARIN    PROC    FAR
RSTAT:    IN      AL,79H      ;read UART status
          AND     AL,02H      ;examine RxRDY
          JZ      RSTAT       ;wait until receiver is ready
          IN      AL,78H      ;read receiver
          AND     AL,7FH      ;ensure 7-bit ASCII code
          RET
CHARIN    ENDP
```

The data byte received is ANDed with 7F to clear the MSB. Because the UART is being used for character transmission, only a 7-bit ASCII code is required.

Character Output

This routine checks to see if the transmitter is ready for a new character. If it is, the ASCII character stored in AL is output to the transmitter.

```
CHAROUT    PROC    FAR
           MOV     AH,AL      ;save character
TSTAT:     IN      AL,79H     ;read UART status
           AND     AL,1       ;examine TxRDY
           JZ      TSTAT      ;wait until transmitter is ready
           MOV     AL,AH      ;get character back
           OUT     78H,AL     ;output to transmitter
           RET
CHAROUT    ENDP
```

An 8251 Application: Video Typewriter

Once the routines to communicate with the UART are in place, we can begin using them in applications. A video typewriter requires that a serial display terminal be connected to the UART that is controlled by CHARIN and CHAROUT. The simple programming loop that follows is used to echo every character received (presumably from the keyboard) to the screen. Line-feed characters are inserted when a carriage return is seen.

```
TVT:    CALL  FAR  PTR CHARIN    ;get keyboard character
        CALL  FAR  PTR CHAROUT   ;send it to screen
        CMP   AL,0DH             ;carriage return?
        JNZ   TVT
        MOV   AL,0AH             ;output a linefeed character
        CALL  FAR  PTR CHAROUT
        JMP   TVT
```

This simple routine can be modified to allow editing and other features, and is left for you to think about on your own.

READCOM1: A Serial I/O Application for the Personal Computer

The personal computer provides serial I/O through its COM1 connector. A mouse or modem commonly uses COM1 to communicate. The single-board computer presented in Chapter 14 is also capable of interfacing with the PC through COM1, which uses a UART similar to the 8251. Port addresses 3F8H through 3FFH are normally reserved for use with COM1 (with COM2 through COM4 using other ranges). Only two of these addresses are needed to perform simple serial I/O through COM1 once its UART has been initialized. These are

3F8H: Data port (receiver/transmitter data registers)

3FDH: Status port (receiver/transmitter ready flags)

Characters received by COM1 can be read by inputting from port 3F8H. Outputting to port 3F8H transmits a character.

Two status bits on port 3FDH must be used to synchronize the I/O operations:

Bit 0: receiver ready when high

Bit 5: transmitter ready when high

These status bits are tested in a manner similar to that used in the CHARIN and CHAROUT routines.

The READCOM1 program that follows reads characters received by COM1 and displays their 8-bit binary equivalents on the display. If a character is a printable ASCII character code, the actual character is output as well. The receiver-ready bit (bit 0 of port 3FDH) is examined through the use of the TEST instruction. Once the UART status has been placed in register AL, the instruction

```
TEST   AL,1
```

is used to check the state of bit 0.

```
;Program READCOM1.ASM: Read and display COM1 data.
;
          .MODEL SMALL
          .DATA
XMSG      DB     'Press any key to exit...',0DH,0AH,'$'
DSTR      DB     ' is '
CDATA     DB     20H         ;reserved for COM1 data
          DB     0DH,0AH,'$'

          .CODE
          .STARTUP
          LEA    DX,XMSG      ;set up pointer to exit message
          MOV    AH,9         ;display string function
          INT    21H          ;DOS call
COM1:     MOV    DX,3FDH      ;set up status port address
MORE:     IN     AL,DX        ;read UART status
          TEST   AL,1         ;has a character been received?
          JNZ    READ         ;yes, go get it
AKEY:     MOV    AH,0BH       ;check keyboard status
          INT    21H
          CMP    AL,0FFH      ;has any key been pressed?
          JNZ    MORE
          JMP    BYE
READ:     MOV    DX,3F8H      ;set up data port address
          IN     AL,DX        ;read UART receiver
          MOV    CDATA,AL     ;save character
          CMP    AL,20H       ;test for printable ASCII
          JC     NONPRT
          CMP    AL,80H
          JC     OKVAL
NONPRT:   MOV    CDATA,20H    ;use a blank when non-printable
OKVAL:    CALL   DISPBIN      ;display data in binary
          LEA    DX,DSTR      ;set up pointer to data string
          MOV    AH,9         ;display string function
          INT    21H          ;DOS call
          JMP    AKEY         ;and repeat
BYE:      .EXIT


DISPBIN PROC NEAR
          MOV    CX,8         ;set up loop counter
NEXT:     SHL    AL,1         ;move bit into Carry flag
          PUSH   AX           ;save number
          JC     ITIS1        ;was the LSB a 1?
          MOV    DL,30H       ;load '0' character
          JMP    SAY01        ;go display it
ITIS1:    MOV    DL,31H       ;load '1' character
SAY01:    MOV    AH,2         ;display character function
          INT    21H          ;DOS call
```

```
            POP    AX            ;get number back
            LOOP   NEXT          ;and repeat
            RET
DISPBIN ENDP

            END
```

READCOM1 executes until any key on the keyboard is pressed. It might be interesting to watch the values returned by a mouse connected to COM1. What patterns are received when the mouse moves left, right, up, and down? What happens when a mouse button is pressed?

Be sure to initialize COM1 to the proper BAUD rate before using READCOM1. This is easily accomplished with the use of the MODE command. For example, to initialize COM1 for 2,400 BAUD, no parity, 8 data bits, and 1 stop bit, use this command:

```
C> MODE COM1:2400,N,8,1
```

This is especially important when using COM1 to communicate with the single-board computer of Chapter 14.

12.9   TROUBLESHOOTING TECHNIQUES

Finding the cause of a faulty I/O device can be tricky. Here are suggestions of some things to try when you encounter an I/O problem.

* Write a short loop that continually accesses the I/O device. This should allow you to use an oscilloscope to look for a stream of pulses on the output of the address decoder. Use something like this to test an 8-bit output port:

```
            MOV    DX,<I/O address>
            SUB    AL,AL
PTEST:      OUT    DX,AL
            INC    AL
            JMP    PTEST
```

In addition to the steady stream of pulses that should appear on the address decoder output, there should be a binary count appearing at the output port. It is easy to see with an oscilloscope whether the waveform periods double (or halve) as you step from bit to bit. This is a good way to check for stuck or crossed outputs.

When checking an input port, use these instructions (or something similar):

```
            MOV    DX,<I/O address>
PTEST:      IN     AL,DX
            JMP    PTEST
```

This loop is good for checking the operation of the address decoder. If possible, combine both loops so that data read from the input port is echoed to the output port.
* Check for easily overlooked mistakes, such as using AD_0 through AD_7 to connect to the I/O device, but not using $IO/\overline{M}$ in the address decoder.
* Verify that the enable signals on the I/O device all go to their active states when accessed.

- For a serial device, examine the serial output for activity. Check for valid transmitter and receiver clocks. If the serial device is connected to a keyboard, press the keys and watch the serial input of the device. Make sure the TTL-to-RS232 driver is working correctly.

Other I/O devices may require you to test the interrupt system, or write a special initialization code to program a peripheral. Keep track of the new software and hardware designs you develop or troubleshoot; this will save you time and effort in the future.

SUMMARY

In this chapter we examined the design and operation of input and output ports. We saw that the 8088 has a smaller port addressing space than memory space. Two types of port addresses (and associated I/O instructions) may be used. One set of port addresses is 1 byte wide, from 00 to FF. The second set of port addresses, which is 2 bytes wide, range from 0000 to FFFF, and must be placed into register DX before use.

The hardware details of full and partial port address decoders were covered, along with examples of actual input and output ports. These ports were expanded into applications dealing with controlled time delays and waveform generation through a digital-to-analog converter. The programming and interfacing requirements of two I/O-based peripherals, the 8255 PPI and 8251 UART, were also covered. Additional peripherals for the 8088 will be covered in the next chapter.

STUDY QUESTIONS

1. Show two ways of reading input port 40H. What instructions are required?
2. Explain the processor bus activity when OUT   20,AL executes. Assume a minmode system.
3. Design a minmode full port address decoder for input/output port B0. You must generate active-low $\overline{RDB0}$ and $\overline{WRB0}$ signals.
4. Design a minmode full port address decoder for input ports B0 through B7. Eight individual port select outputs should be generated.
5. What changes must be made to the designs of Questions 3 and 4 for a maxmode system?
6. Design a minmode partial address decoder for an output port whose binary address is 10X01X1X11111111, where X is a *don't care* bit. What are all possible port address ranges for this decoder?
7. Use the magnitude comparators of Figure 12.5 to design a partial address decoder with selectable port ranges. The base port address is 1011CCCCCCCC0100, where C represents the 8-bit number matched by the comparators.
8. What is the first selectable port address in Question 7? What is the last? How far apart is each port address (in locations)?

9. What are the decoded port address ranges for each circuit in Figure 12.24?

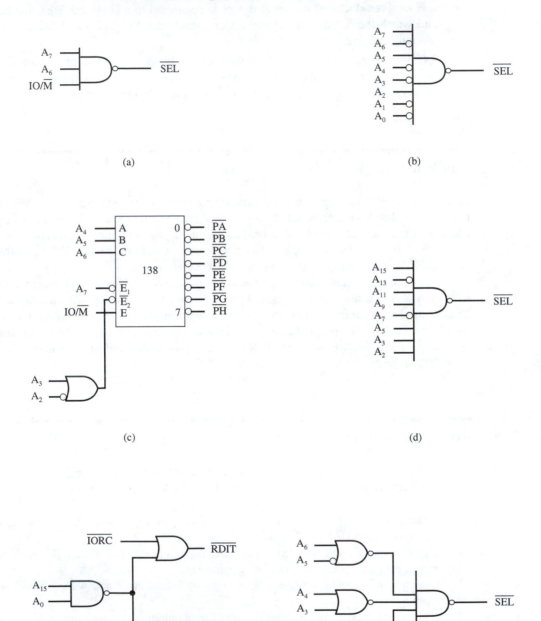

FIGURE 12.24   For Question 9

10. Why are input ports buffered rather than latched?
11. Modify the design of Figure 12.7 so that the port address is 409C.
12. Modify the design of Figure 12.7 so that the port is 16 bits wide (lower 8 at port 9C and upper 8 at 9D).
13. Repeat Questions 11 and 12 for the output port in Figure 12.8.
14. Modify the BINCNT program so that the lower 7 bits of the input port adjust the counting speed, while the most significant bit controls the direction, up or down, of the count.
15. Write a program called KATERPILAR, which will output a sequence of rotating bit patterns to the LEDs of Figure 12.9. The sequence might look like this:

```
XXX-----
-XXX----
--XXX---
---XXX--
----XXX-
-----XXX
```

   The sequence should change direction when all LEDs on the right are lit. Use the input port to control the speed of the display.
16. Modify BINCNT and the design of Figure 12.9 to allow a 16-bit counter to be displayed.
17. The data table in the WAVER routine does not have to be 256 bytes long to get decent looking waveforms. What frequency is possible if only 10 bytes are used to generate a waveform? What about 50 or 100 bytes?
18. Write a subroutine called TONES that generates two different frequency square waves and alternates them. TONES should output 100 cycles of the low-frequency signal and 50 cycles of the high-frequency signal. This should be repeated ten times before TONES returns. The high-frequency signal should have four times the frequency of the low-frequency one.
19. Redesign the port address decoder of Figure 12.11 so that port A is at address 8A. What are the other three port addresses?
20. Determine the mode words for each 8255 configuration:
    a) Mode 0, Ain, Bout, Cin
    b) Mode 1, Aout, Bin
    c) Mode 2, Bin (Mode 0)
21. Redo the BINCNT program by adapting it for use with an 8255 with a base address of 48H. Assign ports A, B, or C as you like. Draw a schematic of your design.
22. Use the VCON procedure to assist you in writing a subroutine called ONEWAVE, which will do the following:
    a) Wait for the analog signal to cross 0 (80H).
    b) Read and store samples from VCON until the signal crosses 0 twice.
    The signal samples should be stored in a data table pointed to by register DI. Do not write more than 1,024 samples into the table.
23. Rewrite VCON so that the $\overline{RD}$ input of the 0804 is always 0, and the $\overline{INT}$ output is fed back to the $\overline{WR}$ input. Note that only a software connection will exist between $\overline{INT}$ and $\overline{WR}$.
24. Sketch the 11-bit transmission code for an ASCII "K." Use even parity and 2 stop bits.

25. How long does the character in Question 24 take to transmit if the BAUD rate is 1,200?

26. Use partial address decoding to put the 8251 of Figure 12.19 into port space at address 40H.

27. Show the instructions needed to program the 8251 for:
    a) Asynchronous mode, 7 data bits, odd parity, X16 clock, 2 stop bits.
    b) Asynchronous mode, 8 data bits, no parity, X1 clock, 1 stop bit.
    c) Synchronous mode, 7 data bits per character.

28. Modify the CHARIN and CHAROUT routines so that bit shift/rotate instructions are used in place of AND to check the status.

29. Modify the CHAROUT routine so that any ASCII code between 01 and 1A will be displayed as a two-character sequence. The first character will always be "^". The second character is found by adding 40H to the original character byte. For example, if the character to display is 03H, CHAROUT should output "^C". All other codes should be directly output.

30. Modify the TVT program so that all characters including the carriage return are stored in a buffer pointed to by register BP. The length of the buffer (not to exceed 255 characters) is returned in CL. When carriage return is hit, or when the buffer is full, call the far routine SCANBUFF.

31. Modify the buffered TVT program so that it detects special control characters. The first is the backspace character (ASCII 08H). If a backspace is received, back up one position in the buffer. Next is Control-C (03H). Reset BP to the beginning of the buffer when this character is received. Last is Control-R (12H). This code causes the entire contents of the current buffer to be displayed on a new line.

32. Write a program that sends the message

    THE QUICK BROWN FOX JUMPS OVER THE LAZY DOG.

    to the COM1 serial port, until 'x' is entered on the keyboard.

CHAPTER 13

Programming with 80x86 Peripherals

OBJECTIVES

In this chapter you will learn about.

- Hardware interrupt handling
- Programmable time delays using an interval counter
- Floating-point coprocessor functions

13.1   INTRODUCTION

The power of a microprocessor can be increased by the use of peripherals designed to implement special functions, functions that may be very difficult to implement via software. A good example of this principle would be in the use of a coprocessor. The coprocessor comes equipped with the ability to perform complex mathematical tasks, such as logarithms, exponentials, and trigonometry. The 8088, although powerful, would require extensive programming to implement these functions, and even then would not compute the results with the same speed. Thus, we see that there are times when we have to make a hardware/software trade-off. In this chapter we will concentrate on applications that employ the use of standard peripherals, designed specifically for the 8088. In each case, we will examine the interfacing requirements of the peripheral, and then see how software is used to control it.

You are encouraged to refer to Appendix C as you read the chapter to get a more detailed look at each peripheral. Section 13.2 covers the 8259 programmable interrupt controller. Section 13.3 shows how accurate time delays can be generated with the 8254 programmable interval timer. Section 13.4 explains the operation of the floating-point unit (80x87 coprocessor). Troubleshooting tips for new peripherals are presented in Section 13.5.

13.2 THE 8259 PROGRAMMABLE INTERRUPT CONTROLLER

We saw in the previous chapter that a microprocessor must be interfaced with an I/O device to communicate with the outside world. Software support is required for each I/O device to ensure its proper operation. For example, the receiver status of a UART must be frequently examined to ensure that no received characters are lost. If a loop is used to test the receiver status, the processor may end up spending a great deal of time waiting for the chance to send the next character. While it is doing this, it cannot do anything else! An efficient solution to this situation is accomplished by adding an interrupt signal to the processor. Whenever a character is received by the UART, the UART will interrupt the processor. A special interrupt service routine will be used to read the UART and process the new character. When the UART is interfaced in this way, the processor is free to execute other codes during the times when the UART has not yet received a character. In this interrupt-driven I/O scheme, the processor accesses the UART only when it has to. This example illustrates the basic differences between **polled** I/O and **interrupt-driven** I/O. For some systems, polling is a good solution. This is especially true when the system is dedicated to doing one task over and over again. When a system is used in a more general way, the processor cannot afford to spend its time constantly polling each I/O device. In this case, interrupts provide a simple way to service all peripherals only when they need the processor's attention.

If we expand the idea of interrupt-driven I/O to an entire system, the number of interrupts required quickly adds up. Separate interrupts may be used for real-time clock/calendars, floppy and hard disk drives, the computer's keyboard, serial and parallel interfaces, video displays, and many other devices. Each device will require its own interrupt handler. The interrupts may also be assigned individual priorities to ensure that they get serviced in a manner desired by the programmer. How is it possible to use this many interrupts on the 8088, which has only two external hardware interrupt inputs (NMI and INTR)? The answer is the 8259 programmable interrupt controller, a special peripheral designed to support eight levels of prioritized hardware interrupts. The 8259 is considered an I/O device on the system bus, and can be written and read like an I/O port. For systems requiring more than eight levels of prioritized interrupts, it is possible to cascade 8259s to obtain up to 64 levels of interrupts. The 8259 is configured after power-on through software, and may be reconfigured at any time. The 8259 is designed to interface with the early 8-bit machines (the 8080 and 8085) and also the 16-bit 8088 and 8086. We will examine only its operation with respect to the 8088.

Figure 13.1 shows a simplified block diagram of the 8259. Eight levels of hardware interrupts are provided by inputs IR_0 through IR_7. These eight inputs go directly to the **interrupt request register** (IRR). This register keeps track of what interrupts have requested service. The 8259 can be programmed to allow level-sensitive or edge-sensitive interrupt inputs.

The output of the IRR feeds the **priority resolver,** which selects the highest priority interrupt from those requesting service. For example, if IR_2 and IR_5 both request service simultaneously, IR_5 will be selected first by the priority resolver.

The output of the priority resolver is used by the *in-service register* (ISR). The ISR indicates which interrupts are being serviced.

All three of these registers communicate with the control logic section, which performs the handshaking with the processor's INTR and $\overline{INTA}$ signals.

Individual interrupts may be disabled by data written to the *interrupt mask register.* Any combination of interrupts can be disabled (or *masked*) without affecting the priority of the remaining interrupts.

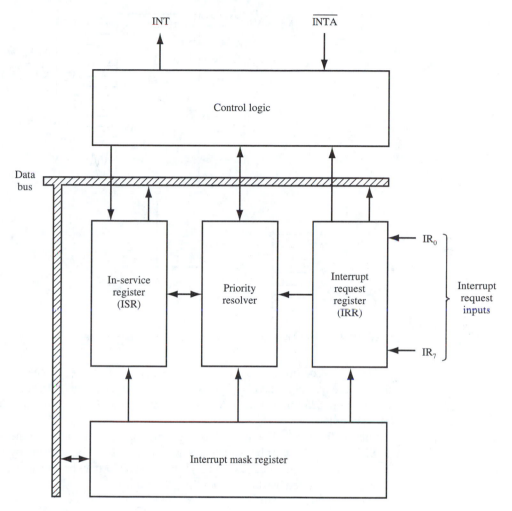

FIGURE 13.1   8259 block diagram

The 8259 can be programmed by writing the appropriate data values to a set of internal registers that are used to control the operation of each of the 8259's functional blocks. Before we see how this is done, let us examine how the 8259 interfaces with the processor.

Interfacing the 8259

The 8259 connects to the system bus like an ordinary I/O device. Figure 13.2 shows an example of the 8259 mapped to I/O ports F8 and F9 in a minmode system. A port-address decoder using an eight-input NAND gate is used to detect port addresses F8 and F9. The processor's A_0 address line connects directly to the 8259 to select internal registers, along with $\overline{RD}$ and $\overline{WR}$. The 8259 takes over the use of the processor's INTR input, which it activates during an interrupt request. Up to eight devices may request an interrupt from the 8259, via interrupt inputs IR_0 through IR_7. These eight lines may be programmed for edge-sensitive or level-sensitive operation, and are prioritized, with IR_7 having the highest priority.

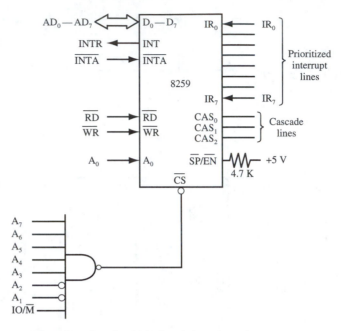

FIGURE 13.2   8259 interfaced to the 8088 in minimum mode

Three cascade lines, CAS_0 through CAS_2, are used to expand the 8259 in a system requiring more than eight levels of interrupts. These signals may be left unconnected when there is a single 8259 in a system. Multiple 8259s require CAS_0 through CAS_2 to be connected in parallel.

Finally, the $\overline{SP/EN}$ line is used to select master/slave operation in the 8259. When a single 8259 is used, it must be operated as a master. This can be selected by placing a 1 on $\overline{SP/EN}$. When two or more 8259s are used, only one 8259 may be a master. The remaining 8259s must be operated as slaves (with $\overline{SP/EN}$ grounded).

The operation of a properly programmed 8259 is as follows:

1. One or more of the interrupting devices connected to the 8259 request an interrupt by activating the appropriate IR input. The corresponding bit in the IRR is set.
2. The 8259 examines the interrupts requested and issues an INTR signal to the 8088 for the highest priority active interrupt input.
3. The processor responds with a pulse on $\overline{INTA}$.
4. The 8259 sets the bit for the highest priority active interrupt in the inservice register and clears the corresponding bit in the interrupt request register (to remove the request).
5. The 8088 outputs a second pulse on $\overline{INTA}$ (as it usually does with *any* interrupt caused by INTR).
6. The 8259 outputs an 8-bit interrupt vector number on the data bus. This number is captured by the processor and used to select the corresponding interrupt service routine address from the vector table.

This process illustrates the usefulness of the 8259. Individual vector numbers are assigned to each of the IR inputs connected to a device. When a particular device issues an

7							0
0	0	0	1	LTIM	0	SNGL	IC4

FIGURE 13.3   Initialization command word 1

interrupt request to the 8259, the 8259 sends the corresponding vector number to the processor.

In the next section we will see how a single 8259 is programmed to provide these vector numbers to the CPU.

Programming the 8259

The port-address decoder shown in Figure 13.2 maps the 8259 to ports F8 and F9. These port addresses must be used to access the 8259 during initialization and also during normal operation. It is important to note that the 8259 can be reconfigured whenever necessary by issuing initialization commands to it. This may be useful to some designers who may wish to modify the interrupt mechanism of their system as necessary.

When a single 8259 is used in a system, operating in master mode, it requires two types of control information from the CPU: **initialization command words (ICW)** and **operation command words (OCW).** Once these words are written to the 8259, it is capable of servicing its eight levels of prioritized interrupts. If more than one 8259 is used, each must be programmed individually.

Initialization is accomplished by sending a sequence of 2–4 bytes (ICWs) to the 8259. Figure 13.3 shows the makeup of the first ICW. Zeros are inserted in bit positions that control the 8259's operation in an 8085 system (bits 2 and 5–7). For the purposes of this discussion, only the bits that directly affect the operation of the 8259 in an 8088 system are explained. IC4 (bit 0), when set, indicates that ICW4 must be read during initialization.

SNGL (bit 1), when set, indicates that only one 8259 is being used in a system. To allow cascaded 8259s, this bit must be cleared (and ICW3 issued).

LTIM (bit 3), when set, indicates that the 8259 should operate the IR inputs in level-sensitive mode. When LTIM is cleared, the eight IR inputs will be edge-sensitive.

Example 13.1:   What is the interpretation of ICW1 when it contains 1AH?

Solution: The bit pattern produced by 1AH is 00011010. This indicates that both LTIM and SNGL are set, and IC4 is cleared. This will select level-sensitive inputs, inform the 8259 that it is the only interrupt controller in the system, and that ICW4 is not required.

Using the circuit of Figure 13.2 as an example, ICW1 must be output to port F8 (A_0 equals 0) to be properly received. The 8259 recognizes ICW1 by the 1 seen in bit position 4. Reception of this control word also causes the 8259 to perform some internal housekeeping.

7							0
T_7	T_6	T_5	T_4	T_3	0	0	0

FIGURE 13.4   Initialization command word 2

Once ICW1 is processed, the 8259 awaits ICW2, which must be output to port F9 (A_0 equals 1). This control word is shown in Figure 13.4, and is used to program the eight interrupt vector numbers that will be associated with the IR inputs. T_3 through T_7 become the five most significant bits of the vector number supplied by the 8259 during an interrupt acknowledge cycle. The lower 3 bits are generated by the IR input that has been selected for service. Figure 13.5 shows the makeup of the interrupt vector number. Do you see the relationship between bits 0–2 and the corresponding IR signal? If IR_0 requests service, the vector number generated will contain three 0s in the lower bits. Suppose that ICW2 is issued with the data byte 68. This indicates that T_7 through T_3 are assigned the values 01101. What are the eight possible interrupt vector numbers, using the information from Figure 13.5? IR_7 will create vector number 6F. IR_0 will create vector number 68. Vector numbers 69 through 6E are generated by IR_1 through IR_6, respectively.

If a single 8259 is used in a system, ICW3 is not needed. For multiple 8259s, ICW3 serves two functions. When an 8259 is used as a master, each bit in ICW3 is used to indicate a slave connected to an IR input. When the 8259 is used as a slave, only the lower 3 bits of ICW3 are used, and they set the slave's cascade number (0 to 7). Figure 13.6 shows two 8259s connected as master and slave. The INT output on the slave device connects to the IR_6 input on the master device. Thus, any requests on interrupt lines INT_7 through INT_{14} will cause IR_6 to be activated on the master. The master will then examine bit 6 in ICW3 to see if it is set (indicating that a slave is connected to IR_6). If so, it will output the cascade number of the slave (110 in this case) on CAS_0 through CAS_2. These cascade bits are received by the slave device, which examines its ICW3 to see if there is a match. The programmer must have previously programmed 110 into the slave's ICW3. If there is a match between the cascade number and ICW3, the slave device will output the appropriate vector number during the second $\overline{INTA}$ pulse.

To get this scheme to work, the first 8259 must have bit 6 set in its ICW3, and the second 8259 must have the bit pattern 110 in the lower 3 bits of its ICW3. Both forms of ICW3 are shown in Figure 13.7.

FIGURE 13.5   Generating the interrupt vector number

Interrupt			Vector number					
	D_7							D_0
IR_7	T_7	T_6	T_5	T_4	T_3	1	1	1
IR_6						1	1	0
IR_5						1	0	1
IR_4						1	0	0
IR_3						0	1	1
IR_2						0	1	0
IR_1						0	0	1
IR_0	T_7	T_6	T_5	T_4	T_3	0	0	0

FIGURE 13.6 Two 8259s cascaded

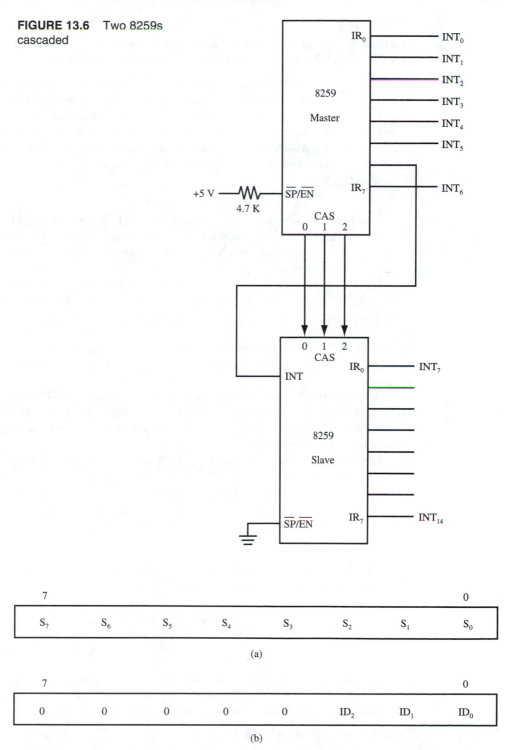

FIGURE 13.7 ICW3 format: (a) master mode; (b) slave mode

Example 13.2:   How many slave devices are required if ICW3 in a master 8259 contains 10010010? What IR inputs are connected to each slave device?

Solution: Because ICW3 contains three 1s, three slave devices are being used. The INT outputs on each slave connect to IR inputs IR_7, IR_4, and IR_1 on the master. Each of the three slaves must have its own ICW3 programmed with a unique cascade code. The three codes are 111 (for IR_7), 100, and 001. All of the slave devices will receive the same cascade code from the master, so using unique codes for each slave will guarantee that only one slave responds during the interrupt acknowledge cycle.

ICW3 must be output to port F9 (A_0 equals 1) to be properly received.

The last initialization command word is ICW4 (also output to port F9). Remember that this word is only needed when bit 0 of ICW1 is set. The format of ICW4 is shown in Figure 13.8. The operation of each bit is as follows:

AEOI (bit 1) is used to program automatic end-of-interrupt mode when high. When AEOI mode is set, the 8259 automatically clears the selected bit in the in-service register. When not operated in AEOI mode, the in-service bit must be cleared manually through software.

M/S (bit 2) is used to set the function of the 8259 when operated in buffered mode. If M/S is set, the 8259 will function as a master. If M/S is cleared, the 8259 will function as a slave.

BUF (bit 3), when set, selects buffered mode. When the 8259 operates in buffered mode, the SP/EN pin becomes an output that can be used to control data buffers connected to the 8259's data bus. When BUF is cleared, SP/EN is used as an input to determine the function (master/slave) of the 8259.

SFNM (bit 4), when set, causes the 8259 to operate in special fully nested mode. This mode is used when multiple 8259s are cascaded. The master must operate in special fully nested mode to support prioritized interrupts in each of its slave 8259s.

Let us look at an example of how a single 8259 can be programmed.

Example 13.3:   Tell what instructions are needed to program a single 8259 to operate as a master and provide the following features:

1. Edge-sensitive interrupts
2. ICW4 needed

7							0
0	0	0	SFNM	BUF	M/S	AEOI	1

FIGURE 13.8   Format of ICW4

3. A base interrupt vector number of 40H
4. No special fully nested mode
5. No buffered mode
6. AEOI mode enabled

Solution: First, the data values for each ICW must be determined. To set SNGL and IC4 and clear LTIM, ICW1 must contain 13H. This takes care of conditions 1 and 2. Placing 40H into ICW2 will program the desired base-interrupt vector of condition 3. This will allow generation of interrupt vectors 40H through 47H. Because SNGL will be set in ICW1, there is no need to write to ICW3 during initialization. The remaining three conditions are met by placing 03H into ICW4.

Initialization can be performed by the instructions shown here:

```
MOV    AL,13H
OUT    0F8H,AL     ;output ICW1
MOV    AL,40H
OUT    0F9H,AL     ;output ICW2
MOV    AL,03H
OUT    0F9H,AL     ;output ICW4
```

A system designed with flexibility in mind may require that the initialization bytes come from a data table placed somewhere in memory. In this case, ICW1's byte could be tested by the initialization routine to see if it needs to output ICW4 (and similar reasoning applies to ICW3).

Once the 8259 has received all of its ICWs, it is ready to begin processing interrupt requests. The exact way the 8259 handles each interrupt is programmed with three OCWs. This information must be output to the 8259 after initialization.

The first OCW is used to mask off selected interrupts by altering the bit pattern in the interrupt mask register. Any of the eight inputs can be disabled by setting its corresponding bit in OCW1.

Example 13.4:   What interrupts are disabled by writing 10100001 into OCW1?

Solution: Interrupts IR_7, IR_5, and IR_0 are masked out by this control word. Interrupt requests on these inputs will be ignored until a different pattern is output to OCW1.

The second OCW is illustrated in Figure 13.9. We saw earlier that the AEOI bit in ICW4 is used to enable/disable *AEOI mode*. AEOI mode supports repetitive interrupts of the same priority by automatically resetting bits in the in-service register. It may be desirable for an interrupt service routine to complete before allowing a second interrupt of the same type or level. In this case, AEOI will be set to 0, and the interrupt service routine is responsible for resetting the in-service register bit for a particular interrupt. This can be done

7							0
R	SL	EOI	0	0	L_2	L_1	L_0

0	0	0
0	0	1
0	1	0
0	1	1
1	0	0
1	0	1
1	1	0
1	1	1

IR level to be acted upon

R	SL	EOI	
0	0	1	Nonspecific EOI command
0	1	1	Specific EOI command*
1	0	1	Rotate on nonspecific EOI command
1	0	0	Rotate in AEOI mode (set)
0	0	0	Rotate in AEOI mode (clear)
1	1	1	Rotate on specific EOI command*
1	1	0	Set priority command*
0	1	0	No operation

*Use $L_0 - L_2$ for specific IR level

FIGURE 13.9   Format of OCW2

by outputting a *specific EOI command* to OCW2. The level of the interrupt being reset must be placed in the lower 3 bits of OCW2. Other features, such as automatic rotation of the interrupt priorities, can also be selected. When many devices in a system have equal priority, rotating interrupt priorities ensures that all devices get serviced in a round-robin fashion.

The third OCW is used to enable/disable *special mask mode*. Remember that OCW1 allows us to mask off individual interrupts. Normally, when an interrupt of a certain priority is processed by the 8259, the in-service register bit for that interrupt is set. This automatically disables additional interrupts of the same and lower level. For example, an IR_3 request will disable further IR_3 through IR_0 requests. In special mask mode, setting a bit in OCW1 will only disable the associated IR input, leaving the lower priority interrupts enabled.

OCW3 is also used to allow the interrupt request and in-service registers to be read by the processor. This is accomplished by writing the necessary pattern into the lower two bits of OCW3 and then reading port F8 (A_0 equals 0).

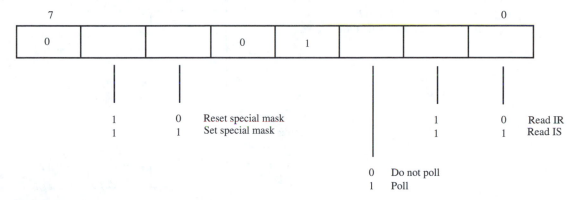

FIGURE 13.10   Format of OCW3

A final function of OCW3 is to enable/disable *poll mode*. In this mode the 8259 does not generate interrupt requests on INT. When read, the 8259 will output the level of the highest priority device requesting service on the lower 3 bits of the data bus. The format of OCW3 is shown in Figure 13.10.

There are times when many of the special features provided by the 8259 will not be needed. However, the introduction to this device should show you that a very complex interrupt system, expandable to 64 inputs, is possible with a handful of 8259s and port-select logic.

13.3   THE 8254 PROGRAMMABLE INTERVAL TIMER

Many applications require the processor to perform an accurate time delay between a set of operations. For example, a microprocessor might be dedicated to reading a custom keypad or driving a multiplexed display. Both applications require a small time delay between repeated input or output operations. A programmer may decide to use a software delay loop, such as:

```
DELAY:  MOV  CX,4000      ;init delay counter
WAIT:   LOOP WAIT         ;LOOP until CX=0
```

The total amount of delay involves 4,000 executions of the LOOP instruction. This time can be estimated by multiplying the processor's clock period by the total number of states required to execute the LOOP instructions. This type of delay loop has two main disadvantages. While the loop is executing, the processor is not able to do anything else, such as execute instructions not related to the loop. Also, the time delay becomes inaccurate if the processor is interrupted. For these reasons some designers (and programmers) prefer to do their timing with hardware. Software is used to program the hardware for a specific time delay. At the end of the time delay, the processor is interrupted. This frees up the processor for other kinds of execution while the hardware is performing the time delay.

FIGURE 13.11   The 8254
programmable interval timer

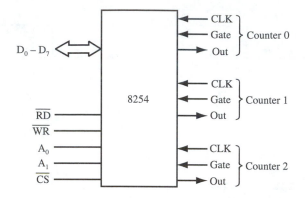

A peripheral designed to implement the type of time delay just described (and many others) is the 8254 programmable interval timer. Figure 13.11 shows the signal groups of the 8254. The 8254 is interfaced through a group of I/O ports. Three internal counters can be programmed in a variety of formats, including, 4-digit BCD or 16-bit binary counting, square-wave generation, and one-shot operation. These formats allow the 8254 to be used for a number of different timing purposes. A short list of these applications includes real-time clocks (and/or calendars), specific time delay generation, frequency synthesis, frequency measurement, and pulse-width modulation.

First let us see how the 8254 is interfaced to the 8088.

Interfacing the 8254

The port-address decoder needed to connect the 8254 to the processor's address bus is shown in Figure 13.12. Here, the 8254 is interfaced with an 8088 operating in maximum mode (indicated by the $\overline{IORC}$ and $\overline{IOWC}$ command signals).

The two 8-input NAND gates in the port-address decoder map the 8254 into four I/O locations, CC80H through CC83H. The first port (CC80H) is used to read and write counter 0. The second two ports (CC81H and CC82H) access counters 1 and 2, respectively. The fourth port (CC83H) is used to control the 8254. Each counter contains CLK and GATE inputs, and one output, OUT. Counters may be cascaded by connecting the OUTput of one to the GATE of the other.

Programming the 8254

Each of the 8254's three counters can be programmed and operated independently of the others. This discussion will concentrate on programming and using counter 0 only.

Counter 0 is a 16-bit synchronous down counter that can be preset to a specific count, and decremented to 0 by pulses on the CLK0 input. Counter 0 may operate in any of six different modes (selected with the use of a control word). Each mode supports four-digit BCD and 16-bit binary counting. Thus, counts may range from 9999 to 0000 BCD or FFFF to 0000 hexadecimal. Programming the counter requires outputting a control word and an initial count to the 8254.

Figure 13.13 shows the bit assignments in the 8254's control word. Two bits (7 and 6) are used to select the counter being programmed. Another two bits (5 and 4) are used to con-

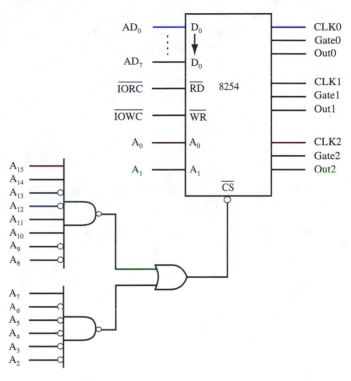

FIGURE 13.12   8254 interfaced to the 8088

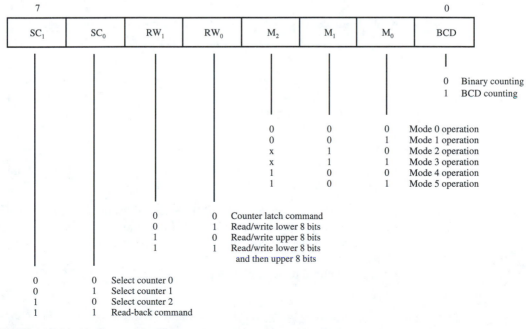

FIGURE 13.13   8254 control word

trol the way the counter is loaded with a new count. Bits 3, 2, and 1 are used to select one of six modes of operation. The least significant bit is used to select BCD or binary counting.

To begin a counting operation, the control word must be output, followed by the 1- or 2-byte initial count. The initial count value output to the 8254 goes into a *count register*. The count register is cleared when the counter is programmed (upon reception of the control word), and transferred to the actual down counter after it gets loaded with the 1- or 2-byte initial count. The counter may be loaded with a new count at any time, without the need for a new control word.

Example 13.5:   What instructions are needed to program counter 0 for BCD counting in mode 4? The initial count is 4788.

Solution: Using the port addresses assigned by the hardware in Figure 13.12, it is necessary to output the control word to port CC83H and the initial count to port CC80H. The control word needed to program counter 0 for BCD counting in mode 4 and a 16-bit initial count is 00111001 (39H). This value must be output to the control port. The initial count is represented by hex-pairs 47H and 88H.

The counter is initialized by the following instructions:

```
MOV    DX,0CC83H
MOV    AL,39H
OUT    DX,AL        ;output control word
MOV    DX,0CC80H
MOV    AL,88H
OUT    DX,AL        ;output lower 8 bits
MOV    AL,47H
OUT    DX,AL        ;output upper 8 bits
```

To load a new BCD count at any time, the last five instructions must be repeated.

Some time delays may be very short, but still require the use of the 8254. If the counter value is less than 256 for binary counting (or 100 for BCD counting), then the counter may be programmed by outputting the lower 8 bits only.

Example 13.6:   What control word is needed to program counter 2 for binary counting in mode 1, with an initial count of A0H?

Solution: The control word required is 10010010 (92H). Because the count register is cleared when the counter is programmed, writing A0H to the lower half results in an initial count of 00A0H. Notice that the control word indicates that only the lower 8 bits are needed to load the counter.

The counter is initialized by the following instructions:

```
MOV    DX,0CC83H
MOV    AL,92H
OUT    DX,AL             ;output control word
```

```
MOV    DX,0CC82H
MOV    AL,0A0H
OUT    DX,AL              ;output initial count
```

It may be necessary to read the state of a counter while counting is in progress. The 8254 allows this to be done three different ways.

The first method employs a read from the input port associated with the counter. This technique will not be accurate unless the counter is temporarily paused by stopping the CLK signal or placing a 0 on the GATE input.

The second technique uses the 8254's *counter latch command*. This command is selected by clearing bits 5 and 4 in the control word for a particular counter. This causes the 8254 to transfer a copy of the selected count register into an output latch. The output latch can then be read at any time by reading the desired 8254 port.

Example 13.7:   What instructions are needed to latch the count in counter 1 and save it in register BX?

Solution: The control word necessary to latch counter 1 is 01000000 (40H). This must be output to the control port (CC83H). The latched count may then be read from port CC81H.

These instructions will read the count and save it in register BX:

```
MOV    DX,0CC83H
MOV    AL,40H
OUT    DX,AL              ;latch counter 1
MOV    DX,0CC81H
IN     AL,DX              ;read lower byte
MOV    BL,AL              ;save lower byte
IN     AL,DX              ;read upper byte
MOV    BH,AL              ;save upper byte
```

Note that if the software fails to read the latched counter value before a new counter latch command is issued for the same counter, no new latching takes place. The count will remain latched and unchanged until it is read.

The third technique uses a *read-back command* to read the state of a counter. The read-back command is issued by setting bits 7 and 6 in the control word. The remaining bits take on new meanings, as shown in Figure 13.14. Bits 1, 2, and 3 are used to select any combination of counters (from none to all three). Bit 4, $\overline{\text{STATUS}}$, causes the status to be latched for any selected counter. Bit 5, $\overline{\text{COUNT}}$, causes the count to be latched in the same fashion. $\overline{\text{STATUS}}$ and $\overline{\text{COUNT}}$ are active low control bits.

7							0
1	1	COUNT	STATUS	CNT_2	CNT_1	CNT_0	0

FIGURE 13.14   8254 read-back command word

Example 13.8:   What control word is needed to latch the count of counters 0 and 2?

Solution: $\overline{\text{COUNT}}$ must be low to latch the count of any counter. CNT_0 and CNT_2 must be high to select counters 0 and 2. The required control word is 11011010 (DAH). Once this control word is output, counter 0 and counter 2 may be read from ports CC80H and CC82H.

Do you see how a single read-back command can be used to eliminate multiple counter latch commands?

When the read-back command is used to read status information, a 1-byte status word is generated for the selected counter. The format of the status word is shown in Figure 13.15. The lower 6 bits indicate the mode and counting scheme currently being used by the counter. NULL COUNT (bit 6) goes low when the new count written to a counter is actually loaded into the counter. This will occur at different times, depending on the mode selected. OUTPUT (bit 7) reflects the current state of the OUT signal for the counter. When counters are assigned and programmed dynamically, it is useful to be able to read the operating parameters by latching the status.

Example 13.9:   What instructions are necessary to latch the count and status of counter 1? The status must be returned in AH and the count in BX. If the status byte indicates that BCD counting is being used, return 01 in CL, otherwise return with CL = 00.

Solution: The command word needed to latch the count and status for counter 1 is 11000100 (C4H). After this byte is output to the control port, the status of counter 1 may be read (from counter 1's input port). Then the 2-byte count may be read.

These instructions will read counter 1's status and count and test for BCD counting:

```
MOV     DX,0CC83H
MOV     AL,0C4H
OUT     DX,AL           ;latch counter 1 data
MOV     DX,0CC81H
IN      AL,DX           ;read status
MOV     AH,AL
IN      AL,DX           ;read lower byte of count
MOV     BL,AL
IN      AL,DX           ;read upper byte of count
MOV     BH,AL
MOV     CL,AH           ;get status back
AND     CL,01H          ;test BCD bit
```

OUTPUT	NULL COUNT	RW_1	RW_0	M_2	M_1	M_0	BCD

FIGURE 13.15   Status word format

The AND instruction is used to test the BCD bit in the status byte. Other bits may be tested in a similar way, using different bit patterns in the immediate byte of the AND instruction.

The remaining discussion is devoted to an explanation of the six modes of operation possible with the 8254.

Mode 0: Interrupt on Terminal Count.   This mode can be used to interrupt the processor after a certain time period has elapsed, or after a number of events have occurred. Its operation is as follows. Initially, the programmer writes the control word specifying mode 0 to the control port of the 8254. This forces the selected counter's OUT signal to go low. Next, the programmer outputs the initial count, which is loaded into the counter during the next falling edge of CLK. Each successive falling edge of CLK will decrement the counter. When the counter gets to 0, OUT will go high. OUT can be used to interrupt the processor (possibly via the 8259 programmable interrupt controller). OUT remains high until a new count (or control word) is issued.

The count can be paused at any time by placing a 0 on the GATE input. Bringing GATE high again resumes counting. If a new count value is output while GATE is low, the next falling edge of CLK will still load it into the counter. When a 2-byte count is output, the first byte output causes the count to terminate and OUT to immediately go low. The second byte output enables the counter to be loaded on the next falling edge of CLK.

Example 13.10:   Show how the 8254 can be used to generate a time delay of 5 ms. A 1-MHz clock is connected to the CLK input of counter 1.

Solution: Figure 13.16 shows the connections to counter 1's input and output signals. GATE is tied high to enable counting all the time. OUT is connected to an IR input on the 8259 programmable interrupt controller. Assuming the IR inputs are edge-sensitive, when OUT goes high at the end of the count, an interrupt will be generated. The trick is to get OUT to go high 5 ms after we start the counter. The period of the 1-MHz clock is 1 μs. Dividing 5 ms by 1 μs gives 5,000, the number of CLK pulses required to get a 5-ms delay. Because one of these CLK pulses will be used to load the counter, the initial counter value must be 4,999. Thus, the counter can operate in binary or BCD mode.

FIGURE 13.16   Generating a 5-ms delay with the 8254

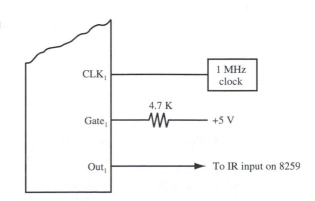

These instructions can be used to implement the 5-ms time delay in mode 0, with BCD counting:

```
MOV    DX,0CC83H
MOV    AL,71H
OUT    DX,AL                    ;counter 1, mode 0, BCD
MOV    DX,0CC81H
MOV    AL,99H
OUT    DX,AL                    ;lower byte of count
MOV    AL,49H
OUT    DX,AL                    ;upper byte of count
```

In general, a count of N requires N + 1 CLK pulses to complete. One CLK pulse is used to load the counter and N are used to count it down to 0. Also note that an initial BCD count of 0000 will require 10,000 CLK pulses to count down to 0000, and an initial binary count of 0000 will require 65,536!

Mode 1: Hardware Retriggerable One-Shot.   In this mode, the 8254 is programmed to output a low-level pulse on OUT for a predetermined length of time. The length of the pulse is obtained by multiplying the CLK period by the initial count value. OUT goes high when the control word is written. The 8254 is triggered by transitions on GATE, and may be retriggered during counting by a high-going pulse on GATE. Triggering the counter causes OUT to go low while the counter decrements to 0.

Example 13.11   Counter 2 is to be programmed to generate a 63-µs pulse when triggered. A 1-MHz clock is connected to CLK. What instructions are needed to use counter 2 as a one-shot?

Solution: The control word for counter 2 must select mode 1, BCD or binary counting, and a load method. Because only 63 CLK pulses will be required for the one-shot pulse, we can get by without having to output a 2-byte count. Only the lower byte value (63 for binary counting, 63H for BCD counting) need be output.

The required instructions are:

```
MOV    DX,0CC83H
MOV    AL,92H
OUT    DX,AL                    ;counter 2, mode 1, binary counting
MOV    DX,0CC82H
MOV    AL,63
OUT    DX,AL                    ;output lower byte of count
```

Mode 2: Rate Generator.   This mode of operation is designed to be *periodic*. Instead of generating a single pulse on OUT, this mode generates a pulse on OUT every N CLK cycle. Thus, the *rate* of output pulses depends on the CLK frequency and the initial count value N. Mode 2 really simulates a modulo-N counter. A 0 on the GATE input can be used to suspend counting.

OUT goes high when the control word is written and remains high until the counter decrements to 1. OUT then goes low for one CLK pulse and the counter is reloaded with the initial count. OUT goes high again at the beginning of the next counting cycle.

Example 13.12: A programmer needs to generate a waveform that goes high once every seven CLK0 cycles. What instructions are required? What does the timing diagram for CLK0 and OUT0 look like? How is the waveform generated?

Solution: Counter 0 must be programmed for mode 2 counting in either binary or BCD. If binary counting is used, these instructions will create the required pulses on OUT0:

```
MOV    DX,0CC83H
MOV    AL,14H
OUT    DX,AL          ;counter 0, mode 2, binary counting
MOV    DX,0CC80H
MOV    AL,7
OUT    DX,AL          ;lower byte of count
```

Mode 2 operation will cause OUT0 to go low once every 7 CLK pulses. An inverter must be used to get the desired output waveform. Figure 13.17 shows the circuit diagram and waveforms for the modulo-7 counter.

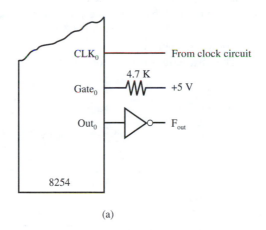

(a)

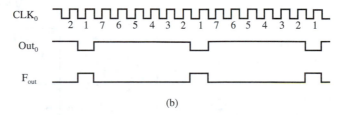

(b)

FIGURE 13.17 Modulo-7 counter: (a) circuit diagram; (b) timing diagram

Because the counters are limited to 16-bit values (either FFFFH or 9999 BCD), what options are available to the programmer who needs to divide by 250,000, or count groups of 10,000 pulses? These kinds of numbers require counters with more bits than those available in the 8254. One simple way to solve this problem is to *cascade two or more counters.* Cascading two counters can result in binary counts of over 4 billion (and BCD counts of up to 100 million). Cascading counters that are operating in different modes can lead to the creation of some interesting and complex waveforms.

Example 13.13:   A 2-MHz clock is available for timing in a system that needs to be interrupted once every 4 seconds. How can two counters be cascaded to obtain this interrupt rate?

Solution: Figure 13.18 shows how counters 0 and 1 are connected to implement the 0.25-Hz interrupt clock. The 2-MHz clock connected to CLK0 will create output pulses on OUT0 when counter 0 is programmed in mode 2. These output pulses serve as the clock for counter 1 (also programmed in mode 2). Dividing 2 MHz by 0.25 Hz gives 8,000,000! This is the count that must be simulated by both counters. Many different counting schemes are possible. One scheme requires that counter 0 be loaded with 50,000 and counter 1 with 160. Note that the product of these two numbers is 8,000,000. Counter 0 will output one pulse for every 50,000 CLK0 pulses. Counter 1 will output one pulse for every 160 CLK1 pulses.

The instructions needed for this interrupt timing circuit are:

```
MOV     DX,0CC83H
MOV     AL,34H
OUT     DX,AL           ;counter 0, mode 2, binary counting
MOV     AL,54H
OUT     DX,AL           ;counter 1, mode 2, binary counting
MOV     DX,0CC80H
MOV     AX,50000
OUT     DX,AL           ;output lower byte of count-0
MOV     AL,AH
```

FIGURE 13.18   Cascading two 8254 counters

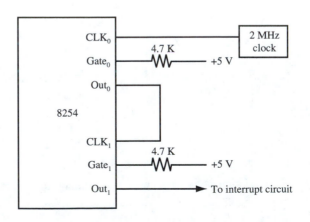

```
OUT     DX,AL           ;output upper byte of count-0
INC     DX              ;point to counter 1
MOV     AL,160
OUT     DX,AL           ;output lower byte of count-1
```

Note the order of the OUT instructions. It is possible to output all control words before sending any counter values. This leads to simpler code and some reuse of registers.

Mode 3: Square Wave Mode.   When a 50 percent duty cycle is required in a timing circuit, mode 3 can be used. The operation of mode 3 is similar to mode 2's rate generation in that it is also periodic. The difference lies in the use of the counter.

When the counter is loaded, the 8254 operating in this mode will decrement it by 2 every CLK pulse. When the counter gets to 0, OUT will change state and the counter will be reloaded. The counter will decrement by 2 again for each CLK pulse. When it reaches 0 a second time, OUT will go back to its original high state. Thus, one complete cycle at OUT requires N clock pulses: N/2 pulses for the first countdown and N/2 pulses for the second. When N is an odd number, OUT will be high for $(N + 1)/2$ CLK pulses and low for $(N - 1)/2$ CLK pulses.

As always, a low on GATE will disable counting.

Example 13.14:   The CLK2 input of the 8254 is connected to a 2.4576-MHz clock (a standard baud-rate generation frequency). OUT2 will be used to drive the transmitter and receiver clock inputs of a UART operating at 2,400 baud with an X16 clock. How must counter 2 be programmed to operate the UART correctly?

Solution: A UART operating at 2,400 baud with an X16 clock requires transmitter and receiver clocks of 38.4 kHz. Dividing 2.4576 MHz by 38.4 kHz gives 64. If counter 2 is programmed for mode 3 with an initial count of 64, the correct frequency will be generated.

The instructions for this application are

```
MOV     DX,0CC83H
MOV     AL,97H
OUT     DX,AL           ;counter 2, mode 3, BCD counting
DEC     DX              ;point to counter 2
MOV     AL,64H
OUT     DX,AL           ;output lower byte of count
```

Mode 4: Software Triggered Strobe.   In this mode of operation, the 8254 generates a low-going pulse on OUT (lasting one CLK pulse) when the counter has decremented to 0. OUT will go low $N + 1$ CLK pulses after the initial count has been written. The extra CLK pulse is needed to load the counter. Only one pulse will be generated on OUT. To get additional pulses, the counter must be reloaded by outputting the initial count value again.

Example 13.15:   A 500-kHz clock is connected to CLK1. The initial count written to counter 1 is 40. How much time expires before OUT1 goes low? How long does OUT1 remain low?

Solution: Counter 1 will be loaded with 40 on the first CLK pulse and decremented to 0 over the next 40 pulses. It will take a total of 41 CLK pulses of time before OUT1 goes low. This corresponds to 82 μs of time. Because OUT1 will remain low for only one CLK period, its duration is 2 μs.

Mode 5: Hardware Triggered Strobe.   This final mode provides the ability for a hardware generated timing pulse. After writing the control word and initial count, OUT will go high. A rising edge on GATE will trigger the 8254 and begin the countdown sequence (after one CLK pulse has been used to load the initial count into the counter). When the counter gets to 0 (after N + 1 CLK pulses), OUT will go low for one CLK pulse. A rising edge on GATE during counting will cause the 8254 to retrigger and begin a new counting sequence (that requires an additional N + 1 pulses to complete).

Example 13.16:   An 8254 will be used to generate a strobe pulse on OUT0 50 CLK cycles after it is triggered by GATE. What instructions are needed to implement this timing need?

Solution: Counter 0 in the 8254 must be programmed for mode 5, and either BCD or binary counting. Because the strobe cannot be issued for 50 CLK cycles, it is necessary to use 49 as the initial count.
    These instructions will program counter 0:

```
MOV   DX,0CC83H
MOV   AL,1AH
OUT   DX,AL        ;counter 0, mode 5, binary counting
MOV   DX,0CC80H
MOV   AL,49
OUT   DX,AL        ;output lower byte of count
```

These six modes of operation provide the programmer (or system designer) with many ways of generating timing pulses, delays, and waveforms. Spend a few moments writing down additional timing applications for the 8254. What modes are needed to implement your applications? What is possible with two or more 8254s?

13.4   THE FLOATING-POINT UNIT (80X87 COPROCESSOR)

Previously we have covered instructions that perform standard mathematical operations, such as multiplication (MUL and IMUL) and division (DIV and IDIV). But the

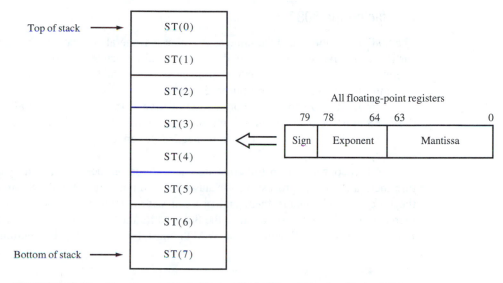

FIGURE 13.19   Floating-point register and stack organization in the FPU

range of values used with these instructions is limited (often to 64-bit integers). The FPU (Floating-Point Unit) improves on this by using 80-bit floating-point numbers, allowing more precision and a huge range of values. A floating-point number, as defined by the IEEE Standards 754 and 854, consists of up to 64 bits of mantissa, a sign bit, and a 15-bit exponent. This 80-bit number does not fit into any of the processor's registers (which are only 16/32 bits wide). For this reason the FPU contains a set of eight internal floating-point registers. The registers are organized into a stack, with ST(0) referring to the register on top of the stack, and ST(7) to the register on the bottom. Figure 13.19 shows this organization. When data is read in from memory, it is *pushed* onto the register stack. Thus, ST(7) becomes ST(6), ST(6) becomes ST(5), and so on, with ST(0) getting replaced by the data from memory. Data written to memory is *popped* off the register stack.

The FPU contains many instructions for manipulating data on the stack. These instructions include the standard add, subtract, multiply, and divide operations (but now with 80-bit precision), as well as logarithms, power functions, square roots, loading useful constants such as pi, and others. The sheer size of the floating-point registers gives the FPU a much higher degree of computational ability over the standard instructions. Computations are performed quickly, because the FPU was specifically designed for floating-point number crunching.

The FPU for the original 8086 and 8088 machines was the 8087. The 80286 and 80386 had their own 80x87 coprocessors as well. Beginning with the 80486, Intel moved the floating-point hardware onto the same chip as the processor, making it internal. There is no need to perform any interfacing anymore. Even so, let us see how the 8087 communicated with the 8088, to get a feel for what goes on between the CPU and the FPU.

Interfacing the 8087

The 8087 is not interfaced the same as the other peripherals we have examined in this (and earlier) chapters. No port-address decoder is used to enable the 8087. Instead, the 8087 operates in parallel with the 8088, sharing access to the address and data buses, and using a handful of signals for control purposes.

Figure 13.20 shows the connections between the 8088 and the 8087. Note that the 8088 must be operated in maximum mode. This allows the 8087 to use $\overline{RQ}/\overline{GT}_0$ to initiate a DMA operation. DMA is needed to enable the 8087 to access system memory.

When coprocessor instructions are encountered on the data bus, the 8087 will capture them and initiate processing. Because they are not part of the 8088's instruction set, the processor will ignore them and proceed with a fetch of the next instruction from memory. If it is necessary to pause the 8088 while the 8087 performs a computation, an FWAIT instruction (floating-point WAIT) must be used. This instruction causes the

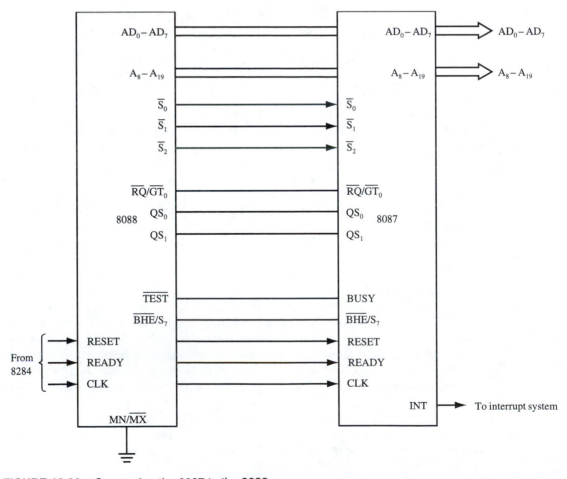

FIGURE 13.20   Connecting the 8087 to the 8088

processor to examine the level of its $\overline{\text{TEST}}$ input. The 8088 will enter a wait state if $\overline{\text{TEST}}$ is high and remain there until $\overline{\text{TEST}}$ goes low. $\overline{\text{TEST}}$ is controlled by the BUSY output of the 8087. If the 8087 is in the middle of a computation, BUSY will be high, which makes $\overline{\text{TEST}}$ high. An FWAIT instruction will then force the processor into a wait state until the 8087 completes execution.

The 8087 may encounter an unexpected result during computation (such as divide by 0, overflow, or precision errors) and generate an interrupt. The INT output must be connected to the system's interrupt circuitry (or directly to INTR if available).

With its shared address and data buses, the 8087 is capable of directly accessing memory to read and write floating-point operands.

FPU Data Types

The FPU implements over 70 different types of floating-point operations on a wide variety of data types. Let us first examine the FPU's data organization, and then proceed to the instruction set.

Seven different data types are available with the FPU. Three of these deal exclusively with positive and negative integers. An integer may be coded as as a *word* integer, a *short* integer, or a *long* integer. Word integers are 16 bits long, with the MSB acting as a sign bit. Two's complement notation is used for storage of negative integers.

Short integers are 32 bits long (with the MSB as sign bit) and long integers occupy 64 bits. All bits except the MSB are used to represent the magnitude of the integer. This leads to three different ranges of integers. Figure 13.21 shows the format of the three integer data types, and their ranges. Integers can be defined within a source file by using certain assembler directives. The assembler will code numbers in the operand field into the format used by the FPU.

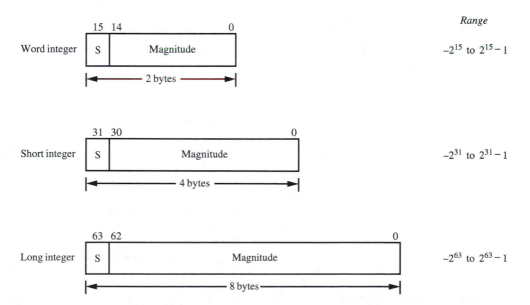

FIGURE 13.21   Integer formats in the FPU

Example 13.17:   The six source lines shown here indicate how the assembler converts decimal numbers into the three integer formats used by the FPU

```
03 E8                          WORDINT   DW    1000
FC 18                                    DW    -1000
00 6A CF C0                    SHORTINT  DD    7000000
FF 95 30 40                              DD    -7000000
00 00 00 6A DE 88 2E 00        LONGINT   DQ    459000000000
FE FE FE 94 20 76 D1 00                  DQ    -459000000000
```

The DW directive is used for word integers, and generates 2 bytes. The DD (define double) and DQ (define quad) directives generate 4 and 8 bytes, respectively. Note the change in the first byte for each group of numbers when the "–" sign is used.

The next data type is *packed BCD*. In this format 10 bytes are used to represent a BCD number. Nine bytes are used to store the magnitude, with each byte holding two BCD digits. The last byte contains only the sign bit. Figure 13.22 shows the organization of this data type. The packed BCD number really represents an integer, with a range roughly equivalent to that of the long integer format.

Example 13.18:   The packed decimal numbers shown here were created by the assembler in response to the DT (define tens) directive.

```
00   00 00 00 00 12 34 56 78 90        PACKBCD   DT    1234567890
80   00 00 00 00 12 34 56 78 90                  DT    -1234567890
```

The effect of the minus sign in the second number is easy to spot by examining the most significant byte of both numbers. Also, the packed decimal numbers are right justified within the 10-byte data area. Familiarity with the format of any data type aids in interpreting its value.

Three data types are devoted to the use of *real* numbers. *Short* real format (single-precision) uses a 32-bit data block composed of 23 magnitude bits, 8 exponent bits, and 1 sign bit. *Long* real numbers (double-precision) use a 64-bit format composed of 52 magni-

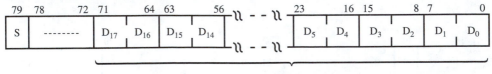

An 18-digit integer

FIGURE 13.22   Packed BCD format

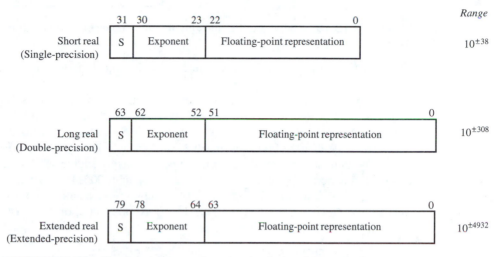

FIGURE 13.23 Short real, long real, and extended real formats

tude bits, 11 exponent bits, and 1 sign bit. Extended-precision reals have 15 bits of exponent and 64 magnitude bits. These formats are shown in Figure 13.23. A brief introduction to floating-point numbers is needed to fully understand these new formats.

Figure 13.24 gives a general equation for the definition and representation of a short real binary number. The first part of the equation generates the sign of the number. A positive number is generated when the sign bit is 0. Negative numbers require the sign bit to be set.

The exponent part of the number is used as a multiplier for the fractional part. The E-127 term moves the decimal point in the fraction to the left (making the number smaller) or to the right (making the number larger) a certain number of places. The E variable is controlled by the 8 exponent bits, which are interpreted as an unsigned number in the range 0 to 255. This gives an exponent range for the short real format of −126 to +128.

The fraction represents the *normalized* binary equivalent of the decimal number. Normalization is a process that converts any binary number into standard format by adjusting the exponent of the number until the mantissa is in the form 1.F. For example, 1010.1111 becomes 1.0101111 with a power-of-2 exponent of 3. Similarly, 0.000011011 becomes 1.1011 with an exponent of −5. Because every number will begin with 1, it is only necessary to store the F part of the number. In the short real format, F is 23 bits wide. Thus, a 24-bit number is actually represented via normalization.

FIGURE 13.24 Equation for short real format representation

$$N = -1^S \times 2^{E-127} \times 1.F$$

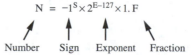

Number   Sign   Exponent   Fraction

Example 13.19:   What are the steps involved in converting 209.8125 into the normalized short real format?

Solution: The first step is to convert 209.8125 into binary. This gives a result of 11010001.1101.

Next, the binary result is normalized by shifting the decimal point seven places to the left. This gives a normalized result of 1.10100011101, with an exponent of +7. Because 23 bits are used to represent the fraction, we end up with a fraction of 10100011101000000000000. Note that the leading 1 (which is always there) is not stored.

The third step creates the required exponent bits by adding 127 to the exponent obtained during normalization (+7). The result is 134, or 10000110.

The final step clears the sign bit so that it represents a positive number. Putting all 32 bits together gives 01000011 01010001 11010000 00000000, or 43 51 D0 00. Checking our conversion with the assembler gives:

```
43 51 D0 00   SHORTREAL   REAL4   209.8125
```

The **REAL4** directive is used to encode a short real number in the format used by the FPU.

The same technique that was used in Example 13.19 will work on long real and extended real formats, the only differences being the number of bits used to represent the exponent (11 or 15) and the fraction (52 or 64). Exponents are generated by adding 1,023 or 16,383 to the exponent found during normalization.

A number stored in real format can be converted back into decimal by reversing the steps outlined in Example 13.19. This can be done without much difficulty in software, and you are encouraged to think about how the conversion could be accomplished. Example 13.20 may give you some ideas.

Example 13.20:   Convert the REAL4 C5 5A 57 00 into its decimal equivalent.

Solution: In binary, we have 11000101 01011010 01010111 00000000.

Grouping the data bits into sign, exponent, and fraction bits gives 1 10001010 10110100101011100000000. The 1 in the sign bit indicates a negative result. The 8-bit pattern in the exponent evaluates to 138. To get the exponent of the normalized fraction, we subtract 127. This gives a normalized exponent of 11. The decimal point in the fraction must be moved eleven places to the right.

The fraction (with the leading 1 added) is 1.10110100101011100000000. Multiplying by the normalized exponent results in 110110100101.0111 (with trailing 0s left off). Converting the integer and fractional parts into decimal (and multiplying by –1) gives –3493.4375.

None of the above steps require an extensive amount of software, with rotate or shift instructions most likely implementing the conversion within the fraction.

Assembler directives REAL8 and REAL10 are used for long real and extended real numbers, respectively.

The instruction set of the FPU consists of a number of functional groups. These groups are comprised of the data transfer, arithmetic, compare, transcendental, constant, and processor control instructions. Each group will be introduced briefly.

FPU Instructions

A wide variety of instructions is available with the FPU. Let us take a brief look at the different groups of instructions used by the FPU.

Data Transfer Instructions. These instructions are used to move data around inside the FPU, and between the FPU and system memory. Source and destination operands are not required on many of the FPU's instructions. When no operands are explicitly included in an instruction, the FPU will use the number (or numbers) located on the top of its floating-point stack. Results are written back onto the stack or into memory. Calculations involving repeated loops are best written in a way that uses the stack to avoid endless delays due to the access time of the system memory.

Source and destination operands are indicated by <src> and <dst>, respectively.

The data transfer instructions are:

```
FLD     <src>       ;load real
FILD    <src>       ;load integer
FBLD    <src>       ;load BCD
FST     <dst>       ;store real
FIST    <dst>       ;store integer
FBSTP   <dst>       ;store BCD and pop
FSTP    <dst>       ;store real and pop
FISTP   <dst>       ;store integer and pop
FXCH    <dst>       ;exchange registers
```

Note that all floating-point instructions begin with F. This helps to differentiate them from ordinary processor instructions.

Arithmetic Instructions. The arithmetic instructions are designed to perform a wide variety of computations on all of the data types supported by the FPU. The hardware performing the operations produces results much faster than an ordinary program could.

The arithmetic instructions are:

```
FADD    <dst>,<src>     ;add real
FADDP   <dst>,<src>     ;add real and pop
FIADD   <src>           ;add integer
FSUB    <dst>,<src>     ;subtract real
FSUBP   <dst>,<src>     ;subtract real and pop
FSUBR   <dst>,<src>     ;subtract real reversed
FSUBRP  <dst>,<src>     ;subtract real reversed and pop
FISUB   <src>           ;subtract integer
FISUBR  <src>           ;subtract integer reversed
FMUL    <dst>,<src>     ;multiply real
FMULP   <dst>,<src>     ;multiply real and pop
FIMUL   <src>           ;multiply integer
FDIV    <dst>,<src>     ;divide real
FDIVP   <dst>,<src>     ;divide real and pop
FDIVR   <dst>,<src>     ;divide real reversed
```

```
FDIVRP    <dst>,<src>      ;divide real reversed and pop
FIDIV     <src>            ;divide integer
FIDIVR    <src>            ;divide integer reversed
FABS                       ;absolute value
FCHS                       ;change sign
FPREM                      ;partial remainder
FPREM1                     ;IEEE standard partial remainder
FRNDINT                    ;round to integer
FSCALE                     ;scale
FSQRT                      ;square root
FXTRACT                    ;extract exponent and fraction
```

Some instructions (like FSQRT and FABS) require no operands, and will perform their indicated operation on the number stored on top of the stack. The programming applications that follow will use some of these mathematical operations and show their various operand forms.

Compare Instructions.   It is useful during a computation to be able to check the results before continuing on to a new set of calculations. The compare instructions allow the programmer to examine the value of the number stored on top of the stack. The compare instructions affect flags stored in the FPU's *status register*. The status register is 16 bits wide, and contains a number of flags and status indicators as indicated in Figure 13.25. One bit indicates that division by 0 has occurred. Another is set when an exponent overflow is detected. A programmer can look for these conditions by testing specific bits within the status register.

The compare instructions are:

```
FCOM     <src>       ;compare real
FCOMP    <src>       ;compare real and pop
FCOMPP               ;compare real and pop twice
```

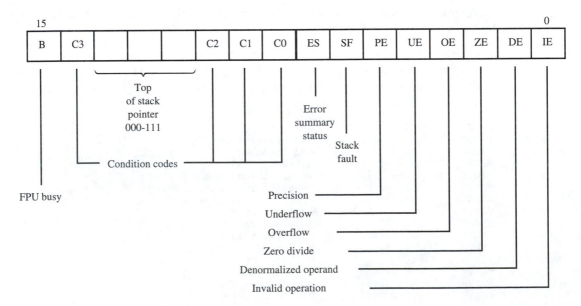

FIGURE 13.25   FPU status word

```
FICOM       <src>       ;compare integer
FICOMP      <src>       ;compare integer and pop
FTST                    ;test top of stack (compare with 0.0)
FUCOM                   ;unordered compare real
FUCOMP                  ;unordered compare real and pop
FUCOMPP                 ;unordered compare real and pop twice
FXAM                    ;examine top of stack
```

Transcendental Instructions.   The transcendental instructions use numbers stored on top of the stack (the first and second stack elements) and write their results back onto the stack. These instructions can be combined with the arithmetic instructions to form other functions that are not implemented on the FPU.

The transcendental instructions are:

```
FSIN        ;calculate sine
FCOS        ;calculate cosine
FSINCOS     ;calculate sine and cosine
FPATAN      ;partial arctangent
FPTAN       ;partial tangent
F2XM1       ;calculate 2^X - 1
FYL2X       ;calculate Y * log2(X)
FYL2XP1     ;calculate Y * log2(X + 1)
```

Constant Instructions.   The constant instructions are used to load the top of the stack with a commonly used number. This helps to speed up execution of routines that would otherwise have to store the constants in memory. All constants are stored with the extended real format.

The constant instructions are:

```
FLDZ        ;load 0.0
FLD1        ;load 1.0
FLDPI       ;load PI (3.14159...)
FLDL2E      ;load log2(E) where E = 2.7182818...
FLDL2T      ;load log2(10)
FLDLG2      ;load log10(2)
FLDLN2      ;load logE(2)
```

Note that some constants are useful for conversion between base-10 and base-2, and also between base-E and base-2.

Processor Control Instructions.   These instructions are useful for controlling the operation of the FPU. Access to the status and control register is provided, and programmer control of error flags and the FPU's internal environment is also available.

The processor control instructions are:

```
FINIT                   ;reset the FPU
FCLEX                   ;clear error flags
FINCSTP                 ;increment stack pointer
FDECSTP                 ;decrement stack pointer
FSTSW       <dst>       ;store status register
FSTSW       AX          ;store status register to AX
FSTCW       <dst>       ;store control register
FLDCW       <src>       ;load control register
FSTENV      <dst>       ;store environment
FLDENV      <src>       ;load environment
FSAVE       <dst>       ;save state
FRSTOR      <src>       ;restore state
```

```
FFREE      <dst>        ;free register
FNOP                    ;no operation
FWAIT                   ;report FPU error
```

FPU Programming Examples

The brief introduction to the FPU's instruction set should not prevent us from examining some simple programming applications. The examples that follow will show the usage and form of a number of different FPU instructions. Remember that these floating-point instructions would need to be implemented in software, through time-consuming subroutines, if the FPU was not available.

Example 13.21:   The FPU instructions shown here convert the floating-point Celsius temperature value stored in CEL into its corresponding Fahrenheit equivalent. The result is saved in FAHR.

```
In current data segment. . .
CEL        REAL4        100.0
C5         REAL4          5.0
C9         REAL4          9.0
C32        REAL4         32.0
FAHR       REAL4          ?
.
.
.
FINIT
FLD    CEL        ;load Celsius temperature
FMUL   C9         ;multiply by 9
FDIV   C5         ;divide by 5
FADD   C32        ;add 32
FST    FAHR       ;save Fahrenheit result
FWAIT
```

The FMUL, FDIV, and FADD instructions have only one operand specified. This means that the FPU will use the floating-point number on top of its internal stack as the second operand in each instruction. Let us take another look at how the floating-point stack is used in a computation.

Example 13.22:   The following FPU instructions compute the length of the hypotenuse of a right triangle.

```
In current data segment...
SIDEA    REAL4      3
SIDEB    REAL4      4
SIDEC    REAL4      ?
.
.
.
FINIT
FLD      SIDEA         ;load side A length
```

```
FMUL    SIDEA       ;square it
FLD     SIDEB       ;load side B length
FMUL    SIDEB       ;square it
FADD                ;add the squares together
FSQRT               ;find the square root
FST     SIDEC       ;store length of side C
FWAIT
```

The floating-point stack is used to save the partial results obtained when the individual side lengths are squared. By the time the FADD instruction executes, the top two numbers on the stack are the squared side lengths. FADD pops both of these numbers, adds them, and pushes the result back onto the stack (for FSQRT to use). Using the floating-point stack instead of external memory locations results in much faster execution.

The next example shows how one of the FPU's built-in constants can be used in a calculation.

Example 13.23:   The area of a circle is equal to πR^2, where π (pi) is roughly 3.14159265, and R is the radius of the circle. The coprocessor routine shown here implements the area formula using the coprocessor's internal **pi** constant.

```
In current data segment. . .
RADIUS    REAL4     5.6
AREA      REAL4     ?
   .
   .
   .
FINIT
FLD     RADIUS      ;load R
FMUL    RADIUS      ;compute R squared
FLDPI               ;load pi constant
FMUL                ;compute area
FST     AREA        ;save result
FWAIT
```

One common limitation of the example programs just examined is that we have no way of viewing the result of each routine because the floating-point result is saved in the standard IEEE format.

One way to view floating-point numbers is to use a program like CodeView, which performs the necessary conversions and displays the floating-point result in a debugging window.

CodeView's MD (Memory Display) and VM (View Memory) commands allow you to specify a display format that controls how a number read from memory is displayed. For example, to display a REAL4 floating-point value, we enter MDR <address>, where address is the memory address to begin displaying data. Let us consider the routine of Example 13.21. Assume the data area begins at offset 000C in the data segment. The VMR 0x000C command will display the floating-point numbers in scientific notation, beginning at offset 000C. The output looks like this:

```
214B:000C   +1.0000000E+002
214B:0010   +5.0000000E+000
214B:0014   +9.0000000E+000
214B:0018   +3.2000000E+001
214B:001C   +2.1200000E+002
```

The result of 212 (at offset 001C) indicates that the program worked correctly.

An option within the VM command can be used to add the associated hexadecimal data for each floating-point number to the display. The new command is VMR 0x000C /R+ and has the following format:

```
214B:000C   00 00 C8 42   +1.0000000E+002
214B:0010   00 00 A0 40   +5.0000000E+000
214B:0014   00 00 10 41   +9.0000000E+000
214B:0018   00 00 00 42   +3.2000000E+001
214B:001C   00 00 54 43   +2.1200000E+002
```

This feature of CodeView is very handy, but we still do not have a way for the program to display a floating-point number.

Another way to view these numbers is to use an FPU procedure to perform the conversion. This procedure is covered next.

SQRT: An FPU Application

The SQRT program contains a set of procedures capable of converting an FPU floating-point number into a 12-digit real number that is displayed on the screen. The floating-point number is converted through repetitive division into its individual digit's values. For example, 371.52 is converted in the following manner:

Divide	Round	Multiply	Subtract
371.52/100 = 3.7152	3	3×100	371.52 − 300
71.52/10 = 7.152	7	7×10	71.52 − 70
1.52/1 = 1.52	1	1×1	1.52 − 1
0.52/0.1 = 5.2	5	5×0.1	0.52 − 0.5
0.02/0.01 = 2.0	2	2×0.01	0.02 − 0.02

The FPCON procedure converts six digits at a time, based on the floating-point number stored in NUM. The first time FPCON is called, the initial divisor value is 100,000. So, numbers as large as 999,999 can be converted with FPCON without making any changes to the program. To convert different-size numbers, the values of DIVI and POW must be changed, as well as the storage requirements for DIGITS.

The DPD procedure converts the integer-based floating-point result values saved in DIGITS into their corresponding ASCII decimal counterparts '0' through '9'.

```
;Program SQRT.ASM: Display floating-point square root using FPU.
;
        .MODEL SMALL
        .586
        .DATA
X       DW      200             ;integer input number
FPX     REAL4   200.0           ;floating point input number
TAI     DB      'The square root of 200 is $'
CRLF    DB      13,10,'$'
DIGITS  REAL4   6 DUP(0.0)      ;leftmost digit storage
```

```
        REAL4   6 DUP(0.0)      ;rightmost digit storage
DIVI    REAL4   1.0E5           ;initial divisor
POW     DW      6               ;number of digits to display
NUM     REAL4   ?               ;FP working number
TEN     REAL4   10.0            ;used in conversion
CW      DW      ?               ;FPU control word

        .CODE
        .STARTUP
        LEA     DX,TAI          ;set up pointer to square root message
        MOV     AH,9            ;display string function
        INT     21H             ;DOS call
        FINIT                   ;initialize FPU
        FLD     FPX             ;load floating-point x value
        FSQRT                   ;calculate square root
        FST     NUM             ;save result
        FWAIT
        CALL    FPDISP          ;display floating point result
        LEA     DX,CRLF         ;set up pointer to newline string
        MOV     AH,9            ;display string function
        INT     21H             ;DOS call
        .EXIT

FPDISP  PROC NEAR
        SUB     SI,SI           ;clear index pointer
        CALL    FPCON           ;convert leftmost digits
        CALL    FPCON           ;convert rightmost digits
        LEA     SI,DIGITS       ;set up pointer to digits
        CALL    DPD             ;display leftmost digits
        MOV     DL,'.'          ;load decimal point character
        MOV     AH,2            ;display character function
        INT     21H             ;DOS call
        CALL    DPD             ;display rightmost digits
        RET
FPDISP  ENDP

FPCON   PROC NEAR
        MOV     CX,POW          ;set up loop counter
DODIG:  FINIT                   ;initialize FPU
        FNSTCW  CW              ;store control word
        OR      CW,0C00H        ;set rounding control to truncate
        FLDCW   CW              ;load control word
        FLD     NUM             ;get current value of number
        FDIV    DIVI            ;divide by multiple of 10
        FRNDINT                 ;round result to integer
        FST     DIGITS[SI]      ;save result in digit buffer
        FLD     DIVI            ;load divisor
        FMUL    DIGITS[SI]      ;multiply by integer result
        FLD     NUM             ;load current value of number
        FSUBR                   ;calculate remainder
        FST     NUM             ;save new value of number
        FLD     DIVI            ;divide divisor by 10
        FDIV    TEN
        FST     DIVI
        FWAIT
        ADD     SI,4            ;advance to next digit
        LOOP    DODIG           ;and repeat
        RET
FPCON   ENDP

DPD     PROC    NEAR
        MOV     CX,POW          ;set up loop counter
OP:     MOV     AH,[SI+3]       ;load exponent information
```

```
          MOV     AL,[SI+2]
          SHL     AX,1                 ;calculate exponent
          SUB     AH,127
          MOV     DL,[SI+2]            ;load initial bits of number
          OR      DL,80H               ;always begin with 1 as MSB
          SUB     DH,DH                ;clear DH
SDL:      SHL     DX,1                 ;double number
          CMP     AH,0                 ;is exponent zero?
          JZ      OP2                  ;if yes, go display digit
          DEC     AH                   ;repeat until exponent equals 0
          JMP     SDL
OP2:      MOV     DL,DH                ;load digit value
          ADD     DL,30H               ;add ASCII bias
          MOV     AH,2                 ;display character function
          INT     21H                  ;DOS call
          ADD     SI,4                 ;advance pointer to next FP number
          LOOP    OP                   ;and repeat
          RET
DPD       ENDP

          END
```

The execution of SQRT results in this output

```
The square root of 200 is 000014.142135
```

Note that leading 0's are not eliminated by the DPD procedure. You may wish to experiment with larger numbers to investigate the accuracy of the conversion process.

Programming Exercise 13.1: Write a program that will display the temperature found with the FPU code from Example 13.21. Use the FPCON procedure from SQRT.ASM to display the result.

Programming Exercise 13.2: Repeat Programming Exercise 13.1 for the other FPU routines covered in Examples 13.22 and 13.23.

Programming Exercise 13.3: Modify the SQRT program so that leading 0's are not displayed.

13.5   TROUBLESHOOTING TECHNIQUES

The peripherals in this chapter are suitably advanced. Although they cover a wide variety of topics, there are many other, even more specialized, peripherals. When you come across a new device and are faced with the challenging task of getting it to work, keep these suggestions in mind:

•   Look over the data manual on the new peripheral. If you do not have a data manual, try searching the Web. Intel has a very useful site, offering downloads of manuals

(80x86 series plus many others) in PDF format. Many other educational institutions post important information on the Web also, as part of class projects.

Skim the figures and captions, look at the register and bit assignments, and read the tables. Read about the hardware signals. Study the timing diagrams. Look at any sample interface designs provided by the manufacturer. Be sure you understand why the signals are used the way they are.

Read about the software architecture of the peripheral. How is it controlled? How do you send data to it, or read data from it? How many different functions does it perform?

- Get the hardware interface working properly. This requires your skill in designing I/O hardware. Some software may be required to fully test the interface.
- If the peripheral has many modes of operation, begin with the simplest. Program the peripheral to operate in this mode to be sure you have control over it. Expand to other modes of operation as you learn more about the device.

Even if all you are doing is modifying someone else's code, written long ago, for a peripheral that is already operational, it is still good to learn as much about the device as possible. This will help you avoid typical problems, such as forgetting to issue the master reset command, even though power was just applied.

SUMMARY

In this chapter we examined the operation of three peripherals designed to complement the operation of the processor. The first peripheral, the 8259 programmable interrupt controller, exhibited many useful features. A range of interrupt vectors can be programmed and then issued via level- or edge-sensitive inputs that are prioritized. Eight levels of prioritized interrupts are available with a single 8259 operating as a master. The 8259 can be cascaded to provide up to 64 levels of interrupts.

The second peripheral examined was the 8254 programmable interval timer. This device contains three independent 16-bit down counters that can be programmed to count in binary or BCD. Six modes of operation are possible, allowing generation of square and pulse waveforms, programmed time delays, and other time-related functions.

Finally, we covered the 80x87 floating-point coprocessor, a device that extends the instruction set of the processor to include operations on a number of different data types, from 32-bit integers to 80-bit real numbers. The 80x87 provides quick execution of complex mathematical operations, eliminating the need for slower—and possibly less accurate—software routines.

STUDY QUESTIONS

1. What are the main differences between polled I/O and interrupt-driven I/O?
2. Redesign the port-address decoder of Figure 13.2 so that the 8259 is mapped to a base port address of 70H. What are all of the port addresses the 8259 will respond to with the new design?

3. What ICWs are needed to program a single 8259 with edge-sensitive inputs and a base interrupt vector of C0H? (ICW4 is not needed.)

4. How many slaves are required if ICW3 in a master 8259 contains 01100101? What devices are connected to each IR input?

5. Draw a schematic of the cascaded 8259 circuit needed in Question 4. Label all interrupt signals in order from lowest to highest priority.

6. A master 8259 has slaves connected to IR2 and IR5. What are the cascade numbers that must be written to each slave?

7. Write the instructions needed to program the 8259 described in Question 2 with the parameters of Question 3.

8. What OCW1 is needed to disable interrupts on IR3 through IR6?

9. A new computer system has five devices that generate interrupts. The devices, in order of highest-to-lowest priority, generate the following signals: DISK, KEYBOARD, TIMER, VIDEO, and IODEV. Show how these signals could be connected to an 8259. What exactly happens if DISK, KEYBOARD, and IODEV interrupts are requested simultaneously?

10. Why does a delay loop become inaccurate if the processor is interrupted? How is this avoided by using the 8254?

11. What instructions are needed to program counter 2 in an 8254 for binary counting in mode 0? The initial count is 3000H. Assume the base port address is B0H.

12. What two ways can be used to load a counter with an initial count of 9F00H?

13. Write a routine that will latch the count of counter 0, store it in register DX, and call TIMEOUT if the count is less than 7.

14. What mode-0 counter value is needed to get a 25-ms time delay? A 1-MHz clock is connected to CLK.

15. What is the longest delay possible with a BCD counter and a 250-kHz clock?

16. Repeat Question 15 for a binary counter.

17. Show how counters 0 and 2 can be cascaded to provide 32-bit counting.

18. Refer to Example 13.13. Name three additional pairs of counts that will produce the 8,000,000 count division.

19. Write a routine to generate a square wave whose frequency, in kiloherts, is specified in BL. BL can take on the values 1 to 99. A 2-MHz clock is available for your use. The 8254 has a base port address of 38H.

20. What are the decimal ranges for each of the FPU's integer formats?

21. Show how the number 1,020.6 is converted into a normalized, short real format.

22. What decimal number is represented by these four assembler-created short real bytes: C3 E5 A9 2B?

23. Show the FPU instructions needed to find the square root of the number stored at LEVEL. Replace LEVEL with the new value.

24. Show how the 80x87 can be used to convert 60 miles/hour into feet/second.

25. Write an 80x87 routine to convert degrees to radians. The input is stored in DEG. Save the result in RAD.

26. Write an 80x87 routine to compute the area of a circle whose diameter is stored in DIAMETER. Save the result in AREA.

27. Name five applications that could make good use of the speed advantage provided by the FPU.

28. Write a routine that will convert a short real number stored at RESULT into an integer part and a fractional part. For example, 80.29 is split into 80 and 0.29. Store the integer part in register AX and the fractional part in BX. Do not use any FPU instructions in your routine.

29. Write the 80x87 instructions needed to evaluate the expression X = A^2 – 5*A*B + B^2.

30. What expression is evaluated by the following section of FPU code?

```
FINIT
FLD    W
FDIV   X
FCHS
FMUL   Y
FLD1
FADD
FSQRT
FST    Z
```

31. Write an 80x87 routine that computes the equivalent resistance of the series-parallel circuit shown in Figure 13.26.

32. Write the FPU code needed to implement the charge equation shown in Figure 13.27. If E = 100 V, find V_C at a time t = 2 ms.

FIGURE 13.26   For Question 31

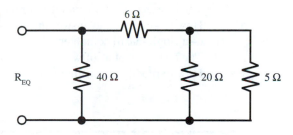

FIGURE 13.27   For Question 32

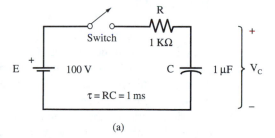

(a)

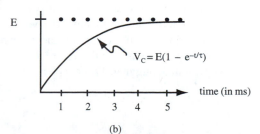

(b)

CHAPTER 14

Building a Working 8088 System

OBJECTIVES

In this chapter you will learn about:

- The main parts of a single-board computer
- The design of custom circuitry for the major sections of the microcomputer system
- How to generate and answer the necessary questions for the design or modification of a single-board computer
- The operation of a software monitor program
- How to modify an existing monitor program by writing additional routines

14.1   INTRODUCTION

This chapter deals exclusively with the design of a custom 8088-based microcomputer system. The system is an ideal project for students wishing to get some hands-on experience, and is also a very educational way of using all of the concepts we have studied so far. In addition, even though we are not designing a stand-alone Pentium system, the act of creating a working microprocessor system is still very rewarding. Do not forget that many projects do not require the horsepower of the Pentium. An 8088 will do a fine job controlling a robot arm, for example, whereas a Pentium would be overkill.

Ideally, we wish to design a system that is easy to build, has a minimal cost, and yet gives the most for the money. The very least we expect the system to do is execute programs written in 8088 code. It is therefore necessary to have some kind of software monitor that will provide us with the ability to enter 8088 code into memory, execute programs, and even aid in debugging. This chapter, then, will consist of two parts. The first part deals with the design of the minimal system, and the second part with the design of a software monitor and the use of its commands.

Pay close attention to the trade-offs that we will be making during the design process. A difficult hardware task can often be performed by cleverly written machine code, and the same goes for the reverse. Do not forget our main goal: to design a *minimal* 8088-based system suitable for custom programming.

Section 14.2 covers the minimum requirements of the system we will design. Section 14.3 describes the design of the system hardware. Section 14.4 contains the parts list for the system. Section 14.5 gives hints on how the system may be constructed. Section 14.6 deals with the design of the software monitor program for the system. Section 14.7 explains a sample session with the single-board computer. Troubleshooting hints are provided in Section 14.8.

14.2   MINIMAL SYSTEM REQUIREMENTS

The requirements of our minimal system are the same as those of any computer system, and consist of four main sections: timing, CPU, memory, and I/O. Because we are the designers building this system for our personal use, it is up to us to answer the following questions:

1. How fast should the CPU clock speed be?
2. How much EPROM memory is needed?
3. How much RAM memory is needed?
4. Should we use static or dynamic RAMs?
5. What kind of I/O should be used—parallel, serial, or both?
6. Do we want interrupt capability?
7. Will future expansion (of memory, I/O, etc.) be required?
8. What kind of software is required?

It should be clear that we have a big task ahead of us. During the design, all of these questions will be answered and the reasons for choosing one answer over another explained. Make sure you understand each step before proceeding to the next one. In this fashion, you should be able to design your *own* computer system, from scratch, and without any outside help.

14.3   DESIGNING THE HARDWARE

In this section, the four main functional components of the system will be designed. In each case, there will be questions to answer regarding specific choices that must be made. You may want to make a list of all important questions as you go.

The Timing Section

The timing section has the main responsibility of providing the CPU with a nicely functioning stable clock. Any type of digital oscillator will work in many cases. It is then necessary to decide on a frequency for the oscillator. Many times, this frequency is the operating frequency of the CPU being used. Microprocessors are commonly available with different clock speeds.

One important factor limiting the clock speed is the speed of the memories being used in the system. A 12-MHz CPU might require RAMs or EPROMs with access times less than 100 ns! In our design, we will use a 10-MHz crystal, together with the 8284 clock generator. This is fast enough to provide very quick execution of programs, while at the same time allowing for use of less expensive RAMs with longer access times (200 ns).

The circuit of Figure 14.1 shows a 10-MHz crystal connected to the 8284 clock generator. The output of the 8284 drives two buffers so that any external loading on the CLK

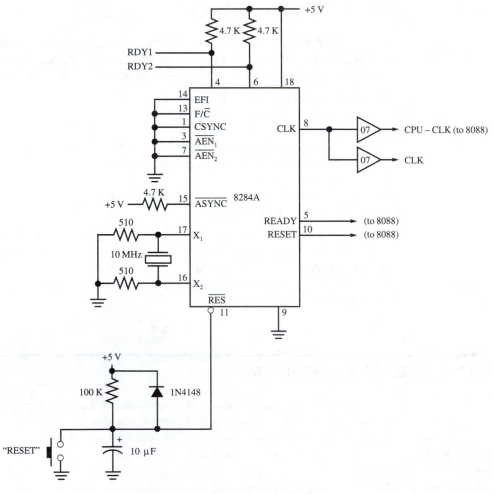

FIGURE 14.1   Clock generator for single-board computer

signal will not affect its operation. One of the outputs, CPU-CLK, is the master CPU clock signal. Because many other circuits might also require the use of this master clock, we make the CLK signal available too. The CPU therefore gets its own clock signal. It is desirable to separate the clock in this fashion to aid in any digital troubleshooting that may need to be done. By making multiple clocks available, it is easier to trace the cause of a missing clock, should that problem occur.

In addition to the clock, the CPU must be provided with a reset pulse upon application of power. It is very important to properly reset the CPU at power-up to ensure that it begins executing its main program correctly. The 8284 has built-in reset circuitry that uses an external R-C network to generate the power-on reset pulse. The values shown in Figure 14.1 (100K ohms and 10 μF) produce a reset pulse of about 1 ms in duration, long enough to satisfy the hardware reset requirement of the CPU and other system devices.

The CPU Section

Once we have a working timing section, we must design the CPU portion of our system. During the design of this section we answer our question about interrupts, and pose a few more important questions. For instance, do we need to buffer the address and data lines? Do we want to give bus-granting capability to an external device? Should the system operate in maxmode or minmode? Take a good look at Figure 14.2 before continuing with the reading.

The figure details the connections we must make to the CPU for it to function in our minimal system. On the right side of the CPU we see the data and address lines. These signals are used in both the memory and I/O sections. The CPU is capable of driving only a few devices by itself (one RAM and one EPROM safely). Because the system we are designing will contain RAM, EPROM, and serial *and* parallel I/O, it is best to buffer the address and data lines. As Figure 14.2 shows, a 74LS244 octal buffer is used to drive address lines A_8 through A_{15}. Address lines A_0 through A_7 are multiplexed together with the eight data lines, requiring an 8282 octal latch (together with the ALE signal from the 8288) to demultiplex and drive the lower byte of the address bus. These sixteen address lines will allow for 64KB of system memory in our design.

The upper four address lines are not used in this system, because we will not be expanding the system memory requirements past 64KB.

The data bus is buffered in two directions by the 8286 bidirectional line driver/receiver. The direction of data in this device is controlled by the DT/$\overline{R}$ output of the 8288.

The schematic of the CPU section shows the 8088's MN/$\overline{MX}$ pin wired to ground. This selects maximum mode operation within the CPU and requires that we use the 8288 bus controller to generate memory and I/O control signals. Do not confuse the description of our system (a *minimal* system) with its mode of operation (*maximum* mode).

The decision to operate the processor in minmode or maxmode depends on a number of factors. If low chip count is necessary, then minmode can be used and the 8288 eliminated. If coprocessor support will be needed in a future expansion of the system, it is best to operate in maxmode from the beginning. The 8288 may also eliminate the need for additional decoding logic in a minimum mode system. Furthermore, bus-granting capability is available only in maxmode. Because no devices in the minimal system use this feature, both $\overline{RQ}/\overline{GT}$ inputs are pulled high. The pullup resistors will not prevent us from connecting a DMA device to either input at a future time.

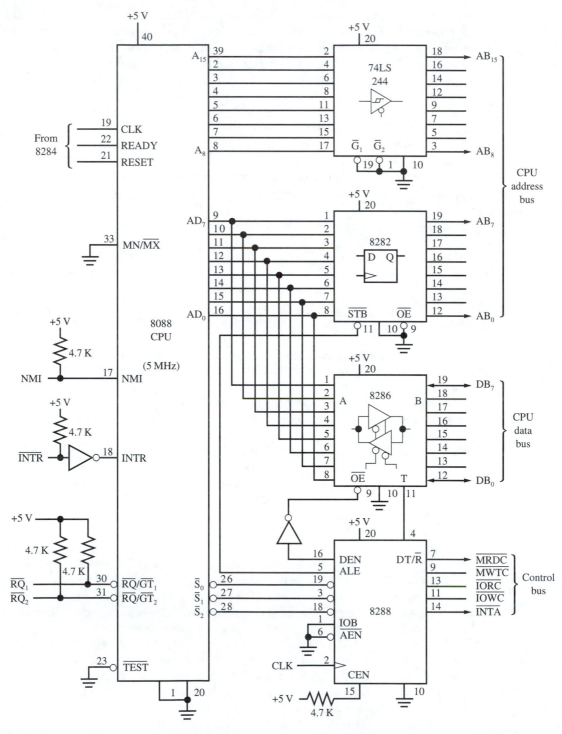

FIGURE 14.2   CPU section of single-board computer

Because our goal is to design a system with a minimum of hardware, extensive interrupt support logic will not be necessary. The processor's two external hardware interrupt inputs should serve our needs adequately.

An inverter is used to make the 8088's high-level INTR interrupt respond to a low-level signal. This technique keeps the INTR input in the inactive state if no devices are connected to $\overline{\text{INTR}}$. NMI is also pulled up to a high level. Remember that NMI is edge-sensitive. When no interrupting device is connected, NMI will remain in the high state, and no interrupt will be requested. If we connect a device to NMI in the future, the pullup resistor will not have an adverse effect on any rising-edge NMI signal that is generated.

Technically, although there are five integrated circuits in the CPU section, a *barebones* system could get by with only the 8088 running in minmode. But this would most likely require the addition of hardware in the future (to drive the buses and/or possibly switch to maxmode). Any unexpected expansion of hardware is a costly, and sometimes impossible, venture. So, although the minimal system already contains a handful of integrated circuits, choosing maxmode for our project leaves the door open for easy expansion in the future.

The Memory Section

A number of questions must be answered before we get involved in the design of our memory section. For instance, how much EPROM memory is needed? How much RAM? Should we use static or dynamic RAM? Should we use full or partial address decoding? Will we allow DMA operations?

The answer to each of these questions will help specify the required hardware for the memory section. If we first consider what *applications* we will be using, the previous questions will almost answer themselves. Our application at this time is educational. We desire an 8088-based system that will run short-machine language programs. Keeping this point in mind, we will now proceed to find answers to our design questions.

A programmer, through experience, can estimate the required amount of machine code needed to perform a desired task. The software monitor that we will need to control our system will have to be placed in the EPROMs of our memory section. One standard 2764 EPROM will provide us with 8,192 bytes of programmable memory. This is more than enough EPROM to implement our software monitor. We will still have space left over in the EPROM in case we want to add more functions to the monitor in the future.

The amount of RAM required also depends on our application. Because we will be using our system to test only short, educational programs, we can get by with a few hundred bytes or so. Because dynamic RAMs are generally used in very large memory systems (64K, 256K, and more), we will not use them because most of the memory would go to waste. Other reasons exist for not choosing dynamic RAMs at this time. They require complex timing and refresh logic, and will also need to be wired very carefully to prevent messy noise problems from occurring. Even if we use a DRAM controller, we will need some external logic to support the controller, which itself could be a very costly item.

For these reasons, we decide to use static RAM. Even though a few hundred bytes will cover our needs, we will use one 6264 static RAM, thus making our RAM memory 8,192 bytes long also. The 6264 is a low-power static RAM, with a pinout almost exactly identical to the 2764 we are using for our EPROM memory. So, by adding only two more integrated circuits (plus a few for control), our memory needs are taken care of.

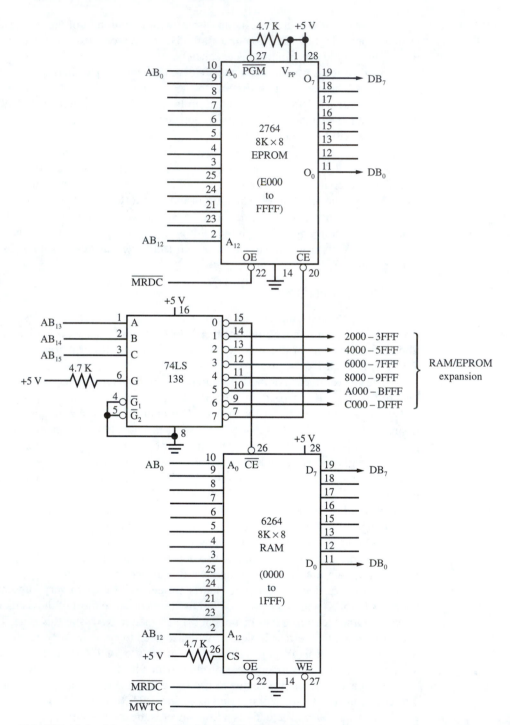

FIGURE 14.3   Memory circuitry for the single-board computer

Figure 14.3 shows how we use a 74LS138 three- to eight-line decoder to perform partial-address decoding for us. Because we are not concerned with future expansion on a large scale, partial-address decoding becomes the cheapest way to generate our addressing signals. Address lines A_{13} through A_{15} are used because they break up the 8088's memory space into convenient ranges (8KB blocks in this case). We completely ignore the state of the upper four address lines (A_{16} through A_{19}). If future expansion beyond the 64KB range is necessary, the upper four address lines must be used to enable the 74LS138.

With A_{13} through A_{15} all low, the 74LS138 decoder will output a 0 on the output connected to the RAM's chip-enable input. With A_0 through A_{12} selecting individual locations within the 6264 RAM, we get an address range of 00000H to 01FFFH. Thus, any time the processor accesses memory in the range 00000H to 01FFFH, the RAM will be enabled. This is a good place for system RAM, because the interrupt vector table must be stored in locations 00000H through 003FFH.

When A_{13} through A_{15} are all high (as they are after a reset causes the initial instruction fetch from FFFF0H), the chip-enable of the 2764 8KB EPROM is pulled low (by the 74LS138). Together with information on the thirteen lower address lines, this maps the EPROM into locations FE000H to FFFFFH. Because partial-address decoding is being used, we can imagine the upper four address lines to be anything we want. This is why we conveniently made them low for the RAM range and high for the EPROM range. Other acceptable RAM ranges are 10000H through 11FFFH, 50000H through 51FFFH, and C0000H through C1FFFH. These are only three more of the sixteen possible RAM address ranges, all of which look identical to the processor. EPROM ranges can be found in a similar manner.

In addition to the RAM and EPROM chip-select signals, the 74LS138 also decodes six additional blocks of addresses. Table 14.1 shows the address range associated with each output of the 74LS138. If additional 8KB RAMs or EPROMs need to be added at a later date, the FREE decode signals can be used to map them into the desired range.

TABLE 14.1 Partially decoded address ranges in the minimal system

74LS138 Output	Decoded Address Range	Use
0	x0000 to x1FFF	Main RAM
1	x2000 to x3FFF	Free
2	x4000 to x5FFF	Free
3	x6000 to x7FFF	Free
4	x8000 to x9FFF	Free
5	xA000 to xBFFF	Free
6	xC000 to xDFFF	Free
7	xE000 to xFFFF	Main EPROM

x = don't care (can be anything from 0 to F)

If we were allowing DMA operations, we might not want the 74LS138 to operate in the same way. The enable inputs of the 74LS138 provide us with a way to disable it (all outputs remain high) during a DMA operation, so that an external device may take over the system.

The $\overline{\text{MRDC}}$ and $\overline{\text{MWTC}}$ signals generated in the CPU section are used to control the transfer of data between the processor and memory.

The Serial Section

The serial section of our computer will contain all hardware required to communicate with the outside world (via an EIA-compatible data terminal). One question that must be answered concerns the baud rate at which we will be transmitting and receiving. A very acceptable speed is 2,400 baud. Speeds higher than this will be too fast to read on the screen, and slower speeds will take too long to read.

Figure 14.4 shows the schematic of the serial section, where an 8251 is used to provide serial communications. The chip-enable input of the 8251 is controlled by output 2 of the 74LS138 port-address decoder. Address lines A_5 through A_7 are used by the 74LS138 to decode eight port-address ranges. The 8251 responds to any I/O accesses to ports 40H through 5FH. The remaining seven groups of port addresses are available for expansion. One of these groups will be used to access an 8255 to provide parallel I/O (as shown in the next section).

The MC14411, together with a 1.8432-MHz crystal, generates the required transmitter and receiver clock frequencies for standard baud rates from 300 to 9,600. A DIP switch or jumper can be used to select one of these rates.

The 8251 communicates with the processor via the 8-bit data bus. $\overline{\text{IORC}}$ and $\overline{\text{IOWC}}$, together with A_0, control read and write operations in the 8251. CLK is provided to take care of the 8251's internal activities, and RESET is used to initialize the 8251 at power-up.

Because no modem is connected, the 8251's $\overline{\text{DSR}}$ and $\overline{\text{CTS}}$ inputs can be grounded. This ensures that the 8251 is always ready to communicate.

Serial data enters and leaves the 8251 on RxD and TxD. These signals are connected to a MAX232CPE, which converts the 8251's TTL signal levels into RS232-compatible voltage levels, and vice versa. Four 22-μF electrolytic capacitors are used to create a ±10V swing on the output of the MAX232. This eliminates the need for an external power supply for these two voltages. The serial-in and serial-out signals from the MAX232 can be wired to a DB25 connector or other suitable connector.

Figure 14.5 shows one way the single-board computer may be connected to a PC. The transmit and receive lines (RS232out and RS232in) on the single-board are connected to the receieve and transmit lines on the PC's COM1 connector. The connecting cable is sometimes referred to as a **null modem,** because it cross-couples the transmit and receive lines for full-duplex communication.

Although any port address in the range 40H through 5FH will activate the 8251, the software (via A_0) uses only ports 40H and 41H. Port 40H is the 8251's *data* port, which is used to read and write to the receiver and transmitter. Port 41H is the 8251's *status* port, which is used by the software to determine when it is safe to access the receiver or transmitter.

If one serial channel is not sufficient for your needs, a second one can be added by interfacing a second 8251. One of the free port-address ranges should be used to enable the

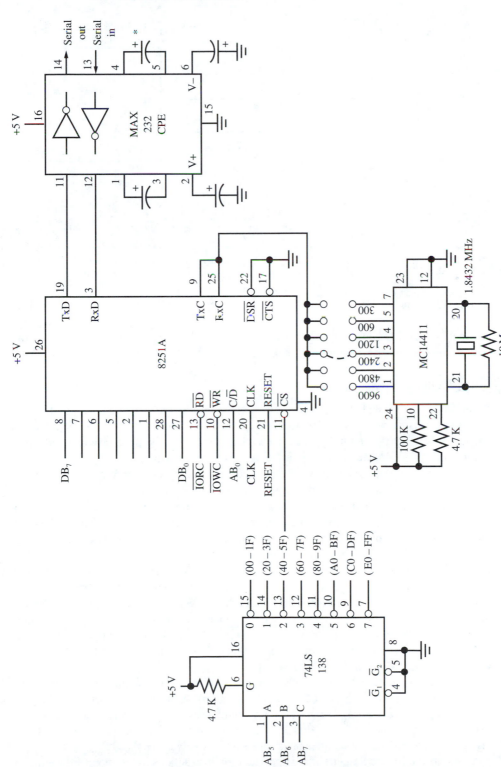

FIGURE 14.4 Serial I/O circuitry for minimal system

* All electrolytic caps are 22 μF/25 V.

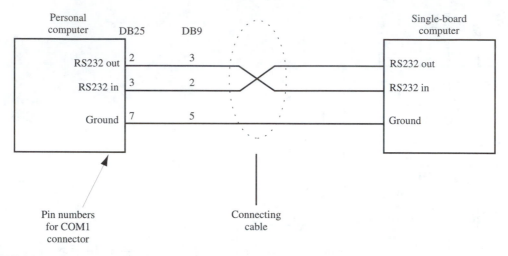

FIGURE 14.5   Serial connection to a personal computer

second 8251. Its baud-rate clock will be supplied by the MC14411, and the other half of
the MAX232 can be used to drive the second set of serial data lines. A second serial
channel is useful for downloading machine code into the minimal system's memory (although this can also be done with a single channel), or for echoing data to a printer.

The Parallel Section

If parallel I/O is needed, simple latching and buffering circuitry can be used to add a single I/O port using one of the free port-address decoder ranges. If more than one port is
needed, it is best to use a multiport device such as the 8255. The 8255 provides three programmable I/O ports and is easily interfaced with the processor.

Figure 14.6 shows how an 8255 is connected in the minimal system. All of the usual
data and I/O signals are connected. Because there are four internal ports in the 8255 (three
for data and one for control), two address lines are required to select one of the four internal ports. A_0 and A_1 are used for this purpose. With the chip-enable input of the 8255
wired to the first output of the 74LS138 port-address decoder, port addresses 00H through
03H can be used to select the 8255.

Figure 14.6 also shows how the 8255 is used to provide the minimal system with
analog I/O capability. This additional circuitry may not be needed in many applications. In
that case, the 8255 merely provides 24 bits of parallel I/O. When analog I/O is a requirement, the circuit of Figure 14.6 provides an acceptable range of analog input and output
voltages. A 1408 8-bit digital-to-analog converter is connected to port A of the 8255
(which must be programmed for output operation). The current output of the 1408 is converted into a ±2.5-V swing by a 741 op-amp.

Port B of the 8255 is used to read the output of an 0804 8-bit analog-to-digital converter (which must be programmed for input operation). A second 741 is used to adjust the
input voltage range of ±2.5 V to the 0- to 5-V swing needed by the 0804. The 0804 is controlled by 2 bits in the 8255's C port. With the 0804 connected in this way, it is possible to
digitize over 8,000 analog samples in 1s (one sample every 125 μs).

Control of the analog circuitry is provided by instructions in the monitor program.

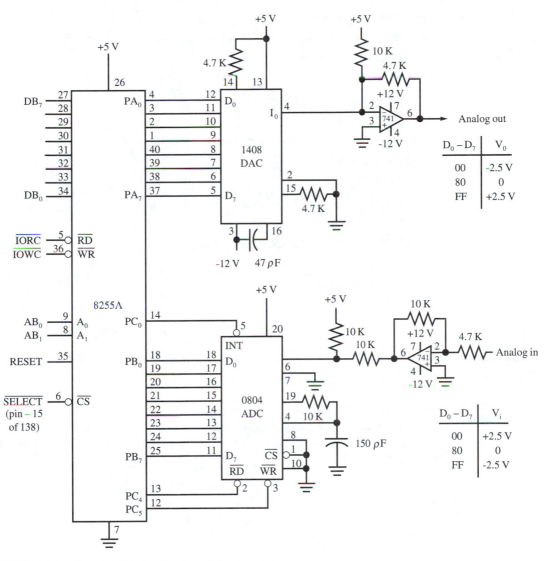

FIGURE 14.6   Parallel I/O circuitry for minimal system

14.4   THE MINIMAL SYSTEM PARTS LIST

Now that we have finished designing the minimal system, we can look back on all the
figures and decide how many ICs we will need to build it. The whole idea behind the
design was to build a working maxmode system with a minimum of parts. The following
list summarizes all the ICs that are needed (excluding the analog I/O circuitry). Pullup
resistors, discrete components, and sockets are not included.

one 8284A clock generator

one 74LS07 buffer

one 74LS04 hex inverter

one 8088 microprocessor

one 74LS244 octal buffer

one 8282 octal latch

one 8286 bidirectional bus driver

one 8288 bus controller

two 74LS138 three- to eight-line decoders

one 2764 8K by 8 EPROM

one 6264 8K by 8 RAM

one 8251 UART

one MC14411 baud-rate generator

one MAX232CPE TTL to RS232 converter

one 10-MHz crystal

one 1.8432-MHz crystal

In short, only fifteen integrated circuits are needed to build a working 8088 maxmode system.

14.5   CONSTRUCTION TIPS

The easiest way to build the minimum system is to wire-wrap it. A printed circuit board may be used, but it would be very complex and most likely double-sided.

The minimum system will work the first time power is applied, if the following points are kept in mind:

1. Keep all wires as short as possible. Long wires pick up noise.
2. Connect 0.1-μF bypass capacitors across +5 V and GND on all ICs.
3. Trim all excess component leads to avoid short circuits.
4. Connect power and ground to all ICs before wiring anything else.
5. Pull all unused TTL inputs to +5 V with 4.7K-ohm resistors.
6. On a copy of the schematic, mark off connections as they are made.
7. Make sure no ICs are plugged in backward before applying power.
8. Use an ohmmeter to check each connection as it is made.
9. Plug in only the clock ICs first. If the clock does not work, neither will the rest of the system.
10. Check that each IC has proper power before beginning any major troubleshooting.

Experience, of course, is the best teacher, but these hints should be enough to get you started. There is nothing like the feeling of building a circuit that works the first time! If it fails to operate properly, do not get discouraged. With your knowledge of TTL, you should be able to track down the source of the problem in no time. You might be surprised that most problems will be due to wrong wiring. Always check your wiring very carefully!

14.6   WRITING THE SOFTWARE MONITOR

Now that we have the system hardware designed, we must tackle the job of writing the system software. Because our goal is to use the system for testing custom 8088 programs, the monitor program must be capable of performing every step that is needed for us to get the new program into memory, edit it if necessary, display it in hexadecimal format, and execute it. This will require the use of a number of monitor commands. The commands available with the monitor program are:

B—set breakpoint

C—clear breakpoint

D—dump memory contents

E—enter new register data

G—go execute a user program

H—display help message

I—input data from port

L—downline load a program

M—move memory

O—output data to port

R—display registers

S—stop processor

T—test analog I/O

X—examine memory

To implement the required monitor commands, we have to write machine language subroutines that perform each function. Many of these routines will perform identical tasks (such as reading an address from the keyboard, converting from hex to ASCII, and outputting data to the display terminal); therefore, we will also need a collection of smaller routines to perform these chores. These routines are called **auxiliary** subroutines, and are summarized in Table 14.2. We will study the operation of each auxiliary subroutine first, and then see how they are used within the command routines.

Keep in mind that the monitor program has been written using simple instructions and addressing modes. Many of the routines presented probably can be simplified. You are encouraged to rewrite them once the computer has been built and the basic monitor program is up and running.

The Auxiliary Routines

Each auxiliary routine is designed to perform a specific function. Enough detail will be provided for you to grasp the overall operation of each routine. Pay attention to the methods used to pass information to/from each routine. It will also be useful for you to make a table showing (for each routine) what registers are used for input, which ones are used for output, and which registers are simply used during computations.

TABLE 14.2   Auxiliary routines

Name	Function
IN_IT	Initialize I/O devices and system tables
BLANK	Send ASCII blank to the display
CRLF	Send ASCII CR (carriage return), LF (line feed) to display
HTOA	Convert hex to ASCII
A_BIAS	Add ASCII bias to hex value
H_OUT	Output four-digit hex number to display
C_OUT	Output ASCII character to display
CH_CASE	Convert ASCII to uppercase
ERROR	Display error message
C_IN	Read ASCII character from keyboard
CHK_SUM	Check sum during downline load
GET_BYT	Read byte value from keyboard
GET_WRD	Read word value from keyboard
CON_V	Convert ASCII to hex
GET_NUM	Read number from keyboard
S_END	Output ASCII message to display
ENVIR	Save system environment
D_ENV	Display system environment
D_FLG	Display processor flag states
DPR	Display processor registers

IN_IT.   This routine initializes the 8251 serial I/O device so that it will be capable of generating waveforms containing 8 data bits, no parity, and 2 stop bits. An x16 clock is also selected. IN_IT also programs the 8255 for port A out and port B in, with port C used for handshaking with the ADC. The DAC is initialized to output 0 V. The monitor's breakpoint flag is reset. The receiver of the 8251 is read to clear any stray character that may be present at power-on. The 8251 is accessed through data port 40H and control/status port 41H. The 8255 is accessed through ports 0 through 3. Equate statements are included in the monitor program to associate these port addresses with labels. For example, the data port of the 8251 is defined like this:

```
S_DATA   EQU   40H
```

The label is used in an instruction to aid in understanding what port is being accessed. The routine looks like this:

```
IN_IT   PROC   NEAR
        MOV    AL,0CEH
        OUT    S_CTRL,AL    ;output 8251 mode word
        MOV    AL,5
```

```
        OUT     S_CTRL,AL    ;output 8251 command word
        MOV     AL,83H
        OUT     AD_CTRL,AL   ;init 8255
        MOV     AL,80H
        OUT     D_AC,AL      ;zero DAC
        MOV     AL,10H
        OUT     AD_STAT,AL
        MOV     AL,30H
        OUT     AD_STAT,AL   ;reset ADC
        MOV     ES:[BR_STAT+RTOP],0  ;clear breakpoint flag
        IN      AL,S_DATA    ;clear 8251 receiver
        RET
IN_IT   ENDP
```

Blank.   This routine outputs an ASCII blank character to the display. It may first appear that a routine dedicated to this simple function is a waste of time (and code). But consider that there may be many routines that need to output blanks during their execution. It is much more convenient for the programmer to simply CALL BLANK than to duplicate the instructions each time.

```
BLANK   PROC    NEAR
        MOV     AL,20H       ;code for ASCII blank
        CALL    C_OUT
        RET
BLANK   ENDP
```

CRLF.   This routine is used to make the display scroll up one line. The codes for carriage return and line feed are output to the display. This routine, like BLANK, is needed often.

```
CRLF    PROC    NEAR
        MOV     AL,13        ;ASCII CR
        CALL    C_OUT
        MOV     AL,10        ;ASCII LF
        CALL    C_OUT
        RET
CRLF    ENDP
```

HTOA.   This routine performs hex-to-ASCII conversion. The data byte contained in register AL is converted into a two-character sequence of ASCII characters and output to the display. For example, if AL contains 3FH, an ASCII "3" and an ASCII "F" are output to the display.

```
HTOA    PROC    NEAR
        PUSH    AX
        SHR     AL,1         ;get upper nibble
        SHR     AL,1
        SHR     AL,1
        SHR     AL,1
        CALL    A_BIAS       ;convert to ASCII and output
        POP     AX           ;get lower nibble
        CALL    A_BIAS       ;convert and output again
        RET
HTOA    ENDP
```

A_BIAS.   This routine converts the lower 4 bits of register AL into a printable ASCII equivalent character. For example, ----0011 becomes 33H, which is "3". ----1011 becomes 42H, which is "B". The character is then output to the display.

```
A_BIAS    PROC    NEAR
          AND     AL,0FH    ;clear upper 4 bits
          ADD     AL,30H    ;add ASCII bias
          CMP     AL,3AH    ;is it A through F?
          JC      NO_7      ;no
          ADD     AL,7      ;yes, correct to alphabetic
NO_7:     CALL    C_OUT     ;output to display
          RET
A_BIAS    ENDP
```

H_OUT.   This routine outputs the four-character ASCII equivalent of the number stored in register DX. For example, if DX contains 3E7CH, the ASCII characters "3", "E", "7", and "C" are output to the display. Notice how HTOA is used to simplify this routine.

```
H_OUT     PROC    NEAR
          MOV     AL,DH     ;do upper byte first
          CALL    HTOA
          MOV     AL,DL     ;then lower byte
          CALL    HTOA
          RET
H_OUT     ENDP
```

C_OUT.   This routine examines the level of the 8251's transmitter-ready flag, and, when it is ready, outputs the character in AL to the transmitter (for viewing on the display).

```
C_OUT     PROC    NEAR
          PUSH    AX
          MOV     AH,AL
CO_S:     IN      AL,S_CTRL   ;get 8251 status
          AND     AL,01H      ;test TRDY
          JZ      CO_S        ;loop until not busy
          MOV     AL,AH
          OUT     S_DATA,AL   ;output character
          POP     AX
          RET
C_OUT     ENDP
```

CH_CASE.   This routine examines the ASCII character in register AL. If the character is lowercase ("a" through "z") it is converted into uppercase ("A" through "Z").

```
CH_CASE   PROC    NEAR
          CMP     AL,'a'      ;test for 'a'...'z' range
          JC      UN_ALPH
          CMP     AL,'z'+1
          JNC     UN_ALPH
          AND     AL,0DFH     ;convert into uppercase
UN_ALPH:  RET
CH_CASE:  ENDP
```

ERROR.   This routine gives an audible beep when an error is detected. A "?" is also displayed on the screen to indicate an error.

```
ERROR     PROC    NEAR
          MOV     AL,'?'      ;output '?'
          CALL    C_OUT
          MOV     AL,7        ;beep terminal (control-G)
```

```
          CALL    C_OUT
          RET
ERROR     ENDP
```

C_IN.   This routine examines the level of the 8251's receiver-ready flag, and, when ready, reads a character from the receiver. The MSB of the byte returned in register AL is always cleared.

```
C_IN      PROC    NEAR
CI_S:     IN      AL,S_CTRL    ;read 8251 status
          AND     AL,02H       ;test RRDY
          JZ      CI_S         ;loop until ready
          IN      AL,S_DATA    ;read new character
          AND     AL,7FH       ;clear MSB
          RET
C_IN      ENDP
```

CHK_SUM.   This routine checks the C_SUM location in the monitor's data table and returns an error message if it is not zero. A nonzero C_SUM means that a program was downline-loaded incorrectly.

```
CHK_SUM   PROC    NEAR
          CMP     ES:[C_SUM+RTOP],0   ;test C_SUM
          JZ      GD_LD               ;OK if zero
          LEA     SI,CSE              ;load address of error message
          CALL    S_END               ;output message
          JMP     GET_COM             ;get a new command
GD_LD:    RET
CHK_SUM   ENDP
```

The error message is stored in the monitor's data area, like this:

```
CSE     DB     '<-Checksum error->$'
```

with the "$" serving as the end-of-string character.

GET_BYT.   This routine is used to read two successive characters from the serial port and convert them into their 8-bit equivalent. For example, if "5" and "C" are received, register AL will contain 5CH on return.

```
GET_BYT   PROC    NEAR
          CALL    C_IN         ;get a character
          CALL    C_OUT        ;echo it to display
          CALL    CON_V        ;convert into binary
          SHL     AL,1         ;move result into upper 4 bits
          SHL     AL,1
          SHL     AL,1
          SHL     AL,1
          MOV     BL,AL        ;save upper half of result
          CALL    C_IN         ;get second character
          CALL    C_OUT        ;echo it
          CALL    CON_V        ;convert to binary
          OR      AL,BL        ;combine with upper half
          ADD     ES:[C_SUM+RTOP],AL   ;add new value to C_SUM
          RET
GET_BYT   ENDP
```

Notice that the received byte is added to C_SUM. This allows GET_BYT to be used in the downline-loading command.

GET_WRD.   This routine calls GET_BYT twice to read a four-character number from the serial port. The number is returned in register AX.

```
GET_WRD    PROC    NEAR
           CALL    GET_BYT    ;get first byte
           MOV     BH,AL      ;save it
           CALL    GET_BYT    ;get second byte
           MOV     AH,BH      ;return result in AX
           RET
GET_WRD    ENDP
```

CON_V.   This routine converts an ASCII character in the range "0" to "9" or "A" to "F" into its corresponding 4-bit binary equivalent. The result is returned in the lower half of register AL.

```
CON_V       PROC    NEAR
            SUB     AL,'0'     ;remove ASCII bias
            CMP     AL,10      ;test 'A' to 'F' range
            JC      NO_SUB7    ;not a letter
            SUB     AL,7       ;remove alpha bias
NO_SUB7:    RET
CON_V       ENDP
```

GET_NUM.   This routine accepts a multidigit hexadecimal number from the keyboard and stores it in register DX. If more than four characters are entered, only the last four will be used in the conversion. A CR or blank will terminate the input. If no characters are entered prior to the CR or blank, register BH will contain "0", otherwise it will contain "1". Any illegal character causes an exit to ERROR.

This short table gives a few examples of GET_NUM at work:

Inputs	Outputs
'3','A','7' <cr>	DX=03A7 BH='1'
'1','2','3','4','5' <cr>	DX=2345 BH='1'
<cr>	DX=0000 BH='0'

This is the actual routine:

```
GET_NUM    PROC    NEAR
TOP_N:     MOV     BH,'0'     ;init BH
           MOV     DX,0       ;clear DX
GT_NM:     CALL    C_IN       ;get a character
           CMP     AL,13      ;is it CR?
           JZ      GT_BYE     ;yes
           CALL    C_OUT      ;no, echo it
           CMP     AL,20H     ;is it a blank?
           JZ      GT_BYE     ;yes
           MOV     BH,'1'     ;no, adjust BH
           CALL    CH_CASE    ;convert to uppercase
           CMP     AL,'0'     ;test '0' to '9' range
           JC      BAD_NM
           CMP     AL,'9'+1
           JC      OK_SUB
```

```
                 CMP    AL,'A'      ;test 'A' to 'F' range
                 JC     BAD_NM
                 CMP    AL,'F'+1
                 JNC    BAD_NM
                 SUB    AL,7        ;remove alpha bias
OK_SUB:          SUB    AL,30H      ;remove ASCII bias
                 SHL    DX,1        ;make room for new nibble
                 SHL    DX,1
                 SHL    DX,1
                 SHL    DX,1
                 ADD    DL,AL       ;adjust result
                 JMP    GT_NM       ;repeat as necessary
GT_BYE:          CLC
                 RET
BAD_NM:          CALL   ERROR
                 JMP    TOP_N
GET_NUM          ENDP
```

SEND. This routine reads ASCII characters from memory and outputs them to the display. The characters are pointed to by register SI. All character strings must terminate with "$". Upon entry, SI must point to the first character in the string.

```
S_END    PROC   NEAR
S_NXT:   MOV    AL,[SI]    ;get a character
         CMP    AL,'$'     ;end of string?
         JZ     S_QWT      ;yes
         CALL   C_OUT      ;no, output to display
         INC    SI         ;point to next character
         JMP    S_NXT      ;repeat
S_QWT:   RET
S_END    ENDP
```

ENVIR. This routine saves the monitor's environment (all registers and flags) when the monitor is reentered from an external source. The usual reentry technique requires an INT 95H instruction at the end of the user routine (see Section 14.7). You may think of ENVIR as the interrupt service routine for INT 95H. ENVIR also reestablishes the monitor's segment registers, which may have been altered by the external software. All registers are stored in memory as 2-byte words (low byte first) beginning at location R_DATA[RTOP] (location 1F80 in the system RAM). The registers are stored in the following order: AX, BX, CX, DX, BP, SI, DI, SP, DS, SS, ES. The flags are stored in locations 1FA1 and 1FA2.

When ENVIR completes execution it *falls into* D_ENV to display the system environment.

```
ENVIR    PROC   NEAR
         PUSHF              ;save flags on stack
         PUSH   AX          ;save AX on stack
         MOV    AX,DS
         PUSH   AX          ;save DS on stack
         MOV    AX,DATA
         ADD    AX,0E00H
         MOV    DS,AX       ;reload monitor DS
         MOV    AX,ES
         PUSH   AX          ;save ES on stack
         MOV    AX,0
         MOV    ES,AX       ;reload monitor ES
         POP    AX          ;pop and save old ES
         MOV    ES:R_DATA[RTOP + 20],AX
```

```
        POP     AX              ;pop and save old DS
        MOV     ES:R_DATA[RTOP + 16],AX
        POP     AX              ;pop and save old AX
        MOV     ES:R_DATA[RTOP+0],AX
        MOV     ES:R_DATA[RTOP+2],BX    ;save all other registers
        MOV     ES:R_DATA[RTOP+4],CX
        MOV     ES:R_DATA[RTOP+6],DX
        MOV     ES:R_DATA[RTOP+8],BP
        MOV     ES:R_DATA[RTOP+10],SI
        MOV     ES:R_DATA[RTOP+12],DI
        POP     AX
        POP     BX
        MOV     ES:R_DATA[RTOP+14],SP   ;save SP
        PUSH    BX
        PUSH    AX
        MOV     ES:R_DATA[RTOP+18],SS   ;save SS
        POP     AX
        MOV     ES:[F_LAGS+RTOP],AX     ;save flags
```

This routine physically appears just before the code for D_ENV. The instruction following the last MOV in ENVIR is the first instruction of D_ENV. The data table containing the stored register values is defined like this:

```
R_DATA  DW   11 DUP(?)
```

where memory space for eleven words (one for each register saved) is reserved. When INT 95H is used at the end of a user routine, it causes ENVIR to execute, which, in turn, saves the final state of each register at the completion of the user routine.

D_ENV.   This routine displays the contents of all CPU registers stored in the R_DATA table by ENVIR. The display format (with sample register values) is as follows:

```
AX:1111  BX:2222  CX:3333  DX:4444
BP:5555  SI:6666  DI:7777  SP:8888
DS:9999  SS:AAAA  ES:BBBB
```

After outputting the value of ES, the routine falls into D_FLG to display the state of the flags.

```
D_ENV:  MOV     SI,0                    ;init pointer to register data
        MOV     CX,11                   ;init loop counter
        CALL    CRLF
T_DR:   MOV     AL,R_LETS[SI]           ;output register name
        CALL    C_OUT
        MOV     AL,R_LETS[SI+1]
        CALL    C_OUT
        MOV     AL,':'                  ;and a ':'
        CALL    C_OUT
        MOV     DX,ES:R_DATA[RTOP+SI]   ;get register data
        CALL    H_OUT                   ;and output it
        CALL    BLANK                   ;2 blanks for spacing
        CALL    BLANK
        ADD     SI,2                    ;point to next register
        MOV     AX,SI                   ;do display formatting
        AND     AL,7
        JNZ     ADJST
        CALL    CRLF
ADJST:  LOOP    T_DR                    ;repeat for all registers
```

D_ENV requires a predefined data table of register names. The table looks like this:

```
R_LETS  DB   'AXBXCXDXBPSIDISPDSSSES'
```

D_FLG. This routine displays the state of each of the five arithmetic flags: sign, zero, auxiliary carry, parity, and carry, in the following format:

```
Flags:    S=1    Z=0    A=0    P=1    C=1
```

Together with the register display of D_ENV, D_FLG provides a method for determining exactly what a user routine has done to the registers and flags during execution.

```
D_FLG:    LEA     SI,FL_MSG    ;output flag message
          CALL    S_END
          MOV     SI,0         ;init pointer to tables
          MOV     CX,5         ;init loop counter
N_FLG:    MOV     AL,F_SYM[SI] ;get a flag name
          CALL    C_OUT        ;display it
          MOV     AL,'='
          CALL    C_OUT        ;display '='
          MOV     AX,ES:[F_LAGS+RTOP]   ;get flag byte
          AND     AL,F_MASK[SI]  ;mask out specific flag
          MOV     AL,'0'         ;adjust AL according to flag state
          JZ      NOT_1
          INC     AL
NOT_1:    CALL    C_OUT        ;display flag state
          CALL    BLANK        ;and output spacing blanks
          CALL    BLANK
          INC     SI           ;point to next flag
          LOOP    N_FLG        ;repeat
          RET
ENVIR     ENDP
```

D_FLG requires three predefined data tables, which are:

```
FL_MSG    DB      13,10,'Flags: $'
F_SYM     DB      'SZAPC'
F_MASK    DB      80H,40H,10H,4,1
```

Note the use of the ENDP statement in the routine. ENVIR, D_ENV, and D_FLG are all contained within the same procedure block.

DPR. This final auxiliary routine is used to enter D_ENV and display the flags, *assuming they have been previously saved.*

```
DPR     PROC    NEAR
        JMP     D_ENV
DPR     ENDP
```

This code is used to display the environment without having to encounter an INT   95H in the user code.

As you study the command routines in the next section, watch how the auxiliary routines are used to simplify the code required to execute a monitor command.

The Monitor Commands

The monitor commands are really the heart of the software monitor. Through the use of the monitor commands, the job of creating new and useful software becomes much easier. We will now examine just how these commands are implemented. Study the methods used to perform I/O with the user, and how decisions are made within the routines. You should be able to gain a very good understanding of the structure of a command routine, and be able to use that knowledge to write *your own* command routine to perform a job that the basic monitor cannot.

The Command Recognizer

To use a command routine, we must be able to get to its starting address in memory (to fetch the first instruction of the command routine). It is much easier and more convenient to enter a single letter command, such as D, G, or R, than to enter a multidigit hexadecimal starting address. The purpose of the command recognizer is to determine which of the monitor commands has been entered by the user, and jump to the command routine for execution. The command recognizer accepts uppercase and lowercase command letters.

Once the command is recognized, the address of the selected command routine is read from a data table, and the routine is jumped to. For this reason, the command routines must jump back to the beginning of the monitor program (GET_COM), and not return as a subroutine would.

The command recognizer is written so that new commands may be easily added with a minimum of change in its code.

```
GET_COM:   MOV    SP,2000H       ;init stack pointer
           CALL   CRLF           ;newline
           MOV    AL,'>'         ;output command prompt
           CALL   C_OUT
           CALL   C_IN           ;get command letter
           CALL   C_OUT          ;echo it
           CALL   CH_CASE        ;convert to uppercase
           CMP    AL,13          ;if CR, start again
           JZ     GET_COM
           MOV    CX,NUM_COM     ;init loop counter
           MOV    SI,0           ;init table pointer
C_TEST:    CMP    AL,COMS[SI]    ;compare user command with table item
           JZ     CHK_SN         ;match
           ADD    SI,2           ;no match, point to next item
           LOOP   C_TEST         ;repeat test
           CALL   ERROR          ;no match at all
           JMP    GET_COM
CHK_SN:    CALL   C_IN           ;command must be followed by CR
           CMP    AL,13          ;or blank
           JZ     DO_JMP
           CALL   C_OUT
           CMP    AL,20H
           JZ     DO_JMP
           CALL   ERROR
           JMP    GET_COM
DO_JMP:    JMP    J_UMPS[SI]     ;fetch command routine address and jump
```

GET_COM uses the COMS data table during the search for a matching command letter. The individual command letters are all followed by a blank, so that when SI is incremented by 2 it will always access the next command letter. This lets GET_COM use SI to point to the correct location within the command routine address table J_UMPS as well.

The data tables used by GET_COM are:

```
NUM_COM    DW    14
COMS       DB    'B C D E G H I L M O R S T X'
J_UMPS     DW    B_RKP, C_BRP, D_UMP, I_NIT
```

```
DW    E_XEC, H_ELP, P_IN, L_OAD
DW    M_OVE, P_OUT, D_REG, S_TOP
DW    T_EST, E_XAM
```

The Command Routines

The command routines are designed to provide the features necessary to load a new program into memory, debug it (find/fix errors), and execute it. Some routines perform operations on the system itself, rather than on the user program. Examples of these types of commands are M (move memory), S (stop processor), and I (input data from port). Study the command routines carefully. Look for ways to improve them once your personal system is up and running.

The B Command: Set Breakpoint. This command is used to specify the address where the user program should break away from its execution and return to the monitor. All registers are saved and displayed upon return.

Breakpoints are very useful when debugging code. To use a breakpoint, select an instruction (and its associated address in memory). The instruction located at the breakpoint address is saved and then replaced by an INT 3 instruction (opcode byte CCH). When the processor gets to the INT 3 instruction (during execution of the user program), a special monitor reentry procedure will be executed, which saves and displays the registers and flags. Then the original instruction is restored and the breakpoint cleared.

If a breakpoint is already set, additional B commands will produce an error message and the breakpoint will not be changed.

The format of the command is B <address>. An example of the B command is:

```
B 110C
```

which will cause a breakpoint at address 110C of the user program.

```
B_RKP:   CALL    GET_NUM                    ;get breakpoint address
         MOV     AL,ES:[BR_STAT+RTOP]       ;check breakpoint status
         CMP     AL,0
         JZ      DO_BP                      ;no breakpoint saved yet
         LEA     SI,BP_AS                   ;breakpoint already saved
         CALL    S_END
         JMP     GET_COM
DO_BP:   MOV     ES:[B_MMA+RTOP],DX         ;save breakpoint address
         MOV     SI,DX
         MOV     AL,ES:[SI]                 ;fetch byte at breakpoint
         MOV     ES:[OP_KODE+RTOP],AL       ;and save it
         MOV     AL,0CCH                    ;insert breakpoint code
         MOV     ES:[SI],AL
         MOV     ES:[BR_STAT+RTOP],1        ;adjust breakpoint status
         LEA     SI,BP_SA                   ;inform user about breakpoint
         CALL    S_END
         JMP     GET_COM
```

B_RKP uses two messages to inform the user of what it has done:

```
BP_AS   DB   'Breakpoint already saved...$'
BP_SA   DB   'Breakpoint saved.$'
```

B_RKP determines which message to send based on the user's actions.

The C Command: Clear Breakpoint.   This command clears the saved breakpoint address. The breakpoint status is adjusted and the user instruction restored. This command must be used before attempting to set a new breakpoint.

The C command has no parameters.

```
C_BRP:  MOV    AL,ES:[BR_STAT+RTOP]    ;check breakpoint status
        CMP    AL,1
        JZ     OP_LD                   ;breakpoint exists
        LEA    SI,BR_ALC               ;no breakpoint, inform user
        CALL   S_END
        JMP    GET_COM
OP_LD:  LEA    SI,BR_CLR               ;tell user breakpoint cleared
        CALL   S_END
        MOV    SI,ES:[B_MMA+RTOP]      ;load breakpoint address
        MOV    AL,ES:[OP_KODE+RTOP]    ;load user instruction byte
        MOV    ES:[SI],AL              ;restore user byte
        MOV    ES:[BR_STAT+RTOP],0     ;clear breakpoint status
        JMP    GET_COM
```

C_BRP uses two predefined messages to inform the user about what it has done:

```
BR_ALC  DB    'Breakpoint already cleared.$'
BR_CLR  DB    'Breakpoint cleared.$'
```

The D Command: Display Memory Contents.   This command displays the contents of memory in hexadecimal format. Each line of the display contains the starting address, followed by 16 bytes of code. A sample line of the display looks like this:

```
1000    F3 1A 29 B3 02 CA 33 F7 88 34 CD 02 10 2A BB C9
```

The first byte (F3H) was read out of memory location 1000, the second byte (1AH) from location 1001, and so on. The last byte on the line (C9H) is read out of location 100F.

The format of the command is: D <starting address> <ending address>. An example of the D command is:

```
D 1000 100F
```

which results in the sample output previously shown.

The routine takes this form:

```
D_UMP:   CALL   GET_NUM   ;get starting address
         PUSH   DX        ;save it on stack
         CALL   BLANK
         CALL   GET_NUM   ;get ending address
         POP    SI        ;retrieve starting address
         SUB    DX,SI     ;compute length of block
         MOV    CX,DX     ;init loop counter
         INC    CX
         MOV    DX,SI     ;go start dump
         JMP    DO_DMP
CHK_ADR: MOV    DX,SI     ;check for address wrap-around
         MOV    AL,DL
         AND    AL,0FH    ;is least-significant digit zero?
         JNZ    NO_ADR    ;no
DO_DMP:  CALL   CRLF      ;yes, output newline and address
         CALL   H_OUT
         CALL   BLANK
NO_ADR:  CALL   BLANK
         MOV    AL,ES:[SI] ;get a byte from memory
```

```
            CALL    HTOA        ;display it
            INC     SI          ;point to next location
            LOOP    CHK_ADR     ;repeat
            JMP     GET_COM
```

The E Command: Enter New Register Data. This command allows the stored register data (in R_DATA) to be changed. All processor registers may be altered. The routine displays the name of each register, its current value, and then gives the user the option of entering a new value. If no new value is entered, the current value is not changed. For example:

```
AX - 1A23?200
```

represents the first line of output from the E command. Register AX currently contains 1A23, but the user has changed this value to 0200. To skip over a register and not change its current value, simply hit the Return key.

The E command has no parameters.

```
I_NIT:  MOV     SI,0                ;init pointer
        MOV     CX,11               ;init loop counter
I_NEX:  CALL    CRLF                ;output register name
        MOV     AL,R_LETS[SI]
        CALL    C_OUT
        MOV     AL,R_LETS[SI+1]
        CALL    C_OUT
        CALL    BLANK
        MOV     AL,'-'              ;and a dash
        CALL    C_OUT
        CALL    BLANK
        MOV     DX,ES:R_DATA[RTOP+SI]   ;and the saved value
        CALL    H_OUT
        MOV     AL,'?'              ;ask for a new value
        CALL    C_OUT
        CALL    GET_NUM             ;read new value
        CMP     BH,'0'              ;change?
        JZ      NUN                 ;no
        MOV     ES:R_DATA[RTOP+SI],DX   ;yes, save new value
NUN:    ADD     SI,2                ;point to next register
        LOOP    I_NEX               ;repeat
        JMP     GET_COM
```

The G Command: Go Execute a User Program. This command loads registers AX, BX, CX, DX, BP, SI, and DI from the monitor's register storage area (R_DATA) and then jumps to the user-supplied address to execute the user routine. Because system RAM is from 00000H to 01FFFH, the user program must be located within the first 8KB of memory. It is suggested that user programs have an ORG of at least 400H so that they do not interfere with the interrupt vector table located in locations 00000H through 003FFH. Also, the monitor reserves a block of system RAM at the high end of free memory (for its internal stack and storage tables). This block begins at address 1F80H. User programs should not utilize RAM above this address either.

The jump to the user program is actually accomplished by pushing the starting address onto the stack and "RETurning" to it.

The format of the command is: G <execution address>. An example of the G command is:

```
G 1000
```

which causes the user program beginning at address 1000H to be executed.

```
E_XEC:  CALL    GET_NUM                       ;get starting address
        SUB     AX, AX                        ;clear AX
        PUSH    AX                            ;let CS=0000 on stack
        PUSH    DX                            ;let IP=DX on stack
        CALL    CRLF
        MOV     AX,ES:R_DATA[RTOP+0]          ;load data registers
        MOV     BX,ES:R_DATA[RTOP+2]
        MOV     CX,ES:R_DATA[RTOP+4]
        MOV     DX,ES:R_DATA[RTOP+6]
        MOV     BP,ES:R_DATA[RTOP+8]
        MOV     SI,ES:R_DATA[RTOP+10]
        MOV     DI,ES:R_DATA[RTOP+12]
        RET                                   ;pop stack to execute user code
```

The H Command: Display Help Message. This command displays a help message showing the syntax of each monitor command. The help message is stored as a long text string (terminated by "$") beginning at H_MSG.

The H command has no parameters. Its routine is:

```
H_ELP:  LEA     SI,H_MSG        ;init pointer to help message
        CALL    S_END           ;display message
        JMP     GET_COM
```

The I Command: Input Data from Port. This command is used to read and display the byte present at a specific input port. The byte read from the specified input port is displayed in square brackets, as in [34].

The format of the command is: I <port address>. An example of the I command is:

```
I 2F
```

which displays the byte seen at input port 2FH.

```
P_IN:   CALL    GET_NUM         ;get input port address
        CALL    CRLF
        MOV     AL,'['          ;display left bracket
        CALL    C_OUT
        IN      AL,DX           ;read input port
        CALL    HTOA            ;display byte
        MOV     AL,']'          ;display right bracket
        CALL    C_OUT
        JMP     GET_COM
```

The L Command: Downline Load a User Program. This command is used to downline load a standard Intel-format HEX file into system RAM. The HEX file is created by the assembler and linker. A sample line from the HEX file might look like this:

```
:11307A006279207468652041535354454D424C4552212A
```

All lines begin with a ":". Embedded within the line of text is a length byte (11), a load address (307A), a record type (00), data bytes (6279 . . .), and a final byte called a **checksum** byte (2A). A downline loader routine must retrieve the information present in the HEX file and use it to load the embedded information into memory. The L command always loads a program into memory beginning at address 00400H and supports four record types:

00—Data record

01—End-of-file record

02—Extended-address record

03—Starting-address record

The HEX file is received by the serial input of the single board (as if someone were quickly entering it through the keyboard). The routine checks the sum of each line as it is received and gives an error message if the checksum is incorrect.

The L command has no parameters.

```
L_OAD:      CALL    CRLF
            MOV     BP,400H   ;program always loads at 400H
NXT_REC:    MOV     ES:[C_SUM + RTOP],0   ;clear checksum
KOLON:      CALL    C_IN      ;wait for ':'
            CALL    C_OUT
            CMP     AL,':'
            JNZ     KOLON
            CALL    GET_BYT   ;get record length
            MOV     CL,AL     ;init loop counter
            MOV     CH,0
            CALL    GET_WRD   ;get load address
            MOV     DI,AX
            CALL    GET_BYT   ;get record type
            CMP     AL,0
            JZ      D_REC     ;record 00
            CMP     AL,1
            JZ      EOF_REC   ;record 01
            CMP     AL,2
            JZ      EA_REC    ;record 02
            CMP     AL,3
            JZ      SA_REC    ;record 03
            LEA     SI,URT    ;record type not valid
            CALL    S_END
            JMP     GET_COM
D_REC:      CALL    GET_BYT               ;now get all data bytes
            MOV     ES:[BP][DI],AL  ;one by one and store them
            INC     DI                    ;in memory
            LOOP    D_REC
            CALL    GET_BYT   ;read checksum
            CALL    CHK_SUM   ;check for correct sum
            JMP     NXT_REC
EOF_REC:    CALL    GET_BYT   ;read checksum
            CALL    CHK_SUM   ;check for correct sum
            JMP     GET_COM
EA_REC:     CALL    GET_WRD   ;get new segment address
            SHL     AX,1      ;shift left 4 bits
            SHL     AX,1
            SHL     AX,1
            SHL     AX,1
            MOV     BP,AX     ;load memory pointer
            CALL    GET_BYT   ;read and check sum
            CALL    CHK_SUM
            JMP     NXT_REC
SA_REC:     CALL    GET_WRD   ;load starting address
            CALL    GET_WRD   ;even though it is ignored
            CALL    GET_BYT
            CALL    CHK_SUM
            JMP     NXT_REC
```

L_OAD requires a predefined error message for acknowledging invalid record types:

```
URT    DB    '<-Unidentified record type->$'
```

The M Command: Move Memory.   This command routine is useful for moving blocks of memory data from one location to another. The data included in the input range is not actually moved anywhere, it is *copied* instead. The starting and ending addresses of the block to be moved (copied) must be specified, along with the starting address of the destination. This command can be used after a downline load to move the user program to a different location in system RAM (or to copy monitor code from EPROM into RAM).

The format of the command is: M <starting address> <ending address> <destination address>. An example of the M command is:

```
M 1200 12FF 1600
```

which copies 256 bytes of RAM from addresses 1200H through 12FFH to memory beginning at address 1600H.

```
M_OVE:   CALL   GET_NUM   ;get starting address
         MOV    SI,DX
         CALL   GET_NUM   ;get ending address
         SUB    DX,SI     ;compute length of block
         MOV    CX,DX     ;init loop counter
         INC    CX
         CALL   GET_NUM   ;get destination address
         MOV    DI,DX
MVIT:    MOV    AL,ES:[SI]  ;read a byte
         MOV    ES:[DI],AL  ;write it at new location
         INC    SI          ;advance pointers
         INC    DI
         LOOP   MVIT        ;repeat
         JMP    GET_COM
```

The O Command: Output Data to Port.   This routine is used to output a data byte to a specific port. The byte and port addresses are specified in the command. This command is useful for testing output-port hardware (such as the serial and parallel I/O sections).

The format of the command is O <output port address> <data>. An example of the O command is:

```
O 7 33
```

which outputs the data byte 33H to port 07H.

```
P_OUT:   CALL   GET_NUM   ;get port address
         PUSH   DX
         MOV    AL,' '
         CALL   C_OUT
         CALL   GET_NUM   ;get output data
         MOV    AL,DL
         POP    DX
         OUT    DX,AL     ;output data to port
         JMP    GET_COM
```

The R Command: Display Registers.   This command is used to display the contents of all processor registers (and flags) saved in the R_DATA storage area. It uses code already contained in the auxiliary routine DPR.

The R command has no parameters.

```
D_REG:  CALL    DPR
        JMP     GET_COM
```

The S Command: Stop the Processor.   This command halts the processor. It may be
useful to examine the level of the 8088's signals when it is halted (or to escape from the
halt state with an interrupt).

```
S_TOP:  LEA     SI,S_MSG
        CALL    S_END
        HLT
```

S_TOP uses a predefined message to indicate that the machine has halted.

```
S_MSG   DB      'Placing 8088 in HALT state.$'
```

The T Command: Test Analog I/O Ports.   This command is used to test the operation of the
analog I/O circuitry (as depicted in Figure 14.6). The user is provided with a choice of two
operations. The first test option exercises the digital-to-analog converter by outputting
values from a data table that generates a sine wave at the analog output. The system must
be RESET to get out of the sine wave routine.

The second test option reads the analog-to-digital converter and echoes the data to
the digital-to-analog converter. The echo test is also terminated by RESET.

The T command has no parameters.

```
T_EST:  LEA     SI,TST_MSG      ;output choice message
        CALL    S_END
GET_RP: CALL    C_IN            ;get user choice
        CALL    C_OUT           ;echo it
        CMP     AL,'1'          ;must be '1' or '2'
        JZ      WAVER
        CMP     AL,'2'
        JZ      EK_0
        CALL    ERROR           ;give an error beep
        JMP     GET_RP
WAVER:  MOV     CX,256          ;init loop counter
        MOV     SI,0            ;init pointer to data
P_IE:   MOV     AL,SINE[SI]     ;get a sine wave sample
        OUT     D_AC,AL         ;output to D/A
        ADD     SI,1            ;point to next data sample
        LOOP    P_IE            ;repeat
        JMP     WAVER           ;go generate another cycle
E_K0:   IN      AL,AD_STAT      ;check for end-of-conversion signal
        AND     AL,01H
        JNZ     EK_0
        MOV     AL,AD_RD        ;do handshaking for A/D read
        OUT     AD_STAT,AL
        IN      AL,A_DC         ;read A/D
        NOT     AL              ;maintain phase relationship
        OUT     D_AC,AL         ;echo data to D/A
        MOV     AL,AD_WR        ;start a new conversion
        OUT     AD_STAT,AL
        MOV     AL,AD_NOM
        OUT     AD_STAT,AL
        JMP     EK_0            ;repeat
```

WAVER makes use of a 256-byte data table containing the digitized samples of a one-
cycle sine wave. The table is as follows:

```
SINE    DB    82H,85H,88H,8BH,8EH,91H,94H,97H
        DB    9BH,9EH,0A1H,0A4H,0A7H,0AAH,0ADH,0AFH
        DB    0B2H,0B5H,0B8H,0BBH,0BEH,0C0H,0C3H,0C6H
        DB    0C8H,0CBH,0CDH,0D0H,0D2H,0D4H,0D7H,0D9H
        DB    0DBH,0DDH,0DFH,0E1H,0E3H,0E5H,0E7H,0E9H
        DB    0EBH,0ECH,0EEH,0EFH,0F1H,0F2H,0F4H,0F5H
        DB    0F6H,0F7H,0F8H,0F9H,0FAH,0FBH,0FBH,0FCH
        DB    0FDH,0FDH,0FEH,0FEH,0FEH,0FEH,0FEH,0FFH
        DB    0FEH,0FEH,0FEH,0FEH,0FEH,0FDH,0FDH,0FCH
        DB    0FBH,0FBH,0FAH,0F9H,0F8H,0F7H,0F6H,0F5H
        DB    0F4H,0F2H,0F1H,0EFH,0EEH,0ECH,0EBH,0F9H
        DB    0E7H,0E5H,0E3H,0E1H,0DFH,0DDH,0DBH,0D9H
        DB    0D7H,0D4H,0D2H,0D0H,0CDH,0CBH,0C8H,0C6H
        DB    0C3H,0C0H,0BEH,0BBH,0B8H,0B5H,0B2H,0AFH
        DB    0ADH,0AAH,0A7H,0A4H,0A1H,9EH,9BH,97H
        DB    94H,91H,8EH,8BH,88H,85H,82H,7EH
        DB    7BH,78H,75H,72H,6FH,6CH,69H,66H
        DB    62H,5FH,5CH,59H,56H,53H,50H,4EH
        DB    4BH,48H,45H,42H,3FH,3DH,3AH,37H
        DB    35H,32H,30H,2DH,2BH,29H,26H,24H
        DB    22H,20H,1EH,1CH,1AH,18H,16H,14H
        DB    12H,11H,0FH,0EH,0CH,0BH,9,8
        DB    7,6,5,4,3,2,2,1,0,0,0,0,0,0,0,0
        DB    0,0,0,0,0,0,0,1,2,2,3,4,5,6,7,8
        DB    9,0BH,0CH,0EH,0FH,11H,12H,14H
        DB    16H,18H,1AH,1CH,1EH,20H,22H,24H
        DB    26H,29H,2BH,2DH,30H,32H,35H,37H
        DB    3AH,3DH,3FH,42H,45H,48H,4BH,4EH
        DB    50H,53H,56H,59H,5CH,5FH,62H,66H
        DB    69H,6CH,6FH,72H,75H,78H,7BH,7FH
```

A predefined message is also required to provide the test choices:

```
TST_MSG    DB    13,10,'1) Send sinewave to D/A converter, or'
           DB    13,10,'2) Echo A/D to D/A'
           DB    13,10,'Choice ?$'
```

The X Command: Examine Memory. This command is used to manually enter a program by hand, or to examine the contents of selected memory locations. The user supplies the starting address in memory. The X command routine displays the address and the contents of each memory location. The user is given the chance to enter a new value or to leave the data unchanged. The format of the X command is: X <starting address>. An example of the X command is:

```
X 1000
```

which generates the following string of addresses:

```
1000    3A?<sp>
1001    29?07<cr>
1002    B6?<cr>
```

The 3A at address 1000 remained unchanged, because a space was entered at the '?' prompt. The 29 at address 1001 was changed to 07 and the <cr> on the third line terminated the X command.

```
E_XAM:    CALL    GET_NUM    ;get starting address
          MOV     SI, DX
DO_LYN:   CALL    CRLF       ;display starting address
          MOV     DX,SI
```

```
              CALL    H_OUT
              CALL    BLANK
              CALL    BLANK
              MOV     AL,ES: [SI]    ;read memory
              CALL    HTOA           ;display byte
              MOV     AL,'?'         ;ask for new data
              CALL    C_OUT
              CALL    GET_NUM
              CMP     BH,'0'         ;change?
              JZ      E_MIP          ;no (must have been CR or SP)
              MOV     ES:[SI],DL     ;yes, replace data
EX_GNA:       INC     SI             ;point to next location
              JMP     DO_LYN         ;repeat
E_MIP:        CMP     AL,20H         ;skip over data?
              JZ      EX_GNA         ;yes
              JMP     GET_COM        ;no, exit
```

The Body of the Monitor

At this point we have covered the creation of all routines (command and auxiliary) that we need inside our monitor. The last step we need to perform is to collect all the routines together and organize them into a source file. The source file must provide an interrupt vector for the INT   3 instruction used by the breakpoint routine. It must also contain code to initialize the stack pointer, and handle breakpoint and reentry (INT 95H) procedures.

The following source code must reside at the beginning of the source file. It is followed by the code of the command and auxiliary routines.

```
         .CODE
;segment usage is as follows
;     CS for machine code
;     DS for rom-based data tables
;     ES for ram-stack operations

         ORG    100H

;this is the starting address of the
;reserved system-ram area.

RTOP   EQU    1F80H

;This is where the re-start (int 95h) and
;breakpoint (int 03h) return vectors are
;generated.

         LEA    BX,RE_STRT
         MOV    CX,0E00H
         MOV    AX,0
         MOV    SI,0
         MOV    DS,AX
         MOV    [SI+254H],BX      ;init restart address (INT 95H)
         MOV    [SI+256H],CX
         LEA    BX,BR_ENTR        ;init breakpoint address (INT 3)
         MOV    [SI+0CH],BX
         MOV    [SI+0EH],CX
         MOV    AX,CS             ;init all segment registers
         ADD    AX,DATA
         MOV    DS,AX
         SUB    AX,AX
         MOV    ES,AX
```

```
        MOV     SS,AX
        MOV     SP,2000H            ;init stack pointer
        CALL    IN_IT              ;init I/O devices
        LEA     SI,HELLO           ;send monitor greeting
        CALL    S_END
        JMP     GET_COM            ;go get first command
;entry point during a breakpoint (via INT  3)
BR_ENTR:
        CALL    ENVIR                    ;save/display environment
        LEA     SI,BR_BAK                ;display breakpoint message
        CALL    S_END
        POP     AX                       ;get breakpoint address from stack
        MOV     ES:[BR_IP+RTOP],AX
        POP     AX
        MOV     ES:[BR_CS+RTOP],AX
        POP     AX
        MOV     DX,ES:[BR_CS+RTOP]       ;display address of breakpoint
        CALL    H_OUT                    ;in CS:IP format
        MOV     AL,':'
        CALL    C_OUT
        MOV     DX,ES:[BR_IP+RTOP]
        DEC     DX
        CALL    H_OUT
        MOV     AL,']'
        CALL    C_OUT
        MOV     SI,ES:[B_MMA+RTOP]       ;restore user code
        MOV     AL,ES:[OP_KODE+RTOP]
        MOV     ES:[SI],AL
        MOV     ES:[BR_STAT+RTOP],0      ;clear breakpoint status
        JMP     GET_COM
;normal entry point (via INT 95H)
;this routine falls into GET_COM
RE_STRT:
        CALL    ENVIR                    ;save/display environment
        LEA     SI,RENTER                ;display reentry message
        CALL    S_END
        POP     AX                       ;get exit address from stack
        MOV     ES:[BR_IP+RTOP],AX
        POP     AX
        MOV     ES:[BR_CS+RTOP],AX
        POP     AX
        MOV     DX,ES:[BR_CS+RTOP]       ;display exit address
        CALL    H_OUT
        MOV     AL,':'
        CALL    C_OUT
        MOV     DX,ES:[BR_IP+RTOP]
        SUB     DX,2
        CALL    H_OUT
        MOV     AL,']'
        CALL    C_OUT
GET_COM: <now include command recognizer code>
```

Two data segments are also used by the monitor software. The first segment, DATA, contains all of the text messages used by the system, other types of data tables (including the sine wave table), and the system equates (for definition of I/O ports).

The second data segment, DATA2, contains all of the reserved storage locations for the monitor.

```
        .DATA
HELLO   DB      13,10,'8088 Monitor, Ver:3.0$'
WR_BY   DB      13,10,'C1997 James L. Antonakos$'
```

```
RENTER    DB        13,10,'Re-entry by external program at [$'
BR_BAK    DB        13,10,'Breakpoint encountered at [$'

;messages from auxiliary and command routines go here also

D_AC      EQU       0
A_DC      EQU       1
AD_STAT   EQU       2
AD_CTRL   EQU       3
AD_RD     EQU       20H
AD_WR     EQU       10H
AD_NOM    EQU       30H
C8255     EQU       83H   ;Aout, Bin, CLin, CHout
S_DATA    EQU       40H
S_STAT    EQU       41H
S_CTRL    EQU       41H
M8251     EQU       0CEH  ;2 stop, no parity, 8 data, *16 clock
C8251     EQU       05H   ;R and T enable only
TRDY      EQU       01H
RRDY      EQU       02H

R_DATA    DW        11 DUP(?)
GO_ADR    DW        ?
BR_STAT   DB        ?
R_MMA     DW        ?     ;breakpoint main-memory address
OP_KODE   DB        ?     ;opcode byte
BR_CS     DW        ?     ;breakpoint CS
BR_IP     DW        ?     ;breakpoint IP
F_LAGS    DW        ?     ;system flags
C_SUM     DB        ?     ;checksum storage
```

The entire source file must be assembled and burned into an EPROM. You should have a copy of the entire source file (SBC-MON.ASM) for your use.

To get the monitor program to start up correctly, a far jump instruction is placed into the EPROM at the location corresponding to address FFFF0H. The code for the jump instruction is EA 00 01 00 0E, which jumps to the first instruction of the monitor program located at CS:IP = 0E00:0100.

You may already have ideas for improving the design of the monitor program (by rewriting existing commands or adding new ones). *But if you plan to build the minimal system, make sure the basic monitor works before you begin changing it!*

14.7   A SAMPLE SESSION WITH THE SINGLE-BOARD COMPUTER

Once you have constructed the minimal system and burned the monitor program into EPROM, you will be ready to test the system. If you are planning on using a PC as the communication console for the single-board computer, it will be necessary to wire both computers together (as previously shown in Figure 14.5), and then execute a *terminal emulation* program on the PC. If no terminal emulator is available, the SBCIO program shown here will do all that is necessary to provide serial I/O between both machines.

```
;Program SBCIO.ASM: Communicate with Single-Board Computer via COM1.
;
        .MODEL SMALL
```

```
        .DATA
XMSG    DB      'SBCIO:    Press Control-X to exit. . .',0DH, 0AH, 0DH, 0AH, '$'

        .CODE
        .STARTUP
        LEA     DX,XMSG         ;set up pointer to exit message
        MOV     AH,9            ;display string function
        INT     21H             ;DOS call
COM1:   MOV     DX,3FDH         ;set up status port address
MORE:   IN      AL,DX           ;read UART status
        TEST    AL,1            ;has a character been received?
        JNZ     READ            ;yes, go get it
AKEY:   MOV     AH,0BH          ;check keyboard status
        INT     21H
        CMP     AL,0FFH         ;has any key been pressed?
        JNZ     COM1
        MOV     AH,7            ;read keyboard function, no echo
        INT     21H             ;DOS call
        CMP     AL,18H          ;is it Control-X?
        JZ      BYE
        PUSH    AX              ;save character
UWAIT:  IN      AL,DX           ;read UART status
        TEST    AL,20H          ;check transmitter ready bit
        JZ      UWAIT
        MOV     DX,3F8H         ;set up data port address
        POP     AX              ;get character back
        OUT     DX,AL           ;output it
        JMP     AKEY
READ:   MOV     DX,3F8H         ;set up data port address
        IN      AL,DX           ;read UART receiver
        MOV     DL,AL           ;load character for output
        MOV     AH,2            ;display character function
        INT     21H             ;DOS call
        JMP     AKEY
BYE:    .EXIT

        END
```

SBCIO continuously checks COM1's UART status, as well as the status of the PC's keyboard. Whenever a key is pressed on the keyboard, it is transmitted to the single-board computer as soon as the transmitter in COM1 is ready. Likewise, if a character is received by COM1's UART, it is immediately output to the display screen (via INT   21H). So, full duplex communication is implemented through the SBCIO program. Please note that COM1's baud rate and other parameters must be set up before executing SBCIO. The following sample session illustrates many of the 8088 single-board computer's features. You may wish to try to duplicate the session on your computer. Keep in mind that all user inputs are in **boldface**:

1. Turn computer on. A greeting should appear:

   ```
   8088 Monitor, Ver3.0
   ```

2. Consider the following test program:

   ```
   1000    03 D8    ADD    BX,AX
   1002    03 D1    ADD    DX,CX
   1004    CD 95    INT    95H
   ```

3. To load the test program into memory, use the X command:

```
>X 1000
1000     C7?03
1001     06?D8
1002     74?03
1003     50?D1
1004     CD?<SP>
1005     00?95
1006     B2?<CR>
>
```

Note: The <sp> skips over the data in location 1004 (which is already correct) and the <CR> terminates the command.

4. Display the program now stored in memory:

```
>D 1000 1005
1000    03 D8 03 D1 CD 95
>
```

5. Load AX through DX with initial data to be passed to the program:

```
>E
AX - FF20?1111
BX - 7E36?2222
CX - 3352?3333
DX - A5E8?4444
     etc.
```

6. Check the register contents:

```
>R
AX:1111   BX:2222   CX:3333   DX:4444
                 etc.
```

7. Execute the program:

```
>G 1000
AX:1111   BX:3333   CX:3333   DX:7777
              etc.
Reentry by external program at [0000:1004]
```

Note: BX now equals BX + AX and DX now equals DX + CX. Also, the INT   95H instruction caused the registers to be saved and displayed upon reentry to the monitor.

8. Place a breakpoint at the beginning of the second instruction:

```
>B 1002
Breakpoint saved.
```

9. Check the user code again:

```
>D 1000 1005
1000    03 D8 CC D1 CD 95
```

Note: The "CC" opcode is the breakpoint interrupt. The user code "03" is now saved in system RAM, to be replaced after the breakpoint is encountered during execution.

10. Execute the program again:

```
>G 1000
AX:1111   BX:4444   CX:3333   DX:7777
              etc.
Breakpoint encountered at [0000:1002]
```

Note: By examining the register contents we see that BX has been changed (by another addition with AX), but CX and DX remain unchanged. This was the intent of placing the breakpoint at address 1002.

11. Check the user code again:

```
>D 1000 1005
1000    03 D8 03 D1 CD 95
```

Note: The breakpoint service routine has restored the original instruction byte at address 1002.

14.8   TROUBLESHOOTING TECHNIQUES

Here are some things we can do if the single-board computer does not work when we turn it on:

*   Feel around the board for hot components. A chip that is incorrectly wired or placed backwards in its socket can get very hot. You may even smell the hot component.
*   Make sure all the ICs have power by measuring with a DMM or oscilloscope. Put the probe right on the pin of the IC, not on the socket lead.
*   Use an oscilloscope to examine the CLK output of the 8284. Push the RESET button to verify that the RESET signal is being generated properly.
*   Look at the address, data, and control lines with an oscilloscope or logic analyzer. Activity is a good sign; there may be something as simple as a missing address or data line, or crossed lines. No activity means the processor is not receiving the right information. By examining the logic analyzer traces, you should be able to determine if the memory and I/O address decoders are working properly, as well as the address and data bus drivers.
*   Verify that the EPROM was burned correctly, and is in the right socket and not switched with the RAM. You should be able to connect a logic analyzer to verify that the processor fetches the first instruction from address FFFF0H. You should also be able to see the first instruction byte come out of the EPROM as well.
*   Examine the TxD output of the 8251. Activity at power on or RESET is a good sign, since the monitor program is designed to output a short greeting to let us know it is alive. If TxD wiggles around, but the serial output of the MAX232 does not, there could be a wiring problem there. Putting the capacitors in backwards is bad for the MAX chip.
*   Check every connection again from a fresh schematic. Many times, a missing connection is found, even when every attempt was made to be careful during construction.
*   Change all the chips, one by one. Look for bent or missing pins when you remove them.
*   When all else fails, tell someone else everything you've done and see if they can suggest anything else.
*   You could also set the project aside for a while to get your mind off it. The problem may present itself to you when you least expect it. You may suddenly remember that you commented out an important I/O routine in the monitor because you were having trouble with the assembler. Silly things like that are really fun to find.

SUMMARY

This chapter dealt with the hardware and software design of a minimal 8088 maxmode single-board computer. The system uses 8KB of EPROM, 8KB of RAM, and has a 2,400-baud serial I/O section. In addition, a parallel device is included to provide analog I/O capability.

Many design ideas were suggested, and reasons for choosing one idea over another were given.

The software monitor program is composed of many auxiliary routines, each of which performs one task. The monitor command routines selectively call the auxiliary routines, resulting in shorter, more logical code.

The software monitor is by no means complete, and may be improved by adding commands or features after the basic system is up and running.

STUDY QUESTIONS

1. Modify the circuitry of Figure 14.2 so that the entire address bus can be tri-stated by a 0 placed on a control signal called $\overline{\text{UNBUS}}$.
2. What changes must be made to the memory section (see Figure 14.3) to allow the use of 16KB 27128 EPROMs? Assume that a total of four 27128s will be used.
3. What then will be the address range for each memory in Question 2?
4. How can the 74LS138 memory-address decoder be disabled?
5. Design a selector circuit that will allow a single switch to select either 300- or 2,400-baud operation in the serial device.
6. Redesign the port-address decoder in Figure 14.4 so that the 74LS138 is enabled only when $\overline{\text{IORC}}$ or $\overline{\text{IOWC}}$ is low.
7. Redesign the port-address decoder to decode port addresses in the range 00?? only (i.e., the upper byte of the address must be 0).
8. Design a parallel circuit (using an 8255) that has LEDs on port B, switches on port A, and a common-cathode seven-segment display on port C. Each LED must turn on when its associated bit is high. An open switch indicates a logic 1 level. The LSB of port C drives segment A of the display. Each successive bit drives the next segment. The base port address is E0.
9. Write the software necessary to initialize the parallel circuit of Question 8. All LEDs should be off and the seven-segment display should be showing a U (for "up and running").
10. Write an auxiliary routine that will display the decimal equivalent (0 to 255) of the contents of register AL.
11. Repeat Question 10 for the value contained in DX. The decimal range is now 0–65,535.
12. Modify C_OUT so that it automatically calls CRLF if register AL contains 0DH on entry.

13. Using the auxiliary routines, complete the following table:

Command	Inputs	Outputs	Registers Used
GET_WRD	–	AX	AX, BH, [GET_BYT]

The routine name GET_BYT in square brackets means that it is called by GET_WRD. Show a table entry for each auxiliary routine.

14. Write an auxiliary routine to accept a multidigit decimal number from the keyboard and place its binary equivalent into register AX. Give an error message if the number is greater than 65,535.

15. What changes must be made to allow the command recognizer to recognize entire names for commands? For example, instead of D, use DUMP or DISP. For G, use EXEC or GO.

16. Show the changes needed to add the command V to the monitor program. The V command needs no inputs. The command routine associated with V is V_ERIFY.

17. Write a command routine called V_ERIFY that will test all available RAM locations. It will display the address, data written, and data read for any locations that fail the test.

18. Write a command routine that will calculate the checksum of RAM and EPROM and display the results in hexadecimal. The checksum is obtained by adding up all bytes in the desired memory range, ignoring carries.

19. Write a command routine to add and subtract two supplied hexadecimal numbers.

20. How can the breakpoint routine be modified to allow for more than one breakpoint at a time?

21. What other formats for displaying memory contents might be useful?

22. Modify the D_UMP routine so that the display is paused if Control-S is entered from the keyboard. A paused display is restarted by pressing any key.

23. Rewrite the I_NIT routine so that a single register may be changed by naming it in the command. For example, E CX will only allow register CX to be updated.

24. What instructions must be added to the E_XEC command routine so that any address outside the range 400H to 1F00H causes an error message?

25. Modify the port-input routine so that the data at the input port is continuously updated until <cr> is entered.

26. Write an interrupt service routine for INT   80H. Modify the monitor's start-up code to initialize the INT   80H vector table entry to point to the D_ISPAT routine. The D_ISPAT interrupt service routine uses the number in register AL to call a specific auxiliary command, according to the following table:

AL Value	Routine Called
00	C_IN
01	C_OUT
02	S_END
10	BLANK
11	CRLF
20	GET_BYT

AL Value	Routine Called
28	GET_WRD
2F	GET_NUM
40	CH_CASE
80	DPR

27. Compute the approximate frequency of the sine wave output during the analog test procedure.

28. Assemble this code fragment using DEBUG. Execute it with the monitor commands. What are the final results?

```
MOV    AX,1234H
MOV    BX,5678H
ADD    AL,BL
XOR    AH,BH
```

29. What monitor commands would you deem nonessential? Explain your reasons for doing so.

30. What monitor commands have not been discussed at all (or mentioned in the study questions)? What are the requirements of any additional commands you can think of? Is there enough room in the 2764 EPROM to include them?

31. How does the SBCIO program poll the UART on COM1?

PART 5

Advanced Topics

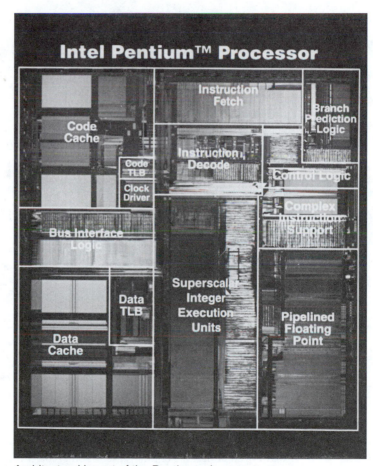

Architectural layout of the Pentium microprocessor.

CHAPTER 15

Hardware Details of the Pentium

OBJECTIVES

In this chapter you will learn about:

- The operation of the Pentium's hardware signals
- The philosophy behind RISC designs
- The various types of bus cycles supported by the Pentium
- The Pentium's pipelined superscalar architecture
- Branch prediction
- The operation of the Pentium's code and data cache
- The Pentium's floating-point unit

15.1   INTRODUCTION

To fully understand the operation of the Pentium, it is necessary to examine the operation of its hardware architecture. Our experience with the hardware architecture of the 8088 and 8086 microprocessors in Chapters 10 through 14 sets the stage for the material presented here. In this chapter, we will see how the Pentium operates from a hardware perspective, a dramatic contrast to the 8088 and 8086.

Learning to design a Pentium-based motherboard for a computer system is *not* the goal of this chapter. Instead, the concentration will be on how the Pentium does what it does. The operation of the U- and V-pipelines, the code and data cache, branch prediction—all of the tricks the designers used to squeeze more performance out of the Pentium—will be explored. The goal is to get a feel for the overall Pentium package and how its various internal components work together.

Section 15.2 presents a detailed discussion of the processor signals. This is followed by an introduction to the techniques used in designing a RISC processor in Section 15.3.

The various types of bus cycles supported by the Pentium are covered in Section 15.4. Superscalar architecture and pipelining are examined next, in Sections 15.5 and 15.6, respectively. Section 15.7 explains how branch prediction is used to keep the instruction pipelines flowing smoothly. A detailed look at the processor's code and data cache is found in Section 15.8. This is followed by a discussion of the redesigned floating-point unit in Section 15.9. The chapter concludes with a set of troubleshooting tips in Section 15.10.

15.2 CPU PIN DESCRIPTIONS

Figure 15.1 shows a top view of the 60-MHz and 66-MHz Pentium packages. The package is a 273-pin PGA (Pin Grid Array). The 60-MHz and 66-MHz Pentiums were the first ones introduced to the computing market.

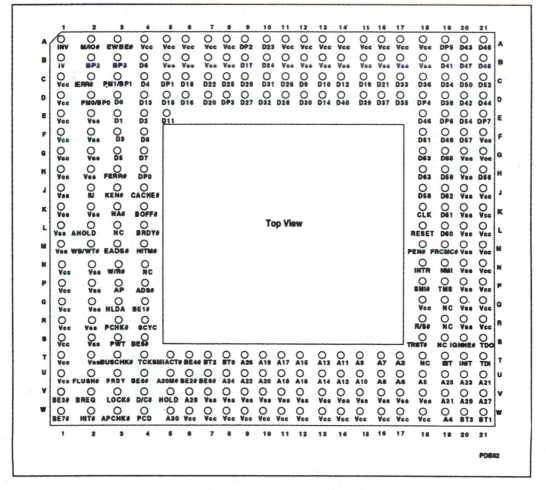

FIGURE 15.1 Pentium® processor (510\60, 567\66) pinout (top view). Reprinted by permission of Intel Corporation. Copyright© 1995 Intel Corporation.

Newer Pentiums with faster clock speeds and other enhancements use a different package, the 296-pin PGA shown in Figure 15.2. These Pentiums allow hardware control over the ratio of bus speed to processor speed, to support motherboard designs that may not be capable of operating at the clock speed of the processor. In addition, the power supply voltage is 3.3 V, compared to 5 V on the earlier 60- and 66-MHz versions. This helps reduce the power used by the processor.

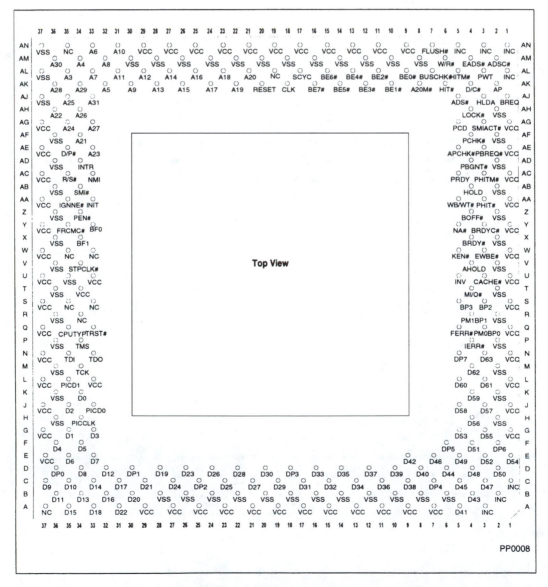

FIGURE 15.2 Pentium® processor (610\75, 735\90, 815\100, 1000\120, 1110\133) pinout (top view). Reprinted by permission of Intel Corporation. Copyright© 1995 Intel Corporation.

Pentium Hardware Signals

$\overline{A20M}$ *(Address 20 Mask).* This input is used to force the Pentium to limit addressable memory to 1MB, to emulate the memory space of the 8086. $\overline{A20M}$ may only be active in real mode.

A_3 *Through A_{31} (Address Lines).* These twenty-nine address lines, together with the byte enable outputs, form the Pentium's 32-bit address bus. A memory space of 4,096MB (4 *gigabytes*) is possible, along with 65,536 I/O ports.

  The address lines are used as inputs during an inquire cycle to read an address *into* the Pentium, for examination by the internal cache.

$\overline{ADS}$ *(Address Strobe).* The $\overline{ADS}$ output, when low, indicates the beginning of a new bus cycle. Signals that define the new bus cycle are valid when $\overline{ADS}$ is active. These signals include the address bus and byte enables, AP, $\overline{CACHE}$, $\overline{LOCK}$, M/$\overline{IO}$, W/$\overline{R}$, D/$\overline{C}$, SCYC, PWT, and PCD.

AHOLD (Address Hold). This input is used to place the Pentium's address bus into a high impedance state so that an inquire cycle can be run. The address used during the inquire cycle is read in by the Pentium when AHOLD is active.

AP (Address Parity). This signal is bidirectional and is used to indicate the even parity of address lines A_5 through A_{31}. AP is used as an output when the Pentium is outputting an address, and as an input during an inquire cycle.

$\overline{APCHK}$ *(Address Parity Check).* This output goes low if the Pentium discovers a parity error on the address lines. External circuitry is responsible for taking the appropriate action if a parity error is encountered.

APICEN (Advanced Programmable Interrupt Controller Enable). This input is used to enable or disable the Pentium's internal APIC interrupt controller circuitry. APICEN is sampled when the processor is reset.

$\overline{BE_0}$ *Through $\overline{BE_7}$.* These 8-byte enable outputs, together with A_3 through A_{31}, make up the 32-bit address output by the Pentium. Each byte enable is used to control a different 8-bit portion of the processor's 64-bit data bus. Table 15.1 shows the purpose of each byte enable.

TABLE 15.1 Byte enable operation

Output	Data Bus Byte Enabled
$\overline{BE_0}$	D_0–D_7
$\overline{BE_1}$	D_8–D_{15}
$\overline{BE_2}$	D_{16}–D_{23}
$\overline{BE_3}$	D_{24}–D_{31}
$\overline{BE_4}$	D_{32}–D_{39}
$\overline{BE_5}$	D_{40}–D_{47}
$\overline{BE_6}$	D_{48}–D_{55}
$\overline{BE_7}$	D_{56}–D_{63}

BF$_0$, BF$_1$ (Bus Frequency).   These inputs, sampled during reset, control the ratio of bus frequency to CPU core frequency. The bus/core ratio is 2/3 when BF$_0$ is high or floating. When low, the ratio becomes 1/2.

$\overline{BOFF}$ *(Backoff).*   This input causes the processor to terminate any bus cycle currently in progress and tristate its buses. Execution of the interrupted bus cycle is restarted when $\overline{BOFF}$ goes high.

PM/BP$_0$, PM/BP$_1$, BP$_2$, and BP$_3$.   The BP (breakpoint) outputs are associated with a set of internal registers called **debug** registers. There are eight debug registers. The first four (DR$_0$ through DR$_3$) are used to store the memory or I/O address of a program *breakpoint*. A breakpoint is a predefined address that the programmer chooses to help determine how a program executes. For example, a breakpoint may be set to address 1000H to see if the program ever reads or writes to that address. The breakpoint outputs go high when the breakpoint address in its respective debug register matches a program-generated address. Two PM (performance monitoring) ouputs are multiplexed with the lower two BP outputs. These signals are active after a reset and indicate the status of two internal performance monitoring counters. Setting the DE (debug extensions) bit in CR4 (Control Register 4) causes these two outputs to change to BP$_0$ and BP$_1$. We will examine the operation of the debug and control registers in Chapter 16.

$\overline{BRDY}$ *(Burst Ready).*   During a read cycle, this input indicates that data is available on the data bus. For write cycles, $\overline{BRDY}$ informs the processor that the output data has been stored. $\overline{BRDY}$ is used for both memory and I/O operations. If $\overline{BRDY}$ is not low when sampled, the Pentium inserts extra clock cycles into the current cycle (*wait* states) to provide extra time for the data transfer.

BREQ (Bus Request).   This output goes low when the Pentium has a pending bus cycle ready to begin. In a multiprocessor system, BREQ may be used to select a processor competing for the system bus.

BT$_0$ through BT$_3$ (Branch Trace).   BT$_0$ through BT$_2$ output the lower 3 bits (A$_0$ through A$_2$) of the target address of the currently executing branch instruction. BT$_3$ indicates the operand size of the current instruction (0 for 16-bit, 1 for 32-bit).

$\overline{BUSCHK}$ *(Bus Check).*   This input, when low, indicates to the Pentium that there was a problem with the last bus cycle. The processor may perform a machine check exception to recover.

$\overline{CACHE}$ *(Cacheability).*   This output indicates whether the data associated with the current bus cycle is being read from or written to the internal cache.

CLK (Clock).   This is the clock input to the processor. CLK must be stable (running at the desired frequency) within 150 ms of power on.

CPUTYP (CPU Type).   This input is used to specify the processor type in a dual-processor system. When low, CPUTYP specifies the primary processor. The dual processor is indicated when CPUTYP is high.

D/$\overline{C}$ (Data/Code).   This output indicates that the current bus cycle is accessing code (D/$\overline{C}$ is low) or data (D/$\overline{C}$ is high).

D/$\overline{P}$ (Dual/Primary). This output indicates the processor type in a dual-processing system. When low, the processor is the primary processor. The dual processor is indicated when high.

D_0 through D_{63} (Data Bus). These signals make up the Pentium's 64-bit bidirectional data bus. The actual data lines in use during a particular bus cycle are indicated by the byte enable ($\overline{BE}$) outputs.

DP_0 through DP_7 (Data Parity). These bidirectional signals are used to indicate the even parity of each data byte on the data bus. DP_0 represents the parity of the lower byte (D_0 through D_7).

$\overline{DPEN}$ (Dual Processing Enable). This signal is an output on the dual processor and an input on the primary processor in a dual-processing system. During reset, $\overline{DPEN}$ is used to initiate dual processing.

$\overline{EADS}$ (External Address Strobe). $\overline{EADS}$ is used to indicate that an external address may be read by the address bus during an inquire cycle.

$\overline{EWBE}$ (External Write Buffer Empty). This input, when low, indicates that the Pentium may proceed with the next cache writethrough operation.

$\overline{FERR}$ (Floating-Point Error). This output indicates that the processor's FPU generated an error. $\overline{FERR}$ is included to maintain compatibility with the error-handling mechanism of MS-DOS.

$\overline{FLUSH}$ (Flush Cache). This input, when low, causes the Pentium to writeback all modified data lines in its internal code and data cache.

$\overline{FRCMC}$ (Functional Redundancy Checking Master/Checker). This input is sampled during a RESET operation to determine whether the Pentium is operating as a master (when high) or a checker (when low). Two Pentiums may be connected in such a way that one is the master and the other is the checker. The checker checks every operation performed by the master to guarantee correct execution. A master/checker pair provides a measure of reliability to critical systems, such as flight control computers.

$\overline{HIT}$ (Inquire Cycle Cache Hit/Miss). This output indicates a cache hit (when low) as the result of an inquire cycle.

$\overline{HITM}$ (Hit/Miss to a Modified Cache Line). This output indicates that a modified line in the cache was hit as the result of an inquire cycle.

HLDA (Bus Hold Acknowledge). The HLDA output goes high (as a result of a HOLD request) to indicate that the Pentium has been placed in a hold state. If the code and data cache contain current instructions and operands, execution continues with no bus activity, only cache accesses.

HOLD (HOLD Bus). If HOLD is high when sampled, the Pentium tri-states its bus signals and activates HLDA. HOLD may be used by another processor that wishes to become the bus master.

IBT (Instruction Branch Taken). This output goes high for one clock cycle whenever the processor takes a branch (a JNZ that jumps, for example).

IERR (Internal Error).   This output, when low, indicates that a parity or redundancy check error has occurred. Parity errors may cause the Pentium to enter shutdown mode (see Section 15.4).

IGNNE (Ignore Numeric Exception).   A zero on this input allows the processor to continue executing floating-point instructions, even if an error is generated.

INIT (Initialization).   INIT is a rising edge-sensitive input that causes the processor to be initialized in the same way as a RESET, except that the internal registers and cache are left unchanged.

INTR (Interrupt Request).   This input, when high, causes the Pentium to initiate interrupt processing. An 8-bit vector number is read on the lower byte of the data bus to select an interrupt service routine. INTR is ignored if the interrupt enable bit in the flag's register is clear.

INV (Invalidation Request).   During an inquire cycle, INV is used to determine what happens to a cache line during a hit. If INV is high, the cache line is invalidated. If INV is low, the line is marked shared. See Section 15.8 for details on shared, and other, cache line states.

IU (U-Pipeline Instruction Complete).   This output goes high for one clock cycle each time an instruction completes in the U pipeline.

IV (V-Pipeline Instruction Complete).   This output goes high for one clock cycle when an instruction completes in the V pipeline.

KEN (Cache Enable).   A zero on this input enables caching of data being read into the processor. If $\overline{KEN}$ is high when sampled, no caching will occur.

LOCK (Bus Lock).   This output goes low to indicate that the current bus cycle is locked and may not be interrupted by another bus master.

M/IO (Memory/Input-Output).   This output indicates the type of bus cycle currently starting. If M/$\overline{IO}$ is high, a memory cycle is beginning; otherwise an IO operation is performed.

NA (Next Address).   This input, when low, indicates that the external memory system is capable of performing a pipelined access (two bus cycles in progress at the same time).

NMI (Nonmaskable Interrupt).   The Pentium responds to this rising edge input by issuing interrupt vector 2. No external interrupt acknowledge cycles are generated.

PBGNT (Private Bus Grant).   This signal is used in a dual-processing system to indicate when private bus arbitration is allowed.

PBREQ (Private Bus Request).   This signal is used to request a private bus operation in a dual-processing system.

PCD (Page Cacheability Disable).   This output indicates the state of CR3's PCD (page cache disable) bit. It is used to control cacheability on a page-by-page basis. More details on PCD will be covered in Chapter 16.

$\overline{PCHK}$ *(Data Parity Check).* This output goes low if the Pentium detects a parity error on the data bus. External hardware must take the appropriate action to handle the error.

$\overline{PEN}$ *(Parity Enable).* If this input is low during the same cycle a parity error is detected, the Pentium will save a copy of the address and control signals in an internal machine check register.

$\overline{PHIT}$ *(Private Hit).* When operating in a dual-processor system, $\overline{PHIT}$ is used to help maintain local cache coherency.

$\overline{PHITM}$ *(Private Modified Hit).* Used in conjunction with $\overline{PHIT}$ to maintain local cache coherency in a dual-processing system.

PICCLK (Programmable Interrupt Controller Clock). This input controls the serial data rate in the internal APIC interrupt controller.

PICD$_0$, PICD$_1$ (Programmable Interrupt Controller Data). These two signals are used to exchange data with the internal APIC interrupt controller.

PRDY. This output is used for debugging purposes. It indicates that the processor has stopped normal execution and is entering a debugging mode called **probe mode,** where it may execute debugging instructions. PRDY goes high in response to activity on R/$\overline{S}$.

PWT (Page Writethrough). This output indicates the status of the cache's writethrough paging bit in CR3.

R/$\overline{S}$. This negative edge-sensitive input places the Pentium in a wait state, and possibly executes instructions in probe mode. It is only used for debugging with the **Intel debug port,** a specific hardware debugging interface.

RESET (Processor Reset). The RESET input causes the Pentium to initialize its registers to known states, invalidate the code and data cache, and fetch its first instruction from address FFFFFFF0H. RESET must be active for at least 1 ms after power on to ensure proper operation.

SCYC (Split Cycle Indication). This output goes high to indicate that two or more locked bus cycles are performing a misaligned transfer. A misaligned transfer involves a 16- or 32-bit operand that is stored at a starting address that is not a multiple of four, or a 64-bit operand that does not begin at an address that is a multiple of eight.

$\overline{SMI}$ *(System Management Interrupt).* This negative edge-sensitive input is used to generate a system management interrupt. System management is used to perform special functions, such as power management.

$\overline{SMIACT}$ *(System Management Interrupt Active).* The $\overline{SMIACT}$ output goes low in response to an $\overline{SMI}$ request, and remains low until the processor leaves system management mode.

$\overline{STPCLK}$ *(Stop Clock).* When low, this input causes the Pentium to stop its internal clock (thus reducing the power consumed).

The next five inputs are all associated with the Intel debug port.

TCK (Test Clock).   This input is used to clock data into and out of the Pentium when performing a specific test procedure called *boundary scan* (IEEE Standard 1149.1).

TDI (Test Data Input).   This input allows serial test data to be input into the Pentium. Data is clocked in on the rising edge of TCK.

TDO (Test Data Output).   Serial test information is clocked out of TDO on the falling edge of TCK.

TMS (Test Mode Select).   This input is used to control the sequencing of boundary scan tests.

$\overline{TRST}$ (Test Reset).   This input, when low, resets the test controller logic.

W/$\overline{R}$ (Write/Read).   This output is used to indicate if the current bus cycle is a read operation (W/$\overline{R}$ is low) or a write operation (W/$\overline{R}$ is high).

WB/$\overline{WT}$ (Writeback/Writethrough).   This input defines the current cache line update protocol as writeback or writethrough. Cache lines may be defined on a line-by-line basis with WB/$\overline{WT}$.

Clearly, the hardware operation of the Pentium is complex, as would be any system designed around it. Only experienced computer engineers, who know all tricks of the trade, can successfully design a high-speed motherboard based on the Pentium.

15.3   RISC CONCEPTS

The last decade has delivered a great deal of change in the area of computer architecture, due to advances in microelectronic manufacturing technology. As more and more logic was crammed into the small space of the silicon wafer, computer architects were able to implement processors with increasingly complex instructions and addressing modes. In general, these processors fall into a category called **CISC,** for Complex Instruction Set Computer. Unfortunately, cramming so much logic into a single package leads to a performance bottleneck, since many CISC instructions require multiple clock cycles to execute.

One solution to the performance bottleneck was the emergence of a new design philosophy called **RISC,** for Reduced Instruction Set Computer. RISC designers chose to make the instruction sets *smaller,* using fewer instructions and simpler addressing modes. This reduced set of operations was easier to implement on silicon, resulting in faster performance. Coupled with other architectural differences and improvements, RISC gained popularity, acceptance, and commercial interest. Video games and laser printers now boast of their internal RISC processors.

In general, a RISC machine is designed with the following goals in mind:

- Reduce accesses to main memory.
- Keep instructions and addressing modes simple.

- Make good use of registers.
- Pipeline everything.
- Utilize the compiler extensively.

Let us briefly examine each goal, with the idea that we will compare the Pentium processor against each goal as we go.

Reduce Accesses to Main Memory

Even with the improvements made in memory technology, processors are much faster than memories. For example, a processor clocked at 100 MHz would like to access memory in 10 ns, the period of its 100-MHz clock. Unfortunately, the memory interfaced to the processor might require 60 ns for an access. So, the processor ends up waiting during each memory access, wasting execution cycles.

To reduce the number of accesses to main memory, RISC designers added instruction and data cache to the design. A **cache** is a special type of high-speed RAM where data and the *address* of the data are stored. Whenever the processor tries to read data from main memory, the cache is examined first. If one of the addresses stored in the cache matches the address being used for the memory read (called a *hit*), the cache will supply the data instead. Cache is commonly ten times faster than main memory, so you can see the advantage of getting data in 10 ns instead of 60. Only when we *miss* (i.e., do not find the required data in the cache), does it take the full access time of 60 ns. But this can only happen once, since a copy of the new data is written into the cache after a miss. The data will be there the next time we need it.

Instruction cache is used to store frequently used instructions, such as those in a short loop. Data cache is used to store frequently used data. Each cache is initially empty and fills as a program executes. The Pentium employs an 8KB instruction cache and an 8KB data cache. We will examine their operation in detail in Section 15.8.

Keep Instructions and Addressing Modes Simple

Computer scientists learned, after studying the operation of many different types of programs, that programmers use only a small subset of the instructions available on the processor they are using. The same is true for the processor's addressing modes.

Implementing fewer instructions and addressing modes on silicon reduces the complexity of the instruction decoder, the addressing logic, and the execution unit. This allows the machine to be clocked at a faster speed, since less work needs to be done each clock period.

Unfortunately, the Pentium processor cannot meet this design goal. In order to remain compatible with the installed software of the entire 80x86 family, the Pentium designers had to keep each and every instruction and addressing mode of the previous machine, the 80486. From this point of view, the Pentium looks like a CISC machine.

Make Good Use of Registers

RISC machines typically have large sets of registers. The number of registers available in a processor can affect performance the same way a memory access does. A complex calculation may require the use of several data values. If the data values all reside in memory during the calculations, many memory accesses must be used to utilize them. If the data values are stored in the internal registers of the processor instead, their access during calculations will be much faster. It is good, then, to have lots of internal registers.

The 8-bit processors of the late 1970s had few registers (two on the 6800 and seven on the 8085) for general purpose use. These registers were also only 8 bits wide. The Pentium has seven general purpose registers (EAX, EBX, etc.), all of which are 32 bits wide. This is four times the internal register storage space of the 8085. The Pentium also contains six 16-bit segment registers, a 32-bit stack pointer, and eight **floating-point registers,** each *80 bits* wide, which are used by the floating-point unit (covered in Section 15.9). So the Pentium has a fairly large set of registers to work with.

Pipeline Everything

Pipelining is a technique used to enable one instruction to complete with each clock cycle. Compare the two three-instruction sequences in Figure 15.3. On a nonpipelined machine, nine clock cycles are needed for the individual fetch, decode, and execute cycles. On a pipelined machine, where fetch, decode, and execute operations are performed *in parallel,* only five cycles are needed to execute the same three instructions. The first instruction requires three cycles to complete. Additional instructions then complete at a rate of one per cycle. As Figure 15.3(b) indicates, during clock cycle 5 we have I_3 completing, I_4 being decoded, and I_5 being fetched. A long sequence of instructions, say 1,000 of them, might require 3,000 clock cycles on a nonpipelined machine, and only 1,002 clock cycles when pipelined (three cycles for the first instruction, and then one cycle each for the remaining 999). So, pipelining results in a tremendous performance gain.

The Pentium employs two types of pipelines. These are the instruction pipelines (U and V pipelines, covered in Section 15.6), and a pipeline that performs special types of bus

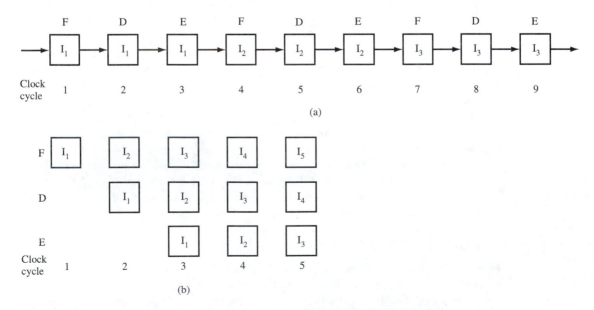

FIGURE 15.3   Execution of three instructions: (a) nonpipelined; (b) pipelined

cycles. The instruction pipelines are five-stage pipelines and are capable of independent operation. The U pipeline also forms the first five stages of the eight-stage floating-point pipeline. Certain types of bus cycles, such as back-to-back burst reads and writes, are possible through pipelined addressing logic.

Furthermore, the Pentium employs a technique called **branch prediction** that helps identify possible interruptions to the normal flow of instructions through the U and V pipelines. By predicting which instructions might branch and change program flow, it is possible to keep a steady stream of instructions flowing into the pipelines. Once again, an attempt is made to keep instructions completing at a rate of one per clock cycle. The Pentium is very like a RISC machine in this respect.

Utilize the Compiler Extensively

When a high-level language program (such as a C program) is compiled, the individual statements within the program's source file are converted into assembly language instructions or groups of instructions. A Pentium compiler, if written properly, can perform many optimizations on the assembly language code to take advantage of the Pentium's architectural advances. For instance, the compiler may reorder certain pairs of instructions to allow them to execute in parallel in the floating-point unit or dual-integer pipelines. Instructions may also be rearranged to take advantage of the Pentium's branch prediction strategy. Other optimizations may involve use of the instruction/data cache, or algorithms to allocate the minimum number of processor registers during parsing of an arithmetic statement. Sometimes it is possible to substitute an instruction (such as MOV EAX,0) with an equivalent instruction (such as SUB EAX,EAX) to reduce the number of clock cycles or the number of bytes of machine code.

Overall, the compiler plays an important role in helping a CISC or RISC machine achieve high performance. As a matter of fact, programs written for earlier 80x86 machines, even the 80486, can be improved by recompiling their sources with a Pentium compiler.

RISC Summary

What we have seen is that the Pentium contains both CISC *and* RISC characteristics. This is due in part to Intel's commitment to support for all 80x86 users. Through pipelining, branch prediction, instruction and data cache, and clever compilation, the Pentium delivers impressive speed. In fact, the Pentium is almost twice as fast as the 80486DX when clocked at identical speeds.

15.4   BUS OPERATIONS

The Pentium processor performs a number of different operations over its address and data buses. Data transfers (both single-cycle and burst transfers), interrupt acknowledge cycles, inquire cycles for examining the internal code and data cache, and I/O operations are just some of the operations possible. In this section we will examine the basic operation and purpose of each type of bus cycle.

TABLE 15.2   Bus cycle encodings

M/$\overline{\text{IO}}$	D/$\overline{\text{C}}$	W/$\overline{\text{R}}$	$\overline{\text{CACHE}}$	$\overline{\text{KEN}}$	Cycle Description
0	0	0	1	X	Interrupt acknowledge
0	0	1	1	X	Special cycle
0	1	0	1	X	I/O read non-cached
0	1	1	1	X	I/O write non-cached
1	0	0	1	X	Code read 8 bytes non-cached
1	0	0	X	1	Code read 8 bytes non-cached
1	0	0	0	0	Code read 32 bytes burst cached
1	1	0	1	X	Memory read up to 8 bytes non-cached
1	1	0	X	1	Memory read up to 8 bytes non-cached
1	1	0	0	0	Memory read 32 bytes burst cached
1	1	1	1	X	Memory write up to 8 bytes non-cached
1	1	1	0	X	32 byte cache writeback burst

Note: X = don't care

Decoding a Bus Cycle

The Pentium bus logic indicates the type of bus cycle currently starting with the use of its cycle definition signals. These are the M/$\overline{\text{IO}}$, D/$\overline{\text{C}}$, W/$\overline{\text{R}}$, $\overline{\text{CACHE}}$, and $\overline{\text{KEN}}$ signals. Table 15.2 shows how these signals define the current bus cycle.

Special cycle bus cycles require additional decoding and use the byte enable outputs for selection. Table 15.3 defines these types of bus cycles.

The memory system must be designed to respond to each type of bus cycle.

Bus Cycle States

There are six possible states the Pentium bus may be in, depending on what type of cycle is being processed. These states are Ti, T1, T2, T12, T2P, and TD. Ti is the idle state and indicates that no bus cycle is currently running. The bus begins in this state after a hardware reset. T1 and T2 are the first and second states of a bus cycle. During T1, a valid address is output on the address lines and $\overline{\text{ADS}}$ is made active (low). Data is read or written during T2, and the $\overline{\text{BDRY}}$ input is examined.

TABLE 15.3   Special bus cycles

$\overline{BE_7}$	$\overline{BE_6}$	$\overline{BE_5}$	$\overline{BE_4}$	$\overline{BE_3}$	$\overline{BE_2}$	$\overline{BE_1}$	$\overline{BE_0}$	Special Bus Cycle
1	1	1	1	1	1	1	0	Shutdown
1	1	1	1	1	1	0	1	Flush cache
1	1	1	1	1	0	1	1	Halt
1	1	1	1	0	1	1	1	Writeback
1	1	1	0	1	1	1	1	Flush acknowledge
1	1	0	1	1	1	1	1	Branch trace message

When a second bus cycle is started before the first one completes, the bus enters the T12 state. Data for the first bus cycle is transferred, and a new address is output on the address lines. State T2P continues the bus cycle started in T12. TD is used to insert a *dead* state between two consecutive cycles (read followed by write or vice versa), in order to give the system bus time to change states.

The bus state controller follows a predefined set of transitions to switch from state to state. Figure 15.4 shows these transitions in the form of a state diagram.

The transitions between states are defined as follows:

0: No bus cycle requested.

1: New bus cycle started. $\overline{\text{ADS}}$ is taken low.

2: Second clock cycle of current bus cycle.

FIGURE 15.4 Bus state transitions

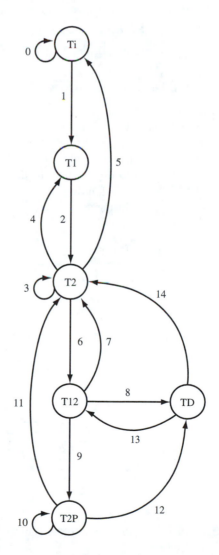

3: Stay in T2 until $\overline{\text{BDRY}}$ is active or new bus cycle is requested.

4: Go back to T1 if a new request is pending.

5: Bus cycle complete; go back to idle state.

6: Begin second bus cycle.

7: Current cycle is finished and no dead clock is needed.

8: A dead clock is needed after the current cycle is finished.

9: Go to T2P to transfer data.

10: Wait in T2P until data is transferred.

11: Current cycle is finished and no dead clock is needed.

12: A dead clock is needed after the current cycle is finished.

13: Begin a pipelined bus cycle if NA is active.

14: No new bus cycle is pending.

Clearly, a lot of thought went into the design of the Pentium's bus controller.

Single-Transfer Cycle

This cycle is used to transfer up to 8 bytes of non-cacheable data between the processor and memory. The byte enable outputs indicate how many bytes will be transferred. The timing for single-transfer read and write operations is shown in Figure 15.5.

The cycle begins during clock cycle T1, when $\overline{\text{ADS}}$ goes low. $\overline{\text{CACHE}}$ is taken high to indicate to external circuitry that the data is not going to, or coming from, the internal cache. If $\overline{\text{BDRY}}$ goes low during the T2 clock cycle, the data will be transferred and the operation completes during clock cycle Ti.

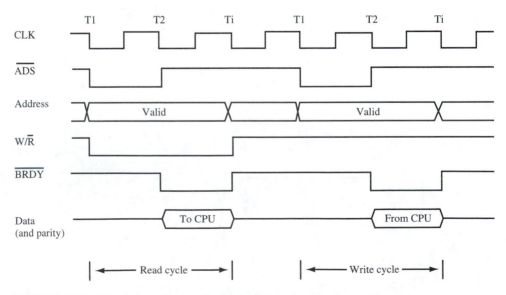

FIGURE 15.5   Single-transfer read/write cycles

If $\overline{\text{BDRY}}$ is not low during T2, additional T2 clock cycles are generated. These extra clock cycles are called *wait states,* and are used to give the memory and I/O systems additional time to complete the read or write request.

Burst Cycles

The Pentium supports burst reads and writes of 32 bytes. The cache uses burst cycles for line loads and writebacks. During a burst operation, a new 8-byte chunk can be transferred every clock cycle. The processor supplies the starting address of the first group of 8 bytes at the beginning of the cycle. The next three groups of 8 bytes are transferred according to the burst order shown in Table 15.4.

The external memory system must generate the remaining three addresses itself, and supply the data in the correct order. Figure 15.6 shows the timing for a cacheable burst read cycle. Note that an 8-byte data chunk is transferred during every T2 clock cycle.

Locked Operations

Many operating system processes depend on what is termed *atomic access* to data stored in memory. An **atomic operation** cannot be broken down into smaller suboperations. The data accessed during the atomic operation often comes in the form of a **semaphore**—a

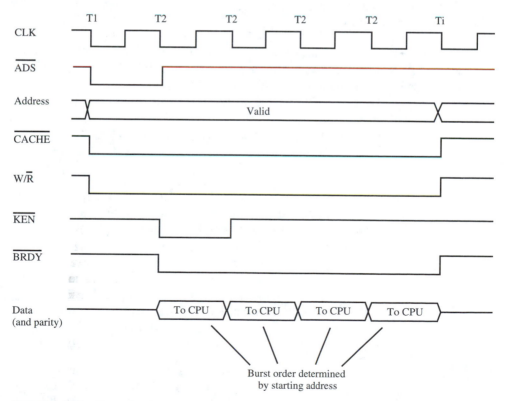

FIGURE 15.6 Burst read cycle

special type of counter variable that must be read, updated, and stored in one single, uninterruptable operation. This requires a read cycle followed by a write cycle. No other devices may take over control of the processor buses while the atomic operation is taking place. This is what it means to "lock" the bus. Some instructions, like XCHG, automatically lock the bus when one of their operands is a memory operand.

BOFF

The $\overline{\text{BOFF}}$ input provides a way for other processors in a multiprocessor system to instantly take over the Pentium's buses. The Pentium samples $\overline{\text{BOFF}}$ every clock cycle, and if low, puts its buses into a high-impedance state, beginning with the next clock cycle. Execution resumes with the interrupted cycle when $\overline{\text{BOFF}}$ is returned high.

Bus Hold

The HOLD input provides a second way for a different bus master to take control of the Pentium's buses. Unlike $\overline{\text{BOFF}}$, HOLD completes the current bus cycle and then tri-states its buses. The HLDA output indicates when the Pentium is in the HOLD state.

Interrupt Acknowledge

The processor runs two interrupt acknowledge cycles in response to an INTR request. Both cycles are locked. External circuitry (such as the 8259A Programmable Interrupt Controller used in the PC) is responsible for supplying an 8-bit interrupt vector number on D_0 through D_7. To maintain hardware compatibility with earlier 80x86 machines, the data on D_0 through D_7 is ignored by the processor during the first interrupt acknowledge cycle and accepted during the second. Figure 15.7 shows the associated timing.

The byte enable outputs are used to indicate which of the two cycles the processor is running. During the first interrupt acknowledge cycle, $\overline{\text{BE}}_4$ is low and the other 7 byte enables are high. During the second cycle, only $\overline{\text{BE}}_0$ is low.

Cache Flush

In response to a zero supplied to the $\overline{\text{FLUSH}}$ input, the Pentium flushes its internal code and data cache by performing writebacks for any lines that have been changed. The lines are then invalidated. When all writebacks complete, the processor runs a **flush acknowledge** cycle to inform external circuitry (such as a second level cache) that its caches are flushed.

TABLE 15.4 Burst transfer order

1st Address	2nd Address	3rd Address	4th Address
0	8	10	18
8	0	18	10
10	18	0	8
18	10	8	0

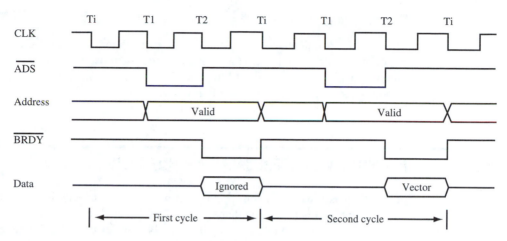

FIGURE 15.7 Interrupt acknowledge cycles

Shutdown

If the Pentium detects an internal parity error, a shutdown cycle is run. Execution is suspended while in shutdown, until the processor receives an NMI, an INIT, or a RESET request. Internal cache data remains unchanged unless the processor is snooped with an inquire cycle or the cache is flushed via $\overline{\text{FLUSH}}$.

HALT

This cycle is similar to shutdown, except that the INTR signal may also be used to resume execution (if interrupts are enabled). This cycle is run when the processor encounters the HLT instruction.

Pipelined Cycles

The Pentium is capable of starting a second bus cycle before the current one is completed. It does so through pipelined read and write logic, in response to a request on the $\overline{\text{NA}}$ input. As always, pipelining improves performance by performing different operations in parallel. Read cycles may be pipelined into write cycles, and vice versa. Writeback operations and locked bus cycles are not pipelined.

Inquire Cycles

Inquire cycles are used to maintain cache coherency in a multiprocessor system. The Pentium processor is able to watch the system bus (address, data, and control signals) in a multiprocessor system. This is called **bus snooping.** If the Pentium detects a memory read/write operation being performed by another CPU, it runs an internal inquire cycle to determine whether the address on the bus is stored in one of its internal caches. If so, the cache may need to be updated. In order to implement bus snooping, it is necessary for the address bus to be bidirectional.

Bus Cycle Summary

It is necessary to design memory systems capable of supporting all of the different kinds of bus cycles that are possible with the Pentium. Good use must be made of the bus cycle definition signals and the byte enable outputs. Not taking advantage of this important part of the Pentium's architecture will result in lower performance.

15.5   THE PENTIUM'S SUPERSCALAR ARCHITECTURE

The Pentium is capable, under special circumstances, of executing two integer or two floating-point instructions simultaneously. This parallel execution is made possible through the Pentium's twin U and V pipelines. Processors capable of parallel instruction execution of multiple instructions are known as *superscalar machines*.

There are four restrictions placed on a pair of integer instructions attempting parallel execution:

- Both must be simple instructions.
- No data dependencies may exist between them.
- Neither instruction may contain both immediate data and a displacement value (such as MOV TABLE[SI],7).
- Prefixed instructions (such as MOV ES:[DI],AL) may only execute in the U pipeline.

A simple instruction does not require microcode control to execute, and generally takes one clock cycle to complete. Some examples are register-to-register MOVs, INC, DEC, and near conditional jumps (JZ, JNZ, etc.). A conditional jump must be the second instruction in the pair.

Some simple instructions may take two or three clock cycles. These are arithmetic and logical instructions (ALU instructions) that use both register and memory operands.

A data dependency between two instructions exists if the second instruction reads an operand written to it by the first instruction (read-after-write dependency), or if both instructions write to the same operand. For example, the instructions

```
ADD   AX,BX
ADD   AX,CX
```

cannot be paired, since both read and write the AX register.

For floating-point instructions, the first instruction of the pair must be one of the following:

```
FLD <single/double>       FLD ST(i)
FADD      FSUB      FMUL      FDIV
FCOM      FUCOM     FTST      FABS
FCHS
```

The second instruction must be FXCH.

As mentioned earlier, the compiler plays an important role in the ordering of instructions during code generation. By looking for the allowable combinations of integer and floating-point instructions, the compiler is able to help keep the Pentium's superscalar architecture running at full speed.

15.6   PIPELINING

As mentioned in Section 15.3, pipelining is a valid technique for improving instruction execution rate. Let us examine the operation of the Pentium's U and V pipelines.

Figure 15.8 shows the names and order of the five-stage U and V instruction pipelines. Specifically, we have:

PF   Prefetch

D1   Instruction Decode

D2   Address Generate

EX   Execute, Cache, and ALU Access

WB   Writeback

Note that each pipeline has its own PF stage, its own D1 stage, and so on. The U pipeline can execute any processor instruction (including the initial stages of the floating-point instructions), whereas the V pipeline only executes *simple* instructions (as defined in Section 15.5). Under ideal conditions, two integer instructions may complete execution every clock cycle, as indicated by Figure 15.9. Recall that a pair of instructions that execute in parallel

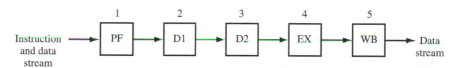

FIGURE 15.8   U and V instruction pipeline stages

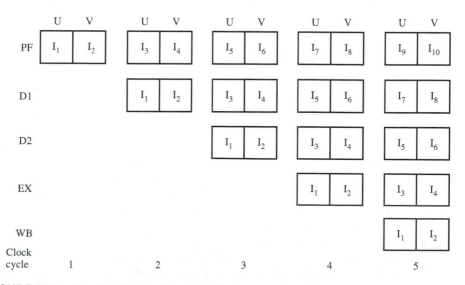

FIGURE 15.9   Pipelined instruction execution

must both be simple instructions and cannot have any data dependencies between them. Examine the state of the pipelines during the fifth clock cycle. The first pair of instructions complete execution in this clock cycle. Four more cycles are needed to complete the other eight instructions that are in various stages of execution.

Instructions are fed into the PF stage after being prefetched from the instruction cache or memory. Decoders in each D1 stage (U and V pipelines) determine if the current pair of instructions can execute together. Instructions that contain a prefix byte require an additional clock cycle in the D1 stage and may only execute in the U pipeline. They may not be paired with any other instruction.

Addresses for operands that reside in memory are calculated in stage D2. In the EX stage, operands are read from the data cache (or memory), ALU operations are performed, and branch predictions for instructions (except conditional branches in the V pipeline) are verified. See Section 15.7 for details on branch prediction.

The final stage, WB, is used to write the results of the completed instruction and verify conditional branch instruction predictions.

When paired instructions reach the EX stage, it is possible that one or the other will stall and require additional cycles to execute. A pipeline stall lowers performance, since no work is done during the stall. Instructions stall for various reasons, most notably when their operands are not found in the data cache. If the instruction in the U pipeline stalls, so does the instruction in the V pipeline. If the V pipeline instruction stalls, the instruction in the U pipeline may continue executing. Both instructions must progress to the WB stage before another pair (or the next single instruction) may enter the EX stage.

15.7   BRANCH PREDICTION

The performance gain realized through pipelining can be reduced by the presence of program transfer instructions in the instruction stream. Instructions such as JMP, CALL, RET, and the conditional jumps change the flow of execution in a program. This is a problem for the instruction pipeline, which is always filled with a group of instructions that occupy sequential locations in memory. Program transfer instructions change the sequence, causing all instructions that entered the pipeline after the program transfer instruction to become invalid. Refer back to Figure 15.3(b). Suppose that instruction I_3 is a conditional jump to instruction I_{50} at some other address in the program (the **target** address). This means that the instructions that entered the pipeline after I_3 (I_4 and I_5) are invalid, since they occur in the program *after* the conditional jump. These instructions must be discarded, or **flushed,** from the pipeline and the new sequence of instructions, beginning with I_{50}, loaded in. This causes **bubbles** in the pipeline, where no work is done as the pipeline stages are reloaded.

To avoid this problem, the Pentium uses a scheme called **dynamic branch prediction.** In this scheme, a prediction is made concerning the branch instruction currently in the pipeline. The prediction will be either taken or not taken. If the prediction turns out to be true, the pipeline will not be flushed, and no clock cycles will be lost. If the prediction turns out to be false, the pipeline is flushed and started over with the correct instruction. It is best if the predictions are true most of the time.

The Pentium accomplishes branch prediction through the use of a **branch target buffer (BTB),** a special cache that stores the instruction and target addresses of any branch instructions that have been encountered in the instruction stream. Along with the addresses for each instruction, the BTB also stores two **history bits** that indicate the execution history of the last two branch instructions. The history bits are initially set to 11 when a new target address is placed into the BTB. Whenever the corresponding branch instruction is encountered, the history bits are updated, as indicated in Figure 15.10. Note that repeated failures to take a branch cause the history bits to become 00 and the prediction to become not taken.

The BTB uses the history bits to predict whether the branch is taken or not taken. As long as they are not both zero, the prediction will be taken. If a new branch instruction is encountered (no target address in the BTB), the prediction is not taken.

Two 32-byte prefetch buffers work with the BTB and the D1 stage of the U and V pipelines to keep a steady stream of instructions flowing into the pipelines. One buffer prefetches instructions from the current program address. The other buffer, activated when the BTB predicts "taken," will prefetch instructions from the target address. The BTB is accessed with the address of the branch instruction during the D1 stage. If found and the BTB's prediction is "taken," the second prefetch buffer becomes active and no cycles are lost. Incorrect predictions, or correct predictions with the wrong target address, cause the pipelines to be flushed. This wastes three clock cycles in the U pipeline and four in the V pipeline.

In accordance with good programming practice, conditional jumps are used to perform loops. Since we generally require multiple passes through a loop, we would want to begin predicting "taken" right away. The Pentium sets the history bits to 11 for a new entry, so the number of lost clock cycles will be minimized by the dynamic branch prediction mechanism.

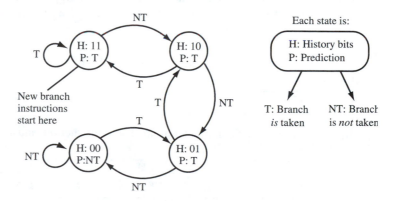

FIGURE 15.10   *Dynamic branch prediction operation*

15.8   THE INSTRUCTION AND DATA CACHES

The Pentium processor employs two separate internal cache memories, one for instructions and the other for data. In this section, we will examine the benefits of using caches, their internal organization and operation, and methods of maintaining valid data in both cache and main memory.

Cache Background

Which is faster: 10-ns memory or 70-ns memory? The answer is obvious and at the heart of the performance goal for a system using cache memory. Cache is a special type of high-speed RAM (access time of 10 ns or less) that is used to help speed up accesses to memory and reduce traffic on the processor's busses.

As indicated by Figure 15.11, an on-chip cache is used to feed instructions and data to the CPU's pipeline. When an instruction or data operand is required from main memory, the on-chip cache will be searched first. If the instruction or data is found in the cache (a *hit*), a copy is sent to the pipeline very quickly, usually within one clock cycle.

When the required instruction or data is not found in the internal cache (a *miss*), the processor is forced to go to external memory. An **external cache** (also called a *second-level cache*) is examined next. If there is another miss, or there is no external cache, the main memory is accessed. A copy of the instruction or data from main memory is written to the cache so that it will be there if needed again.

How does all this activity speed things up? Consider a system that has an internal cache access time of 10 ns and a main memory access time of 70 ns. There is no external cache. The time for a hit is 10 ns. The time for a miss is 80 ns, since both cache and main memory are accessed during a miss. So, the average access time will be between 10 ns and 80 ns, depending on how many hits there are. A *hit ratio* specifies the percentage of hits to total cache accesses. For example, a hit ratio of 0.9 means that nine times out of ten the cache contains the requested information. Thus, the hit ratio

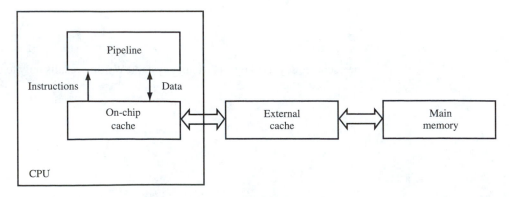

FIGURE 15.11   Using cache in a microcomputer system

affects the average access time. To predict the average access time, use the following equation:

$$T_{acc} = HitRatio \times T_{cache} + (1 - HitRatio) \times (T_{cache} + T_{ram})$$

Continuing with our example, a hit ratio of 0.9 results in an access time of 17 ns (9 ns for hits, 8 ns for misses). The hit ratio is governed by many factors, including the size of the program, the type and amount of data used by the program, and the addressing activity during execution.

Two characteristics of a running program pave the way for a performance improvement when cache is used. First, when we access a memory location, there is a good chance we will access it again in the future. Second, when we access one location, there is a good possibility that we will access the *next* location also. In general, these accesses maintain a *locality of reference* in a small range of memory locations. Consider the following loop of instructions:

```
          MOV    CX,1000
          SUB    AX,AX
NEXT:     ADD    AX,[SI]
          MOV    [SI],AX
          INC    SI
          LOOP   NEXT
```

The code for the four instructions that make up the body of the loop will be executed 1,000 times. If the cache is initially empty, each instruction fetch will generate a miss in the cache, causing the instruction to be read from main memory. The first pass through the loop fills the cache with the code for the entire loop. The next 999 passes will all generate hits for each instruction fetch. This will speed things up tremendously.

When a miss occurs, the cache reads a copy of a *group* of locations from main memory. This group is called a **line of data.** This prepares the cache for hits in case there are more localized references made (to data within the line). Our loop example shows the benefit of this process. The first instruction, MOV   CX,1000, causes a miss. When its code is copied from main memory, so are the bytes for the next few instructions (as many as will fit within the line). So, after fetching the first instruction, the rest of the loop is already in the cache, before we have even finished the first pass!

In addition to the instruction fetches, the loop example also contains accesses to data operands stored in memory via ADD   AX,[SI] and MOV   [SI],AX. Once again, during reads, a miss will cause a line of data to be read from main memory. So each miss reads in data for the next few passes through the loop, guaranteeing fast data access.

But what about the data writes caused by the MOV   [SI],AX instruction? Should they be written just to the cache (10 ns) or to the cache and main memory (70 ns)? This depends on the cache policy used by a particular system. Writing results only to the cache are called **writeback;** writing to the cache and to main memory is called **writethrough.** The writeback policy results in fast writes at the expense of out-of-date main memory data. Writethrough maintains valid data in main memory, but requires long write times, which slows down execution.

Out-of-date main memory data in the writeback policy is eventually updated with the correct data from the cache. This happens when the cache is full and a line must be replaced with a new line of data coming in. If the line being replaced was written to, it must be copied back to main memory. So during a miss, it may be necessary to access main memory twice: once to store the updated line being replaced, and a second time to read the new line of data to satisfy the miss.

When the cache is full and a line must be replaced, the victim line may be chosen any number of ways. One algorithm used to pick a victim is called **LRU (least recently used)**. LRU is based on the idea that the cache entry least recently used is not likely to be used in the future, so it may be replaced. One or more bits are added to the cache entry to support the LRU algorithm. These bits are updated during hits and examined when a victim must be chosen.

Cache Organization

How is it possible that a cache that contains numerous entries can search them so quickly and report a hit if a match is found? This has to do with the organization of the cache. The address and data that make up each cache entry may be organized in different hardware configurations. There are three main designs: direct-mapped, fully associative, and set-associative. A direct-mapped cache uses a portion of the incoming physical address to select an entry. A *tag* stored with the entry is compared with the remaining address bits. A match represents a hit. This cache is illustrated in Figure 15.12(a). Note that there are 128 entries in the cache, which has a line size of 32 bytes. The lower 5 bits of the physical address are not needed, since all the bytes accessed by them are stored in a single line. The next 7 address bits (called *index bits*) are used to select one of 128 entries in the cache. The tag at the selected entry is compared with the remaining 20 upper address bits.

Fully associative cache, shown in Figure 15.12(b), uses larger tags and does not select an entry based on index bits, as direct-mapped cache does. Instead, the upper address

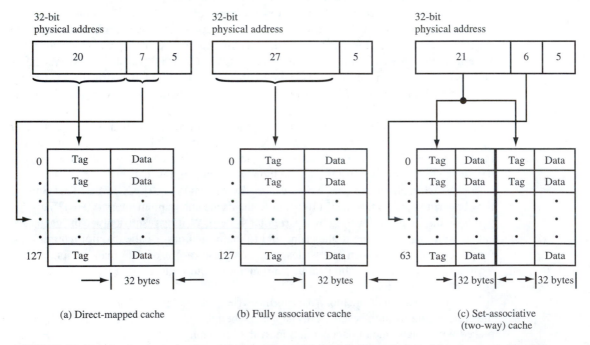

(a) Direct-mapped cache   (b) Fully associative cache   (c) Set-associative (two-way) cache

FIGURE 15.12   Cache organization: (a) direct-mapped; (b) fully associative; (c) set-associative (two-way)

bits of the incoming physical address are compared with *every* tag in the cache. This requires more extensive hardware than the direct-mapped cache. However, the fully associative cache is capable of storing data in more flexible ways than a direct-mapped cache. For instance, in the direct-mapped cache there can be only one data entry with an index value of 7 at any time. In the fully associative cache, multiple tags may contain addresses that correspond to an index of 7 (although index bits are not used). This helps to increase the hit ratio over that of the direct-mapped cache when the same data is accessed frequently.

The compromise between the direct-mapped and fully associative designs is the **set-associative cache.** The entries are divided into sets containing two, four, eight, or more entries. Two entries per set is referred to as *two-way set associative cache.* Each entry in a set has its own tag. A particular set is selected through the use of index bits. The remaining upper address bits are compared with each tag in the set at the same time. The tag comparators are smaller than those used in the fully associative cache, since they do not have to compare as many bits. Figure 15.12(c) shows how a two-way set associative cache is organized.

The set-associative cache combines the simple tag comparator hardware of the direct-mapped cache with the flexible data caching capability of the fully associative cache. This is the type of cache used by the Pentium.

The Data and Instruction Cache

The Pentium's data and instruction caches are both organized as two-way set-associative caches with 128 sets. This gives 256 entries per cache. There are 32 bytes in a line (64 bytes per set), resulting in 8KB of storage per cache. The data and instruction caches may be accessed simultaneously. An LRU algorithm is used to select victims in each cache.

Each entry in a set has its own tag. Figure 15.13 shows the internal structure of each cache. The tags in the data cache are **triple ported,** meaning that they can be accessed from three different places at the same time. Two of these are the U and V pipelines, which access the data cache to read/write instruction operands. The third port is used for a special operation called bus **snooping.** Bus snooping is used to help maintain consistent data in a multiprocessor system where each processor has a separate cache.

In addition, each entry in the data cache can be configured for writethrough or writeback.

FIGURE 15.13 Structure of 8KB instruction and data cache

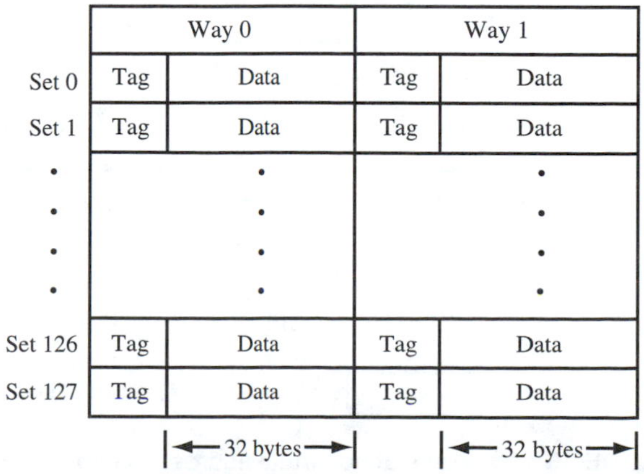

	Way 0		Way 1	
Set 0	Tag	Data	Tag	Data
Set 1	Tag	Data	Tag	Data
•		•		•
•		•		•
•		•		•
•		•		•
Set 126	Tag	Data	Tag	Data
Set 127	Tag	Data	Tag	Data

|←— 32 bytes —→| |←— 32 bytes —→|

The instruction cache is write protected to prevent self-modifying code from changing the executing program. Tags in the instruction cache are also triple ported, with two ports used for split-line accesses (reading the upper half of one line and the lower half of the next line at the same time). The third port is for bus snooping.

Parity bits are used in each cache to help maintain data integrity. Each tag has its own parity bit, as does every byte in the data cache. There is one parity bit for every 8 bytes of data (a quarter of a line) in the instruction cache.

The Translation Lookaside Buffers

The instruction and data caches contain **TLBs (Translation Lookaside Buffers)** that translate *virtual* (also called *logical* or *linear*) *addresses* into physical addresses. Physical addresses are used to access the cache because the same address is used to access main memory.

The TLBs are caches themselves. The data cache contains two TLBs. The first is four-way set-associative with 64 entries. This TLB translates addresses for 4KB pages of main memory. Figure 15.14 shows how this TLB is used to translate a virtual address into a physical address.

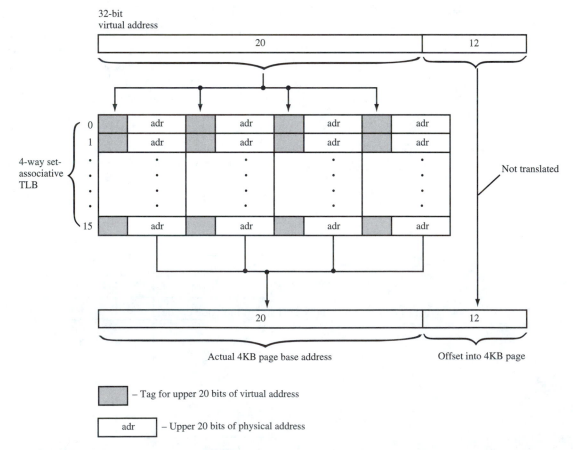

FIGURE 15.14   Translating virtual addresses into physical addresses with a TLB

The lower 12 bits of the physical address are the same as the lower 12 bits of the virtual address. The upper 20 bits of the virtual address are checked against four tags and translated into the upper 20 physical address bits during a hit. Since this translation must be done very quickly, the TLB is kept small (only 64 entries).

The second TLB used by the data cache is four-way set-associative with eight entries, and is used to handle 4MB pages. Both TLBs are parity protected and dual ported.

The instruction cache uses a single four-way set-associative TLB with 32 entries. Both 4KB and 4MB page addresses are supported, with the 4MB pages cached in 4KB chunks. Once again, parity bits are used on the tags and data to maintain data integrity.

Entries are replaced in all three TLBs through use of a 3-bit LRU counter stored with each set.

Cache Coherency in a Multiprocessor System

When multiple processors are used in a single system, there needs to be a mechanism whereby all processors agree on the contents of shared cache information. For example, two or more processors may utilize data from the same main memory location, X. Each processor will maintain a copy of the data for location X in its own data cache. So after each processor has changed the value of the data item, it is possible to have different (incoherent) values of X's data in each cache, as shown in Figure 15.15. When this happens, which value of X's data should actually be written back into main memory?

The Pentium's mechanism for maintaining *cache coherency* in its data cache is called **MESI (Modified/Exclusive/Shared/Invalid).** MESI is a cache-consistence protocol that uses 2 bits stored with each line of data to keep track of the state of the cache line. The four states are modified, exclusive, shared, and invalid. Specifically, each state is defined as follows:

Modified: The current line has been modified (does not match main memory) and is only available in a single cache.

Exclusive: The current line has not been modified (matches main memory) and is only available in a single cache. Writing to this line changes its state to modified.

Shared: Copies of the current line may exist in more than one cache. A write to this line causes a writethrough to main memory and may invalidate the copies in the other cache.

Invalid: The current line is empty. A read from this line will generate a miss. A write will cause a writethrough to main memory.

Only the shared and invalid states are used in the code cache.

The MESI protocol requires the Pentium to monitor all accesses to main memory in a multiprocessor system. This is called **bus snooping.** Referring back to Figure 15.15, if Processor 3 writes its local copy of X (30) back to memory, the memory write cycle will be detected by the other three processors. Each processor will then run an internal *inquire* cycle to determine whether its data cache contains the address of X. Processors 1 and 2 will update their cache based on their individual MESI states.

Inquire cycles examine the code cache as well. Recall that the code and data cache tags are triple ported, with one port dedicated to bus snooping. The Pentium's address lines are used as inputs during an inquire cycle to accomplish bus snooping.

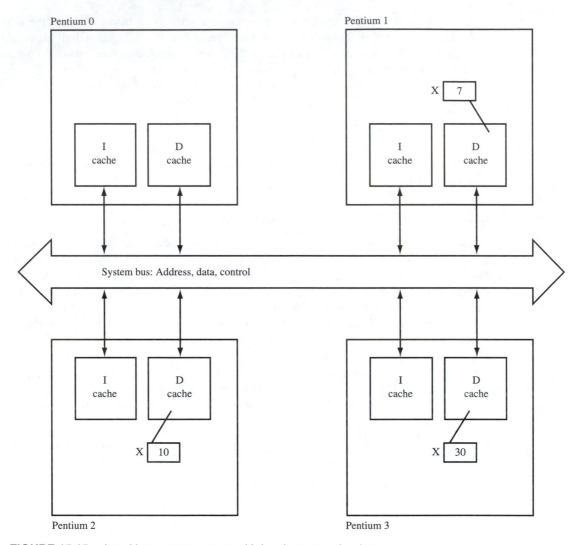

FIGURE 15.15   A multiprocessor system with incoherent cache data

Cache Instructions

Three instructions are provided to allow the programmer some control over the operation of the cache. These instructions are INVD (Invalidate cache), INVLPG (Invalidate TLB entry), and WBINVD (Write back and invalidate cache). INVD effectively erases all information in the data cache (by marking it all invalid). Any values not previously written back will be lost when INVD executes. This problem can be avoided by using WBINVD, which first writes back any updated cache entries, and then invalidates them. INVLPG invalidates the TLB entry associated with a supplied memory operand. This may be necessary when implementing a paged memory system.

Clearly these instructions are meant to be used by system programmers. Indiscriminately wiping out the cache during execution generally leads to trouble.

Cache Summary

In this section we examined how data and code cache may be used to enhance processing speed. The operation of the Pentium's 8KB two-way set-associative code and data caches was discussed. We also examined the operation of a translation lookaside buffer, the different cache policies (such as writethrough and writeback), and the MESI protocol for maintaining cache consistency.

In conclusion, it is important to note that all of these cache operations are performed automatically by the Pentium. No programming code is needed to get the cache to work.

15.9   THE FLOATING-POINT UNIT

Like the 80486 processor, the Pentium contains an on-chip floating-point unit. Prior to the 80486, an external coprocessor was used to perform floating-point operations for the 8086, 80286, and 80386. These 80x87 coprocessors, illustrated in Table 15.5, shared address and data signals, as well as control lines, with the processor. The simple fact that these coprocessors were *outside* the CPU increased the time required to perform a floating-point operation, due to synchronization issues. Moving the coprocessor onto the same chip as the processor, like the 80486 and Pentium, allows faster communication and quicker execution.

The Pentium's FPU is totally redesigned over that used in the 80486. Many floating-point instructions now require fewer clock cycles than previous 80x87 units, and new algorithms provide additional speed increases. Table 15.6 shows the improvement in the floating-point multiply instruction, FMUL, in each generation. Clearly, the Pentium's FPU leaves little room for improvement.

The FPU achieves its quick speed through a pipeline containing eight stages. The first five stages make up the U pipeline, which processes integer instructions. The opera-

TABLE 15.5   Coprocessor family

Processor	Coprocessor
8086/88	8087
80286	80287
80386	80387
80486	Internal FPU
Pentium	Internal FPU

TABLE 15.6   FMUL instruction performance

Coprocessor	Minimum Clock Cycles Required
8087	130
80287	130
80387	29
80486 FPU	16
Pentium FPU	1

tion of the fifth stage is different, however, and feeds the remaining three stages that complete the floating-point pipeline. Recall that the fifth stage of the U pipeline was WB (Write Back). When configured for a floating-point operation, the fifth stage becomes X1, the first floating-point execution stage. Figure 15.16 shows the eight-stage FPU pipeline. The stages, and their functions, are as follows:

PF	Prefetch
D1	Instruction decode
D2	Address generation
EX	Memory and register read, floating-point data converted into memory format, memory write
X1	Floating-point execute, stage one. Memory data converted into floating-point format, write operand to floating-point register file, bypass 1 (send data back to EX stage)
X2	Floating-point execute stage two
WF	Round floating-point result and write to floating-point register file, bypass 2 (send data back to EX stage)
ER	Error reporting, update status word

Note the use of two different **bypass** connections. Bypass 1 connects the output of the X1 stage to the input of the EX stage. This is done to allow a floating-point register write operation in the X1 stage to bypass the floating-point register file and send the results to the instruction performing a floating-point register read in the EX stage. For example, the two instructions:

```
FLD     ST
FMUL    ST
```

both use the ST register as an operand. The result produced by FLD must be available before the FMUL instruction can proceed. Using bypass 1, the data from FLD (in stage X1) is made available as an input operand to FMUL (in stage EX) before it is written into the floating-point register file. This prevents the pipeline from stalling while waiting for the register file to provide a copy of ST. Overall, we get a decrease in the number of clock cycles required to perform many different combinations of instructions. This technique is also called *forwarding*.

The second type of bypass takes place between the WF and EX stages, where the result of an arithmetic instruction (in stage WF) is made available to the next instruction fetching operands in the EX stage.

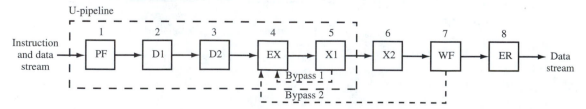

FIGURE 15.16   FPU pipeline stages

The floating-point register file contains the eight 80-bit floating-point registers, ST(0) through ST(7). The read and write sections are dual ported, to allow two reads or two writes to take place simultaneously. As indicated by Figure 15.17, the two read ports send data into the EX stage (which reads register operands). The two write ports receive data from the X1 and WF stages of the pipeline. Both bypasses are shown to illustrate how results are fed back from X1 and WF to EX.

The Pentium is capable of executing two floating-point instructions at the same time under special circumstances. To do so, the first instruction of the pair must be one of the following:

```
FLD <single/double>          FLD ST(i)
FADD        FSUB        FMUL        FDIV
FCOM        FUCOM       FTST        FABS
FCHS
```

The second instruction must be FXCH. Since the U pipeline makes up the first five stages of the FPU pipeline, the FXCH instruction executes in the V pipeline. Since both instructions execute in parallel, the FXCH instruction literally requires *zero* clock cycles! Normally FXCH requires one clock cycle when issued by itself.

Since many of the floating-point instructions use the operand on top of the floating-point stack, there is a good possibility of a stall when two instructions require the top-of-stack operand. Using FXCH as the second instruction of a legal pair helps avoid this problem.

If the second floating-point instruction in a pair is not FXCH, both floating-point instructions are executed one at a time. If the second instruction is not a floating-point instruction, it cannot be paired with the first instruction because the FPU uses both pipelines (U and V) when fetching a 64-bit operand from memory.

Overall, the Pentium's FPU has many features that contribute to fast floating-point execution.

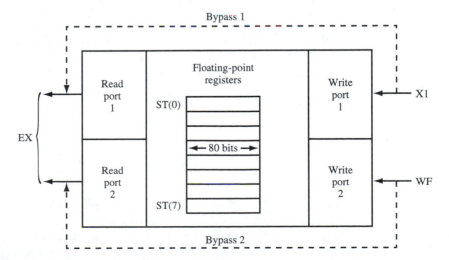

FIGURE 15.17   FPU register file

15.10   TROUBLESHOOTING TECHNIQUES

The advanced nature of the Pentium microprocessor requires us to think differently about the nature of computing. As we have already seen, exotic techniques such as branch prediction, pipelining, and superscalar processing have paved the way for improved performance. Let us take a quick look at some other improvements from Intel:

*   Intel has added MMX Technology to its line of Pentium processors (Pentium, Pentium Pro, and Pentium II). A total of fifty-seven new instructions enhance the processor's ability to manipulate audio, graphic, and video data. Intel accomplished this major architectural addition by *reusing* the 80-bit floating-point registers in the FPU. Using a method called **SIMD** (single instruction multiple data), one MMX instruction is capable of operating on 64 bits of data stored in an FPU register.

*   The Pentium Pro processor (and the newer Pentium II) employ a technique called *speculative execution.* In this technique, multiple instructions are fetched and executed, possibly out of order, in order to keep the pipeline busy. The results of each instruction are speculative until the processor determines that they are needed (based on the result of branch instructions and other program variables). Overall, a high level of parallelism is maintained.

*   First used in the Pentium Pro, a new bus technology called Dual Independent Bus architecture utilizes two data buses to transfer data between the processor and main memory (including the level-2 cache). One bus is for main memory, the second for the level-2 cache. The buses may be used independently or in parallel, significantly improving the bus performance over that of a single-bus machine.

*   The five-stage Pentium pipeline was redesigned for the Pentium Pro into a *superpipelined* fourteen-stage pipeline. By adding more stages, less logic can be used in each stage, which allows the pipeline to be clocked at a higher speed. This is easily verified by the 200- and 300-MHz processors currently available. Although there are drawbacks to superpipelining, such as bigger branch penalties during an incorrect prediction, its benefits are well worth the price.

Every aspect of computing must be studied in order to fully understand how to develop improved methods, software, and hardware for increased performance. Spend some time trying to think of an improvement of your own, as if you are designing a new microprocessor. Look through recent issues of computer architecture journals, or search the Web for information. You will find that a lot of people are thinking about improvements, too.

SUMMARY

In this chapter we examined the architectural organization and operation of the Pentium microprocessor. The Pentium's 64-bit data bus, internal code and data caches, twin integer pipelines, on-chip FPU, and branch prediction hardware combine to produce impressive execution speed. In addition, the MESI cache coherency protocol allows the Pentium to operate in parallel with other Pentiums in a multiprocessing system. In short, the Pentium utilizes a new architectural direction while maintaining software compatibility with the rest of the 80x86 family.

STUDY QUESTIONS

1. How is a 32-bit address output by the Pentium?
2. Why are the Pentium's address lines bidirectional?
3. What type of error checking is performed on the address and data lines?
4. How is instruction execution possible when the processor buses are in a HOLD state?
5. What processor signals are used to define the type of bus cycle currently running?
6. What signals are associated with the internal cache?
7. What are the Pentium's hardware interrupt signals?
8. Explain the differences between a CISC processor and a RISC processor.
9. How are accesses to main memory reduced when a cache is used?
10. How does an instruction pipeline speed up execution?
11. What is branch prediction?
12. How does a compiler affect the performance of the machine code?
13. How does the Pentium indicate the type of bus cycle it is running?
14. What bus cycle states are directly reachable from state T2?
15. How many bytes are read/written during a single-transfer bus cycle? How many for a burst transfer?
16. What are burst transfers used for?
17. What are locked bus cycles used for? How does the Pentium indicate that the current bus cycle is locked?
18. Is it possible for two bus cycles to be in progress at the same time?
19. What is the main difference between the shutdown and halt bus cycles with regard to the internal state of the processor?
20. Why is the Pentium a superscalar processor?
21. Is ADD DATA[BP],3 a simple instruction?
22. What type of dependency exists between these two instructions?

```
ADD   BX,CX
SUB   DX,BX
```

23. Name the five stages used by the U and V pipelines.
24. Are there any differences between the U and V pipelines? If so, what are they?
25. What is a pipeline stall? What role does the data cache play in a stall?
26. How is it possible, through instruction pairing, to execute ten instructions in only nine clock cycles?
27. What are the history bits used for in a BTB entry?
28. The instruction JMP [BP] may be correctly predicted each time it is encountered in a program, but still cause stalls in the pipeline. Why is this?
29. What is the BTB's prediction for a conditional jump the first time it is encountered?
30. What pipeline stage plays a role in branch prediction?

31. What is a cache hit?
32. What happens during a cache miss?
33. What is the average access time for a system that contains 10-ns cache and 80-ns RAM if the hit ratio is 0.95?
34. What does locality of reference mean? How does it apply to an executing program?
35. Describe the structure of the Pentium's 8KB two-way set-associative cache.
36. Explain the code and data cache activity during execution of this loop of code:

```
          MOV    CX,20
TOP: CMP    AL,[SI]
          ADC    BL,[SI]
          INC    SI
          LOOP   TOP
```

37. What is the difference between writethrough and writeback?
38. What is second-level cache?
39. What does MESI stand for?
40. What is bus snooping?
41. What is an inquire cycle?
42. List the reasons why the Pentium's FPU is faster than the 80486's FPU.
43. Which pipeline (U or V) is part of the FPU pipeline?
44. How many stages does the FPU pipeline have? What are they?
45. What is the purpose of each bypass in the FPU pipeline?
46. Why is the floating-point register file dual ported?
47. Which pairs of instructions may execute in parallel?
    a) FMUL   b) FXCH   c) FADD   d) FSUB
       FXCH   FSUB   DEC BX   FDIV
48. When FXCH is executed in parallel with another floating-point instruction, how many clock cycles are required for its execution? Explain.

CHAPTER 16

Protected-Mode Operation

OBJECTIVES

In this chapter you will learn about:

- The way memory is utilized in protected mode
- Segmented addressing
- Virtual addressing
- Paged memory management
- Multitasking
- How memory and I/O are protected
- Protected-mode interrupt operation
- Virtual-8086 mode

16.1   INTRODUCTION

The architecture of the Pentium's **protected mode** is significantly different from that of **real mode.** In real mode, addresses are generated by shifting 16-bit segment registers to the left, and adding a 16-bit offset to create a 20-bit physical address. As we will see, in protected-mode memory addresses are generated in a totally different way. Segment registers are now called *segment selectors,* and point to a structure called a **segment descriptor.** The segment descriptor contains addressing and control information which is used to control how a 32-bit linear address is generated. These addresses may then be further translated by a **paging** mechanism before emerging as a physical address.

A number of additional registers are available in protected mode. These registers are shown in Figure 16.1. The five control registers in Figure 16.1(a) control how memory and cache are used and how the FPU is handled, and provide information on the current execution state.

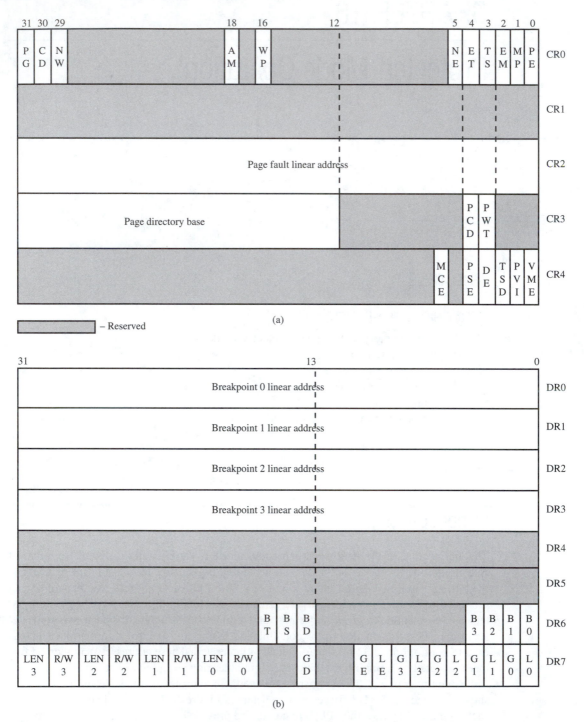

FIGURE 16.1 Additional protected-mode registers: (a) control registers; (b) debug registers

CR0 contains many important control and status bits. Their functions are as follows:

PG: Paging. Enables paging when set.

CD: Cache Disable. Disables cache writes when set.

NW: Not Writethrough. Disables cache writethrough operations when set.

AM: Alignment Mask. Allows alignment checking when set.

WP: Write Protect. Enforces supervisor-level write protection when set.

NE: Numeric Error. Allows floating-point errors to be reported when set.

ET: Extension Type. Reserved.

TS: Task Switch. Set when a task switch occurs.

EM: Emulation. Indicates the presence of a coprocessor. Should be zero on the Pentium, which has an internal FPU.

MP: Monitor Coprocessor. Must be set to run 80286 and 80386 programs on the Pentium.

CR2 contains the 32-bit linear address that generated the most recent page fault.

CR3 contains the base address of the current Page Directory, which is used to support paging.

CR4 has 6 bits whose operation is as follows:

VME: Virtual-8086 Mode Extensions. When set, enables emulation of a virtual interrupt flag.

PVI: Protected Mode Virtual Interrupts. When set, allows a virtual interrupt flag to be maintained in protected mode.

TSD: Time Stamp Disable. Used to make the RDTSC instruction privileged.

DE: Debugging Extensions. Enables I/O breakpoints when set.

PSE: Allows 4MB pages when set.

MCE: Enables the machine check exception.

Figure 16.1(b) shows the eight debug registers. These registers are used to support program debugging by indicating the address at which a program breakpoint was generated, as well as the size of the breakpoint data or instruction, whether it was a read or write request, and what kind of bus cycle (instruction fetch, data, or I/O access) to generate breakpoints on. Both control and debug registers may be loaded or saved using the MOV instruction.

Additional protected-mode registers are used to support interrupts and tasks, and these will be covered as well, including the GDTR (Global Descriptor Table Register), LDTR (Local Descriptor Table Register), IDTR (Interrupt Descriptor Table Register), and TR (Task Register). Privileged instructions that operate on these new registers are present in protected mode, and these will be examined also. Briefly, these new instructions are as follows:

ARPL	Adjust requested privilege level
CLTS	Clear task switched flag

CPUID	CPU identification
LAR	Load access rights
LGDT	Load global descriptor table register
LIDT	Load interrupt descriptor table register
LLDT	Load local descriptor table register
LMSW	Load machine status word
LSL	Load segment limit
LTR	Load task register
MOV	Move data to/from control register
RDTSC	Read from time stamp counter
SGDT	Store global descriptor table register
SIDT	Store interrupt descriptor table register
SLDT	Store local descriptor table register
SMSW	Store machine status word
STR	Store task register
VERR	Verify segment for reading
VERW	Verify segment for writing

Section 16.2 describes how segmented memory is accessed in protected mode. This is followed by a discussion of the Pentium's virtual address paging mechanism in Section 16.3. Protection methods are covered in Section 16.4. Section 16.5 introduces the powerful multitasking capability of protected mode (through the use of task state segments). Exceptions and interrupts are explained in Section 16.6, followed by protected-mode input and output in Section 16.7. A third mode of execution, virtual-8086 mode, is described in Section 16.8. An example protected-mode application appears in Section 16.9. The chapter concludes with some protected-mode troubleshooting techniques in Section 16.10.

16.2   SEGMENTATION

Segmented memory is utilized by protected mode to allow tasks to have their own separate memory spaces, which are protected from access by other tasks. In this section we will examine the operation of segmented memory.

Selectors

As previously mentioned, the segment registers we are familiar with from real mode have a different function in protected mode. Figure 16.2 shows the format of a segment selector. **Segment selectors** contain a 13-bit index field that is used to select one of 8,192 segment **descriptors** that reside either in the global descriptor table (GDT) or the local descriptor

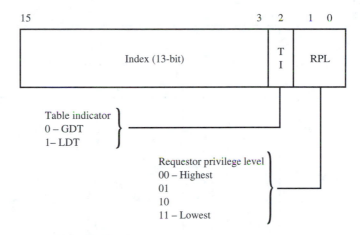

FIGURE 16.2 Segment selector

table (LDT). There is only one GDT in protected-mode. Protected mode tasks, however, may each have their own LDT. The TI bit in the segment selector picks the appropriate descriptor table during translation.

The GDT is located in memory through use of the GDTR. The GDTR is initialized with the LGDT (load global descriptor table register) instruction. LGDT loads 6 bytes of data from a source memory operand into the GDTR.

Local descriptor tables are referenced through the LDTR, which is initialized through use of the LLDT (load local descriptor table register) instruction. LLDT requires a word-size register or memory operand, which represents the index of the LDT in the GDT.

To obtain a copy of the current GDTR or LDTR, use the SGDT (store global descriptor table register) or SLDT (store local descriptor table register) instructions. SGDT requires a 6-byte destination operand in memory. The destination for SLDT is a word-size register or memory operand.

Two requestor privilege level (RPL) bits are used in protection checks to determine if access to the segment is allowed. Selectors may be loaded into any of the six segment registers (CS, DS, ES, FS, GS, and SS). A selector that has an index value of zero and points to the GDT is called a *null selector*. This selector value is reserved to provide a method of initializing segment registers, since any access using a null selector generates an exception.

Segment Descriptors

A selector points to one of 8,192 segment descriptors stored in the GDT or LDT. The structure of a segment descriptor is shown in Figure 16.3. The segment descriptor contains a 32-bit base address that specifies the beginning of the segment of memory controlled by the descriptor. The size of the segment is indicated by a 20-bit limit field and the state of the G (granularity) bit. When G is clear, the limit bits represent a segment size up to 64KB. Any attempt to access a memory location outside the limit generates an exception.

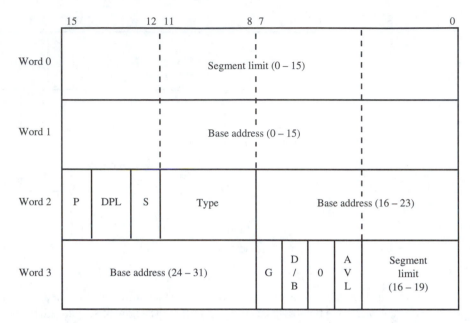

FIGURE 16.3   Segment descriptor format

When the granularity bit is set, the limit bits represent the number of 4KB pages contained in the segment. This allows the size of a segment to be from 4KB to 4GB!

Two descriptor privilege level (DPL) bits specify the privilege level required to access the segment. An attempt by a less privileged task to use the segment results in an exception.

The remaining bits are defined as follows:

P: Indicates whether the segment is present in memory. A segment-not-present exception is generated if this bit is clear when the segment descriptor is accessed.

S: When set, indicates that the segment is a system segment. When clear, the segment is a code or data segment.

D/B: For code segments, D/B controls the default operand and address size (16 bits when D/B is clear versus 32 bits when set). For data segments, D/B controls how the stack is manipulated (via SP or ESP with 16- or 32-bit pushes/pops).

AVL: Available to the programmer.

Type: This 4-bit field determines what kind of segment descriptor is being used. Table 16.1 shows the various types of segment descriptors that may be used by applications. Table 16.2 shows the different system segments that are available.

Generating a Linear Address

When a valid descriptor is in place in the GDT or LDT, the linear address associated with it is generated by the process shown in Figure 16.4. The 13-bit index from the segment selector points to a segment descriptor in the GDT or LDT. The 32-bit base address

TABLE 16.1 Segment descriptor types

Type	Description	
0	Read-only	
1	Read-only, accessed	
2	Read/write	
3	Read/write, accessed	Data Descriptors
4	Read-only, expand down	
5	Read-only, expand down, accessed	
6	Read/write, expand down	
7	Read/write, expand down, accessed	
8	Execute-only	
9	Execute-only, accessed	
A	Execute/read	
B	Execute/read, accessed	Code Descriptors
C	Execute-only, conforming	
D	Execute-only, conforming, accessed	
E	Execute/road-only, conforming	
F	Execute/read-only, conforming, accessed	

from the segment is added to the 32-bit offset to create the linear address. This address is compared with the limit information to check for illegal memory references. The segment limit of a selector can be examined with the LSL (load segment limit) instruction. The segment limit of the segment specified by the source operand is loaded into a destination register.

TABLE 16.2 System segment descriptor types

Type	Description
0	Reserved
1	Available 16-bit TSS
2	LDT
3	Busy 16-bit TSS
4	16-bit call gate
5	Task gate
6	16-bit interrupt gate
7	16-bit trap gate
8	Reserved
9	Available 32-bit TSS
A	Reserved
B	Busy 32-bit TSS
C	32-bit call gate
D	Reserved
E	32-bit interrupt gate
F	32-bit trap gate

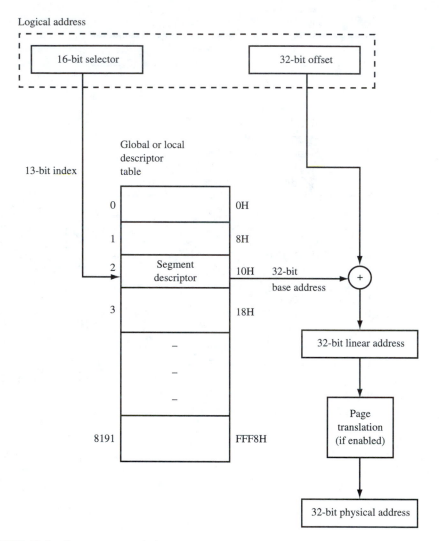

FIGURE 16.4   Segment translation

If paging is not used, the 32-bit linear address is the same as the 32-bit physical address output on the address lines. Otherwise, the paging hardware converts the linear address into a physical address (see Section 16.3 for details).

Privilege Levels

The RPL and DPL bits found in the segment selector and descriptor are used to perform protection checks each time an address is generated. These protection checks are based on a four-level privilege hierarchy, with level 0 being the highest level of privilege, and level 3 the lowest. Programs execute with a particular level of privilege, and therefore make memory requests (and other types of requests, such as interrupt and subroutine requests, or task switches) based on their privilege level. The privilege level of the currently executing program is called the current privilege level (CPL). The lower 2 bits of the CS register specify the CPL of the program. The CPL is compared with the RPL and DPL during address generation to enforce protection. In general, a less privileged program may not access higher privileged segments. Intel refers to the four privilege levels as *rings of protection.* As shown in Figure 16.5, a typical operating system might use privilege level 0 (the highest) for private OS functions, level 1 for OS services available to applications, level 2 for device drivers, and level 3 for application programs. This allows the operating system to have control over what code and data structures are available to software running on the system. Any protection violation will cause an exception that may be serviced by the operating system. So, programs will not be able to get away with illegal memory references (overwriting important operating system data tables) or function calls (like calling a function that changes privilege levels).

A number of instructions are provided to manipulate and examine protections. These are ARPL (adjust requested privilege level), LAR (load access rights), VERR (verify segment for reading), and VERW (verify segment for writing). ARPL is used to adjust the privilege level of a selector by comparing privilege levels of source and destination operands. LAR loads a copy of the access rights of a selector containing a source operand into a destination register. VERR and VERW both compare the current privilege level with that of the source operand selector. If read or write access is allowed, the zero flag is set.

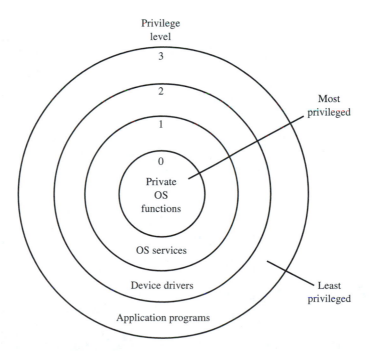

FIGURE 16.5 Rings of protection

16.3   PAGING

The Pentium supports translation of *virtual (linear) addresses* into physical addresses through the use of special tables that map portions of the virtual address into actual physical memory locations. Figure 16.6 illustrates the general process. Physical memory is divided into fixed-size **page frames** of 4KB each. 32-bit virtual (linear) addresses generated by a running task select entries in the systems page directory and page table, which translate the upper 20 bits of the virtual address into the actual physical address where a page frame is located. The lower 12 bits of the virtual address are not translated and point to one of 4,096 byte locations within a page frame. Page translation allows the physical memory used by a system to be much smaller than the linear addressing space. For instance, the Pentium's 4GB linear addressing space may be mapped to a physical memory of only 16MB. This does not pose a problem, as evidenced by Windows, which runs quite nicely with 16MB of RAM.

As Figure 16.6 shows, the pages used by a program do not need to be stored consecutively. A program's code and data may be spread out all over physical memory, and even moved around (with help from the hard disk) while the program is executing! This helps to explain why the linear addresses are also called virtual addresses, since they have no relation to the actual physical memory address used, except for the lower 12 bits.

Paging is enabled when the PG bit in CR0 is set. This is a requirement for running multiple tasks in virtual-8086 mode. In addition, many operating systems employ a memory management technique called **demand paging,** which requires the kind of address translation described here.

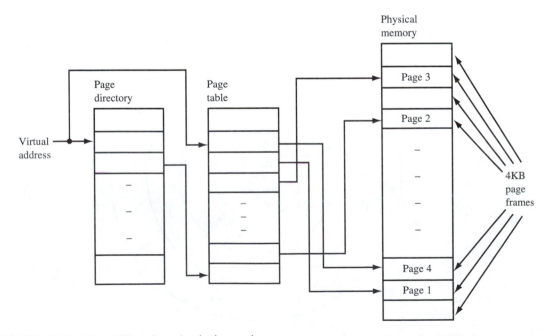

FIGURE 16.6   Virtual (linear) to physical mapping

Page Directories and Page Tables

Figure 16.7 shows how a 32-bit virtual address is translated into a physical address. The upper 10 bits of the virtual address select one of 1,024 entries in the **page directory.** The base address of the page directory is stored in the **page directory base register (PDBR).** Each entry in the page directory is 4 bytes wide and contains the base address of a **page table.** The next 10 bits from the virtual address select one of 1,024 entries in the page table pointed to by the page directory entry. This entry is also 4 bytes wide and contains the base address of the actual physical memory page frame. This address is combined with the lower 12 bits of the virtual address to access the desired location in memory. Note that the page directory and page table are themselves also 4KB page frames stored in memory.

Let us examine an actual virtual-to-physical address translation. Figure 16.8 outlines the translation process. The virtual address 801C3400H is broken up into three parts. The upper 10 bits contain the value of 200H (when reinterpreted). This is the offset of the page directory entry. The PDE contains the base address 000E4000H. This is the base address of the page table.

The next 10 bits of the virtual address have a value of 1C3H, which becomes the offset into the page table. The address stored in the PTE (00028000H) is the base address of the physical memory page frame. This address is combined with the lower 12 bits of the virtual address (400H), giving a physical address of 00028400H.

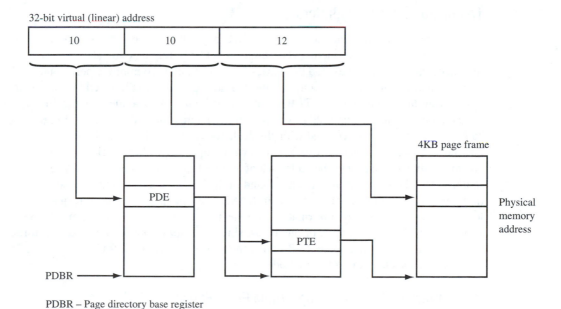

PDBR – Page directory base register
PDE – Page directory entry
PTE – Page table entry

FIGURE 16.7 Translating page addresses

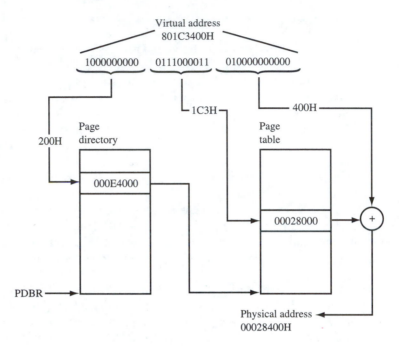

FIGURE 16.8   Actual virtual-to-physical address translation

Translation Lookaside Buffers

Since the page directory and page table are 4KB pages themselves, it would be very inefficient to have to access them every time an address requires translation, since two memory reads are needed to read the entries from each table. To improve performance, the internal instruction and data caches of the Pentium contain small, special caches called **translation lookaside buffers (TLBs)** that automatically translate the upper 20 bits of the virtual address into the upper 20 physical memory address bits. The TLBs are needed because the cache must be accessed with physical, not virtual, addresses. The TLB can translate the virtual address *and* access the cache with it in a single clock cycle. Since the TLBs are small, they contain only the addresses of a few of the most recently used pages. If the required translation information is not found in the TLB, the processor accesses the page directory and page table entries stored in RAM. The operating system is responsible for loading the new translation information into the TLB. Prior to doing this it may be necessary to invalidate the contents of the TLBs, since they may contain out-of-date information. The INVLPG (invalidate TLB entry) instruction is provided for this purpose, and may only be executed in protected mode.

Page Directory Entry and Page Table Entry Formats

Figure 16.9 indicates the formats of page directory and page table entries. The upper 20 bits of each entry specify the base address of a page frame. In the PDE, this address is the base address of a page table. In the PTE, this address is the base address of the physical memory page frame.

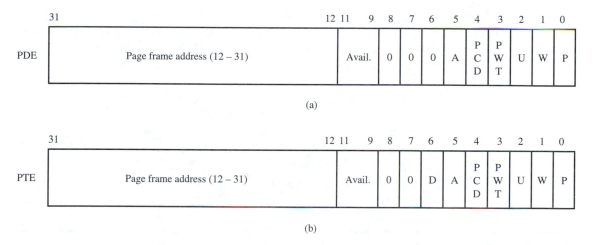

FIGURE 16.9 Format of (a) page directory entry; (b) page table entry

Three bits are available for the programmer to use for any purpose, such as counting the number of times the entry is accessed. The remaining bits in each entry are defined as follows:

D: Dirty. This bit is set if a write has been performed to the page pointed to by the PTE. Dirty bits are used to determine if the page should be written back to hard disk when the page is swapped out (to make room for a new page coming in).

A: Accessed. This bit is set if a read or write was performed to the page selected by the PDE and PTE. This bit is used by the operating system to help choose a victim page to swap out when all pages are in use and a new page must be loaded into RAM. A page that has been accessed is less likely to be swapped out than a page that has not been accessed.

PCD: Cache Disable. This bit determines whether the current memory access is cached.

PWT: Writethrough. This bit enables writethrough operations between the cache and memory.

U: User. This bit is used when performing protection checks on the current memory address.

W: Writeable. This bit determines whether the page may be written to and is also used in protection checks.

P: Present. This bit indicates whether the page is actually stored in memory. In a demand-paging system, when a new page is needed, one of two conditions may be true:

- There is a free page frame available.
- No page frames are available.

If a page frame is available, the new page is simply copied into memory at the appropriate address, the TLBs are updated, and the P-bit is set to indicate that the page is in memory.

If no free pages exist, a victim page must be chosen to make room for the new page. The P-bit of the victim's PTE is cleared, to show that the page has been swapped out. The page may be copied back to hard disk (as required by the dirty bit) before the new page is read in.

The Pentium uses the P-bit to generate a *page fault* when attempting to access a page that is not in memory. One characteristic of a demand-paging system is that pages are only brought into memory when needed. Page faults are used to load a page into memory the first time it is needed, and to reload it if it has been swapped out.

It is interesting to note that, using demand-paged virtual memory management, all or part of many different programs may be stored in many different locations in physical memory. Page faults are used to bring in other parts (pages) of the programs as needed. Performance depends on how many pages a program is allowed to have in memory at one time.

16.4   PROTECTION

Consider a multiuser operating system based on the Pentium. Each user is capable of executing programs and using system resources, such as the hard disk, printer, and other hardware supported by the system. Now, imagine what might happen if one user's program goes out of control due to an unforeseen bug and begins writing over important operating system data structures stored in memory, or even the code and data of programs being executed by the other users. The system will most likely grind to a halt and require a complete reboot. Even worse, the problem may go undetected for a long time, causing additional bugs that may be hard to find when the initial problem is eventually discovered. This situation must not occur. Through the use of certain protection mechanisms, this catastrophe can be prevented, and possibly even corrected before any damage occurs.

The Pentium provides protection for segmented and paged memory accesses. Protection is accomplished by comparing privilege levels during address translation. One task can be prevented from accessing code and data of another task, or even performing a task switch.

Protecting Segmented Accesses

Prior to any memory access using segment selectors, the Pentium performs five different checks. These checks are as follows:

- Type check
- Limit check
- Addressable domain check
- Procedure entry point check
- Privileged instruction check

Any violation of these protection checks results in an exception.

Type checking is used to determine whether the current memory access (read/write) is allowed. For example, a memory write is not allowed on a read-only data segment. It may also be illegal to read from an execute-only segment. The types of accesses allowed are based on individual bits in the data and code segment descriptors. These bits are the writeable bit (data segment descriptor) and the readable bit (code segment descriptor).

Limit checking uses the twenty limit bits stored in the segment descriptor to guarantee that addresses outside the range of the segment are not generated. The granularity bit determines how the limit bits are interpreted. When the granularity is zero, the limit bits specify the total number of byte addresses that may be used. For example, if the limit bits have been set to 01FFFH, no addresses above xxxx1FFFH may be generated.

When the granularity bit is one, the limit bits specify the number of 4KB pages used by the segment. Thus, a limit value of 00400H represents a segment size of 1,024 4KB pages (a total segment size of 4MB). Any attempt to generate an address above the addressable limit results in a general protection violation exception.

The addressable domain of a task is a function of the task's CPL. A CPL of 0 is the highest privilege level, and a task with a CPL of 0 may thus access data operands in segments with any privilege level. As the CPL changes to lower privilege levels (1, 2, or 3), only segments with the same, or lower, privilege level may be accessed. For example, a task with a CPL of 2 may access data segments with privilege levels of 2 or 3 only. A task with a CPL of 3 may only access data segments with a privilege level of 3. The privilege level of a segment is specified by the two DPL bits stored in its descriptor.

The procedure entry point check is performed through the use of a **call gate.** Call gates are used to control the transfer of execution between procedures of different privilege levels. The structure of a call gate is shown in Figure 16.10. The call gate has a structure similar to that used for a descriptor. The P (present) bit and DPL bits indicate the presence of the segment in memory and its privilege level. The DWORD count specifies the number of double-words to transfer from the caller's stack to the stack in the new procedure, if there is a privilege change. The calling task's CPL is compared with the DPL of the call gate to determine whether control may be transferred to the procedure entry point specified in the call gate. Call gates are used by JMP and CALL instructions and may only reside in the GDT and LDT.

Finally, some instructions are privileged and may only be executed when the CPL is 0. These instructions include:

CLTS	HLT	INVD	INVLPG	WBINVD
LGDT	LLDT	LIDT	LMSW	LTR
MOV to/from CR		MOV to/from DR		

A general protection violation exception is generated if an attempt is made to execute any of these instructions with a CPL greater than 0.

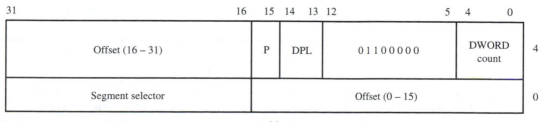

(a)

FIGURE 16.10   Structure of a call gate

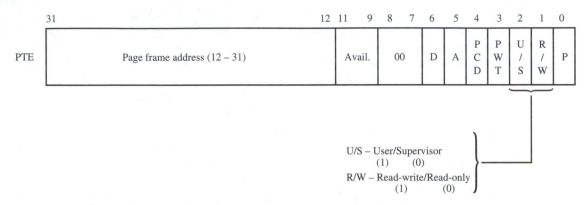

FIGURE 16.11   Page table entry protection bits

Page-Level Protection

Protection for memory pages is performed after the protection checks for segmented address generation, and consists of two checks:

*   Type check (reads and writes)
*   Addressable domain check (via privilege levels)

The page directory and page table entries (PDE and PTE) contain two bits that are used to perform these two checks. Figure 16.11 shows the format of a PDE or PTE. The two protection bits are U/S (user/supervisor) and R/W (read-write/read-only). Pages are marked as user (U/S equals 1) or supervisor (U/S is 0). A task is running at the supervisor level if the CPL is 0, 1, or 2. A task running at the user level (CPL equals 3) may only access a user page. A task running at supervisor level may access any page.

When R/W is 0 the page is a read-only page. A user-level task may only perform reads from the page. No user-level writes are allowed. A supervisor-level task may also read the page. If write-protection is disabled (through the WP bit in CR0) a supervisor task may write to the read-only page.

When R/W is a 1, the page is available for reads and writes. A user-level task may not read or write a supervisor-level page. If write-protection is enabled, the Pentium will catch any write to a user or supervisor page (via the page fault exception).

Combining segment-level and page-level protections adds a large measure of security and reliability to a Pentium-based system.

16.5   MULTITASKING

One of the most significant features of protected mode is its ability to support execution of multiple programs (called *tasks*) simultaneously. In actuality, only one task is ever running at one point in time, since there is only one Pentium to execute on. But the ability to switch from task to task at very high speeds gives the impression that many tasks are all running at the

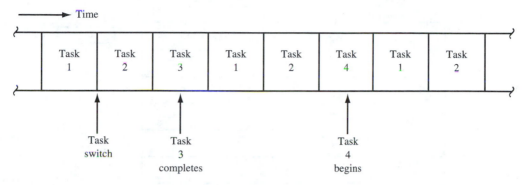

FIGURE 16.12 Running multiple tasks simultaneously

same time. This is illustrated in Figure 16.12. Note that each task executes for a period of time (called a **time slice**) and then a *task switch* is used to switch from one task to the next. Rapidly switching from task to task gives the impression that all tasks are running at the same time.

The Task State Segment

During a task switch, the contents of all processor registers, as well as other information, are saved for the task being suspended and new information is loaded for the next task. This information is not saved on the stack as you might expect, but in a special memory structure called the **task state segment (TSS).** The structure of a 32-bit TSS is shown in Figure 16.13.

The TSS contains storage areas for all of the Pentium's 32-bit registers and 16-bit segment selectors, plus additional storage for the stack pointers and segment selectors for each protection level stack.

When a task is created, the task's LDT selector (offset 60H), PDBR (offset 1CH), protection level stacks, T-bit, and I/O permission bit map are filled in. During a task switch, these items are read but not changed. Only the register portion (offset 20H through 5CH) is modified during a task switch, being overwritten by the current contents of each register. These values are read during a task switch that restarts a suspended task.

TSS Descriptors

As with any segment, the TSS utilizes a descriptor that defines the various characteristics the segment will exhibit. Figure 16.14 illustrates the format of the TSS descriptor.

The individual bit fields are defined as follows:

Base address: 32-bit segment base address

Segment limit: 20-bit segment size limit

G: Granularity

AVL: Segment available

P: Segment present

DPL: Descriptor privilege level

B: Busy

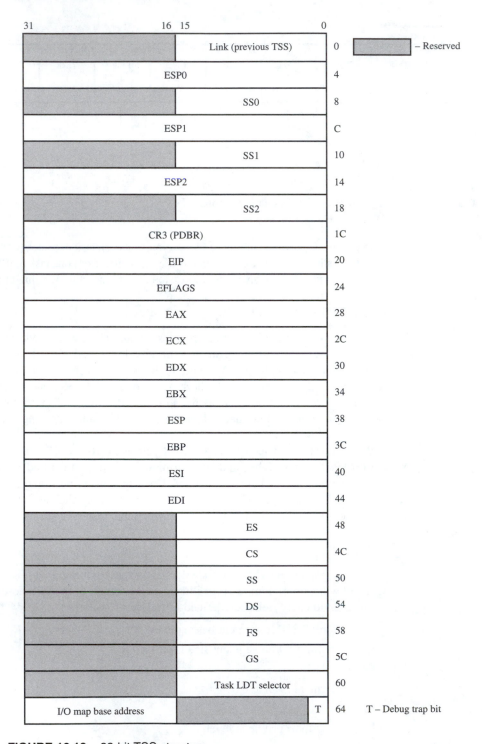

FIGURE 16.13   32-bit TSS structure

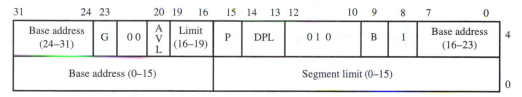

FIGURE 16.14 TSS descriptor

The granularity bit determines how the limit field is interpreted (size in bytes or size in 4KB chunks). When G is clear, the limit field represents a segment size from 1 byte to 1MB. When G is set, the segment size goes from 4KB to 4GB (4,096MB, in chunks of 4KB). If the segment is available for use, the AVL bit will be set. The present bit indicates whether the segment is actually in memory or not (possibly having been swapped out during a page fault). The 2-bit DPL field indicates the privilege level of the segment and is used in protection checking. The busy bit indicates that the task is currently running or waiting to run when high.

The Task Register

TSS descriptors may only be loaded into the GDT. When multiple TSS descriptors exist in the GDT, the TSS currently in use is accessed through the use of the **task register.** The task register is used as an index pointer into the GDT to locate a TSS descriptor. The format of the task register is shown in Figure 16.15.

The task register contains two parts: a visible portion accessible by the programmer, and an invisible portion that is automatically loaded with information from the associated TSS descriptor. The task register may be loaded with a new TSS selector with the LTR (Load Task Register) instruction. LTR requires a 16-bit register or memory operand and may only be executed in protected mode with a CPL of 0. Initially, the task register is loaded with the first protected-mode task to execute via LTR. Then, during a task switch, the task register is changed to reflect the new TSS being used.

The visible portion of the task register may be read with the STR (Store Task Register) instruction. Only the 16-bit selector portion visible to the programmer may be stored. STR may only be executed in protected mode.

Task Gates

Since the TSS descriptor contains two DPL bits that specify the privilege level of the segment, a task switch may result in a privilege violation if the new task has a lower priority

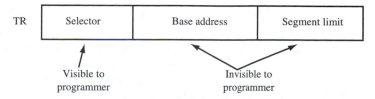

FIGURE 16.15 Task register format

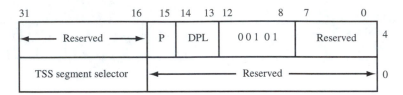

FIGURE 16.16   Format of a task gate

than the currently executing task. In addition, it may be necessary for an interrupt or exception to cause a task switch to a segment containing the handler code. The Pentium provides **task gates** as an additional way to facilitate task switching. Task gates may be stored in the LDT of a task, or in the IDT (see Section 16.6). The format of a task gate is illustrated in Figure 16.16.

Task gates allow a single busy bit to be used for a segment (the one contained in its TSS descriptor). Even though many different tasks might have access to a segment through their respective task gates, only one TSS descriptor is required for the segment. For example, suppose that a TSS descriptor points to the handler code for the divide error exception. Since many different tasks may generate this exception, each will require a task gate in its LDT to access the TSS descriptor of the divide error handler.

Task Switching

Switching from one task to another is accomplished in four different ways:

- The current task JMPs or CALLs a TSS descriptor.
- The current task JMPs or CALLs a task gate.
- The current task executes an IRET when the NT flag is set.
- An interrupt or exception selects a task gate.

When a task switch is called for, the following steps take place:

1. The new TSS descriptor or task gate must have sufficient privilege to allow a task switch. The DPL, CPL, and RPL values are compared before any further processing takes place. Interrupts and exceptions do not force protection checking.
2. The new TSS descriptor must have its present bit set and have a valid limit field.
3. The state of the current task (also called its *context*) is saved. This involves copying the contents of all processor registers into the TSS for the current task.
4. The task register is loaded with the selector of the new TSS descriptor.
5. The busy bit in the new TSS descriptor is set, as is the TS bit in CR0.
6. The state of the new task is loaded from its TSS and execution is resumed.

The selector for the old task's TSS descriptor is copied into the TSS of the new task to facilitate a return when tasks are nested (NT flag is set). An IRET instruction checks the NT flag to determine if the previous TSS selector may be used during a task switch.

The TS bit in CR0 that is set during a task switch may be cleared by executing the CLTS (clear task switched flag) instruction.

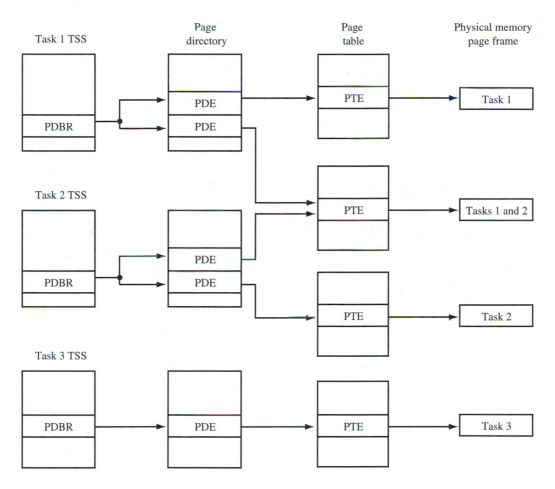

PDBR – Page directory base register
PDE – Page directory entry
PTE – Page table entry

FIGURE 16.17   Logical to physical address mapping in multiple tasks

Task Addressing Space

If paging is not enabled, the linear addresses generated by a task are the same as the physical addresses sent to the memory system. When paging is enabled it is possible for each
task to have its own separate, protected addressing space, through the use of the PDBR
stored within each TSS. As Figure 16.17 indicates, tasks may map their logical addresses
into different, or overlapping, physical memory spaces. Overlapping (or *shared*) physical
addresses are useful for providing the same information to many different tasks (such as
the contents of DOS's interrupt vector table).

16.6   EXCEPTIONS AND INTERRUPTS

In this section, we will examine the operation of interrupts and exceptions in the Pentium's protected mode. An interrupt is generated in response to a hardware request on the INTR or NMI inputs, whereas an exception is generated during the course of execution. For example, divide error is an exception generated when the Pentium's DIV or DIVI instructions are executed with a divisor operand of 0.

Let us examine how interrupts and exceptions are supported.

The Interrupt Descriptor Table

Real mode uses a 1KB interrupt vector table (IVT) beginning at address 00000H. Each 4-byte entry in the IVT consists of a CS:IP pair that specifies the address of the first instruction in the interrupt service routine. An 8-bit vector number is shifted 2 bits to the left to form an index into the IVT.

Protected mode relies on an *interrupt descriptor table* (IDT) to support interrupts and exceptions. The IDT comprises 8-byte gate descriptors for task, trap, or interrupt gates. The IDT has a maximum size of 256 descriptors. The size of the IDT is controlled by a 16-bit limit value stored in the **interrupt table descriptor register (ITDR)**. This 48-bit register contains the 32-bit base address for the IDT and the 16-bit size limit. Figure 16.18 shows how the IDTR is used to locate the IDT, which can be placed anywhere in physical memory.

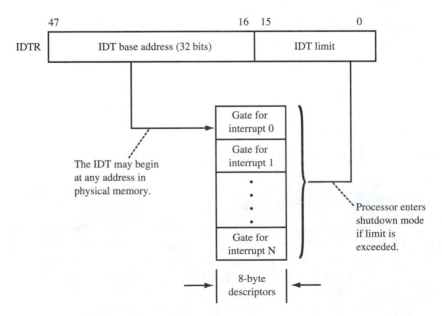

FIGURE 16.18   Using the IDTR to access the IDT

The 8-bit vector number for the currently recognized interrupt is shifted 3 bits to the left and used as an index into the IDT. Thus, vector 10H accesses the descriptor at offset 80H in the IDT.

The LIDT (Load Interrupt Descriptor Table Register) and SIDT (Store Interrupt Descriptor Table Register) instructions are used in conjunction with the IDTR. Each instruction uses a single operand that specifies the address of a 48-bit memory word. This 6-byte word is used to change the location of the IDT, or find its current address. LIDT may only be executed when the CPL is 0. SIDT can be executed anytime.

READIDTR: Reading and Displaying the IDTR

The READIDTR program uses the SIDT instruction to store a copy of the contents of the IDTR in memory, where it is then examined and converted into printable form.

```
;Program READIDTR.ASM: Read and display contents of IDTR.
;
        .MODEL SMALL
        .586
        .DATA
IDTRA   DW      ?          ;storage for limit bits
IDTRD   DD      ?          ;storage for IDT base address
MSG     DB      'The IDTR contains '
        DB      8 DUP(?)
        DB      ':'
        DB      4 DUP(?)
DBASE   DB      0DH,0AH,'$'
HEXTAB  DB      '0123456789ABCDEF'

        .CODE
        .STARTUP
        SIDT    IDTRA             ;store IDTR
        MOV     AX,IDTRA          ;load limit bits
        LEA     SI,DBASE-1        ;set up conversion pointer
        CALL    CONV              ;convert limit bits
        DEC     SI                ;skip over ':'
        MOV     EAX,IDTRB         ;load IDT base address
        CALL    CONV              ;convert 32-bit address
        CALL    CONV
        LEA     DX,MSG            ;display results
        MOV     AH,9
        INT     21H
        .EXIT

CONV    PROC    NEAR
        MOV     CX,4              ;prepare for 4 passes
DIGIT:  MOV     DI,AX             ;get a copy of input
        AND     DI,000FH          ;mask out offset
        MOV     BL,HEXTAB[DI]     ;load corresponding ASCII code
        MOV     [SI],BL           ;save it in buffer
        DEC     SI                ;adjust pointer to buffer
        SHR     EAX,4             ;get next hex digit
        LOOP    DIGIT             ;and repeat
        RET
CONV    ENDP

        END
```

When READIDTR is executed from DOS, the output looks like this:

`The IDTR contains 00000000:03FF`

If, however, you start Windows and run READIDTR from inside a DOS shell, you get a different result:

`The IDTR contains 8000DA70:02FF`

This indicates that Windows, which runs in protected mode, has changed the nature of the underlying interrupt system.

IDT Descriptors

There are three descriptors that may be used within the IDT: task gates, trap gates, and interrupt gates. The format of each descriptor is shown in Figure 16.19.

The P-bit in each descriptor stands for present, and indicates whether the segment is present in memory. The 2-bit DPL field specifies the descriptor privilege level (0 is the highest). The 32-bit offset points to the first instruction in the handler's code segment. The

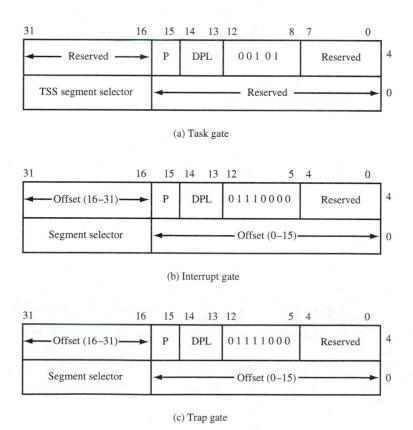

FIGURE 16.19   IDT descriptors

segment selector points to an executable segment selector in the GDT or LDT for interrupt and trap gates. The TSS segment selector for a task gate points to a TSS descriptor in the GDT. Interrupt and trap gates operate like a CALL to a call gate. Task gates operate like a CALL to a task gate. Once again, there may be up to 256 descriptors in the IDT. When fewer interrupts/exceptions are required, the limit field of the IDTR is used to specify the addressable limit within the IDT. The Pentium will enter shutdown mode if the limit is exceeded.

Interrupt and Exception Descriptions

Table 16.3 summarizes the protected-mode interrupts and exceptions available on the Pentium. Descriptions of these interrupts and exceptions are as follows:

Vector 0: Divide Error. This exception is generated when the divisor is 0 in a DIV or DIVI instruction.

Vector 1: Debug Exception. This exception has multiple uses during debugging. It is used for single-stepping, instruction, data, and task switch breakpoints.

Vector 2: NMI Interrupt. This interrupt is generated upon recognition of a rising edge on the Pentium's NMI input. NMI cannot be disabled through software (as the INTR interrupt can).

Vector 3: Breakpoint. This 1-byte instruction (opcode CCH) can be used to trigger a debugging routine by replacing the first byte of an instruction (in RAM) with the CCH opcode. The breakpoint handler is responsible for replacing the original byte of the instruction modified when the breakpoint was set up.

TABLE 16.3 Protected-mode interrupts and exceptions

Vector	Description	Error Code
0	Divide error	No
1	Debug exception	No
2	NMI interrupt	No
3	Breakpoint	No
4	Overflow	No
5	Bounds check	No
6	Invalid opcode	No
7	Device not available	No
8	Double fault	Yes, 0
10	Invalid TSS	Yes
11	Segment not present	Yes
12	Stack fault	Yes
13	General protection	Yes
14	Page fault	Yes (special format)
16	Floating-point error	No
17	Alignment check	Yes, 0
18	Machine check	Depends on CPU model
19–31	Reserved	—
32–255	Maskable interrupts	No

Vector 4: Overflow.   This exception is called when the INTO instruction is executed with the overflow flag set. Recall that the overflow flag is modified in accordance with the signed results of arithmetic and logical instructions.

Vector 5: Bounds Check.   The BOUNDS instruction calls this exception when it detects an array subscript out of range.

Vector 6: Invalid Opcode.   Any opcode not recognized by the instruction decoder generates this exception. In addition, using the wrong operand size in an instruction (presumably via self-modifying code), or using the LOCK prefix with the wrong instructions also causes an invalid opcode exception.

Opcodes D6H and F1H do not generate this exception, even though they have no designed function. They are reserved by Intel.

Vector 7: Device Not Available.   Two bits in CR0 (EM and MP) are used to control when, if at all, an ESC or WAIT instruction generates this exception. On earlier 80x86 machines, this exception was used to indicate that there was no external floating-point coprocessor interfaced to the CPU.

Vector 8: Double Fault.   When two exceptions occur in sequence (the second one detected while the first is being processed), there are some combinations that cause a double fault to be signaled. Double-fault exceptions are reserved for the most severe sequences, such as a page fault (Vector 14) followed by a second page fault. Interrupts and exceptions are classified into three categories: *benign* interrupts and exceptions, *contributory* exceptions, and *page faults.* Table 16.4 shows the respective classifications by vector.

Whether a double-fault interrupt is generated depends on the classification of both exceptions in the sequence, as indicated by Table 16.5. If any other exception is signaled during processing of a double fault, the processor enters shutdown mode.

TABLE 16.4   Interrupt/exception classifications

Class	Vector	Description
Benign interrupts and exceptions	1	Debug exceptions
	2	NMI interrupt
	3	Breakpoint
	4	Overflow
	5	Bounds check
	6	Invalid opcode
	7	Device not available
	16	Floating-point error
Contributory exceptions	0	Divide error
	10	Invalid TSS
	11	Segment not present
	12	Stack fault
	13	General protection
Page faults	14	Page fault

TABLE 16.5 Generating a double fault

		Second Exception		
		Benign	Contributory	Page Fault
First Exception	Benign	No	No	No
	Contributory	No	Yes	No
	Page fault	No	Yes	Yes

Vector 9: Reserved.   This vector was previously used to signal a page fault during transfer of a 387 coprocessor operand. It is not available on the Pentium.

Vector 10: Invalid TSS.   This exception is generated when a problem is detected with a new TSS during a task switch. Depending on the error, the exception may be signaled before or after the task switch.

Vector 11: Segment Not Present.   This exception is generated when the present bit in the current descriptor is clear. This indicates that the segment is not in memory and must be reloaded (from the hard drive). This exception is useful for virtual memory implementation.

Vector 12: Stack Fault.   A stack fault is signaled when the limit of the SS selector is reached during execution of stack-based instructions PUSH, POP, ENTER, and LEAVE. Instructions that use SS as a segment override, or that use the BP register to reference memory, will also generate a stack fault if the limit is reached.

A stack fault is also generated when the present bit of a new descriptor for SS is clear.

Vector 13: General Protection.   This exception is the result of many different conditions that may arise. All of the following will cause a general protection exception:

*   Exceeding the segment limit with CS, DS, ES, FS, or GS.
*   Reading from an execute-only code segment.
*   Writing to a read-only data or code segment.
*   Loading a segment register with an inappropriate segment selector (such as loading DS with an execute-only segment).
*   Switching to a busy task.
*   Privilege violations.
*   Exceeding the instruction length limit.
*   Loading CR0 with improper PE/PG combination.
*   Using the wrong interrupt handler when leaving virtual-8086 mode.

Windows users know that this is a bad exception to get when running a program inside Windows. Usually, Windows must be restarted to get things back to normal after a general protection exception.

Vector 14: Page Fault.   A page fault exception is generated when the processor attempts to access a page that is not in memory. A page fault occurs to let the operating system know that the page must be loaded into memory. The present bit in the page directory or page table entry, when clear, indicates the page is not in memory.

A page fault exception is also generated when page-level privilege is violated (the privilege is not high enough).

The Pentium pushes a 32-bit error code onto the stack of the page fault exception handler, whose meaning is described in Figure 16.20.

The error code allows the operating system to respond to the page fault in various ways. For example, the P-bit determines whether the page must be read in from the hard disk.

Vector 16: Floating-Point Error.   This exception is generated if the NE bit is set in CR0 and an FPU instruction causes an error. The handler for floating-point error may adjust the operation of the FPU accordingly (via the FPU's control word).

Vector 17: Alignment Check.   This exception is generated when a memory operand larger than 1 byte begins at an odd address, or at an even address that is not the proper multiple of 2, 4, 8, etc. For example, a word beginning at address 1001H (must be 1000H or 1002H instead), or a double-word beginning at address 1001H, 1002H, or 1003H (1000H and 1004H are acceptable) generate this exception.

To enable alignment checking, the AM bit (in CR0) must be set. Then, if the AC flag is set and the CPL is 3, alignment check will be generated.

Vector 18: Machine Check.   This exception may or may not exist, depending on the model of the CPU. The CPUID instruction returns a bit that indicates the status of this exception (available, not available).

Additional Interrupt and Exception Descriptions

The remainder of the usable interrupts in Table 16.3 (32 through 255) are called **maskable interrupts.** These interrupts may be generated internally by software (via INT 32 through INT 255) or externally through an 8-bit vector number supplied with a hardware INTR request. Since the INTR signal may be masked by the interrupt enable flag, these hardware interrupts may be masked.

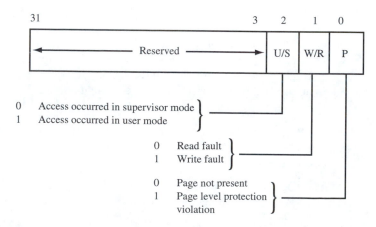

FIGURE 16.20   Page fault error code

16.7   INPUT/OUTPUT

The Pentium provides protection for input and output operations that take place in protected mode or virtual-8086 mode. When an IN or OUT instruction executes, the processor
checks the task's CPL against the IOPL (Input/Output Privilege Level) bits stored in the
EFLAGS register. If the CPL is less than or equal to the IOPL, the operation is allowed. If
CPL is greater than IOPL, a general protection violation exception is generated.

It is important to protect I/O operations in an operating system, since many hardware
features depend on proper I/O settings. For example, an operating system might employ a
counter/timer peripheral mapped to a few I/O ports that control the rate of special timing
interrupts (such as a multitasking time slice interrupt). If no protection is employed, any
user program may change the timer interrupt at will, causing havoc for the rest of the users.
It is better to restrict I/O operations to privileged users, and force user programs to request
I/O operations from the operating system. The operating system can then decide which requests to honor to keep things running smoothly.

When the CPL is greater than the IOPL, or when the processor is operating in vir-
tual-8086 mode (see Section 16.8), I/O operations are allowed on a port-by-port basis via
permission bits stored in the **I/O permission bit map** section of the task's TSS. The offset
of the I/O permission bit map within the TSS is stored in the I/O map base section of the
TSS. Each byte in the bit map stores permission bits for eight consecutive ports. The
sample bit map in Figure 16.21 indicates that access to ports 3, 7, and 12 is not allowed.
Any attempt to read or write these three ports will cause a general protection violation

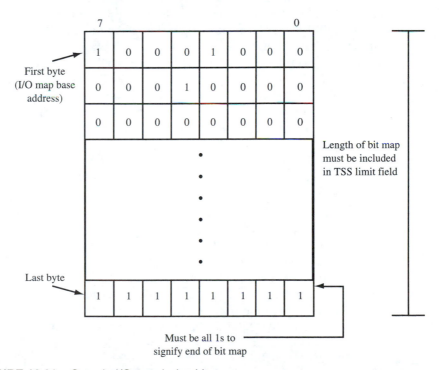

FIGURE 16.21   Sample I/O permission bit map

exception. The size of the bit map may vary according to the number of ports that must be protected.

16- and 32-bit ports must have two or four consecutive zeros in their associated bit map positions to be allowed access.

16.8   VIRTUAL-8086 MODE

Virtual-8086 mode is the last of the three main operating modes of the Pentium processor. Virtual-8086 mode is entered from protected mode when the VM bit of the flag's register is set, and executes programs written for the 8086 (and 8088) microprocessor. Multiple virtual-8086 programs may execute simultaneously (via the multitasking capabilities of protected mode) on *virtual machines*. A virtual machine comprises the hardware and software required to implement a particular task. Each virtual machine has its own 1MB addressing space and set of processor registers. The 1MB addressing space may be located anywhere in memory through the use of the Pentium's paging mechanism. The processor registers are maintained through their respective entries in the virtual task's TSS. As in real mode, register sizes default to 16 bits. Override prefixes may be used to allow 32-bit registers and addressing modes, with the usual restrictions. Figure 16.22 illustrates the concept of virtual machines running on the Pentium.

Address generation in virtual-8086 mode is like that of real mode. A 16-bit segment register is shifted 4 bits to the left and added to a 16-bit offset to form the desired effective address. Unlike the 8086 processor, which wraps addresses around the end of a segment from FFFFH to 0000H, a virtual-8086 task retains the carryout of the 20-bit effective address, and thus accesses a larger real-mode addressing space using 21-bit addresses. The actual range is from 000000H to 10FFEFH (FFFF0H plus FFFFH). These 21 bits are part of the 32-bit linear address used within the virtual-8086 task, and may be translated/paged to any physical address in the 4GB range of system memory.

All real-mode instructions are also available in virtual-8086 mode, though there are some differences in execution from that of the 8086. In short:

- Pentium instructions require fewer clock cycles than the 8086.
- CS:IP points to the DIV instruction, and not the following instruction, during an exception.
- Divide exceptions are not generated for IDIV quotients that equal 80H or 8000H.
- Undefined 8086 opcodes that represent valid Pentium instructions are executed and do not generate an invalid-opcode exception.
- The value of the SP pushed with PUSH SP is the value before it is decremented, instead of the value after it is decremented.
- Shift/rotate counts are limited to 31 bits.
- The Pentium generates an exception if a data access or instruction fetch crosses the end of a segment (offset FFFFH).
- Bits 12 through 15 of the flag register contain different values. Bit 15 is clear, and bits 12 through 15 are set according to the NT and IOPL states. On the 8086, these bits were undefined.

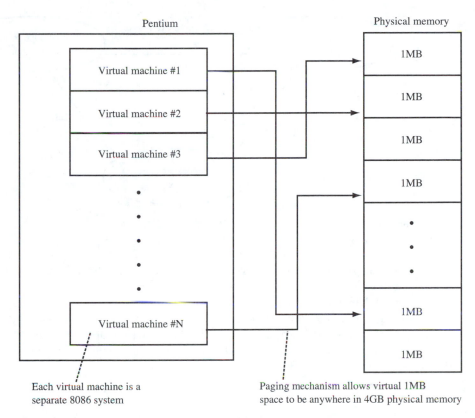

FIGURE 16.22   Concept of a virtual machine

These differences are slight, and should not interfere with the normal operation of programs originally written for the 8086.

Virtual-8086 mode tasks always execute with a privilege level of 3 (the lowest), and may be entered/exited in a number of ways. A virtual-8086 task may be initiated by a task switch, which loads a new 32-bit TSS that has the VM bit set in its copy of the flag register. Also, a procedure with a CPL of 0 (highest priority) may execute an IRET instruction that pops a 1 into the VM bit of the flag register. This indicates that the calling procedure was a virtual-8086 task, and causes the processor to reenter virtual-8086 mode.

To exit virtual-8086 mode, an interrupt or exception must be generated. If the interrupt or exception causes a task switch, the system may exit virtual-8086 mode if the new TSS is a 32-bit TSS and its copy of the VM bit is clear, or if the new TSS is only a 16-bit TSS.

If the interrupt or exception calls a procedure with a priority of 0, the processor will also exit virtual-8086 mode. Figure 16.23 shows the possible ways to enter and leave virtual-8086 mode.

As Figure 16.23 shows, interrupts and exceptions cause the processor to switch between a virtual-8086 task and a virtual-8086 *monitor task*. The monitor task is itself a protected-mode task, and is responsible for initialization, interrupt and exception handling, and I/O for the running virtual-8086 task. There may be literally hundreds or

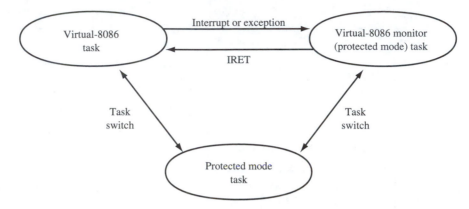

FIGURE 16.23   Entering and leaving virtual-8086 mode

thousands of virtual-8086 tasks running simultaneously. Each requires the support of a monitor task.

A handful of instructions will cause a general protection violation if executed in virtual-8086 mode with an IOPL less than 3 (recall that the CPL is always 3 in virtual-8086 mode). These instructions are CLI, STI, PUSHF, POPF, INT, and IRET. Intel calls these instructions **sensitive instructions.** Sensitive instructions may need special handling by the virtual-8086 monitor.

16.9   A PROTECTED-MODE APPLICATION

In this final section, we will examine a program named PVIEW that has been designed to be executed in protected mode. Recall that in Section 8.11 the DPMISTAT program used function 1687H of the multiplex interrupt (INT   2FH) to detect the presence of a DPMI host (such as the one provided in the Windows environment).

If a DPMI host is present, function 1687H returns a *mode switch entry point* in registers ES:DI. Performing a CALL to the entry point places the Pentium into protected mode! The PVIEW program presented here takes advantage of the capabilities provided by the DPMI host to execute some privileged instructions, such as SGDT and STR, and display the results of the execution.

A number of new instructions are included in PVIEW. These are CPUID (CPU identification), RDTSC (read time stamp counter), and SMSW (store machine status word). The CPUID instruction requires a zero or one in EAX prior to execution, and returns information such as model number and family. The RDTSC instruction reads the current value of an internal 64-bit time stamp counter that is updated every clock cycle. The SMSW instruction saves a copy of the lower 16 bits of CR0. These bits are known as the **machine status word** and can be loaded with new data using the LMSW (load machine status word) instruction. Do not play with the bits in the machine status word unless you are looking for unexpected results.

Look over PVIEW.ASM. Can you locate the protected instructions?

```
;Program PVIEW.ASM: View selected protected-mode information.
;
                .MODEL SMALL
                .586P
                .DATA

NODPMI  DB      'DPMI host missing.',0DH,0AH,'$'
NOMEM   DB      'Cannot allocate memory.',0DH,0AH,'$'
NOPM    DB      'Cannot switch to protected mode.',0DH,0AH,'$'

DPMI    DW      ?,?
HEXTAB  DB      '0123456789ABCDEF'
TSCLO   DD      ?
TSCHI   DD      ?
MSW     DW      ?
CPUMOD  DW      ?
TASKREG DW      ?
GDT     DD      ?
GDTB    DW      ?
IDT     DD      ?
IDTB    DW      ?
LDT     DW      ?

TSMSG   DB      'Time Stamp Counter : '
        DB      8 DUP(?)
        DB      ':'
        DB      8 DUP(?)
TSEND   DB      0DH,0AH
        DB      'Machine Status Word : '
        DB      4 dup(?)
MSWB    DB      0DH,0AH
        DB      'CPU identification : '
        DB      4 DUP(?)
CID     DB      0DH,0AH
        DB      'Task Register : '
        DB      4 DUP(?)
TRB     DB      0DH,0AH
        DB      'Global Descriptor Table Register : '
        DB      4 DUP(?)
G2      DB      4 DUP(?)
        DB      ':'
        DB      4 DUP(?)
G1      DB      0DH,0AH
        DB      'Interrupt Descriptor Table Register : '
        DB      4 DUP(?)
I2      DB      4 DUP(?)
        DB      ':'
        DB      4 DUP(?)
I1      DB      0DH,0AH
        DB      'Local Descriptor Table Register : '
        DB      4 DUP(?)
LI      DB      0DH,0AH, '$'

        .CODE
        .STARTUP
        MOV     BX,DS           ;find size of code and data
        MOV     AX,ES           ;in paragraphs
        SUB     BX,AX
        MOV     AX,SP           ;find stack size in paragraphs
```

```
                SHR     AX,4
                ADD     BX,AX               ;find total paragraphs
                MOV     AH,4AH              ;adjust allocation
                INT     21H
                MOV     AX,1687H            ;get DPMI entry point
                INT     2FH
                OR      AX,AX               ;host present?
                JNZ     NOHOST              ;not there
                MOV     DPMI[0],DI          ;save entry address
                MOV     DPMI[2],ES
                OR      SI,SI               ;Any allocation needed?
                JZ      GOPM                ;No, so continue
                MOV     BX,SI
                MOV     AH,48H              ;try to allocate memory
                INT     21H
                JC      SORRY               ;not enough memory
GOPM:           MOV     ES,AX
                MOV     AX,0                ;choose 16-bit application
                CALL    DWORD PTR DPMI      ;switch to protected mode
                JNC     PMODE

                LEA     DX,NOPM             ;display could not switch message
                JMP     ERROR
NOHOST:  LEA     DX,NODPMI           ;display no DPMI host message
                JMP     ERROR
SORRY:   LEA     DX,NOMEM            ;display not enough memory message
ERROR:   MOV     AH,9
                INT     21H
                JMP     BYE                 ;and exit

PMODE:   RDTSC                       ;otherwise, do some stuff in protected mode!
                MOV     TSCLO,EAX           ;save time stamp counter
                MOV     TSCHI,EDX
                SMSW    MSW                 ;save machine status word
                MOV     EAX,1
                CPUID                       ;get cpu identification
                MOV     CPUMOD,AX
                STR     TASKREG             ;save task register
                SGDT    GDT                 ;save GDTR
                SIDT    IDT                 ;save IDTR
                SLDT    LDT                 ;save LDTR

                MOV     EAX,TSCLO           ;convert time stamp counter
                LEA     SI,TSEND
                CALL    CONV8
                MOV     EAX,TSCHI
                CALL    CONV8
                MOV     AX,MSW              ;convert machine status word
                LEA     SI,MSWB
                CALL    CONV4
                MOV     AX,CPUMOD           ;convert cpuid results
                LEA     SI,CID
                CALL    CONV4
                MOV     AX,TASKREG          ;convert task register
                LEA     SI,TRB
                CALL    CONV4
                MOV     AX,WORD PTR GDT     ;convert 6-byte GDTR
                LEA     SI,G1
                CALL    CONV4
                MOV     AX,WORD PTR GDT[2]
                CALL    CONV4
                MOV     AX,GDTB
```

```
            LEA     SI,G2
            CALL    CONV4
            MOV     AX,WORD PTR IDT    ;convert 6-byte IDTR
            LEA     SI,I1
            CALL    CONV4
            MOV     AX,WORD PTR IDT[2]
            CALL    CONV4
            MOV     AX,IDTB
            LEA     SI,I2
            CALL    CONV4
            MOV     AX,LDT              ;convert LDTR
            LEA     SI,L1
            CALL    CONV4
            LEA     DX,TSMSG            ;view results
            MOV     AH,9
            INT     21H
BYE:        .EXIT

CONV4       PROC    NEAR
            DEC     SI                 ;adjust ascii pointer
            MOV     CX,4               ;prepare for 4 passes
DIGIT:      MOV     DI,AX              ;get a copy of input
            AND     DI,000FH           ;mask out offset
            MOV     BL,HEXTAB[DI]      ;load corresponding ASCII code
            MOV     [SI],BL            ;save it in buffer
            DEC     SI                 ;adjust buffer pointer
            SHR     EAX,4              ;get next hex digit
            LOOP    DIGIT              ;and repeat
            RET
CONV4       ENDP

CONV8       PROC    NEAR
            CALL    CONV4              ;convert lower 4 digits
            INC     SI                 ;adjust ascii pointer
            CALL    CONV4              ;convert upper 4 digits
            RET
CONV8       ENDP

            END
```

Try executing PVIEW from DOS. You should get the error message:

```
DPMI host missing.
```

Instead, start up Windows and double-click the DOS-prompt icon in the Main group. Then try running PVIEW. The results should now look like this:

```
Time Stamp Counter : 000000B1:2F3F6733
Machine Status Word : 0013
CPU identification : 0525
Task Register : 0018
Global Descriptor Table Register : 800517D8:010F
Interrupt Descriptor Table Register : 80BAB000:02FF
Local Descriptor Table Register : 00A8
```

Keep in mind that much of this information is privileged and only accessible through protected mode.

Many other functions are provided by the DPMI host, from memory management to interrupt servicing. These functions are best used by programmers skilled in operating system design. Even so, just taking a peek inside protected mode can be the beginning of greater things to come.

16.10   TROUBLESHOOTING TECHNIQUES

The complexity of protected mode poses a significant challenge to the programmer. Fortunately, there are many tools that aid in the development of protected-mode applications. These include compilers (capable of generating optimized Pentium code), code profilers (that determine where your program is spending its time), debuggers, disassemblers, and many good papers and books.

On-line documentation on many aspects of protected-mode programming can be found on Intel's Web site, and in many other locations. Protected-mode programming is a favorite of computer game designers, who are always searching for new ways to increase speed, improve graphics, and control action.

To begin experimenting with protected-mode programming, download any of the DOS Extender packages available over the Web, or purchase your own. A DOS Extender allows your real-mode program to execute in protected mode, breaking the 640KB DOS barrier and opening up the entire 4GB addressing space of the processor. DOS Extenders come with example programs, the most basic being one that switches back and forth from real mode to protected mode. Be prepared to invest a good deal of time in learning how to run the processor in protected mode, and do not be surprised if your PC crashes during development.

SUMMARY

In this chapter we have examined the features of the Pentium's powerful protected mode. These features included memory management through segmentation and paging, protection mechanisms for memory and I/O, interrupts, exceptions, and multitasking.

We finished with an examination of the Pentium's virtual-8086 mode, which emulates the operation of an 8086 machine, and a short trip into protected mode.

STUDY QUESTIONS

1. What are the additional registers available in protected mode?
2. What is a selector? What is its purpose?
3. What information is contained in a segment descriptor?
4. What types of segment descriptors are available?
5. Explain how a linear address is generated using segment selectors and descriptors.
6. How is paging enabled?
7. How is a virtual address translated into a physical address?
8. How many bytes of physical memory can be accessed using a single page directory entry?

9. How many bytes of physical memory can be accessed using all the entries in a page directory?
10. Describe the operation of demand paging.
11. What are the dirty and accessed bits used for?
12. What is a page fault? When is a page fault generated?
13. Why bother with protection mechanisms in a multiuser operating system? Why are they needed?
14. How does the Pentium protect one task's memory space from being overwritten (or even read) by another task?
15. What two bits in a segment descriptor play a role in protection?
16. The limit bits in a segment selector have the value 047FFH. How many bytes does the segment contain if:
    a) the granularity bit is 0?
    b) the granularity bit is 1?
17. What is the purpose of a call gate?
18. What two instructions may specify a call gate to initiate a task switch?
19. What type of protection is provided by a call gate?
20. How is page-level protection implemented?
21. Can a user-level task access a supervisor page?
22. Can a supervisor-level task write to a read-only page when write-protection is disabled?
23. What is the purpose of a TSS? What does it contain?
24. How is the current TSS located?
25. What instructions are associated with the task register? Can they be executed in real mode?
26. What are the four ways to initiate a task switch?
27. What is a task gate? How does it differ from a TSS descriptor?
28. What controls the addressing space of a task?
29. What are the IDTR and IDT? How are they related?
30. The IDTR contains the value 4E00800000FFH. What is the base address of the IDT? What is the last address?
31. What kind of descriptors may be used in the IDT?
32. Give an example of a double fault exception.
33. What value does the error code have to the operating system during a page fault exception?
34. The page fault error code is 0006H. What does this mean?
35. What is a nonmaskable interrupt? How many are there?
36. What is a maskable interrupt? How many are there?
37. What vector has an offset of C0H in the IDT?
38. What two methods are used to restrict access to I/O ports?
39. How many bytes are needed in the I/O permission bit map to protect access to ports 00H through FFH?
40. What is a virtual machine?
41. Why are so many virtual-8086 tasks possible?
42. What is required by a virtual-8086 task?

43. What is a virtual-8086 monitor?
44. What is the address range allowed for a virtual-8086 task?
45. Can virtual-8086 mode be entered from real mode?
46. How is virtual-8086 mode entered and exited?
47. What are some of the differences in instruction execution between virtual-8086 mode and the original 8086?
48. Which instructions are sensitive to IOPL in virtual-8086 mode?

APPENDIX A

8088 Data Sheets

8088
8-BIT HMOS MICROPROCESSOR
8088/8088-2

- 8-Bit Data Bus Interface
- 16-Bit Internal Architecture
- Direct Addressing Capability to 1 Mbyte of Memory
- Direct Software Compatibility with 8086 CPU
- 14-Word by 16-Bit Register Set with Symmetrical Operations
- 24 Operand Addressing Modes

- Byte, Word, and Block Operations
- 8-Bit and 16-Bit Signed and Unsigned Arithmetic in Binary or Decimal, Including Multiply and Divide
- Two Clock Rates:
  — 5 MHz for 8088
  — 8 MHz for 8088-2
- Available in EXPRESS
  — Standard Temperature Range
  — Extended Temperature Range

The Intel® 8088 is a high performance microprocessor implemented in N-channel, depletion load, silicon gate technology (HMOS-II), and packaged in a 40-pin CERDIP package. The processor has attributes of both 8- and 16-bit microprocessors. It is directly compatible with 8086 software and 8080/8085 hardware and peripherals.

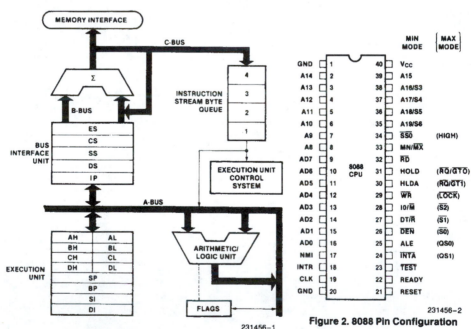

Figure 1. 8088 CPU Functional Block Diagram

Figure 2. 8088 Pin Configuration

8088

Table 1. Pin Description

The following pin function descriptions are for 8088 systems in either minimum or maximum mode. The "local bus" in these descriptions is the direct multiplexed bus interface connection to the 8088 (without regard to additional bus buffers).

Symbol	Pin No.	Type	Name and Function
AD7–AD0	9–16	I/O	**ADDRESS DATA BUS:** These lines constitute the time multiplexed memory/IO address (T1) and data (T2, T3, Tw, T4) bus. These lines are active HIGH and float to 3-state OFF during interrupt acknowledge and local bus "hold acknowledge".
A15–A8	2–8, 39	O	**ADDRESS BUS:** These lines provide address bits 8 through 15 for the entire bus cycle (T1–T4). These lines do not have to be latched by ALE to remain valid. A15–A8 are active HIGH and float to 3-state OFF during interrupt acknowledge and local bus "hold acknowledge".
A19/S6, A18/S5, A17/S4, A16/S3	35–38	O	**ADDRESS/STATUS:** During T1, these are the four most significant address lines for memory operations. During I/O operations, these lines are LOW. During memory and I/O operations, status information is available on these lines during T2, T3, Tw, and T4. S6 is always low. The status of the interrupt enable flag bit (S5) is updated at the beginning of each clock cycle. S4 and S3 are encoded as shown. This information indicates which segment register is presently being used for data accessing. These lines float to 3-state OFF during local bus "hold acknowledge".

S4		S3	Characteristics
0 (LOW)		0	Alternate Data
0		1	Stack
1 (HIGH)		0	Code or None
1		1	Data
S6 is 0 (LOW)			

Symbol	Pin No.	Type	Name and Function
RD̄	32	O	**READ:** Read strobe indicates that the processor is performing a memory or I/O read cycle, depending on the state of the IO/M̄ pin or S2. This signal is used to read devices which reside on the 8088 local bus. RD̄ is active LOW during T2, T3 and Tw of any read cycle, and is guaranteed to remain HIGH in T2 until the 8088 local bus has floated. This signal floats to 3-state OFF in "hold acknowledge".
READY	22	I	**READY:** is the acknowledgement from the addressed memory or I/O device that it will complete the data transfer. The RDY signal from memory or I/O is synchronized by the 8284 clock generator to form READY. This signal is active HIGH. The 8088 READY input is not synchronized. Correct operation is not guaranteed if the set up and hold times are not met.
INTR	18	I	**INTERRUPT REQUEST:** is a level triggered input which is sampled during the last clock cycle of each instruction to determine if the processor should enter into an interrupt acknowledge operation. A subroutine is vectored to via an interrupt vector lookup table located in system memory. It can be internally masked by software resetting the interrupt enable bit. INTR is internally synchronized. This signal is active HIGH.
TEST	23	I	**TEST:** input is examined by the "wait for test" instruction. If the TEST input is LOW, execution continues, otherwise the processor waits in an "idle" state. This input is synchronized internally during each clock cycle on the leading edge of CLK.

621

Table 1. Pin Description (Continued)

Symbol	Pin No.	Type	Name and Function
NMI	17	I	**NON-MASKABLE INTERRUPT:** is an edge triggered input which causes a type 2 interrupt. A subroutine is vectored to via an interrupt vector lookup table located in system memory. NMI is not maskable internally by software. A transition from a LOW to HIGH initiates the interrupt at the end of the current instruction. This input is internally synchronized.
RESET	21	I	**RESET:** causes the processor to immediately terminate its present activity. The signal must be active HIGH for at least four clock cycles. It restarts execution, as described in the instruction set description, when RESET returns LOW. RESET is internally synchronized.
CLK	19	I	**CLOCK:** provides the basic timing for the processor and bus controller. It is asymmetric with a 33% duty cycle to provide optimized internal timing.
V_{CC}	40		**V_{CC}:** is the +5V ±10% power supply pin.
GND	1, 20		**GND:** are the ground pins.
MN/$\overline{MX}$	33	I	**MINIMUM/MAXIMUM:** indicates what mode the processor is to operate in. The two modes are discussed in the following sections.

The following pin function descriptions are for the 8088 minimum mode (i.e., MN/$\overline{MX}$ = V_{CC}). Only the pin functions which are unique to minimum mode are described; all other pin functions are as described above.

Symbol	Pin No.	Type	Name and Function
IO/$\overline{M}$	28	O	**STATUS LINE:** is an inverted maximum mode $\overline{S2}$. It is used to distinguish a memory access from an I/O access. IO/$\overline{M}$ becomes valid in the T4 preceding a bus cycle and remains valid until the final T4 of the cycle (I/O = HIGH, M = LOW). IO/$\overline{M}$ floats to 3-state OFF in local bus "hold acknowledge".
$\overline{WR}$	29	O	**WRITE:** strobe indicates that the processor is performing a write memory or write I/O cycle, depending on the state of the IO/$\overline{M}$ signal. WR is active for T2, T3, and Tw of any write cycle. It is active LOW, and floats to 3-state OFF in local bus "hold acknowledge".
$\overline{INTA}$	24	O	**INTA:** is used as a read strobe for interrupt acknowledge cycles. It is active LOW during T2, T3, and Tw of each interrupt acknowledge cycle.
ALE	25	O	**ADDRESS LATCH ENABLE:** is provided by the processor to latch the address into an address latch. It is a HIGH pulse active during clock low of T1 of any bus cycle. Note that ALE is never floated.
DT/$\overline{R}$	27	O	**DATA TRANSMIT/RECEIVE:** is needed in a minimum system that desires to use a data bus transceiver. It is used to control the direction of data flow through the transceiver. Logically, DT/$\overline{R}$ is equivalent to $\overline{S1}$ in the maximum mode, and its timing is the same as for IO/$\overline{M}$ (T = HIGH, R = LOW). This signal floats to 3-state OFF in local "hold acknowledge".
$\overline{DEN}$	26	O	**DATA ENABLE:** is provided as an output enable for the data bus transceiver in a minimum system which uses the transceiver. $\overline{DEN}$ is active LOW during each memory and I/O access, and for $\overline{INTA}$ cycles. For a read or $\overline{INTA}$ cycle, it is active from the middle of T2 until the middle of T4, while for a write cycle, it is active from the beginning of T2 until the middle of T4. $\overline{DEN}$ floats to 3-state OFF during local bus "hold acknowledge".

8088

Table 1. Pin Description (Continued)

Symbol	Pin No.	Type	Name and Function
HOLD, HLDA	31, 30	I, O	**HOLD:** indicates that another master is requesting a local bus "hold". To be acknowledged, HOLD must be active HIGH. The processor receiving the "hold" request will issue HLDA (HIGH) as an acknowledgement, in the middle of a T4 or Ti clock cycle. Simultaneous with the issuance of HLDA the processor will float the local bus and control lines. After HOLD is detected as being LOW, the processor lowers HLDA, and when the processor needs to run another cycle, it will again drive the local bus and control lines. HOLD and HLDA have internal pull-up resistors. Hold is not an asynchronous input. External synchronization should be provided if the system cannot otherwise guarantee the set up time.
$\overline{SSO}$	34	O	**STATUS LINE:** is logically equivalent to $\overline{S0}$ in the maximum mode. The combination of $\overline{SSO}$, IO/$\overline{M}$ and DT/$\overline{R}$ allows the system to completely decode the current bus cycle status.

IO/$\overline{M}$	DT/$\overline{R}$	$\overline{SSO}$	Characteristics
1(HIGH)	0	0	Interrupt Acknowledge
1	0	1	Read I/O Port
1	1	0	Write I/O Port
1	1	1	Halt
0(LOW)	0	0	Code Access
0	0	1	Read Memory
0	1	0	Write Memory
0	1	1	Passive

The following pin function descriptions are for the 8088/8288 system in maximum mode (i.e., MN/$\overline{MX}$ = GND). Only the pin functions which are unique to maximum mode are described; all other pin functions are as described above.

Symbol	Pin No.	Type	Name and Function
$\overline{S2}$, $\overline{S1}$, $\overline{S0}$	26–28	O	**STATUS:** is active during clock high of T4, T1, and T2, and is returned to the passive state (1,1,1) during T3 or during Tw when READY is HIGH. This status is used by the 8288 bus controller to generate all memory and I/O access control signals. Any change by $\overline{S2}$, $\overline{S1}$, or $\overline{S0}$ during T4 is used to indicate the beginning of a bus cycle, and the return to the passive state in T3 and Tw is used to indicate the end of a bus cycle. These signals float to 3-state OFF during "hold acknowledge". During the first clock cycle after RESET becomes active, these signals are active HIGH. After this first clock, they float to 3-state OFF.

$\overline{S2}$	$\overline{S1}$	$\overline{S0}$	Characteristics
0(LOW)	0	0	Interrupt Acknowledge
0	0	1	Read I/O Port
0	1	0	Write I/O Port
0	1	1	Halt
1(HIGH)	0	0	Code Access
1	0	1	Read Memory
1	1	0	Write Memory
1	1	1	Passive

623

Table 1. Pin Description (Continued)

Symbol	Pin No.	Type	Name and Function
R̄Q̄/ḠT̄0, R̄Q̄/ḠT̄1	30, 31	I/O	**REQUEST/GRANT:** pins are used by other local bus masters to force the processor to release the local bus at the end of the processor's current bus cycle. Each pin is bidirectional with R̄Q̄/ḠT̄0 having higher priority than R̄Q̄/ḠT̄1. R̄Q̄/ḠT̄ has an internal pull-up resistor, so may be left unconnected. The request/grant sequence is as follows (See Figure 8):
			1. A pulse of one CLK wide from another local bus master indicates a local bus request ("hold") to the 8088 (pulse 1).
			2. During a T4 or TI clock cycle, a pulse one clock wide from the 8088 to the requesting master (pulse 2), indicates that the 8088 has allowed the local bus to float and that it will enter the "hold acknowledge" state at the next CLK. The CPU's bus interface unit is disconnected logically from the local bus during "hold acknowledge". The same rules as for HOLD/HOLDA apply as for when the bus is released.
			3. A pulse one CLK wide from the requesting master indicates to the 8088 (pulse 3) that the "hold" request is about to end and that the 8088 can reclaim the local bus at the next CLK. The CPU then enters T4.
			Each master-master exchange of the local bus is a sequence of three pulses. There must be one idle CLK cycle after each bus exchange. Pulses are active LOW.
			If the request is made while the CPU is performing a memory cycle, it will release the local bus during T4 of the cycle when all the following conditions are met:
			1. Request occurs on or before T2. 2. Current cycle is not the low bit of a word. 3. Current cycle is not the first acknowledge of an interrupt acknowledge sequence. 4. A locked instruction is not currently executing.
			If the local bus is idle when the request is made the two possible events will follow:
			1. Local bus will be released during the next clock. 2. A memory cycle will start within 3 clocks. Now the four rules for a currently active memory cycle apply with condition number 1 already satisfied.
L̄O̅C̄K̄	29	O	**LOCK:** indicates that other system bus masters are not to gain control of the system bus while L̄O̅C̄K̄ is active (LOW). The L̄O̅C̄K̄ signal is activated by the "LOCK" prefix instruction and remains active until the completion of the next instruction. This signal is active LOW, and floats to 3-state off in "hold acknowledge".
QS1, QS0	24, 25	O	**QUEUE STATUS:** provide status to allow external tracking of the internal 8088 instruction queue. The queue status is valid during the CLK cycle after which the queue operation is performed.

QS1	QS0	Characteristics
0(LOW)	0	No Operation
0	1	First Byte of Opcode from Queue
1(HIGH)	0	Empty the Queue
1	1	Subsequent Byte from Queue

| — | 34 | O | Pin 34 is always high in the maximum mode. |

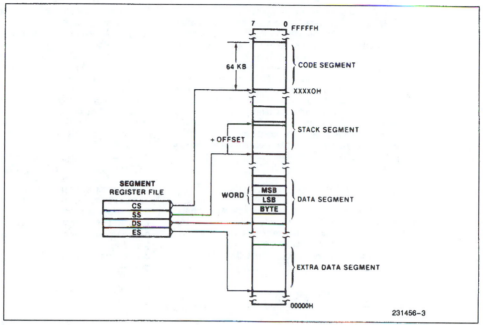

Figure 3. Memory Organization

FUNCTIONAL DESCRIPTION

Memory Organization

The processor provides a 20-bit address to memory which locates the byte being referenced. The memory is organized as a linear array of up to 1 million bytes, addressed as 00000(H) to FFFFF(H). The memory is logically divided into code, data, extra data, and stack segments of up to 64K bytes each, with each segment falling on 16-byte boundaries (See Figure 3).

All memory references are made relative to base addresses contained in high speed segment registers. The segment types were chosen based on the ad-

dressing needs of programs. The segment register to be selected is automatically chosen according to the rules of the following table. All information in one segment type share the same logical attributes (e.g. code or data). By structuring memory into relocatable areas of similar characteristics and by automatically selecting segment registers, programs are shorter, faster, and more structured.

Word (16-bit) operands can be located on even or odd address boundaries. For address and data operands, the least significant byte of the word is stored in the lower valued address location and the most significant byte in the next higher address location. The BIU will automatically execute two fetch or write cycles for 16-bit operands.

Memory Reference Used	Segment Register Used	Segment Selection Rule
Instructions	CODE (CS)	Automatic with all instruction prefetch.
Stack	STACK (SS)	All stack pushes and pops. Memory references relative to BP base register except data references.
Local Data	DATA (DS)	Data references when: relative to stack, destination of string operation, or explicitly overridden.
External (Global) Data	EXTRA (ES)	Destination of string operations: Explicitly selected using a segment override.

8088

Certain locations in memory are reserved for specific CPU operations (See Figure 4). Locations from addresses FFFF0H through FFFFFH are reserved for operations including a jump to the initial system initialization routine. Following RESET, the CPU will always begin execution at location FFFF0H where the jump must be located. Locations 00000H through 003FFH are reserved for interrupt operations. Four-byte pointers consisting of a 16-bit segment address and a 16-bit offset address direct program flow to one of the 256 possible interrupt service routines. The pointer elements are assumed to have been stored at their respective places in reserved memory prior to the occurrence of interrupts.

Minimum and Maximum Modes

The requirements for supporting minimum and maximum 8088 systems are sufficiently different that they cannot be done efficiently with 40 uniquely defined pins. Consequently, the 8088 is equipped with a strap pin (MN/$\overline{\text{MX}}$) which defines the system con-

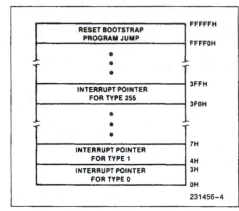

Figure 4. Reserved Memory Locations

figuration. The definition of a certain subset of the pins changes, dependent on the condition of the strap pin. When the MN/$\overline{\text{MX}}$ pin is strapped to GND, the 8088 defines pins 24 through 31 and 34 in maximum mode. When the MN/$\overline{\text{MX}}$ pin is strapped to V$_{CC}$, the 8088 generates bus control signals itself on pins 24 through 31 and 34.

The minimum mode 8088 can be used with either a multiplexed or demultiplexed bus. The multiplexed bus configuration is compatible with the MCS-85™ multiplexed bus peripherals. This configuration (See Figure 5) provides the user with a minimum chip count system. This architecture provides the 8088 processing power in a highly integrated form.

The demultiplexed mode requires one latch (for 64K addressability) or two latches (for a full megabyte of addressing). A third latch can be used for buffering if the address bus loading requires it. A transceiver can also be used if data bus buffering is required (See Figure 6). The 8088 provides $\overline{\text{DEN}}$ and DT/$\overline{\text{R}}$ to control the transceiver, and ALE to latch the addresses. This configuration of the minimum mode provides the standard demultiplexed bus structure with heavy bus buffering and relaxed bus timing requirements.

The maximum mode employs the 8288 bus controller (See Figure 7). The 8288 decodes status lines $\overline{\text{S0}}$, $\overline{\text{S1}}$, and $\overline{\text{S2}}$, and provides the system with all bus control signals. Moving the bus control to the 8288 provides better source and sink current capability to the control lines, and frees the 8088 pins for extended large system features. Hardware lock, queue status, and two request/grant interfaces are provided by the 8088 in maximum mode. These features allow co-processors in local bus and remote bus configurations.

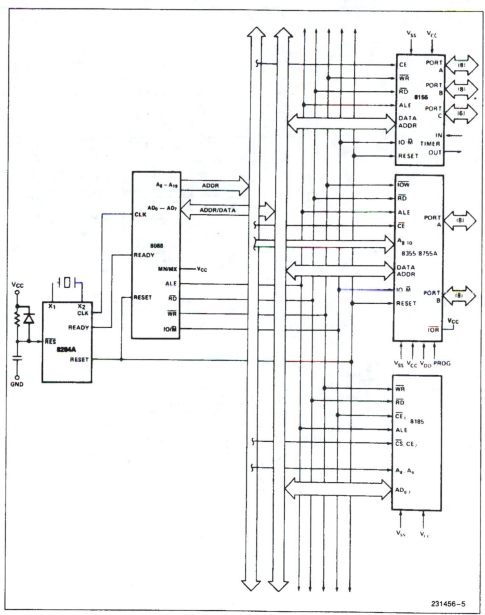

Figure 5. Multiplexed Bus Configuration

231456–5

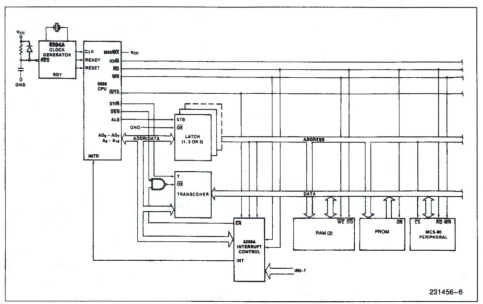

Figure 6. Demultiplexed Bus Configuration

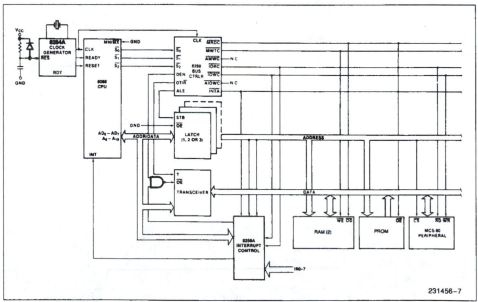

Figure 7. Fully Buffered System Using Bus Controller

8088

Bus Operation

The 8088 address/data bus is broken into three parts—the lower eight address/data bits (AD0–AD7), the middle eight address bits (A8–A15), and the upper four address bits (A16–A19). The address/data bits and the highest four address bits are time multiplexed. This technique provides the most efficient use of pins on the processor, permitting the use of a standard 40 lead package. The middle eight address bits are not multiplexed, i.e. they remain val-

id throughout each bus cycle. In addition, the bus can be demultiplexed at the processor with a single address latch if a standard, non-multiplexed bus is desired for the system.

Each processor bus cycle consists of at least four CLK cycles. These are referred to as T1, T2, T3, and T4 (See Figure 8). The address is emitted from the processor during T1 and data transfer occurs on the bus during T3 and T4. T2 is used primarily for chang-

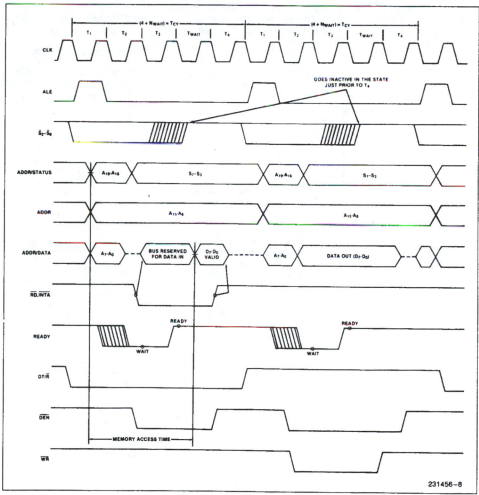

Figure 8. Basic System Timing

231456-8

629

ing the direction of the bus during read operations. In the event that a "NOT READY" indication is given by the addressed device, "wait" states (Tw) are inserted between T3 and T4. Each inserted "wait" state is of the same duration as a CLK cycle. Periods can occur between 8088 driven bus cycles. These are referred to as "idle" states (Ti), or inactive CLK cycles. The processor uses these cycles for internal housekeeping.

During T1 of any bus cycle, the ALE (address latch enable) signal is emitted (by either the processor or the 8288 bus controller, depending on the MN/$\overline{MX}$ strap). At the trailing edge of this pulse, a valid address and certain status information for the cycle may be latched.

Status bits $\overline{S0}$, $\overline{S1}$, and $\overline{S2}$ are used by the bus controller, in maximum mode, to identify the type of bus transaction according to the following table:

$\overline{S2}$	$\overline{S1}$	$\overline{S0}$	Characteristics
0(LOW)	0	0	Interrupt Acknowledge
0	0	1	Read I/O
0	1	0	Write I/O
0	1	1	Halt
1(HIGH)	0	0	Instruction Fetch
1	0	1	Read Data from Memory
1	1	0	Write Data to Memory
1	1	1	Passive (No Bus Cycle)

Status bits S3 through S6 are multiplexed with high order address bits and are therefore valid during T2 through T4. S3 and S4 indicate which segment register was used for this bus cycle in forming the address according to the following table:

S_4	S_3	Characteristics
0(LOW)	0	Alternate Data (Extra Segment)
0	1	Stack
1(HIGH)	0	Code or None
1	1	Data

S5 is a reflection of the PSW interrupt enable bit. S6 is always equal to 0.

I/O Addressing

In the 8088, I/O operations can address up to a maximum of 64K I/O registers. The I/O address appears in the same format as the memory address on bus lines A15–A0. The address lines A19–A16 are zero in I/O operations. The variable I/O instructions, which use register DX as a pointer, have full address capability, while the direct I/O instructions directly address one or two of the 256 I/O byte locations in page 0 of the I/O address space. I/O ports are addressed in the same manner as memory locations.

Designers familiar with the 8085 or upgrading an 8085 design should note that the 8085 addresses I/O with an 8-bit address on both halves of the 16-bit address bus. The 8088 uses a full 16-bit address on its lower 16 address lines.

EXTERNAL INTERFACE

Processor Reset and Initialization

Processor initialization or start up is accomplished with activation (HIGH) of the RESET pin. The 8088 RESET is required to be HIGH for greater than four clock cycles. The 8088 will terminate operations on the high-going edge of RESET and will remain dormant as long as RESET is HIGH. The low-going transition of RESET triggers an internal reset sequence for approximately 7 clock cycles. After this interval the 8088 operates normally, beginning with the instruction in absolute locations FFFF0H (See Figure 4). The RESET input is internally synchronized to the processor clock. At initialization, the HIGH to LOW transition of RESET must occur no sooner than 50 μs after power up, to allow complete initialization of the 8088.

NMI asserted prior to the 2nd clock after the end of RESET will not be honored. If NMI is asserted after that point and during the internal reset sequence, the processor may execute one instruction before responding to the interrupt. A hold request active immediately after RESET will be honored before the first instruction fetch.

All 3-state outputs float to 3-state OFF during RESET. Status is active in the idle state for the first clock after RESET becomes active and then floats to 3-state OFF. ALE and HLDA are driven low.

Interrupt Operations

Interrupt operations fall into two classes: software or hardware initiated. The software initiated interrupts and software aspects of hardware interrupts are specified in the instruction set description in the iAPX 88 book or the iAPX 86,88 User's Manual. Hardware interrupts can be classified as nonmaskable or maskable.

Interrupts result in a transfer of control to a new program location. A 256 element table containing address pointers to the interrupt service program locations resides in absolute locations 0 through 3FFH (See Figure 4), which are reserved for this purpose. Each element in the table is 4 bytes in size and corresponds to an interrupt "type." An interrupting device supplies an 8-bit type number, during the interrupt acknowledge sequence, which is used to vector through the appropriate element to the new interrupt service program location.

Non-Maskable Interrupt (NMI)

The processor provides a single non-maskable interrupt (NMI) pin which has higher priority than the maskable interrupt request (INTR) pin. A typical use would be to activate a power failure routine. The NMI is edge-triggered on a LOW to HIGH transition. The activation of this pin causes a type 2 interrupt.

NMI is required to have a duration in the HIGH state of greater than two clock cycles, but is not required to be synchronized to the clock. Any higher going transition of NMI is latched on-chip and will be serviced at the end of the current instruction or between whole moves (2 bytes in the case of word moves) of a block type instruction. Worst case response to NMI would be for multiply, divide, and variable shift instructions. There is no specification on the occurrence of the low-going edge; it may occur before, during, or after the servicing of NMI. Another high-going edge triggers another response if it occurs after the start of the NMI procedure. The signal must be free of logical spikes in general and be free of bounces on the low-going edge to avoid triggering extraneous responses.

Maskable Interrupt (INTR)

The 8088 provides a single interrupt request input (INTR) which can be masked internally by software with the resetting of the interrupt enable (IF) flag bit. The interrupt request signal is level triggered. It is internally synchronized during each clock cycle on the high-going edge of CLK. To be responded to, INTR must be present (HIGH) during the clock period preceding the end of the current instruction or the end of a whole move for a block type instruction. During interrupt response sequence, further interrupts are disabled. The enable bit is reset as part of the response to any interrupt (INTR, NMI, software interrupt, or single step), although the FLAGS register which is automatically pushed onto the stack reflects the state of the processor prior to the interrupt. Until the old FLAGS register is restored, the

enable bit will be zero unless specifically set by an instruction.

During the response sequence (See Figure 9), the processor executes two successive (back to back) interrupt acknowledge cycles. The 8088 emits the LOCK signal (maximum mode only) from T2 of the first bus cycle until T2 of the second. A local bus "hold" request will not be honored until the end of the second bus cycle. In the second bus cycle, a byte is fetched from the external interrupt system (e.g., 8259A PIC) which identifies the source (type) of the interrupt. This byte is multiplied by four and used as a pointer into the interrupt vector lookup table. An INTR signal left HIGH will be continually responded to within the limitations of the enable bit and sample period. The interrupt return instruction includes a flags pop which returns the status of the original interrupt enable bit when it restores the flags.

HALT

When a software HALT instruction is executed, the processor indicates that it is entering the HALT state in one of two ways, depending upon which mode is strapped. In minimum mode, the processor issues ALE, delayed by one clock cycle, to allow the system to latch the halt status. Halt status is available on IO/$\overline{M}$, DT/$\overline{R}$, and $\overline{SSO}$. In maximum mode, the processor issues appropriate HALT status on $\overline{S2}$, $\overline{S1}$, and $\overline{S0}$, and the 8288 bus controller issues one ALE. The 8088 will not leave the HALT state when a local bus hold is entered while in HALT. In this case, the processor reissues the HALT indicator at the end of the local bus hold. An interrupt request or RESET will force the 8088 out of the HALT state.

Read/Modify/Write (Semaphore) Operations via LOCK

The LOCK status information is provided by the processor when consecutive bus cycles are required during the execution of an instruction. This allows the processor to perform read/modify/write operations on memory (via the "exchange register with memory" instruction), without another system bus master receiving intervening memory cycles. This is useful in multiprocessor system configurations to accomplish "test and set lock" operations. The $\overline{LOCK}$ signal is activated (LOW) in the clock cycle following decoding of the LOCK prefix instruction. It is deactivated at the end of the last bus cycle of the instruction following the LOCK prefix. While $\overline{LOCK}$ is active, a request on a $\overline{RQ}/\overline{GT}$ pin will be recorded, and then honored at the end of the LOCK.

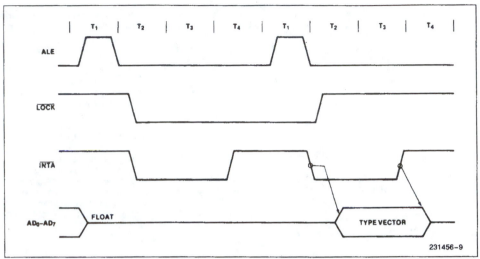

Figure 9. Interrupt Acknowledge Sequence

External Synchronization via $\overline{TEST}$

As an alternative to interrupts, the 8088 provides a single software-testable input pin ($\overline{TEST}$). This input is utilized by executing a WAIT instruction. The single WAIT instruction is repeatedly executed until the $\overline{TEST}$ input goes active (LOW). The execution of WAIT does not consume bus cycles once the queue is full.

If a local bus request occurs during WAIT execution, the 8088 3-states all output drivers. If interrupts are enabled, the 8088 will recognize interrupts and process them. The WAIT instruction is then refetched, and reexecuted.

Basic System Timing

In minimum mode, the MN/$\overline{MX}$ pin is strapped to V_{CC} and the processor emits bus control signals compatible with the 8085 bus structure. In maximum mode, the MN/$\overline{MX}$ pin is strapped to GND and the processor emits coded status information which the 8288 bus controller uses to generate MULTIBUS compatible bus control signals.

System Timing—Minimum System

(See Figure 8)

The read cycle begins in T1 with the assertion of the address latch enable (ALE) signal. The trailing (low

going) edge of this signal is used to latch the address information, which is valid on the address/data bus (AD0–AD7) at this time, into the 8282/8283 latch. Address lines A8 through A15 do not need to be latched because they remain valid throughout the bus cycle. From T1 to T4 the IO/$\overline{M}$ signal indicates a memory or I/O operation. At T2 the address is removed from the address/data bus and the bus goes to a high impedance state. The read control signal is also asserted at T2. The read ($\overline{RD}$) signal causes the addressed device to enable its data bus drivers to the local bus. Some time later, valid data will be available on the bus and the addressed device will drive the READY line HIGH. When the processor returns the read signal to a HIGH level, the addressed device will again 3-state its bus drivers. If a transceiver is required to buffer the 8088 local bus, signals DT/$\overline{R}$ and $\overline{DEN}$ are provided by the 8088.

A write cycle also begins with the assertion of ALE and the emission of the address. The IO/$\overline{M}$ signal is again asserted to indicate a memory or I/O write operation. In T2, immediately following the address emission, the processor emits the data to be written into the addressed location. This data remains valid until at least the middle of T4. During T2, T3, and Tw, the processor asserts the write control signal. The write ($\overline{WR}$) signal becomes active at the beginning of T2, as opposed to the read, which is delayed somewhat into T2 to provide time for the bus to float.

The basic difference between the interrupt acknowledge cycle and a read cycle is that the interrupt acknowledge ($\overline{\text{INTA}}$) signal is asserted in place of the read ($\overline{\text{RD}}$) signal and the address bus is floated. (See Figure 9) In the second of two successive $\overline{\text{INTA}}$ cycles, a byte of information is read from the data bus, as supplied by the interrupt system logic (i.e. 8259A priority interrupt controller). This byte identifies the source (type) of the interrupt. It is multiplied by four and used as a pointer into the interrupt vector lookup table, as described earlier.

Bus Timing—Medium Complexity Systems

(See Figure 10)

For medium complexity systems, the MN/$\overline{\text{MX}}$ pin is connected to GND and the 8288 bus controller is added to the system, as well as a latch for latching the system address, and a transceiver to allow for bus loading greater than the 8088 is capable of handling. Signals ALE, $\overline{\text{DEN}}$, and DT/$\overline{\text{R}}$ are generated by the 8288 instead of the processor in this configuration, although their timing remains relatively the same. The 8088 status outputs ($\overline{\text{S2}}$, $\overline{\text{S1}}$, and $\overline{\text{S0}}$) provide type of cycle information and become 8288 inputs. This bus cycle information specifies read (code, data, or I/O), write (data or I/O), interrupt acknowledge, or software halt. The 8288 thus issues control signals specifying memory read or write, I/O read or write, or interrupt acknowledge. The 8288 provides two types of write strobes, normal and advanced, to be applied as required. The normal write strobes have data valid at the leading edge of write. The advanced write strobes have the same timing as read strobes, and hence, data is not valid at the leading edge of write. The transceiver receives the usual T and $\overline{\text{OE}}$ inputs from the 8288's DT/$\overline{\text{R}}$ and $\overline{\text{DEN}}$ outputs.

The pointer into the interrupt vector table, which is passed during the second $\overline{\text{INTA}}$ cycle, can derive from an 8259A located on either the local bus or the system bus. If the master 8289A priority interrupt controller is positioned on the local bus, a TTL gate is required to disable the transceiver when reading from the master 8259A during the interrupt acknowledge sequence and software "poll".

The 8088 Compared to the 8086

The 8088 CPU is an 8-bit processor designed around the 8086 internal structure. Most internal functions of the 8088 are identical to the equivalent 8086 functions. The 8088 handles the external bus the same way the 8086 does with the distinction of handling only 8 bits at a time. Sixteen-bit operands are fetched or written in two consecutive bus cycles. Both processors will appear identical to the software engineer, with the exception of execution time. The internal register structure is identical and all instructions have the same end result. The differences between the 8088 and 8086 are outlined below. The engineer who is unfamiliar with the 8086 is referred to the iAPX 86, 88 User's Manual, Chapters 2 and 4, for function description and instruction set information. Internally, there are three differences between the 8088 and the 8086. All changes are related to the 8-bit bus interface.

- The queue length is 4 bytes in the 8088, whereas the 8086 queue contains 6 bytes, or three words. The queue was shortened to prevent overuse of the bus by the BIU when prefetching instructions. This was required because of the additional time necessary to fetch instructions 8 bits at a time.

- To further optimize the queue, the prefetching algorithm was changed. The 8088 BIU will fetch a new instruction to load into the queue each time there is a 1 byte hole (space available) in the queue. The 8086 waits until a 2-byte space is available.

- The internal execution time of the instruction set is affected by the 8-bit interface. All 16-bit fetches and writes from/to memory take an additional four clock cycles. The CPU is also limited by the speed of instruction fetches. This latter problem only occurs when a series of simple operations occur. When the more sophisticated instructions of the 8088 are being used, the queue has time to fill and the execution proceeds as fast as the execution unit will allow.

The 8088 and 8086 are completely software compatible by virtue of their identical execution units. Software that is system dependent may not be completely transferable, but software that is not system dependent will operate equally as well on an 8088 and an 8086.

The hardware interface of the 8088 contains the major differences between the two CPUs. The pin assignments are nearly identical, however, with the following functional changes:

- A8–A15—These pins are only address outputs on the 8088. These address lines are latched internally and remain valid throughout a bus cycle in a manner similar to the 8085 upper address lines.

- $\overline{\text{BHE}}$ has no meaning on the 8088 and has been eliminated.

8088

- $\overline{\text{SSO}}$ provides the $\overline{\text{SO}}$ status information in the minimum mode. This output occurs on pin 34 in minimum mode only. DT/$\overline{\text{R}}$, IO/$\overline{\text{M}}$, and $\overline{\text{SSO}}$ provide the complete bus status in minimum mode.

- IO/$\overline{\text{M}}$ has been inverted to be compatible with the MCS-85 bus structure.

- ALE is delayed by one clock cycle in the minimum mode when entering HALT, to allow the status to be latched with ALE.

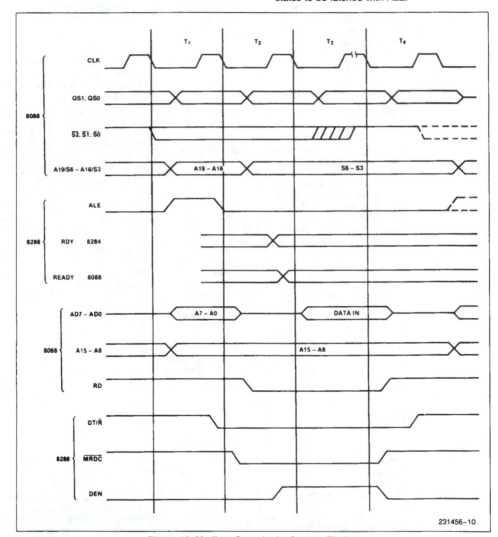

231456–10

Figure 10. Medium Complexity System Timing

APPENDIX B

Instruction Set and Execution Times

8088

8086/8088 Instruction Set Summary

Mnemonic and Description	Instruction Code			
DATA TRANSFER				
MOV = Move:	76543210	76543210	76543210	76543210
Register/Memory to/from Register	100010dw	mod reg r/m		
Immediate to Register/Memory	1100011w	mod 0 0 0 r/m	data	data if w = 1
Immediate to Register	1011w reg	data	data if w = 1	
Memory to Accumulator	1010000w	addr-low	addr-high	
Accumulator to Memory	1010001w	addr-low	addr-high	
Register/Memory to Segment Register	10001110	mod 0 reg r/m		
Segment Register to Register/Memory	10001100	mod 0 reg r/m		
PUSH = Push:				
Register/Memory	11111111	mod 1 1 0 r/m		
Register	01010 reg			
Segment Register	000 reg 1 1 0			
POP = Pop:				
Register/Memory	10001111	mod 0 0 0 r/m		
Register	01011 reg			
Segment Register	000 reg 1 1 1			
XCHG = Exchange:				
Register/Memory with Register	1000011w	mod reg r/m		
Register with Accumulator	10010 reg			
IN = Input from:				
Fixed Port	1110010w	port		
Variable Port	1110110w			
OUT = Output to:				
Fixed Port	1110011w	port		
Variable Port	1110111w			
XLAT = Translate Byte to AL	11010111			
LEA = Load EA to Register	10001101	mod reg r/m		
LDS = Load Pointer to DS	11000101	mod reg r/m		
LES = Load Pointer to ES	11000100	mod reg r/m		
LAHF = Load AH with Flags	10011111			
SAHF = Store AH into Flags	10011110			
PUSHF = Push Flags	10011100			
POPF = Pop Flags	10011101			

Reprinted by permission of Intel Corporation. Copyright/Intel Corporation 1991.

8086/8088 Instruction Set Summary (Continued)

Mnemonic and Description	Instruction Code			
ARITHMETIC	76543210	76543210	76543210	76543210
ADD = Add:				
Reg./Memory with Register to Either	000000dw	mod reg r/m		
Immediate to Register/Memory	100000sw	mod 0 0 0 r/m	data	data if s:w = 01
Immediate to Accumulator	0000010w	data	data if w = 1	
ADC = Add with Carry:				
Reg./Memory with Register to Either	000100dw	mod reg r/m		
Immediate to Register/Memory	100000sw	mod 0 1 0 r/m	data	data if s:w = 01
Immediate to Accumulator	0001010w	data	data if w = 1	
INC = Increment:				
Register/Memory	1111111w	mod 0 0 0 r/m		
Register	0 1 0 0 0 reg			
AAA = ASCII Adjust for Add	00110111			
BAA = Decimal Adjust for Add	00100111			
SUB = Subtract:				
Reg./Memory and Register to Either	001010dw	mod reg r/m		
Immediate from Register/Memory	100000sw	mod 1 0 1 r/m	data	data if s:w = 01
Immediate from Accumulator	0010110w	data	data if w = 1	
SSB = Subtract with Borrow				
Reg./Memory and Register to Either	000110dw	mod reg r/m		
Immediate from Register/Memory	100000sw	mod 0 1 1 r/m	data	data if s:w = 01
Immediate from Accumulator	000111w	data	data if w = 1	
DEC = Decrement:				
Register/memory	1111111w	mod 0 0 1 r/m		
Register	0 1 0 0 1 reg			
NEG = Change sign	1111011w	mod 0 1 1 r/m		
CMP = Compare:				
Register/Memory and Register	001110dw	mod reg r/m		
Immediate with Register/Memory	100000sw	mod 1 1 1 r/m	data	data if s:w = 01
Immediate with Accumulator	0011110w	data	data if w = 1	
AAS = ASCII Adjust for Subtract	00111111			
DAS = Decimal Adjust for Subtract	00101111			
MUL = Multiply (Unsigned)	1111011w	mod 1 0 0 r/m		
IMUL = Integer Multiply (Signed)	1111011w	mod 1 0 1 r/m		
AAM = ASCII Adjust for Multiply	11010100	00001010		
DIV = Divide (Unsigned)	1111011w	mod 1 1 0 r/m		
IDIV = Integer Divide (Signed)	1111011w	mod 1 1 1 r/m		
AAD = ASCII Adjust for Divide	11010101	00001010		
CBW = Convert Byte to Word	10011000			
CWD = Convert Word to Double Word	10011001			

8086/8088 Instruction Set Summary (Continued)

Mnemonic and Description	Instruction Code			
LOGIC	76543210	76543210	76543210	76543210
NOT = Invert	1111011w	mod 0 1 0 r/m		
SHL/SAL = Shift Logical/Arithmetic Left	110100vw	mod 1 0 0 r/m		
SHR = Shift Logical Right	110100vw	mod 1 0 1 r/m		
SAR = Shift Arithmetic Right	110100vw	mod 1 1 1 r/m		
ROL = Rotate Left	110100vw	mod 0 0 0 r/m		
ROR = Rotate Right	110100vw	mod 0 0 1 r/m		
RCL = Rotate Through Carry Flag Left	110100vw	mod 0 1 0 r/m		
RCR = Rotate Through Carry Right	110100vw	mod 0 1 1 r/m		
AND = And:				
Reg./Memory and Register to Either	001000dw	mod reg r/m		
Immediate to Register/Memory	1000000w	mod 1 0 0 r/m	data	data if w = 1
Immediate to Accumulator	0010010w	data	data if w = 1	
TEST = And Function to Flags. No Result:				
Register/Memory and Register	1000010w	mod reg r/m		
Immediate Data and Register/Memory	1111011w	mod 0 0 0 r/m	data	data if w = 1
Immediate Data and Accumulator	1010100w	data	data if w = 1	
OR = Or:				
Reg./Memory and Register to Either	000010dw	mod reg r/m		
Immediate to Register/Memory	1000000w	mod 0 0 1 r/m	data	data if w = 1
Immediate to Accumulator	0000110w	data	data if w = 1	
XOR = Exclusive or:				
Reg./Memory and Register to Either	001100dw	mod reg r/m		
Immediate to Register/Memory	1000000w	mod 1 1 0 r/m	data	data if w = 1
Immediate to Accumulator	0011010w	data	data if w = 1	
STRING MANIPULATION				
REP = Repeat	1111001z			
MOVS = Move Byte/Word	1010010w			
CMPS = Compare Byte/Word	1010011w			
SCAS = Scan Byte/Word	1010111w			
LODS = Load Byte/Wd to AL/AX	1010110w			
STOS = Stor Byte/Wd from AL/A	1010101w			
CONTROL TRANSFER				
CALL = Call:				
Direct Within Segment	11101000	disp-low	disp-high	
Indirect Within Segment	11111111	mod 0 1 0 r/m		
Direct Intersegment	10011010	offset-low	offset-high	
		seg-low	seg-high	
Indirect Intersegment	11111111	mod 0 1 1 r/m		

8088

8086/8088 Instruction Set Summary (Continued)

Mnemonic and Description	Instruction Code		
JMP = Unconditional Jump:	7 6 5 4 3 2 1 0	7 6 5 4 3 2 1 0	7 6 5 4 3 2 1 0
Direct Within Segment	1 1 1 0 1 0 0 1	disp-low	disp-high
Direct Within Segment-Short	1 1 1 0 1 0 1 1	disp	
Indirect Within Segment	1 1 1 1 1 1 1 1	mod 1 0 0 r/m	
Direct Intersegment	1 1 1 0 1 0 1 0	offset-low	offset-high
		seg-low	seg-high
Indirect Intersegment	1 1 1 1 1 1 1 1	mod 1 0 1 r/m	
RET = Return from CALL:			
Within Segment	1 1 0 0 0 0 1 1		
Within Seg Adding Immed to SP	1 1 0 0 0 0 1 0	data-low	data-high
Intersegment	1 1 0 0 1 0 1 1		
Intersegment Adding Immediate to SP	1 1 0 0 1 0 1 0	data-low	data-high
JE/JZ = Jump on Equal/Zero	0 1 1 1 0 1 0 0	disp	
JL/JNGE = Jump on Less/Not Greater or Equal	0 1 1 1 1 1 0 0	disp	
JLE/JNG = Jump on Less or Equal/ Not Greater	0 1 1 1 1 1 1 0	disp	
JB/JNAE = Jump on Below/Not Above or Equal	0 1 1 1 0 0 1 0	disp	
JBE/JNA = Jump on Below or Equal/ Not Above	0 1 1 1 0 1 1 0	disp	
JP/JPE = Jump on Parity/Parity Even	0 1 1 1 1 0 1 0	disp	
JO = Jump on Overflow	0 1 1 1 0 0 0 0	disp	
JS = Jump on Sign	0 1 1 1 1 0 0 0	disp	
JNE/JNZ = Jump on Not Equal/Not Zero	0 1 1 1 0 1 0 1	disp	
JNL/JGE = Jump on Not Less/Greater or Equal	0 1 1 1 1 1 0 1	disp	
JNLE/JG = Jump on Not Less or Equal/ Greater	0 1 1 1 1 1 1 1	disp	
JNB/JAE = Jump on Not Below/Above or Equal	0 1 1 1 0 0 1 1	disp	
JNBE/JA = Jump on Not Below or Equal/Above	0 1 1 1 0 1 1 1	disp	
JNP/JPO = Jump on Not Par/Par Odd	0 1 1 1 1 0 1 1	disp	
JNO = Jump on Not Overflow	0 1 1 1 0 0 0 1	disp	
JNS = Jump on Not Sign	0 1 1 1 1 0 0 1	disp	
LOOP = Loop CX Times	1 1 1 0 0 0 1 0	disp	
LOOPZ/LOOPE = Loop While Zero/Equal	1 1 1 0 0 0 0 1	disp	
LOOPNZ/LOOPNE = Loop While Not Zero/Equal	1 1 1 0 0 0 0 0	disp	
JCXZ = Jump on CX Zero	1 1 1 0 0 0 1 1	disp	
INT = Interrupt			
Type Specified	1 1 0 0 1 1 0 1	type	
Type 3	1 1 0 0 1 1 0 0		
INTO = Interrupt on Overflow	1 1 0 0 1 1 1 0		
IRET = Interrupt Return	1 1 0 0 1 1 1 1		

639

8086/8088 Instruction Set Summary (Continued)

Mnemonic and Description	Instruction Code	
	76543210	76543210
PROCESSOR CONTROL		
CLC = Clear Carry	11111000	
CMC = Complement Carry	11110101	
STC = Set Carry	11111001	
CLD = Clear Direction	11111100	
STD = Set Direction	11111101	
CLI = Clear Interrupt	11111010	
STI = Set Interrupt	11111011	
HLT = Halt	11110100	
WAIT = Wait	10011011	
ESC = Escape (to External Device)	11011xxx	mod x x x r/m
LOCK = Bus Lock Prefix	11110000	

NOTES:
AL = 8-bit accumulator
AX = 16-bit accumulator
CX = Count register
DS = Data segment
ES = Extra segment
Above/below refers to unsigned value
Greater = more positive:
Less = less positive (more negative) signed values
if d = 1 then "to" reg; if d = 0 then "from" reg
if w = 1 then word instruction; if w = 0 then byte
    instruction
if mod = 11 then r/m is treated as a REG field
if mod = 00 then DISP = 0*, disp-low and disp-high are
    absent
if mod = 01 then DISP = disp-low sign-extended to
    16 bits, disp-high is absent
if mod = 10 then DISP = disp-high; disp-low
if r/m = 000 then EA = (BX) + (SI) + DISP
if r/m = 001 then EA = (BX) + (DI) + DISP
if r/m = 010 then EA = (BP) + (SI) + DISP
if r/m = 011 then EA = (BP) + (DI) + DISP
if r/m = 100 then EA = (SI) + DISP
if r/m = 101 then EA = (DI) + DISP
if r/m = 110 then EA = (BP) + DISP*
if r/m = 111 then EA = (BX) + DISP
DISP follows 2nd byte of instruction (before data if required)

*except if mod = 00 and r/m = then EA = disp-high:
disp-low.
if s:w = 01 then 16 bits of immediate data form the operand
if s:w = 11 then an immediate data byte is sign extended
    to form the 16-bit operand
if v = 0 then "count" = 1; if v = 1 then "count" in (CL)
    register
x = don't care
z is used for string primitives for comparison with ZF FLAG
SEGMENT OVERRIDE PREFIX

    0 0 1 reg 1 1 0

REG is assigned according to the following table:

16-Bit (w = 1)		8-Bit (w = 0)		Segment	
000	AX	000	AL	00	ES
001	CX	001	CL	01	CS
010	DX	010	DL	10	SS
011	BX	011	BL	11	DS
100	SP	100	AH		
101	BP	101	CH		
110	SI	110	DH		
111	DI	111	BH		

Instructions which reference the flag register file as a 16-bit
object use the symbol FLAGS to represent the file:
FLAGS =
X:X:X:X:(OF):(DF):(IF):(TF):(SF):(ZF):X:(AF):X:(PF):X:(CF)

Mnemonics © Intel, 1978

DATA SHEET REVISION REVIEW

The following list represents key differences between this and the -005 data sheet. Please review this summary carefully.

1. The Intel® 8088 implementation technology (HMOS) has been changed to (HMOS-II).

INSTRUCTION EXECUTION TIMES

The following addressing-mode abbreviations are used throughout this section:

Reg—Register   **Mem**—Memory

Imm—Immediate   **Acc**—Accumulator

Arithmetic Instructions

Instruction	Clock Cycles	Transfers
AAA	4	
AAD	60	
AAM	83	
AAS	4	
ADC		
Reg to Reg	3	
Mem to Reg	9 + EA	1
Reg to Mem	16 + EA	2
Imm to Reg	4	
Imm to Mem	17 + EA	2
Imm to Acc	4	
ADD		
Reg to Reg	3	
Mem to Reg	9 + EA	1
Reg to Mem	16 + EA	2
Imm to Reg	4	
Imm to Mem	17 + EA	2
Imm to Acc	4	
CBW	2	
CWD	5	
DAA	4	
DAS	4	
DEC		
8-bit	3	
16-bit	2	
Mem	15 + EA	2
DIV		
8-bit Reg	80–90	
16-bit Reg	144–162	
8-bit Mem	86–96 + EA	1
16-bit Mem	150–168 + EA	1
IDIV		
8-bit Reg	101–112	
16-bit Reg	165–184	
8-bit Mem	107–118 + EA	1
16-bit Mem	171–190 + EA	1

Instruction	Clock Cycles	Transfer
IMUL		
8-bit Reg	80–98	
16-bit Reg	128–154	
8-bit Mem	86–104 + EA	1
16-bit Mem	134–160 + EA	1
INC		
8-bit	3	
16-bit	2	
Mem	15 + EA	2
MUL		
8-bit Reg	70–77	
16-bit Reg	118–133	
8-bit Mem	76–83 + EA	1
16-bit Mem	124–139 + EA	1
NEG		
Reg	3	
Mem	16 + EA	2
SBB		
Reg from Reg	3	
Mem from Reg	9 + EA	1
Reg from Mem	16 + EA	2
Imm from Reg	4	
Imm from Mem	17 + EA	2
Imm from Acc	4	
SUB		
Reg from Reg	3	
Mem from Reg	9 + EA	1
Reg from Mem	16 + EA	2
Imm from Reg	4	
Imm from Mem	17 + EA	2
Imm from Acc	4	

Logical Instructions

Instruction	Clock Cycles	Transfers
AND		
Reg to Reg	3	
Mem to Reg	9 + EA	1
Reg to Mem	16 + EA	2
Imm to Reg	4	
Imm to Mem	17 + EA	2
Imm to Acc	4	
CMP		
Reg to Reg	3	
Mem to Reg	9 + EA	1
Reg to Mem	9 + EA	1
Imm to Reg	4	
Imm to Mem	10 + EA	1
Imm to Acc	4	
NOT		
Reg	3	
Mem	16 + EA	2
OR		
Reg to Reg	3	
Mem to Reg	9 + EA	1
Reg to Mem	16 + EA	2
Imm to Reg	4	
Imm to Mem	17 + EA	2
Imm to Acc	4	
RCL		
1-bit Reg	2	
Multi-bit Reg	8 + 4/bit	
1-bit Mem	15 + EA	2
Multi-bit Mem	20 + EA + 4/bit	2
RCR		
1-bit Reg	2	
Multi-bit Reg	8 + 4/bit	
1-bit Mem	15 + EA	2
Multi-bit Mem	20 + EA + 4/bit	2
ROL		
1-bit Reg	2	
Multi-bit Reg	8 + 4/bit	
1-bit Mem	15 + EA	2
Multi-bit Mem	20 + EA + 4/bit	2
ROR		
1-bit Reg	2	
Multi-bit Reg	8 + 4/bit	
1-bit Mem	15 + EA	2
Multi-bit Mem	20 + EA + 4/bit	2

Instruction	Clock Cycles	Transfers
SAL/SHL		
1-bit Reg	2	
Multi-bit Reg	8 + 4/bit	
1-bit Mem	15 + EA	2
Multi-bit Mem	20 + EA + 4/bit	2
SAR		
1-bit Reg	2	
Multi-bit Reg	8 + 4/bit	
1-bit Mem	15 + EA	2
Multi-bit Mem	20 + EA + 4/bit	2
SHR		
1-bit Reg	2	
Multi-bit Reg	8 + 4/bit	
1-bit Mem	15 + EA	2
Multi-bit Mem	20 + EA + 4/bit	2
XOR		
Reg to Reg	3	
Mem to Reg	9 + EA	1
Reg to Mem	16 + EA	2
Imm to Reg	4	
Imm to Mem	17 + EA	2
Imm to Acc	4	

String Instructions

Instruction	Clock Cycles	Transfers
CMPS/CMPSB/CMPSW		
Not Repeated	22	2
Repeated	9 + 22/rep	2/rep
LDS	16 + EA	2
LEA	2 + EA	
LES	16 + EA	2
LODS/LODSB/LODSW		
Not Repeated	12	1
Repeated	9 + 13/rep	1/rep
MOVS/MOVSB/MOVSW		
Not Repeated	18	2
Repeated	9 + 17/rep	2/rep
SCAS/SCASB/SCASW		
Not Repeated	15	1
Repeated	9 + 15/rep	1/rep
STOS/STOSB/STOSW		
Not Repeated	11	1
Repeated	9 + 10/rep	1/rep

Loop and Jump Instructions

Instruction	Clock Cycles	Transfers
JA/JNBE	16/4	
JAE/JNB/JNC	16/4	
JB/JNAE/JC	16/4	
JBE/JNA	16/4	
JCXZ	18/6	
JE/JZ	16/4	
JG/JNLE	16/4	
JGE/JNL	16/4	
JL/JNGE	16/4	
JLE/JNG	16/4	
JMP		
Intrasegment direct short	15	
Intrasegment direct	15	
Intersegment direct	15	
Intrasegment Mem-indirect	18 + EA	1
Intrasegment Reg-indirect	11	
Intersegment indirect	24 + EA	2
JNE/JNZ	16/4	
JNO	16/4	
JNP/JPO	16/4	
JNS	16/4	
JO	16/4	
JP/JPE	16/4	
JS	16/4	
LOOP	17/5	
LOOPE/LOOPZ	18/6	
LOOPNE/LOOPNZ	19/5A	

x/y – x cycles when jump is not taken.
      y cycles when jump is taken.

Data Transfer Instructions

Instruction	Clock Cycles	Transfers
IN		
Fixed-port	10	1
Variable-port	8	1
LAHF	4	
MOV		
Acc to Mem	10	1
Mem to Acc	10	1
Reg to Reg	2	
Mem to Reg	8 + EA	1
Reg to Mem	9 + EAs	1
Imm to Reg	4	
Imm to Mem	10 + EA	1
Reg to SS, DS, or ES	2	
Mem to SS, DS, or ES	8 + EA	1
Segment Reg to Reg	2	
Segment Reg to Mem	9 + EA	1
OUT		
Fixed-port	10	1
Variable-port	8	1
POP		
Reg	8	1
SS, DS, or ES	8	1
Mem	16 + EA	2
POPF	8	1
PUSH		
Reg	11	1
Segment Reg	10	1
Mem	16 + EA	2
PUSHF	10	1
SAHF	4	
XCHG		
Reg with Acc	3	
Reg with Mem	17 + EA	2
Reg with Reg	4	
XLAT/XLATB	11	1

Subroutine and Interrupt Instructions

Instruction	Clock Cycles	Transfers
CALL		
Intrasegment direct	19	1
Intersegment direct	28	2
Intrasegment Mem-indirect	21 + EA	2
Intrasegment Reg-indirect	16	1
Intersegment indirect	37 + EA	4
INT		
Type-3	52	5
Not Type-3	51	5
INTO		
Taken	53	5
Not Taken	4	
INTR	61	7
IRET	24	3
NMI	50	5
RET		
Intrasegment	8	1
Intrasegment with pop	12	1
Intersegment	18	2
Intersegment with pop	17	2

Note: INTR and NMI are included for timing purposes only.

Processor Control Instructions

Instruction	Clock Cycles	Transfers
CLC	2	
CLD	2	
CLI	2	
CMC	2	
ESC		
Reg	2	
Mem	8 + EA	1
HLT	2	
LOCK	2	
NOP	3	
REP	2	
REPE/REPZ	2	
REPNE/REPNZ	2	
STC	2	
STD	2	
STI	2	
TEST		
Reg with Reg	3	
Mem with Reg	9 + EA	1
Imm with Acc	4	
Imm with Reg	5	
Imm with Mem	11 + EA	
WAIT	3 + 5n	

EFFECTIVE ADDRESS (EA) CALCULATION TIMES

Addressing Mode	Clock Cycles
Displacement	6
Base or Index	5
Disp + Base or Index	9
Base + Index	
BP + DI, BX + SI	7
BP + SI, BX + DI	8
Disp + Base + Index	
BP + DI + Disp	11
BX + SI + Disp	11
BP + SI + Disp	12
BX + DI + Disp	12

Notes: Add two clock cycles for segment override (e.g., MOV AL,**ES:**[100]). Add four clock cycles for each word transfer to/from memory.

APPENDIX C

8088 Peripheral Data Sheets

8259A
PROGRAMMABLE INTERRUPT CONTROLLER
(8259A/8259A-2)

- 8086, 8088 Compatible
- MCS-80®, MCS-85® Compatible
- Eight-Level Priority Controller
- Expandable to 64 Levels
- Programmable Interrupt Modes
- Individual Request Mask Capability

- Single +5V Supply (No Clocks)
- Available in 28-Pin DIP and 28-Lead PLCC Package
  (See Packaging Spec., Order #231369)
- Available in EXPRESS
  — Standard Temperature Range
  — Extended Temperature Range

The Intel 8259A Programmable Interrupt Controller handles up to eight vectored priority interrupts for the CPU. It is cascadable for up to 64 vectored priority interrupts without additional circuitry. It is packaged in a 28-pin DIP, uses NMOS technology and requires a single +5V supply. Circuitry is static, requiring no clock input.

The 8259A is designed to minimize the software and real time overhead in handling multi-level priority interrupts. It has several modes, permitting optimization for a variety of system requirements.

The 8259A is fully upward compatible with the Intel 8259. Software originally written for the 8259 will operate the 8259A in all 8259 equivalent modes (MCS-80/85, Non-Buffered, Edge Triggered).

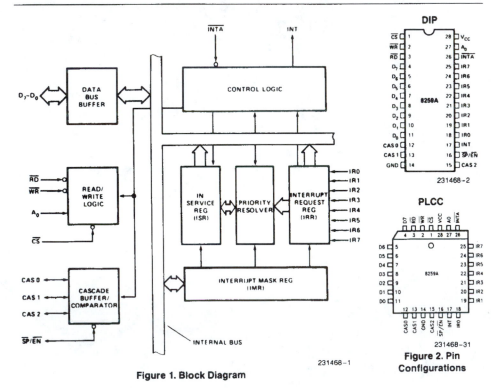

231468-1

Figure 1. Block Diagram

Figure 2. Pin Configurations

231468-2

231468-31

Table 1. Pin Description

Symbol	Pin No.	Type	Name and Function
V_{CC}	28	I	**SUPPLY:** + 5V Supply.
GND	14	I	**GROUND**
$\overline{CS}$	1	I	**CHIP SELECT:** A low on this pin enables $\overline{RD}$ and $\overline{WR}$ communication between the CPU and the 8259A. INTA functions are independent of CS.
$\overline{WR}$	2	I	**WRITE:** A low on this pin when CS is low enables the 8259A to accept command words from the CPU.
$\overline{RD}$	3	I	**READ:** A low on this pin when CS is low enables the 8259A to release status onto the data bus for the CPU.
D_7-D_0	4-11	I/O	**BIDIRECTIONAL DATA BUS:** Control, status and interrupt-vector information is transferred via this bus.
CAS_0-CAS_2	12, 13, 15	I/O	**CASCADE LINES:** The CAS lines form a private 8259A bus to control a multiple 8259A structure. These pins are outputs for a master 8259A and inputs for a slave 8259A.
$\overline{SP}/\overline{EN}$	16	I/O	**SLAVE PROGRAM/ENABLE BUFFER:** This is a dual function pin. When in the Buffered Mode it can be used as an output to control buffer transceivers (EN). When not in the buffered mode it is used as an input to designate a master (SP = 1) or slave (SP = 0).
INT	17	O	**INTERRUPT:** This pin goes high whenever a valid interrupt request is asserted. It is used to interrupt the CPU, thus it is connected to the CPU's interrupt pin.
IR_0-IR_7	18-25	I	**INTERRUPT REQUESTS:** Asynchronous inputs. An interrupt request is executed by raising an IR input (low to high), and holding it high until it is acknowledged (Edge Triggered Mode), or just by a high level on an IR input (Level Triggered Mode).
$\overline{INTA}$	26	I	**INTERRUPT ACKNOWLEDGE:** This pin is used to enable 8259A interrupt-vector data onto the data bus by a sequence of interrupt acknowledge pulses issued by the CPU.
A_0	27	I	**AO ADDRESS LINE:** This pin acts in conjunction with the $\overline{CS}$, $\overline{WR}$, and $\overline{RD}$ pins. It is used by the 8259A to decipher various Command Words the CPU writes and status the CPU wishes to read. It is typically connected to the CPU A0 address line (A1 for 8086, 8088).

8259A

FUNCTIONAL DESCRIPTION

Interrupts in Microcomputer Systems

Microcomputer system design requires that I.O devices such as keyboards, displays, sensors and other components receive servicing in a an efficient manner so that large amounts of the total system tasks can be assumed by the microcomputer with little or no effect on throughput.

The most common method of servicing such devices is the *Polled* approach. This is where the processor must test each device in sequence and in effect "ask" each one if it needs servicing. It is easy to see that a large portion of the main program is looping through this continuous polling cycle and that such a method would have a serious detrimental effect on system throughput, thus limiting the tasks that could be assumed by the microcomputer and reducing the cost effectiveness of using such devices.

A more desirable method would be one that would allow the microprocessor to be executing its main program and only stop to service peripheral devices when it is told to do so by the device itself. In effect, the method would provide an external asynchronous input that would inform the processor that it should complete whatever instruction that is currently being executed and fetch a new routine that will service the requesting device. Once this servicing is complete, however, the processor would resume exactly where it left off.

This method is called *Interrupt*. It is easy to see that system throughput would drastically increase, and thus more tasks could be assumed by the microcomputer to further enhance its cost effectiveness.

The Programmable Interrupt Controller (PIC) functions as an overall manager in an Interrupt-Driven system environment. It accepts requests from the peripheral equipment, determines which of the incoming requests is of the highest importance (priority), ascertains whether the incoming request has a higher priority value than the level currently being serviced, and issues an interrupt to the CPU based on this determination.

Each peripheral device or structure usually has a special program or "routine" that is associated with its specific functional or operational requirements; this is referred to as a "service routine". The PIC, after issuing an Interrupt to the CPU, must somehow input information into the CPU that can "point" the Program Counter to the service routine associated with the requesting device. This "pointer" is an address in a vectoring table and will often be referred to, in this document, as vectoring data.

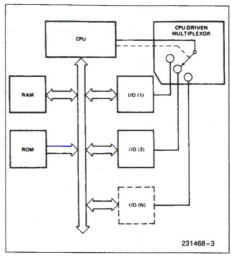

Figure 3a. Polled Method

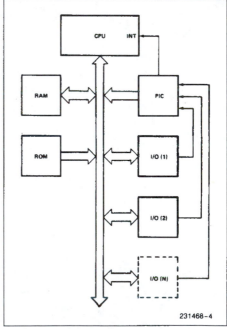

Figure 3b. Interrupt Method

8259A

The 8259A is a device specifically designed for use in real time, interrupt driven microcomputer systems. It manages eight levels or requests and has built-in features for expandability to other 8259A's (up to 64 levels). It is programmed by the system's software as an I/O peripheral. A selection of priority modes is available to the programmer so that the manner in which the requests are processed by the 8259A can be configured to match his system requirements. The priority modes can be changed or reconfigured dynamically at any time during the main program. This means that the complete interrupt structure can be defined as required, based on the total system environment.

INTERRUPT REQUEST REGISTER (IRR) AND IN-SERVICE REGISTER (ISR)

The interrupts at the IR input lines are handled by two registers in cascade, the Interrupt Request Register (IRR) and the In-Service (ISR). The IRR is used to store all the interrupt levels which are requesting service; and the ISR is used to store all the interrupt levels which are being serviced.

PRIORITY RESOLVER

This logic block determines the priorites of the bits set in the IRR. The highest priority is selected and strobed into the corresponding bit of the ISR during INTA pulse.

INTERRUPT MASK REGISTER (IMR)

The IMR stores the bits which mask the interrupt lines to be masked. The IMR operates on the IRR. Masking of a higher priority input will not affect the interrupt request lines of lower quality.

INT (INTERRUPT)

This output goes directly to the CPU interrupt input. The V_{OH} level on this line is designed to be fully compatible with the 8080A, 8085A and 8086 input levels.

INTA (INTERRUPT ACKNOWLEDGE)

INTA pulses will cause the 8259A to release vectoring information onto the data bus. The format of this data depends on the system mode (μPM) of the 8259A.

DATA BUS BUFFER

This 3-state, bidirectional 8-bit buffer is used to interface the 8259A to the system Data Bus. Control words and status information are transferred through the Data Bus Buffer.

READ/WRITE CONTROL LOGIC

The function of this block is to accept OUTput commands from the CPU. It contains the Initialization Command Word (ICW) registers and Operation Command Word (OCW) registers which store the various control formats for device operation. This function block also allows the status of the 8259A to be transferred onto the Data Bus.

CS (CHIP SELECT)

A LOW on this input enables the 8259A. No reading or writing of the chip will occur unless the device is selected.

WR (WRITE)

A LOW on this input enables the CPU to write control words (ICWs and OCWs) to the 8259A.

RD (READ)

A LOW on this input enables the 8259A to send the status of the Interrupt Request Register (IRR), In Service Register (ISR), the Interrupt Mask Register (IMR), or the Interrupt level onto the Data Bus.

A0

This input signal is used in conjunction with WR and RD signals to write commands into the various command registers, as well as reading the various status registers of the chip. This line can be tied directly to one of the address lines.

653

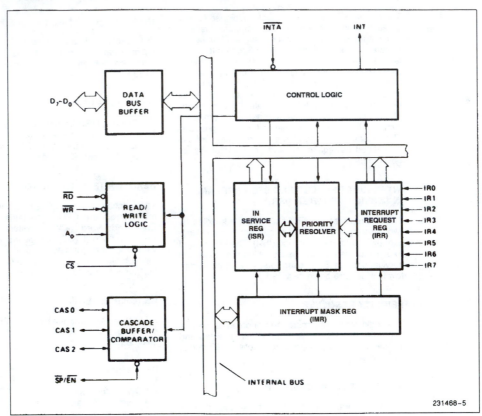

Figure 4a. 8259A Block Diagram

231468-5

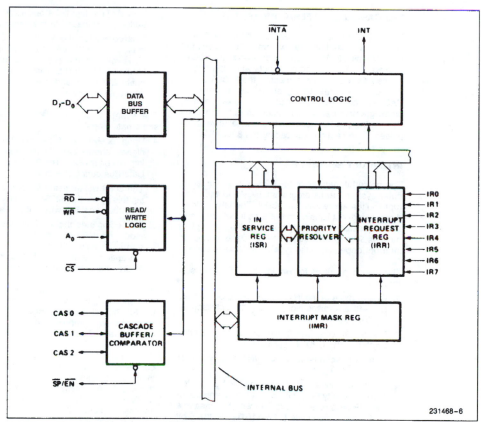

Figure 4b. 8259A Block Diagram

231468-6

8259A

THE CASCADE BUFFER/COMPARATOR

This function block stores and compares the IDs of all 8259A's used in the system. The associated three I/O pins (CAS0-2) are outputs when the 8259A is used as a master and are inputs when the 8259A is used as a slave. As a master, the 8259A sends the ID of the interrupting slave device onto the CAS0-2 lines. The slave thus selected will send its preprogrammed subroutine address onto the Data Bus during the next one or two consecutive INTA pulses. (See section "Cascading the 8259A".)

INTERRUPT SEQUENCE

The powerful features of the 8259A in a microcomputer system are its programmability and the interrupt routine addressing capability. The latter allows direct or indirect jumping to the specific interrupt routine requested without any polling of the interrupting devices. The normal sequence of events during an interrupt depends on the type of CPU being used.

The events occur as follows in an MCS-80/85 system:

1. One or more of the INTERRUPT REQUEST lines (IR7–0) are raised high, setting the corresponding IRR bit(s).

2. The 8259A evaluates these requests, and sends an INT to the CPU, if appropriate.

3. The CPU acknowledges the INT and responds with an INTA pulse.

4. Upon receiving an INTA from the CPU group, the highest priority ISR bit is set, and the corresponding IRR bit is reset. The 8259A will also release a CALL instruction code (11001101) onto the 8-bit Data Bus through its D7–0 pins.

5. This CALL instruction will initiate two more INTA pulses to be sent to the 8259A from the CPU group.

6. These two INTA pulses allow the 8259A to release its preprogrammed subroutine address onto the Data Bus. The lower 8-bit address is re-

leased at the first INTA pulse and the higher 8-bit address is released at the second INTA pulse.

7. This completes the 3-byte CALL instruction released by the 8259A. In the AEOI mode the ISR bit is reset at the end of the third INTA pulse. Otherwise, the ISR bit remains set until an appropriate EOI command is issued at the end of the interrupt sequence.

The events occuring in an 8086 system are the same until step 4.

4. Upon receiving an INTA from the CPU group, the highest priority ISR bit is set and the corresponding IRR bit is reset. The 8259A does not drive the Data Bus during this cycle.

5. The 8086 will initiate a second INTA pulse. During this pulse, the 8259A releases an 8-bit pointer onto the Data Bus where it is read by the CPU.

6. This completes the interrupt cycle. In the AEOI mode the ISR bit is reset at the end of the second INTA pulse. Otherwise, the ISR bit remains set until an appropriate EOI command is issued at the end of the interrupt subroutine.

If no interrupt request is present at step 4 of either sequence (i.e., the request was too short in duration) the 8259A will issue an interrupt level 7. Both the vectoring bytes and the CAS lines will look like an interrupt level 7 was requested.

When the 8259A PIC receives an interrupt, INT becomes active and an interrupt acknowledge cycle is started. If a higher priority interrupt occurs between the two INTA pulses, the INT line goes inactive immediately after the second INTA pulse. After an unspecified amount of time the INT line is activated again to signify the higher priority interrupt waiting for service. This inactive time is not specified and can vary between parts. The designer should be aware of this consideration when designing a system which uses the 8259A. It is recommended that proper asynchronous design techniques be followed.

656

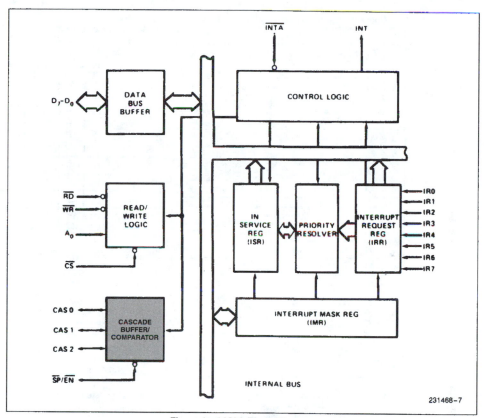

Figure 4c. 8259A Block Diagram

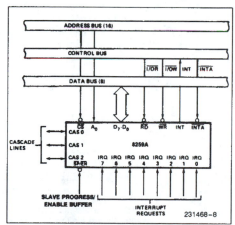

Figure 5. 8259A Interface to Standard System Bus

INTERRUPT SEQUENCE OUTPUTS

MCS-80®, MCS-85®

This sequence is timed by three $\overline{\text{INTA}}$ pulses. During the first $\overline{\text{INTA}}$ pulse the CALL opcode is enabled onto the data bus.

Content of First Interrupt Vector Byte

	D7	D6	D5	D4	D3	D2	D1	D0
CALL CODE	1	1	0	0	1	1	0	1

During the second $\overline{\text{INTA}}$ pulse the lower address of the appropriate service routine is enabled onto the data bus. When Interval = 4 bits A_5-A_7 are programmed, while A_0-A_4 are automatically inserted by the 8259A. When Interval = 8 only A_6 and A_7 are programmed, while A_0-A_5 are automatically inserted.

8259A

Content of Second Interrupt Vector Byte

IR	Interval = 4							
	D7	D6	D5	D4	D3	D2	D1	D0
7	A7	A6	A5	1	1	1	0	0
6	A7	A6	A5	1	1	0	0	0
5	A7	A6	A5	1	0	1	0	0
4	A7	A6	A5	1	0	0	0	0
3	A7	A6	A5	0	1	1	0	0
2	A7	A6	A5	0	1	0	0	0
1	A7	A6	A5	0	0	1	0	0
0	A7	A6	A5	0	0	0	0	0

IR	Interval = 8							
	D7	D6	D5	D4	D3	D2	D1	D0
7	A7	A6	1	1	1	0	0	0
6	A7	A6	1	1	0	0	0	0
5	A7	A6	1	0	1	0	0	0
4	A7	A6	1	0	0	0	0	0
3	A7	A6	0	1	1	0	0	0
2	A7	A6	0	1	0	0	0	0
1	A7	A6	0	0	1	0	0	0
0	A7	A6	0	0	0	0	0	0

During the third $\overline{INTA}$ pulse the higher address of the appropriate service routine, which was programmed as byte 2 of the initialization sequence (A_8–A_{15}), is enabled onto the bus.

Content of Third Interrupt Vector Byte

D7	D6	D5	D4	D3	D2	D1	D0
A15	A14	A13	A12	A11	A10	A9	A8

8086, 8088

8086 mode is similar to MCS-80 mode except that only two Interrupt Acknowledge cycles are issued by the processor and no CALL opcode is sent to the processor. The first interrupt acknowledge cycle is similar to that of MCS-80, 85 systems in that the 8259A uses it to internally freeze the state of the interrupts for priority resolution and as a master it issues the interrupt code on the cascade lines at the end of the $\overline{INTA}$ pulse. On this first cycle it does not issue any data to the processor and leaves its data bus buffers disabled. On the second interrupt acknowledge cycle in 8086 mode the master (or slave if so programmed) will send a byte of data to the processor with the acknowledged interrupt code

composed as follows (note the state of the ADI mode control is ignored and A_5–A_{11} are unused in 8086 mode):

Content of Interrupt Vector Byte for 8086 System Mode

	D7	D6	D5	D4	D3	D2	D1	D0
IR7	T7	T6	T5	T4	T3	1	1	1
IR6	T7	T6	T5	T4	T3	1	1	0
IR5	T7	T6	T5	T4	T3	1	0	1
IR4	T7	T6	T5	T4	T3	1	0	0
IR3	T7	T6	T5	T4	T3	0	1	1
IR2	T7	T6	T5	T4	T3	0	1	0
IR1	T7	T6	T5	T4	T3	0	0	1
IR0	T7	T6	T5	T4	T3	0	0	0

PROGRAMMING THE 8259A

The 8259A accepts two types of command words generated by the CPU:

1. *Initialization Command Words (ICWs):* Before normal operation can begin, each 8259A in the system must be brought to a starting point—by a sequence of 2 to 4 bytes timed by $\overline{WR}$ pulses.

2. *Operation Command Words (OCWs):* These are the command words which command the 8259A to operate in various interrupt modes. These modes are:

   a. Fully nested mode

   b. Rotating priority mode

   c. Special mask mode

   d. Polled mode

The OCWs can be written into the 8259A anytime after initialization.

INITIALIZATION COMMAND WORDS (ICWS)

General

Whenever a command is issued with A0 = 0 and D4 = 1, this is interpreted as Initialization Command Word 1 (ICW1). ICW1 starts the intiitalization sequence during which the following automatically occur.

a. The edge sense circuit is reset, which means that following initialization, an interrupt request (IR) input must make a low-to-high transistion to generate an interrupt.

658

b. The Interrupt Mask Register is cleared.

c. IR7 input is assigned priority 7.

d. The slave mode address is set to 7.

e. Special Mask Mode is cleared and Status Read is set to IRR.

f. If IC4 = 0, then all functions selected in ICW4 are set to zero. (Non-Buffered mode*, no Auto-EOI, MCS-80, 85 system).

*NOTE:

Master/Slave in ICW4 is only used in the buffered mode.

Initialization Command Words 1 and 2 (ICW1, ICW2)

A_5–A_{15}: *Page starting address of service routines.* In an MCS 80/85 system, the 8 request levels will generate CALLs to 8 locations equally spaced in memory. These can be programmed to be spaced at intervals of 4 or 8 memory locations, thus the 8 routines will occupy a page of 32 or 64 bytes, respectively.

The address format is 2 bytes long (A_0–A_{15}). When the routine interval is 4, A_0–A_4 are automatically inserted by the 8259A, while A_5–A_{15} are programmed externally. When the routine interval is 8, A_0–A_5 are automatically inserted by the 8259A, while A_6–A_{15} are programmed externally.

The 8-byte interval will maintain compatibility with current software, while the 4-byte interval is best for a compact jump table.

In an 8086 system A_{15}–A_{11} are inserted in the five most significant bits of the vectoring byte and the 8259A sets the three least significant bits according to the interrupt level. A_{10}–A_5 are ignored and ADI (Address interval) has no effect.

LTIM: If LTIM = 1, then the 8259A will operate in the level interrupt mode. Edge detect logic on the interrupt inputs will be disabled.

ADI: CALL address interval. ADI = 1 then interval = 4; ADI = 0 then interval = 8.

SNGL: Single. Means that this is the only 8259A in the system. If SNGL = 1 no ICW3 will be issued.

IC4: If this bit is set—ICW4 has to be read. If ICW4 is not needed, set IC4 = 0.

Initialization Command Word 3 (ICW3)

This word is read only when there is more than one 8259A in the system and cascading is used, in which

case SNGL = 0. It will load the 8-bit slave register. The functions of this register are:

a. In the master mode (either when SP = 1, or in buffered mode when M/S = 1 in ICW4) a "1" is set for each slave in the system. The master then will release byte 1 of the call sequence (for MCS-80/85 system) and will enable the corresponding slave to release bytes 2 and 3 (for 8086 only byte 2) through the cascade lines.

b. In the slave mode (either when $\overline{SP}$ = 0, or if BUF = 1 and M/S = 0 in ICW4) bits 2–0 identify the slave. The slave compares its cascade input with these bits and, if they are equal, bytes 2 and 3 of the call sequence (or just byte 2 for 8086) are released by it on the Data Bus.

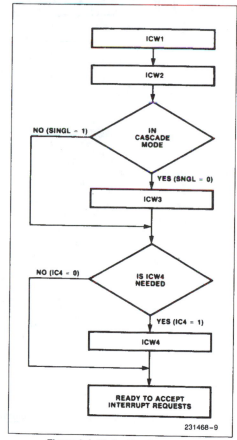

Figure 6. Initialization Sequence

231468-9

659

8259A

Initialization Command Word 4 (ICW4)

SFNM: If SFNM = 1 the special fully nested mode is programmed.

BUF: If BUF = 1 the buffered mode is programmed. In buffered mode SP/EN becomes an enable output and the master/slave determination is by M/S.

M/S: If buffered mode is selected: M/S = 1 means the 8259A is programmed to be a master, M/S = 0 means the 8259A is programmed to be a slave. If BUF = 0, M/S has no function.

AEOI: If AEOI = 1 the automatic end of interrupt mode is programmed.

μPM: Microprocessor mode: μPM = 0 sets the 8259A for MCS-80, 85 system operation, μPM = 1 sets the 8259A for 8086 system operation.

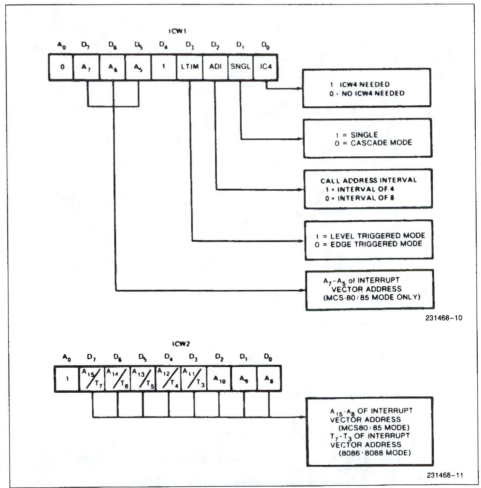

Figure 7. Initialization Command Word Format

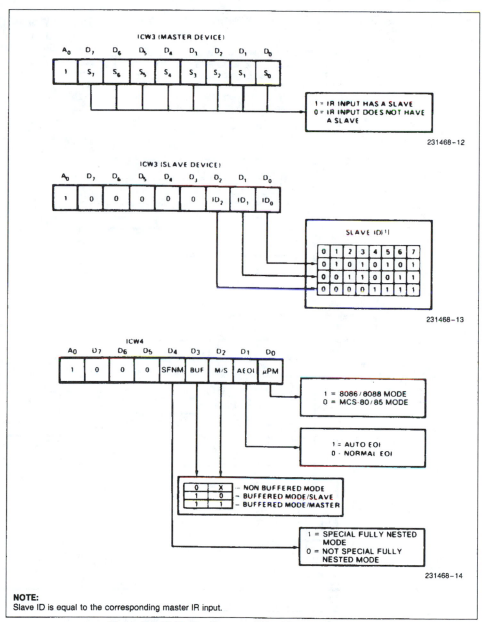

NOTE:
Slave ID is equal to the corresponding master IR input.

Figure 7. Initialization Command Word Format (Continued)

8259A

OPERATION COMMAND WORDS (OCWS)

After the Initialization Command Words (ICWs) are programmed into the 8259A, the chip is ready to accept interrupt requests at its input lines. However, during the 8259A operation, a selection of algorithms can command the 8259A to operate in various modes through the Operation Command Words (OCWs).

Operation Control Words (OCWs)

OCW1

A0	D7	D6	D5	D4	D3	D2	D1	D0
1	M7	M6	M5	M4	M3	M2	M1	M0

OCW2

0	R	SL	EOI	0	0	L2	L1	L0

OCW3

0	0	ESMM	SMM	0	1	P	RR	RIS

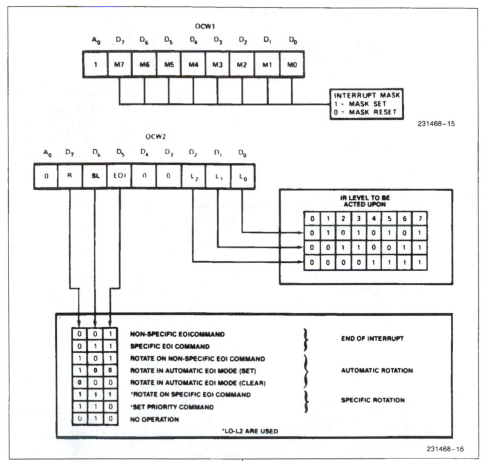

Figure 8. Operation Command Word Format

662

Operation Control Word 1 (OCW1)

OCW1 sets and clears the mask bits in the interrupt Mask Register (IMR). M_7–M_0 represent the eight mask bits. M = 1 indicates the channel is masked (inhibited), M = 0 indicates the channel is enabled.

Operation Control Word 2 (OCW2)

R, SL, EOI—These three bits control the Rotate and End of Interrupt modes and combinations of the two. A chart of these combinations can be found on the Operation Command Word Format.

L_2, L_1, L_0—These bits determine the interrupt level acted upon when the SL bit is active.

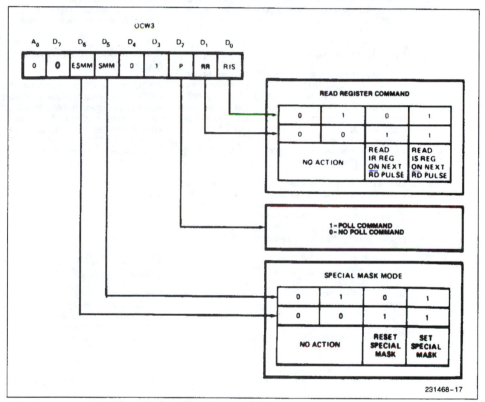

Figure 8. Operation Command Word Format (Continued)

8259A

Operation Control Word 3 (OCW3)

ESMM—Enable Special Mask Mode. When this bit is set to 1 it enables the SMM bit to set or reset the Special Mask Mode. When ESMM = 0 the SMM bit becomes a "don't care".

SMM—Special Mask Mode. If ESMM = 1 and SMM = 1 the 8259A will enter Special Mask Mode. If ESMM = 1 and SMM = 0 the 8259A will revert to normal mask mode. When ESMM = 0, SMM has no effect.

Fully Nested Mode

This mode is entered after initialization unless another mode is programmed. The interrupt requests are ordered in priority from 0 through 7 (0 highest). When an interrupt is acknowledged the highest priority request is determined and its vector placed on the bus. Additionally, a bit of the Interrupt Service register (ISO-7) is set. This bit remains set until the microprocessor issues an End of Interrupt (EOI) command immediately before returning from the service routine, or if AEOI (Automatic End of Interrupt) bit is set, until the trailing edge of the last INTA. While the IS bit is set, all further interrupts of the same or lower priority are inhibited, while higher levels will generate an interrupt (which will be acknowledged only if the microprocessor internal Interupt enable flip-flop has been re-enabled through software).

After the initialization sequence, IR0 has the highest prioirity and IR7 the lowest. Priorities can be changed, as will be explained, in the rotating priority mode.

End of Interrupt (EOI)

The In Service (IS) bit can be reset either automatically following the trailing edge of the last in sequence INTA pulse (when AEOI bit in ICW1 is set) or by a command word that must be issued to the 8259A before returning from a service routine (EOI command). An EOI command must be issued twice if in the Cascade mode, once for the master and once for the corresponding slave.

There are two forms of EOI command: Specific and Non-Specific. When the 8259A is operated in modes which perserve the fully nested structure, it can determine which IS bit to reset on EOI. When a Non-Specific EOI command is issued the 8259A will automatically reset the highest IS bit of those that are set, since in the fully nested mode the highest IS level was necessarily the last level acknowledged and serviced. A non-specific EOI can be issued with OCW2 (EOI = 1, SL = 0, R = 0).

When a mode is used which may disturb the fully nested structure, the 8259A may no longer be able to determine the last level acknowledged. In this case a Specific End of Interrupt must be issued which includes as part of the command the IS level to be reset. A specific EOI can be issued with OCW2 (EOI = 1, SL = 1, R = 0, and L0–L2 is the binary level of the IS bit to be reset).

It should be noted that an IS bit that is masked by an IMR bit will not be cleared by a non-specific EOI if the 8259A is in the Special Mask Mode.

Automatic End of Interrupt (AEOI) Mode

If AEOI = 1 in ICW4, then the 8259A will operate in AEOI mode continuously until reprogrammed by ICW4. in this mode the 8259A will automatically perform a non-specific EOI operation at the trailing edge of the last interrupt acknowledge pulse (third pulse in MCS-80/85, second in 8086). Note that from a system standpoint, this mode should be used only when a nested multilevel interrupt structure is not required within a single 8259A.

The AEOI mode can only be used in a master 8259A and not a slave. 8259As with a copyright date of 1985 or later will operate in the AEOI mode as a master or a slave.

Automatic Rotation (Equal Priority Devices)

In some applications there are a number of interrupting devices of equal priority. In this mode a device, after being serviced, receives the lowest priority, so a device requesting an interrupt will have to wait, in the worst case until each of 7 other devices are serviced at most *once*. For example, if the priority and "in service" status is:

Before Rotate (IR4 the highest prioirity requiring service)

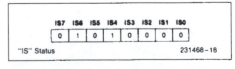

"IS" Status   231468-18

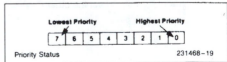

Priority Status   231468-19

664

After Rotate (IR4 was serviced, all other priorities rotated correspondingly)

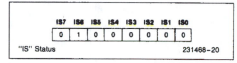

"IS" Status   231468-20

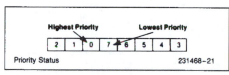

Priority Status   231468-21

There are two ways to accomplish Automatic Rotation using OCW2, the Rotation on Non-Specific EOI Command (R = 1, SL = 0, EOI = 1) and the Rotate in Automatic EOI Mode which is set by (R = 1, SL = 0, EOI = 0) and cleared by (R = 0, SL = 0, EOI = 0).

Specific Rotation (Specific Priority)

The programmer can change priorities by programming the bottom priority and thus fixing all other priorities; i.e., if IR5 is programmed as the bottom priority device, then IR6 will have the highest one.

The Set Priority command is issued in OCW2 where: R = 1, SL = 1, L0–L2 is the binary priority level code of the bottom priority device.

Observe that in this mode internal status is updated by software control during OCW2. However, it is independent of the End of Interrupt (EOI) command (also executed by OCW2). Priority changes can be executed during an EOI command by using the Rotate on Specific EOI command in OCW2 (R = 1, SL = 1, EOI = 1 and L0–L2 = IR level to receive bottom priority).

Interrupt Masks

Each Interrupt Request input can bem masked individually by the Interrupt Mask Register (IMR) programmed through OCW1. Each bit in the IMR masks one interrupt channel if it is set (1). Bit 0 masks IR0, Bit 1 masks IR1 and so forth. Masking an IR channel does not affect the other channels operation.

Special Mask Mode

Some applications may require an interrupt service routine to dynamically alter the system priority structure during its execution under software control. For example, the routine may wish to inhibit lower priority requests for a portion of its execution but enable some of them for another portion.

The difficulty here is that if an Interrupt Request is acknowledged and an End of Interrupt command did not reset its IS bit (i.e., while executing a service routine), the 8259A would have inhibited all lower priority requests with no easy way for the routine to enable them.

That is where the Special Mask Mode comes in. In the special Mask Mode, when a mask bit is set in OCW1, it inhibits further interrupts at that level *and* enables interrupts from *all other* levels (lower as well as higher) that are not masked.

Thus, any interrupts may be selectively enabled by loading the mask register.

The special Mask Mode is set by OWC3 where: SSMM = 1, SMM = 1, and cleared where SSMM = 1, SMM = 0.

Poll Command

In Poll mode the INT output functions as it normally does. The microprocessor should ignore this output. This can be accomplished either by not connecting the INT output or by masking interrupts within the microprocessor, thereby disabling its interrupt input. Service to devices is achieved by software using a Poll command.

The Poll command is issued by setting P = '1'' in OCW3. The 8259A treats the next $\overline{RD}$ pulse to the 8259A (i.e., $\overline{RD}$ = 0, $\overline{CS}$ = 0) as an interrupt acknowledge, sets the appropriate IS bit if there is a request, and reads the priority level. Interrupt is frozen from $\overline{WR}$ to $\overline{RD}$.

The word enabled onto the data bus during $\overline{RD}$ is:

D7	D6	D5	D4	D3	D2	D1	D0
I	—	—	—	—	W2	W1	W0

W0–W2: Binary code of the highest priority level requesting service.

I: Equal to "1" if there is an interrupt.

This mode is useful if there is a routine command common to several levels so that the $\overline{INTA}$ sequence is not needed (saves ROM space). Another application is to use the poll mode to expand the number of priority levels to more than 64.

Reading the 8259A Status

The input status of several internal registers can be read to update the user information on the system.

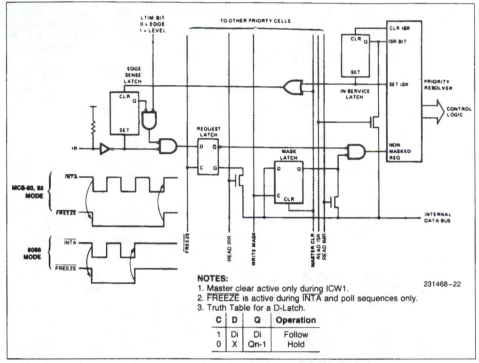

NOTES:
1. Master clear active only during ICW1.
2. FREEZE is active during INTA and poll sequences only.
3. Truth Table for a D-Latch.

C	D	Q	Operation
1	Di	Di	Follow
0	X	Qn-1	Hold

231468–22

Figure 9. Priority Cell—Simplified Logic Diagram

The following registers can be read via OCW3 (IRR and ISR or OCW1 [IMR]).

Interrupt Request Register (IRR): 8-bit register which contains the levels requesting an interrupt to be acknowledged. The highest request level is reset from the IRR when an interrupt is acknowledged. (Not affected by IMR.)

In-Service Register (ISR): 8-bit register which contains the priority levels that are being serviced. The ISR is updated when an End of Interrupt Command is issued.

Interrupt Mask Register: 8-bit register which contains the interrupt request lines which are masked.

The IRR can be read when, prior to the RD pulse, a Read Register Command is issued with OCW3 (RR = 1, RIS = 0.)

The ISR can be read, when, prior to the RD pulse, a Read Register Command is issued with OCW3 (RR = 1, RIS = 1.)

There is no need to write an OCW3 before every status read operation, as long as the status read corresponds with the previous one; i.e., the 8259A "remembers" whether the IRR or ISR has been previously selected by the OCW3. This is not true when poll is used.

After initialization the 8259A is set to IRR.

For reading the IMR, no OCW3 is needed. The output data bus will contain the IMR whenever RD is active and A0 = 1 (OCW1).

Polling overrides status read when P = 1, RR = 1 in OCW3.

Edge and Level Triggered Modes

This mode is programmed using bit 3 in ICW1.

If LTIM = '0', an interrupt request will be recognized by a low to high transition on an IR input. The IR input can remain high without generating another interrupt.

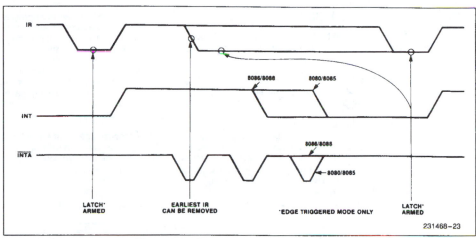

Figure 10. IR Triggering Timing Requirements

If LTIM = '1', an interrupt request will be recognized by a 'high' level on IR input, and there is no need for an edge detection. The interrupt request must be removed before the EOI command is issued or the CPU interrupts is enabled to prevent a second interrupt from occurring.

The priority cell diagram shows a conceptual circuit of the level sensitive and edge sensitive input circuitry of the 8259A. Be sure to note that the request latch is a transparent D type latch.

In both the edge and level triggered modes the IR inputs must remain high until after the falling edge of the first INTA. If the IR input goes low before this time a DEFAULT IR7 will occur when the CPU acknowledges the interrupt. This can be a useful safeguard for detecting interrupts caused by spurious noise glitches on the IR inputs. To implement this feature the IR7 routine is used for "clean up" simply executing a return instruction, thus ignoring the interrupt. If IR7 is needed for other purposes a default IR7 can still be detected by reading the ISR. A normal IR7 interrupt will set the corresponding ISR bit, a default IR7 won't. If a default IR7 routine occurs during a normal IR7 routine, however, the ISR will remain set. In this case it is necessary to keep track of whether or not the IR7 routine was previously entered. If another IR7 occurs it is a default.

The Special Fully Nest Mode

This mode will be used in the case of a big system where cascading is used, and the priority has to be conserved within each slave. In this case the fully nested mode will be programmed to the master (us-

ing ICW4). This mode is similar to the normal nested mode with the following exceptions:

a. When an interrupt request from a certain slave is in service this slave is not locked out from the master's priority logic and further interrupt requests from higher priority IR's within the slave will be recognized by the master and will initiate interrupts to the processor. (In the normal nested mode a slave is masked out when its request is in service and no higher requests from the same slave can be serviced.)

b. When exiting the Interrupt Service routine the software has to check whether the interrupt serviced was the only one from that slave. This is done by sending a non-specific End of Interrupt (EOI) command to the slave and then reading its In-Service register and checking for zero. If it is empty, a non-specific EOI can be sent to the master too. If not, no EOI should be sent.

Buffered Mode

When the 8259A is used in a large system where bus driving buffers are required on the data bus and the cascading mode is used, there exists the problem of enabling buffers.

The buffered mode will structure the 8259A to send an enable signal on SP/EN to enable the buffers. In this mode, whenever the 8259A's data bus outputs are enabled, the SP/EN output becomes active.

This modification forces the use of software programming to determine whether the 8259A is a master or a slave. Bit 3 in ICW4 programs the buffered mode, and bit 2 in ICW4 determines whether it is a master or a slave.

CASCADE MODE

The 8259A can be easily interconnected in a system of one master with up to eight slaves to handle up to 64 priority levels.

The master controls the slaves through the 3 line cascade bus. The cascade bus acts like chip selects to the slaves during the INTA sequence.

In a cascade configuration, the slave interrupt outputs are connected to the master interrupt request inputs. When a slave request line is activated and afterwards acknowledged, the master will enable the corresponding slave to release the device routine address during bytes 2 and 3 of INTA. (Byte 2 only for 8086/8088).

The cascade bus lines are normally low and will contain the slave address code from the trailing edge of the first INTA pulse to the trailing edge of the third pulse. Each 8259A in the system must follow a separate initialization sequence and can be programmed to work in a different mode. An EOI command must be issued twice: once for the master and once for the corresponding slave. An address decoder is required to activate the Chip Select (CS) input of each 8259A.

The cascade lines of the Master 8259A are activated only for slave inputs, non-slave inputs leave the cascade line inactive (low).

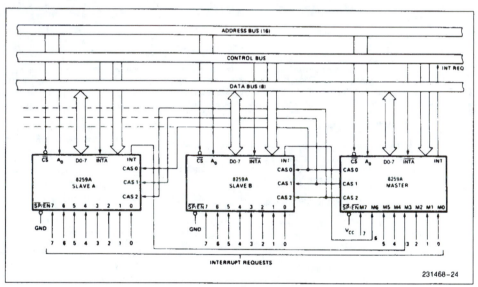

Figure 11. Cascading the 8259A

8254
PROGRAMMABLE INTERVAL TIMER

- ■ **Compatible with All Intel and Most Other Microprocessors**
- ■ **Handles Inputs from DC to 10 MHz**
  - — 5 MHz 8254-5
  - — 8 MHz 8254
  - — 10 MHz 8254-2
- ■ **Status Read-Back Command**

- ■ **Six Programmable Counter Modes**
- ■ **Three Independent 16-Bit Counters**
- ■ **Binary or BCD Counting**
- ■ **Single +5V Supply**
- ■ **Available in EXPRESS**
  - **— Standard Temperature Range**

The Intel® 8254 is a counter/timer device designed to solve the common timing control problems in microcomputer system design. It provides three independent 16-bit counters, each capable of handling clock inputs up to 10 MHz. All modes are software programmable. The 8254 is a superset of the 8253.

The 8254 uses HMOS technology and comes in a 24-pin plastic or CERDIP package.

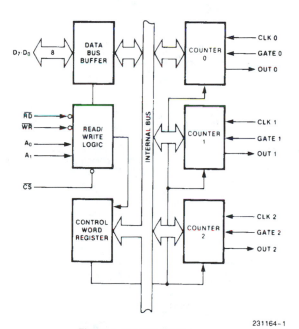

231164-1

Figure 1. 8254 Block Diagram

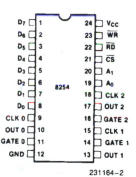

231164-2

Figure 2. Pin Configuration

Table 1. Pin Description

Symbol	Pin No.	Type	Name and Function
D₇–D₀	1–8	I/O	**DATA:** Bi-directional three state data bus lines, connected to system data bus.
CLK 0	9	I	**CLOCK 0:** Clock input of Counter 0.
OUT 0	10	O	**OUTPUT 0:** Output of Counter 0.
GATE 0	11	I	**GATE 0:** Gate input of Counter 0.
GND	12		**GROUND:** Power supply connection.
V_CC	24		**POWER:** +5V power supply connection.
$\overline{WR}$	23	I	**WRITE CONTROL:** This input is low during CPU write operations.
$\overline{RD}$	22	I	**READ CONTROL:** This input is low during CPU read operations.
$\overline{CS}$	21	I	**CHIP SELECT:** A low on this input enables the 8254 to respond to $\overline{RD}$ and $\overline{WR}$ signals. $\overline{RD}$ and $\overline{WR}$ are ignored otherwise.
A₁, A₀	20–19	I	**ADDRESS:** Used to select one of the three Counters or the Control Word Register for read or write operations. Normally connected to the system address bus.

A₁	A₀	Selects
0	0	Counter 0
0	1	Counter 1
1	0	Counter 2
1	1	Control Word Register

Symbol	Pin No.	Type	Name and Function
CLK 2	18	I	**CLOCK 2:** Clock input of Counter 2.
OUT 2	17	O	**OUT 2:** Output of Counter 2.
GATE 2	16	I	**GATE 2:** Gate input of Counter 2.
CLK 1	15	I	**CLOCK 1:** Clock input of Counter 1.
GATE 1	14	I	**GATE 1:** Gate input of Counter 1.
OUT 1	13	O	**OUT 1:** Output of Counter 1.

FUNCTIONAL DESCRIPTION

General

The 8254 is a programmable interval timer/counter designed for use with Intel microcomputer systems. It is a general purpose, multi-timing element that can be treated as an array of I/O ports in the system software.

The 8254 solves one of the most common problems in any microcomputer system, the generation of accurate time delays under software control. Instead of setting up timing loops in software, the programmer configures the 8254 to match his requirements and programs one of the counters for the desired delay. After the desired delay, the 8254 will interrupt the CPU. Software overhead is minimal and variable length delays can easily be accommodated.

Some of the other counter/timer functions common to microcomputers which can be implemented with the 8254 are:

- Real time clock
- Event-counter
- Digital one-shot
- Programmable rate generator
- Square wave generator
- Binary rate multiplier
- Complex waveform generator
- Complex motor controller

Block Diagram

DATA BUS BUFFER

This 3-state, bi-directional, 8-bit buffer is used to interface the 8254 to the system bus (see Figure 3).

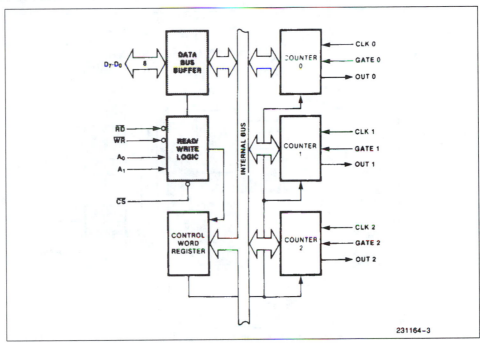

Figure 3. Block Diagram Showing Data Bus Buffer and Read/Write Logic Functions

READ/WRITE LOGIC

The Read/Write Logic accepts inputs from the system bus and generates control signals for the other functional blocks of the 8254. A_1 and A_0 select one of the three counters or the Control Word Register to be read from/written into. A "low" on the $\overline{RD}$ input tells the 8254 that the CPU is reading one of the counters. A "low" on the $\overline{WR}$ input tells the 8254 that the CPU is writing either a Control Word or an initial count. Both $\overline{RD}$ and $\overline{WR}$ are qualified by $\overline{CS}$; $\overline{RD}$ and $\overline{WR}$ are ignored unless the 8254 has been selected by holding $\overline{CS}$ low.

CONTROL WORD REGISTER

The Control Word Register (see Figure 4) is selected by the Read/Write Logic when $A_1,A_0 = 11$. If the CPU then does a write operation to the 8254, the data is stored in the Control Word Register and is interpreted as a Control Word used to define the operation of the Counters.

The Control Word Register can only be written to; status information is available with the Read-Back Command.

COUNTER 0, COUNTER 1, COUNTER 2

These three functional blocks are identical in operation, so only a single Counter will be described. The internal block diagram of a single counter is shown in Figure 5.

The Counters are fully independent. Each Counter may operate in a different Mode.

The Control Word Register is shown in the figure; it is not part of the Counter itself, but its contents determine how the Counter operates.

The status register, shown in Figure 5, when latched, contains the current contents of the Control Word Register and status of the output and null count flag. (See detailed explanation of the Read-Back command.)

The actual counter is labelled CE (for "Counting Element"). It is a 16-bit presettable synchronous down counter.

OL_M and OL_L are two 8-bit latches. OL stands for "Output Latch"; the subscripts M and L stand for "Most significant byte" and "Least significant byte"

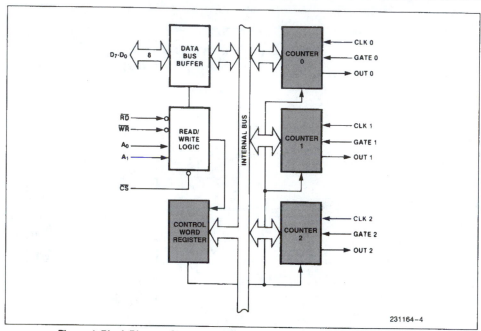

Figure 4. Block Diagram Showing Control Word Register and Counter Functions

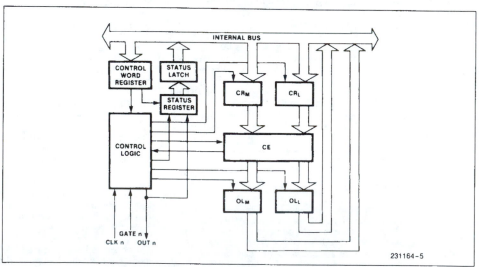

Figure 5. Internal Block Diagram of a Counter

8254

respectively. Both are normally referred to as one unit and called just OL. These latches normally "follow" the CE, but if a suitable Counter Latch Command is sent to the 8254, the latches "latch" the present count until read by the CPU and then return to "following" the CE. One latch at a time is enabled by the counter's Control Logic to drive the internal bus. This is how the 16-bit Counter communicates over the 8-bit internal bus. Note that the CE itself cannot be read; whenever you read the count, it is the OL that is being read.

Similarly, there are two 8-bit registers called CR_M and CR_L (for "Count Register"). Both are normally referred to as one unit and called just CR. When a new count is written to the Counter, the count is stored in the CR and later transferred to the CE. The Control Logic allows one register at a time to be loaded from the internal bus. Both bytes are transferred to the CE simultaneously. CR_M and CR_L are cleared when the Counter is programmed. In this way, if the Counter has been programmed for one byte counts (either most significant byte only or least significant byte only) the other byte will be zero. Note that the CE cannot be written into; whenever a count is written, it is written into the CR.

The Control Logic is also shown in the diagram. CLK n, GATE n, and OUT n are all connected to the outside world through the Control Logic.

8254 SYSTEM INTERFACE

The 8254 is a component of the Intel Microcomputer Systems and interfaces in the same manner as all other peripherals of the family. It is treated by the system's software as an array of peripheral I/O ports; three are counters and the fourth is a control register for MODE programming.

Basically, the select inputs A_0, A_1 connect to the A_0, A_1 address bus signals of the CPU. The CS can be derived directly from the address bus using a linear select method. Or it can be connected to the output of a decoder, such as an Intel 8205 for larger systems.

OPERATIONAL DESCRIPTION

General

After power-up, the state of the 8254 is undefined. The Mode, count value, and output of all Counters are undefined.

How each Counter operates is determined when it is programmed. Each Counter must be programmed before it can be used. Unused counters need not be programmed.

Programming the 8254

Counters are programmed by writing a Control Word and then an initial count.

The Control Words are written into the Control Word Register, which is selected when $A_1, A_0 = 11$. The Control Word itself specifies which Counter is being programmed.

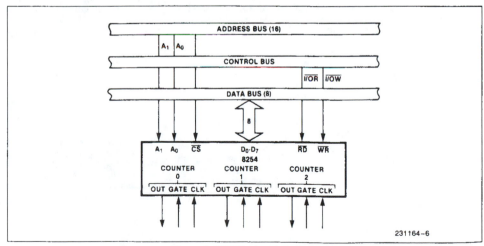

Figure 6. 8254 System Interface

673

8254

Control Word Format

$A_1, A_0 = 11$ $\overline{CS} = 0$ $\overline{RD} = 1$ $\overline{WR} = 0$

D7	D6	D5	D4	D3	D2	D1	D0
SC1	SC0	RW1	RW0	M2	M1	M0	BCD

SC—Select Counter

SC1	SC0	
0	0	Select Counter 0
0	1	Select Counter 1
1	0	Select Counter 2
1	1	Read-Back Command (see Read Operations)

M—Mode

M2	M1	M0	
0	0	0	Mode 0
0	0	1	Mode 1
X	1	0	Mode 2
X	1	1	Mode 3
1	0	0	Mode 4
1	0	1	Mode 5

RW—Read/Write

RW1	RW0	
0	0	Counter Latch Command (see Read Operations)
0	1	Read/Write least significant byte only
1	0	Read/Write most significant byte only
1	1	Read/Write least significant byte first, then most significant byte

BCD

0	Binary Counter 16-bits
1	Binary Coded Decimal (BCD) Counter (4 Decades)

NOTE:
Don't care bits (X) should be 0 to insure compatibility with future Intel products.

Figure 7. Control Word Format

By contrast, initial counts are written into the Counters, not the Control Word Register. The A_1, A_0 inputs are used to select the Counter to be written into. The format of the initial count is determined by the Control Word used.

Write Operations

The programming procedure for the 8254 is very flexible. Only two conventions need to be remembered:

1) For each Counter, the Control Word must be written before the initial count is written.

2) The initial count must follow the count format specified in the Control Word (least significant byte only, most significant byte only, or least significant byte and then most significant byte).

Since the Control Word Register and the three Counters have separate addresses (selected by the A_1, A_0 inputs), and each Control Word specifies the Counter it applies to (SC0, SC1 bits), no special instruction sequence is required. Any programming sequence that follows the conventions in Figure 7 is acceptable.

A new initial count may be written to a Counter at any time without affecting the Counter's programmed Mode in any way. Counting will be affected as described in the Mode definitions. The new count must follow the programmed count format.

If a Counter is programmed to read/write two-byte counts, the following precaution applies: A program must not transfer control between writing the first and second byte to another routine which also writes into that same Counter. Otherwise, the Counter will be loaded with an incorrect count.

674

	A₁	A₀		A₁	A₀
Control Word—Counter 0	1	1	Control Word—Counter 2	1	1
LSB of count—Counter 0	0	0	Control Word—Counter 1	1	1
MSB of count—Counter 0	0	0	Control Word—Counter 0	1	1
Control Word—Counter 1	1	1	LSB of count—Counter 2	1	0
LSB of count—Counter 1	0	1	MSB of count—Counter 2	1	0
MSB of count—Counter 1	0	1	LSB of count—Counter 1	0	1
Control Word—Counter 2	1	1	MSB of count—Counter 1	0	1
LSB of count—Counter 2	1	0	LSB of count—Counter 0	0	0
MSB of count—Counter 2	1	0	MSB of count—Counter 0	0	0
	A₁	**A₀**		**A₁**	**A₀**
Control Word—Counter 0	1	1	Control Word—Counter 1	1	1
Control Word—Counter 1	1	1	Control Word—Counter 0	1	1
Control Word—Counter 2	1	1	LSB of count—Counter 1	0	1
LSB of count—Counter 2	1	0	Control Word—Counter 2	1	1
LSB of count—Counter 1	0	1	LSB of count—Counter 0	0	0
LSB of count—Counter 0	0	0	MSB of count—Counter 1	0	1
MSB of count—Counter 0	0	0	LSB of count—Counter 2	1	0
MSB of count—Counter 1	0	1	MSB of count—Counter 0	0	0
MSB of count—Counter 2	1	0	MSB of count—Counter 2	1	0

NOTE:
In all four examples, all Counters are programmed to read/write two-byte counts. These are only four of many possible programming sequences.

Figure 8. A Few Possible Programming Sequences

Read Operations

It is often desirable to read the value of a Counter without disturbing the count in progress. This is easily done in the 8254.

There are three possible methods for reading the counters: a simple read operation, the Counter Latch Command, and the Read-Back Command. Each is explained below. The first method is to perform a simple read operation. To read the Counter, which is selected with the A1, A0 inputs, the CLK input of the selected Counter must be inhibited by using either the GATE input or external logic. Otherwise, the count may be in the process of changing when it is read, giving an undefined result.

COUNTER LATCH COMMAND

The second method uses the "Counter Latch Command". Like a Control Word, this command is written to the Control Word Register, which is selected when $A_1,A_0 = 11$. Also like a Control Word, the SC0, SC1 bits select one of the three Counters, but two other bits, D5 and D4, distinguish this command from a Control Word.

$A_1,A_0 = 11; CS = 0; RD = 1; WR = 0$

D₇	D₆	D₅	D₄	D₃	D₂	D₁	D₀
SC1	SC0	0	0	X	X	X	X

SC1,SC0—specify counter to be latched

SC1	SC0	Counter
0	0	0
0	1	1
1	0	2
1	1	Read-Back Command

D5,D4—00 designates Counter Latch Command

X—don't care

NOTE:
Don't care bits (X) should be 0 to insure compatibility with future Intel products.

Figure 9. Counter Latching Command Format

675

The selected Counter's output latch (OL) latches the count at the time the Counter Latch Command is received. This count is held in the latch until it is read by the CPU (or until the Counter is reprogrammed). The count is then unlatched automatically and the OL returns to "following" the counting element (CE). This allows reading the contents of the Counters "on the fly" without affecting counting in progress. Multiple Counter Latch Commands may be used to latch more than one Counter. Each latched Counter's OL holds its count until it is read. Counter Latch Commands do not affect the programmed Mode of the Counter in any way.

If a Counter is latched and then, some time later, latched again before the count is read, the second Counter Latch Command is ignored. The count read will be the count at the time the first Counter Latch Command was issued.

With either method, the count must be read according to the programmed format; specifically, if the Counter is programmed for two byte counts, two bytes must be read. The two bytes do not have to be read one right after the other; read or write or programming operations of other Counters may be inserted between them.

Another feature of the 8254 is that reads and writes of the same Counter may be interleaved; for example, if the Counter is programmed for two byte counts, the following sequence is valid.

1) Read least significant byte.
2) Write new least significant byte.
3) Read most significant byte.
4) Write new most significant byte.

If a Counter is programmed to read/write two-byte counts, the following precaution applies: A program must not transfer control between reading the first and second byte to another routine which also reads from that same Counter. Otherwise, an incorrect count will be read.

READ-BACK COMMAND

The third method uses the Read-Back Command. This command allows the user to check the count value, programmed Mode, and current states of the OUT pin and Null Count flag of the selected counter(s).

The command is written into the Control Word Register and has the format shown in Figure 10. The command applies to the counters selected by setting their corresponding bits D3, D2, D1 = 1.

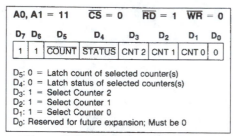

Figure 10. Read-Back Command Format

The read-back command may be used to latch multiple counter output latches (OL) by setting the COUNT bit D5 = 0 and selecting the desired counter(s). This single command is functionally equivalent to several counter latch commands, one for each counter latched. Each counter's latched count is held until it is read (or the counter is reprogrammed). The counter is automatically unlatched when read, but other counters remain latched until they are read. If multiple count read-back commands are issued to the same counter without reading the count, all but the first are ignored; i.e., the count which will be read is the count at the time the first read-back command was issued.

The read-back command may also be used to latch status information of selected counter(s) by setting STATUS bit D4 = 0. Status must be latched to be read; status of a counter is accessed by a read from that counter.

The counter status format is shown in Figure 11. Bits D5 through D0 contain the counter's programmed Mode exactly as written in the last Mode Control Word. OUTPUT bit D7 contains the current state of the OUT pin. This allows the user to monitor the counter's output via software, possibly eliminating some hardware from a system.

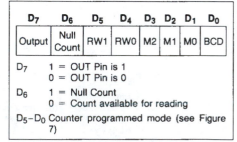

Figure 11. Status Byte

8254

NULL COUNT bit D6 indicates when the last count written to the counter register (CR) has been loaded into the counting element (CE). The exact time this happens depends on the Mode of the counter and is described in the Mode Definitions, but until the count is loaded into the counting element (CE), it can't be read from the counter. If the count is latched or read before this time, the count value will not reflect the new count just written. The operation of Null Count is shown in Figure 12.

This Action	Causes
A. Write to the control word register;[1]	Null Count = 1
B. Write to the count register (CR);[2]	Null Count = 1
C. New Count is loaded into CE (CR → CE);	Null Count = 0

NOTE:
1. Only the counter specified by the control word will have its Null Count set to 1. Null count bits of other counters are unaffected.
2. If the counter is programmed for two-byte counts (least significant byte then most significant byte) Null Count goes to 1 when the second byte is written.

Figure 12. Null Count Operation

If multiple status latch operations of the counter(s) are performed without reading the status, all but the first are ignored; i.e., the status that will be read is the status of the counter at the time the first status read-back command was issued.

Both count and status of the selected counter(s) may be latched simultaneously by setting both $\overline{COUNT}$ and $\overline{STATUS}$ bits D5,D4 = 0. This is functionally the same as issuing two separate read-back commands at once, and the above discussions apply here also. Specifically, if multiple count and/or status read-back commands are issued to the same counter(s) without any intervening reads, all but the first are ignored. This is illustrated in Figure 13.

If both count and status of a counter are latched, the first read operation of that counter will return latched status, regardless of which was latched first. The next one or two reads (depending on whether the counter is programmed for one or two type counts) return latched count. Subsequent reads return unlatched count.

$\overline{CS}$	$\overline{RD}$	$\overline{WR}$	A_1	A_0	
0	1	0	0	0	Write into Counter 0
0	1	0	0	1	Write into Counter 1
0	1	0	1	0	Write into Counter 2
0	1	0	1	1	Write Control Word
0	0	1	0	0	Read from Counter 0
0	0	1	0	1	Read from Counter 1
0	0	1	1	0	Read from Counter 2
0	0	1	1	1	No-Operation (3-State)
1	X	X	X	X	No-Operation (3-State)
0	1	1	X	X	No-Operation (3-State)

Figure 14. Read/Write Operations Summary

Command								Description	Result
D_7	D_6	D_5	D_4	D_3	D_2	D_1	D_0		
1	1	0	0	0	0	1	0	Read back count and status of Counter 0	Count and status latched for Counter 0
1	1	1	0	0	1	0	0	Read back status of Counter 1	Status latched for Counter 1
1	1	1	0	1	1	0	0	Read back status of Counters 2, 1	Status latched for Counter 2, but not Counter 1
1	1	0	1	1	0	0	0	Read back count of Counter 2	Count latched for Counter 2
1	1	0	0	0	1	0	0	Read back count and status of Counter 1	Count latched for Counter 1, but not status
1	1	1	0	0	0	1	0	Read back status of Counter 1	Command ignored, status already latched for Counter 1

Figure 13. Read-Back Command Example

8254

Mode Definitions

The following are defined for use in describing the operation of the 8254.

CLK Pulse:	a rising edge, then a falling edge, in that order, of a Counter's CLK input.
Trigger:	a rising edge of a Counter's GATE input.
Counter loading:	the transfer of a count from the CR to the CE (refer to the "Functional Description")

MODE 0: INTERRUPT ON TERMINAL COUNT

Mode 0 is typically used for event counting. After the Control Word is written, OUT is initially low, and will remain low until the Counter reaches zero. OUT then goes high and remains high until a new count or a new Mode 0 Control Word is written into the Counter.

GATE = 1 enables counting; GATE = 0 disables counting. GATE has no effect on OUT.

After the Control Word and initial count are written to a Counter, the initial count will be loaded on the next CLK pulse. This CLK pulse does not decrement the count, so for an initial count of N, OUT does not go high until N + 1 CLK pulses after the initial count is written.

If a new count is written to the Counter, it will be loaded on the next CLK pulse and counting will continue from the new count. If a two-byte count is written, the following happens:

1) Writing the first byte disables counting. OUT is set low immediately (no clock pulse required)

2) Writing the second byte allows the new count to be loaded on the next CLK pulse.

This allows the counting sequence to be synchronized by software. Again, OUT does not go high until N + 1 CLK pulses after the new count of N is written.

If an initial count is written while GATE = 0, it will still be loaded on the next CLK pulse. When GATE goes high, OUT will go high N CLK pulses later; no CLK pulse is needed to load the Counter as this has already been done.

MODE 1: HARDWARE RETRIGGERABLE ONE-SHOT

OUT will be initially high. OUT will go low on the CLK pulse following a trigger to begin the one-shot pulse, and will remain low until the Counter reaches zero.

OUT will then go high and remain high until the CLK pulse after the next trigger.

After writing the Control Word and initial count, the Counter is armed. A trigger results in loading the Counter and setting OUT low on the next CLK pulse, thus starting the one-shot pulse. An initial count of N will result in a one-shot pulse N CLK cycles in duration. The one-shot is retriggerable, hence OUT will remain low for N CLK pulses after any trigger. The one-shot pulse can be repeated without rewriting the same count into the counter. GATE has no effect on OUT.

If a new count is written to the Counter during a one-shot pulse, the current one-shot is not affected unless the counter is retriggered. In that case, the Counter is loaded with the new count and the one-shot pulse continues until the new count expires.

MODE 2: RATE GENERATOR

This Mode functions like a divide-by-N counter. It is typically used to generate a Real Time Clock interrupt. OUT will initially be high. When the initial count has decremented to 1, OUT goes low for one CLK pulse. OUT then goes high again, the Counter reloads the initial count and the process is repeated. Mode 2 is periodic; the same sequence is repeated indefinitely. For an initial count of N, the sequence repeats every N CLK cycles.

GATE = 1 enables counting; GATE = 0 disables counting. If GATE goes low during an output pulse, OUT is set high immediately. A trigger reloads the Counter with the initial count on the next CLK pulse; OUT goes low N CLK pulses after the trigger. Thus the GATE input can be used to synchronize the Counter.

After writing a Control Word and initial count, the Counter will be loaded on the next CLK pulse. OUT goes low N CLK Pulses after the initial count is written. This allows the Counter to be synchronized by software also.

Writing a new count while counting does not affect the current counting sequence. If a trigger is received after writing a new count but before the end of the current period, the Counter will be loaded with the new count on the next CLK pulse and counting will continue from the new count. Otherwise, the new count will be loaded at the end of the current counting cycle. In mode 2, a COUNT of 1 is illegal.

MODE 3: SQUARE WAVE MODE

Mode 3 is typically used for Baud rate generation. Mode 3 is similar to Mode 2 except for the duty cycle of OUT. OUT will initially be high. When half the

678

intel

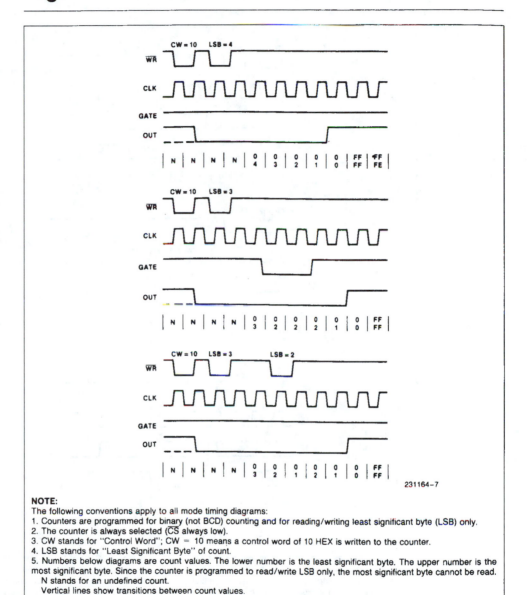

231164-7

NOTE:
The following conventions apply to all mode timing diagrams:
1. Counters are programmed for binary (not BCD) counting and for reading/writing least significant byte (LSB) only.
2. The counter is always selected (CS always low).
3. CW stands for "Control Word"; CW = 10 means a control word of 10 HEX is written to the counter.
4. LSB stands for "Least Significant Byte" of count.
5. Numbers below diagrams are count values. The lower number is the least significant byte. The upper number is the most significant byte. Since the counter is programmed to read/write LSB only, the most significant byte cannot be read.
   N stands for an undefined count.
   Vertical lines show transitions between count values.

Figure 15. Mode 0

679

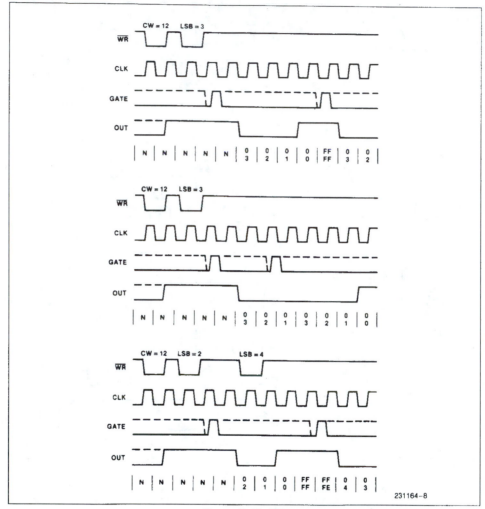

Figure 16. Mode 1

initial count has expired, OUT goes low for the remainder of the count. Mode 3 is periodic; the sequence above is repeated indefinitely. An initial count of N results in a square wave with a period of N CLK cycles.

GATE = 1 enables counting; GATE = 0 disables counting. If GATE goes low while OUT is low, OUT is set high immediately; no CLK pulse is required. A trigger reloads the Counter with the initial count on the next CLK pulse. Thus the GATE input can be used to synchronize the Counter.

After writing a Control Word and initial count, the Counter will be loaded on the next CLK pulse. This allows the Counter to be synchronized by software also.

Writing a new count while counting does not affect the current counting sequence. If a trigger is received after writing a new count but before the end of the current half-cycle of the square wave, the Counter will be loaded with the new count on the next CLK pulse and counting will continue from the

intel

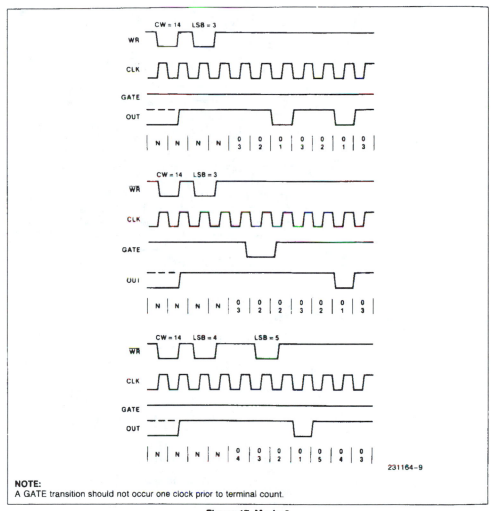

NOTE:
A GATE transition should not occur one clock prior to terminal count.

Figure 17. Mode 2

new count. Otherwise, the new count will be loaded at the end of the current half-cycle.

Mode 3 is implemented as follows:

Even counts: OUT is initially high. The initial count is loaded on one CLK pulse and then is decremented by two on succeeding CLK pulses. When the count expires OUT changes value and the Counter is reloaded with the initial count. The above process is repeated indefinitely.

Odd counts: OUT is initially high. The initial count minus one (an even number) is loaded on one CLK pulse and then is decremented by two on succeeding CLK pulses. One CLK pulse *after* the count expires, OUT goes low and the Counter is reloaded with the initial count minus one. Succeeding CLK pulses decrement the count by two. When the count expires, OUT goes high again and the Counter is reloaded with the initial count minus one. The above process is repeated indefinitely. So for odd counts, OUT will be high for $(N + 1)/2$ counts and low for $(N - 1)/2$ counts.

681

intel

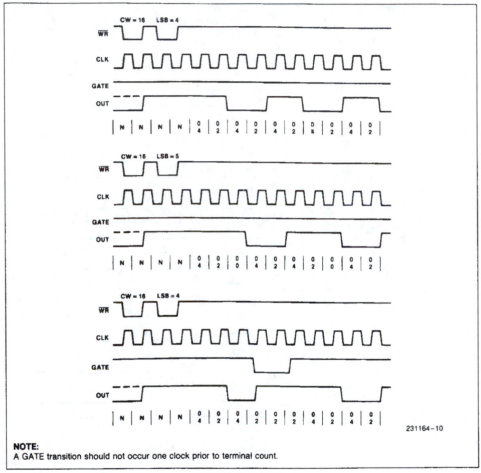

NOTE:
A GATE transition should not occur one clock prior to terminal count.

Figure 18. Mode 3

231164-10

MODE 4: SOFTWARE TRIGGERED STROBE

OUT will be initially high. When the initial count expires, OUT will go low for one CLK pulse and then go high again. The counting sequence is "triggered" by writing the initial count.

GATE = 1 enables counting; GATE = 0 disables counting. GATE has no effect on OUT.

After writing a Control Word and initial count, the Counter will be loaded on the next CLK pulse. This CLK pulse does not decrement the count, so for an initial count of N, OUT does not strobe low until N + 1 CLK pulses after the initial count is written.

If a new count is written during counting, it will be loaded on the next CLK pulse and counting will continue from the new count. If a two-byte count is written, the following happens:

1) Writing the first byte has no effect on counting.
2) Writing the second byte allows the new count to be loaded on the next CLK pulse.

This allows the sequence to be "retriggered" by software. OUT strobes low N + 1 CLK pulses after the new count of N is written.

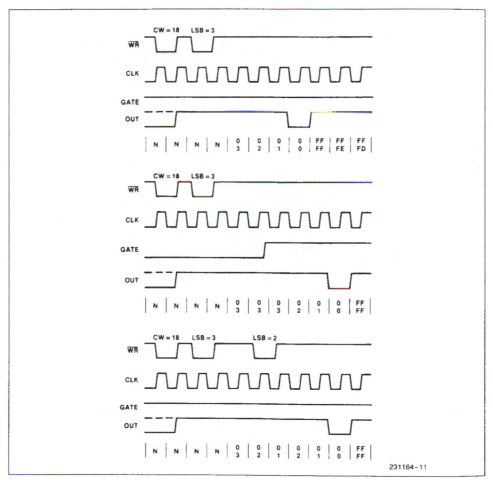

Figure 19. Mode 4

231164–11

683

8254

MODE 5: HARDWARE TRIGGERED STROBE (RETRIGGERABLE)

OUT will initially be high. Counting is triggered by a rising edge of GATE. When the initial count has expired, OUT will go low for one CLK pulse and then go high again.

After writing the Control Word and initial count, the counter will not be loaded until the CLK pulse after a trigger. This CLK pulse does not decrement the count, so for an initial count of N, OUT does not strobe low until N + 1 CLK pulses after a trigger.

A trigger results in the Counter being loaded with the initial count on the next CLK pulse. The counting sequence is retriggerable. OUT will not strobe low for N + 1 CLK pulses after any trigger. GATE has no effect on OUT.

If a new count is written during counting, the current counting sequence will not be affected. If a trigger occurs after the new count is written but before the current count expires, the Counter will be loaded with the new count on the next CLK pulse and counting will continue from there.

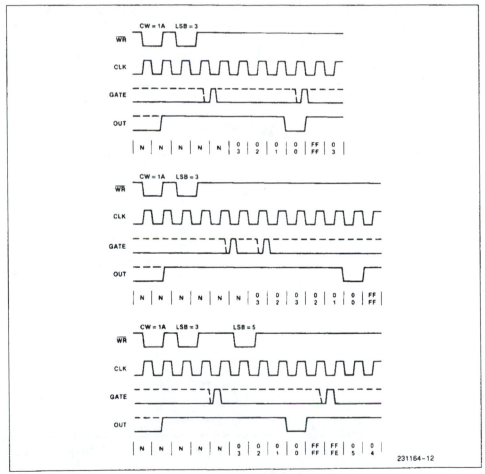

Figure 20. Mode 5

684

8254

Signal Status Modes	Low Or Going Low	Rising	High
0	Disables Counting	— —	Enables Counting
1	— —	1) Initiates Counting 2) Resets Output after Next Clock	— —
2	1) Disables Counting 2) Sets Output Immediately High	Initiates Counting	Enables Counting
3	1) Disables Counting 2) Sets Output Immediately High	Initiates Counting	Enables Counting
4	Disables Counting	— —	Enables Counting
5	— —	Initiates Counting	— —

Figure 21. Gate Pin Operations Summary

Mode	Min Count	Max Count
0	1	0
1	1	0
2	2	0
3	2	0
4	1	0
5	1	0

NOTE:
0 is equivalent to 2^{16} for binary counting and 10^4 for BCD counting.

Figure 22. Minimum and Maximum Initial Counts

Operation Common to All Modes

PROGRAMMING

When a Control Word is written to a Counter, all Control Logic is immediately reset and OUT goes to a known initial state; no CLK pulses are required for this.

GATE

The GATE input is always sampled on the rising edge of CLK. In Modes 0, 2, 3, and 4 the GATE input is level sensitive, and the logic level is sampled on the rising edge of CLK. In Modes 1, 2, 3, and 5 the GATE input is rising-edge sensitive. In these Modes, a rising edge of GATE (trigger) sets an edge-sensitive flip-flop in the Counter. This flip-flop is then sampled on the next rising edge of CLK; the flip-flop is reset immediately after it is sampled. In this way, a trigger will be detected no matter when it occurs—a high logic level does not have to be maintained until the next rising edge of CLK. Note that in Modes 2 and 3, the GATE input is both edge- and level-sensitive. In Modes 2 and 3, if a CLK source other than the system clock is used, GATE should be pulsed immediately following WR of a new count value.

COUNTER

New counts are loaded and Counters are decremented on the falling edge of CLK.

The largest possible initial count is 0; this is equivalent to 2^{16} for binary counting and 10^4 for BCD counting.

The Counter does not stop when it reaches zero. In Modes 0, 1, 4, and 5 the Counter "wraps around" to the highest count, either FFFF hex for binary counting or 9999 for BCD counting, and continues counting. Modes 2 and 3 are periodic; the Counter reloads itself with the initial count and continues counting from there.

8254

ABSOLUTE MAXIMUM RATINGS*

Ambient Temperature Under Bias ......0°C to 70°C
Storage Temperature .... −65°C to +150°C
Voltage on Any Pin with
  Respect to Ground............. −0.5V to +7V
Power Dissipation 1W

D.C. CHARACTERISTICS $T_A = 0°C$ to $70°C$, $V_{CC} = 5V \pm 10\%$

Symbol	Parameter	Min	Max	Units	Test Conditions
V_{IL}	Input Low Voltage	−0.5	0.8	V	
V_{IH}	Input High Voltage	2.0	$V_{CC} + 0.5V$	V	
V_{OL}	Output Low Voltage		0.45	V	$I_{OL} = 2.0$ mA
V_{OH}	Output High Voltage	2.4		V	$I_{OH} = -400 \mu A$
I_{IL}	Input Load Current		±10	μA	$V_{IN} = V_{CC}$ to 0V
I_{OFL}	Output Float Leakage		±10	μA	$V_{OUT} = V_{CC}$ to 0.45V
I_{CC}	V_{CC} Supply Current		170	mA	
C_{IN}	Input Capacitance		10	pF	$f_c = 1$ MHz
$C_{I/0}$	I/O Capacitance		20	pF	Unmeasured pins returned to V_{SS}[4]

A.C. CHARACTERISTICS $T_A = 0°C$ to $70°C$, $V_{CC} = 5V \pm 10\%$, GND = 0V

Bus Parameters[1]

READ CYCLE

Symbol	Parameter	8254-5		8254		8254-2		Unit
		Min	Max	Min	Max	Min	Max	
t_{AR}	Address Stable Before $\overline{RD}$ ↓	45		45		30		ns
t_{SR}	$\overline{CS}$ Stable Before $\overline{RD}$ ↓	0		0		0		ns
t_{RA}	Address Hold Time After $\overline{RD}$ ↑	0		0		0		ns
t_{RR}	$\overline{RD}$ Pulse Width	150		150		95		ns
t_{RD}	Data Delay from $\overline{RD}$ ↓		120		120		85	ns
t_{AD}	Data Delay from Address		220		220		185	ns
t_{DF}	$\overline{RD}$ ↑ to Data Floating	5	90	5	90	5	65	ns
t_{RV}	Command Recovery Time	200		200		165		ns

NOTE:
1. AC timings measured at $V_{OH} = 2.0V$, $V_{OL} = 0.8V$.

8254

A.C. CHARACTERISTICS T_A = 0°C to 70°C, V_{CC} = 5V ± 10%, GND = 0V (Continued)

WRITE CYCLE

Symbol	Parameter	8254-5		8254		8254-2		Unit
		Min	Max	Min	Max	Min	Max	
t_{AW}	Address Stable Before $\overline{WR}$ ↓	0		0		0		ns
t_{SW}	$\overline{CS}$ Stable Before $\overline{WR}$ ↓	0		0		0		ns
$t_{\overline{WA}}$	Address Hold Time After $\overline{WR}$ ↓	0		0		0		ns
t_{WW}	$\overline{WR}$ Pulse Width	150		150		95		ns
t_{DW}	Data Setup Time Before $\overline{WR}$ ↑	120		120		95		ns
t_{WD}	Data Hold Time After $\overline{WR}$ ↑	0		0		0		ns
t_{RV}	Command Recovery Time	200		200		165		ns

CLOCK AND GATE

Symbol	Parameter	8254-5		8254		8254-2		Unit
		Min	Max	Min	Max	Min	Max	
t_{CLK}	Clock Period	200	DC	125	DC	100	DC	ns
t_{PWH}	High Pulse Width	60[3]		60[3]		30[3]		ns
t_{PWL}	Low Pulse Width	60[3]		60[3]		50[3]		ns
t_R	Clock Rise Time		25		25		25	ns
t_F	Clock Fall Time		25		25		25	ns
t_{GW}	Gate Width High	50		50		50		ns
t_{GL}	Gate Width Low	50		50		50		ns
t_{GS}	Gate Setup Time to CLK ↑	50		50		40		ns
t_{GH}	Gate Setup Time After CLK ↑	50[2]		50[2]		50[2]		ns
t_{OD}	Output Delay from CLK ↓		150		150		100	ns
t_{ODG}	Output Delay from Gate ↓		120		120		100	ns
t_{WC}	CLK Delay for Loading ↓	0	55	0	55	0	55	ns
t_{WG}	Gate Delay for Sampling	−5	50	−5	50	−5	40	ns
t_{WO}	OUT Delay from Mode Write		260		260		240	ns
t_{CL}	CLK Set Up for Count Latch	−40	45	−40	45	−40	40	ns

NOTES:
2. In Modes 1 and 5 triggers are sampled on each rising clock edge. A second trigger within 120 ns (70 ns for the 8254-2) of the rising clock edge may not be detected.
3. Low-going glitches that violate t_{PWH}, t_{PWL} may cause errors requiring counter reprogramming.
4. Sampled, not 100% tested. T_A = 25°C.
5. If CLK present at TWC min then Count equals N + 2 CLK pulses, TWC max equals Count N + 1 CLK pulse. TWC min to TWC max, count will be either N + 1 or N + 2 CLK pulses.
6. In Modes 1 and 5, if GATE is present when writing a new Count value, at TWG min Counter will not be triggered, at TWG max Counter will be triggered.
7. If CLK present when writing a Counter Latch or ReadBack Command, at TCL min CLK will be reflected in count value latched, at TCL max CLK will not be reflected in the count value latched.

8087
MATH COPROCESSOR

- Adds Arithmetic, Trigonometric, Exponential, and Logarithmic Instructions to the Standard 8086/8088 and 80186/80188 Instruction Set for All Data Types
- CPU/8087 Supports 7 Data Types: 16-, 32-, 64-Bit Integers, 32-, 64-, 80-Bit Floating Point, and 18-Digit BCD Operands
- Compatible with IEEE Floating Point Standard 754

- Available in 5 MHz (8087), 8 MHz (8087-2) and 10 MHz (8087-1): 8 MHz 80186/80188 System Operation Supported with the 8087-1
- Adds 8 x 80-Bit Individually Addressable Register Stack to the 8086/8088 and 80186/80188 Architecture
- 7 Built-In Exception Handling Functions
- MULTIBUS® System Compatible Interface

The Intel 8087 Math CoProcessor is an extension to the Intel 8086/8088 microprocessor architecture. When combined with the 8086/8088 microprocessor, the 8087 dramatically increases the processing speed of computer applications which utilize mathematical operations such as CAM, numeric controllers, CAD or graphics.

The 8087 Math CoProcessor adds 68 mnemonics to the 8086 microprocessor instruction set. Specific 8087 math operations include logarithmic, arithmetic, exponential, and trigonometric functions. The 8087 supports integer, floating point and BCD data formats, and fully conforms to the ANSI/IEEE floating point standard.

The 8087 is fabricated with HMOS III technology and packaged in a 40-pin cerdip package.

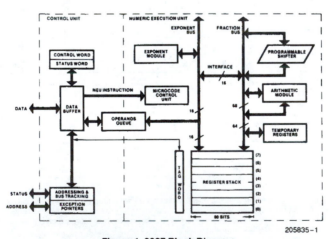

205835-1

Figure 1. 8087 Block Diagram

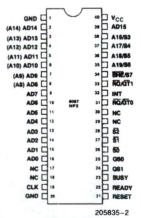

205835-2

Figure 2. 8087 Pin Configuration

Table 1. 8087 Pin Description

Symbol	Type	Name and Function
AD15–AD0	I/O	**ADDRESS DATA:** These lines constitute the time multiplexed memory address (T_1) and data (T_2, T_3, T_W, T_4) bus. A0 is analogous to the $\overline{BHE}$ for the lower byte of the data bus, pins D7–D0. It is LOW during T_1 when a byte is to be transferred on the lower portion of the bus in memory operations. Eight-bit oriented devices tied to the lower half of the bus would normally use A0 to condition chip select functions. These lines are active HIGH. They are input/output lines for 8087-driven bus cycles and are inputs which the 8087 monitors when the CPU is in control of the bus. A15–A8 do not require an address latch in an 8088/8087 or 80188/8087. The 8087 will supply an address for the T_1–T_4 period.
A19/S6, A18/S5, A17/S4, A16/S3	I/O	**ADDRESS MEMORY:** During T_1 these are the four most significant address lines for memory operations. During memory operations, status information is available on these lines during T_2, T_3, T_W, and T_4. For 8087-controlled bus cycles, S6, S4, and S3 are reserved and currently one (HIGH), while S5 is always LOW. These lines are inputs which the 8087 monitors when the CPU is in control of the bus.
BHE/S7	I/O	**BUS HIGH ENABLE:** During T_1 the bus high enable signed ($\overline{BHE}$) should be used to enable data onto the most significant half of the data bus, pins D15–D8. Eight-bit-oriented devices tied to the upper half of the bus would normally use $\overline{BHE}$ to condition chip select functions. $\overline{BHE}$ is LOW during T_1 for read and write cycles when a byte is to be transferred on the high portion of the bus. The S7 status information is available during T_2, T_3, T_W, and T_4. The signal is active LOW. S7 is an input which the 8087 monitors during the CPU-controlled bus cycles.
$\overline{S2}$, $\overline{S1}$, $\overline{S0}$	I/O	**STATUS:** For 8087-driven, these status lines are encoded as follows: $\overline{S2}$ $\overline{S1}$ $\overline{S0}$ 0 (LOW) X X Unused 1 (HIGH) 0 0 Unused 1 0 1 Read Memory 1 1 0 Write Memory 1 1 1 Passive Status is driven active during T_4, remains valid during T_1 and T_2, and is returned to the passive state (1, 1, 1) during T_3 or during T_W when READY is HIGH. This status is used by the 8288 Bus Controller (or the 82188 Integrated Bus Controller with an 80186/80188 CPU) to generate all memory access control signals. Any change in $\overline{S2}$, $\overline{S1}$, or $\overline{S0}$ during T_4 is used to indicate the beginning of a bus cycle, and the return to the passive state in T_3 or T_W is used to indicate the end of a bus cycle. These signals are monitored by the 8087 when the CPU is in control of the bus.
$\overline{RQ}/\overline{GT0}$	I/O	**REQUEST/GRANT:** This request/grant pin is used by the 8087 to gain control of the local bus from the CPU for operand transfers or on behalf of another bus master. It must be connected to one of the two processor request/grant pins. The request/grant sequence on this pin is as follows: 1. A pulse one clock wide is passed to the CPU to indicate a local bus request by either the 8087 or the master connected to the 8087 $\overline{RQ}/\overline{GT}1$ pin. 2. The 8087 waits for the grant pulse and when it is received will either initiate bus transfer activity in the clock cycle following the grant or pass the grant out on the $\overline{RQ}/\overline{GT}1$ pin in this clock if the initial request was for another bus master. 3. The 8087 will generate a release pulse to the CPU one clock cycle after the completion of the last 8087 bus cycle or on receipt of the release pulse from the bus master on $\overline{RQ}/\overline{GT}1$. For 80186/80188 systems the same sequence applies except $\overline{RQ}/\overline{GT}$ signals are converted to appropriate HOLD, HLDA signals by the 82188 Integrated Bus Controller. This is to conform with 80186/80188's HOLD, HLDA bus exchange protocol. Refer to the 82188 data sheet for further information.

Table 1. 8087 Pin Description (Continued)

Symbol	Type	Name and Function
RQ/GT1	I/O	**REQUEST/GRANT:** This request/grant pin is used by another local bus master to force the 8087 to request the local bus. If the 8087 is not in control of the bus when the request is made the request/grant sequence is passed through the 8087 on the RQ/GT0 pin one cycle later. Subsequent grant and release pulses are also passed through the 8087 with a two and one clock delay, respectively, for resynchronization. RQ/GT1 has an internal pullup resistor, and so may be left unconnected. If the 8087 has control of the bus the request/grant sequence is as follows: 1. A pulse 1 CLK wide from another local bus master indicates a local bus request to the 8087 (pulse 1). 2. During the 8087's next T_4 or T_1 a pulse 1 CLK wide from the 8087 to the requesting master (pulse 2) indicates that the 8087 has allowed the local bus to float and that it will enter the "RQ/GT acknowledge" state at the next CLK. The 8087's control unit is disconnected logically from the local bus during "RQ/GT acknowledge." 3. A pulse 1 CLK wide from the requesting master indicates to the 8087 (pulse 3) that the "RQ/GT" request is about to end and that the 8087 can reclaim the local bus at the next CLK. Each master-master exchange of the local bus is a sequence of 3 pulses. There must be one dead CLK cycle after each bus exchange. Pulses are active LOW. For 80186/80188 system, the RQ/GT1 line may be connected to the 82188 Integrated Bus Controller. In this case, a third processor with a HOLD, HLDA bus exchange system may acquire the bus from the 8087. For this configuration, RQ/GT1 will only be used if the 8087 is the bus master. Refer to 82188 data sheet for further information.
QS1, QS0	I	**QS1, QS0:** QS1 and QS0 provide the 8087 with status to allow tracking of the CPU instruction queue. QS1 QS0 0 (LOW) 0 No Operation 0 1 First Byte of Op Code from Queue 1 (HIGH) 0 Empty the Queue 1 1 Subsequent Byte from Queue
INT	O	**INTERRUPT:** This line is used to indicate that an unmasked exception has occurred during numeric instruction execution when 8087 interrupts are enabled. This signal is typically routed to an 8259A for 8086/8088 systems and to INT0 for 80186/80188 systems. INT is active HIGH.
BUSY	O	**BUSY:** This signal indicates that the 8087 NEU is executing a numeric instruction. It is connected to the CPU's TEST pin to provide synchronization. In the case of an unmasked exception BUSY remains active until the exception is cleared. BUSY is active HIGH.
READY	I	**READY:** READY is the acknowledgement from the addressed memory device that it will complete the data transfer. The RDY signal from memory is synchronized by the 8284A Clock Generator to form READY for 8086 systems. For 80186/80188 systems, RDY is synchronized by the 82188 Integrated Bus Controller to form READY. This signal is active HIGH.
RESET	I	**RESET:** RESET causes the processor to immediately terminate its present activity. The signal must be active HIGH for at least four clock cycles. RESET is internally synchronized.
CLK	I	**CLOCK:** The clock provides the basic timing for the processor and bus controller. It is asymmetric with a 33% duty cycle to provide optimized internal timing.
V_{CC}		**POWER:** V_{CC} is the +5V power supply pin.
GND		**GROUND:** GND are the ground pins.

NOTE:
For the pin descriptions of the 8086, 8088, 80186 and 80188 CPUs, reference the respective data sheets (8086, 8088, 80186, 80188).

APPLICATION AREAS

The 8087 provides functions meant specifically for high performance numeric processing requirements. Trigonometric, logarithmic, and exponential functions are built into the coprocessor hardware. These functions are essential in scientific, engineering, navigational, or military applications.

The 8087 also has capabilities meant for business or commercial computing. An 8087 can process Binary Coded Decimal (BCD) numbers up to 18 digits without roundoff errors. It can also perform arithmetic on integers as large as 64 bits $\pm 10^{18}$).

PROGRAMMING LANGUAGE SUPPORT

Programs for the 8087 can be written in Intel's high-level languages for 8086/8088 and 80186/80188 Systems; ASM-86 (the 8086, 8088 assembly language), PL/M-86, FORTRAN-86, and PASCAL-86.

RELATED INFORMATION

For 8086, 8088, 80186 or 80188 details, refer to the respective data sheets. For 80186 or 80188 systems, also refer to the 82188 Integrated Bus Controller data sheet.

FUNCTIONAL DESCRIPTION

The 8087 Math CoProcessor's architecture is designed for high performance numeric computing in conjunction with general purpose processing.

The 8087 is a numeric processor extension that provides arithmetic and logical instruction support for a variety of numeric data types. It also executes numerous built-in transcendental functions (e.g., tangent and log functions). The 8087 executes instructions as a coprocessor to a maximum mode CPU. It effectively extends the register and instruction set of the system and adds several new data types as well. Figure 3 presents the registers of the CPU + 8087. Table 2 shows the range of data types supported by the 8087. The 8087 is treated as an extension to the CPU, providing register, data types, control, and instruction capabilities at the hardware level. At the programmer's level the CPU and the 8087 are viewed as a single unified processor.

System Configuration

As a coprocessor to an 8086 or 8088, the 8087 is wired in parallel with the CPU as shown in Figure 4. Figure 5 shows the 80186/80188 system configuration. The CPU's status (S0–S2) and queue status lines (QS0–QS1) enable the 8087 to monitor and decode instructions in synchronization with the CPU and without any CPU overhead. For 80186/80188 systems, the queue status signals of the 80186/80188 are synchronized to 8087 requirements by the 8288 Integrated Bus Controller. Once started, the 8087 can process in parallel with, and independent of, the host CPU. For resynchronization, the 8087's BUSY signal informs the CPU that the 8087 is executing an instruction and the CPU WAIT instruction tests this signal to insure that the 8087 is ready to execute subsequent instructions. The 8087 can interrupt the CPU when it detects an error or exception. The 8087's interrupt request line is typically routed to the CPU through an 8259A Programmable Interrupt Controller for 8086, 8088 systems and INT0 for 80186/80188.

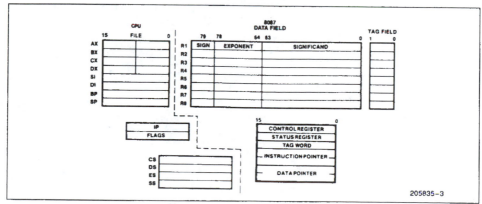

Figure 3. CPU + 8087 Architecture

691

The 8087 uses one of the request/grant lines of the 8086/8088 architecture (typically $\overline{RQ}/\overline{GT}0$) to obtain control of the local bus for data transfers. The other request/grant line is available for general system use (for instance by an I/O processor in LOCAL mode). A bus master can also be connected to the 8087's $\overline{RQ}/\overline{GT}1$ line. In this configuration the 8087 will pass the request/grant handshake signals between the CPU and the attached master when the 8087 is not in control of the bus and will relinquish the bus to the master directly when the 8087 is in control. In this way two additional masters can be configured in an 8086/8088 system; one will share the 8086/8088 bus with the 8087 on a first-come first-served basis, and the second will be guaranteed to be higher in priority than the 8087.

For 80186/80188 systems, $\overline{RQ}/\overline{GT}0$ and $\overline{RQ}/\overline{GT}1$ are connected to the corresponding inputs of the 82188 Integrated Bus Controller. Because the 80186/80188 has a HOLD, HLDA bus exchange protocol, an interface is needed which will translate $\overline{RQ}/\overline{GT}$ signals to corresponding HOLD, HLDA signals and vice versa. One of the functions of the 82188 IBC is to provide this translation. $\overline{RQ}/\overline{GT}0$ is translated to HOLD, HLDA signals which are then directly connected to the 80186/80188. The $\overline{RQ}/\overline{GT}1$ line is also translated into HOLD, HLDA signals (referred to as SYSHOLD, SYSHLDA signals) by the 82188 IBC. This allows a third processor (using a HOLD, HLDA bus exchange protocol) to gain control of the bus.

Unlike an 8086/8087 system, $\overline{RQ}/\overline{GT}$ is only used when the 8087 has bus control. If the third processor requests the bus when the current bus master is the 80186/80188, the 82188 IBC will directly pass the request onto the 80186/80188 without going through the 8087. The third processor has the highest bus priority in the system. If the 8087 requests the bus while the third processor has bus control, the grant pulse will not be issued until the third processor releases the bus (using SYSHOLD). In this configuration, the third processor has the highest priority, the 8087 has the next highest, and the 80186/80188 has the lowest bus priority.

Bus Operation

The 8087 bus structure, operation and timing are identical to all other processors in the 8086/8088 series (maximum mode configuration). The address is time multiplexed with the data on the first 16/8 lines of the address/data bus. A16 through A19 are time multiplexed with four status lines S3–S6. S3, S4 and S6 are always one (HIGH) for 8087-driven bus cycles while S5 is always zero (LOW). When the 8087 is monitoring CPU bus cycles (passive mode) S6 is also monitored by the 8087 to differentiate 8086/8088 activity from that of a local I/O processor or any other local bus master. (The 8086/8088 must be the only processor on the local bus to drive S6 LOW). S7 is multiplexed with and has the same value as $\overline{BHE}$ for all 8087 bus cycles.

Table 2. 8087 Data Types

Data Formats	Range	Precision	Most Significant Byte								
			7 0	7 0	7 0	7 0	7 0	7 0	7 0	7 0	7 0
Word Integer	10^4	16 Bits	I_{15} I_0 Two's Complement								
Short Integer	10^9	32 Bits	I_{31} I_0 Two's Complement								
Long Integer	10^{18}	64 Bits	I_{63} I_0 Two's Complement								
Packed BCD	10^{18}	18 Digits	S — $D_{17}D_{16}$ D_1 D_0								
Short Real	$10^{\pm38}$	24 Bits	S E_7 E_0 F_1 F_{23} F_0 Implicit								
Long Real	$10^{\pm308}$	53 Bits	S E_{10} E_0 F_1 F_{52} F_0 Implicit								
Temporary Real	$10^{\pm4932}$	64 Bits	S E_{14} E_0 F_0 F_{63}								

Integer: I
Packed BCD: $(-1)^S(D_{17}...D_0)$
Real: $(-1)^S(2^{E-Bias})(F_0 \bullet F_1...)$
bias = 127 for Short Real
       1023 for Long Real
       16383 for Temp Real

The first three status lines, $\overline{S0}$–$\overline{S2}$, are used with an 8288 bus controller or 82188 Integrated Bus Controller to determine the type of bus cycle being run:

$\overline{S2}$	$\overline{S1}$	$\overline{S0}$	
0	X	X	Unused
1	0	0	Unused
1	0	1	Memory Data Read
1	1	0	Memory Data Write
1	1	1	Passive (no bus cycle)

Programming Interface

The 8087 includes the standard 8086, 8088 instruction set for general data manipulation and program control. It also includes 68 numeric instructions for extended precision integer, floating point, trigonometric, logarithmic, and exponential functions. Sample execution times for several 8087 functions are shown in Table 3. Overall performance is up to 100 times that of an 8086 processor for numeric instructions.

Any instruction executed by the 8087 is the combined result of the CPU and 8087 activity. The CPU and the 8087 have specialized functions and registers providing fast concurrent operation. The CPU controls overall program execution while the 8087 uses the coprocessor interface to recognize and perform numeric operations.

Table 2 lists the seven data types the 8087 supports and presents the format for each type. Internally, the 8087 holds all numbers in the temporary real format. Load and store instructions automatically convert operands represented in memory as 16-, 32-, or 64-bit integers, 32- or 64-bit floating point numbers or 18-digit packed BCD numbers into temporary real format and vice versa. The 8087 also provides the capability to control round off, underflow, and overflow errors in each calculation.

Computations in the 8087 use the processor's register stack. These eight 80-bit registers provide the equivalent capacity of 20 32-bit registers. The 8087 register set can be accessed as a stack, with instructions operating on the top one or two stack elements, or as a fixed register set, with instructions operating on explicitly designated registers.

Table 5 lists the 8087's instructions by class. All appear as ESCAPE instructions to the host. Assembly language programs are written in ASM-86, the 8086, 8088 assembly language.

Table 3. Execution Times for Selected 8086/8087 Numeric Instructions and Corresponding 8086 Emulation

Floating Point Instruction	Approximate Execution Time (μs)	
	8086/8087 (8 MHz Clock)	8086 Emulation
Add/Subtract	10.6	1000
Multiply (Single Precision)	11.9	1000
Multiply (Extended Precision)	16.9	1312
Divide	24.4	2000
Compare	~5.6	812
Load (Double Precision)	~6.3	1062
Store (Double Precision)	13.1	750
Square Root	22.5	12250
Tangent	56.3	8125
Exponentiation	62.5	10687

NUMERIC PROCESSOR EXTENSION ARCHITECTURE

As shown in Figure 1, the 8087 is internally divided into two processing elements, the control unit (CU) and the numeric execution unit (NEU). The NEU executes all numeric instructions, while the CU receives and decodes instructions, reads and writes memory operands and executes 8087 control instructions. The two elements are able to operate independently of one another, allowing the CU to maintain synchronization with the CPU while the NEU is busy processing a numeric instruction.

Control Unit

The CU keeps the 8087 operating in synchronization with its host CPU. 8087 instructions are intermixed with CPU instructions in a single instruction stream. The CPU fetches all instructions from memory; by monitoring the status ($\overline{S0}$–$\overline{S2}$, S6) emitted by the CPU, the control unit determines when an instruction is being fetched. The CPU monitors the data bus in parallel with the CPU to obtain instructions that pertain to the 8087.

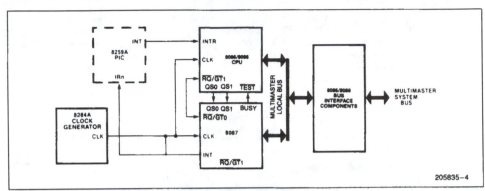

Figure 4. 8086/8087, 8088/8087 System Configuration

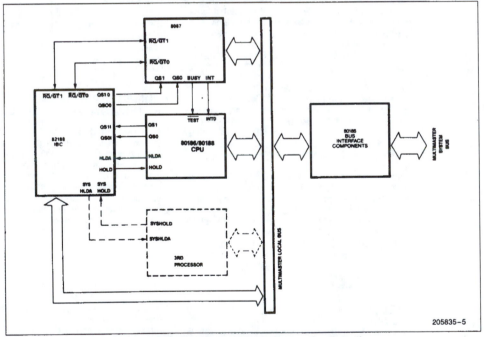

Figure 5. 80186/8087, 80188/8087 System Configuration

694

The CU maintains an instruction queue that is identical to the queue in the host CPU. The CU automatically determines if the CPU is an 8086/80186 or an 8088/80188 immediately after reset (by monitoring the BHE/S7 line) and matches its queue length accordingly. By monitoring the CPU's queue status lines (QS0, QS1), the CU obtains and decodes instructions from the queue in synchronization with the CPU.

A numeric instruction appears as an ESCAPE instruction to the CPU. Both the CPU and 8087 decode and execute the ESCAPE instruction together. The 8087 only recognizes the numeric instructions shown in Table 5. The start of a numeric operation is accomplished when the CPU executes the ESCAPE instruction. The instruction may or may not identify a memory operand.

The CPU does, however, distinguish between ESC instructions that reference memory and those that do not. If the instruction refers to a memory operand, the CPU calculates the operand's address using any one of its available addressing modes, and then performs a "dummy read" of the word at that location. (Any location within the 1M byte address space is allowed.) This is a normal read cycle except that the CPU ignores the data it receives. If the ESC instruction does not contain a memory reference (e.g. an 8087 stack operation), the CPU simply proceeds to the next instruction.

An 8087 instruction can have one of three memory reference options: (1) not reference memory; (2) load an operand word from memory into the 8087; or (3) store an operand word from the 8087 into memory. If no memory reference is required, the 8087 simply executes its instruction. If a memory reference is required, the CU uses a "dummy read" cycle initiated by the CPU to capture and save the address that the CPU places on the bus. If the instruction is a load, the CU additionally captures the data word when it becomes available on the local data bus. If data required is longer than one word, the CU immediately obtains the bus from the CPU using the request/grant protocol and reads the rest of the information in consecutive bus cycles. In a store operation, the CU captures and saves the store address as in a load, and ignores the data word that follows in the "dummy read" cycle. When the 8087 is ready to perform the store, the CU obtains the bus from the CPU and writes the operand starting at the specified address.

Numeric Execution Unit

The NEU executes all instructions that involve the register stack; these include arithmetic, logical, transcendental, constant and data transfer instructions. The data path in the NEU is 84 bits wide (68 fractions bits, 15 exponent bits and a sign bit) which allows internal operand transfers to be performed at very high speeds.

When the NEU begins executing an instruction, it activates the 8087 BUSY signal. This signal can be used in conjunction with the CPU WAIT instruction to resynchronize both processors when the NEU has completed its current instruction.

Register Set

The CPU+8087 register set is shown in Figure 3. Each of the eight data registers in the 8087's register stack is 80 bits and is divided into "fields" corresponding to the 8087's temporary real data type.

At a given point in time the TOP field in the control word identifies the current top-of-stack register. A "push" operation decrements TOP by 1 and loads a value into the new top register. A "pop" operation stores the value from the current top register and then increments TOP by 1. Like CPU stacks in memory, the 8087 register stack grows "down" toward lower-addressed registers.

Instructions may address the data registers either implicitly or explicitly. Many instructions operate on the register at the top of the stack. These instructions implicitly address the register pointed to by the TOP. Other instructions allow the programmer to explicitly specify the register which is to be used. Explicit register addressing is "top-relative."

Status Word

The status word shown in Figure 6 reflects the overall state of the 8087; it may be stored in memory and then inspected by CPU code. The status word is a 16-bit register divided into fields as shown in Figure 6. The busy bit (bit 15) indicates whether the NEU is either executing an instruction or has an interrupt request pending (B = 1), or is idle (B = 0). Several instructions which store and manipulate the status word are executed exclusively by the CU, and these do not set the busy bit themselves.

8087

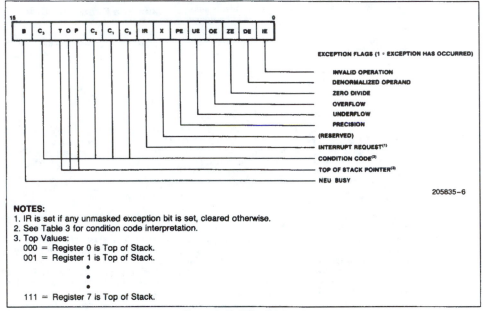

205835-6

NOTES:
1. IR is set if any unmasked exception bit is set, cleared otherwise.
2. See Table 3 for condition code interpretation.
3. Top Values:
   000 = Register 0 is Top of Stack.
   001 = Register 1 is Top of Stack.
       ·
       ·
       ·
   111 = Register 7 is Top of Stack.

Figure 6. 8087 Status Word

The four numeric condition code bits (C_0–C_3) are similar to flags in a CPU: various instructions update these bits to reflect the outcome of the 8087 operations. The effect of these instructions on the condition code bits is summarized in Table 4.

Bits 14–12 of the status word point to the 8087 register that is the current top-of-stack (TOP) as described above.

Bit 7 is the interrupt request bit. This bit is set if any unmasked exception bit is set and cleared otherwise.

Bits 5–0 are set to indicate that the NEU has detected an exception while executing an instruction.

Tag Word

The tag word marks the content of each register as shown in Figure 7. The principal function of the tag word is to optimize the 8087's performance. The tag word can be used, however, to interpret the contents of 8087 registers.

Instruction and Data Pointers

The instruction and data pointers (see Figure 8) are provided for user-written error handlers. Whenever the 8087 executes a math instruction, the CU saves the instruction address, the operand address (if present) and the instruction opcode. 8087 instructions can store this data into memory.

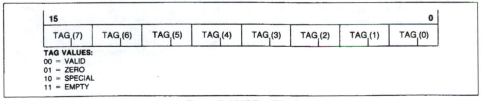

Figure 7. 8087 Tag Word

Table 4a. Condition Code Interpretation

Instruction Type	C_3	C_2	C_1	C_0	Interpretation
Compare, Test	0	0	X	0	ST > Source or 0 (FTST)
	0	0	X	1	ST < Source or 0 (FTST)
	1	0	X	0	ST = Source or 0 (FTST)
	1	1	X	1	ST is not comparable
Remainder	Q_1	0	Q_0	Q_2	Complete reduction with three low bits of quotient (See Table 4b)
	U	1	U	U	Incomplete Reduction
Examine	0	0	0	0	Valid, positive unnormalized
	0	0	0	1	Invalid, positive, exponent = 0
	0	0	1	0	Valid, negative, unnormalized
	0	0	1	1	Invalid, negative, exponent = 0
	0	1	0	0	Valid, positive, normalized
	0	1	0	1	Infinity, positive
	0	1	1	0	Valid, negative, normalized
	0	1	1	1	Infinity, negative
	1	0	0	0	Zero, positive
	1	0	0	1	Empty
	1	0	1	0	Zero, negative
	1	0	1	1	Empty
	1	1	0	0	Invalid, positive, exponent = 0
	1	1	0	1	Empty
	1	1	1	0	Invalid, negative, exponent = 0
	1	1	1	1	Empty

NOTES:
1. ST = Top of stack
2. X = value is not affected by instruction
3. U = value is undefined following instruction
4. Q_n = Quotient bit n

Table 4b. Condition Code Interpretation after FPREM Instruction As a Function of Divided Value

Dividend Range	Q_2	Q_1	Q_0
Dividend < 2 * Modulus	C_3[1]	C_1[1]	Q_0
Dividend < 4 * Modulus	C_3[1]	Q_1	Q_0
Dividend ≥ 4 * Modulus	Q_2	Q_1	Q_0

NOTE:
1. Previous value of indicated bit, not affected by FPREM instruction execution.

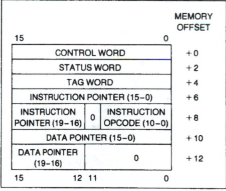

Figure 8. 8087 Instruction and Data Pointer Image in Memory

8087

Control Word

The 8087 provides several processing options which are selected by loading a word from memory into the control word. Figure 9 shows the format and encoding of the fields in the control word.

The low order byte of this control word configures 8087 interrupts and exception masking. Bits 5–0 of the control word contain individual masks for each of the six exceptions that the 8087 recognizes and bit 7 contains a general mask bit for all 8087 interrupts. The high order byte of the control word configures the 8087 operating mode including precision, rounding, and infinity controls. The precision control bits (bits 9–8) can be used to set the 8087 internal operating precision at less than the default of temporary real precision. This can be useful in providing compatibility with earlier generation arithmetic processors of smaller precision than the 8087. The rounding control bits (bits 11–10) provide for directed rounding and true chop as well as the unbiased round to nearest mode specified in the proposed IEEE standard. Control over closure of the number space at infinity is also provided (either affine closure, $\pm\infty$, or projective closure, ∞, is treated as unsigned, may be specified).

Exception Handling

The 8087 detects six different exception conditions that can occur during instruction execution. Any or all exceptions will cause an interrupt if unmasked and interrupts are enabled.

If interrupts are disabled the 8087 will simply continue execution regardless of whether the host clears the exception. If a specific exception class is masked and that exception occurs, however, the 8087 will post the exception in the status register and perform an on-chip default exception handling procedure, thereby allowing processing to continue. The exceptions that the 8087 detects are the following:

1. INVALID OPERATION: Stack overflow, stack underflow, indeterminate form (0/0, $\infty - \infty$, etc.) or the use of a Non-Number (NAN) as an operand. An exponent value is reserved and any bit pattern with this value in the exponent field is termed a Non-Number and causes this exception. If this exception is masked, the 8087's default response is to generate a specific NAN called INDEFINITE, or to propagate already existing NANs as the calculation result.

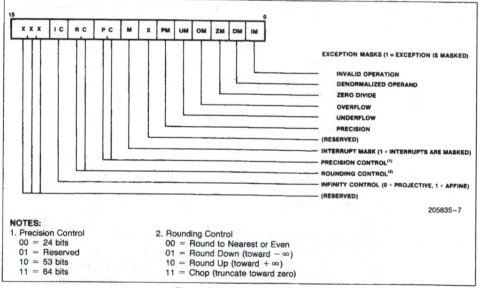

Figure 9. 8087 Control Word

2. OVERFLOW: The result is too large in magnitude to fit the specified format. The 8087 will generate an encoding for infinity if this exception is masked.

3. ZERO DIVISOR: The divisor is zero while the dividend is a non-infinite, non-zero number. Again, the 8087 will generate an encoding for infinity if this exception is masked.

4. UNDERFLOW: The result is non-zero but too small in magnitude to fit in the specified format. If this exception is masked the 8087 will denormal-ize (shift right) the fraction until the exponent is in range. This process is called gradual underflow.

5. DENORMALIZED OPERAND: At least one of the operands or the result is denormalized; it has the smallest exponent but a non-zero significand. Normal processing continues if this exception is masked off.

6. INEXACT RESULT: If the true result is not exactly representable in the specified format, the result is rounded according to the rounding mode, and this flag is set. If this exception is masked, processing will simply continue.

ABSOLUTE MAXIMUM RATINGS*

Ambient Temperature Under Bias ......0°C to 70°C
Storage Temperature .... −65°C to +150°C
Voltage on Any Pin with
  Respect to Ground.............. −1.0V to +7V
Power Dissipation.......................3.0 Watt

NOTICE: This is a production data sheet. The specifications are subject to change without notice.

*WARNING: Stressing the device beyond the "Absolute Maximum Ratings" may cause permanent damage. These are stress ratings only. Operation beyond the "Operating Conditions" is not recommended and extended exposure beyond the "Operating Conditions" may affect device reliability.

D.C. CHARACTERISTICS T_A = 0°C to 70°C, V_{CC} = 5V ±5%

Symbol	Parameter	Min	Max	Units	Test Conditions
V_{IL}	Input Low Voltage	−0.5	0.8	V	
V_{IH}	Input High Voltage	2.0	V_{CC} + 0.5	V	
V_{OL}	Output Low Voltage (Note 8)		0.45	V	I_{OL} = 2.5 mA
V_{OH}	Output High Voltage	2.4		V	I_{OH} = −400 µA
I_{CC}	Power Supply Current		475	mA	T_A = 25°C
I_{LI}	Input Leakage Current		±10	µA	0V ≤ V_{IN} ≤ V_{CC}
I_{LO}	Output Leakage Current		±10	µA	T_A = 25°C
V_{CL}	Clock Input Low Voltage	−0.5	0.6	V	
V_{CH}	Clock Input High Voltage	3.9	V_{CC} + 1.0	V	
C_{IN}	Capacitance of Inputs		10	pF	fc = 1 MHz
C_{IO}	Capacitance of I/O Buffer (AD0–15, A_{16}–A_{19}, BHE, S2–S0, RQ/GT) and CLK		15	pF	fc = 1 MHz
C_{OUT}	Capacitance of Outputs BUSY INT		10	pF	fc = 1 MHz

A.C. CHARACTERISTICS $T_A = 0°C$ to $70°C$, $V_{CC} = 5V \pm 5\%$

TIMING REQUIREMENTS

Symbol	Parameter	8087 Min	8087 Max	8087-2 Min	8087-2 Max	8087-1 (See Note 7) Min	8087-1 (See Note 7) Max	Units	Test Conditions
TCLCL	CLK Cycle Period	200	500	125	500	100	500	ns	
TCLCH	CLK Low Time	118		68		53		ns	
TCHCL	CLK High Time	69		44		39		ns	
TCH1CH2	CLK Rise Time		10		10		15	ns	From 1.0V to 3.5V
TCL2CL2	CLK Fall Time		10		10		15	ns	From 3.5V to 1.0V
TDVCL	Data In Setup Time	30		20		15		ns	
TCLDX	Data In Hold Time	10		10		10		ns	
TRYHCH	READY Setup Time	118		68		53		ns	
TCHRYX	READY Hold Time	30		20		5		ns	
TRYLCL	READY Inactive to CLK (Note 6)	−8		−8		−10		ns	
TGVCH	RQ/GT Setup Time (Note 8)	30		15		15		ns	
TCHGX	RQ/GT Hold Time	40		30		20		ns	
TQVCL	QS0−1 Setup Time (Note 8)	30		30		30		ns	
TCLQX	QS0−1 Hold Time	10		10		5		ns	
TSACH	Status Active Setup Time	30		30		30		ns	
TSNCL	Status Inactive Setup Time	30		30		30		ns	
TILIH	Input Rise Time (Except CLK)		20		20		20	ns	From 0.8V to 2.0V
TIHIL	Input Fall Time (Except CLK)		12		12		15	ns	From 2.0V to 0.8V

TIMING RESPONSES

Symbol	Parameter	8087 Min	8087 Max	8087-2 Min	8087-2 Max	8087-1 (See Note 7) Min	8087-1 (See Note 7) Max	Units	Test Conditions
TCLML	Command Active Delay (Notes 1, 2)	10/0	35/70	10/0	35/70	10/0	35/70	ns	$C_L = 20-100$ pF for all 8087 Outputs (in addition to 8087 self-load)
TCLMH	Command Inactive Delay (Notes 1, 2)	10/0	35/55	10/0	35/55	10/0	35/70	ns	
TRYHSH	Ready Active to Status Passive (Note 5)		110		65		45	ns	
TCHSV	Status Active Delay	10	110	10	60	10	45	ns	
TCLSH	Status Inactive Delay	10	130	10	70	10	55	ns	
TCLAV	Address Valid Delay	10	110	10	60	10	55	ns	
TCLAX	Address Hold Time	10		10		10		ns	

8087

A.C. CHARACTERISTICS $T_A = 0°C$ to $70°C$, $V_{CC} = 5V \pm 5\%$ (Continued)

TIMING RESPONSES (Continued)

Symbol	Parameter	8087		8087-2		8087-1 (See Note 7)		Units	Test Conditions
		Min	Max	Min	Max	Min	Max		
TCLAZ	Address Float Delay	TCLAX	80	TCLAX	50	TCLAX	45	ns	$C_L = 20-100$ pF for all 8087 Outputs (in addition to 8087 self-load)
TSVLH	Status Valid to ALE High (Notes 1, 2)		15/30		15/30		15/30	ns	
TCLLH	CLK Low to ALE Valid (Notes 1, 2)		15/30		15/30		15/30	ns	
TCHLL	ALE Inactive Delay (Notes 1, 2)		15/30		15/30		15/30	ns	
TCLDV	Data Valid Delay	10	110	10	60	10	50	ns	
TCHDX	Status Hold Time	10		10		10	45	ns	
TCLDOX	Data Hold Time	10		10		10		ns	
TCVNV	Control Active Delay (Notes 1, 3)	5	45	5	45	5	45	ns	
TCVNX	Control Inactive Delay (Notes 1, 3)	10	45	10	45	10	45	ns	
TCHBV	BUSY and INT Valid Delay	10	150	10	85	10	65	ns	
TCHDTL	Direction Control Active Delay (Notes 1, 3)		50		50		50	ns	
TCHDTH	Direction Control Inactive Delay (Notes 1, 3)		30		30		30	ns	
TSVDTV	STATUS to DT/$\overline{R}$ Delay (Notes 1, 4)	0	30	0	30	0	30	ns	
TCLDTV	DT/$\overline{R}$ Active Delay (Notes 1, 4)	0	55	0	55	0	55	ns	
TCHDNV	$\overline{DEN}$ Active Delay (Notes 1, 4)	0	55	0	55	0	55	ns	
TCHDNX	$\overline{DEN}$ Inactive Delay (Notes 1, 4)	5	55	5	55	5	55	ns	
TCLGL	$\overline{RQ}$/$\overline{GT}$ Active Delay (Note 8)	0	85	0	50	0	38	ns	$C_L = 40$ pF (in addition to 8087 self-load)
TCLGH	$\overline{RQ}$/$\overline{GT}$ Inactive Delay	0	85	0	50	0	45	ns	
TOLOH	Output Rise Time		20		20		15	ns	From 0.8V to 2.0V
TOHOL	Output Fall Time		12		12		12	ns	From 2.0V to 0.8V

NOTES:
1. Signal at 8284A, 8288, or 82188 shown for reference only.
2. 8288 timing/82188 timing.
3. 8288 timing.
4. 82188 timing.
5. Applies only to T_3 and wait states.
6. Applies only to T_2 state (8 ns into T_3).
7. IMPORTANT SYSTEM CONSIDERATION: Some 8087-1 timing parameters are constrained relative to the corresponding 8086-1 specifications. Therefore, 8086-1 systems incorporating the 8087-1 should be designed with the 8087-1 specifications.
8. Changes since last revision.

A.C. TESTING INPUT, OUTPUT WAVEFORM

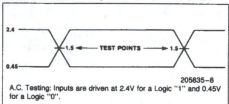

205835-8

A.C. Testing: Inputs are driven at 2.4V for a Logic "1" and 0.45V
for a Logic "0".

A.C. TESTING LOAD CIRCUIT

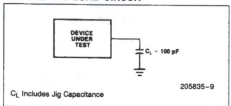

C_L Includes Jig Capacitance

205835-9

WAVEFORMS

MASTER MODE (with 8288 references)

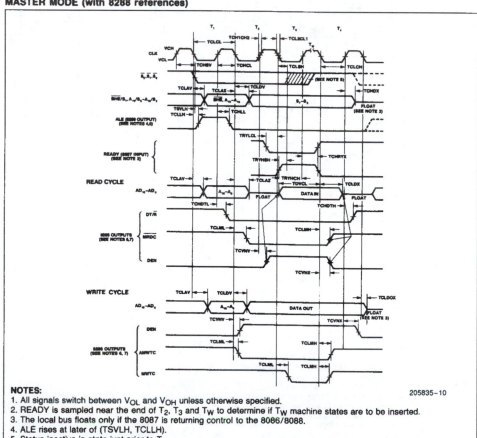

205835-10

NOTES:
1. All signals switch between V_{OL} and V_{OH} unless otherwise specified.
2. READY is sampled near the end of T_2, T_3 and T_W to determine if T_W machine states are to be inserted.
3. The local bus floats only if the 8087 is returning control to the 8086/8088.
4. ALE rises at later of (TSVLH, TCLLH).
5. Status inactive in state just prior to T_4.
6. Signals at 8284A or 8288 are shown for reference only.
7. The issuance of 8288 command and control signals (MRDC, (MWTC, AMWC, and DEN) lags the active high 8288 CEN.
8. All timing measurements are made at 1.5V unless otherwise noted.

8087

WAVEFORMS (Continued)

MASTER MODE (with 82188 references)

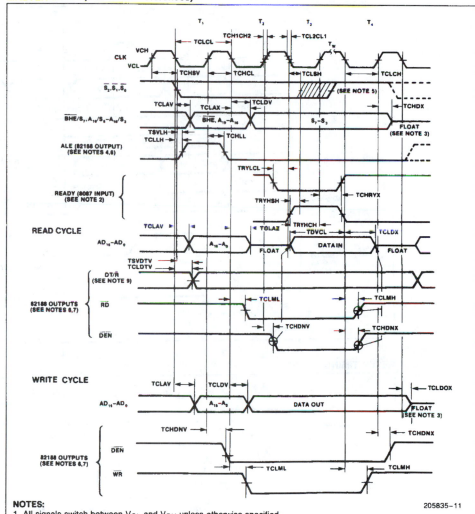

205835-11

NOTES:
1. All signals switch between V_{OL} and V_{OH} unless otherwise specified.
2. READY is sampled near the end of T_2, T_3 and T_W to determine if T_W machine states are to be inserted.
3. The local bus floats only if the 8087 is returning control to the 80186/80188.
4. ALE rises at later of (TSVLH, TCLLH).
5. Status inactive in state just prior to T_4.
6. Signals at 8284A or 82188 are shown for reference only.
7. The issuance of 8288 command and control signals ($\overline{MRDC}$, ($\overline{MWTC}$, $\overline{AMWC}$, and DEN) lags the active high 8288 CEN.
8. All timing measurements are made at 1.5V unless otherwise noted.
9. DT/$\overline{R}$ becomes valid at the later of (TSVDTV, TCLDTV).

8087

WAVEFORMS (Continued)

PASSIVE MODE

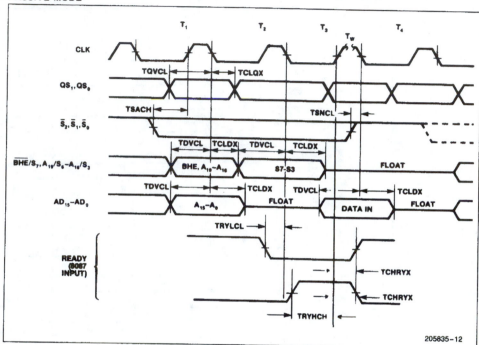

205835–12

RESET TIMING

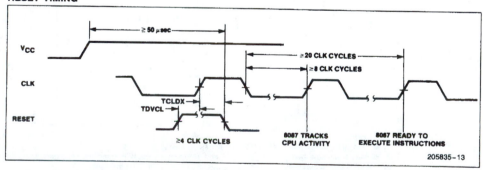

205835–13

704

WAVEFORMS (Continued)

REQUEST/GRANT₀ TIMING

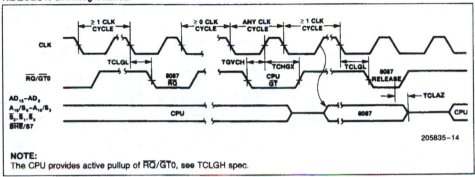

205835-14

NOTE:
The CPU provides active pullup of RQ/GT0, see TCLGH spec.

REQUEST/GRANT₁ TIMING

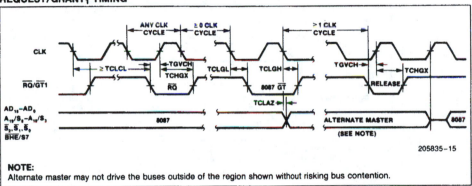

205835-15

NOTE:
Alternate master may not drive the buses outside of the region shown without risking bus contention.

BUSY AND INTERRUPT TIMING

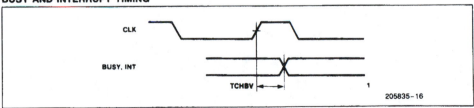

205835-16

Table 5. 8087 Extensions to the 86/186 Instructions Sets

Data Transfer	Encoding		Optional 8,16 Bit Displacement	Clock Count Range			
FLD = LOAD	MF =			**32 Bit Real** 00	**32 Bit Integer** 01	**64 Bit Real** 10	**16 Bit Integer** 11
Integer/Real Memory to ST(0)	ESCAPE MF 1	MOD 0 0 0 R/M	DISP	38–56 +EA	52–60 +EA	40–60 +EA	46–54 +EA
Long Integer Memory to ST(0)	ESCAPE 1 1 1	MOD 1 0 1 R/M	DISP	60–68 +EA			
Temporary Real Memory to ST(0)	ESCAPE 0 1 1	MOD 1 0 1 R/M	DISP	53–65 +EA			
BCD Memory to ST(0)	ESCAPE 1 1 1	MOD 1 0 0 R/M	DISP	290–310 +EA			
ST(i) to ST(0)	ESCAPE 0 0 1	1 1 0 0 0 ST(i)		17–22			
FST = STORE							
ST(0) to Integer/Real Memory	ESCAPE MF 1	MOD 0 1 0 R/M	DISP	84–90 +EA	82–92 +EA	96–104 +EA	80–90 +EA
ST(0) to ST(i)	ESCAPE 1 0 1	1 1 0 1 0 ST(i)		15–22			
FSTP = STORE AND POP							
ST(0) to Integer/Real Memory	ESCAPE MF 1	MOD 0 1 1 R/M	DISP	86–92 +EA	84–94 +EA	98–106 +EA	82–92 +EA
ST(0) to Long Integer Memory	ESCAPE 1 1 1	MOD 1 1 1 R/M	DISP	94–105 +EA			
ST(0) to Temporary Real Memory	ESCAPE 0 1 1	MOD 1 1 1 R/M	DISP	52–58 +EA			
ST(0) to BCD Memory	ESCAPE 1 1 1	MOD 1 1 0 R/M	DISP	520–540 +EA			
ST(0) to ST(i)	ESCAPE 1 0 1	1 1 0 1 1 ST(i)		17–24			
FXCH = Exchange ST(i) and ST(0)	ESCAPE 0 0 1	1 1 0 0 1 ST(i)		10–15			

Comparison

	Encoding		Optional 8,16 Bit Displacement	32 Bit Real	32 Bit Integer	64 Bit Real	16 Bit Integer
FCOM = Compare							
Integer/Real Memory to ST(0)	ESCAPE MF 0	MOD 0 1 0 R/M	DISP	60–70 +EA	78–91 +EA	65–75 +EA	72–86 +EA
ST(i) to ST(0)	ESCAPE 0 0 0	1 1 0 1 0 ST(i)		40–50			
FCOMP = Compare and Pop							
Integer/Real Memory to ST(0)	ESCAPE MF 0	MOD 0 1 1 R/M	DISP	63–73 +EA	80–93 +EA	67–77 +EA	74–88 +EA
ST(i) to ST(0)	ESCAPE 0 0 0	1 1 0 1 1 ST(i)		45–52			
FCOMPP = Compare ST(1) to ST(0) and Pop Twice	ESCAPE 1 1 0	1 1 0 1 1 0 0 1		45–55			
FTST = Test ST(0)	ESCAPE 0 0 1	1 1 1 0 0 1 0 0		38–48			
FXAM = Examine ST(0)	ESCAPE 0 0 1	1 1 1 0 0 1 0 1		12–23			

205835–17

Table 5. 8087 Extensions to the 86/186 Instructions Sets (Continued)

Constants	Instruction Encoding	Optional 8,16 Bit Displacement	32 Bit Real (00)	32 Bit Integer (01)	64 Bit Real (10)	16 Bit Integer (11)
	MF =					
FLDZ = LOAD + 0.0 into ST(0)	ESCAPE 0 0 1 | 1 1 1 0 1 1 1 0			11–17		
FLD1 = LOAD + 1.0 into ST(0)	ESCAPE 0 0 1 | 1 1 1 0 1 0 0 0			15–21		
FLDPI = LOAD π into ST(0)	ESCAPE 0 0 1 | 1 1 1 0 1 0 1 1			16–22		
FLDL2T = LOAD $\log_2$ 10 into ST(0)	ESCAPE 0 0 1 | 1 1 1 0 1 0 0 1			16–22		
FLDL2E = LOAD $\log_2$ e into ST(0)	ESCAPE 0 0 1 | 1 1 1 0 1 0 1 0			15–21		
FLDLG2 = LOAD $\log_{10}$ 2 into ST(0)	ESCAPE 0 0 1 | 1 1 1 0 1 1 0 0			18–24		
FLDLN2 = LOAD $\log_e$2 into ST(0)	ESCAPE 0 0 1 | 1 1 1 0 1 1 0 1			17–23		

Arithmetic

FADD = Addition

	Instruction Encoding	Optional 8,16 Bit Displacement	32 Bit Real (00)	32 Bit Integer (01)	64 Bit Real (10)	16 Bit Integer (11)
Integer/Real Memory with ST(0)	ESCAPE MF 0 | MOD 0 0 0 R/M	DISP	90–120 + EA	108–143 + EA	95–125 + EA	102–137 + EA
ST(i) and ST(0)	ESCAPE d P 0 | 1 1 0 0 0 ST(i)		70–100 (Note 1)			

FSUB = Subtraction

	Instruction Encoding	Optional 8,16 Bit Displacement	32 Bit Real (00)	32 Bit Integer (01)	64 Bit Real (10)	16 Bit Integer (11)
Integer/Real Memory with ST(0)	ESCAPE MF 0 | MOD 1 0 R R/M	DISP	90–120 + EA	108–143 + EA	95–125 + EA	102–137 + EA
ST(i) and ST(0)	ESCAPE d P 0 | 1 1 1 0 R R/M		70–100 (Note 1)			

FMUL = Multiplication

	Instruction Encoding	Optional 8,16 Bit Displacement	32 Bit Real (00)	32 Bit Integer (01)	64 Bit Real (10)	16 Bit Integer (11)
Integer/Real Memory with ST(0)	ESCAPE MF 0 | MOD 0 0 1 R/M	DISP	110–125 + EA	130–144 + EA	112–168 + EA	124–138 + EA
ST(i) and ST(0)	ESCAPE d P 0 | 1 1 0 0 1 R/M		90–145 (Note 1)			

FDIV = Division

	Instruction Encoding	Optional 8,16 Bit Displacement	32 Bit Real (00)	32 Bit Integer (01)	64 Bit Real (10)	16 Bit Integer (11)
Integer/Real Memory with ST(0)	ESCAPE MF 0 | MOD 1 1 R R/M	DISP	215–225 + EA	230–243 + EA	220–230 + EA	224–238 + EA
ST(i) and ST(0)	ESCAPE d P 0 | 1 1 1 1 R R/M		193–203 (Note 1)			

	Instruction Encoding		Clock Count Range			
FSQRT = Square Root of ST(0)	ESCAPE 0 0 1 | 1 1 1 1 1 0 1 0			180–186		
FSCALE = Scale ST(0) by ST(1)	ESCAPE 0 0 1 | 1 1 1 1 1 1 0 1			32–38		
FPREM = Partial Remainder of ST(0) ÷ ST(1)	ESCAPE 0 0 1 | 1 1 1 1 1 0 0 0			15–190		
FRNDINT = Round ST(0) to Integer	ESCAPE 0 0 1 | 1 1 1 1 1 1 0 0			16–50		

205835–18

Table 5. 8087 Extensions to the 86/186 Instructions Sets (Continued)

			Optional 8,16 Bit Displacement	Clock Count Range
FXTRACT = Extract Components of St(0)	ESCAPE 0 0 1	1 1 1 1 0 1 0 0		27–55
FABS = Absolute Value of ST(0)	ESCAPE 0 0 1	1 1 1 0 0 0 0 1		10–17
FCHS = Change Sign of ST(0)	ESCAPE 0 0 1	1 1 1 0 0 0 0 0		10–17

Transcendental

			Optional 8,16 Bit Displacement	Clock Count Range
FPTAN = Partial Tangent of ST(0)	ESCAPE 0 0 1	1 1 1 1 0 0 1 0		30–540
FPATAN = Partial Arctangent of ST(0) ÷ ST(1)	ESCAPE 0 0 1	1 1 1 1 0 0 1 1		250–800
F2XM1 = $2^{ST(0)}$ -1	ESCAPE 0 0 1	1 1 1 1 0 0 0 0		310–630
FYL2X = ST(1) · $\log_2$ |ST(0)|	ESCAPE 0 0 1	1 1 1 1 0 0 0 1		900–1100
FYL2XP1 = ST(1) · $\log_2$ [ST(0) +1]	ESCAPE 0 0 1	1 1 1 1 1 0 0 1		700–1000

Processor Control

			Optional 8,16 Bit Displacement	Clock Count Range
FINIT = Initialized 8087	ESCAPE 0 1 1	1 1 1 0 0 0 1 1		2–8
FENI = Enable Interrupts	ESCAPE 0 1 1	1 1 1 0 0 0 0 0		2–8
FDISI = Disable Interrupts	ESCAPE 0 1 1	1 1 1 0 0 0 0 1		2–8
FLDCW = Load Control Word	ESCAPE 0 0 1	MOD 1 0 1 R/M	DISP	7–14 + EA
FSTCW = Store Control Word	ESCAPE 0 0 1	MOD 1 1 1 R/M	DISP	12–18 + EA
FSTSW = Store Status Word	ESCAPE 1 0 1	MOD 1 1 1 R/M	DISP	12–18 + EA
FCLEX = Clear Exceptions	ESCAPE 0 1 1	1 1 1 0 0 0 1 0		2–8
FSTENV = Store Environment	ESCAPE 0 0 1	MOD 1 1 0 R/M	DISP	40–50 + EA
FLDENV = Load Environment	ESCAPE 0 0 1	MOD 1 0 0 R/M	DISP	35–45 + EA
FSAVE = Save State	ESCAPE 1 0 1	MOD 1 1 0 R/M	DISP	197 – 207 + EA
FRSTOR = Restore State	ESCAPE 1 0 1	MOD 1 0 0 R/M	DISP	197 – 207 + EA
FINCSTP = Increment Stack Pointer	ESCAPE 0 0 1	1 1 1 1 0 1 1 1		6–12
FDECSTP = Decrement Stack Pointer	ESCAPE 0 0 1	1 1 1 1 0 1 1 0		6–12

205835–19

708

Table 5. 8087 Extensions to the 86/186 Instructions Sets (Continued)

			Clock Count Range
FFREE = Free ST(i)	ESCAPE 1 0 1	1 1 0 0 0 ST(i)	9–16
FNOP = No Operation	ESCAPE 0 0 1	1 1 0 1 0 0 0 0	10–16
FWAIT = CPU Wait for 8087	1 0 0 1 1 0 1 1		3 + 5n*
			205835–20

*n = number of times CPU examines TEST line before 8087 lowers BUSY.

NOTES:
1. if mod = 00 then DISP = 0*, disp-low and disp-high are absent
   if mod = 01 then DISP = disp-low sign-extended to 16-bits, disp-high is absent
   if mod = 10 then DISP = disp-high; disp-low
   if mod = 11 then r/m is treated as an ST(i) field
2. if r/m = 000 then EA = (BX) + (SI) + DISP
   if r/m = 001 then EA = (BX) + (DI) + DISP
   if r/m = 010 then EA = (BP) + (SI) + DISP
   if r/m = 011 then EA = (BP) + (DI) + DISP
   if r/m = 100 then EA = (SI) + DISP
   if r/m = 101 then EA = (DI) + DISP
   if r/m = 110 then EA = (BP) + DISP
   if r/m = 111 then EA = (BX) + DISP
   *except if mod = 000 and r/m = 110 then EA = disp-high; disp-low.
3. MF = Memory Format
      00–32-bit Real
      01–32-bit Integer
      10–64-bit Real
      11–16-bit Integer
4. ST(0) = Current stack top
   ST(i) = i^{th} register below stack top
5. d = Destination
      0—Destination is ST(0)
      1—Destination is ST(i)
6. P = Pop
      0—No pop
      1—Pop ST(0)
7. R = Reverse: When d = 1 reverse the sense of R
      0—Destination (op) Source
      1—Source (op) Destination
8. For **FSQRT:** $-0 \leq ST(0) \leq +\infty$
   For **FSCALE:** $-2^{15} \leq ST(1) < +2^{15}$ and ST(1) integer
   For **F2XM1:** $0 \leq ST(0) \leq 2^{-1}$
   For **FYL2X:** $0 < ST(0) < \infty$
                     $-\infty < ST(1) < +\infty$
   For **FYL2XP1:** $0 \leq |ST(0)| < (2 - \sqrt{2})/2$
                     $-\infty < ST(1) < \infty$
   For **FPTAN:** $0 \leq ST(0) \leq \pi/4$
   For **FPATAN:** $0 \leq ST(0) < ST(1) < +\infty$

APPENDIX D

A Review of Number Systems and Binary Arithmetic

When working with microprocessors, you will find it necessary to have a good grasp of the binary and hexadecimal number systems, and the arithmetic associated with them. This appendix is intended as a review (or a brief introduction) to this material.

DECIMAL VERSUS BINARY

A decimal number is composed of one or more digits chosen from a set of ten digits {0, 1, 2, 3, 4, 5, 6, 7, 8, 9}. Each digit in a decimal number has an associated *weight* that is used to give the digit meaning. For example, the decimal number 357 contains three 100s, five 10s, and seven 1s. The weight of the digit 3 is 100, the weight of the digit 5 is 10, and the weight of the digit 7 is 1. The weights of each digit in a decimal number are related to its *base*. Decimal numbers are base-10 numbers. Thus, the weights are all multiples of 10. Look at our example decimal number again:

3	5	7 —digits
10^2	10^1	10^0—weights as powers of 10
100	10	1 —actual weight value
300	50	7 —components of number

Notice that the weights are all powers of 10 beginning with 0. The components of the number are found by multiplying each digit value by its respective weight. The number itself is found by adding the individual components together. This technique applies to numbers in *any* base.

Now, a binary number is a number composed of digits (called **bits**) chosen from a set of only **two** digits {0, 1}. Base-2 is used for binary numbers because there are only two legal digits in a binary number. This means that the weights of the bits in a binary number will all be multiples of two! Consider the binary number 10110. The associated weights are as follows:

1	0	1	1	0 —bits
2^4	2^3	2^2	2^1	2^0—weights as powers of 2
16	8	4	2	1 —actual weight value
16		4	2	—components of number

The components are again found by multiplying each bit in the number by its respective power of 2. The individual components add up to 22. Thus, 10110 binary equals 22 decimal. We now have a technique for determining the value of any binary number.

Going from one base to another requires a *conversion*. As we just saw, going from 10110 to 22 required us to perform a **binary-to-decimal** conversion. How do we go the other way? For example, what binary number represents the decimal number 37? This requires a **decimal-to-binary** conversion. One way to do this conversion is as follows:

$$37 / 2 = 18 \text{ with 1 left over}$$
$$18 / 2 = 9 \text{ with 0 left over}$$
$$9 / 2 = 4 \text{ with 1 left over}$$
$$4 / 2 = 2 \text{ with 0 left over}$$
$$2 / 2 = 1 \text{ with 0 left over}$$
$$1 / 2 = 0 \text{ with 1 left over}$$

The number is repeatedly divided by 2 and the remainder recorded. When we get to *1 / 2 = 0 with 1 left over* we are done dividing. The binary result is found by reading the remainder bits from the bottom up. So, 37 decimal equals 100101 binary. To check, we use the binary-to-decimal conversion technique:

1	0	0	1	0	1
32	16	8	4	2	1
32			4		1

Adding 32, 4, and 1 gives 37, the original number!

Here are some common binary and decimal numbers:

Binary	Decimal
1010	10
1111	15
1100100	100
10000000	128
11111111	255
1111101000	1000

Clearly, it helps to have a good understanding of the various powers of 2 to perform conversions. The first 20 powers of 2 are as follows:

$$2^0 = 1 \qquad 2^5 = 32$$
$$2^1 = 2 \qquad 2^6 = 64$$
$$2^2 = 4 \qquad 2^7 = 128$$
$$2^3 = 8 \qquad 2^8 = 256$$
$$2^4 = 16 \qquad 2^9 = 512$$

$$2^{10} = 1024 \qquad 2^{15} = 32768$$
$$2^{11} = 2048 \qquad 2^{16} = 65536$$
$$2^{12} = 4096 \qquad 2^{17} = 131072$$
$$2^{13} = 8192 \qquad 2^{18} = 262144$$
$$2^{14} = 16384 \qquad 2^{19} = 524288$$

THE HEXADECIMAL NUMBER SYSTEM

It is not easy to remember large binary numbers. For instance, examine the 20-bit binary number shown here for 5 seconds. Then close your eyes and try to repeat it:

10101111011010110011

Were you able to do it? Most people cannot, because their short-term memory is not capable of storing so many bits of information. It is possible, however, to remember shorter sequences of characters. Try this character sequence:

A F 6 B 3

Were you able to remember it? Those who have difficulty with the 20-bit example can usually do the 5-character example easily. Here is where the trick comes in: The 20-bit binary number *is the same* as the 5-character sequence. This is because AF6B3 represents a **hexadecimal** number (base 16). Here is how the examples relate to each other:

* Separate the number into groups of 4 bits each

    1010 1111 0110 1011 0011

* Find the individual decimal equivalents of each group

    1010 1111 0110 1011 0011
     10   15   6   11   3

* Replace each 10...15 value with its A...F equivalent

    1010 1111 0110 1011 0011
     10   15   6   11   3
      A   F   6   B   3

This technique makes it possible to work with large binary numbers through their hexadecimal equivalents.

The word **hexadecimal** refers to "6" and "10," or "16." This number system contains numbers composed of digits and letters chosen from the set {0, 1, 2, 3, 4, 5, 6, 7, 8, 9, A, B, C, D, E, F}. The decimal number 10 is represented in hexadecimal as **A.** Letters B, C, D, E, and F are equivalent to 11 through 15, respectively. Each digit or letter in a hexadecimal number represents 4 binary bits (as we have seen). The binary patterns associated with the 16 hexadecimal symbols are as follows:

0 = 0 0 0 0	4 = 0 1 0 0	8 = 1 0 0 0	C = 1 1 0 0
1 = 0 0 0 1	5 = 0 1 0 1	9 = 1 0 0 1	D = 1 1 0 1
2 = 0 0 1 0	6 = 0 1 1 0	A = 1 0 1 0	E = 1 1 1 0
3 = 0 0 1 1	7 = 0 1 1 1	B = 1 0 1 1	F = 1 1 1 1

It is much more convenient (and easier to remember) to use 2 hexadecimal symbols than 8 binary bits. For instance, 3EH (the H stands for Hex) means 00111110B (B for Binary). Larger binary numbers prove this point even better (as our AF6B3H example showed). Because microprocessor address and data lines commonly utilize 8, 16, or even 32 bits of data, the 2-, 4-, or 8-symbol hexadecimal equivalents are easier to deal with.

BINARY ADDITION

Once a decimal number has been translated into binary, what can we do with it? Usually, a number is loaded into a computer so that it can be manipulated. When two or more numbers are input, we often need to find their sum. So, it is necessary for the microprocessor to know how to add in binary. The rules for binary addition of two bits are short and straightforward:

$$
\begin{array}{ccccc}
 & & & & 1 \\
0 & 0 & 1 & 1 & +\,1 \\
+\,0 & +\,1 & +\,0 & +\,1 & +\,1 \\
\hline
0 & 1 & 1 & 10 & 11
\end{array}
$$

Because the binary number system allows only the two symbols 0 and 1, the sum of 1 + 1 cannot be 2! So, it is 0 with *1 to carry* into the next column of bits.

Likewise, the sum of 1 + 1 + 1 is not 3, but 1 with 1 to carry. These are all the rules we have to remember. Let us use them to add two 8-bit numbers:

```
      1 1 1 1 1    —carry bits
    1 0 0 1 0 1 1 0—first number  = 150 (or 96H)
  + 0 0 1 1 1 0 1 1—second number = 59 (or 3BH)
  ─────────────────
    1 1 0 1 0 0 0 1—result = 209 (or D1H)
```

This is the type of addition performed by microprocessors.

BINARY SUBTRACTION

Oddly enough, we use binary *addition* to perform subtraction in binary. For example, subtracting these two numbers:

$$
\begin{array}{r}
50 \\
-\,18 \\
\hline
\end{array}
$$

is the same as **adding** these two numbers:

$$50$$
$$+ \, -18$$

So, we need a method for converting positive 18 into negative 18. We use a technique called *2's complement* to perform this conversion. To find the 2's complement of 18, we do the following:

```
0 0 0 1 0 0 1 0—positive 18 in 8-bit binary
1 1 1 0 1 1 0 1—complement of each bit
+               1—add 1
   _____
1 1 1 0 1 1 1 0—the 2's complement of 18
```

It is not possible to distinguish a 2's complement binary number from an ordinary binary number. Two's complement refers to the way in which we *interpret* a binary number. The –18 representation 11101110 must be interpreted as a **signed** binary number. A signed binary number uses its MSB as a *sign* bit, where 0 means positive and 1 means negative. So, we have:

0 0 0 1 0 0 1 0   +18, sign bit is 0

1 1 1 0 1 1 1 0   –18, sign bit is 1

The value of the sign bit in these two numbers supports our interpretation.

Getting back to the original problem, we can now subtract 18 from 50 by adding the 2's complement of 18 (which is –18) to 50:

```
   1 1 1 1 1 1      —carry bits
   0 0 1 1 0 0 1 0—  50
+  1 1 1 0 1 1 1 0— –18
   _____
   0 0 1 0 0 0 0 0—  32
```

The carry out of the MSB is ignored. Note that the MSB of the result is 0, indicating a positive result.

This brief discussion should have familiarized you with the types of numbers encountered when dealing with microprocessors. Now, when you encounter instructions like:

```
CODES    DW    14H, 3EA9H, 2200

         MOV   AL,4CH
         AND   BL,10110101B
```

you will have an idea what the numbers mean.

APPENDIX E

BIOS Function Calls

These interrupt references are designed to remind you what registers are used for a particular interrupt function. Refer to the actual coverage of each function for details.

VIDEO SERVICES: SELECTED FUNCTIONS

INT 10H, Function 00H: Set video mode
    Input: AH = 00H
               AL = Video mode (0 to 7)
    Output: None

INT 10H, Function 01H: Set cursor type
    Input: AH = 01H
               CH = Top scan line
                      Bits 0 to 4: Start line of cursor
                      Bits 5 to 6: 00, Normal
                                01, Cursor not displayed
                      Bit 7: Reserved
               CL = Bottom scan line
                      Bits 0 to 4: End line of cursor
                      Bits 5 to 7: Reserved
    Output: None

INT 10H, Function 02H: Set cursor position
    Input: AH = 02H
               BH = Display page number (0 to 7)
               DH = Row (first is 0)
               DL = Column (first is 0)
    Output: None

INT 10H, Function 03H: Read cursor position
        Input: AH = 03H
                    BH = Display page number (0 to 7)
        Output: AX = 00H
                    CH = Starting cursor scan line
                    CL = Ending cursor scan line
                    DH = Row
                    DL = Column

INT 10H, Function 05H: Select new video page
        Input: AH = 05H
                    AL = Page number (0 to 7)
        Output: None

INT 10H, Function 06H: Scroll page up
        Input: AH = 06H
                    AL = Scroll distance in rows (0 for all)
                    BH = Attribute to use on blanked lines
                    CH = Top row
                    CL = Leftmost column
                    DH = Bottom row
                    DL = Rightmost column
        Output: None

INT 10H, Function 07H: Scroll page down
        Input: AH = 07H
                    AL = Scroll distance in rows (0 for all)
                    BH = Attribute to use on blanked lines
                    CH = Top row
                    CL = Leftmost column
                    DH = Bottom row
                    DL = Rightmost column
        Output: None

INT 10H, Function 08H: Read character/attribute from screen
        Input: AH = 08H
                    BH = Display page (0 to 7)
        Output: AH = Attribute
                    AL = Character

INT 10H, Function 09H: Write character/attribute to screen
        Input: AH = 09H
                    AL = Character to write
                    BH = Display page (0 to 7)
                    BL = Attribute
                    CX = Number of characters
        Output: None

INT 10H, Function 0AH: Write character only to screen
    Input: AH = 0AH
            AL = Character to write
            BH = Display page (0 to 7)
            CX = Number of characters
    Output: None

INT 10H, Function 0FH: Read video status
    Input: AH = 0FH
    Output: AH = Number of columns
            AL = Video mode
            BH = Display page

DISK SERVICES: SELECTED FUNCTIONS

INT 13H, Function 00H: Reset disk system
    Input: AH = 00H
            DL = Drive number (0 or 1)
    Output: AH = 0: No error
                    else Error code (see following list)
            CF = 0: No error
                 1: Error
            Error code: 00H, No error
                      01H, Invalid function request
                      02H, Address mark not found
                      03H, Write protect error
                      04H, Sector not found
                      06H, Disk change line active
                      08H, DMA overrun on operation
                      09H, Data boundary error
                      0CH, Media type not found
                      10H, Uncorrectable ECC or CRC error
                      20H, General controller failure
                      40H, Seek operation failed
                      80H, Timeout

INT 13H, Function 01H: Read disk status
    Input: AH = 01H
            DL = Drive number (0 or 1)
    Output: AH = 0: No error
                    else Error code (as in Function 00H)
            CF = 0: No error
                 1: Error

INT 13H, Function 02H: Read disk sectors
    Input: AH = 02H

AL = Number of sectors (1 to 18)
CH = Track number (0 to 79)
CL = Sector number (1 to 18)
DH = Head number (0 or 1)
DL = Drive number (0 or 1)
ES:BX = Pointer to buffer

Output: AH = 0: No error
            else Error code (as in Function 00H)
       AL = Number of sectors read
       CF = 0: No error
            1: Error

INT 13H, Function 03H: Write disk sectors

Input: AH = 03H
       AL = Number of sectors (1 to 18)
       CH = Track number (0 to 79)
       CL = Sector number (1 to 18)
       DH = Head number (0 or 1)
       DL = Drive number (0 or 1)
       ES:BX = Pointer to buffer

Output: AH = 0: No error
            else Error code (as in Function 00H)
       AL = Number of sectors written
       CF = 0: No error
            1: Error

KEYBOARD FUNCTIONS

INT 16H, Function 00H: Read keyboard input

Input: AH = 00H
Output: AH = Scan code or character ID for special character
       AL = ASCII code or other translation of character

INT 16H, Function 01H: Read keyboard status

Input: AH = 01H
Output: AH = Scan code or character ID for special character
       AL = ASCII code or other translation of character
       ZF = 0: Character is ready
            1: No character is available

INT 16H, Function 02H: Return shift flag status

Input: AH = 02H
Output: AL = Current shift status
            Bit 0: Right Shift
            Bit 1: Left Shift
            Bit 2: Ctrl

Bit 3: Alt
Bit 4: Scroll Lock
Bit 5: Num Lock
Bit 6: Caps Lock
Bit 7: Insert

INT 16H, Function 03H: Set typematic rate and delay
    Input: AH = 03H
              AL = 05H
              BH = Delay in milliseconds
                    00H: 250, 01H: 500, 02H: 750, 03H: 1000
              BL = Typematic rate in characters per second
                    00H: 30.0, 08H: 15.0,10H: 7.5, 18H: 3.7
                    01H: 26.7, 09H: 13.3,11H: 6.7, 19H: 3.3
                    02H: 24.0, 0AH: 12.0,12H: 6.0,1AH: 3.0
                    03H: 21.8, 0BH: 10.9,13H: 5.5,1BH: 2.7
                    04H: 20.0, 0CH: 10.0,14H: 5.0,1CH: 2.5
                    05H: 18.5, 0DH: 9.2,15H: 4.6,1DH: 2.3
                    06H: 17.1, 0EH: 8.6,16H: 4.3,1EH: 2.1
                    07H: 16.0, 0FH: 8.0,17H: 4.0,1FH: 2.0

INT 16H, Function 05H: Store key data
    Input: AH = 05H
              CL = ASCII character
              CH = Scan code
    Output: AL = 0: No error
                   1: Keyboard buffer full
              CF = 0: No error
                   1: Buffer full

PARALLEL PRINTER FUNCTIONS

INT 17H, Function 00H: Print character
    Input: AH = 00H
              AL = Character to print
              DX = Printer number (0, 1, or 2)
    Output: AH = Printer status
                   Bit 0: Time out
                   Bit 1: Reserved
                   Bit 2: Reserved
                   Bit 3: I/O error
                   Bit 4: Printer selected
                   Bit 5: Out of paper
                   Bit 6: Acknowledgment from printer
                   Bit 7: Printer not busy

INT 17H, Function 01H: Initialize printer
    Input: AH = 01H
                 DX = Printer number (0, 1, or 2)
    Output: AH = Printer status
                 (same bit assignments as Function 00H)

INT 17H, Function 02H: Read printer status
    Input: AH = 02H
                 DX = Printer number (0, 1, or 2)
    Output: AH = Printer status
                 (same bit assignments as Function 00H)

APPENDIX F

DOS Function Calls

These interrupt references are designed to remind you what registers are used for a particular interrupt function. Refer to the actual coverage of each function for details.

SELECTED DOS FUNCTIONS

INT 21H, Function 01H: Keyboard input
    Input: AH = 01H
    Output: AL = Keyboard character

INT 21H, Function 02H: Display output
    Input: AH = 02H
               DL = Character to output
    Output: None

INT 21H, Function 05H: Printer output
    Input: AH = 05H
               DL = Character to output
    Output: None

INT 21H, Function 08H: Console input without echo
    Input: AH = 08H
    Output: AL = Character from console

INT 21H, Function 09H: Display string
    Input: AH = 09H
               DS:DX = Pointer to character string
    Output: None

INT 21H, Function 0AH: Buffered keyboard input
    Input: AH = 0AH
                 DS:DX = Pointer to input buffer
    Output: None

INT 21H, Function 0BH: Check standard input status
    Input: AH = 0BH
    Output: AL = 0FFH: Character available
                       00H: No character available

INT 21H, Function 19H: Get current drive
    Input: AH = 19H
    Output: AL = Drive ID

INT 21H, Function 1AH: Set disk transfer area
    Input: AH = 1AH
                 DS:DX = Pointer to DTA
    Output: None

INT 21H, Function 1BH: Get current drive information
    Input: AH = 1BH
    Output: AL = Sectors per allocation unit
                 CX = Bytes per sector
                 DX = Number of allocation units
                 DS:BX = Pointer to FAT ID byte

INT 21H, Function 1CH: Get drive information
    Input: AH = 1CH
                 DL = Drive ID
    Output: AL = Sectors per allocation unit
                 CX = Bytes per sector
                 DX = Number of allocation units
                 DS:BX = Pointer to FAT ID byte

INT 21H, Function 25H: Set interrupt vector
    Input: AH = 25H
                 AL = Interrupt number
                 DX = IP of interrupt routine
                 DS = CS of interrupt routine
    Output: None

INT 21H, Function 2AH: Get date
    Input: AH = 2AH
    Output: AL = Day of the week
                 CX = Year
                 DH = Month
                 DL = Day

INT 21H, Function 2BH: Set date
    Input: AH = 2BH
                CX = Year
                DH = Month
                DL = Day
    Output: AL = 00H: Date is valid
                     0FFH: Date is invalid

INT 21H, Function 2CH: Get time
    Input: AH = 2CH
    Output: CH = Hour
                CL = Minutes
                DH = Seconds
                DL = 100ths of seconds

INT 21H, Function 2DH: Set time
    Input: AH = 2DH
                CH = Hour
                CL = Minutes
                DH = Seconds
                DL = 100ths of seconds
    Output: AL = 00H: Time is valid
                     0FFH: Time is invalid

INT 21H, Function 30H: Get DOS version number
    Input: AH = 30H
    Output: AL = Major version number
                AH = Minor version number
                BX = 0000H
                CX = 0000H

INT 21H, Function 31H: Terminate and remain resident
    Input: AH = 31H
                AL = Return code
                DX = Memory size in paragraphs
    Output: None

INT 21H, Function 35H: Get interrupt vector
    Input: AH = 35H
                AL = Interrupt number
    Output: BX = IP of interrupt routine
                ES = CS of interrupt routine

INT 21H, Function 39H: Create subdirectory
    Input: AH = 39H
                DS:DX = Pointer to ASCIIZ string
    Output: CF = 0: No error
                     1: Error, AX contains error code

INT 21H, Function 3AH: Delete subdirectory
        Input: AH = 3AH
                  DS:DX = Pointer to ASCIIZ string
        Output: CF = 0: No error
                        1: Error, AX contains error code

INT 21H, Function 3BH: Change current directory
        Input: AH = 3BH
                  DS:DX = Pointer to ASCIIZ string
        Output: CF = 0: No error
                        1: Error, AX contains error code

INT 21H, Function 3CH: Create file
        Input: AH = 3CH
                  DS:DX = Pointer to ASCIIZ string
                  CX = File attributes
        Output: CF = 0: No error, AX contains file handle
                        1: Error, AX contains error code

INT 21H, Function 3DH: Open file with handle
        Input: AH = 3DH
                  DS:DX = Pointer to ASCIIZ string
                  AL = Access code
        Output: CF = 0: No error, AX contains file handle
                        1: Error, AX contains error code

INT 21H, Function 3EH: Close file with handle
        Input: AH = 3EH
                  BX = File handle
        Output: CF = 0: No error
                        1: Error, AX contains error code

INT 21H, Function 3FH: Read from file
        Input: AH = 3FH
                  BX = File handle
                  DS:DX = Pointer to buffer
                  CX = Number of bytes to read
        Output: CF = 0: No error, AX equals number of bytes read
                        1: Error, AX contains error code

INT 21H, Function 40H: Write to file
        Input: AH = 40H
                  BX = File handle
                  DS:DX = Pointer to buffer
                  CX = Number of bytes to write
        Output: CF = 0: No error, AX equals number of bytes written
                        1: Error, AX contains error code

INT 21H, Function 41H: Delete file
Input: AH = 41H
            DS:DX = Pointer to ASCIIZ string
Output: CF = 0: No error
                 1: Error, AX contains error code

INT 21H, Function 42H: Position file pointer
Input: AH = 42H
            AL = Method code
            BX = File handle
            CX = Upper 16 bits of offset
            DX = Lower 16 bits of offset
Output: CF = 0: No error, DX = Upper 16 bits of new pointer
                        AX = Lower 16 bits of new pointer
                 1: Error, AX contains error code

INT 21H, Function 43H: Get or set file attributes
Input: AH = 43H
            AL = 0: Get
                 1: Set
            CX = Attribute
            DS:DX = Pointer to ASCIIZ string
Output: CF = 0: No error, CX contains old attribute
                 1: Error, AX contains error code

INT 21H, Function 47H: Get current directory
Input: AH = 47H
            DS:SI = Pointer to buffer
            DL = Drive number
Output: CF = 0: No error, DS:SI contains path name
                 1: Error, AX contains error code

INT 21H, Function 48H: Allocate memory
Input: AH = 48H
            BX = Number of paragraphs requested
Output: CF = 0: No error, AX contains allocation segment
                 1: Error, AX contains error code, BX contains
                        largest block of paragraphs available

INT 21H, Function 49H: Free allocated memory
Input: AH = 49H
            ES = Segment of block to free
Output: CF = 0: No error
                 1: Error, AX contains error code

INT 21H, Function 4AH: Modify allocated memory
Input: AH = 4AH

ES = Segment of block

BX = New size of block in paragraphs

Output: CF = 0: No error

CF = 1: Error, AX contains error code, BX contains
largest paragraph size possible

INT 21H, Function 4CH: Terminate program

Input: AH = 4CH

AL = Return code

Output: None

INT 21H, Function 4EH: Find file

Input: AH = 4EH

CX = Attribute to use during search

DS:DX = Pointer to ASCIIZ string

Output: AX = Return code (same definitions as error code)

INT 21H, Function 56H: Rename file

Input: AH = 56H

DS:DX = Pointer to ASCIIZ string (old name)

ES:DI = Pointer to ASCIIZ string (new name)

Output: CF = 0: No error

1: Error, AX contains error code

INT 21H, Function 57H: Get or set file date and time

Input: AH = 57H

AL = 0: Get

1: Set

BX = File handle (if setting)

CX = Time

DX = Date

Output: AX = Return code (same definitions as error code)

CX = Time

DX = Date

APPENDIX G

Assembler Reference

Microsoft MASM Version 6.11 contains updated software capable of processing Pentium instructions. Machine codes and instruction cycle counts are generated by MASM for all instructions on each processor beginning with the 8086.

The combined assembler and linker is the program ML. To see the format of ML's command line, and the possible option switches, use the DOS command:

```
ML /?
```

or

```
ML /help
```

You will get an output that looks like this:

```
ML [ /options ] filelist [ /link linkoptions ]
```

/AT Enable tiny model (.COM file)	/nologo Suppress copyright message
/Bl<linker> Use alternate linker	/Sa Maximize source listing
/c Assemble without linking	**/Sc Generate timings in listing**
/Cp Preserve case of user identifiers	/Sf Generate first pass listing
/Cu Map all identifiers to uppercase	/Sl<width> Set line width
/Cx Preserve case in publics, externs	/Sn Suppress symbol-table listing
/coff generate COFF format object file	/Sp<length> Set page length
/D<name>[=text] Define text macro	/Ss<string> Set subtitle
/EP Output preprocessed listing to stdout	/St<string> Set title
/F <hex> Set stack size (bytes)	/Sx List false conditionals
/Fe<file> Name executable	/Ta<file> Assemble non-.ASM file
/Fl[file] Generate listing	/w Same as /W0 /WX
/Fm[file] Generate map	/WX Treat warnings as errors
/Fo<file> Name object file	/W<number> Set warning level
/FPi Generate 80x87 emulator encoding	/X Ignore INCLUDE environment path
/Fr[file] Generate limited browser info	/Zd Add line number debug info
/FR[file] Generate full browser info	/Zf Make all symbols public
/G<c|d|z> Use Pascal, C, or Stdcall calls	/Zi Add symbolic debug info
/H<number> Set max external name length	/Zm Enable MASM 5.10 compatibility
/I<name> Add include path	/Zp[n] Set structure alignment
/link <linker options and libraries>	/Zs Perform syntax check only

The more commonly used options are highlighted in bold. For example, when a new program is being written and debugged, it is helpful to generate a list file to aid in debugging. The list file will identify any errors that show up.

To assemble the file PROG.ASM and create a list file, use this command line:

```
ML /Fl PROG.ASM
```

The ML program will assemble and link (if no errors are found) the PROG.ASM file, creating PROG.LST, PROG.OBJ, and PROG.EXE. It is important to use the correct case in the option switches, as ML command options are case sensitive.

To generate instruction clock counts in the list file, use this command line:

```
ML /Fl /Sc PROG.ASM
```

Clock counts are useful when determining the execution time of a program. It may be necessary to assemble a source file, but not link it. This situation may occur when a multi-source program is being written by a group of individuals, and all the modules are not available yet. Or, the program may be in the debug stage and not require an executable file yet. To assemble without linking, use this command line:

```
ML /c /Fl PROG.ASM
```

which generates list and object files but no .EXE file. To create an .EXE file, you must use the LINK utility provided with MASM. LINK has its own set of command line parameters (use LINK /? to see them) and is normally called automatically by ML. When the /c option is used, ML does not call LINK. To create PROG.EXE from PROG.OBJ, use this LINK command:

```
LINK PROG,,;
```

which converts the contents of PROG.OBJ into PROG.EXE. To link more than one object file, use + signs between their names, as in:

```
LINK PA + PB + PC,,;
```

The linker will name the .EXE file after the first file in the object name list.

The following is a list of MASM's reserved words used throughout the text:

ASSUME	assume definition
BYTE	byte (as in BYTE PTR)
.CODE	begin code segment
.DATA	begin data segment
DB	define byte
DD	define double-word
DQ	define quadword
DS	define storage
DUP	duplicate
DW	define word
ELSE	else statement
END	end program

ENDIF	end if statement
ENDM	end macro
ENDP	end procedure
ENDS	end segment
EQU	equate
.EXIT	generate exit code
EXTRN	external reference
FAR	far reference
IF	if statement
MACRO	define macro
.MODEL	model type
NEAR	near reference
OFFSET	offset
ORG	origin
PARA	paragraph
PROC	define procedure
PTR	pointer
PUBLIC	public reference
SEG	locate sement
SEGMENT	define segment
.STARTUP	generate startup code
WORD	word (as in WORD PTR)

It might be useful to design a program *template* to assist you when writing new programs. A program template consists of the minimal instructions most programs need to function, such as those that set up the segment registers and others that return to DOS. The program template shown here is a good beginning template:

```
;Program -------.ASM: <Brief description goes here>
;
        .MODEL SMALL
        .586
        .DATA
;
;Put your application data here
;
        .CODE
        .STARTUP
;
;Put your main application code here
;
        .EXIT
;
;Put your application procedures here
;
        END
```

A different assembler, called TASM, may also be used. TASM (by Borland International) uses different command line options than ML does. To see a display of TASM's options, just enter TASM with no parameters. To generate list and object files use:

```
TASM /L PROG
```

TASM does not require the .ASM extension.

To create PROG.EXE you must use Borland's TLINK utility. This linker operates the same way LINK does.

APPENDIX H

DEBUG and Codeview Reference

The DEBUG and Codeview utilities are discussed in this appendix. DEBUG comes with DOS and allows the user instruction-level control over an .EXE or .COM program. Codeview does the same, with a fancier environment (multiple color windows) and with the addition of being able to handle newer 80386, 80486, and Pentium instructions. Codeview is supplied with Microsoft MASM.

Let us examine DEBUG first.

USING DEBUG TO EXECUTE AN 80X86 PROGRAM

What Is DEBUG?

DEBUG is a utility program that allows a user to load an 80X86 program into memory and execute it step by step. DEBUG displays the contents of all processor registers after each instruction executes, allowing the user to determine if the code is performing the desired task. DEBUG only displays the 16-bit portion of the general purpose registers. Codeview is capable of displaying the entire 32 bits. DEBUG is a very useful debugging tool. We will use DEBUG to step through a number of simple programs, gaining familiarity with DEBUG's commands as we do so. DEBUG contains commands that can display and modify memory, assemble instructions, disassemble code already placed into memory, trace through single or multiple instructions, load registers with data, and do much more.

DEBUG loads into memory like any other program, in the first available slot. The memory space used by DEBUG for the user program begins *after* the end of DEBUG's code. If an .EXE or .COM file were specified, DEBUG would load the program according to accepted DOS conventions.

Getting Started

The best way to get familiar with DEBUG is to work through some examples with it. The first example we will use is this three-instruction sequence:

```
MOV   AL,7
MOV   BH,2
ADD   AL,BH
```

The first instruction places the number 7 into register AL. The second instruction places 2 into register BH. These two registers are added together in the third instruction, with the results going into AL. With DEBUG we will be able to type in the instructions as they appear. DEBUG will automatically assemble them and place their respective codes into memory. We will then be able to examine the results of each instruction by tracing through the instructions one at a time.

The first step is to invoke DEBUG. This is done with a simple command, printed here in boldface. Make sure your floppy or hard disk has a copy of DEBUG.COM installed on it, and that it is in your current directory. At the DOS command prompt, enter:

C> debug <cr>

The expression <cr> indicates that you should hit the return key. Since DEBUG is a .COM file, DOS will load it into memory and execute it. DEBUG uses a minus sign as its command prompt, so you should see a "–" appear on your display.

To get a list of all commands available with DEBUG, enter a question mark at DE-BUG's command prompt and press <cr>. The command summary appears like this:

```
-?<cr>
assemble            A [address]
compare             C range address
dump                D [range]
enter               E address [list]
fill                F range list
go                  G [=address] [addresses]
hex                 H value1 value2
input               I port
load                L [address] [drive] [firstsector] [number]
move                M range address
name                N [pathname] [arglist]
output              O port byte
proceed             P [=address] [number]
quit                Q
register            R [register]
search              S range list
trace               T [=address] [value]
unassemble          U [range]
write               W [address] [drive] [firstsector] [number]
allocate expanded memory        XA [#pages]
deallocate expanded memory      XD [handle]
map expanded memory pages       XM [Lpage] [Ppage] [handle]
display expanded memory status  XS
-
```

We will begin with a small group of these commands to get the feel for how DEBUG is used.

To enter the three instructions we wish to execute, we need to use the *assemble* command. Entering the command letter "a" at the prompt should result in a display similar to this:

```
-a<cr>
1539:0100
```

This is the familiar CS:IP format used throughout the text. Instead of using the actual effective address, DEBUG shows us the CS value and the IP value. The 1539 address will most likely be different on your machine, because it is probably not configured like the one used to generate the DEBUG examples. The second address, 0100, should be the same. This is DEBUG telling us that it will place the first instruction into the code segment at offset 0100H. Bear in mind that DEBUG interprets *all* numbers as hexadecimal numbers.

When DEBUG begins execution, it sets the value of each processor register to a default value. The segment registers (CS, DS, ES, and SS) are all set to the beginning of free memory. This accounts for the 1539 address in the CS register at the beginning of the **a** command. The instruction pointer is always set to address 0100 and all other processor registers are set to 0000 with the exception of the stack pointer (SP), which is set to address FFEE. In addition, all processor *flags* are cleared. It is useful to know how the processor registers are initialized. This will become clear as we proceed through the example.

At this point you can enter the three instructions. If you make a typing mistake, use the backspace key to correct your errors before hitting return. You should see something similar to this on your display:

```
-a<cr>
1539:0100 mov al,7<cr>
1539:0102 mov bh,2<cr>
1539:0104 add al,bh<cr>
1539:0106 <cr>
-
```

Notice that the IP address changes after each instruction. The fourth instruction, if there were one, would begin at address 0106. Because we are entering only three instructions, hitting a return on the fourth line will terminate the assemble command and get us back to the command prompt.

To examine the code that was generated by each instruction we use the *unassemble* command. Unassemble will show us the addresses, opcodes, and instruction mnemonics for the three instructions we have just entered. If unassemble is used without parameters it will show the next 20H bytes and their corresponding 80X86 instruction equivalents. We can shorten this display by using a range parameter, like this:

```
-u 100 104<cr>
```

This command tells DEBUG to unassemble the bytes between addresses 0100 and 0104. The resulting display looks like this:

```
1539:0100 B007    MOV   AL,07
1539:0102 B702    MOV   BH,02
1539:0104 00F8    ADD   AL,BH
```

Here, we can see that each instruction entered required 2 bytes of machine code.

To begin execution we should examine the contents of each register. Then we will know which registers change as we step through the program. DEBUG's *register* command can be used to display (and modify) any of the processor's registers. To display their contents, simply enter "r" and return. You should get a display similar to this:

```
AX=0000  BX=0000  CX=0000  DX=0000  SP=FFEE  BP=0000  SI=0000  DI=0000
DS=1539  ES=1539  SS=1539  CS=1539  IP=0100    NV UP EI PL NZ NA PO NC
1539:0100 B007          MOV  AL,07
```

Spend a few moments looking at the value displayed for each register. Note that all general purpose registers have been loaded with 0000. Also, the CS, DS, SS, and ES registers have all been initialized to the 1539 address we have seen earlier. IP points to address 0100, where we placed the first instruction. The end of the second line indicates the state of the flags. Table H.1 explains the meaning of each flag code. We can see that currently there is no carry, odd parity, no auxiliary carry, not zero, and plus indicated, along with enabled interrupts, up direction (for use with string operations), and no overflow. It is sometimes important to watch the changes in flags as we step through a program.

The last line of the display shows the instruction that will be executed next. Because we have not executed any instructions yet, this is our first instruction! To execute a single instruction we use DEBUG's *trace* command, abbreviated "t," and get the following display:

```
-t <cr>
AX=0007  BX=0000  CX=0000  DX=0000  SP=FFEE  BP=0000  SI=0000  DI=0000
DS=1539  ES=1539  SS=1539  CS=1539  IP=0102    NV UP EI PL NZ NA PO NC
1539:0102 B702          MOV    BH,02
```

By comparing this display with the previous one, we can determine that only the lower byte of AX has been changed (it now contains 07H). No other registers except IP have been affected. No condition codes have been changed. Isn't that what MOV   AL,7 should do?

Tracing through the next two instructions should look something like this:

```
-t <cr>
AX=0007  BX=0200  CX=0000  DX=0000  SP=FFEE  BP=0000  SI=0000  DI=0000
DS=1539  ES=1539  SS=1539  CS=1539  IP=0104    NV UP EI PL NZ NA PO NC
1539:0104 00F8          ADD    AL,BH
```

TABLE H.1   Flag codes

Flag	Set	Clear
Overflow	OV	NV
Direction	DN	UP
Interrupt	EI	DI
Sign	NG	PL
Zero	ZR	NZ
Aux. Carry	AC	NA
Parity	PE	PO
Carry	CY	NC

```
-t <cr>
AX=0009  BX=0200  CX=0000  DX=0000  SP=FFEE  BP=0000  SI=0000  DI=0000
DS=1539  ES=1539  SS=1539  CS=1539  IP=0106  NV UP EI PL NZ NA PE NC
1539:0106 8B0EDF47      MOV    CX,[47DF]
```

As expected, the final value in AL is 09H (the sum of 7 and 2). The flag display indicates that the result of the ADD instruction changed the parity flag. All other flags kept their states.

As a point of interest, look at the instruction located at address 0106. Where did it come from? We did not enter this code during any point of our exercise. Nonetheless, DEBUG thinks this is the next instruction to be executed. The reason for this is that each of the PC's 640KB comes up in a random pattern when power is first applied. DEBUG will try to interpret these random patterns as valid 80x86 instructions, as it is doing with the MOV   CX instruction.

To return to DOS, exit DEBUG by entering "q" (for *quit)* at the command prompt.

You, or your instructor, may find it necessary to save all of the work you perform while using DEBUG. This is easily accomplished by pressing the **Control-PrintScreen** button on the keyboard. This will cause all characters entered from the keyboard, and all characters output to the screen, to be *echoed* to the printer, which must be turned on and loaded with paper. This will produce a hard copy of your work and will allow you to scan the results of each instruction by watching how the registers change.

To stop echoing text to your printer, press the **Control-PrintScreen** button again. You may wish to use this feature of DOS as you work through the next few examples.

Example H.1:   What is the final value in AL after this sequence of instructions executes?

```
MOV    AL,27H
MOV    BL,37H
ADD    AL,BL
DAA
```

Use DEBUG to enter these four instructions and trace their execution. What are the machine codes for each instruction? What is the final value in AL? What changes must be made when entering the instructions with the assemble command?

Solution: The purpose of this example was to use the properties of the DAA instruction. When we add 27 to 37 we expect to get 64, the correct *decimal* answer. Using DEBUG to trace through the ADD instruction shows that AL contains 5EH. The next trace command executes the DAA instruction, which corrects the value in AL to 64H, the correct packed-decimal result. You should also see that the auxiliary carry flag has been set as a result of the DAA instruction.

The machine codes for each instruction are as follows:

```
B027    MOV    AL,27
B337    MOV    BL,37
00D8    ADD    AL,BL
27      DAA
```

The machine codes can be obtained in two ways. During a trace, the machine codes and mnemonics for each instruction are displayed after the register list. The unassemble command can also be used. Try **u 100 106** for this example and see what you get.

Also, notice that the "H" was left off the 27 and 37 operands, because DEBUG expects all numbers to be in hexadecimal form. If you include the "H" by accident, DEBUG will display an error message and wait for you to reenter the instruction.

Example H.2:   The following sequence of DEBUG instructions converts an integer stored in AX from a miles-per-hour (mph) value into a feet-per-second (fps) value. For example, 60 mph equals 88 fps. The conversion is accomplished by first multiplying AX by 5280 (14A0H) and then dividing the result by 3600 (0E10H).

```
MOV   BX,14A0
MUL   BX
MOV   BX,E10
DIV   BX
```

Add the necessary instructions to convert 60 mph into 88 fps. Show that the results are correct at each step in the conversion.

Solution: Because DEBUG works with hexadecimal numbers only, we must first convert 60 into hex. This gives us 3CH. Register AX must be loaded with this value prior to execution. The resulting code is as follows:

```
MOV    AX,3C
MOV    BX,14A0
MUL    BX
MOV    BX,E10
DIV    BX
```

When the instructions are entered with the assemble command and executed with trace, the contents of registers AX, BX, and DX are as follows:

```
MOV    AX,3C     ->    AX=003C   BX=0000   DX=0000
MOV    BX,14A0   ->    AX=003C   BX=14A0   DX=0000
MUL    BX        ->    AX=D580   BX=14A0   DX=0004
MOV    BX,E10    ->    AX=D580   BX=0E10   DX=0004
DIV    BX        ->    AX=0058   BX=0E10   DX=0000
```

Notice that after the MUL   BX instruction, AX contains D580H and DX contains 0004H. Remember that when a 16-bit register is used in MUL, the result of the multiplication is 32 bits wide and is saved in registers DX and AX (with DX holding the upper 16 bits). Thus, the product is 4D580H, which is 316,800 decimal. Check for yourself that 316,800 equals 60 times 5,280.

After the DIV   BX, the final result in AX is 0058H. When converted into decimal, we get 88.

Example H.3:   Give a sequence of instructions that will add up all integers from 1 to 10. The sum should be found in register BL.

Solution: One way to find the required sum is to add each integer from 1 to 10 to BL one at a time, like this:

```
SUB    BL,BL     ;set result to zero
ADD    BL,1      ;add 1 to BL
ADD    BL,2      ;add 2 to BL
```

```
.
.
.
ADD    BL,9      ;add 9 to BL
ADD    BL,0A     ;add 10 to BL
```

Perhaps a better solution can be found by using a *loop*. A loop is a technique used to repeat a group of instructions any number of times. These instructions are called *loop instructions* in general, and in this example the loop instructions need to be executed ten times. Examine the following DEBUG session:

```
-a<cr>
19DC:0100 SUB BL,BL<cr>
19DC:0102 MOV AL,0A<cr>
19DC:0104 ADD BL,AL<cr>
19DC:0106 DEC AL<cr>
19DC:0108 JNZ 104<cr>
19DC:010A <cr>
-
```

Register AL is used as the *loop counter*. The loop counter is decremented once each time the loop is executed. The JNZ instruction causes the processor to jump back to address 104 until AL gets to 0. A long sequence of trace commands will show how the sum builds in BL as AL keeps getting smaller. To keep from having to enter the trace commands so many times, you might try:

```
-t 1e<cr>
```

This will cause DEBUG to trace 1EH (or 30 decimal) instructions. The last two traces should be similar to these:

```
AX=0001  BX=0037  CX=0000  DX=0000  SP=FFEE  BP=0000  SI=0000  DI=0000
DS=19DC  ES=19DC  SS=19DC  CS=19DC  IP=0106    NV UP EI PL NZ NA PO NC
19DC:0106 FEC8          DEC    AL
-t

AX=0000  BX=0037  CX=0000  DX=0000  SP=FFEE  BP=0000  SI=0000  DI=0000
DS=19DC  ES=19DC  SS=19DC  CS=19DC  IP=0108    NV UP EI PL ZR NA PE NC
19DC:0108 75FA          JNZ    0104
-
```

Notice how the zero flag changes (from **NZ** to **ZR**) when AX goes from 0001 to 0000. Because the JNZ command watches the state of the zero flag, execution will now continue at address 10A. The final result in register BL is 37H, which is 55 decimal, the correct sum of all the integers between 1 and 10.

Calling DOS Functions from DEBUG

Now that you have a little experience using DEBUG to execute simple sequences of instructions, we can move on to more complicated applications. We will make use of three new DEBUG commands: *enter, dump,* and *proceed*. We will also use a DOS function call through INT 21H. This is a very versatile DOS function, capable of performing many different operations. For our first example, we will use the display-string function of INT 21H. This function is selected by placing 9 into register AH, the register used by INT 21H to determine which of its many functions have been selected. Display-string requires

that the address of the first byte of the text string be placed into DS:DX before using INT
21H. This means that the string must be contained in the data segment pointed to by DS,
and have an offset equal to the value in DX. Use the assemble command of DEBUG to
enter these instructions:

```
MOV   AH,9
MOV   DX,200
INT   21
```

Notice that we do not initialize the DS register. Remember that DEBUG automatically sets
DS, CS, ES, and SS to the same value. This guarantees that our text string will be placed
into the current data segment area. The offset value of 200H used to initialize DX is not a
special value, it just happened to be a round number. Because the machine codes for the
instructions are being placed into memory around address 100H, 200H seemed a good
place to put the text string.

We can enter the text string two ways. First, we will make use of a new DEBUG
command called *enter*. Enter allows memory to be modified on a byte-by-byte basis, beginning at the address specified in the instruction. To load the text string "Hello!$" into
memory at address 200H enter **e 200** and each individual ASCII byte for every character in
the text string. You will get a display similar to this:

```
-e200 <cr>
1539:0200   66.48    6F.65    75.6C    6E.6C    64.6F    0D.21    0A.24    00.<cr>
```

The first pair of numbers are the actual address where the string is being loaded. The next
number (66H) is the byte stored at location 200H. DEBUG follows it with a period and waits
for you to enter the new byte value. The new value of 48H is entered, followed by the spacebar. Hitting the spacebar without entering any new value will skip over the location without
changing its value. The display shows that 7 new bytes were entered. The last byte displayed
(00) is not followed by anything because Return was hit to terminate the enter command. The
7 bytes entered are the ASCII values for the characters in the "Hello!$" string. The "$" character must be at the end of a text string for display-string to know where the string ends.

We can check our work with another new command: *dump*. Dump displays the bytes
stored in a range of memory loacations. To verify that the text string has been properly
stored, do the following:

```
-d200 20f <cr>
1539:0200 48 65 6C 6C 6F 21 24 00-F6 38 53 79 6E 74 61 78 Hello!$..8Syntax
```

The dump command displays the starting address, followed by 16 bytes of data read from
memory. The final part of each line of a dump display are the ASCII-equivalent characters for
each of the 16 bytes. The dump display clearly shows that we entered the string correctly.

A second technique that can be used to enter strings uses the assemble command. To
place a different string at address 400H, do this:

```
-a400<cr>
1539:0400 db 'Try this string too...$' <cr>
1539:0417 <cr>
```

The "db" directive stands for *define-byte,* and causes DEBUG to look up the ASCII values of any characters surrounded by single quotes. You could easily enter numeric values
with db as well. If you count all of the characters in the new string, including the blanks,
you should get 23. This indicates that locations 400H through 416H will be loaded with

the corresponding ASCII byte values. The next possible location to put anything in is 417H, which DEBUG is already indicating.

Getting back to the example at hand, we have placed a text string into memory at address 200H, and entered the instructions necessary to INT 21H's display-string function. Use the trace command until it gets to the INT 21 instruction. You should see that AH contains 09 and DX contains 0200. Unfortunately, there may be hundreds of instructions involved in the execution of INT 21H. It would be a waste of time to trace through every one of them. It would be nice if INT 21 could be treated as a single instruction by DEBUG, with all instructions of INT 21, including the final RETurn, executing by entering a single DEBUG command. Fortunately for us, DEBUG does have such a command: *proceed!* Proceed causes all INT, CALL, and REP instructions to be treated as single instructions. So, if a subroutine contains forty-five instructions, using the proceed command when the CALL instruction shows up in the trace will cause DEBUG to execute all forty-five instructions, and show the contents of each register *upon return from the subroutine!* We can use proceed to see what happens when we call INT 21. Your last DEBUG trace should look something like this:

```
AX=0900  BX=0000  CX=0000  DX=0200  SP=FFEE  BP=0000  SI=0000  DI=0000
DS=1539  ES=1539  SS=1539  CS=1539  IP=0105    NV UP EI PL NZ NA PO NC
1539:0105 CD21           INT    21
```

If the proceed command is now used, the resulting display becomes:

```
-p<cr>
Hello!
AX=0924  BX=0000  CX=0000  DX=0200  SP=FFEE  BP=0000  SI=0000  DI=0000
DS=1539  ES=1539  SS=1539  CS=1539  IP=0107    NV UP EI PL NZ NA PO NC
1539:0107 6C            DB      6C
```

You can see that the "Hello" string appeared on the screen in the current cursor location, and that DEBUG will get its next instruction from 1539:0107. Obviously, INT 21H must have done its job, or the string would not have appeared. The proceed command is very useful for tracing programs that involve DOS function calls.

Example H.4: What must be done to display the second string, which was entered with the assemble command and the DB directive?

Solution: Because the string was placed at address 400H, the MOV DX,200 instruction must be changed to MOV DX,400. Then the entire sequence of instructions is executed again. DEBUG updates the IP register after each instruction executes, so it will be necessary to load IP with 100 again (if the first instruction is at 100). The register command can be used to do this. The following steps will set IP back to 100:

```
-r ip<cr>
IP 0107
:100<cr>
-
```

The instructions for the second string can now be traced as those for the first were, with the proceed command used when INT 21 shows up again.

We will finish this section with one final example using two more DOS function calls. INT   21H can also read the computer's time and date if the appropriate value is placed into AH. To read the time, AH must contain 2CH. To read the date, AH must contain 2AH. The time and date are returned in various registers, as Example H.5 shows.

Example H.5:   INT   21H requires no register setup before it is called when we are reading only the time or date. The time is returned in the following way: CH contains hours, CL contains minutes, and DH seconds. Hundredths of seconds are returned in DL. The date is returned with AL containing the day of the week, CX the year (1980 to 2099 only), DH the month, and DL the day. Can you determine the time and date from these DEBUG trace displays?

```
Time trace:
AX=2C00  BX=0000  CX=0F1C  DX=0235  SP=FFEE  BP=0000  SI=0000  DI=0000
DS=1539  ES=1539  SS=1539  CS=1539  IP=0104    NV UP EI PL NZ NA PO NC
Date trace:
AX=2A02  BX=0000  CX=07CD  DX=0417  SP=FFEE  BP=0000  SI=0000  DI=0000
DS=1539  ES=1539  SS=1539  CS=1539  IP=0108    NV UP EI PL NZ NA PO NC
```

Solution: The values contained in CX and DX in the time trace indicate that the computer's time was 15:28:02 when INT   21H was called. The values in CX and DX in the date trace show the date to be 4/23/97. The day of the week stored in AL is 2, corresponding to Tuesday. Sunday is indicated by 0.

RUNNING .EXE AND .COM PROGRAMS WITH DEBUG

In this section we will examine one more command: *go*. DEBUG can automatically load an .EXE or .COM file into memory, performing all necessary relocation and initialization. The great advantage here is that we do not have to enter the program by hand. We can also use DEBUG to examine existing programs, executing portions of them to determine how they work. Microcode exploration with DEBUG can be a tremendous learning experience.

To load a program with DEBUG, include the name of the program in the command line. For example:

```
C> debug hello.exe<cr>
```

will cause DEBUG to load the HELLO.EXE code into memory. Once done, we can dump, unassemble, trace, or modify the code as we see fit. If we need only to execute the program, we issue the go command, and see the following display:

```
-g<cr>
Hello!
Program terminated normally
-
```

A nice advantage of using DEBUG to execute a newly developed program in this manner is that the program exits to DEBUG, not DOS, when completed. If we desire, we can now examine the stack area to see what values were pushed onto the stack. Or, if the program performed calculations and saved the results in the data segment, we can use dump to examine the results and verify their correctness. An example of this technique involves the data summing program FINDAVE. Ten numbers are added and averaged. The results are then saved in the data segment (via the SUM and AVERAGE variables). The list file is included here to give an indication of what the program looks like and some idea as to where the results can be found.

```
                           ;Program FINDAVE.ASM: Find the average of ten words.
                           ;
                                           .MODEL SMALL
0000                                       .DATA
0000 0A                    COUNT    DB     10
0001 0000                  SUM      DW     ?
0003 0000                  AVERAGE  DW     ?
0005 0064 00C8 012C 0190   VALUES   DW     100,200,300,400,500
     01F4
000F 03E8 07D0 0BB8                 DW     1000,2000,3000,4000,5000
     0FA0 1388

0000                               .CODE
                                   .STARTUP
0017 8D 36 0005 R                  LEA     SI,VALUES    ;set up pointer to data
001B B8 0000                       MOV     AX,0         ;set initial sum to zero
001E 8A 0E 0000 R                  MOV     CL,COUNT     ;set up loop counter
0022 03 04            ADDLOOP:     ADD     AX,[SI]      ;add new data item to sum
0024 83 C6 02                      ADD     SI,2         ;point to next data item
0027 FE C9                         DEC     CL           ;decrement loop counter
0029 75 F7                         JNZ     ADDLOOP      ;jump if CL is not zero
002B A3 0001 R                     MOV     SUM,AX       ;save sum
002E 99                            CWD                  ;convert sum to double-word
002F BB 000A                       MOV     BX,10        ;prepare for division by ten
0032 F7 FB                         IDIV    BX           ;divide to find average
0034 A3 0003 R                     MOV     AVERAGE,AX   ;save average
                                   .EXIT

                                   END
```

By examining the list file, we can see that the sum will be stored at address 0001 and the average at address 0003 within the data segment. After execution by DEBUG, we can dump these locations to view the results. We must first use the unassemble command to find out where DEBUG located the data segment.

```
C> debug findave.exe<cr>
-g<cr>

Program terminated normally
-u<cr>
2B60:0000 BA632B     MOV     DX,2B63
2B60:0003 8EDA       MOV     DS,DX
2B60:0005 8CD3       MOV     BX,SS
2B60:0007 2BDA       SUB     BX,DX
2B60:0009 D1E3       SHL     BX,1
2B60:000B D1E3       SHL     BX,1
2B60:000D D1E3       SHL     BX,1
```

```
2B60:000F  D1E3        SHL      BX,1
2B60:0011  FA          CLI
2B60:0012  8ED2        MOV      SS,DX
2B60:0014  03E3        ADD      SP,BX
2B60:0016  FB          STI
2B60:0017  8D361100    LEA      SI,[0011]
2B60:001B  B80000      MOV      AX,0000
2B60:001E  8A0E0C00    MOV      CL,[000C]
```

The unassembly shows that the data segment begins at 2B63:0000. However, the address of VALUES loaded into SI is no longer 0005 as shown in the list file, but 0011. The linker relocated the entire data segment when it created FINDAVE.EXE from FINDAVE.OBJ. Further use of the u command will show that SUM and AVERAGE have the new offset addresses 000D and 000F, respectively. Now we can use the dump command to examine the results.

```
-d 2B63:0 1f
2B63:0000  0A 00 F7 FB A3 0F 00 B4-4C CD 21 00 0A 74 40 72    ........L.!..t@r
2B63:0010  06 64 00 C8 00 2C 01 90-01 F4 01 E8 03 D0 07 B8    .d...,..........
-q
```

The sum stored at address 000D is 4074H. This equates to 16,500, the correct sum for the ten data items included in the program. The average stored at address 000F is 0672H, which is the proper average of 1650. Thus, we see that DEBUG can be used to verify the operation of an .EXE file and let us know if the program is working properly.

Using Script Files with DEBUG

There are times when a new programming exercise presents a significant challenge and requires many DEBUG sessions to finally get it right. It is not difficult to imagine that typing in the same group of instructions over and over quickly leads to frustration, even when the last attempt provides the correct solution. One way to avoid having to duplicate the same work in a series of DEBUG sessions while a programming exercise is being developed is to use a *script* file. A script file contains all of the DEBUG commands and statements you might enter during a session, and is created using an ordinary text editor. For example, consider the following script file:

```
a
mov al,5
mov cl,6
mul cl
sub al,12
mov bl,al

r
t 5
q
```

This script file contains ten lines. The first line contains DEBUG's **a** command. This will put DEBUG into assembly mode. The next five lines are the instructions we wish to assemble and place into memory. Line seven is a blank line and is important because it gets DEBUG out of assembly mode. Line eight contains DEBUG's **r** command. This will display the current register contents prior to execution. Line nine uses DEBUG's trace command to show the execution results of the five instructions entered earlier. Finally, line ten allows us to quit DEBUG from within the script file.

To use the script file, create it with a text editor and save it as FILENAME.SCR. Then, use the following command to allow DEBUG to access the script file:

```
DEBUG < FILENAME.SCR
```

The < symbol is a DOS function that allows the standard input device (the keyboard in this case) to be *redirected*. This means that instead of DEBUG's waiting for a keystroke for each new command or statement, it will simply read it from the script file. So to DEBUG, using a script file is not any different from simply typing in all the statements really fast.

Once again, to get a printout of your DEBUG session, use the **Control-PrintScreen** function to toggle the printer prior to using the script file.

The script file previously discussed is used to solve the following equation:

$$BL = (6 \times AL) - 18$$

Note that the 18 in the equation is a 12 in instruction line five, because DEBUG uses only hexadecimal numbers.

In the script file, AL is initially loaded with the number 5. The equation predicts BL to have a value of 12. Examine the following DEBUG session (which was generated by the script file) to verify that the given instructions performed the calculation correctly.

```
-a
1CE2:0100 mov al,5
1CE2:0102 mov cl,6
1CE2:0104 mul cl
1CE2:0106 sub al,12
1CE2:0108 mov bl,al
1CE2:010A
-r
AX=0000  BX=0000  CX=0000  DX=0000  SP=FFEE  BP=0000  SI=0000  DI=0000
DS=1CE2  ES=1CE2  SS=1CE2  CS=1CE2  IP=0100   NV UP EI PL NZ NA PO NC
1CE2:0100 B005          MOV     AL,05
-t 5

AX=0005  BX=0000  CX=0000  DX=0000  SP=FFEE  BP=0000  SI=0000  DI=0000
DS=1CE2  ES=1CE2  SS=1CE2  CS=1CE2  IP=0102   NV UP EI PL NZ NA PO NC
1CE2:0102 B106          MOV     CL,06

AX=0005  BX=0000  CX=0006  DX=0000  SP=FFEE  BP=0000  SI=0000  DI=0000
DS=1CE2  ES=1CE2  SS=1CE2  CS=1CE2  IP=0104   NV UP EI PL NZ NA PO NC
1CE2:0104 F6E1          MUL     CL

AX=001E  BX=0000  CX=0006  DX=0000  SP=FFEE  BP=0000  SI=0000  DI=0000
DS=1CE2  ES=1CE2  SS=1CE2  CS=1CE2  IP=0106   NV UP EI PL NZ NA PO NC
1CE2:0106 2C12          SUB     AL,12

AX=000C  BX=0000  CX=0006  DX=0000  SP=FFEE  BP=0000  SI=0000  DI=0000
DS=1CE2  ES=1CE2  SS=1CE2  CS=1CE2  IP=0108   NV UP EI PL NZ NA PE NC
1CE2:0108 88C3          MOV     BL,AL

AX=000C  BX=000C  CX=0006  DX=0000  SP=FFEE  BP=0000  SI=0000  DI=0000
DS=1CE2  ES=1CE2  SS=1CE2  CS=1CE2  IP=010A   NV UP EI PL NZ NA PE NC
1CE2:010A 48            DEC     AX
-q
```

The final value in register BL is 0CH, which is the correct result.

A BRIEF LOOK AT CODEVIEW

Codeview is a fancier version of DEBUG that allows user control over the execution of a program being developed or examined. Codeview displays its information in different windows, such as the source window, register window, command window, etc. To switch from one window to another just press the F6 button.

The source window can display source code in the original form of the source file or in machine language and mnemonics. Pressing F3 switches the display format.

The register window displays the hexadecimal contents of all processor registers, both 16- and 32-bit. The state of each arithmetic flag is also displayed, along with the address and data of the last data access. When a program is traced instruction by instruction, the contents of any registers that change during execution are highlighted. This makes it easy to follow the progress of the program. The contents of any window can be sent to the printer or to a file (called CODEVIEW.LST by default). The register window looks like this when printed:

```
AX = 0000
BX = 0000
CX = 0000
DX = 0000
SP = 0000
BP = 0000
SI = 0000
DI = 0000
DS = 2B71
ES = 2B71
SS = 2B81
CS = 2B81
IP = 0010
FL = 0200

NV UP EI PL
NZ NA PO NC
```

A command window is provided to enable the user to enter debugging commands, in a fashion similar to that of DEBUG. Some of the more useful commands are as follows:

A	Assemble
BC	Breakpoint Clear
BD	Breakpoint Disable
BE	Breakpoint Enable
BL	Breakpoint List
BP	Breakpoint Set
E	Animate
G	Go
H	Help
I	Port Input
K	Stack Trace
L	Restart
MC	Memory Compare

MD	Memory Dump
ME	Memory Enter
MF	Memory Fill
MM	Memory Move
MS	Memory Search
N	Radix
O	Options
O	Port Output
P	Program Step
Q	Quit
R	Register
T	Trace
U	Unassemble
VM	View Memory
X	Examine Symbols

Help on every command is available on line. For example, the command:

```
H VM
```

displays the help information that describes the VM command.

There are other windows that allow the user to watch the contents of a variable change as the program executes, and see a display of the contents of a block of memory.

The assembler places special symbolic debugging information into an .EXE file when it is assembled with the /Zi option, as in:

```
ML /Zi FINDAVE.ASM
```

This symbolic information helps Codeview keep track of things at the source-file level. You must specify the name of an .EXE file when starting Codeview (as a command line parameter):

```
CV FINDAVE
```

Codeview does not require the .EXE extension.

Codeview is also capable of handling not just the 32-bit registers (EAX, EBX, etc.) but the newer 80386 and 80486 instructions as well. Recall that DEBUG is only capable of displaying the 16-bit register sizes, and cannot assemble or unassemble instructions other than those of the 8086.

Codeview allows data operands to be displayed in their intended format. For example, a 4-byte integer and a 4-byte real number must be interpreted and displayed differently. This is easily accomplished with the MD (or VM) command. A format specifier is used to control the data format. These specifiers are:

A	ASCII characters
B	Byte
C	Code

I	Integer (2 bytes)
IU	Integer unsigned (2 bytes)
IX	Integer hexadecimal (2 bytes)
L	Long (4-byte decimal)
LU	Long unsigned (4 byte)
LX	Long hexadecimal (4 bytes)
R	Real (4-byte floating point)
RL	Real long (8-byte floating point)
RT	Real ten-byte floating point

Altogether, Codeview offers a significant improvement over the debugging capabilities of DEBUG. It is worth the time invested in learning how to use all of Codeview's features.

APPENDIX I

Extended Keyboard Scan Codes

00H		20H	ALT-D	40H	F6	60H	CTRL-F3
01H		21H	ALT-F	41H	F7	61H	CTRL-F4
02H		22H	ALT-G	42H	F8	62H	CTRL-F5
03H		23H	ALT-H	43H	F9	63H	CTRL-F6
04H		24H	ALT-J	44H	F10	64H	CTRL-F7
05H		25H	ALT-K	45H		65H	CTRL-F8
06H		26H	ALT-L	46H		66H	CTRL-F9
07H		27H		47H	Home	67H	CTRL-F10
08H		28H		48H	Up Arrow	68H	ALT-F1
09H		29H		49H	Page Up	69H	ALT-F2
0AH		2AH		4AH		6AH	ALT-F3
0BH		2BH		4BH	Left Arrow	6BH	ALT-F4
0CH		2CH	ALT-Z	4CH		6CH	ALT-F5
0DH		2DH	ALT-X	4DH	Right Arrow	6DH	ALT-F6
0EH		2EH	ALT-C	4EH		6EH	ALT-F7
0FH	Shift-Tab	2FH	ALT-V	4FH	End	6FH	ALT-F8
10H	ALT-Q	30H	ALT-B	50H	Down Arrow	70H	ALT-F9
11H	ALT-W	31H	ALT-N	51H	Page Down	71H	ALT-F10
12H	ALT-E	32H	ALT-M	52H	Insert	72H	CTRL-PRINT SCREEN
13H	ALT-R	33H		53H	Delete	73H	CTRL-LEFT ARROW
14H	ALT-T	34H		54H	Shift-F1	74H	CTRL-RIGHT ARROW
15H	ALT-Y	35H		55H	Shift-F2	75H	CTRL-END
16H	ALT-U	36H		56H	Shift-F3	76H	CTRL-PAGE DOWN
17H	ALT-I	37H		57H	Shift-F4	77H	CTRL-HOME
18H	ALT-O	38H		58H	Shift-F5	78H	ALT-1
19H	ALT-P	39H		59H	Shift-F6	79H	ALT-2
1AH		3AH		5AH	Shift-F7	7AH	ALT-3
1BH		3BH	F1	5BH	Shift-F8	7BH	ALT-4
1CH		3CH	F2	5CH	Shift-F9	7CH	ALT-5
1DH		3DH	F3	5DH	Shift-F10	7DH	ALT-6
1EH	ALT-A	3EH	F4	5EH	CTRL-F1	7EH	ALT-7
1FH	ALT-S	3FH	F5	5FH	CTRL-F2	7FH	ALT-8

80H	ALT-9
81H	ALT-0
82H	ALT-HYPHEN
83H	ALT-EQUAL
84H	CTRL-PGUP

Solutions and Answers to Selected Odd-Numbered Study Questions

Chapter 1: Microprocessor-Based Systems

1.1: Microwave oven, digital television, heart monitors, FAX machines, UPC scanners, cash registers, automobiles, cameras, sporting equipment, answering machines.

1.3: In all three cases, timing signals are required to ensure that data is transferred at the correct time. Specifically, the Serial I/O section uses timing signals to generate the required BAUD rate clock. The Memory section uses timing signals to control access to dynamic RAMs and to insert wait states as necessary (to allow for chip access times). The Interrupt section uses timing signals to latch interrupt requests and to service certain types of interrupts (for real-time clocks/calendars) at regular intervals.

1.7: A cash register without record keeping would require less RAM, because all transactions do not have to be stored.

1.9: Doors: any opening/closing. Windows: any opening, closing, breaking. Elevators: all switches (up, down, open, close, floor number). Timing interrupts to keep the elevator from getting stuck on any floor and to close the door after a sufficient delay.

1.11: Downloading of main program, status information, serial number of part being assembled, commands from host computer.

1.13: The connectors on the plug-in card allow access to all of the required processor signals.

1.15: One CPU to run the game program, one CPU to control graphics, one CPU to generate sounds and perform user I/O.

1.17: Image processing, system modeling, database search, 3-D graphics, and code breaking.

1.19: The new CPU must execute all instructions from earlier models. Unused bit patterns from the previous instruction set can be implemented as additional instructions on the new machine. Hardware must be added to decode and execute additional instructions.

1.21: The AND instruction takes the longest, because it must both read and write the memory location pointed to by SI.

1.23: DEBUG statements assume that input numbers are hexadecimal.

Chapter 2: An Introduction to the 80x86 Microprocessor Family

2.1: AX, BX, CX, DX, SI, DI, BP. AX is used for multiplication and division. CX is used for loops. SI and DI are used for string operations.

2.3: (a) 03E00 + 1F20 = 05D20
(b) 04000 + 3FFE = 07FFE
(c) 24000 + FFFF = 33FFF

2.5: No. The U pipeline is used for FPU operations as well as integer instructions.

2.7: A segment is a 64KB block of memory beginning on any 16-byte boundary (paragraph).

2.9: Intel byte swapping refers to the way 16-bit numbers are stored in memory. The lower 8 bits are stored in location N, the upper 8 bits in location N+1.

2.11: The processor allows for 11 different addressing modes.

2.13: Register addressing is fast because the physical register hardware is contained within the CPU.

2.15: MOV   AX,1000H places the immediate value 1000H into register AX. MOV AX,[1000H] reads the word stored at location 1000H and places its value into AX.

2.17: An interrupt is an event that temporarily changes execution to service a request.

2.19: The processor supports 256 interrupts.

2.21: The 80286 supports a 16MB physical memory space, protected mode operation, virtual memory, and multitasking.

2.23: A page fault occurs when the processor attempts to access a memory page that has not been loaded into memory.

2.25: Physical memory has a greater effect on the number of page faults. The more physical memory, the smaller the chance that a page will not have been previously loaded.

2.27: Applications may include imaging, world-wide databases, real-time simulation of weather, chemical reactions, and astronomical experiments.

2.29: The 486 uses an 8KB cache memory, has a redesigned internal architecture resulting in fewer clock cycles per instruction, and contains an on-chip coprocessor.

2.31: The Pentium differs from other 80x86 CPUs in that it was designed with RISC techniques in mind, such as branch prediction and dual integer pipelines.

Chapter 3: 80x86 Instructions, Part 1

3.1: ORG is used to set the initial program counter for the program being assembled. DB is used to define (reserve) a byte of storage. DW is used to define a word of storage. END is used to tell the assembler that it has reached the end of the source file.

3.3: Object (.obj) and list (.lst) files.

3.5: 16-bit numbers are stored in two consecutive memory locations with the lower byte going into the first location and the upper byte into the second. Thus, the 2 bytes are swapped.

3.7: A linker is used to combine multiple object files into a single object file.

3.9: (a) register, (b) immediate, (c) indexed, (d) based, (e) indexed, (f) relative, (g) implied.

3.11: Upon completion, AX contains the original contents of BX. BX contains the original contents of CX, and CX contains the orginal contents of AX.

3.13: (a) 9C0A, (b) B29C, (c) 78B2

3.15: LDS loads a string pointer using the DS register; LES uses the ES register. LDS should be used to load source string addresses. LES should be used for destination strings.

3.21: The simplified segment directives are .DATA and .CODE.

3.23: (a) EAX is a base register.
(b) EBX is a base register. EAX is an index register.
(c) EAX is a base register. EBX is an index register.
(d) ESI is a base register. EDI is an index register.

3.25: Scale values of 1, 2, 4, or 8 may be used.

3.27: The port address must be placed in register DX, which is used as an operand in IN and OUT.

3.29: (a) EBX contains 00003000H.
(b) EBX contains 00003000H.

3.31: The PUSHA/POPA instructions can be used to save/restore all general purpose registers via the stack.

Chapter 4: 80x86 Instructions, Part 2

4.1: AH is unchanged. AL contains the sum of AL and AH (46H).

4.3: After the INC instruction executes, location 2000 will contain 01 and location 2001 will contain 50.

4.5: The zero flag is cleared.

4.7:
```
MOV   BX,25
MUL   BX
```

4.9: DX contains 0000. AX contains FDE8.

4.11: CBW converts 30H into 0030H, and 98H into FF98H.

4.13: DX contains 8365H.

4.15: AL contains 6AH.

4.17: ROR BYTE PTR [1000H],1

4.19: JZ

4.21: (a) no, (b) no, (c) yes, (d) no, (e) cannot determine without OF status, (f) cannot determine without OF status.

4.25: SP for near call: 27FEH. SP for far call: 27FCH.

4.27: First the flags are pushed onto the stack. Then the trace and interrupt-enable flags are cleared. Then the CS and IP registers are pushed. The interrupt vector table is accessed for the address of INT 16H's ISR. Then the CS and IP registers are loaded with their new addresses and execution resumes.

4.29:
```
ADD   AL,BL
ADD   AL,DL
MOV   CL,AL
```

4.31: Dividing by 8 is the same as multiplying by 0.125.
```
MOV   BL,8
DIV   BL
```

4.33: MUL multiplies two unsigned 16-bit numbers. Each can be as large as 65535. IMUL multiplies two signed 16-bit numbers. The largest positive number that can be used is 32767.

4.35: AND BX,23F9H

4.39: To pop the stack and return to the correct place in the calling program.

```
4.41:  MUL    BL
       MOV    DX,0
       MOV    CX,2
       DIV    CX
4.43:  MOV    AL,STATUS
       TEST   AL,80H
       JZ     ROUT1
       TEST   AL,0CH
       JZ     ROUT3
       TEST   AL,20H
       JNZ    ROUT2
       AND    AL,3
       CMP    AL,2
       JZ     ROUT4
4.45:        MOV    AL,COUNT
             CLC
             INC    AL
             DAA
             CMP    AL,60H
             JNZ    DONE
             SUB    AL,AL
       DONE: MOV    COUNT,AL
```

Chapter 5: Interrupt Processing

5.1: The processor's environment is the information contained in all of its internal registers. Saving the environment during interrupt processing ensures that the state of the machine prior to the interrupt will be correctly restored.

5.3: INTR is enabled with the STI instruction and disabled with the CLI instruction.

```
5.7:  MOV    DI,94H
      MOV    [DI],9AE2H
      MOV    [DI+2],3C0H
```

5.9: 280H divided by 4 gives 0A0H. The address range is for INT   0A0H.

5.11: After 1st instruction: AX = E03F.
      After 2nd instruction: AX = E040.
      After 3rd instruction: AX = DF40.
      After 4th instruction: AX = 20BF.

5.13: 2.5 ms divided by 0.2 μs gives 12,250. An average of 12,250 instructions can be executed after the power failure.

5.15: 1FH.

5.17: C0, C8, D0, D8, E0, E8, F0, F8.

5.19: Yes, because NMITIME is called every 16.67 ms. It is possible that interrupts would be lost, resulting in inaccurate timekeeping.

5.21: ISR20H will select the DIV   BL instruction. The resulting values in AX is 03C0.

5.23: The new program counter will be 0480:1234.

5.25: The results of each instruction can be examined, effectively running the program in slow motion and watching all the results.

5.27: INTO will not generate an interrupt. Trace is enabled. Interrupts are enabled.

5.29:
```
T7:     PUSH BX
        PUSH DI
        MOV  BL,7
        IMUL BL
        CMP  AX,8400H
        JLE  EXIT
        PUSH AX
        CALL FAR PTR OVERSCAN
        POP  AX
EXIT:   POP  DI
        POP  BX
        IRET
```

Chapter 6: An Introduction to Programming the 80x86

6.1: Writing programs in pseudocode is often easier than writing programs in assembly language. For many programmers, a statement like IF CHAR = 'A' THEN ACOUNT = ACOUNT + 1 is easier to come up with than the equivalent assembly language statements.

6.3: The DS register must be loaded with the segment address of the source strings, and ES must be loaded with the segment address of the destination strings.

6.5: Add an additional byte to the BCD string to save a signed exponent.

6.7: Fractional values can be more accurately represented in BCD. Consider the example value 0.7. The binary representation of 0.7 repeats over and over, whereas the BCD representation is 07 with an appropriate exponent.

6.9: Add an additional byte for use as a sign byte.

6.11: Field terminators are useful for searching purposes. Looking for a field terminator is one way of finding the end of a record, or the beginning of the next record.

Chapter 7: Programming with DOS and BIOS Function Calls

7.1: The BIOS keyboard interrupt does not echo the key to the display.

7.3: PASSWORD looks for the carriage return code (0DH) to detect the end of the secret password.

7.5: 25 lines times 80 characters/line equals 2,000 characters. Each character requires 2 bytes of storage, giving 4,000 bytes of screen memory.

7.7: (a) 27H, (b) 42H, (c) 81H, (d) 64H.

7.9: 1.19 MHz/2380 = 500 Hz.

7.11: The printer status should be read and checked for errors before outputting a character to the printer. This helps avoid lost printer data.

7.13: The ES register is automatically initialized by DOS to the proper segment address when the ECHO program loads.

7.15: BSORT must scan for a '−' sign before any number and save the 2's complement of the input number when found. Then the proper conditional jump instruction must be used when comparing numbers.

Chapter 8: Advanced Programming Applications

8.1: The PUBLIC assembler directive is used to make the value of a symbol available to other link modules. The EXTRN directive is used to access the value of an external symbol.

8.3: A nested macro is a macro definition that contains a second macro definition within itself.

8.7: DEBUG can display the interrupt vector for INT 21H by using the following command: D 0:84 L 4

8.11: DOS manages memory by assigning blocks of paragraphs beginning at the first free segment address.

8.13: One way to access mouse data is through INT 33H. Mouse information is available when a mouse is connected and an appropriate mouse driver is loaded.

8.15: The size of the resident code (in paragraphs) must be determined prior to the exit to DOS through DOS INT 21H, Function 31H.

8.17: The BTXT (board text) string contains the printable text of the tic tac toe board. Nine locations within the BTXT string are modified during game play with the user/computer moves. These locations are specified in the BPOS (board position) array.

8.19: The stack in a C program is used to hold variable values, return addresses, and pointers to other stack areas.

Chapter 9: Using Disks and Files

9.1: 5.25-inch diskettes have fewer tracks (40 versus 80) and may also have fewer sectors/track (9/15 versus 9/18).

9.3: Hard drives spin at a faster rate than floppies.

9.5: The BIOS parameter block contains information about the physical makeup of a disk and is located in the boot sector.

9.7: A cluster chain indicates all clusters associated with a specific file. Each cluster in the chain contains an associated FAT entry that points to the next (or last) cluster in the cluster chain.

9.9: A directory entry uses 32 bytes. If 14 sectors are available for directory information, a total of 14 * 512/32 = 224 files may be stored.

9.11: 12H is 00010010 binary. The two high bits indicate that the attribute is for a hidden directory.

9.13: An ASCIIZ string is a string of characters describing a file name. It must end with a 00 byte. The source statement FILEABC DB 'ABC.DAT',0 is the required ASCIIZ string for the file ABC.DAT.

9.15: Assuming a valid ASCIIZ string for SPIN.ASM exists at the label FILE1, use the following instructions to open SPIN.ASM for read/write access:

```
Lea   DX,FILE1
MOV   AL,2          ;read/write access
MOV   AH,3DH
INT   21H
```

9.17: File pointers may be moved in the following ways: (1) absolute byte offset from the beginning of the file, (2) relative byte offset from the current position, and (3) absolute byte offset from the end of the file.

Chapter 10: Hardware Details of the 8088

10.1: Some CPU signals are different (ALE vs. QS_0 for example). Minmode does not allow bus request/grant and does not support a coprocessor.

10.5: 10 MHz/3 = 3.333 MHz.

10.9: The initial instruction fetch is from address FFFF0.

10.11: HOLD/HLDA or RQ/GT.

10.13: The processor will automatically be interrupted each time interrupts are enabled.

10.15: A 74LS374 or two 74LS75s.

10.17: The address 3A8C4 is represented as follows:

A19	18	17	16	15	14	13	12	11	10	9	8	7	6	5	4	3	2	1	A0
0	0	1	1	1	0	1	0	1	0	0	0	1	1	0	0	0	1	0	0

10.19: Demultiplexing takes additional clock cycles.

10.23: T_1

10.25: The bus activity on the entire system will be very busy. The bus activity of each processor will have gaps in processing as the other processor alternately takes over.

System Bus	-- busy --	-- busy --	-- busy --	-- ...
CPU #1 Bus	-- busy --	-- idle --	-- busy --	-- ...
CPU #2 Bus	-- idle --	-- busy --	-- idle --	-- ...

10.27: An 8-bit interrupt vector number.

Chapter 11: Memory System Design

11.1: AD_0 through AD_7 act as a bidirectional data bus. They also supply the lower byte of the address bus and receive an 8-bit vector number during INTA.

11.3: Minmode: IO/M, RD, WR.
Maxmode: MRDC, MWTC.

11.5: 4(250 ns) = 1,000 ns = 1 μs.

11.7: 2^{17} = 128KB. Seventeen address lines are needed (A_0 through A_{16}).
2^{21} = 2MB. Twenty-one address lines are needed (A_0 through A_{20}).
Note: The 8088 cannot access 2MB of memory.

11.9: Address line A_{11} is used to select EPROM #1 when low and EPROM #2 when high. Address lines A_{12} through A_{19} must be used in the address decoder.

11.13: (a) 2A000–2AFFF
(b) 54000–57FFF
(c) 40000–5FFFF
(d) E0000–EFFFF

11.15: A: 60000–63FFF
B: 64000–67FFF
C: 68000–6BFFF
D: 6C000–6FFFF
E: 70000–73FFF
F: 74000–77FFF
G: 78000–7BFFF
H: 7C000–7FFFF

11.17: Advantages: less hardware needed, easier to design. Disadvantages: future expansion requires additional hardware, addresses may conflict or select more than one device at a time.

11.23: Each address line provides a row address bit and a column address bit. These two bits can be in four different states.

11.25: The interrupt service routine could maintain a counter (for RAS refreshing). Each time the interrupt occurs, the counter could be written to the DRAM address circuitry to cause the refresh.

11.27: Waiting in hold acknowledge.

Chapter 12: I/O System Design

12.1: First way: IN   AL,40H

      Second way: MOV   DX,40H

               IN   AL,DX

12.5: Use IORC and IOWC instead of RD, WR, and IO/M.

12.9: (a) C0 – FF

     (b) 1 0 x 0 0 x 0 0 <---> 80, 84, A0, A4

     (c) 0 C B A 0 1 x x

        PA: 04–07

        PB: 14–17

        PC: 24–27

        PD: 34–37

        PE: 44–47

        PF: 54–57

        PG: 64–67

        PH: 74–77

     (d) 1 x 0 x 1 x 1 x 0 x 1 x 1 1 x x

        There are 128 different addresses. Here are a few of them: 8A2C, 8A2D, 8A2E, 8A3C, 8A6C, 8B2C, 8E2C, 9A2C, and CA2C.

     (e) 8001, plus 16383 more that have their LSB and MSB set.

     (f) x 0 1 0 0 1 x x <---> 24 - 27, A4 - A7

12.15:
```
KP:     MOV   AL,0E0H
        MOV   CX,6
ALOP:   OUT   9CH,AL
        CALL  DELAY
        SHR   AL,1
        LOOP  ALOP
        MOV   AL,0EH
        MOV   CX,4
BLOP:   OUT   9CH,AL
        CALL  DELAY
        SHL   AL,1
        LOOP  BLOP
        JMP   KP
DELAY:  PUSH  AX
        PUSH  CX
```

```
            IN    AL,9CH
            MOV   CH,AL
            MOV   CL,1
    WAIT:   NOP
            NOP
            LOOP  WAIT
            POP   CX
            POP   AX
            RET
```

12.17: Assuming no DELAY cycles, 10-byte waveforms require 10*17 = 170 cycles. With a clock speed of 5 MHz, this works out to 34 μs/cycle, or 29.4 KHz. Fifty-byte waveforms require 50*17 = 850 cycles, giving a frequency of 5,882 Hz; 100-byte waveforms require 100*17 = 1,700 cycles, giving a frequency of 2,941 Hz.

12.19: This cannot be done, because 8A does not begin on a 4-port boundary (such as 88 or 8C).

12.21: Assume 8255 is programmed for Aout, Bin.

```
    BINCNT:   MOV    AL,0
    DISPCNT:  OUT    48H,AL
              CALL   DELAY
              INC    AL
              JMP    DISPCNT
    DELAY:    MOV    BL,AL
              IN     AL,49H
              MOV    CH,AL
              MOV    CL,1
    WAIT:     NOP
              NOP
              LOOP   WAIT
              MOV    AL,BL
              RET
```

12.25: 11/1,200 = 9.17 ms.

12.27: (a) MOV AL,0DAH
            OUT DX,AL
       (b) MOV AL,4DH
            OUT DX,AL
       (c) MOV AL,48H
            OUT DX,AL

12.29:
```
    CHAROUT:  MOV    AH,AL
    TSTAT:    IN     AL,79H
              AND    AL,1
              JNZ    TSTAT
              MOV    AL,AH
              CMP    AL,1
              JC     COUT
              CMP    AL,1BH
              JNC    COUT
              PUSH   AX
              MOV    AL,'^'
              CALL   CHAROUT
              POP    AX
              ADD    AL,40H
```

```
                  JMP     CHAROUT
        COUT:     OUT     78H,AL
                  RET
```

Chapter 13: Programming Peripherals

13.1: The CPU cannot perform other processing while polling.

13.3: ICW1: 12, ICW2: C0, ICW3: 00 (although not needed).

13.7:
```
MOV AL,12H
OUT 70H,AL
MOV AL,0C0H
OUT 71H,AL
```

13.11:
```
MOV AL,0B0H
OUT 0B3H,AL
SUB AL,AL
OUT 0B2H,AL
MOV AL,30H
OUT 0B2H,AL
```

13.13:
```
LATCH0: MOV AL,0
        MOV DX,0CC83H
        OUT DX,AL
        SUB DX,3
        IN  AL,DX
        MOV DL,AL
        IN  AL,DX
        MOV DH,AL
        CMP DX,7
        JNZ EXIT
        CALL TIMEOUT
EXIT:   RET
```

13.15: $10,000/250K = 40$ ms.

13.19:
```
SWAVE: MOV AL,36H
       OUT 3CH,AL
       SUB DX,DX
       MOV AX,2000
       SUB BH,BH
       DIV BX
       OUT 38H,AL
       MOV AL,AH
       OUT 38H,AL
       RET
```

13.21: $1,020.6 = 1111111100.1001$

Normalized with exponent of $9 = 1.1111111001001$

23-bit mantissa = .11111110010011001100110

Exponent = $127 + 9 = 136 = 10001000$

Sign bit = 0

01000100	01111111	00100110	01100110
44	7F	26	66

13.23:
```
FLD   LEVEL
FSQRT
FST   LEVEL
```

13.25:
```
PI      DD    3.14159265
C180    DD    180.0
DTR:    FLD   DEG
        FMUL  PI
        FDIV  C180
        FST   RAD
        RET
```

13.27: Image processing, matrix calculations, fft (fast Fourier transform), spreadsheets, signal processing, graphics (rotation, scaling).

13.29:
```
NEG5    DD    -5.0
EQUN:   FLD   A
        FMUL  A
        FLD   B
        FMUL  B
        FLD   NEG5
        FMUL  A
        FMUL  B
        FADD
        FADD
        FST   X
        RET
```

Chapter 14: Building a Working 8088 System

14.3: Use 138 outputs 0 (0000-3FFF), 2 (4000-7FFF), 4 (8000-BFFF), and 6 (C000-FFFF) for EPROM CS inputs.

14.9:
```
INIT:   MOV AL,90H
        OUT 0E3H,AL
        SUB AL,AL
        OUT 0E1H,AL
        MOV AL,3CH
        OUT 0E2H,AL
        RET
```

14.11:
```
DXOUT:  MOV   AX,DX
        SUB   DX,DX
        MOV   BX,10000
        CALL  DOUT
        MOV   BX,1000
        CALL  DOUT
        MOV   BX,100
        CALL  DOUT
        MOV   BX,10
        CALL  DOUT
        ADD   AL,30H
        CALL  C_OUT
        RET
DOUT:   DIV   BX
        ADD   AL,30H
        CALL  C_OUT
        MOV   AX,DX
        SUB   DX,DX
        RET
```

14.15: Store commands in a data table (i.e., db 'DUMPDISPGO EXEC...'). Compare input text with each four-letter group in the table. If a match is found, look up corresponding routine address in a second data table (i.e., dw D_UMP, D_UMP, E_XEC, E_XEC,...). Read this address and jump to it.

14.17:
```
V_ERIFY: MOV  DI,500H
   NEXT: MOV  BL,0ABH
         CALL TESTLOC
         MOV  BL,65H
         CALL TESTLOC
         INC  DI
         CMP  DI,1F80H
         JNZ  NEXT
         JMP  GET_COM
TESTLOC: MOV  ES:[DI],BL
         MOV  AL,ES:[DI]
         CMP  AL,BL
         JZ   OK
         PUSH AX
         PUSH BX
         CALL CRLF
         MOV  DX,DI
         CALL H_OUT
         CALL BLANK
         POP  BX
         MOV  AL,BL
         CALL HTOA
         CALL BLANK
         POP  AX
         CALL HTOA
     OK: RET
```

14.19:
```
HEXPM: CALL GET_BYT
       MOV  DL,AL
       CALL GET_BYT
       MOV  AL,DH
       CALL CRLF
       MOV  AL,DH
       ADD  DL
       CALL HTOA
       CALL BLANK
       MOV  AL,DL
       SUB  AL,DH
       CALL HTOA
       JMP  GET_COM
```

14.21: It might be useful to include the segment address in the dump, as in 01CE:0240 3E 43 12 06 0F E5..., and printing the ASCII equivalent characters after the 16 bytes would also be a good addition.

14.23: Turn I_NEX code (up to NUN) into a subroutine. Use SI to pass the index of the R_LETS and R_DATA tables. Rewrite I_NIT so that I_NEX is called 11 times if no register name is specified. Otherwise, search R_LETS for the supplied register name and adjust SI accordingly.

14.25: Add PIN2: to the MOV AL,'[' instruction. Insert these instructions between CALL C_OUT and JMP GET_COM:

```
IN   AL,S_CTRL
AND  AL,2
JZ   PIN2
IN   AL,S_DATA
AND  AL,7FH
CMP  AL,0DH
JNZ  PIN2
```

14.27:
```
P_IE: MOV AL,SINE[SI]    8 + 9
      OUT  D_AC,AL          10
      ADD  SI,1             4
      LOOP P_IE             17
```

These 48 cycles are repeated 256 times, giving a total estimate of 12,288 cycles. The period is approximately 12,288/3.333 MHz = 3.68 ms. The frequency is approximately 271 Hz.

Chapter 15: Hardware Details of the Pentium

15.1: A_3 through A_{31}, and the eight-byte enable outputs, are used to output a 32-bit address. The byte enable outputs simulate the operation of A_0, A_1, and A_2.

15.3: Parity checking is performed on address and data lines.

15.5: Processor signals D/C, M/IO, W/R, CACHE, and KEN define the type of bus cycle currently running. The byte enable outputs specify special cycles in some cases.

15.7: INTR, NMI, and SMI.

15.9: The cache, in many cases, is able to supply a stored copy of the requested memory data, without the need for a main memory access.

15.11: Branch prediction is a technique used to predict the outcome of a conditional branch instruction. When the prediction is correct, instructions are kept flowing smoothly through the instruction pipelines.

15.13: The Pentium indicates the current bus cycle type through its D/C, M/IO, W/R, CACHE, and KEN signals.

15.15: Single transfer: up to eight bytes. Burst transfer: 32 bytes.

15.17: Locked bus cycles are used to restrict access to shared variables, such as semaphores. The current bus cycle is locked if the LOCK output is active.

15.21: No, ADD   DATA[BP],3 is not a simple instruction. It contains both a displacement value and an immediate operand.

15.23: PF, D1, D2, EX, and WB.

15.25: A pipeline stall is one or more clock cycles in which nothing happens in the instruction pipeline. Stalls are caused by dependencies between instructions and also by long main memory accesses. Having valid copies of requested data in cache reduces the number of stalls by reducing main memory accesses.

15.27: The history bits in a BTB entry are used to indicate results of the most recent branches (taken, not taken).

15.29: The BTB predicts "not taken" for a new conditional jump.

15.31: A cache hit is a cache access that results in a tag match.

15.37: Writethrough: writes to the cache also update main memory. Writeback: writes to the cache do not update main memory.

15.39: MESI stands for Modified Exclusive Shared Invalid, a protocol used to help maintain cache coherency.

15.41: An inquire cycle is used to determine if the cache contains a copy of a shared data item.

15.43: The U pipeline makes up the first five stages of the FPU pipeline.

Chapter 16: Protected Mode Operation

16.1: The additional protected mode registers are the control registers, the debug registers, the GDTR, LDTR, IDTR, and TR.

16.3: A segment descriptor controls the base address, length, and access permissions of a segment of memory.

16.5: A logical address uses a segment selector and a 32-bit offset to access memory. The segment selector points to one of 8,192 segment descriptors in the GDT or LDT, which supplies a 32-bit linear base address to combine with the 32-bit offset.

16.7: A virtual address is translated into a physical address through two levels of page tables, a PDE and a PTE. The upper 20 bits of the virtual address are replaced by 20 bits from the PTE, giving a new, 32-bit physical address.

16.9: A page directory contains 1,024 entries. Each entry can access 1,024 PTEs. Each PTE represents 4KB of memory. So, 1,024 times 1,024 times 4KB gives 4 gigabytes, the entire addressing space of the processor.

16.11: Dirty bit: indicates if page has been written to. Accessed bit: indicates if a read or write has been performed on a page.

16.15: The RPL bits in the segment selector play a role in protected accesses.

16.17: The purpose of a call gate is to protect access to the entry point of a procedure.

16.21: A user-level task may not access a supervisor page.

16.23: The TSS is used to support task switching. The TSS contains a copy of each processor register for the task it is associated with.

16.25: The LTR and STR instructions are used with the task register. They may only be executed in protected mode.

16.29: The IDTR is the interrupt descriptor table register. The IDT is the interrupt descriptor table. The IDTR is used to locate the base address of the IDT in memory. The IDT stores interrupt descriptors.

16.31: Task gates, interrupt gates, and trap gates may be used in the IDT.

16.33: The page fault error code provides valuable information that may be used to correct the page fault (swap in a new page, choose a victim).

16.37: Divide C0H by eight (the number of bytes in a descriptor). This gives vector number 18H.

16.39: 32 bytes are required in the I/O permission bit map to protect ports 0 through FFH.

16.43: A virtual-8086 monitor is a protected mode program that watches over a running virtual-8086 mode task, providing services such as interrupt handling.

16.45: No, virtual-8086 mode may not be entered from real mode.

INDEX

Function 00H: Mouse reset and status, 307
Function 01H: Show mouse cursor, 307
Function 02H: Hide mouse cursor, 308
Function 03H: Get button status and mouse position, 308
Intel debug port, 553
Interrupt, 6, 154, 293
    Breakpoint, 163
    Divide error, 39, 161
    INTR, 165
    NMI, 163
    Overflow, 165
    Single step, 162
Interrupt acknowledge cycle, 159, 395
Interrupt chain, 311
Interrupt descriptor table, 602
Interrupt descriptor table register, 602
Interrupt enable flag, 141, 385
Interrupt gate, 602
Interrupt instructions, 138
Interrupt pointer table, 40
Interrupt request, 154
Interrupt service routine, 138, 168, 384
Interrupt vector table, 40, 141, 155, 385
Intersegment transfer, 131
Intrasegment transfer, 131
ISA bus, 397
I/O addressing space, 434
I/O permission bitmap, 609
Iterative formula, 208

Jump, 131

Label, 27
Least recently used algorithm, 570
Lexical nesting level, 143
LIB utility, 280

Library file, 59
LIFO, 224
Line of data, 569
Linear address, 572, 586, 590
LINK utility, 14
Linked list, 221
Linker, 59
List file, 55
Little endian format, 88
Local bus, 398
Local descriptor table, 585
Local descriptor table register, 585
LOCAL directive, 284
Local variable, 284
Locality of reference, 569
Logical address, 572
Logical instructions, 118
Loop instructions, 136

Machine code, 55
Machine language, 13
Machine status word, 612
Macro, 281
MACRO directive, 282
Macro assembler, 14
Maskable interrupt, 608
Masking, 385
Master device, 427
Maximum mode, 41, 381
ML utility, 14
Megabyte, 20, 36, 379
Memory, 4
Memory address decoder, 407
Memory management, 301
Memory-mapped I/O, 427
Memory read cycle, 406, 560
Memory write cycle, 406, 560
MESI protocol, 573
Microcontroller, 7
Microprocessor based system, 4
Minimum mode, 41, 379
MMX technology, 578
Mode-set flag, 446
MODEL directive, 59

Module, 59, 180
Monitor task, 611
Motherboard, 10
Mouse, 307
Multiplexed address/data bus, 41
Multitasking, 43, 296, 596

NEAR directive, 132
Nested loop, 137
Nested macro, 284
Nested procedures, 145
Nil, 221
Node, 221
Nonmaskable interrupt, 39, 154, 384
Nonvolatile memory, 6, 163
Normalization, 205, 491
Null modem, 512
Null selector, 585

Object code library, 280
Object file, 55
OFFSET directive, 92
Open loop system, 211
Operand size prefix, 23
Operation command word, 469
ORG directive, 56
Output port, 71, 440
Overflow, 135
Overhead, 291

Packed decimal, 115
Page, 44
Page directory, 591
Page directory base register, 591
Page fault, 44, 594, 606
Page frame, 590
Page table, 591
Paging, 590
Paragraph, 302
Parallel data, 233
Parallel I/O, 445
Parallel processing, 7